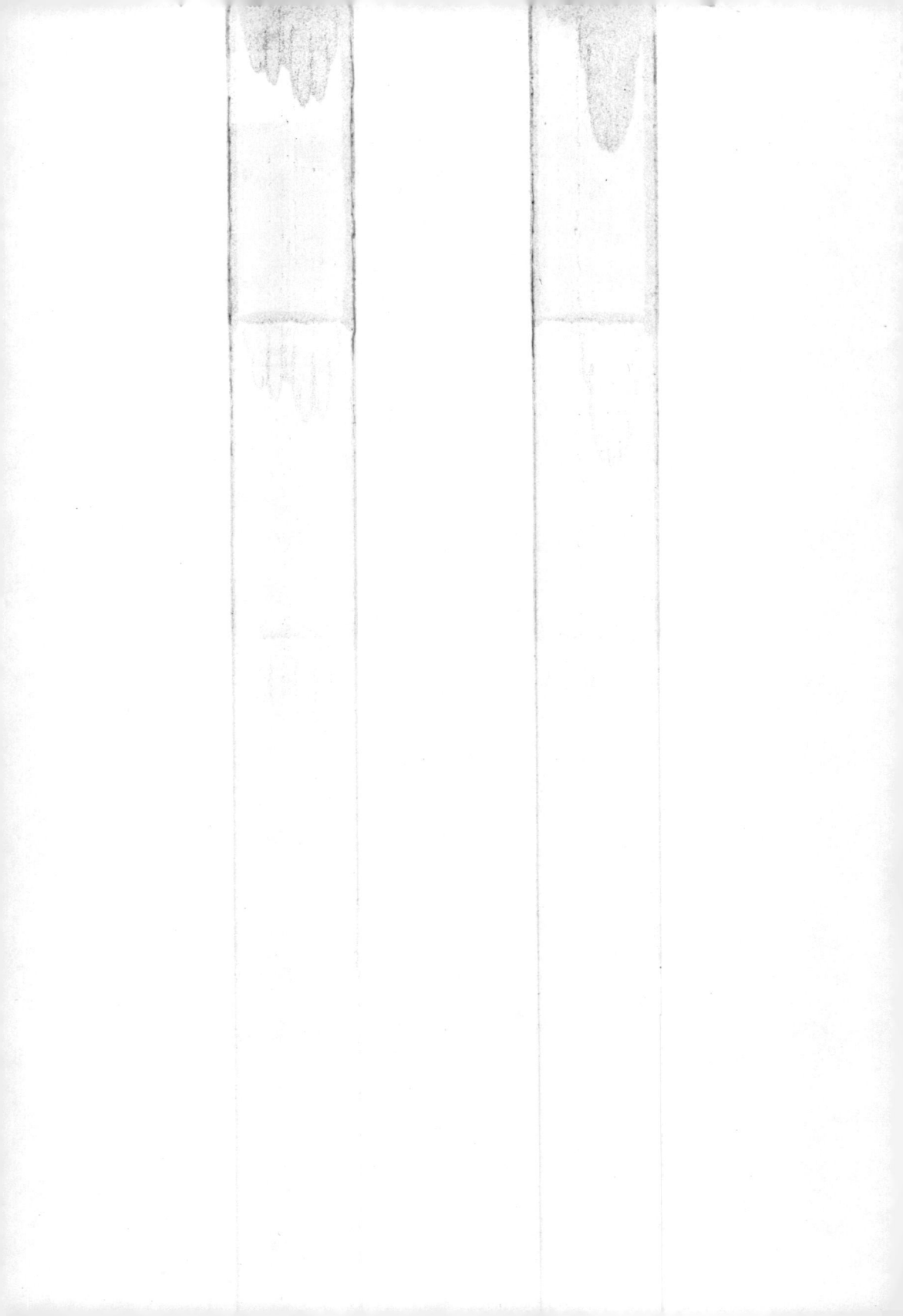

Polarization Phenomena in Nuclear Physics-1980
(Fifth International Symposium, Santa Fe)

AIP Conference Proceedings
Series Editor: Hugh C. Wolfe
Number 69

Polarization Phenomena in Nuclear Physics-1980
(Fifth International Symposium, Santa Fe)

Part 1

Editors
G.G. Ohlsen, Ronald E. Brown, Nelson Jarmie,
W.W. McNaughton and G.M. Hale
Los Alamos Scientific Laboratory

American Institute of Physics
New York   1981

Copying fees: The code at the bottom of the first page of each article in this volume gives the fee for each copy of the article made beyond the free copying permitted under the 1978 US Copyright Law. (See also the statement following "Copyright" below). This fee can be paid to the American Institute of Physics through the Copyright Clearance Center, Inc., Box 765, Schenectady, N.Y. 12301.

Copyright © 1981 American Institute of Physics

Individual readers of this volume and non-profit libraries, acting for them, are permitted to make fair use of the material in it, such as copying an article for use in teaching or research. Permission is granted to quote from this volume in scientific work with the customary acknowledgment of the source. To reprint a figure, table or other excerpt requires the consent of one of the original authors and notification to AIP. Republication or systematic or multiple reproduction of any material in this volume is permitted only under license from AIP. Address inquiries to Series Editor, AIP Conference Proceedings, AIP.

L.C. Catalog Card No. 81-65107
ISBN 0-88318-168-1
DOE CONF- 800824

SUMMARY TABLE OF CONTENTS

Part 1

Summary Table of Contents.	v
Detailed Table of Contents	vi
Sponsors and Staff	xxxxii
Introduction	xxxxv
Opening Address.	1
Section 1: Nucleon-Nucleon and High Energy Polarization Phenomena.	10
Section 2: Polarization in Nuclear Structure Physics.	203
Author Index	753

Part 2

Summary Table of Contents.	v
Detailed Table of Contents	vi
Section 3: Polarized Beam and Target Technology, Techniques, and Formalism.	769
Section 4: Polarization Studies with Heavy Ions, New Developments in Related Fields, Exotica, and Applications	1000
Section 5: Polarization Effects in Light Nuclei	1131
Section 6: Basic Symmetries	1139
Conference Summary	1491
Conference Participants.	1502
Author Index	1521

DETAILED TABLE OF CONTENTS

Part 1

Sponsors and Staff. xxxxii
Introduction. xxxxv

      OPENING ADDRESS   1
  The International Polarization Symposia 1960-1980:
  An Overview. 2
    L. C. Biedenharn

  SECTION 1: NUCLEON-NUCLEON AND HIGH ENERGY
      POLARIZATION PHENOMENA   10

1.A Spin Correlation Measurements in Nucleon-Nucleon
   Scattering at High Energies. 11
    D. G. Crabb

1.B Evidence for Dibaryon Resonances in Nucleon-Nucleon
   Scattering 31
    J. B. Roberts

1.C Polarization Phenomena in (p,π) Reactions. 62
    J. A. Niskanen

1.D The Interpretation of Recent Measurements of np and
   pp Cross Sections. 84
    J. A. Edgington

1.E Polarization Analyzing Power Measurements in Coherent
   Pion Production by Protons 93
    E. G. Auld

1.RR Rapporteur's Report: Nucleon-Nucleon and High Energy
   Polarization Phenomena 111
    L. C. Northcliffe

1.1 High-Precision np Polarization Data Between 13.5 and
   16.9 MeV and the Paris-Potential Predictions 114
    J. Côté, M. Lacombe, B. Loiseau, P. Pirès,
    R. de Tourreil, and R. Vinh Mau

1.2 The Influence of Multiple Scattering Corrections on
   High Accuracy Neutron-Proton Analyzing Power Data
   Measured at 16.9 MeV 117
    W. Tornow and R. L. Walter

1.3	Polarization Transfer in n-p Scattering at 50 MeV H. L. Woolverton, J. C. Hiebert, L. C. Northcliffe, M. J. Marolda, S. Nath, and W. F. Woodward	120
1.4	The Measurement of A_{nn} for Free N-P Scattering for Neutron Energies 300-665 MeV. T. S. Bhatia, G. Glass, J. C. Hiebert, L. C. Northcliffe, W. B. Tippens, B. E. Bonner, J. E. Simmons, C. L. Hollas, C. R. Newsom, R. D. Ransome, and P. J. Riley	123
1.5	Measurement of the Free Neutron-Proton Analyzing Power from 300 to 800 MeV C. R. Newsom, C. L. Hollas, R. D. Ransome, P. J. Riley, B. E. Bonner, J. J. Jarmer, M. W. McNaughton, J. E. Simmons, T. S. Bhatia, G. Glass, J. C. Hiebert, L. C. Northcliffe, and W. B. Tippens	126
1.6	A Measurement of the Analyzing Power for n-p Elastic Scattering at 800 MeV for $\theta_{c.m.} > 100°$. C. L. Hollas, P. J. Riley, C. R. Newsom, R. D. Ransome, B. E. Bonner, J. E. Simmons, T. S. Bhatia, G. Glass, J. C. Hiebert, L. C. Northcliffe, and W. B. Tippens	129
1.7	Measurement of K_{NN}, K_{SS}, K_{SL}, and K_{LL} in $\vec{n}p \to \vec{p}n$ at 800 MeV in the CEX Region R. D. Ransome, C. L. Hollas, P. J. Riley, B. E. Bonner, W. R. Gibbs, M. W. McNaughton, J. E. Simmons, T. S. Bhatia, G. Glass, J. C. Hiebert, L. C. Northcliffe, and W. B. Tippens	132
1.8	Quasi-Free $\vec{p}$ + n Analyzing Powers at 800 MeV. M. Barlett, G. W. Hoffmann, J. McGill, B. Bonner, B. Hoistad, and G. S. Blanpied	135
1.9	Analyzing Power of Proton-Proton Scattering at 9.88 MeV. M. D. Barker, P. C. Colby, W. Haeberli, and J. Ulbricht	137
1.10	Polarization in Proton-Proton Scattering at 52 MeV and Effect of the Electromagnetic LS Force. K. Imai, T. Matsusue, H. Shimizu, J. Shirai, K. Nisimura, K. Hatanaka, and T. Saito	140
1.11	(withdrawn)	

1.12 Measurement of the Spin Correlation Parameters A_{ookk}, A_{ooks} and A_{ooss} in the p-p Elastic Scattering between 400 and 600 MeV. 143
E. Aprile, R. Hausammann, E. Heer, R. Hess, S. Jaccard, C. Lechanoine-LeLuc, W. Leo, Y. Onel, S. Mango, and D. Rapin

1.13 Asymmetries for Elastic pp Scattering in the Coulomb Interference Region at 800 MeV. 146
G. Pauletta, G. S. Adams, G. Igo, J. B. McClelland, A. T. M. Wang, C. A. Whitten, Jr., A. Wriekat, M. Gazzaly, and B. Hoistad

1.14 Measurement of D_{NN}, D_{SS}, D_{LS} in pp → pp at 800 MeV. . 149
M. W. McNaughton, B. E. Bonner, W. D. Cornelius, E. W. Hoffman, O. B. van Dyck, R. L. York, R. D. Ransome, C. L. Hollas, P. J. Riley, K. Toshioka, H. Spinka, P. R. Bevington, and H. B. Willard

1.15 Energy Dependence of A_{nn} (90°) in pp Elastic Scattering from 1.10 to 2.75 GeV/c. 152
H. E. Miettinen, D. A. Bell, J. A. Buchanan, M. M. Calkin, J. M. Clement, W. H. Dragoset, M. Furić, K. A. Johns, J. D. Lesikar, T. A. Mulera, G. S. Mutchler, G. C. Phillips, J. B. Roberts, and S. E. Turpin

1.16 Spin-Dependent Forces in pp→pp at 6 GeV/c 155
S. Wakaizumi and M. Sawamoto

1.17 Inclusive Scattering of Protons on Helium and Nickel at 500 MeV 158
G. Roy, L. G. Greeniaus, G. A. Moss, D. A. Hutcheon, R. Liljestrand, R. M. Woloshyn, D. Boal, A. W. Stetz, K. Aniol, A. Willis, N. Willis, and R. McCamis

1.18 Measurement of Polarization Parameters in pp → π^+d at Intermediate Energies. 161
E. Aprile, R. Hausammann, E. Heer, R. Hess, C. Lechanoine-LeLuc, W. Leo, Y. Onel, D. Rapin, J. M. Cameron, S. Mango, and S. Jaccard

1.19 Differential Cross Section for Pion Production in the Reaction pp→πd between 500 and 600 MeV. 163
J. Hoftiezer, Ch. Weddigen, B. Favier, S. Jaccard, P. Chatelain, F. Foroughi, and C. Nussbaum

1.20 The Measurement of K_{NN} and K_{LL} in $\vec{p}p \to \vec{n}X$ at 800 MeV. ... 166
T. S. Bhatia, G. Glass, J. C. Hiebert, L. C. Northcliffe, W. B. Tippens, C. L. Hollas, C. R. Newsom, R. D. Ransome, P. J. Riley, G. P. Pepin, B. E. Bonner, and J. E. Simmons

1.21 Polarized Cross Section for the Reaction $p\uparrow\uparrow p \to p\pi^+n$. ... 169
A. D. Hancock, R. W. Hackenburg, E. V. Hungerford, B. W. Mayes, L. S. Pinsky, J. C. Allred, T. M. Williams, S. D. Baker, J. A. Buchanan, J. M. Clement, M. Copel, D. M. Judd, G. S. Mutchler, G. P. Pepin, E. A. Umland, G. C. Phillips, M. McNaughton, and C. Hwang

1.22 Calculation of Asymmetries and Cross Sections for the Reaction $p\uparrow p \to p\pi^+n$ 172
E. A. Umland, I. M. Duck, and G. S. Mutchler

1.23 Deuteron Photodisintegration Induced by Monochromatic Linearly Polarized Gamma-Rays 175
W. Del Bianco, L. Federici, G. Giordano, G. Matone, G. Pasquariello, P. Picozza, R. Caloi, L. Casano, L. Ingrosso, M. P. De Pascale, M. Mattioli, E. Poldi, C. Schaerf, P. Pelfer, D. Prosperi, S. Frullani, B. Girolami, and H. Jeremie

1.24 Phase Shift Analyses of np and pn Elastic Scattering between 10-750 MeV. 177
J. Bystricky, C. Lechanoine-LeLuc, and F. Lehar

1.25 Phase Shift Analyses of pp Elastic Scattering between 10-750 MeV. 178
J. Bystricky, C. Lechanoine-LeLuc, and F. Lehar

1.26 Effect of Recent np A_y, A_{yy} Measurements on Scattering Analyses below 850 MeV 179
R. A. Arndt and B. J. VerWest

1.27 Coulomb Corrections in Proton-Proton Observables. ... 182
W. Plessas

1.28 The Effect of the Coulomb Distortion on Proton-Proton Observables. 185
W. Plessas and L. Mathelitsch

1.29	Search for the Dibaryon Bandhead. . . . . R. Abegg, J. M. Cameron, D. A. Hutcheon, R. P. Liljestrand, W. J. McDonald, C. A. Miller, L. E. Antonuk, C. E. Stronach, and J. R. Tinsley	188
1.30	Some Evidence for the Existence of the $^1S_0$- Dibaryon Resonances . . . . V. V. Komarov, A. M. Popova, and Yu. V. Popov	191
1.31	The 3F3 Diproton in πNN Dynamics. . . . . M. Araki, T. Ueda, and Y. Koike	194
1.32	Elastic Widths of $N\overline{N}$-Resonances from a Momentun-Space One-Boson-Exchange Potential. . . . . L. Heins, K. Holinde, and D. Schütte	197
1.33	Effect of Non-Iterative Isobar Diagrams on NN-Scattering Data . . . . K. Holinde, R. Machleidt, A. Faessler, H. Müther, and M. R. Anastasio	200

SECTION 2: POLARIZATION IN NUCLEAR STRUCTURE PHYSICS   203

2.A	Elastic Scattering Potentials for Nucleons and Composite Particles . . . . W. J. Thompson	204
2.B	Polarization Effects in Inelastic Proton Scattering at Intermediate Energies . . . . A. D. Bacher	220
2.C	Reaction Mechanism Studies with Polarized Ions. . . . K. Yagi	254
2.D	Survey of Direct Reaction Studies with Polarized Tritons . . . . E. R. Flynn	270
2.E	Polarization Effects in ^{3}He Induced Reactions . . . . S. Roman	282
2.F	Neutron Density Distribution Studies Using Intermediate Energy Polarized Protons . . . . L. Ray	295
2.G	Polarized Neutron Capture Studies in the Giant Resonance Region. . . . . H. R. Weller	308

2.H	Polarized Proton Radiative Capture Studies of Giant Resonances. K. A. Snover	321
2.I	Polarization Transfer in Inelastic Scattering . . . . J. M. Moss	334
2.J	Advances in Neutron Polarization Studies. . . . . . . R. L. Walter	344
2.K	DWBA Analysis of $(\vec{p},d)$: A Spectacular Failure. . . . J. R. Shepard, E. Rost, and P. D. Kunz	361
2.RR	Rapporteur's Report: Polarization in Nuclear Structure Physics . . . . R. N. Boyd	373
2.1	Polarization Effects in the Small-Angle Scattering of Fast Neutrons by Bismuth . . . . M. Ahmed and F. W. K. Firk	389
2.2	Analyzing Power of Lanthanum Using 7.65 MeV Neutrons. . . . . G. Schleussner, J. W. Hammer, K. W. Hoffmann, D. Kollewe, W. Kratschmer, and E. Speller	392
2.3	Analyzing Power of Uranium-238 Using 7.65 MeV Neutrons. . . . . J. W. Hammer, G. Schleussner, K. W. Hoffmann, D. Kollewe, W. Kratschmer, and E. Speller	395
2.4	Scattering of Polarized Neutrons from ^{54}Fe and ^{65}Cu at 10 MeV. . . . . C. E. Floyd, P. P. Guss, K. Murphy, R. C. Byrd, S. A. Wender, R. L. Walter, and T. B. Clegg	398
2.5	Neutron Scattering from ^{58}Ni and ^{208}Pb at 10 MeV. . . P. P. Guss, G. Tungate, C. E. Floyd, E. Woye, K. Murphy, R. C. Byrd, R. L. Walter, and T. B. Clegg	401
2.6	The Analyzing Power for Elastic Scattering of 9.9, 11.9 and 13.9 MeV Neutrons from Ca. . . . . W. Tornow, E. Woye, G. Mack, C. E. Floyd, K. Murphy, P. P. Guss, S. A. Wender, R. C. Byrd, R. L. Walter, and T. B. Clegg	404

2.7	The Scattering of Polarized Neutrons from ^{9}Be between 9 and 15 MeV. . . . . *C. E. Floyd, P. P. Guss, R. C. Byrd, K. Murphy, S. A. Wender, W. Tornow, R. L. Walter, T. B. Clegg, and W. J. Thompson*	407
2.8	Optical-Model and Coupled-Channel Predictions in Comparison to n-^{12}C Analyzing Power Data. . . . . . . *E. Woye, W. Tornow, and G. Mack*	410
2.9	A Program of Systematic Measurement and Analysis for the Low-Energy Nucleon Optical-Model Potential. . *R. L. Walter, R. C. Byrd, T. B. Clegg, E. J. Ludwig, and W. J. Thompson*	413
2.10	Polarized Proton Beam Studies of Very Narrow Resonances. . . . . *J. F. Wilkerson, W. J. Thompson, E. J. Ludwig, and T. B. Clegg*	415
2.11	The Depth of the Imaginary Potential Around A = 105 Determined by Proton Scattering . . . . *A. Feigel, E. Finckh, B. Rowedder, K. Rüskamp, H. Scheuring, U. Schneidereit, and P. Tröger*	418
2.12	Investigation of the Reaction Mechanism in the Elastic Scattering of Polarized Protons on ^{27}Al . . . *W. Kretschmer, J. Jordan, H. Löh, and W. Stach*	421
2.13	Proton Optical Potential for the Elastic Scattering on Molybdenum Isotopes. . . . . *W. Kretschmer, E. Heitz, J. Jordan, H. Löh, W. Schuster, W. Stach, R. Stingl, P. Urbainsky, and M. B. Wango*	424
2.14	Optical Potential for p+^{208}Pb Scattering below the Coulomb-Barrier . . . . *W. Kretschmer, K. H. Frank, E. Heitz, J. Jordan, H. Löh, W. Schuster, K. Spitzer, W. Stach, P. Urbainsky, and M. B. Wango*	427
2.15	Polarized Proton Scattering from Se Isotopes. . . . . *R. L. Varner, J. F. Wilkerson, W. J. Thompson, Y. Tagishi, E. J. Ludwig, T. B. Clegg, and B. L. Burks*	430

2.16 Analysis of Elastic Scattering with an L-Dependent Optical Potential 432
    P. J. van Hall, R. S. Mackintosh, A. M. Kobos, and W. H. L. Moonen

2.17 Nuclear Structure Effects in Polarized Proton Scattering. 434
    P. J. van Hall, W. H. L. Moonen, and S. D. Wassenaar

2.18 Analyzing Power in Proton-Nucleus Elastic Scattering in the Small Angle Region at 65 MeV 437
    T. Matsusue, K. Imai, H. Shimizu, J. Shirai, K. Nisimura, K. Hatanaka, and T. Saito

2.19 The Shell Effect Observed in the Spin Orbit Part of the Optical Potential. 440
    H. Sakaguchi, M. Nakamura, K. Hatanaka, T. Noro, F. Ohtani, H. Sakamoto, and S. Kobayashi

2.20 65 MeV Polarized Proton Elastic Scattering and the Effective Two-Body Interaction Range. 445
    H. Sakaguchi, M. Nakamura, K. Hatanaka, A. Goto, T. Noro, F. Ohtani, H. Sakamoto, and S. Kobayashi

2.21 Even-Odd Effect Observed in the Elastic Scattering of Polarized Protons at 65 MeV. 448
    M. Nakamura, H. Sakaguchi, K. Hatanaka, T. Noro, F. Ohtani, H. Sakamoto, and S. Kobayashi

2.22 The Isospin Dependence of the 65 MeV Proton Optical Potential in the f-p Shell Nuclei 451
    T. Noro, K. Hatanaka, H. Sakaguchi, M. Nakamura, H. Sakamoto, F. Ohtani, and S. Kobayashi

2.23 Elastic Scattering of Polarized Protons at 200 to 500 MeV. 454
    D. A. Hutcheon, J. M. Cameron, R. P. Liljestrand, P. Kitching, C. A. Miller, W. J. McDonald, D. M. Sheppard, W. C. Olsen, G. C. Neilson, H. S. Sherif, R. N. MacDonald, G. M. Stinson, D. K. McDaniels, J. R. Tinsley, L. W. Swensen, P. Schwandt, C. E. Stronach, and L. Ray

2.24   The Spin Dependence of Intermediate-Energy Proton-
        Nucleus Elastic Scattering.   457
            P. Schwandt, A. D. Bacher, W. W. Jacobs,
            H. -O. Meyer, and S. E. Vigdor

2.25   Spin "Length" Scales and Polarization in p-
        Nucleus Scattering.   461
            J. A. McNeil

2.26   Origin of the Deuteron Imaginary Spin-Dependent
        Potential   464
            W. H. Wong and P. A. Quin

2.27   A Test of Collective Effects in Se(d,d)Se   467
            Y. Tagishi, B. L. Burks, T. B. Clegg,
            E. J. Ludwig, R. L. Varner, J. F.
            Wilkerson, and W. J. Thompson

2.28   Phase Shift Analysis of ^{58}Ni(d,d$_o$) at 12 MeV.   470
            A. Lindner

2.29   Mass Dependence of the T_R Tensor Potential for
        Elastic Scattering of 20 MeV Deuterons.   473
            R. Frick, H. Clement, G. Graw, P.
            Schiemenz, N. Seichert, and Sun Tsu-Hsun

2.30   On the Optical Potential for Deuteron Scattering
        at E_d = 20 MeV.   476
            H. Clement

2.31   Measurements of A_{xx} and A_{yy} for Elastic Scattering
        of 56 MeV Deuterons from ^{28}Si, ^{64}Ni and ^{144}Sm 478
            K. Hatanaka, M. Nakamura, K. Imai,
            T. Noro, H. Shimizu, H. Sakamoto, J.
            Shirai, T. Matsusue, and K. Nisimura

2.32   Features in the Analyzing Powers in Deuteron
        Elastic Scattering Near 80 MeV.   481
            E. J. Stephenson, C. C. Foster,
            P. Schwandt, and D. A. Goldberg

2.33   Measurements of A_y and A_{yy} for 80 MeV Deuteron
        Elastic Scattering on ^{58}Ni and ^{208}Pb.   484
            E. J. Stephenson, J. C. Collins, C. C.
            Foster, D. L. Friesel, J. R. Hall, W. W.
            Jacobs, W. P. Jones, S. Kailas, M.
            Kaitchuck, P. Schwandt, W. W. Daehnick,
            and D. A. Goldberg

2.34   Complex $\vec{L}\cdot\vec{S}$ Term in Global Optical Model Potentials
        for Elastic Deuteron Scattering   487
            W. W. Daehnick

2.35	The Deuteron Optical Potential. . . . . G. H. Rawitscher and S. N. Mukherjee	490
2.36	New Representation of Analyzing Powers in Elastic Scattering of Polarized Deuterons . . . . H. Ohnishi, M. Tanifuji, and H. Noya	493
2.37	Polarized Triton Scattering from ^{26}Mg, ^{27}Al and ^{28}Si at 17 MeV. . . . . R. A. Hardekopf, Ronald E. Brown, F. D. Correll, G. G. Ohlsen, and P. Schwandt	496
2.38	Imaginary Spin-Orbit Optical Potential in Triton Elastic Scattering on ^{9}Be . . . . J. Meyer and E. Elbaz	499
2.39	Analyses of the Scattering of Polarized Helions from ^{32}S. . . . . J. M. Barnwell, N. M. Clarke, and R. J. Griffiths	502
2.40	Elastic and Inelastic Scattering of Polarized Protons through Isobaric Analog Resonances in ^{207}Bi and ^{209}Bi . . . . N. L. Back, H. C. Bhang, J. G. Cramer, T. A. Trainor, and R. Von Lintig	505
2.41	Giant Resonance Analysis of the ^{54}Fe($\vec{p},p'$) Reaction . P. J. van Hall, S. D. Wassenaar, and J. P. M. G. Melssen	508
2.42	Depolarization in the Inelastic Scattering of Protons from Copper . . . . W. G. Weitkamp, T. A. Trainor, I. Halpern, H. Bhang, and S. K. Lamoreaux	511
2.43	Scattering of Polarized Protons from ^{64,66,68,70}Zn. . P. J. van Hall, J. F. A. G. Ruyl, J. Krabbenborg, W. H. L. Moonen, and H. Offermans	514
2.44	Study of ^{68,64}Ni and ^{86,88}Sr by Scattering of Polarized Protons . . . . S. D. Wassenaar, P. J. van Hall, S. S. Klein, G. J. Nijgh, O. J. Poppema, W. F. Feix, J. H. Polane, and J. F. J. Dautzenberg	517

2.45 Scattering of Polarized Protons from ^{110,112,114}Cd and ^{115}In 520
S. D. Wassenaar, J. F. J. Dautzenberg,
J. H. Polane, P. J. van Hall, S. S. Klein,
G. J. Nijgh, and O. J. Poppema

2.46 Scattering of Polarized Protons from Even Tin Isotopes. 523
S. D. Wassenaar, P. J. van Hall, S. S.
Klein, G. J. Nijgh, O. J. Poppema,
J. H. Polane, and J. F. J. Dautzenberg

2.47 Microscopic Analysis of Polarized Proton Scattering from Tin. 526
S. D. Wassenaar and P. J. van Hall

2.48 One- and Two-Step Analysis of ^{14}N($\vec{p},p'$)^{14}N(2.31 MeV) Reaction at 21.0 MeV and the Effective Interaction. . 529
Y. Aoki, K. Nagano, Y. Toba,
S. Kunori, and K. Yagi

2.49 The ^{89}Y($\vec{p},p'$) Reaction at 21.1 MeV. 532
J. P. M. G. Melssen, P. J. van Hall,
S. D. Wassenaar, O. J. Poppema, S. S.
Klein, and G. J. Nijgh

2.50 Spin Flip Asymmetry in the Inelastic Scattering of Protons on ^{12}C at Energies from 22.0 to 29.0 MeV . . 535
T. Fujisawa, N. Kishida, T. Kubo,
T. Hasegawa, M. Sekiguchi, N. Ueda,
M. Yasue, Y. Wakuta, and A. Nagao

2.51 Microscopic Analysis of the ^{54}Fe($\vec{p},p'$) Reaction Between 20 and 30 MeV 538
P. J. van Hall, J. P. M. G. Melssen,
and S. D. Wassenaar

2.52 New Evidences for an Imaginary Spin-Orbit Potential in the Inelastic Scattering of Polarized Protons from ^{12}C and ^{16}O. 541
R. de Swiniarski and Dinh-Lien Pham

2.53 Measurements of Analyzing Powers for 6$^-$ States in ^{28}Si and ^{24}Mg by Inelastic Scattering of 65 MeV Polarized Protons 544
K. Hosono, N. Matsuoka, T. Saito, K.
Hatanaka, M. Kondo, T. Noro, H. Shimizu,
S. Kato, K. Okada, K. Ogino, and Y. Kadota

2.54 Analyzing Powers for ^{12}C($\vec{p},p'$)^{12}C at Intermediate Energies. 547
  J. R. Comfort, C. C. Foster, C. D.
  Goodman, D. W. Miller, G. L. Moake,
  P. Schwandt, J. R. Rapaport, and
  R. E. Segel

2.55 New Aspects of the TRIUMF ($\vec{p},\pi$) Program 550
  G. J. Lolos, E. L. Mathie, P. L.
  Walden, E. G. Auld, G. Jones, and
  R. B. Taylor

2.56 ^{90,92}Zr($\vec{p},p'$) Reactions at 800 MeV. 554
  F. T. Baker, C. Glashausser, A. Scott,
  G. Adams, M. Grimm, G. Hoffmann, G. Igo,
  W. G. Love, J. Moss, V. Penumetcha,
  W. Swenson, and B. E. Wood

2.57 Nuclear Information from the Wolfenstein Parameters in Inelastic Proton Nucleus Scattering. 556
  E. Bleszynski, M. Bleszynski, and
  Ch. A. Whitten, Jr.

2.58 Analyzing Powers of the Continuum Spectra (I): 65 MeV Polarized Protons on ^{12}C, ^{28}Si, ^{45}Sc, ^{58}Ni, ^{93}Nb, ^{165}Ho, ^{166}Er and ^{209}Bi. 559
  H. Sakai, K. Hosono, N. Matsuoka,
  S. Nagamachi, K. Okada, K. Maeda,
  and H. Shimizu

2.59 Analyzing Powers of the Continuum Spectra (II): 56 MeV Polarized Deuterons on ^{58}Ni, ^{93}Nb and ^{209}Bi. . 562
  H. Sakai, N. Matsuoka, K. Hatanaka,
  K. Okada, and H. Shimizu

2.60 Analyzing Power in the Continuum in Light-Ion Induced Reactions 565
  H. Lenske, T. Tamura, and T. Udagawa

2.61 Quasi-Elastic ^{40}Ca($\vec{p},2p$) Scattering at 200 MeV at TRIUMF 568
  P. Kitching, L. Antonuk, C. A. Miller,
  D. A. Hutcheon, W. J. McDonald, W. C.
  Olsen, G. C. Neilson, G. M. Stinson,
  and A. W. Stetz

2.62 The Nuclear Quadrupole-Quadrupole Interaction in the Inelastic Scattering of Tensor Polarized Deuterons 571
  H. Clement, R. Frick, G. Graw, F. D.
  Santos, P. Schiemenz, N. Seichert, and
  Sun Tsu-Hsun

2.63 Measurement of Quadrupole Moments and P3-Terms by
      Inelastic Scattering of Vector Polarized Deuterons. .   573
         H. Clement, R. Frick, G. Graw, F. Merz,
         P. Schiemenz, N. Seichert, and Sun
         Tsu-Hsun

2.64 Inelastic Scattering of Vector Polarized Deuterons
      from Samarium Isotopes.   576
         H. Clement, R. Frick, G. Graw, I.
         Oelrich, H. J. Scheerer, P. Schiemenz,
         N. Seichert, and Sun Tsu-Hsun

2.65 Octupole-Quadrupole Coupling Observed in the Excita-
      tion of 3^- States with Polarized Deuterons.   579
         H. Clement, R. Frick, G. Graw, P.
         Schiemenz, and N. Seichert

2.66 The Polarization of Break-Up Protons from Vector
      Polarized Deuteron Induced Reaction   581
         M. Nakamura, H. Sakaguchi, K. Imai,
         T. Noro, H. Shimizu, H. Sakamoto,
         S. Kobayashi, S. Kato, N. Matsuoka,
         and K. Hatanaka

2.67 On Deformed Tensor Potential for Inelastic
      Deuteron Scattering   584
         Jacques Raynal

2.68 Interactions of Polarized ^{3}He Particles with ^{24}Mg . .   587
         F. Entezami, A. K. Basak, O. Karban,
         P. M. Lewis, and S. Roman

2.69 The Scattering of Polarized ^{3}He by Oxygen Isotopes. .   590
         P. M. Lewis, O. Karban, A. K. Basak,
         E. C. Pollacco, and S. Roman

2.70 Analyzing Powers of (^{3}He,t) Reactions on Light
      Nuclei.   593
         A. K. Basak, O. Karban, P. M. Lewis,
         G. C. Morrison, and S. Roman

2.71 Intermediate Structure in the Giant E1 Resonance
      of ^{20}Ne Studied by Polarized Proton Capture   596
         P. M. Kurjan, G. A. Fisher, J. R.
         Calarco, and S. S. Hanna

2.72 Polarized Proton Capture to the First Excited
      State of ^{60}Ni   598
         K. Sparks, J. D. Turner, N. R.
         Roberson, D. R. Tilley, and
         H. R. Weller

2.73 Decay Mechanisms of the Giant E1 Resonance in ^{90}Zr Studied by Polarized Proton Capture. 600
J. R. Calarco, P. M. Kurjan, G. A. Fisher, and S. S. Hanna

2.74 Polarized Proton Capture to the First Excited State in ^{31}P. 602
C. Fitzpatrick, C. P. Cameron, Hideo Kitazawa, N. R. Roberson, D. R. Tilley, and H. R. Weller

2.75 Polarized Proton Capture on ^{88}Sr. 605
R. D. Ledford, C. Cameron, M. Potokar, N. R. Roberson, D. R. Tilley, and H. R. Weller

2.76 The Unique Extraction of E2 Strength in Polarized Proton Capture Reactions. 608
J. Sowinski and D. G. Mavis

2.77 Analyzing Powers and Cross Sections for (p,d) Reactions on Nuclei of N=50-82. 611
K. Nagano, Y. Aoki, H. Iida, S. Kunori, Y. Toba, and K. Yagi

2.78 Analyzing Powers for (p,d) Reactions on ^{208}Pb and ^{142,144}Nd Exciting Neutron-Hole States. 614
Y. Toba, K. Nagano, Y. Aoki, S. Kunori, and K. Yagi

2.79 The Reaction ^{56}Fe($\vec{p}$,d) at 24.6 MeV. 617
J. H. Polane and P. J. van Hall

2.80 Two-Step Processes in the Reaction ^{58}Ni($\vec{p}$,d) at 24.6 MeV 620
J. H. Polane, P. J. van Hall, O. J. Poppema, S. S. Klein, G. J. Nijgh, S. D. Wassenaar, J. F. J. Dautzenberg, and W. Feix

2.81 Effects of the Deuteron D State on the Polarization of the Residual Nuclear State 623
N. Kishida, H. Ohnuma, J. Kasagi, T. Kubo, and M. Yasue

2.82 Measurement of the Analyzing Powers for the Fragmented $f_{7/2}$ States by the ^{58}Ni($\vec{p}$,d)^{57}Ni Reaction at 65 MeV. 626
M. Fujiwara, Y. Fujita, S. Morinobu, I. Katayama, T. Yamazaki, H. Ikegami, and K. Imai

2.83 The (p,d) Reactions on A=12-94 Nuclei by 65 MeV
      Polarized Protons 629
           K. Hosono, M. Kondo, T. Saito, N.
           Matsuoka, S. Nagamachi, T. Noro,
           H. Shimizu, S. Kato, K. Okada, K.
           Ogino, and Y. Kadota

2.84 Spin Determination of Deep Hole States from
      ($\vec{p}$,d) Reactions 632
           J. Kasagi, G. M. Crawley, S. Gales,
           E. Gerlic, D. Friesel, and A. Bacher

2.85 ($\vec{p}$,d) Analyzing-Power Measurements at 94 MeV. 635
           D. W. Miller, W. W. Jacobs, D. W.
           Devins, and W. P. Jones

2.86 Polarization Transfer in the Reaction ^{56}Fe($\vec{d}$,p)^{57}Fe
      by In-Beam Mössbauer Measurements 638
           B. J. vom Feld, Th. Müller, C. Günther,
           H. Hübel, and H. Paetz gen. Schieck

2.87 Determination of j-Mixing in ^{53}Cr(d,p)^{54}Cr from
      Tensor Analyzing Power Measurements 641
           J. E. Kammeraad, J. A. Bieszk,
           L. D. Knutson, and W. Haeberli

2.88 Tensor Analyzing Power of the Reaction ^{64}Ni(d,p)^{65}Ni. 644
           K. Rüskamp, W. Drenckhahn, A. Feigel,
           E. Finckh, G. Gademann, and M. Wangler

2.89 Measurement and DWBA Analysis of the ^{78}Kr($\vec{d}$,p)^{79}Kr
      Reaction. 647
           B. L. Burks, R. R. Cadmus, Jr.,
           T. B. Clegg, and E. J. Ludwig

2.90 An Investigation of Configuration Mixing in ^{210}Bi
      Using Polarized Deuterons 650
           C. A. Gossett, L. D. Knutson,
           and P. A. Quin

2.91 Analyzing Power of the ^{12}C(d,p) and ^{12}C(d,n)
      Reaction. 653
           W. Drenckhahn, A. Feigel, E. Finckh,
           G. Gademann, K. Rüskamp, M. Wangler,
           and L. Zemło

2.92 Vector Analyzing Powers for the ^{12}C($\vec{d}$,n)^{13}N,
      ^{9}Be($\vec{d}$,n)^{10}B and ^{28}Si(d,n)^{29}P Reactions. 656
           F. D. Brooks, P. M. Lister, J. M.
           Nelson, and K. S. Dhuga

2.93 Spectroscopy of ^{62}Ni from the ^{61}Ni(d,p)^{62}Ni
      Reaction. 659
           O. Karban, A. K. Basak, F. Entezami,
           and S. Roman

2.94 Spin-Tensor Interaction in Polarization Transfer
      Reactions 662
           J. W. Hugg and S. S. Hanna

2.95 Spectroscopy of ^{145}Sm at $E_x \leq 3.2$ MeV via the
      $(\vec{d},p)$ Reaction at $E_d = 19$ MeV 664
           Sun Tsu-Hsun, H. Clement, R. Frick,
           G. Graw, F. Merz, F. Riess, P. Schiemenz,
           and N. Seichert

2.96 Inelastic Transfer in $(\vec{d},p)$ Reactions from ^{28}Si
      and ^{54}Cr 667
           N. Seichert, H. Clement, R. Frick,
           G. Graw, P. Schiemenz, and Sun Tsu-Hsun

2.97 D-State Effects in (d,p), (d,t) and (d,^{3}He)
      Reactions at $E_d = 22$ MeV. 670
           N. Seichert, H. Clement, R. Frick,
           G. Graw, S. Roman, F. D. Santos,
           P. Schiemenz, and Sun Tsu-Hsun

2.98 Contributions of Spin-Dependent Effects to the
      Vector Analyzing Power and Proton Polarization for
      $\ell_n = 0$ (d,p) Reaction 673
           T. Hasegawa, N. Ueda, T. Kubo, N.
           Kishida, H. Ohnuma, T. Fujisawa,
           T. Wada, and K. Iwatani

2.99 On Some New Effects of the Deuteron D State
      Observed in Low Energy (p,d) and (d,p)
      Reactions 676
           H. Ohnuma

2.100 Mixed-j Transfer in ^{55}Mn$(\vec{d},t)^{54}$Mn 679
           J. A. Cameron, E. Habib, and
           A. A. Pilt

2.101 Core-Coupled States Excited in the ^{208}Pb$(\vec{d},t)^{207}$Pb
      Reaction. 682
           E. Sugarbaker, W. P. Alford, R. N.
           Boyd, J. Cameron, E. Flynn, and
           J. Sunier

2.102 Configuration Mixing of Particle-Hole States in A = 16 Nuclei Studied by the $^{17}O(\vec{d},t)^{16}O$ and $^{17}O(\vec{d},\tau)^{16}N$ Reaction. 685
G. Mairle, K. T. Knöpfle, H. Riedesel, K. Schindler, G. J. Wagner, V. Bechtold, and L. Friedrich

2.103 Tensor Analyzing Power of (d,^{3}He) Reaction and the D-State of ^{3}He. 688
F. Entezami, K. S. Dhuga, O. Karban, J. M. Nelson, and S. Roman

2.104 Spin Determination of Deeply-Bound Hole States from $(\vec{d},^3He)$ Reactions. 691
A. Stuirbrink, K. T. Knöpfle, G. Mairle, H. Riedesel, K. Schindler, G. J. Wagner, V. Bechtold, and L. Friedrich

2.105 The (t,d) Reaction on the Ni Isotopes with Polarized Tritons 694
E. R. Flynn, J. A. Cizewski, Ronald E. Brown, R. A. Hardekopf, and J. W. Sunier

2.106 Measurements of Masses and Spins of Neutron Rich Nuclei by the $(\vec{t},\alpha)$ Reaction. 697
F. Ajzenberg-Selove, E. R. Flynn, Ronald E. Brown, J. A. Cizewski, and J. W. Sunier

2.107 The $^{194,196,198}Pt(t,\alpha)^{193,195,197}Ir$ Reactions with Polarized Tritons 700
J. A. Cizewski, E. R. Flynn, J. W. Sunier, Ronald E. Brown, and D. G. Burke

2.108 Application of ($\vec{^3He}$,d) and ($\vec{^3He}$,α) Reactions in Spectroscopy 703
O. Karban, A. K. Basak, G. C. Morrison, J. M. Nelson, and S. Roman

2.109 Reaction Mechanism Studies with Polarized ^{3}He 706
O. Karban, A. K. Basak, and S. Roman

2.110 Unnatural Parity Transition in (p,t) Reaction 709
Y. Toba, Y. Aoki, H. Iida, S. Kunori, K. Nagano, and K. Yagi

2.111 Multistep Processes in the Reaction ^{116}Sn$(\vec{p},t)$
       at 25.1 MeV   712
       W. F. Feix, J. H. Polane, and
       P. J. van Hall

2.112 Reaction Mechanisms of the ^{208}Pb(p,t)^{206}Pb
       (3^+ Ex=1.34 MeV) Reaction   715
       M. Igarashi and K.-I. Kubo

2.113 Cross Section and Analyzing Power in the
       ^{206}Pb$(\vec{t},p)$ ^{208}Pb(4^-) Reaction   718
       W. P. Alford, R. N. Boyd, E.
       Sugarbaker, F. deBoer, Ronald
       E. Brown, and E. R. Flynn

2.114 Spin-Parity Determinations from the
       ^{42}Ca$(\vec{d},\alpha)^{40}$K Reaction Near $0°$ 721
       Shang Ren-cheng, J. A. Kuehner,
       A. A. Pilt, M. A. M. Shahabuddin,
       and A. Trudel

2.115 Investigation of the Reaction ^{14}N$(d,\alpha)^{12}$C
       with Vector Polarized Deuterons   724
       W. Kretschmer, G. Pröbstle, and
       W. Stach

2.116 Investigation of the Reaction ^{24}Mg$(d,\alpha)^{22}$Na
       with Vector Polarized Deuterons   727
       W. Kretschmer, E. Heitz, C.
       Glashausser, A. B. Robbins,
       J. Duder, and D. Melnik

2.117 Tensor Analyzing Powers in the Reaction
       ^{24}Mg$(d,\alpha)^{22}$Na   730
       W. Kretschmer, E. Heitz, C.
       Glashausser, A. B. Robbins,
       J. C. Duder, and D. Melnik

2.118 Investigation of the Reaction ^{27}Al$(d,\alpha)^{25}$Mg
       with Vector Polarized Deuterons   733
       W. Kretschmer, E. Heitz, C.
       Glashausser, A. B. Robbins,
       J. C. Duder, and D. Melnik

2.119 Tensor Analyzing Power Measurements for (d,α)
       on s-d Shell Nuclei   736
       Y. Tagishi, T. B. Clegg, E. J.
       Ludwig, S. A. Tonsfeldt, and
       J. F. Wilkerson

2.120 Spectroscopy of Stretched Configurations with
the $(\vec{d},\alpha)$ Reaction at 52 MeV 739
G. Mairle, Liu Ken Pao, K. T.
Knöpfle, H. Riedesel, K. Schindler,
G. J. Wagner, V. Bechtold, J. Bialy,
and L. Friedrich

2.121 The ^{46,48}Ti$(\vec{p},\alpha)^{43,45}$Sc Reaction 742
R. N. Boyd, S. L. Blatt, T. R.
Donoghue, H. J. Hausman,
E. Sugarbaker, and S. E. Vigdor

2.122 Asymmetries in (^{3}He,^{7}Be) Cross Sections and
Spin-Orbit Coupling 745
P. Lezoch, H.-J. Trost, Md. A.
Rahman, and U. Strohbusch

2.123 Photonuclear Reactions with Linearly Polarized
Photons 747
K. Wienhard, K. Ackermann,
K. Bangert, U. E. P. Berg,
C. Bläsing, K. Kobras, W.
Naatz, D. Rück, R. K. M.
Schneider, and R. Stock

2.124 Antisymmetrization Effects in Deuteron-Nucleus
Elastic Scattering 750
J. A. Tostevin, M. H. Lopes, and
R. C. Johnson

Author Index 753

Part 2

SECTION 3: POLARIZED BEAM AND TARGET TECHNOLOGY,
TECHNIQUES, AND FORMALISM   769

3.A Polarized Targets in Nuclear and High Energy
Physics 770
W. Heeringa

3.B Polarized Electron Sources 785
G. Baum

3.C New Technique for the Production of Polarized
^{3}He Ions 797
R. J. Slobodrian

3.D	Lasers in the Production of Polarized Beams and Targets . . . . *D. E. Murnick and M. S. Feld*	804
3.E	Absolute Polarization Standards at Medium and High Energies . . . . *M. W. McNaughton*	818
3.F	Survey of Methods for Rapid Spin Reversal . . . . . . *J. L. McKibben*	830
3.G	New Developments in Polarized Ion Sources . . . . . . *W. Grüebler and P. A. Schmelzbach*	848
3.RR	Rapporteur's Report: Polarized Beam and Target Technology, Techniques, and Formalism . . . . *P. A. Quin*	866
3.1	A Polarized Neutron Beam at LAMPF . . . . *T. S. Bhatia, G. Glass, J. C. Hiebert, L. C. Northcliffe, W. B. Tippens, B. E. Bonner, J. E. Simmons, C. L. Hollas, C. R. Newsom, R. D. Ransome, and P. J. Riley*	871
3.2	The Measurement of K_{NN}, K_{LL} in $\vec{p}d \to \vec{n}X$ and $\vec{p}\,^9Be \to \vec{n}X$ at 800 MeV. . . . . *P. J. Riley, C. L. Hollas, C. R. Newsom, R. D. Ransome, B. E. Bonner, J. E. Simmons, T. S. Bhatia, G. Glass, J. C. Hiebert, L. C. Northcliffe, and W. B. Tippens*	874
3.3	The Wisconsin Colliding-Beam Negative Polarized-Ion Source. . . . . *W. Haeberli, M. D. Barker, G. Caskey, C. A. Gossett, D. G. Mavis, P. A. Quin, J. Sowinski, and T. Wise*	877
3.4	New Type Polarized H$^-$ Ion Source for the KEK 12 GeV Synchrotron. . . . . *Y. Mori, K. Ito, A. Takagi, C. Kubota, and S. Fukumoto*	879
3.5	The Duoplasmatron Emission Aperture: A Very Critical Parameter in Lamb-Shift Sources. . . . . . . *W. Arnold, H. Berg, and G. Clausnitzer*	882

3.6   An Upgrading Program for the TUNL Lamb-Shift Polarized Source. 884
       T. B. Clegg, S. M. Mitchell, R. L. Varner, W. R. Wylie, S. A. Wender, C. E. Floyd, R. K. Murphy, and M. E. Wright

3.7   Improvements in the LAMPF Lamb-Shift Polarized Source. 887
       E. P. Chamberlin, R. L. York, H. E. Williams, and E. L. Rios

3.8   The University of Manitoba Polarized D⁻ and H⁻ Source 890
       S. Oh, M. de Jong, J. Bruckshaw, I. Gusdal, F. Konopasek, A. McIlwain, and R. Pogson

3.9   The Munich Polarized Ion Source 893
       P. Schiemenz, D. Ehrlich, R. Frick, G. Graw, and U. Meyer-Berkhout

3.10   Operating Experience and Cesium Recycling on the LASL Polarized Triton Source. 896
       R. A. Hardekopf

3.11   A High Efficiency Ionizer for the ETH Polarized Ion Source. 899
       P. A. Schmelzbach, W. Grüebler, V. König, and B. Jenny

3.12   Acceleration of the Polarized Proton and Deuteron at the INS SF Cyclotron. 901
       N. Ueda, S. Yamada, T. Hasegawa, M. Sekiguchi, T. Fujisawa, and S. Motonaga

3.13   S.I.N. Upgraded Polarized Beams 904
       S. Jaccard and H. Einenkel

3.14   Performance of the IPCR Polarized Proton and Deuteron Source 906
       T. Fujisawa, S. Motonaga, N. Ueda, S. Yamada, and T. Hasegawa

3.15   Polarized Proton Source Improvements at the ZGS . . . 909
       P. F. Schultz, E. F. Parker, and J. J. Madsen

3.16	Development of Polarized D^+ and $^3He^{++}$ Ion Sources . . W. C. Hardy, R. G. Green, G. W. Guest, O. Karban, W. B. Powell, and S. Roman	913
3.17	Polarized Heavy Ion Beams (^{6,7}Li, ^{23}Na) at the Heidelberg EN-Tandem. . . . . P. Egelhof, B. Bauer, R. Böttger, S. Kossionides, K.-H. Möbius, Z. Moroz, D. Presinger, R. Schuch, E. Steffens, G. Tungate, W. Dreves, I. Koenig, and D. Fick	916
3.18	Production of Vector Polarized Nuclei by $^7\vec{Li}$ Beams. . I. Koenig, D. Fick, R. Böttger, P. Egelhof, H. Ingwersen, S. Kossionides, K.-H. Möbius, D. Presinger, and E. Steffens	919
3.19	Polarization of Atomic Transitions in Ionized ^{14}N After Ion Beam Surface Interaction at Grazing Incidence . . . . H. Winter	922
3.20	On the Production of a Polarized ^{23}Na Beam by Optical Pumping . . . . W. Dreves, W. Broermann, M. Elbel, W. Kamke, D. Fick, and E. Steffens	925
3.21	Experimental Results for the LADON Photon Beam at Frascati . . . . L. Federici, G. Giordano, G. Matone, G. Pasquariello, P. Picozza, R. Caloi, L. Casano, M. P. De Pascale, M. Mattioli, E. Poldi, C. Schaerf, M. Vanni, P. Pelfer, D. Prosperi, S. Frullani, and B. Girolami	928
3.22	A Target of Polarized Hydrogen by Storage of Atoms in a Coated Pyrex Vessel. . . . . M. D. Barker, G. Caskey, C. A. Gossett, W. Haeberli, D. G. Mavis, P. A. Quin, S. Riedhauser, J. Sowinski, and J. Ulbricht	931
3.23	Dynamic Nuclear Polarization of Irradiated Targets. . M. L. Seely, M. R. Bergstrom, S. K. Dhawan, R. A. Fong-Tom, V. W. Hughes, R. F. Oppenheim, K. P. Schüler, P. A. Souder, K. Kondo, S. Miyashita, I. Nakano, S. J. St. Lorant, Y.-N. Guo, and A. Winnacker	933

3.24 Proposal for a Pulsed Lithium Beam Polarized
Target. 936
    J. E. Clendenin and K. P. Schüler

3.25 A Polarized ^{6}Li Target to Study Parity Violation. . . 939
    C. A. Gagliardi, A. R. Davis, G. T.
    Garvey, R. D. McKeown, B. Myslek-
    Laurikainen, R. G. H. Robertson,
    S. J. Freedman, and T. J. Bowles

3.26 Polarized Gas Jet Targets 941
    J. S. Dunham, C. S. Galovich,
    S. W. Wissink, D. G. Mavis,
    and S. S. Hanna

3.27 Cryogenic Polarized Target Facility for Spin- Spin
Neutron Total Cross Section Measurements. 944
    C. R. Gould, D. G. Haase, T. B.
    Clegg, W. J. Thompson, and R. L.
    Walter

3.28 Polarized Targets and Polarized Low Energy
Neutrons at WNR 947
    P. P. J. Delheij, G. L. Morgan,
    and P. Lisowski

3.29 QPAN - A Quick Polarization Analyzer for Neutrons . . 949
    C. L. Hollas, C. R. Newsom, R. D.
    Ransome, P. J. Riley, B. E. Bonner,
    J. G. Boissevain, J. E. Simmons, T. S.
    Bhatia, G. Glass, J. C. Hiebert,
    L. C. Northcliffe, and W. B. Tippens

3.30 Polarization of Elastic Proton-Silicon Scattering . . 952
    S. Kato, S. Kobayashi, H. Sakaguchi,
    M. Nakamura, K. Imai, T. Noro,
    F. Otani, H. Shimizu, H. Sakamoto,
    H. Ogawa, N. Matsuoka, and
    K. Hatanaka

3.31 Proton-Carbon Analyzing Power between 140 and
560 MeV 955
    E. Aprile, C. Eisenegger, R.
    Hausammann, E. Heer, R. Hess,
    C. Lechanoine-LeLuc, W. Leo,
    Y. Onel, and D. Rapin

3.32 A Polarimeter for Protons between 300 and 800 MeV . . 956
    R. D. Ransome, C. L. Hollas, S. J.
    Greene, B. E. Bonner, M. W.
    McNaughton, C. L. Morris, and H. A.
    Thiessen

3.33 Measurement of the pC and p^{9}Be Analyzing Power at
800 MeV   958
   R. D. Ransome, C. L. Hollas, P. J.
   Riley, B. E. Bonner, W. D. Cornelius,
   E. W. Hoffman, M. W. McNaughton,
   O. B. Van Dyck, R. L. York, K.
   Toshioka, and P. R. Bevington

3.34 A Deuteron Vector and Tensor Polarimeter for
E_d=4-17 MeV   961
   S. A. Tonsfeldt, T. B. Clegg,
   E. J. Ludwig, and J. F. Wilkerson

3.35 Analyzing Powers of ^{6}Li($\vec{d}$,α)^{4}He at 0.4 and 0.6
MeV with a New Polarized ^{6}Li Target   964
   J. Ulbricht, R. Beckmann, and
   U. Holm

3.36 A Vertical-Type High-Purity Ge Detector System
for Deuteron Tensor Analyzing Power Measurements. . .   967
   N. Matsuoka, K. Hatanaka, T. Saito,
   K. Hosono, A. Shimizu, and M. Kondo

3.37 Polarization Monitors for Polarized Heavy Ions. . . .   970
   P. Zupranski, Z. Moroz, R. Böttger,
   P. Egelhof, K.-H. Möbius, G. Tungate,
   E. Steffens, W. Dreves, I. Koenig,
   and D. Fick

3.38 Resonant Depolarisation of Polarised H$^-$ Ions
in the University of Manitoba Cyclotron   973
   M. de Jong, S. Oh, J. Birchall,
   I. Gusdal, A. McIlwain, and
   J. S. C. McKee

3.39 Polarization Eigenvector in Synchrotrons Equipped
with Siberian Snakes.   976
   A. Turrin

3.40 Depolarization of Alkali Beams During Ionization
on A Surface Ionizer.   979
   R. Böttger, B. Bauer, P. Egelhof,
   K.-H. Möbius, Z. Moroz, E. Steffens,
   G. Tungate, W. Dreves, I. Koenig,
   and D. Fick

3.41 Polarization of On-Line Separated Isotopes Using
Surface Scattering and Low Temperature Implantation .   982
   L. Vanneste, H. Pattyn, D.
   Vandeplassche, J. Geenen,
   C. Nuytten, and E. Van Walle

3.42　Techniques with H⁰ Produced from Polarized H⁻
Beams 985
　　　*Olin B. van Dyck, Andrew J. Jason, and
　　　Daniel W. Hudgings*

3.43　Use of Neutron Time-of-Flight Techniques in (p,n)
A_y Measurements 988
　　　*R. C. Byrd, K. Murphy, and R. L.
　　　Walter*

3.44　Determination of the Electric Quadrupole Moment
of ^{7}Li by Coulomb Scattering of an Aligned ^{7}Li
Beam. 991
　　　*P. Egelhof, R. Böttger, W. Dreves,
　　　K.-H. Möbius, E. Steffens, G. Tungate,
　　　P. Zupranski, D. Fick, and F. Roesel*

3.45　0° Polarization Transfer Measurements Using
Charged Particles 994
　　　W. D. Cornelius and R. L. York

3.46　Relations Between Cartesian and Spherical
Tensors 997
　　　J. M. Normand and J. Raynal

SECTION 4:　POLARIZATION STUDIES WITH HEAVY IONS,
　　　　　　　NEW DEVELOPMENTS IN RELATED FIELDS,
　　　　　　　EXOTICA, AND APPLICATIONS　　　　1000

4.A　Shape Effects in the Interaction of Aligned
Heavy Ions. 1001
　　　E. Steffens

4.B　Spin Polarization of Products in Heavy Ion Reactions
Observed via β-Ray Asymmetry Measurements 1016
　　　N. Takahashi

4.C　Polarization in Heavy Ion Reactions 1027
　　　G. Graw

4.D　Polarization Effects in Atomic Collision Physics
Related to Nuclear Spin Polarized Ion Beams . . . 1037
　　　H. J. Andrä

4.E　A Review of Muon Spin Rotation with Selected
Applications in Solid State Physics and
Chemistry 1057
　　　D. W. Cooke

4.F	Physics with Muons Stopped in Polarized Nuclear Targets . . . . N. C. Mukhopadhyay and A. Hintermann	1068
4.RR	Rapporteur's Report: Polarization Studies with Heavy Ions, New Developments in Related Fields, Exotica, and Applications . . . . P. Zupranski	1080
4.1	Tensor Analyzing Powers of $\vec{^6Li}$ on $^{12,13}C$ at $E(^6Li) = 9$ MeV. . . . . J. Meyer and E. Elbaz	1083
4.2	Elastic Scattering of Vector Polarized ^6Li on ^6Li . . V. G. Avrigeanu, P. Egelhof, E. Steffens, K. Bethge, G. Gruber, W. Dreves, and D. Fick	1085
4.3	Elastic and Inelastic Scattering of Vector Polarized ^6Li on ^{28}Si at 22.8 MeV . . . . Dinh-Lien Pham	1088
4.4	Imaginary Spin-Orbit Optical Potential in $\vec{^6Li}$ Elastic Scattering. . . . . J. Meyer and E. Elbaz	1091
4.5	Polarization and Scattering of ^6Li on ^{12}C . . . . . Ahmed Osman and M. T. Youssef	1094
4.6	Shape Effects in the ^7Li-^{51}V Reaction Cross Section . . . . K.-H. Möbius, R. Böttger, P. Egelhof, Z. Moroz, E. Steffens, G. Tungate, W. Dreves, I. Koenig, and D. Fick	1096
4.7	The Elastic Scattering of Tensor Polarized ^7Li on ^{58}Ni . . . . G. Tungate, R. Böttger, P. Egelhof, K.-H. Möbius, Z. Moroz, E. Steffens, W. Dreves, I. Koenig, and D. Fick	1099
4.8	Elastic and Inelastic Scattering of Vector Polarized ^7Li on ^{58}Ni . . . . Z. Moroz, R. Böttger, P. Egelhof, K.-H. Möbius, G. Tungate, E. Steffens, W. Dreves, I. Koenig, and D. Fick	1102

4.9   Q-Value Dependence of the Vector Analyzing Power in the ^{58}Ni($^{7}\vec{\text{Li}}$, ^{6}Li)^{59}Ni Reaction 1105
         G. Tungate, R. Böttger, P. Egelhof,
         K.-H. Möbius, Z. Moroz, E. Steffens,
         W. Dreves, D. Fick, D. M. Brink, and
         T. F. Hill

4.10   Analysis of Elastic Scattering of Polarized ^{7}Li on ^{58}Ni within the Diffraction Model. 1108
         Z. Moroz, R. Böttger, P. Egelhof,
         T. F. Hill, K.-H. Möbius, G.
         Tungate, E. Steffens, W. Dreves,
         and D. Fick

4.11   Spin Flip Probability of ^{13}C in the Inelastic Scattering of ^{13}C by ^{12}C at E(^{13}C) = 87 MeV 1111
         M. Tanaka, J. Kawa, T. Fukuda,
         S. Nakayama, I. Miura, and
         H. Ogata

4.12   Spin Polarization of ^{12}B in Heavy Ion Reaction ^{27}Al(^{14}N, ^{12}B)$^{+}$. 1114
         N. Takahashi, K. H. Tanaka,
         Y. Nojiri, T. Minamisono, and
         K. Sugimoto

4.13   Measurements of m-State Populations Via γ-Recoil. 1117
         H. G. Bohlen, M. Clover, G. Ingold,
         H. Lettau, H. Ossenbrink, and
         W. von Oertzen

4.14   Study of the ^{16}O(^{16}O, ^{12}C)^{20}Ne Reaction Mechanism by Polarization Measurements. 1120
         F. Pougheon, P. Roussel, M. Bernas,
         F. Diaf, B. Fabbro, F. Naulin,
         E. Plagnol, and G. Rotbard

4.15   Analysing Powers of Heavy Ion Induced Transfer Reactions 1123
         F. D. Santos and A. M. Gonçalves

4.16   Atomic-Excitation Effects in Nuclear Reactions Studied with Polarized Beams. 1126
         W. J. Thompson, J. F. Wilkerson,
         T. B. Clegg, J. M. Feagin,
         E. J. Ludwig, and E. Merzbacher

4.17   Magnetic Moments of Mirror Nuclei Measured by the Polarized Beam-NMR Method 1129
         J. W. Hugg and S. S. Hanna

	SECTION 5: POLARIZATION EFFECTS IN LIGHT NUCLEI	1131
5.A	Low Energy Three Nucleon Experiment and Theory. . . . W. M. Kloet	1132
5.B	Microscopic Calculations of Polarization Observables in Light Nuclei . . . . R. J. Philpott	1144
5.C	Intermediate and High Energy Polarization Experiments Involving the 3-6 Nucleon Systems . . . . G. J. Igo	1157
5.D	Measurement of Tensor Polarization in Pion-Deuteron Elastic Scattering . . . . R. J. Holt	1180
5.RR	Rapporteur's Report: Polarization Effects in Light Nuclei. . . . . H. E. Conzett	1195
5.1	Constraints on πd Phase Parameters from Polarization Measurements . . . . J. Arvieux and A. S. Rinat	1201
5.2	Polarization Observables in π-d Scattering. . . . . . G. H. Lamot and C. Fayard	1202
5.3	The Reaction $^2\text{H}(\vec{p},t)\pi^+$ at 470 and 500 MeV . . . . . . R. Abegg, D. K. Hasell, W. T. H. van Oers, J. M. Cameron, L. G. Greeniaus, D. A. Hutcheon, C. A. Miller, G. A. Moss, R. P. Liljestrand, H. Wilson, A. W. Stetz, M. B. Epstein, and D. J. Margaziotis	1205
5.4	Study of the Three Nucleon System via p-d Elastic Scattering. . . . . V. König, W. Grüebler, R. E. White, P. A. Schmelzbach, B. Jenny, F. Sperisen, C. Schweizer, and P. Doleschall	1208
5.5	Polarization Transfer Measurements of p-d Scattering as a Test of Faddeev Calculations. . . . . F. Sperisen, W. Grüebler, V. König, K. Elsener, C. Schweizer, B. Jenny, P. A. Schmelzbach, and P. Doleschall	1211

5.6 Determination of the Asymptotic D- to S-State
     Normalization of the Deuteron Wave Function 1214
        W. Grüebler, V. König, P. A.
        Schmelzbach, B. Jenny, and
        F. Sperisen

5.7 Tensor and Vector Asymmetries in p-$\vec{D}$ Elastic
     Scattering at 800 MeV 1217
        M. Bleszynski, J. Carroll, M.
        Haji-Saeid, G. Igo, J. B.
        McClelland, A. Sagle, C. L.
        Morris, R. Klem, T. Joyce,
        Y. Makdisi, M. Marshak,
        B. Mossberg, E. A. Peterson,
        K. Ruddick, and J. Whittaker

5.8 Analyzing Power for pp and pd Elastic Scattering
     in the Coulomb-Nuclear Interference Region at
     800 MeV 1220
        J. B. McClelland, M. Bleszynski,
        G. Igo, F. Irom, and C. A.
        Whitten, Jr.

5.9 Measurement of Polarization Transfer in the
     $H(\vec{d},\vec{p})D$ Reaction at 0° 1223
        W. D. Cornelius, R. L. York,
        L. C. Northcliffe, and J. C.
        Hiebert

5.10 Vector and Tensor Analyzing Powers for Proton-
     Deuteron Elastic and Inelastic Scattering 1226
        M. Sawada, S. Seki, M. Ishikawa,
        K. Furuno, Y. Nagashima, J.
        Schimizu, and J. Sanada

5.11 Analyzing Powers in the Quasi-Free Scattering
     Region of the Reactions $^2H(\vec{p}, 2p)n$ and $^2H(\vec{p}, pn)p$
     at 65 MeV 1229
        H. Shimizu, K. Imai, T. Matsusue,
        J. Shirai, R. Takashima, K.
        Nisimura, K. Hatanaka, T. Saito,
        and A. Okihana

5.12 Proton Analyzing Power Measurement in the
     $D(\vec{p},2p)n$ Reaction 1232
        F. Foroughi, P. Chatelain, H.
        Vuillème, and Ch. Nussbaum

5.13 Analyzing Power Measurement for the Reaction
$\vec{p} + d \to {}^3He + mm$ 1234
M. Gazzaly, M. Hajisaeid, G. Igo,
F. Irom, G. Pauletta, A. Rahbar,
A. T. Wang, C. A. Whitten, Jr.,
and H. A. Thiessen

5.14 Deuteron Breakup in Collinear Geometry 1237
J. P. Svenne, J. Birchall, M. de Jong,
N. T. Okumusoḡlu, and J. S. C. McKee

5.15 Analyzing-Power Formalism for Three-Body Final
States 1240
G. G. Ohlsen, Ronald E. Brown,
F. D. Correll, and R. A. Hardekopf

5.16 Analyzing Powers for the Three-Nucleon Breakup
Reaction ${}^1H(\vec{d},p)pn$ at 16 MeV 1243
Ronald E. Brown, G. G. Ohlsen,
F. D. Correll, R. A. Hardekopf,
and Nelson Jarmie

5.17 Faddeev Calculations for the Breakup Reaction
${}^1H(\vec{d},p)pn$ at 16 MeV 1246
F. D. Correll, Ronald E. Brown,
G. G. Ohlsen, R. A. Hardekopf,
Nelson Jarmie, and P. Doleschall

5.18 Tensor Analyzing Powers in the ${}^1H(\vec{d},pp)n$ Reaction
at 16 MeV   I. The Symmetric, Constant-Relative-
Energy Configurations 1249
F. D. Correll, G. G. Ohlsen,
Ronald E. Brown, R. A. Hardekopf,
Nelson Jarmie, P. Schwandt, and
P. Doleschall

5.19 Tensor Analyzing Powers in the ${}^1H(\vec{d},pp)n$ Reaction
at 16 MeV   II. The Collinear Configurations 1252
F. D. Correll, Ronald E. Brown,
G. G. Ohlsen, R. A. Hardekopf,
Nelson Jarmie, J. M. Lambert,
P. A. Treado, I. Slaus, P. Schwandt,
and P. Doleschall

5.20 Analyzing Powers of Proton Continuum Spectra from
Deuteron Break-Up at 56 MeV 1255
N. Matsuoka, M. Kondo, K. Hosono,
T. Saito, A. Shimizu, K. Hatanaka,
S. Nagamachi, T. Itahashi, H.
Sakaguchi, and F. Ohtani

5.21 Analyzing Power of the ^{3}H$(\vec{p},\gamma)^4$He Reaction Near
      0.9 MeV 1258
      K. Krämer, W. Arnold, H. Berg,
      E. Huttel, and G. Clausnitzer

5.22 Analyzing Power of the Elastic n-^{3}He Scattering . . . 1260
      H. O. Klages, H. Dobiasch,
      R. Fischer, B. Haesner, W.
      Heeringa, P. Schwarz, J.
      Wilczynski, and B. Zeitnitz

5.23 Analyzing Powers of ^{3}He$(\vec{p},p)^3$He Elastic Scattering
      Between 30 and 50 MeV 1263
      J. Birchall, W. T. H. van Oers,
      H. E. Conzett, P. von Rossen,
      R. M. Larimer, J. Watson, and
      Ronald E. Brown

5.24 Analyzing Powers of $^3\vec{\text{He}}(p,p)^3$He Elastic Scattering
      Between 20 and 35 MeV 1266
      R. H. McCamis, P. J. T. Verheijen,
      G. Maughan, N. T. Okumusoglu,
      W. T. H. van Oers, J. M. Daniels,
      and A. D. May

5.25 Cluster-Model Study of the Excited States of
      ^{4}He-^{3}H$(p,n)^3$He Reaction 1269
      H. Furutani

5.26 Vector Analyzing Power for the ^{2}H$(d,n)^3$He Reaction . . 1272
      P. P. Guss, K. Murphy, R. C. Byrd,
      C. E. Floyd, S. A. Wender, R. L.
      Walter, T. B. Clegg, and W. R. Wylie

5.27 Small Angle Neutron Polarization for ^{2}H$(\vec{d},\vec{n})^3$He
      at E_d=8 MeV 1275
      W. Tornow, E. Woye, G. Mack,
      R. L. Walter, P. P. Guss, and
      R. C. Byrd

5.28 Polarization Transfer in the D$(\vec{d},\vec{n})$ Breakup
      Reaction at 50 MeV 1278
      R. L. York, J. C. Hiebert, L. C.
      Northcliffe, and H. L. Woolverton

5.29 Polarization Measurement and Analysis in n-^{4}He
      Scattering at 50 MeV 1281
      R. L. York, J. C. Hiebert, L. C.
      Northcliffe, and H. L. Woolverton

5.30	Spin Dependence in the $^3\vec{\text{He}}$+d Three-Body Break-Up Reaction . . . . *Nazmi T. Okumuşoğlu, A. K. Basak, and C. O. Blyth*	1284
5.31	Asymmetries from the $^4\text{He}(\vec{p},2p)^3\text{H}$ Reaction at 250 and 500 MeV Using Polarized Protons . . . . *M. B. Epstein, D. J. Margaziotis, R. Abegg, D. K. Hasell, W. T. H. van Oers, J. M. Cameron, G. A. Moss, L. G. Greeniaus, and A. W. Stetz*	1287
5.32	Measurement of A_y and A_{yy} in $^4\text{He}(\vec{d},d)^4\text{He}$ Scattering at 12.6 MeV . . . . *J. Birchall, N. T. Okumuşoğlu, M. de Jong, M. S. A. L. Al-Ghazi, and J. S. C. McKee*	1290
5.33	None, One or Two A_{yy}=1 Points in d-α Scattering around E_d=35 MeV . . . . *B. Jenny, W. Grüebler, C. Schweizer, V. König, and P. A. Schmelzbach*	1293
5.34	A Complete Set of d-α Scattering Experiments for the Unique Determination of the M-Matrix . . . . . . *K. Elsener, F. Sperisen, B. Jenny, W. Grüebler, V. König, P. A. Schmelzbach, C. Schweizer, and J. Ulbricht*	1296
5.35	Polarization Observables from the Kinematically Complete Measurement of the Break-Up Reaction $^4\text{He}(\vec{d},p\alpha)n$ at E_d = 18 MeV . . . . *H. Oswald, W. Burgmer, D. Gola, C. Heinrich, H. J. Helten, H. Paetz gen. Schieck, and Y. Koike*	1299
5.36	Vector Analyzing Power at ^{5}He FSI in the d-α Breakup Reaction -The Dependence on the Interacting Pair Angle- . . . . *Y. Koike*	1302
5.37	The Tensor Polarization of $^5\text{Li}(3/2^-)$ Determined by Angular Correlation Experiments in the Reaction α+d→α+p+n . . . . *U. Berghaus, H. Brückmann, P. Lara, and K. Wick*	1305
5.38	Analyzing Power of $^3\vec{\text{He}}(^3\text{He},p)^5\text{Li}$ at 14 MeV and the Reaction Mechanism . . . . *R. Beckmann, D. Fröhling, U. Holm, and H.-G. Körber*	1308

5.39 Studies of $\vec{n}$-^{6}Li Scattering and the Structure of
^{7}Li 1311
    Y. H. Chiu and F. W. K. Firk

5.40 Elastic Scattering of Polarized Tritons by
Helium-4 1314
    Nelson Jarmie, F. D. Correll,
    Ronald E. Brown, R. A. Hardekopf,
    and G. G. Ohlsen

5.41 Measurement of the Depolarization Parameter D in
Elastic ^{9}Be(p,p) Scattering 1317
    G. Roy, L. G. Greeniaus, D. P. Gurd,
    D. A. Hutcheon, R. Liljestrand,
    C. A. Miller, G. A. Moss, H. S. Sherif,
    J. Soukup, G. M. Stinson, H. Wilson,
    and R. Abegg

5.42 Discrepancies in the Polarization of Elastically
Scattered Neutrons from ^{12}C Around 16 MeV 1320
    W. Tornow, E. Woye, and
    R. L. Walter

5.43 The Analyzing Power $A_y(\theta)$ for ^{12}C($\vec{n},n_{0,1}$)^{12}C
from 9 to 16.8 MeV 1323
    E. Woye, W. Tornow, G. Mack,
    C. E. Floyd, P. P. Guss, R. K.
    Murphy, R. C. Byrd, S. A. Wender,
    R. L. Walter, T. B. Clegg, and
    W. Wylie

5.44 Observation of $K_y^{y'}(0°)$ for ^{15}N($\vec{p},\vec{n}$)^{15}O from
12.5 to 16.5 MeV 1326
    R. L. Walter, R. K. Murphy, P. P.
    Guss, C. E. Floyd, R. C. Byrd,
    S. A. Wender, and T. B. Clegg

5.45 Resonances in the Low Energy Elastic Scattering
of Polarized Protons on ^{16}O 1329
    W. Kretschmer and E. Renner

5.46 Study of ^{21}Na by ^{20}Ne ($\vec{p},p_0$) and ^{20}Ne ($\vec{p},p_1$) 1332
    M. Fernandez, G. Murillo, J. Ramirez,
    O. Avila, S. E. Darden, M. C. Rozak,
    J. L. Foster, B. P. Hichwa, and
    P. L. Jolivette

5.47 Validity of a Simple Model for Polarization
Transfer in Deuteron-Stripping Reactions 1335
    L. C. Northcliffe, W. D. Cornelius,
    R. L. York, and J. C. Hiebert

5.48   Asymmetry in the Reaction $\vec{p}+d\to t+\pi^+$ at 0.8 GeV 1338
       Kamal K. Seth, H. Nann, S.
       Iversen, and M. Kaletka

       SECTION 6: BASIC SYMMETRIES   1339

6.A   Parity Violation in the Strong Interaction.   1340
       W. Haeberli

6.B   Parity Violation in the np System at Low Energy . . .   1358
       A. B. McDonald

6.C   Polarization Techniques and Experimental Tests
       of Fundamental Symmetries in Nuclear Physics.   1367
       E. G. Adelberger

6.D   Symmetry Tests in Atomic Physics.   1384
       W. L. Williams

6.E   Parity Violation in Polarized Electron Scattering . .   1400
       C. Y. Prescott

6.F   Symmetry and Coulomb Corrections in Light
       Nuclear Systems   1413
       H. Zankel

6.G   Large Deviations from the Polarization-Analyzing
       Power Equality and Implied Breakdown of Time
       Reversal Invariance   1422
       H. E. Conzett

6.RR   Rapporteur's Report: Basic Symmetries   1429
       S. E. Vigdor

6.1   An Experiment to Measure Parity Violation in the
       $^2H(\gamma,n)^1H$ Reaction.   1436
       E. D. Earle, A. B. McDonald,
       and J. W. Knowles

6.2   Parity Violation in pp Scattering at 45 MeV:
       Discussion of Systematic Error Sources.   1439
       R. Henneck, Ch. Jacquemart,
       J. Lang, M. Simonius, R. Balzer,
       W. Haeberli, S. Jaccard, W.
       Reichart, and Ch. Weddigen

6.3   Test of Parity Conservation in pp Scattering
       at 46 MeV   1442
       P. von Rossen, U. von Rossen,
       and H. E. Conzett

6.4 Calculation of the Parity Violating Asymmetry A_z
     in $\vec{p}$-^{4}He Scattering 1446
         Th. Roser and M. Simonius

6.5 Parity Violation in Proton-Nucleon Scattering at
     6 GeV/C 1449
         E. C. Swallow, J. D. Bowman, C. M.
         Hoffman, R. E. Mischke, D. E. Nagle,
         J. M. Potter, R. L. Talaga, N.
         Lockyer, T. A. Romanowski, D. M. Alde,
         and D. R. Moffett

6.6 Concerning Tests of Time Reversal Invariance via
     the Polarization-Analyzing Power Equality 1452
         Homer E. Conzett

6.7 An Experimental Test of Charge Symmetry in n-p
     Scattering. 1455
         S. E. Vigdor, A. D. Bacher,
         D. DuPlantis, W. W. Jacobs, H.-O.
         Meyer, G. L. Moake, P. Schwandt,
         E. J. Stephenson, L. D. Knutson,
         P. A. Quin, J. Sowinski, B. P.
         Hichwa, and P. L. Jolivette

6.8 No Evidence for Charge Symmetry Violation from
     Differences of the ^{2}H(d,$\vec{p}$) and ^{2}H(d,$\vec{n}$) Polarization . 1458
         H. M. Hofmann and W. Zahn

6.9 Measurement of Polarization Transfer in the
     Charge-Symmetric D($\vec{d}$,$\vec{n}$)^{3}He and D($\vec{d}$,$\vec{p}$)T Reactions. . . 1461
         R. L. York, W. D. Cornelius,
         L. C. Northcliffe, and J. C.
         Hiebert

6.10 Analyzing Powers in the Reactions ^{2}H($\vec{d}$,^{3}H)p and
     ^{2}H($\vec{d}$,^{3}He)n at 56 MeV and Charge Symmetry. 1464
         K. Nisimura, K. Imai, T.
         Matsusue, H. Shimizu, R.
         Takashima, K. Hatanaka,
         T. Saito, and A. Okihana

6.11 Polarization Transfer Coefficient Measurements
     for the ^{3}He($\vec{d}$,$\vec{p}$)^{4}He Reaction. 1466
         R. Detomo, Jr., H. W. Clark,
         T. C. Rinckel, J. C. Brown,
         and T. R. Donoghue

6.12 Charge Symmetric Reactions $T(\vec{d},n)^4He$ and $^3He(\vec{d},p)^4He$ below 6 MeV 1469
H. W. Clark, R. Detomo, Jr., L. J. Dries, T. C. Rinckel, J. C. Brown, and T. R. Donoghue

6.13 Comparisons of Polarizations and Analyzing Powers for the $^{11}B(p,n)^{11}C$ and $^{13}C(p,n)^{13}N$ Reactions 1472
K. Murphy, R. C. Byrd, P. P. Guss, C. E. Floyd, R. L. Walter, S. A. Wender, and T. B. Clegg

6.14 Status of Comparisons Between Polarization and Analyzing Power in (p,n) Reactions. 1475
R. C. Byrd and R. L. Walter

6.15 Analyzing Powers for (p,n) Reactions on Light Nuclei. 1478
K. Murphy, R. C. Byrd, C. E. Floyd, P. P. Guss, R. L. Walter, S. A. Wender, and T. B. Clegg

6.16 Lane Model Constraints on Nucleon-Nucleus Scattering Potentials 1481
R. C. Byrd, R. L. Walter, and S. R. Cotanch

6.17 Independence of Permutation Properties of Observables on Reaction Mechanism 1484
A. M. Yasnogorodsky

6.18 A Comparison of Neutron-Deuteron and Proton-Deuteron Analysing Powers at 14 MeV 1488
A. Chisholm, J. C. Duder, and R. Garrett

CONFERENCE SUMMARY 1491
Summary of the Fifth International Symposium on Polarization Phenomena in Nuclear Physics 1492
S. S. Hanna

Conference Participants. 1502

Author Index 1521

SPONSORS AND STAFF

Conference Host
    Los Alamos Scientific Laboratory of the University of California

Sponsors
    International Union of Pure and Applied Physics
    United States Department of Energy
    United States National Science Foundation
    American Institute of Physics

Industrial Sponsor
    National Electrostatics Corporation

International Advisory Committee
    G. G. Ohlsen (Chairman), LASL, Los Alamos
    E. G. Adelberger, Univ. of Washington, Seattle
    D. V. Bugg, Queen Mary College, London
    G. Clausnitzer, Justus Liebig Univ., Giessen
    T. B. Clegg, Univ. of North Carolina, Chapel Hill
    H. E. Conzett, LBL, Berkeley
    P. Doleschall, Central Research Institute, Budapest
    T. R. Donoghue, Ohio State Univ., Columbus
    J. C. Duder, Univ. of Auckland, Auckland
    D. Fick, Philipps Univ., Marburg
    C. Glashausser, Rutgers Univ., New Brunswick
    G. Graw, Univ. München, Munich
    W. E. Grüebler, ETH, Zürich
    W. Haeberli, Univ. of Wisconsin, Madison
    S. S. Hanna, Stanford Univ., Stanford
    A. D. Krisch, Univ. of Michigan, Ann Arbor
    J. Kuehner, McMaster Univ., Hamilton
    F. Lehar, CEN, Saclay
    J. S. C. McKee, Univ. of Manitoba, Winnipeg
    M. J. Moravcsik, Univ. of Oregon, Eugene
    O. F. Nemets, Ukrainian Academy of Sciences, Kiev
    L. C. Northcliffe, Texas A & M Univ., College Station
    H. Paetz gen Schieck, Univ. Köln, Cologne
    G. R. Plattner, Univ. Basel, Basel
    B. A. Robson, Australian National Univ., Canberra
    S. Roman, Univ. of Birmingham, Birmingham
    J. Sanada, Univ. of Tsukuba, Ibaraki
    G. R. Satchler, ORNL, Oak Ridge
    P. Schwandt, Indiana Univ., Bloomington
    P. Signell, Michigan State Univ., East Lansing
    M. Simonius, ETH, Zürich
    E. Steffens, Max Planck Institute, Heidelberg
    W. T. H. van Oers, Univ. of Manitoba, Winnipeg
    R. L. Walter, Duke Univ., Durham
    H. B. Willard, Case Western Reserve Univ., Cleveland

Local Organizing Committee
- G. G. Ohlsen (Chairman)
- B. E. Bonner
- R. E. Brown
- E. R. Flynn
- G. M. Hale
- R. A. Hardekopf
- N. Jarmie
- N. S. P. King
- P. W. Lisowski
- M. W. McNaughton
- J. M. Moss
- J. E. Simmons

Contributed-Paper Selection
- B. E. Bonner
- R. E. Brown
- E. R. Flynn
- G. M. Hale
- R. A. Hardekopf
- J. M. Moss

Scientific Program
- G. M. Hale (Chairman)
- B. E. Bonner
- N. S. P. King
- J. M. Moss
- G. G. Ohlsen
- J. E. Simmons

Scientific Secretary
- M. W. McNaughton
- Student aides: D. Cremans, E. Milner, J. Riley, W. B. Tippens

Transportation and LASL Tour
- P. W. Lisowski

Convention-Center Facilities
- R. A. Hardekopf (Chairman)
- R. E. Brown
- G. M. Hale
- N. Jarmie
- M. W. McNaughton
- G. G. Ohlsen
- J. E. Simmons
- Student aides: D. Cremans, E. Milner, J. Riley, G. E. Simmons, W. B. Tippens

Welcome Speakers
- H. H. Barschall, Univ. of Wisconsin, on behalf of the International Union of Pure and Applied Physics
- L. Rosen, LAMPF Director, on behalf of the Los Alamos Scientific Laboratory
- J. D. Rogers, State Senator, on behalf of the State of New Mexico

Session Chairmen
- F. P. Brady
- G. Clausnitzer
- S. E. Darden
- T. R. Donoghue
- L. Grenacs
- P. W. Keaton
- F. Lehar
- G. G. Ohlsen
- G. Roy
- J. Sanada
- F. D. Santos
- I. Slaus
- K. Sugimoto
- W. T. H. van Oers

Discussion Leaders
- B. E. Bonner
- T. B. Clegg
- D. Fick
- C. Glashausser
- J. S. C. McKee
- M. Simonius

Administration
- J. Elder
- E. T. Jurney
- M. Peacock
- M. J. Phillips
- L. S. Robinson
- B. H. Talley
- B. Thompson
- P. V. Ungnade
- R. F. Warner
- M. S. Wooten

INTRODUCTION

The Fifth International Symposium on Polarization Phenomena in Nuclear Physics was held in the Sweeney Convention Center in Santa Fe, New Mexico, August 11-15, 1980. There were 331 registrants, which was a record attendance for symposia in this series. This popularity was surely due in part to the cultural and historical interest of the Santa Fe area and the beauty of mountainous Northern New Mexico. Past symposia in the series had been held at Basel, Karlsruhe, Madison, and Zürich.

A broad spectrum of topics was covered during the symposium, with subjects that ranged from the fundamental to the practical and from questions of nuclear structure and reaction mechanisms appropriate to the low and medium energy region to questions of particle physics at high energies. To arrange this program, which consisted of 42 invited talks, six discussion sessions, and six rapporteur talks, a local program committee considered a large volume of input from an international advisory committee. It was inevitable that the final program could not reflect all the suggestions received by the program committee, and decisions not to include a topic or speaker were only arrived at after considerable deliberation.

All manuscripts were submitted to the conference in camera-ready format and have not been proof-read in detail by the editors. However, all contributed papers were refereed for content and relevance by a local committee, and 285 contributions were accepted. The few that were rejected were either irrelevant to the subject matter of the conference or involved gross violations of format or length requirements.

There was no oral presentation of contributed papers in the plenary sessions; however, their subject matter was considered in the disucssion sessions and reported to the symposium by the rapporteurs. Many contributors took advantage of the opportunity to display posters, and about 170 contributions were so displayed. This conference marked the first use of poster sessions at a Polarization Symposium. Our arrangement permitted good exposure of the posters throughout the meeting and was successful and well received.

A highlight of the conference was the tour of facilities at the Los Alamos Scientific Laboratory followed by an outdoor barbque dinner near the Indian ruins at Bandelier National Monument. Also enjoyed was an unusually successful banquet followed by a performance by Flamenco dancers. A visit to Las Golondrinas, a living museum of an old Spanish village, was also on the social program, and several participants took advantage of the opportunity to attend the highly popular Santa Fe Opera or the Indian Market immediately following the conference. We are only sorry that crowded conditions in the hotels prevented some who wished to stay in Santa Fe after the conference from doing so.

We express deep gratitude to all who worked so hard to make the Fifth Symposium a success. Credit is given to many in the Sponsors and Staff section. We also thank Margaret Biava and

the staff of Discover Santa Fe, Inc., for their expert help with
the social program and transportation, and Charlotte Hall and the
staff at the Sweeney Convention Center for excellent service throughout the symposium. The following organizations provided us with
much needed financial aid: The International Union of Pure and
Applied Physics, the U. S. Department of Energy, the U. S. National
Science Foundation, the National Electrostatics Corporation, and
P and MP Divisions of the Los Alamos Scientific Laboratory.

These Proceedings of the Fifth International Symposium on
Polarization Phenomena in Nuclear Physics are organized by subject
matter into two volumes and into six sections. Within a section,
the order is as follows: first, the invited papers, including
discussions; second, the rapporteur's report on the relevant discussion session; and third, the contributed papers. The numbering
system used in the advance copy of the contributed papers distributed at the conference has been retained in the Table of Contents
for use as a secondary reference--several authors have referred to
contributed papers by this means. A complete author index is included at the rear of each volume.

In order to help establish the site of the 1985 conference,
we undertook to solicit opinion and comment as to the most appropriate available location. Considerable input was received, but
throughout the conference a consensus appeared to develop that
the next conference should take place in Japan. Professor H. H.
Barschall kindly consented to address the conference on this subject
and to obtain informal ratification of this choice. The attendees
overwhelmingly endorsed the proposal.

Finally, I would like to make a few personal remarks to my
colleagues in polarization physics. This conference marks the
termination of my 20-year career in nuclear physics, which was begun at the University of Texas at Austin, then continued at the
Australian National University in Canberra, and has continued for
the last 15 years at the Los Alamos Scientific Laboratory. In the
past several years, I have become increasingly interested in financial investments with a view toward self-preservation in the face
of an increasingly inflationary economy. I have decided that my
full attention is now required to maintain and expand these ac-
tivities. I have enjoyed my association with many of you, and with
polarization physics in general, and I hope to continue to have
contacts with many of my good friends in physics in the years ahead.

                                          Gerald G. Ohlsen
                                          711 Central Avenue
                                          Los Alamos, New Mexico   87544

                                          December, 1980

OPENING ADDRESS

The INTERNATIONAL POLARIZATION SYMPOSIA 1960-1980: AN OVERVIEW

L.C. Biedenharn*
Physics Department, Duke University
Durham, North Carolina 27706 U.S.A.

INTRODUCTION

This is the fifth of an ongoing series of symposia on "Polarization Phenomena in Nuclear Physics" which were begun in Basel in 1960 by the late Professor Paul Huber. It is my privilege to respond to the request of the organizing committee to survey these past four conferences. My credentials for this are it seems only two: I spoke at the initial Basel conference, and I also had the opportunity to join in an early review paper with the founder, Paul Huber; my only other possible qualification is the impartiality born of having not been active in the field recently.

What I would like to do is to survey in broad strokes the problems, accomplishments, and trends in the field as exemplified by the four past conferencs:

Basel (July 1960)(P.Huber and K.P.Meyer)
Karlsruhe(Sept.1965)(P.Huber and H.Schopper)
Madison(Sept.1970)(H.H.Barschall and W. Haeberli)
Zurich(Aug.1975)(W.Grüebler and V.König)

(The names cited are the editors of the conference proceedings).[1-4]

My task is made much easier by the fine survey[5] Barschall gave at the fourth conference, which included an overview of the conferences at that time.

First the statistics:

TABLE 1

SYMPOSIUM	ATTENDANCE	CONTRIBUTED PAPERS		PROCEEDINGS	EXP/THEORY RATIO
		EXPERIMENTAL	THEORETICAL	PAGES	
1	174	42	9	436	5:1
2	200	94	13	535	7:1
3	239	130	38	930	3½:1
4	247	177	51	930	3½:1
5	331	219	66	1600	3½:1

*Supported in part by the National Science Foundation.

0094-243X/81/690002-8$1.50 Copyright 1981 American Institute of Physics

One sees from this that the conferences grew rapidly but have now levelled off; that experimental work dominates (as it should--physics is based on experimental reality); that theory is a healthy 25% or so.

It is also not too hard to characterize broadly the individual conferences. The first conference, as noted by Fleischmann,[6] was devoted largely to discussing <u>experimental possibilities</u> for the polarization equipment then under construction at a half-dozen laboratories. The theory presented at the conference was largely on the nucleon-nucleon scattering problem, and on polarization in direct reactions as a useful experimental technique. The second conference was even more heavily experimental as the polarization equipment was perfected, tested, and began to be exploited. The third and fourth conferences are typified by the flood of new data, and theory made a come-back via increasingly detailed analysis.

Barschall noted that there was a trend toward theoretical work being done by the experimenters themselves, and he worried that "full-time theorists" might be leaving the field. In my view the possibility that experimenters be their own theorists is one of the healthiest features of polarization physics; contrast this with high-energy physics with its enormous teams of increasingly narrow gauge specialists (including, sadly, grant-swingers and hype-artists). Surely this latter is the more unhealthy trend!

What can one say as to the accomplishments? Physics progresses primarily through new experimental tools and techniques, and from this viewpoint the existence of dozens of laboratories with almost miraculous polarization facilities is a genuine triumph for our field. Polarization studies have proven to be absolutely essential in unravelling nuclear structure--the primary aim of our efforts being, as Huber[7] put it, paraphrasing Goethe, to learn "was die Kerne im Innersten zusammenhält".

Let me survey now the principal topics of the conferences. These divide into three broad categories:
1. *Fundamental Problems*
    Symmetries:Parity conservation,Time reversal
        invariance,Charge symmetry
    Nucleon-Nucleon interaction.
2. *Nuclear Spectroscopy*
    Spins, parities, electromagnetic moments,
        Giant-resonances, IAS;
    Theoretical tools for analysis: optical
        potentials, DWBA,multistep and coupled
        channel codes, D-state effects.
3. *Polarized Beam and Target Technology*

What sort of trends can one see in the treatment of these topics? (I exclude the third topic as outside my competence.)

(a) One sees right away that the trend, noted by Barschall[5], for authors of experimental papers to be also authors of the theoretical analysis, has continued, abetted by the development (and availability) of *large* computer programs. (There are many examples of this in the papers of Section 2.) Moreover there is a trend to experimental-theoretical collaboration in designing the experiment as well as in carrying out the analysis afterward.(Papers 5.18, 5.19 are a good example.)

(b) There is also a trend toward using polarization as a *qualitative* tool to obtain quantitative information. This is particularly clear in determining J-values. Earlier methods using empirical effects seen in angular correlation studies have been completely eliminated in favor of analyzing power(A_y) measurements, as discussed in recent reviews[8,9]. (Papers 2.114, 2.118 are good examples.)

(c) It has been emphasized, repeatedly, from the beginning of these conferences that polarization depends very sensitively on differences between amplitudes, and exploiting this feature has been a continuing trend. (The collaborations mentioned in (a) aims at using just this feature.) An outstanding example of the sensitivity of polarization to fine details is in the Stanford and TUNL studies of Giant Resonance phenomena with polarized beams. A recent TUNL result[10] detects a giant quadrupole (GQR) contribution of only 0.15%(in cross-section) to a 19 Mev giant dipole resonance (GDR) in ^{41}Ca--this is spectacular sensitivity in a region where orders of magnitude (almost) reigned previously.

(d) As experimental tools become more highly perfected there is a trend toward elegant experiments using conceptually simple ideas(like the electron beam interference experiment in the Aharanov-Bohm effect). Polarization physics has its share here too. At the last conference there were two contributions[11,12] on a novel way to measure unnatural spin-parity states, a method which appealed very much to me. Since the published theory[13] behind the experiment was not very intuitive(involving "Clebscherei" to use the current jargon), I hope I will be forgiven if I present here a more directly physical proof.

The experiment considers a ($\vec{d},\alpha$) reaction using a (tensor) polarized deuteron beam on a spin-zero target, and the object is to determine the spin-parity of the final excited states. Let us consider, as a definite example, the reaction: $^{12}C(\vec{d},\alpha)^{10}B*$ at say, 15 MeV beam energy.

Now the theoretical discussions at our conferences have used quite often, the simple(but basic) symmetry first stated by Aage Bohr[14,15]. Consider any two-body nuclear reaction for which both parity and angular momentum are conserved, and for which the initial and final particles have definite(but arbitrary) spin projections along the y-axis(Madison convention, i.e., the normal to the reaction plane). *Then the reaction plane is a plane of reflection symmetry.* Since a reflection in the ($\vec{k}_{in}$, $\vec{k}_{out}$) plane (that is, $\hat{y}\to-\hat{y}$) is equivalent to a parity operation:$(\hat{x},\hat{y},\hat{z}\to(-\hat{x},-\hat{y},-\hat{z})$ followed by a rotation of π around the y-axis ($\hat{x}\to-\hat{x},\hat{z}\to-\hat{z}$) we easily obtain:

$$\Pi_{initial}\exp(i\pi S_i) = \Pi_{final}\exp(i\pi S_f), \quad (1)$$

where Π_i (Π_f) is the product of the initial(final) intrinsic parities and S_i (S_f) the sum of the initial(final) spin projections.

The key idea in this novel version of the $(\vec{d},\alpha)$ reaction is to consider *forward scattering*(collinear momenta). Observe now that the Bohr symmetry remains valid, but now *all planes through the collinear momentum axis ($\hat{z}$-axis) are planes of reflection symmetry*, provided that (according to our initial assumptions on spin projections) the deuteron is in a pure m=0 eigenstate, *measured now along the $\hat{z}$-axis*.

Using once again the fact that reflection in a plane is equivalent to a parity-operation followed by a rotation of π around the y-axis (or any axis perpendicular to the collinear axis) we find:

$$\Pi_i \exp i\pi\Sigma J_i = \Pi_f \exp(i\pi\Sigma J_f), \quad (2)$$

where ΣJ_i (ΣJ_f) is the sum of the initial(final) spins (not spin projections!). For the example $^{12}C(\vec{d},\alpha)^{10}B*$ we get:

$$\Pi(^{10}B*) = (-1)^{J(^{10}B*)+1} \quad (3)$$

which is the "unnatural parity" condition stated.

The sole difference between equations (1) and (2) is that in the latter case we rotate a state *having spin projection m=0 along the $\hat{z}$-axis* by π around the $\hat{y}$-axis. (This uses the general result: $R_y(\pi)|J,M\rangle = (-1)^{J-M}|J,-M\rangle$).

(e) As a final trend in the conferences, let us note a decrease in interest in the nucleon-nucleon interaction *per se*.[16] This, I believe, is symptomatic of something deeper: *the complete change in the conceptions of hadron structure that has occurred over the span of these conferences*. When the conferences began in 1960, the proton had been found a few years earlier to have a size--but this was viewed, unproblematically, as just the meson cloud to be expected around a fundmental point nucleon. The facts are now different: we know now that the size of hadrons comes entirely from quark constituents(the meson cloud was, and is, untenable). This makes a profound difference in attitude, for now the nucleon-nucleon interaction is not elementary at all, no more elementary than the inter-atomic potential of two hydrogen atoms. In a very real sense the use of the nucleon-nucleon interaction in nuclear structure physics is a close analog to the use of the optical model in nuclear reaction analyses; both are heuristic and both necessarily have inherent limitations. In neither case can one expect to treat too fine detail without revising the model, just as polarization calculations with optical potentials of composite particles (say Li^7) are found to have inherent inaccuracies.

The program of calculating the nucleon-nucleon interaction by range (that is, pion exchange at large distances, then ω and ρ exchange,...) is changed fundamentally.[17] Hadrons are now to be viewed as "bags" containing quarks, and the bags are <u>not</u> small (radius ~1 Fermi). At close distances the bags are touching and meson exchange becomes meaningless.

This makes a profound change in the way one views the origin of the hadronic spin-spin and spin-orbit coupling. To appreciate this, consider the qualitative problem of the spin-orbit interaction in *nuclear matter*. In uniform nuclear matter, as Fermi pointed out[18], there is *no* $\underline{L} \cdot \underline{S}$ interaction (since there is no intrinsic way to define $\underline{L}$, i.e., no distinguished origin). But the existence of a nuclear surface does define an intrinsic direction (a gradient) and hence (using also the momentum vector) an $\underline{L} \cdot \underline{S}$ interaction, involving the radial derivative of the potential(matter distribution). Thus we can see, as nuclear physicists, that it is the *surface* of the hadron bag that will eventually be the key to sub-nuclear structure!

At the fourth conference, Krisch[19] gave a preliminary report on 6 GeV polarized beam-polarized target nucleon-nucleon scattering; at these energies asymptotic trends appear to have set in. These experiments have been completed in the interim, and the results[20] are unexpected and exciting. At 1 GeV momentum transfer(where the two bags are touching) there are *large* spin-spin effects, and for Ann the NP results differ both in sign and magnitude (factor of 3) from the PP results. This effect was predicted by no model, and contradicts some models(Regge exchange). There are indications these effects can be explained by a model where the bag surface is region of vacuum

($q\bar{q}$) pairs.[21,22]

Taking the quark-bag concept seriously, and recognizing the analogy to 1950's nuclear fluid models, one might say, with some justice, that particle physics (high-energy physics) is but a new version of nuclear physics at a different scale.

Recognizing these trends in no way diminishes the importance of nuclear polarization physics, and the importance of our attempts to understand the nucleus.

I look forward to learning more new information, and more clever experimental techniques at this meeting.

ACKNOWLEDGEMENTS

The favor of discussions with Professors T.B. Clegg, R.L. Walter and H.R. Weller of TUNL, and with Dr. Michael Danos of the National Bureau of Standards are gratefully acknowledged.

FOOTNOTES AND REFERENCES

1. "Proceedings of the International Symposium on Polarization Phenomena of Nucleons,"(eds. P.Huber and K.P. Meyer), Helv.Phys.Acta, Supplementum VI,(Birkhäuser Verlag, Basel) 1961.
2. "Proceedings of the Second International Symposium on Polarization Phenomena of Nucleons," (eds. P.Huber and H. Schopper),(Birkhäuser Verlag,Basel)1966.
3. "Polarization Phenomena in Nuclear Reactions," (eds. H.H. Barschall and W. Haeberli), (Univ. of Wisconsin Press, Madison) 1971.
4. "Proceedings of the Fourth International Symposium on Polarization Phenomena in Nuclear Reactions," (eds. W. Gruebler and V. König) Experentia Supplementum 25, (Birkhäuser Verlag, Basel) 1976.
5. H.H. Barshcall, ref.(4) above, p.427ff.
6. R. Fleischmann, ref.(3) above, p.397.
7. P. Huber, ref.(2) above, p.15.
8. Charles Glashausser, "Nuclear Physics with Polarized Beams," Ann. Rev. Nucl. Part. Sci. 1979, $\underline{29}$:33-68.
9. W.J. Thompson and Thomas B. Clegg, "Physics with Polarized Nuclei," Phys. Today, 32-39, February,1979.
10. H.R. Weller, invited paper to be presented at this conference.
11. D.O. Boerma,W. Gruebler,V. König, P.A. Schmelzbach, and R. Risler, ref.(4)above, 693-694.
12. J.A. Kuehner, P.W. Green, G.D. Jones, D.T. Petty, J.Szucs and H.R. Weller, ref.(4)above,695-696.
13. J.A. Kuehner, P.W. Green, G.D. Jones, and D.T. Petty, P.R.L. $\underline{35}$, 423-426(1975).

14. A. Bohr, Nucl. Phys. $\underline{10}$,486 (1959).
15. H.A. Weidenmüller, ref.(2)above, p.224 cites another interesting application of the Bohr symmetry.
16. There is only one contributed paper on the nucleon-nucleon interaction at this conference. This is paper 1.1 on the Paris NN potential; this potential is partly phenomenological, partly dispersion-theoretic and not directly or solely based on boson exchange.
17. J.J.de Swart, G. Austen, $\underline{P}$.J. Mulders and T.A. Rijken, "Quarks, the NN and N$\overline{\text{N}}$ Interaction and Nuclear Physics", International Symposium on Few Particle Problems in Nucl. Phys." Dubna(U.S.S.R.)June 5-8(1979), to be published. We wish to thank Professor Johann de Swart for correspondence on this topic.
18. E. Fermi, N. Cimento Supplementum, Ser.10,$\underline{2}$, 91ff, (1955).
19. A.D. Krisch, ref.(4)above, 41-47.
20. D.G. Crabb, et al., P.R.L. $\underline{43}$, 983-986,(1979).
21. Michael Danos, private communication.
22. A gluon-theoretic quark interchange model[S.J. Brodsky, C.E. Carlson, and H. Lipkin, Phys. Rev. $\underline{D20}$, 2278-2288,(1979)]also fits the data. In nuclear physics language this model is *in effect* (not a priori!) non-local, non-relativistic, spin-isospin exchange in the Wigner SU4 super multiplet symmetry. We wish to thank Professor Harry Lipkin for sending us (p)reprints on this model.

DISCUSSION

MUKHOPADHYAY: While thanking the speaker for his concise summary of the past achievements of this conference series, I would like to take issue with him on his optimism that the theories of the internal structure of hadrons (quark model, bag, etc.) have already replaced the N-N interaction problem as the fundamental one (in the sense that the Van der Waals force can be derived from QED). QCD and bag model cannot, at the moment, treat the two-nucleon problem and, in particular, it cannot describe a system like the deuteron. Amongst the pathologies in the bag model, one does not know how to account for the special role of pions in the PCAC. So, this may be the future, (in fact, it must be) but we are not there yet. Should the polarization phenomena discussed here have clear bearing on this, those bearings would be very interesting to explore.

BIEDENHARN: You are quite right that we are very far from being able to calculate the nucleon-nucleon interaction from first principles using quarks. However, some results along these lines have already been obtained. The three most successful NN interactions are the Paris potential, the potential found by the Bonn group and the Nijmegen potential. The Nijmegen potential has the best X^2; for this potential there are nonets of PS, S and V mesons. The quark-bag model has contributed to the determination of the NN potential in several ways: (a) a resolution to the long-standing puzzle of the scalar mesons; (b) form factors for the coupling constants (vertices); and (c) di-baryon resonances (still tentative).
[A discussion of the use of the bag model in the Nijmegen NN calculations is in the paper of de Swart et al., Ref. 17, above.]

ADELBERGER: It is premature to conclude that QCD has solved the N-N force. Malcolm Harvey has recently looked at the N-N force from the point of view of a six-quark system. He doesn't get any significant attraction!

BIEDENHARN: I do not know of the calculation to which you refer, but I am rather skeptical of the result you stated since, I believe, it contradicts the work of de Tar, Jaffe, Shatz, among others, on the properties of NN (six quark) systems. You are correct though, that such calculations are only at the very beginning stage.

SECTION 1

NUCLEON-NUCLEON AND HIGH ENERGY

POLARIZATION PHENOMENA

SPIN CORRELATION MEASUREMENTS IN NUCLEON-NUCLEON SCATTERING AT HIGH ENERGY

D.G. Crabb
University of Michigan, Ann Arbor, Michigan 48109

ABSTRACT

Recent high energy measurements of spin correlation parameters are reviewed and discussed in terms of recent theoretical models.

INTRODUCTION

In this review I shall consider recent measurements of spin correlation parameters above 3 GeV/c. I have defined high energy to start at 3 GeV/c so that I can discuss an energy region which would not otherwise be covered at this conference. The region below 3 GeV/c is well covered: a large number of papers have been contributed with much new data in the range 500 MeV/c to 2.75 GeV/c and two review papers also discuss interesting phenomena in this region.

The advent of accelerated beams of polarized protons together with the availability of polarized targets has allowed rather precise measurements of the initial spin state correlation parameters such as A_{NN} and A_{LL} in elastic scattering. This has superseded the more traditional (and much less precise) measurements of the final state spin correlation parameters C_{NN}, C_{LL} etc. Of course the initial and final state parameters are equal by time reversal invariance.

At higher energies the large loss of event rate in the rescattering of both final state particles makes impossible any meaningful measurements of C_{NN} etc. Thus at high energies the only possible spin correlation measurements are in the initial state using a polarized beam and target. The only machine ever to accelerate polarized protons to high energy was the ZGS at Argonne National Laboratory which was shut down in October 1979. All the measurements I shall discuss were done at the ZGS.

SPIN PARAMETERS

Convention

I shall use the Ann Arbor Convention[1] for spin parameters. Here A_{NN}, for example, refers to a scattering asymmetry measurement in which the spins are aligned in the initial state whereas C_{NN} is obtained from rescattering the final state nucleons. The convention is not universally accepted and some groups, for historical reasons, still prefer to use C_{NN} instead of A_{NN}.

Amplitudes

I shall refer to the usual set^2 of s channel helicity amplitudes and associated t channel exchange amplitudes.

$$\begin{array}{ll}
\underline{\text{s channel}} & \underline{\text{t channel}} \\
\varphi_1 = \langle ++|++\rangle & N_o = \tfrac{1}{2}(\varphi_1+\varphi_3) \\
\varphi_2 = \langle --|++\rangle & N_1 = \varphi_5 \\
\varphi_3 = \langle +-|+-\rangle & N_2 = \tfrac{1}{2}(\varphi_4-\varphi_2) \\
\varphi_4 = \langle +-|-+\rangle & U_o = \tfrac{1}{2}(\varphi_1-\varphi_3) \\
\varphi_5 = \langle ++|+-\rangle & U_2 = \tfrac{1}{2}(\varphi_4+\varphi_2)
\end{array}$$

Observables

$$\sigma = \tfrac{1}{2}\left(|\varphi_1|^2+|\varphi_2|^2+|\varphi_3|^2+|\varphi_4|^2+4|\varphi_5|^2\right)$$

$$\sigma A = -\text{Im}\left(\varphi_1+\varphi_2+\varphi_3-\varphi_4\right)\varphi_5^*$$

$$\sigma A_{NN} = \text{Re}\left(\varphi_1\varphi_2^*-\varphi_3\varphi_4^*+2|\varphi_5|^2\right)$$

$$\sigma A_{LL} = \tfrac{1}{2}\left(-|\varphi_1|^2-|\varphi_2|^2+|\varphi_3|^2+|\varphi_4|^2\right)$$

$$\sigma A_{SS} = \text{Re}\left(\varphi_1\varphi_2^*+\varphi_3\varphi_4^*\right)$$

$$\sigma A_{SL} = \text{Re}\left(\varphi_1+\varphi_2-\varphi_3+\varphi_4\right)\varphi_5^*$$

Measurements

A spin correlation parameter A_{II} is defined in terms of pure initial spin cross sections

$$A_{II} = \frac{\left.\frac{d\sigma}{dt}\right]_{\uparrow\uparrow}+\left.\frac{d\sigma}{dt}\right]_{\downarrow\downarrow}-\left.\frac{d\sigma}{dt}\right]_{\uparrow\downarrow}-\left.\frac{d\sigma}{dt}\right]_{\downarrow\uparrow}}{\left.\frac{d\sigma}{dt}\right]_{\uparrow\uparrow}+\left.\frac{d\sigma}{dt}\right]_{\downarrow\downarrow}+\left.\frac{d\sigma}{dt}\right]_{\uparrow\downarrow}+\left.\frac{d\sigma}{dt}\right]_{\downarrow\uparrow}}$$

but since in an actual experiment the only thing which is changed for each measurement is a spin direction

$$A_{II} = \frac{1}{P_B P_T}\frac{N_{\uparrow\uparrow}+N_{\downarrow\downarrow}-N_{\uparrow\downarrow}-N_{\downarrow\uparrow}}{N_{\uparrow\uparrow}+N_{\downarrow\downarrow}+N_{\uparrow\downarrow}+N_{\downarrow\uparrow}}$$

where P_B and P_T are the beam and target polarizations.

Another parameter which is used is the ratio of the spin parallel cross section to the spin antiparallel cross sections, r_{II}, where

$$r_{II} = \frac{\left.\frac{d\sigma}{dt}\right]_{\uparrow\uparrow} + \left.\frac{d\sigma}{dt}\right]_{\downarrow\downarrow}}{\left.\frac{d\sigma}{dt}\right]_{\uparrow\downarrow} + \left.\frac{d\sigma}{dt}\right]_{\downarrow\uparrow}} = \frac{1 + A_{II}}{1 - A_{II}}$$

SPIN CORRELATION MEASUREMENTS IN ELASTIC SCATTERING

Small Momentum Transfer

One of the original motivations for a high energy polarized beam was for a detailed study of the dynamics of the nucleon-nucleon interaction. A major part of this effort was to obtain a complete amplitude analysis for p-p elastic scattering at small t. The various spin parameters contain different combinations of the five complex amplitudes and a measurement of a sufficient number of them enables the analysis to be made. However the structure is so rich that measurements are necessary over a range of momentum transfer and energy. Unfortunately some of the spin parameters are very difficult and time consuming to measure so a detailed study was undertaken only at 6 GeV/c and 11.75 GeV/c in order to get some idea of the energy dependence. Prior to this a large amount of data on the analysing power and a small amount on the depolarization parameter D_{NN} had allowed some limits to be put on the amplitudes.

The program was carried out mainly by the Argonne group of Yokosawa et al. and a sufficient amount of data has been analyzed to allow a reasonable description of the amplitudes. The result of such an analysis by Berger et al.[3] is shown in Fig. 1 for $-t = .4(GeV/c)^2$. Similar analyses have been carried out by Kroll et al.[4] and Wakaizumi and Sawamoto[5]. There seems to be general agreement among the analyses though they differ in details. This again is a reflection of the complicated nature of the nucleon-nucleon interaction and the fact that the errors on the measurements still leave room for maneuver.

Wakaizumi and Sawamoto contributed a paper to this conference with details of their analysis and Fig. 2 shows the fits to the various spin parameters together with the currently available data. The curves match the overall trend of the data but details of the structure are generally missed.

Experiments at small momentum transfer

One of the last spin correlation experiments to be done at the ZGS at 6 GeV/c was a measurement of A_{NN} in pp elastic scattering over the momentum transfer squared, $P_\perp^2$, range 1.0-2.4 $(GeV/c)^2$. Although A_{NN} had been measured earlier the data for $P_\perp^2 > 1.0$ was sparse with large errors and did not extend to $90°$ cm. It was felt necessary to fill in this gap before the end of the ZGS.

The apparatus[6] is shown in Fig. 3. The polarized beam came in from the left and entered the hydrogen target of the polarimeter. The polarimeter used two double arm spectrometers to measure the left-right scattering asymmetry in pp elastic scattering to find the beam polarization P_B where

$$P_B = \frac{1}{A} \frac{L-R}{L+R}$$

A is the previously measured analyzing power and L(R) is the total number of elastic scatters to the left (right).

After the polarimeter, the beam entered the polarized proton target. Elastic scatters from the polarized protons were detected in the two spectrometer arms F and B. Tight constraints on angle and momentum allowed the detection of a clean elastic signal.

The results are presented in Fig. 4. Considerable structure is apparent and a noticeable feature is the rapid rise to $A_{NN} \approx$ 12% at 90 cm. The predictions of three theoretical models are shown. The curves from Kroll et al. (KLS) and Field and Stevens[7] (FS) represent the data quite well up to at least $P_\perp^2 = 1.0 (GeV/c)^2$. The Regge Pole model of Field and Stevens is interesting because it is a prediction from several years ago before most of the spin parameters had been measured.

A somewhat neglected area before the polarized beam came along was the study of spin effects in pn scattering. The use of a polarized proton beam with a liquid deuterium target allows easier and more precise measurements of pn scattering than was possible beforehand. An early measurement by the Argonne EMS group[8] of the pn analyzing power at 2,3,4 and 6 GeV/c showed some surprising results (Fig. 5). It had been expected that either the pn analyzing power would be equal to the pp analyzing power or mirror symmetric with it depending on whether a geometric or Regge Pole approach was used. Clearly it was neither and led to a rapid reappraisal among theorists.

As a further test of the models and probe of the pn system a measurement of A_{NN} was undertaken by the Michigan-Argonne and Rice University groups. The simplest method, at least for the experimenters, was to obtain polarized deuterons from the ZGS and then to use the polarized neutron in the deuteron to interact in the polarized proton target. Using 12 GeV/c deuterons meant that the np interaction was at 6 GeV/c. The apparatus used was essentially the same as shown in Fig. 3 except that a neutron detector was used in the F arm of the spectrometer[9]. In this experiment one nucleon in the deuteron was used in the scattering while the other one continued on relatively unaffected by the interaction. The polarized neutron was allowed to interact in the polarized proton target while its paired proton continued on. The neutron polarization was measured in the polarimeter using the polarized proton in the deuteron for pp elastic scattering. The proton and neutron polarizations are equal.

In the time available it was possible to measure two data

points at $P_\perp^2 = 0.8$ and 1.0 $(GeV/c)^2$ and they are shown in Fig. 6. A_{NN} in pn elastic scattering for this P^2 region is negative with a magnitude of ~20%, twice as large and with the opposite sign to the pp case. The predictions of Berger et al. and Field and Stevens are shown. Again it is interesting that the older model predictions are nicely in agreement with the data while the more recent one with the benefit of a much greater body of spin measurements fails rather badly.

The data from the Rice University experiment is at lower $P_\perp^2$ and unfortunately was not completely analyzed in time for this conference[10].

At present the analyzing power for pn scattering has been measured at a number of energies over a large angular range but no data, except that discussed above, exist for other spin parameters.

Experiments at large momentum transfer

While considerable effort was going into disentangling the amplitudes at small momentum transfer, some groups were engaged in extending spin measurements out to large momentum transfer. Here the hope was that the interaction might be simpler to describe; indeed for pp scattering at $90°$ cm the spin flip amplitude ϕ_5 vanishes and $\phi_3 = -\phi_4$ from symmetry considerations. Further, by going to sufficiently large momentum transfer one might enter a hard scattering region where the interaction takes place between the constituents of the nucleons.

In particular, during the past few years the Michigan-Argonne group has been responsible for pushing the measurements of A_{NN} to increasingly large values of $P_\perp^2$. The apparatus used was essentially that shown in Fig. 3 and the results of this series of experiments[11] are shown in Fig. 7. The graph shows what must be familiar to many people by now, the dramatic rise of A_{NN} to a level of about 60%, close to the limits of momentum transfer available at the ZGS. This was a totally unexpected result and was subsequently interpreted as the onset of a hard scattering region where the spin structure of the nucleon constituents was being probed. In terms of r_{NN} (the ratio of the spin parallel to spin antiparallel cross sections defined earlier) a value of 60% for A_{NN} means $r_{NN} = 4$.

Further investigations of this effect were conducted by measuring A_{NN} for $90°$ cm scattering as a function of beam momentum. Again dramatic structure was seen and is shown in Fig. 8. In the region 4-8 GeV/c $A_{NN} \approx 10\%$; below 4 GeV/c it rises rapidly to a value of about 60%. There seems to be further structure below 2 GeV/c which has recently been investigated in more detail.[12] Above 8 GeV/c A_{NN} also rises rapidly to a level of 60%. The point at 12.75 was the highest energy at which the ZGS could be operated.

The two sets of data are combined in Fig. 9, and plotted as $r_{NN} = \sigma_{\uparrow\uparrow}/\sigma_{\uparrow\downarrow}$. Some additional data at $P_\perp^2 = 4.5$ and 5.09 are included in the 11.75 GeV/c points. The 90^0 cm data are plotted against the equivalent $P_\perp^2$ value. It is clear from the figure that both sets of data have the same structure at high $P_\perp^2$, suggesting that the pure spin cross sections may depend only on $P_\perp^2$ in the hard scattering region. It should be pointed out that if the figure had been plotted against the four momentum transfer squared the two sets of data would not have overlapped.

Inspired by these results and subsequent models which tried to explain them and make predictions for other spin variables, the Argonne group of Auer et al.[13] set up to measure A_{LL} at high $P_\perp^2$ at 11.75 GeV/c.

Their apparatus is shown in Fig. 10. The longitudinally polarized beam was incident on the longitudinally polarized target. The momentum and angle of the forward going particle was measured by the spectrometer magnet and planes of multi-wire proportional chambers (MWPC). A large Cerenkov counter was used for particle identification. The recoil particle was detected by MWPC's and scintillation counters. About 35% of the data have been analyzed and the results are shown in Fig. 11 together with the A_{NN} data. It is evident that as the 90 cm scattering angle is approached there is a sudden change in the structure of A_{LL}. The two sets of data should help to establish the credibility of theoretical models.

Theoretical Interpretation
If the data do indicate the onset of a hard scattering region where the scattering of the constituents is important then the ideas of QCD can be applied to try and understand the spin interactions of the constituents. QCD generally has been applied to processes at large s and t and it is not clear whether the spin data discussed above is in a region of applicability. However in the past, spin data at low energies has signalled changes which have only become evident at higher energies in spin averaged parameters (e.g., structure in the analyzing power at low energies could be related to the emergence of structure in the spin averaged cross sections at higher energies). An indication that the ideas of QCD might be applied comes from the fact that the momentum transfer region corresponds to that where the quark counting rule for fixed angle scattering cross sections applies. Here

$$\frac{d\sigma}{dt}\left\{ A+B \rightarrow C+D \right\} = \frac{1}{s^{n-2}} f\left(\frac{t}{s}\right)$$

with $n = n_A+n_B+n_C+n_D$, the minimum number of fundamental constituents of the composite particles. For pp elastic scattering $n_A=n_B=n_C=n_D=3$ and $n=12$.

$$\frac{d\sigma}{dt}\left(p+p \to p+p\right) \approx s^{-10} f(\theta_{cm})$$

An overall fit to the available data gives $s^{-9.7} f(\theta_{cm})$.

The first attempts at fitting to the data were by Farrar et al.[14] and Brodsky et al.[15] who proposed the quark interchange model (QIM). In this simple picture quarks are interchanged between nucleons in a helicity conserving interaction which is independent of the helicity of the exchanged and spectator quarks. The model makes some specific predictions which can be tested.

The requirements of quark helicity conservation means that the amplitudes ϕ_5 and ϕ_2 are zero for all scattering angles θ_{cm} for pp and np scattering. This has the consequence

$$A = A_{SL} = 0$$
$$A_{NN} = -A_{SS} \quad \text{for all } \theta_{cm}$$

In addition for pp scattering at 90 cm symmetry requirements have

$$\phi_5(90°) = 0$$
$$\phi_3(90°) = -\phi_4(90°)$$

which leads to the model independent sum rule

$$A_{NN}(90°) - A_{SS}(90°) - A_{LL}(90°) = 1$$

Therefore tests for the validity of QIM involve the measurement of A and A_{SL} for $\theta_{cm} \neq 90°$ and require

$$A = A_{SL} = 0$$

Note that for the 11.75 GeV/c data $P_\perp^2 = 5.1$ (GeV/c)^2 corresponds to $90°$ cm

Another test is to measure directly A_{NN} and A_{SS} to check for $A_{NN} = -A_{SS}$ or to use the 90° sum rule and measure A_{LL} and either A_{NN} or A_{SS}.

From the data described above $A_{SS} = -0.6\pm.12 \approx -A_{NN}$ which is in agreement with the idea of quark helicity conservation.

However the simple predictions for the QIM models are

$$A_{NN}^{pp} = \frac{1}{3}$$

$$A_{LL}^{pp} = A_{SS}^{pp} = -\frac{1}{3}$$

$$A_{NN}^{np} = -.44$$

$$A_{LL}^{np} = A_{SS}^{np} = .44$$

Clearly these are not in agreement with the data. The simple QIM models all obtain these results but use different approaches to generate large A_{NN}^{pp} values. Generally this means the introduction of a nonpertubative component to interfere with the main QIM process which will die away with increasing $P_\perp^2$. They also seem to require that in these processes $\phi_2 \neq 0$.

A number of such solutions have been proposed and the predictions of two of them are shown in Figs. 12 and 13 along with the data. The models were generated to fit the A_{NN}^{pp} data but do not agree very well with the A_{LL}^{pp} data.

An interesting attempt to justify the validity of the QIM approach was made by Wolters[16] who used a different mechanism for the quark interchange, namely backward quark-quark scattering via one gluon exchange. The other concepts are retained. Using A_{NN}^{pp} as input the following predictions are made for 12 GeV/c at 90°cm.

$$A_{NN}^{np} = -0.22 \pm .22$$

$$A_{LL}^{np} = 0.57 \pm .22$$

These have not been experimentally verified. It was pointed out by Brodsky et al.[15] that the fixed angle cross section has fluctuations around the s^{-10} prediction and that for the 90° cm cross section one of the more noticeable fluctuations is around 13 GeV/c. It was suggested that this might be related to the large asymmetry in A_{NN}.

Wolters has quantified this and, in his model, A_{NN} and the cross section are related such that

$$s^{10} \frac{d\sigma^{pp}}{dt} \propto \left(1 + r_{NN}^{pp}\right) \propto \frac{1}{1 - A_{NN}^{pp}}$$

so that fluctuations in $s^{10} \frac{d\sigma^{pp}}{dt}$ vs. s should reflect structure in A_{NN}^{pp}. Further at 90°cm, $s = 4(P_\perp^2 + m_p^2)$ and $s^{10}(d\sigma^{pp}/dt)(90°)$ can be plotted vs. $P_\perp^2$. The result is shown in Fig. 14. The prediction has been normalized to the A_{NN} point at $P_\perp^2 = 5.1$. Certainly the coincidence of data and prediction at high $P_\perp^2$ is interesting. Above $P_\perp^2 = 3.5(\text{GeV}/c)^2$ there is agreement, below there is complete disagreement. This is taken to mean that some form of QIM can be applied above $P_\perp^2 \sim 3.5(\text{GeV}/c)^2$. Interestingly the prediction shows $A_{NN} \sim 1/3$ at $P_\perp^2 = 10(\text{GeV}/c)^2$, the value predicted by the simple QIM model.

Finally I should mention an alternative approach, the massive quark model of Preparata and Soffer[17]. This has been promoted as the theory to replace QCD and the predictions for the spin variables are shown in Fig. 15. It appears to fare no better than the QIM approach.

CONCLUSIONS

I have been able to review only a small part of the considerable body of spin data which has accumulated in the past few years. The ZGS and its high energy polarized beam opened many new areas of study which have contributed greatly to our basic understanding of fundamental processes. The structure of the amplitudes at 6 GeV/c are quite well understood and there is more data to come.

Paradoxically the surprises have occurred at the low energy limit and high energy limit for ZGS beams. I have not mentioned the issue of the structure in the $\Delta\sigma_L$ measurements around 1.5 GeV/c and whether it is due to dibaryon resonances because it is a common topic at this conference and will be reviewed by Jay Roberts in the next talk. However I'm sure there will be considerable activity in the future at the lower energy machines to resolve this point.

At the highest ZGS energy and available momentum transfer dramatic structure was seen. This has been linked to the scattering of the constituents of the nucleons. However a simple approach does not explain the data. A number of predictions have been made on various spin correlation parameters but the expectation is that the simple approach is more likely to apply at higher momentum transfer. These ideas should be tested in the near future because the higher P^2 region will be accessible when the 26 GeV/c polarized beam at the Brookhaven AGS starts up in two or three years time.

In addition there are plans for a polarized beam at KEK and a 100-300 GeV/c polarized beam derived from Λ^0 decay at Fermilab, so the future for spin physics looks bright.

REFERENCES

1. Proceedings of Conference on Higher Energy Polarized Proton Beams, Ann Arbor 1977, A.D. Krisch and A.J. Salthouse eds., (AIP, New York 1977), p. 142.
2. F. Halzen and G.H. Thomas, Phys. Rev. D10, 344 (1974).
3. E.L. Berger, A.C. Irving and C. Sorenson, Phys. Rev. D17, 2971 (1978).

4. P. Kroll, E. Leader, and W. Von Schlippe, J. Phys. G. $\underline{5}$, 1179 (1979).
5. S. Wakaizuni and M. Sawamoto, Contributed Paper 1.16 at this conference.
6. A. Lin et al., Phys. Lett. $\underline{74B}$, 273 (1978).
7. R.D. Field and P.R. Stevens, Argonne National Laboratory Report ANL-HEP-CP-75-73 (1975).
8. R.E. Diebold et al., Phys. REv. Lett. $\underline{35}$, 632 (1975).
9. D.G. Crabb et al., Phys. Rev. Lett. $\underline{43}$, 983 (1979).
10. H.E. Miettinen, private communication.
11. D.G. Crabb et al., Phys. Rev. Lett. $\underline{41}$, 1257 (1978).
12. H.E. Miettinen et al., Contributed paper 1.15 to this conference.
13. I.P. Auer et al., Argonne National Laboratory Report ANL-HEP-CP-80-38 (1980).
14. G.R. Farrar et al., Phys. Rev. $\underline{D20}$, 202 (1979).
15. S.J. Brodsky et al., Phys. Rev. $\underline{D20}$, 2278 (1979).
16. G.F. Wolters, Phys. Rev. Lett. $\underline{45}$, 776 (1980).
17. G. Preparata and J. Soffer. Phys. Lett. $\underline{86B}$ 304 (1980).

(Editor's Note: Figures follow. References to this conference are given as paper numbers and are cross-indexed to page numbers in the Table of Contents.)

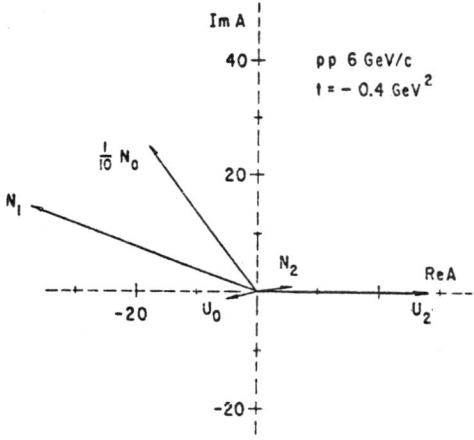

Fig. 1. Amplitude structure from Berger et al.[3]

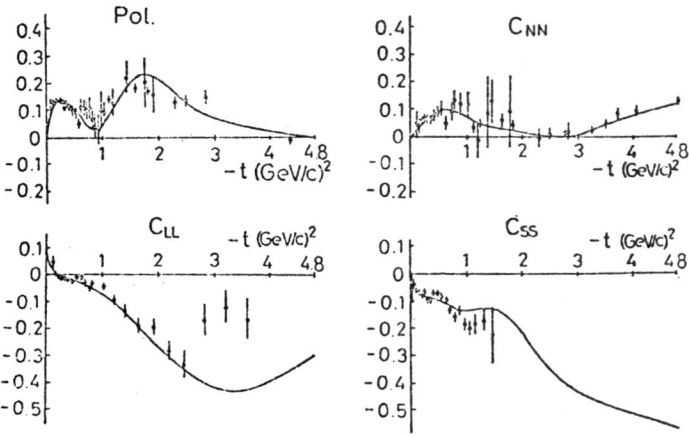

Fig. 2. Spin parameter fits of Wakaizuni and Sawamoto[5]

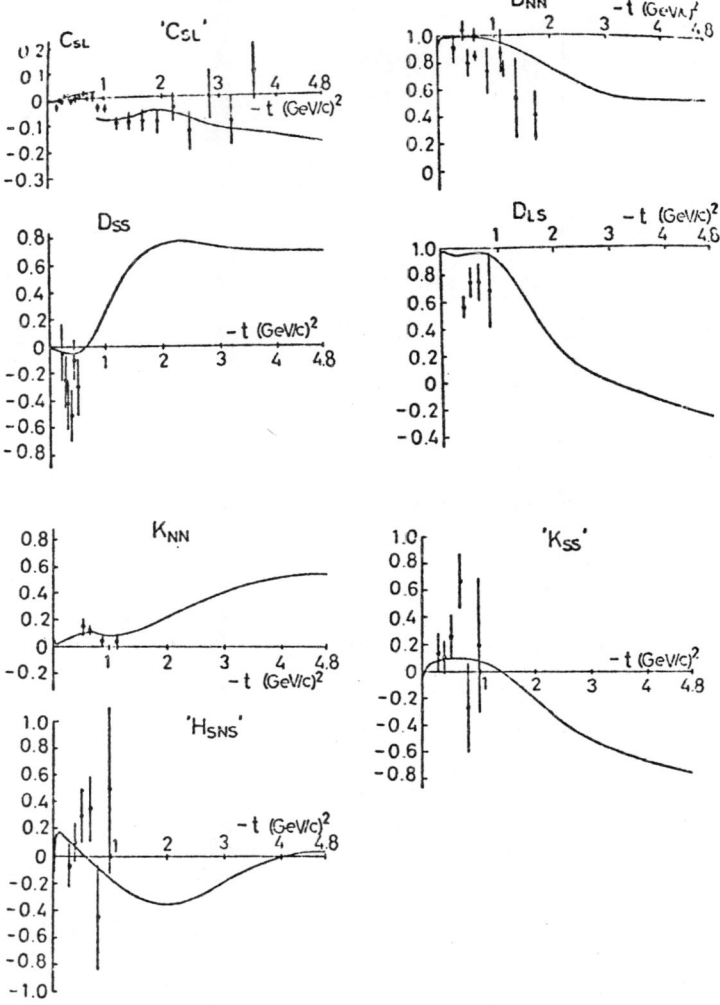

Fig. 2. Continued

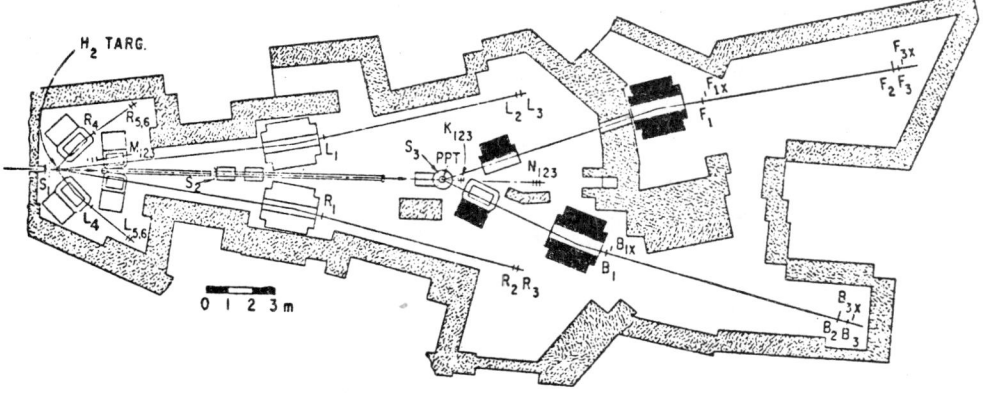

Fig. 3. Layout of the A_{NN} experiment[6]

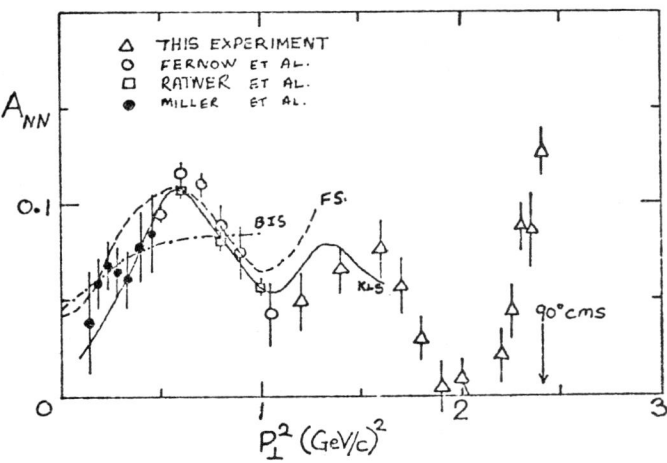

Fig. 4. A_{NN} for pp elastic scattering. The predictions of Berger et al.[3] (BIS), Field and Stevens[7] (FS) and Kroll et al.[4] (KLS) are shown.

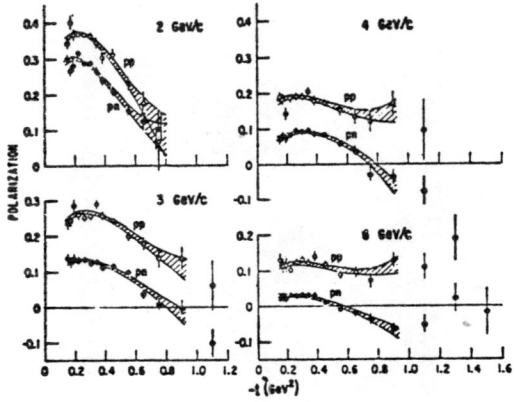

Fig. 5. Analyzing power in pp and pn elastic scattering

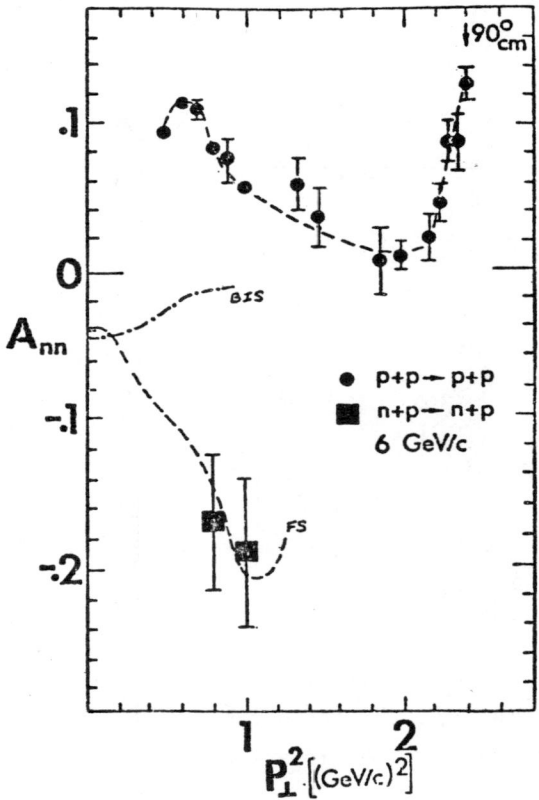

Fig. 6. A_{NN} in np elastic scattering. The predictions of Berger et al.[3] (BIS) and Field and Stevens[7] (FS) are shown. A_{NN} for pp elastic scattering is shown for comparison.

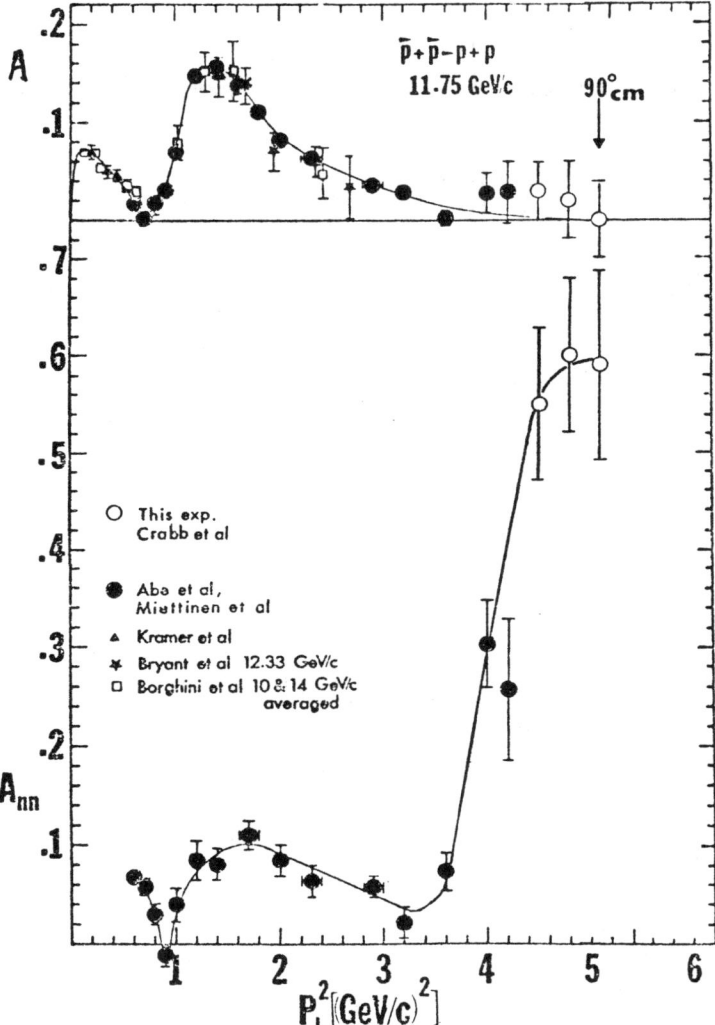

Fig. 7. A and A_{NN} for pp elastic scattering at 11.75 GeV/c

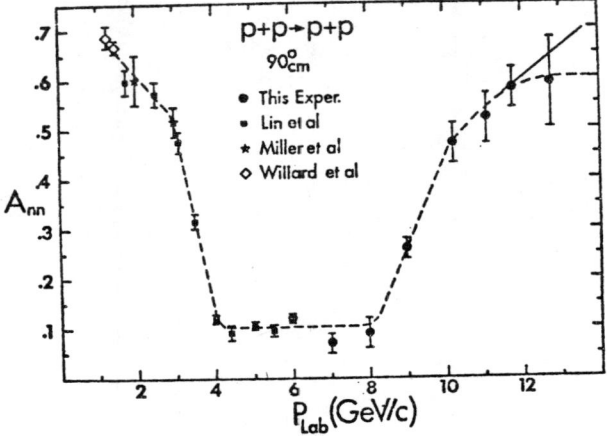

Fig. 8. A_{NN} for pp elastic scattering at $90°$ cm as a function of incident momentum

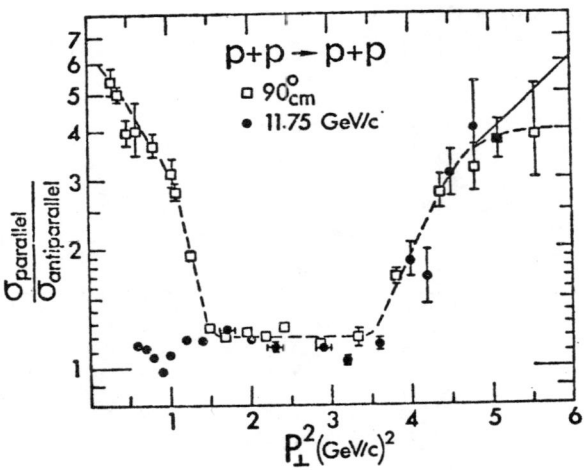

Fig. 9. The cross section ratio $r_{NN} = \dfrac{\sigma_{parallel}}{\sigma_{antiparallel}}$ of the fixed angle ($90°$ cm) and fixed incident momentum (11.75 GeV/c) plotted against the equivalent $P_\perp^2$

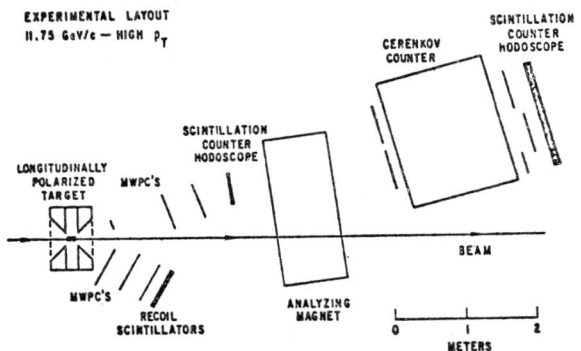

Fig. 10. Layout of the A_{LL} experiment[13]

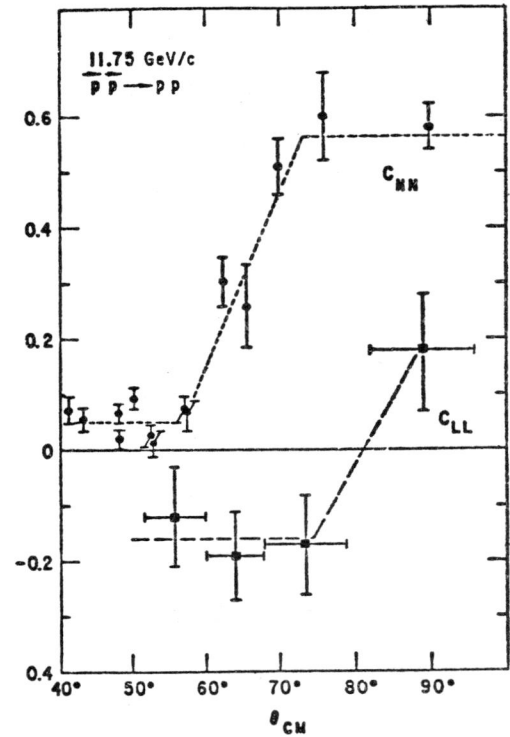

Fig. 11. A_{LL} for pp elastic scattering plotted against cm scattering angle. The A_{NN} data is shown for comparison.

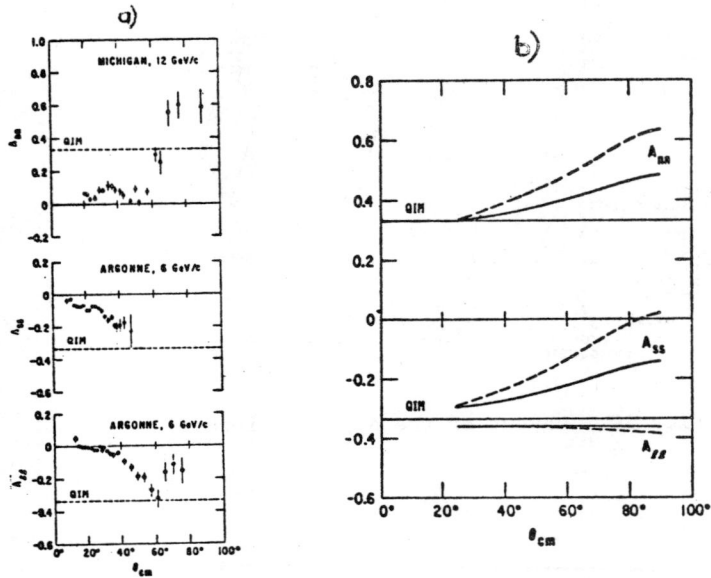

Fig. 12 a) The simple QIM predictions of Farrar et al.[14]
b) The effects of adding instantons to the simple picture

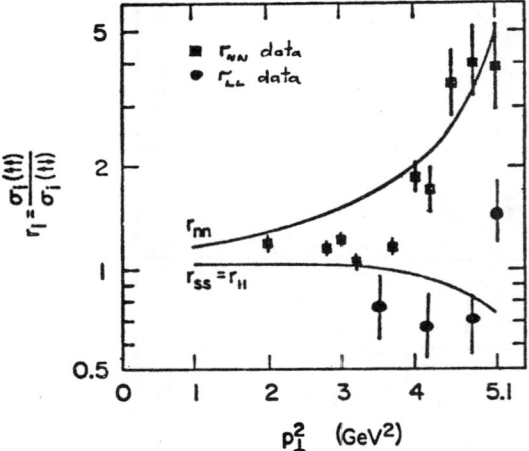

Fig. 13. The QIM predictions of Brodsky et al.[15] compared with the data

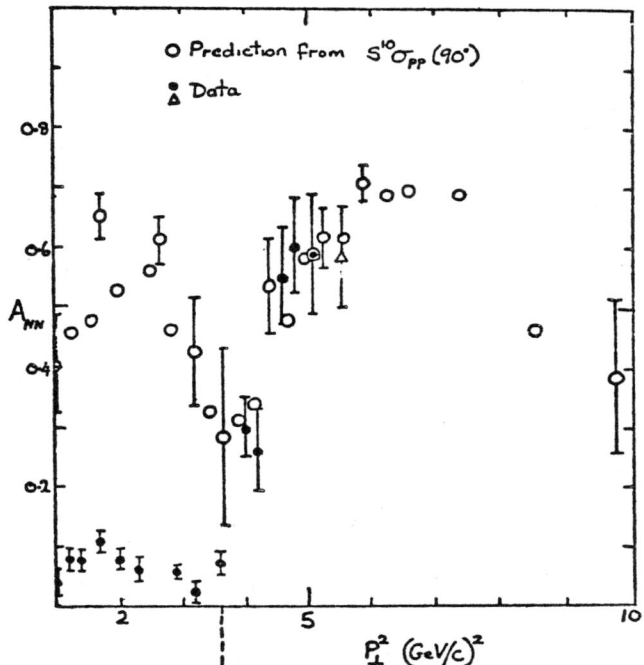

Fig. 14. A prediction of Wolters[16] compared with the data for A_{NN}[9] showing that QIM might be valid above $P_\perp^2 \sim 4 (\text{GeV/c})^2$

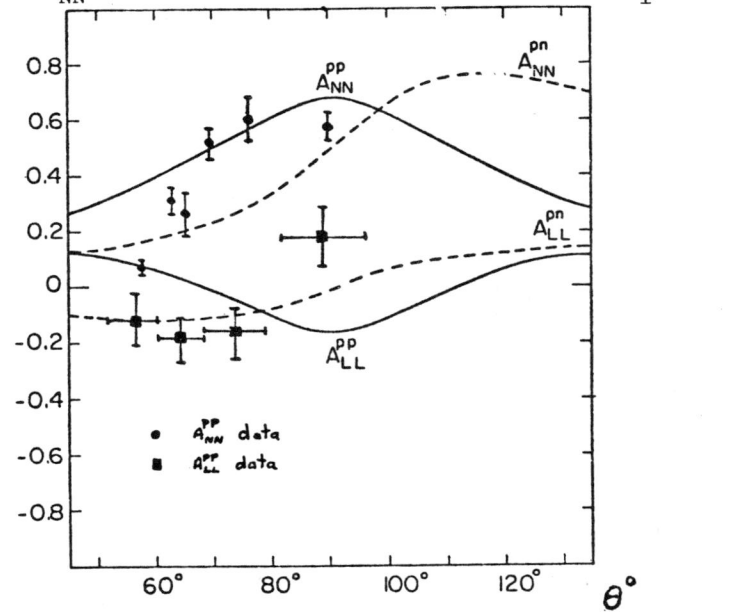

Fig. 15. Spin parameter predictions of the massive quark model of Preparata and Soffer[17] compared with the data

DISCUSSION

BIEDENHARN: There are two comments I would like to make concerning the very nice results presented by Dr. Crabb in his talk. The first concerns the relation he noted between changes in slope of the differential cross section (vs t) and the polarization parameters. This kind of diffraction connection is familiar in these conferences as the "derivative rule" and has been discussed by Miller (p. 410, 2nd Conf.) and by Darden and Haeberli (p. 224, 4th Conf.).

The second remark concerns the theoretical calculations of the coefficients A_{nn}, ..., by the QCD theorists (Farrar, Brodsky, ...). Although the verbal justification for the calculations is quite impressively fancy, what is <u>actually</u> calculated in the "quark interchange model" is the crudest sort of overlap between (NR) SU6 wave functions using the (SU4) product operator of spin- and isospin-exchange. The numerical results agree surprisingly well, but the actual calculation is not at all reliable, being effectively both nonrelativistic and highly nonlocal.

LEHAR: Structure in the parameter C_{NN} can be observed also at lower energies ($\sim$ 140 MeV). The energy dependence of the C_{NN} was plotted by Hess (University of Geneva), and we can see the $C_{NN} \simeq 1$ at 140 MeV.

KLOET: You showed very nice data for np scattering, while the experiment that is really done is dp scattering. How are these np data obtained given the dp data?

CRABB: By using the nucleons in the deuteron for np and pp scattering; one nucleon in the deuteron scatters from the target proton while the other continues on relatively unaffected by the interaction.

EVIDENCE FOR DIBARYON RESONANCES IN NUCLEON-NUCLEON SCATTERING

J. B. Roberts
T. W. Bonner Nuclear Laboratories and Physics Department
Rice University, Houston, TX 77001

ABSTRACT

There has been a revival of interest in the subject of nucleon-nucleon resonances in the past 3 to 4 years largely generated by experimental results from the polarized beam program at the Argonne ZGS. Evidence from experimental results and phase shift and phenomenological analyses incorporating these results regarding the existence of these resonances is summarized.

INTRODUCTION

From its name one might surmise that a dibaryon is an elementary particle with baryon number B=2. Such particles are exotic states in traditional flavor SU(3), which reproduces the mass spectrum of known hadrons very well. Dibaryons were thought not to exist through the 1960's and most of the 1970's. The prejudice against dibaryons was based on experimental data and the desire for the hadron spectrum to obey the simplest possible symmetry scheme. There was no dynamical reason for six quark objects not to exist: the mechanism of binding and confining quarks to form hadrons is still not well understood. Presently several models predict a spectrum of dibaryons in a bag[1,2] or string[3] picture; in other models there are N-Δ or Δ-Δ states bound by meson exchange.[4] We shall primarily discuss evidence for I=1 dibaryons, since the data for proton-proton scattering is much more copious than that for neutron-proton.

THE I=1 SYSTEM

The major reason for the prejudice against the existence of dibaryons was the absence of any structure in the proton-proton total cross section. Whereas the meson-nucleon total cross sections exibit structure due to the s-channel formation of the low lying N^* resonances, the p-p cross section falls up to about 400 MeV, where the scattering is almost purely elastic, rises smoothly by about 20 mb, and then flattens out around 800 MeV (Fig. 1a).[5] The difference between the elastic (Fig. 1b) and total cross sections above 400 MeV is attributed to the onset of pion production, especially the channel pp→nΔ^{++}. It was thus generally believed that there was no structure in nucleon-nucleon scattering other than that due to the NΔ, $\Delta\Delta$, NN*, and N^*N^* thresholds, NΔ being the only one producing noticable energy dependence. There was, however, to the more critical eye structure in several observables centering around 800 MeV (Fig. 2). The total p-p elastic cross section has a broad peak in this vicinity (Fig. 2a).[6] The maximum polarization (-t∼.1-.2) in p-p elastic scattering peaks about 700 MeV[7] (Fig. 2b) and the ratio of the real to imaginary parts of the spin averaged forward scattering amplitude has a zero-crossing at about the same energy[8] (Fig. 2c). Such was the state of affairs in the mid 1970's when systematic measurements of spin observables in p-p scattering were begun in the 1-3 GeV/c energy range.

The measurement[9] of the longitudinal spin-dependent total cross section difference $\Delta\sigma_L$ made by the Argonne polarized target group (Fig. 3) show remarkable structure in the 1-2 GeV/c range, considerably more than the spin-averaged total cross section does. The measurements of the total cross section differences in transverse spin states[10] made by the Michigan and Rice-Houston groups also shows significant structure in this energy range (Fig. 4). The $\Delta\sigma_L$ data has a striking peak at about 1.2 GeV/c (singlet) and an equally striking dip at 1.5 GeV/c (triplet). The $\Delta\sigma_T$ data shows peaks at 1.2 and 2.0 GeV/c (singlet), although due to systematic errors the height of the peak at 1.2 GeV/c is uncertain to within a factor of 1.5-2. (All of these measurements are now being repeated in finer energy steps from 400-800 MeV at LAMPF with hopefully smaller systematic errors. In addition, the lower energy region is being extensively studied at TRIUMPF). The importance of the structure at 2.0 GeV/c is emphasized if $\Delta\sigma_T$ is multiplied by K_{cm}^2 (Fig. 5), giving an energy independent weight to all the phase shifts. A dispersion analysis[11] indicates a loop in the amplitude ϕ_2 at this momentum which might be interpreted as evidence for a singlet ($^1G_4$) dibaryon resonance. However, this energy is too high for a reliable phase shift analysis; therefore, this structure will not be discussed further.

After the appearance of these data, there was considerable theoretical and phenomenological activity regarding the existence or nonexistence of dibaryon resonances. The phase-shift analysis of Hoshizaki[12] (Fig. 6a) indicated counter clockwise loops in the Argand diagrams for the $^1D_2$ and $^3F_3$ partial waves. (The "nonresonant background" has been subtracted in this analysis). Grein and Kroll[11] have used the imaginary parts of the three spin dependent forward scattering amplitudes gotten from measurements of σ_{TOT}, $\Delta\sigma_L$, $\Delta\sigma_T$, phase shifts at lower energies, and some assumptions about the high energy behavior of $\Delta\sigma_L$ and $\Delta\sigma_T$ to get the real parts of the three forward scattering amplitudes via dispersion relations.

They find a resonant-like structure at 1.5 GeV/c but not at 1.2 GeV/c (Fig. 6b). On the other hand, analyses by Hollas[13], Arndt[14], and Minami[15] argued against the necessity of resonances. Hollas was able to fit the $\Delta\sigma_T$ and $\Delta\sigma_L$ data using arguments from the early work of Mandelstam[16], after separating the total elastic and inelastic cross sections into singlet and triplet parts (Fig. 7). Also the phase shift analysis of Arndt showed no loops in the Argand plots for the $^1D_2$ and $^3F_3$ phase shifts at that time.[14]

Since then the Argonne PPT group has measured C_{LL} at 90° C.M. and 73° C.M. at eleven energies between 1.0 and 3.0 GeV/c[17] (Fig. 8). C_{LL} at 90° shows a dip around 1.2 GeV/c and a peak near 1.5. This would indicate dominance of singlet and triplet partial waves at these respective momenta. Also note that the peak just above 2 GeV/c is absent in the 73° data. Since $P_4(\cos\theta)$ has a zero near this angle, this is perhaps evidence of structure caused by activity in the above mentioned $^1G_4$ partial wave. The Argonne PPT group has gotten the full angular distribution of C_{LL} at these energies, which will shortly be submitted for publication.[18] These data should significantly constrain the phase shift analysis. The recent A and A_{nn} data of the Rice group[7] is shown in Figs. 9 and 10. The energy dependence of A_{nn} (90°) is shown in Fig. 11. There is a striking peak showing triplet dominance at ∼700 MeV, consistent with older data which had large error bars. Fig. 12 shows the D_{NN}, D_{SS}, and D_{LS} data of the LASL-UT-CWRU-TAMU collaboration reported at this conference. These very nice data cover almost the full angular range at 800 MeV, and should place significant constraints on the phase shift analysis. Note that the recent phase shift analysis of Arndt[19] fits all three variables very well, whereas his 1979 analysis[14] gives a much poorer fit. The Argand diagrams for the $^1D_2$ and $^3F_3$ partial waves from this recent energy dependent analysis (Fig. 13) going up to 850 MeV both show loops. This behavior is distinctly different from the 1979 analysis, and in fact a K-matrix calculation reported by Arndt at this conference shows striking $^1D_2$ and $^3F_3$ poles when the scattering amplitude is extrapolated into the complex plane. Finally, there is new data on $\Delta\sigma_T$, pp elastic, and pp→dπ from S.I.N. and TRIUMPF presented at this conference which may affect these analyses significantly.

So far we have only discussed the effects on the phase shift analysis of data on total cross sections and elastic scattering. In this energy region various inelastic thresholds are crossed which can give sharp energy-dependent structure to total cross sections and through unitarity possibly to elastic scattering as well.[16,20] Thus the study of the energy dependent behavior of the phase shifts in a coupled channel analysis using all available information on inelasticities seems important. Such an analysis

has been performed by Edwards and Thomas[20] and independently by Arndt as reported at this conference. Some of the results of Edwards and Thomas are shown in Fig. 14a-d for a coupled channel analysis using only elastic and $n\Delta^{++}$ channels. In the first case (14a) the pp→pp is constrained to fit the Arndt phase shifts, the second, Hoshizaki's (14b). The two fits give different values for the $N\Delta \to N\Delta$ phase shift δ_2.
Nonetheless both fits give a similar loop in the Argand plot for the $^1D_2$ partial wave, and a pole in the K-matrix when extrapolated to the complex plane (14c-d). The analysis has been extended to include the $d\pi^+$ channel and extended up in energy to investigate the $^3F_3$ around 800 MeV. In this 3-channel analysis K-matrix poles were found in the $^1D_2$ and $^3F_3$ amplitudes.[21] Similar results were reported here by Arndt. Kloet and Silbar[22] have calculated the p-p elastic phase shifts in a unitary dynamical model using π, ρ, σ, σ', and ω exchanges, for different values of the various coupling constants. In particular, the short range forces (ρ and ω exchange) were varied in strength over a wide range. For "reasonable" values of the coupling constants the $^1D_2$ and $^3F_3$ amplitudes show counterclockwise rotation in the Argand diagrams similar to Arndt's recent phase shift analysis (Fig. 15). This result is not surprising since this dynamical model was constructed to reproduce Arndt's phase shifts.

Umland and Duck have calculated cross sections and single spin asymmetries for pp→pnπ$^+$ at 800 MeV (reported by Umland). Results for a single production and decay angle of the Δ are shown in Fig. 16a-b. The fit to the cross section (this data reported here by Hancock) is improved by adding s-channel $^3F_3$ and $^1D_2$ dibaryon amplitudes to the π and ρ exchanges. However, the fit to the asymmetry data is improved much more dramatically by this addition. On the other hand, even when the couplings are adjusted to fit the cross sections roughly, meson exchange alone gives asymmetries which do not resemble the data at all.

None of these analyses or the data used as input conclusively prove that I=1 dibaryons resonances exist; indeed, some of the analyses were begun with the opposite intent. However, there is increasing experimental evidence for structure in nucleon-nucleon cross sections and spin observables, and increasing evidence from theoretical analyses for counter clockwise rotation of Argand plots and for poles in the complex plane in various partial waves.

THE I=0 SYSTEM

Because of time constraints we shall comment only briefly on the I=0 system, where the data is, in general, much sparser due to

the greater difficulty of making neutron-proton measurements and
subsequently subtracting the I=1 parts. Neutron-proton total
cross section measurements[23] are shown in Fig. 17. We remember
that the n-p system does not have a strong N-Δ threshold, and
therefore, the total cross section does not reach a maximum until
above P_{lab} = 2 GeV/c, distinctly different from p-p (Fig. 1).
Although the n-p cross section rises much slower than p-p, there
is a shoulder around 1.5 GeV/c, which is made more visible by the
absence of the strong N-Δ threshold present in p-p.
Measurements of the Argonne group of $\Delta\sigma_L(I=0)$[24] are shown
in Fig. 18. These data are obtained by measuring $\Delta\sigma_L(pd)$,
then subtracting $\Delta\sigma_L(I=1)$ after attempting to take into
account affects due to screening, rescattering, and Fermi motion
inside the deuteron, very difficult procedures. Nonetheless,
there is a clear peak at 1.5 GeV/c, which has been interpreted as
evidence for a $^1F_3$ dibaryon resonance,[25] which may plausibly
exist if the $^3F_3$ does. Data on $\Delta\sigma_T(pd)$ taken by the
Rice group immediately before the ZGS shutdown, which is presently
under analysis, should help to resolve this question, since the
same singlet enhancements should appear in $\Delta\sigma_T$ and
$\Delta\sigma_L$. However, the analysis is subject to the above
mentioned difficulties in addition to a considerable uncertainty
in the knowledge of the deuteron polarization in the target.
Japanese groups[4] also have found an anomalous peak in the proton
polarization in the photodisintegration of the deuteron at
$\sqrt{s} \sim 2400$ MeV. They are able to fit the data by adding a
Breit-Wigner-type amplitude for a Δ-Δ bound state with this
mass (Fig. 19a), whereas without any resonances they are unable to
account for the large polarization (Fig. 19b). Thus there is some
evidence for structure in the n-p system not associated with
inelastic thresholds, but data is sufficiently sparse in this
energy range that we are far from having a reliable phase shift
analysis.

EPILOG

Finally let us briefly discuss one additional topic. There
has been some recent speculation and evidence from lower energy
accelerators, in particular, TRIUMPF and LAMPF, that some of the
lowest energy data from Argonne may be in error. In particular,
at 1.2 GeV/c, the earliest $\Delta\sigma_T$ point has been found to be
low by subsequent yet to be published measurements at Argonne and
LAMPF, and similarly the $\Delta\sigma_L$ point may be somewhat in error,
both possibly due to unknown depolarization of the beam in the ZGS
and the difficulties of handling the low momentum beams at the
ZGS. For this reason, the experiments now in progress at TRIUMF,
S.I.N., and LAMPF are particularly important; hopefully high

quality data will be obtained which will help settle the question of the existence of dibaryon resonances. My personal belief is that despite quantitative errors in the early data, the qualitative results are correct, i.e., there are structures at 1.1-1.2 GeV/c and at 1.4-1.5 GeV/c associated with singlet and triplet enhancements, respectively.

I very much appreciate the hospitality shown by the organizers of this conference, particularly Dr. B. Bonner and Dr. G. Ohlsen. I am particularly grateful to Dr. A. Yokosawa for making available data and other information relevant to this subject. I am also grateful to Dr. G. Thomas, Dr. M. Johnson, Professor E. Lomon, and Professor I. Duck for illuminating discussions.

REFERENCES

1. R. L. Jaffe, Phys. Rev. Lett. $\underline{38}$, 195 (1977).
   P. J. G. Mulders, et al., Phys. Rev. Lett. $\underline{40}$, 1543 (1978).
   A. Th. M. Aerts, Phys. Rev. $\underline{D17}$, 260 (1978).
2. E. Lomon, "Unification of the Quark Model and Hadron Field Theory," (1980) unpublished.
3. S. Ishida and M. Oda, Prog. Theor. Phys. $\underline{61}$, 1401 (1979) and references therein.
4. T. Kamae, et al., Phys. Rev. Lett. $\underline{38}$, 471 (1977).
   H. Ikeda, et al., Phys. Rev. Lett. $\underline{42}$, 1321 (1978).
5. G. Giacomelli, Prog. in Nucl. Phys. $\underline{12}$, 214 (1971).
   P. Schwaller, et al., Nucl. Phys. $\underline{A316}$, 317 (1979).
6. The integrated data of B. A. Ryan, et al., Phys. Rev. $\underline{3}$, 1 (1971) compiled by the Argonne PPT group is used.
7. M. G. Albrow, et al., Nucl. Phys. $\underline{B23}$, 445 (1970).
   D. A. Bell, et al., Phys. Lett. (1980).
8. U. Amaldi, et. al., Ann. Rev. Nucl. Sci. $\underline{26}$, 385 (1976).
9. I. P. Auer, et al., Phys. Lett. $\underline{67B}$, 113 (1977); ibid $\underline{70B}$, 475, (1977); I. P. Auer, et al., Phys. Rev. Lett. $\underline{41}$, $\overline{354}$ (1978).
10. W. deBoer, et al., Phys. Rev. Lett. $\underline{34}$, 558 (1975).
    E. K. Biegert, et al., Phys. Lett. $\underline{73B}$, 235 (1978).
11. W. Grein and P. Kroll, Nucl. Phys. $\underline{B137}$, 173 (1978).
12. N. Hoshizaki, Prog. Theor. Phys. $\underline{57}$, 1099 (1977); $\underline{58}$, 716 (1977); $\underline{60}$, 1796 (1978); $\underline{61}$, 129 (1979).
13. C. L. Hollas, Phys. Rev. Lett. $\underline{44}$, 1186 (1980).
14. R. A. Arndt, private communication, 1978.
15. S. Minami, Phys. Rev. $\underline{D18}$, 3273 (1978).
16. S. Mandelstam, Proc. Royal Society London, $\underline{A244}$, 491 (1958).
17. I. P. Auer, et al., Phys. Rev. Lett. $\underline{41}$, 1436 (1978).
18. A. Yokosawa, private communication.
19. R. A. Arndt, private communication, 1979.

20. B. J. Edwards and G. H. Thomas, "Inelastic Thresholds and Dibaryon Resonances" ANL-HEP-PR-80-130, June 1980.
21. G. Thomas, private communication.
22. W. M. Kloet and R. R. Silbar, "Effects of Heavy Meson Exchange on the $^1D_2$ and $^3F_3$ N-N Partial Waves and the Question of Dibaryon Resonances" RU-80-219 (1980).
23. T. J. Devlin, et al., Phys. Rev. $\underline{D8}$, 136 (1973).
24. CERN Courier, Sept. 1980, p.252.
25. A. Yokosawa, Physics Reports $\underline{64}$, Sept. 1980.

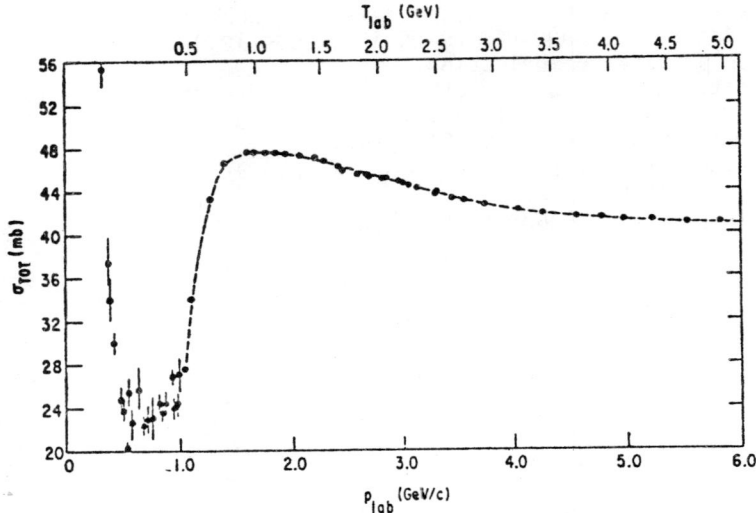

Fig. 1a. Proton-proton total cross section.

Fig. 1b. Total elastic p-p cross section.

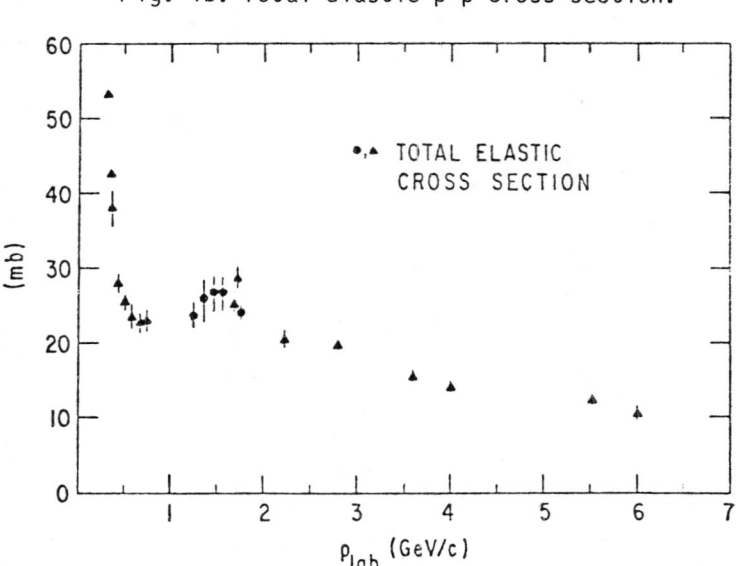

Fig. 2a.

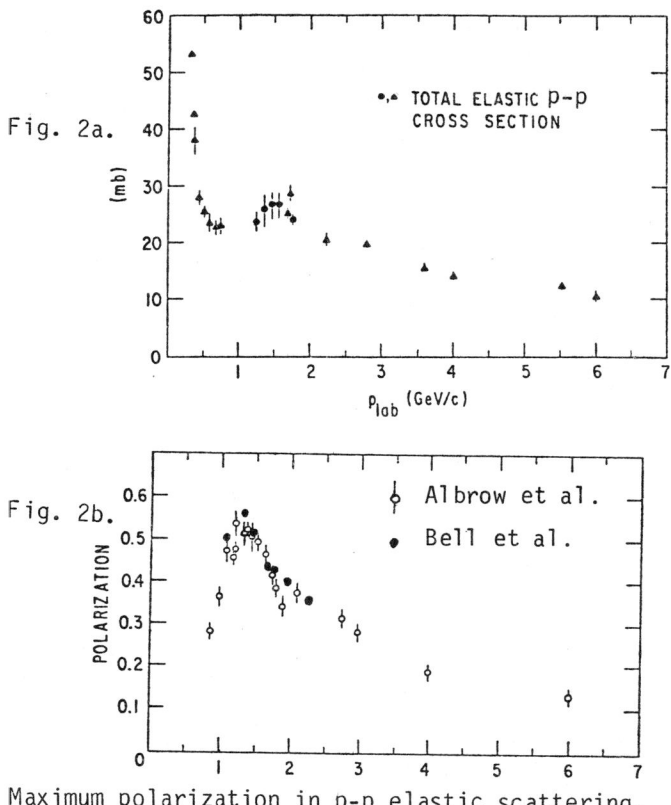

Fig. 2b.

Maximum polarization in p-p elastic scattering.

Fig. 2c.

Re/Im p-p forward scattering amplitude.

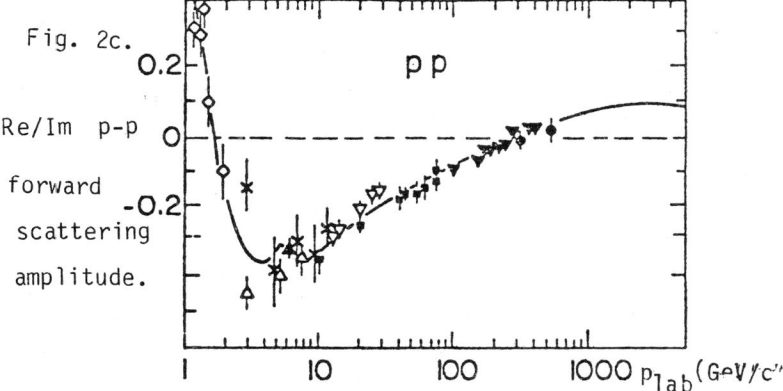

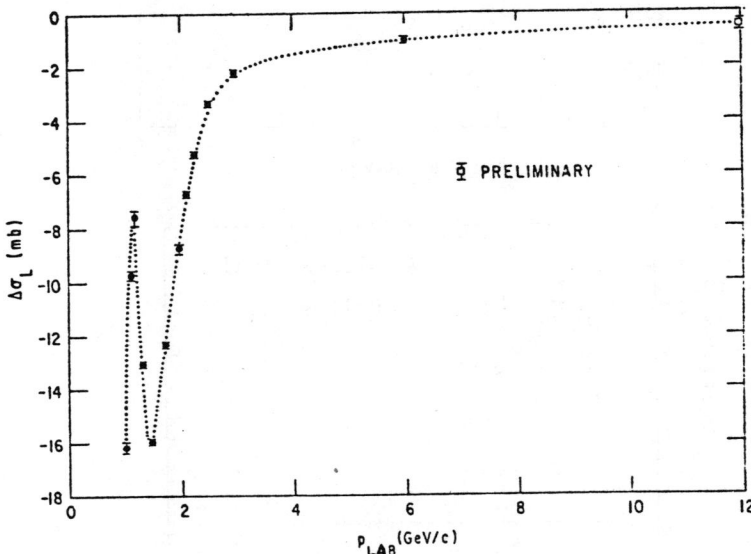

Fig. 3. Total cross section difference, $\Delta\sigma_L$, 1-12 GeV/c.

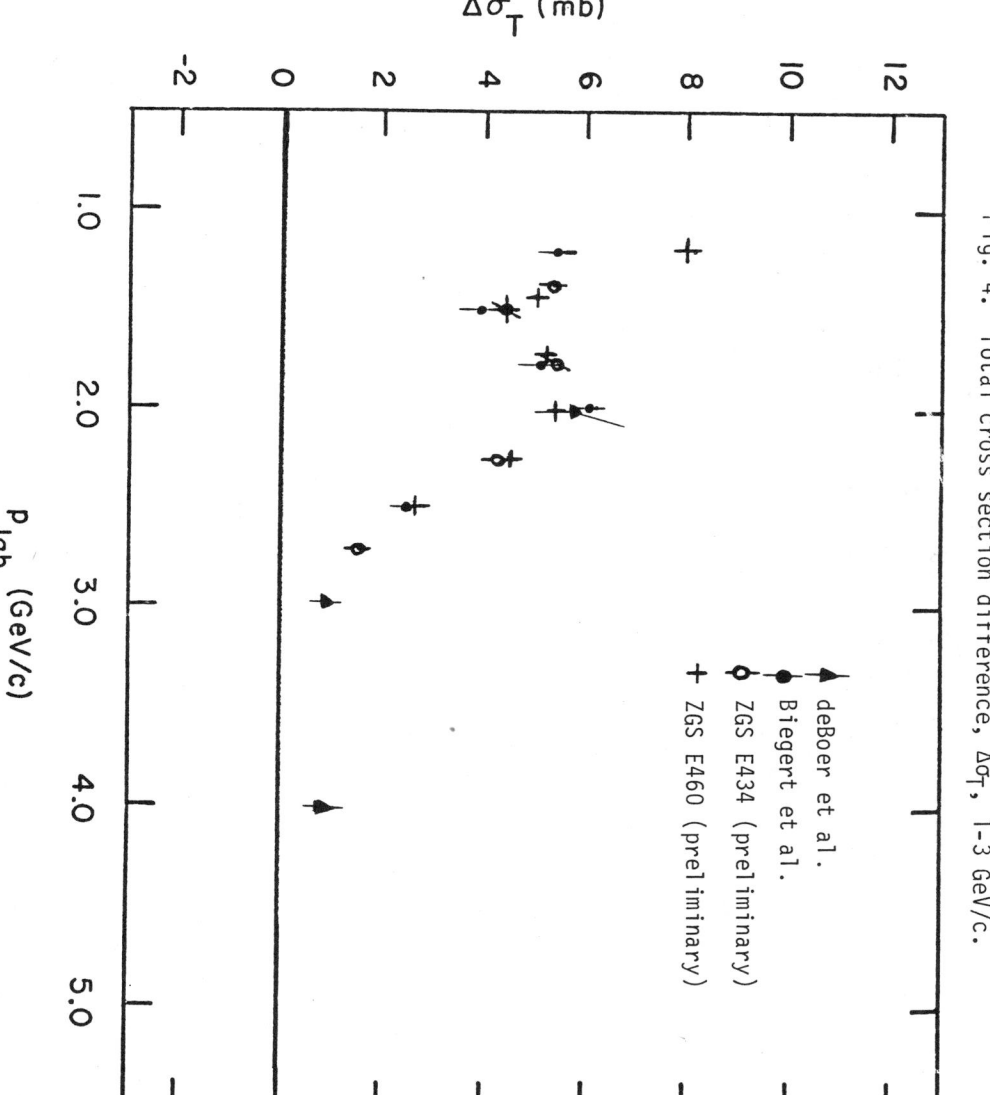

Fig. 4. Total cross section difference, $\Delta\sigma_T$, 1-3 GeV/c.

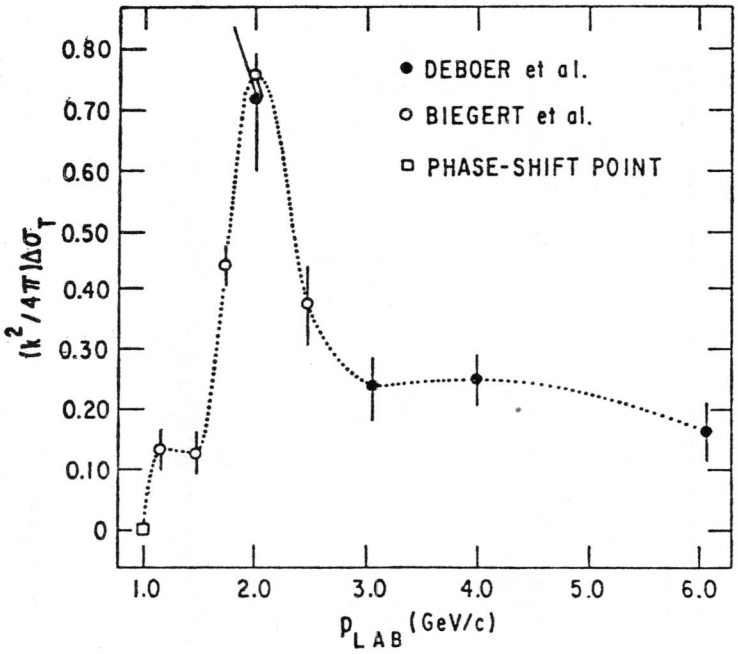

Fig. 5. $K^2_{cm}/4\pi \; \Delta\sigma_T$.

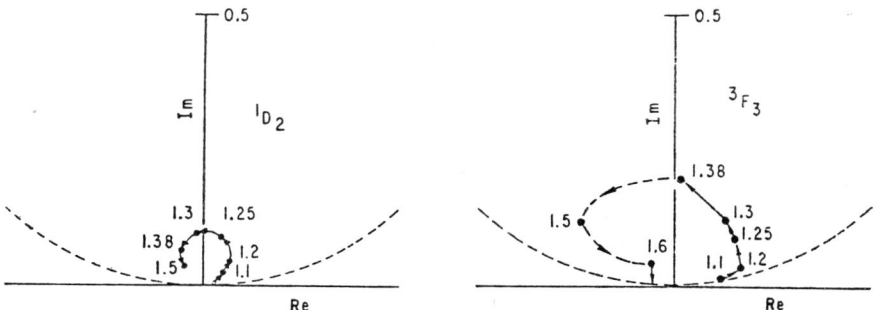

Fig. 6. $^1D_2$ and $^3F_3$ phase shifts from analysis of Hoshizaki.

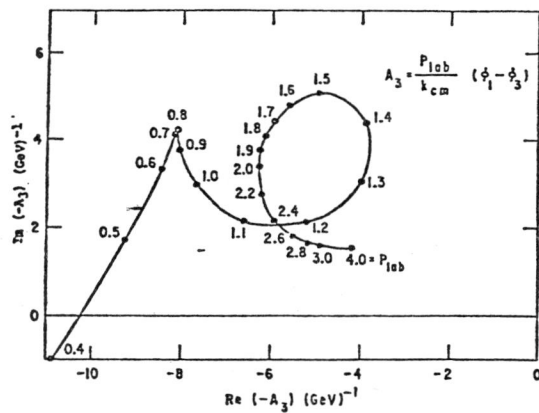

Fig. 7. Argand diagram from analysis of Grein and Kroll.

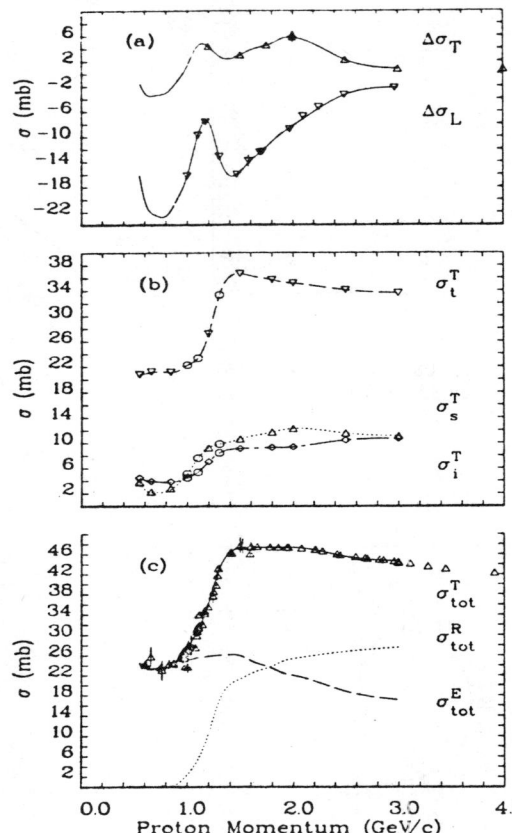

Fig. 8. The momentum dependence of (a) the cross-section differences $\Delta\sigma_T$ and $\Delta\sigma_L$ (the data are from Refs. 1 and 2; the curves are described in the text); (b) the singlet (σ_s^T), triplet (σ_t^T), and triplet-interference (σ_i^T) cross sections, as described in the text; (c) the spin-averaged total cross sections, σ_{tot}^T (triangles), σ_{tot}^E (dotted line), and σ_{tot}^R (dashed line).

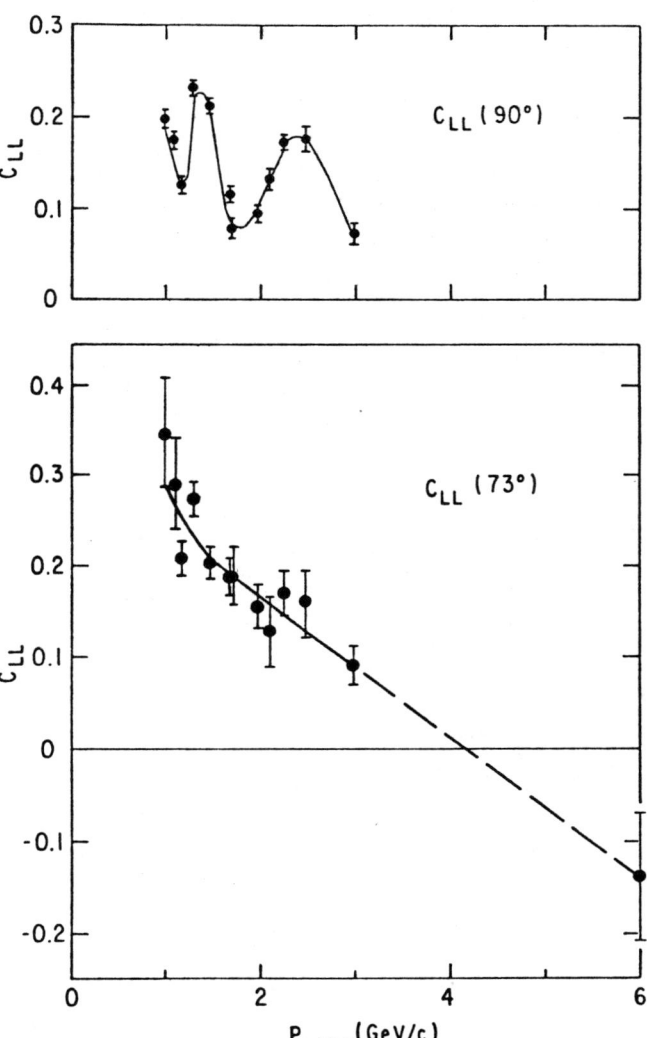

Fig. 9. C_{LL} for p-p elastic at two C.M. angles.

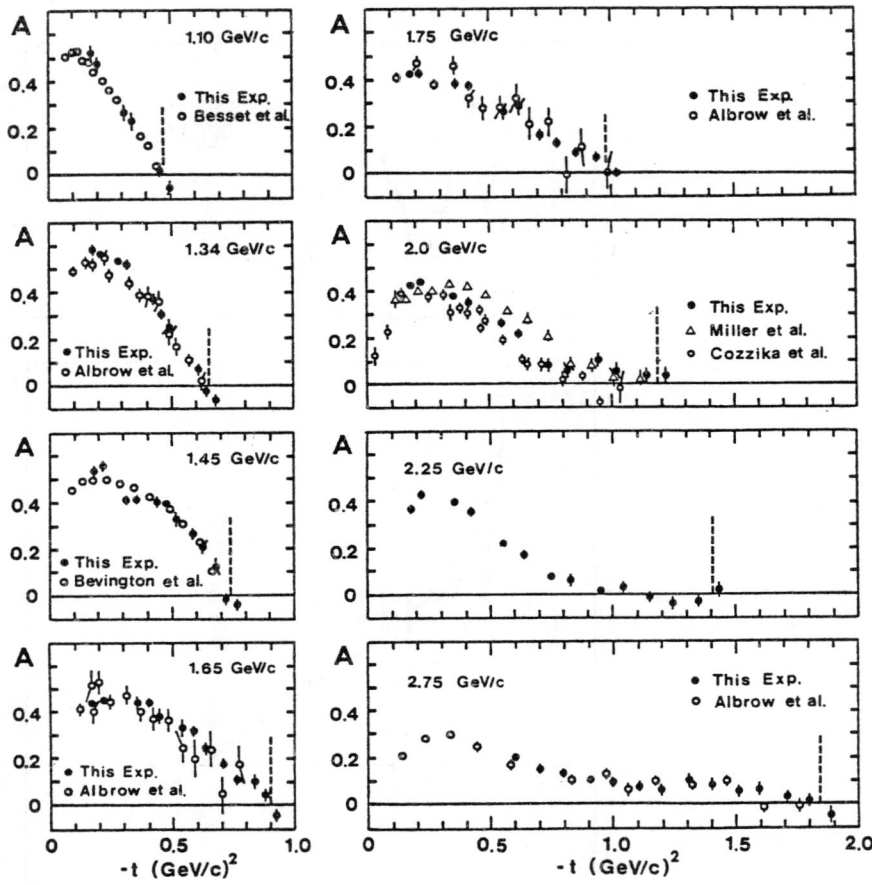

Fig. 10. p-p polarization, 1.1-2.75 GeV/c.

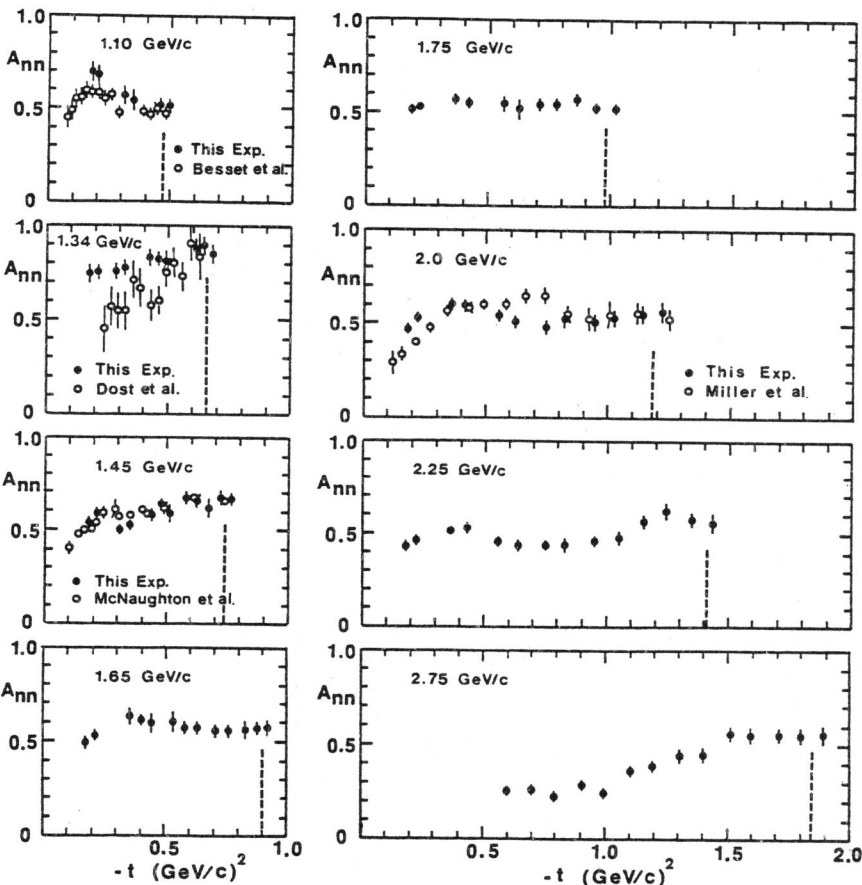

Fig. 11. A_{nn} for p-p elastic, 1.1-2.75 GeV/c.

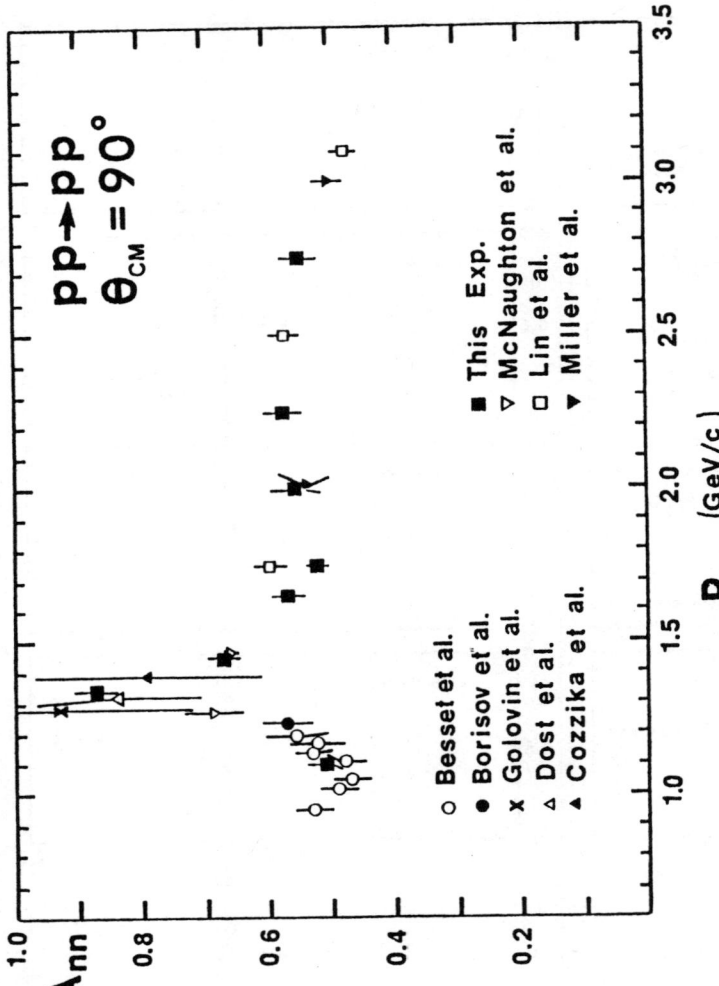

Fig. 12. A_{nn} for p-p elastic at 90° C.M.

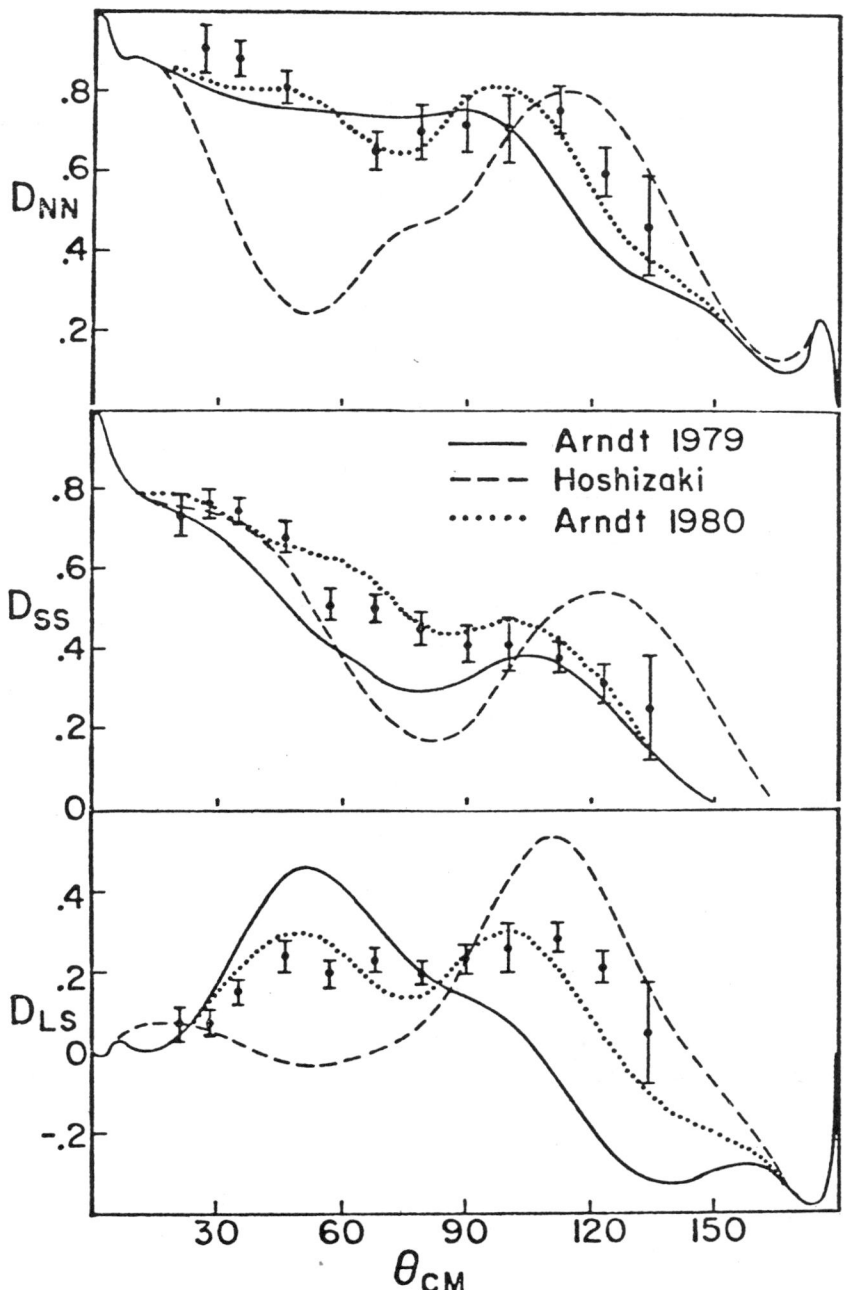

Fig. 13 D_{NN}, D_{SS}, amd D_{LS} for p-p elastic scattering at 800 MeV.

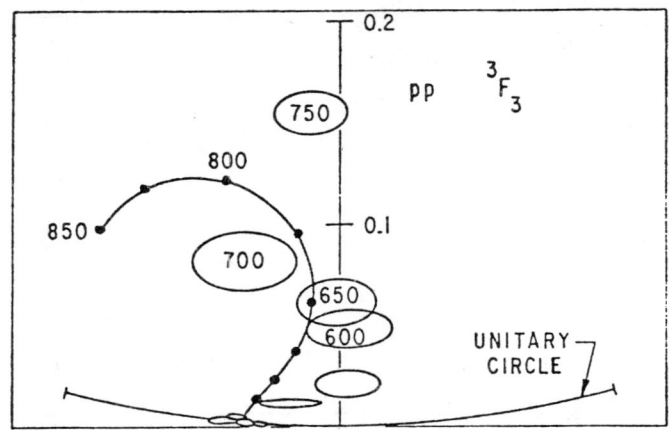
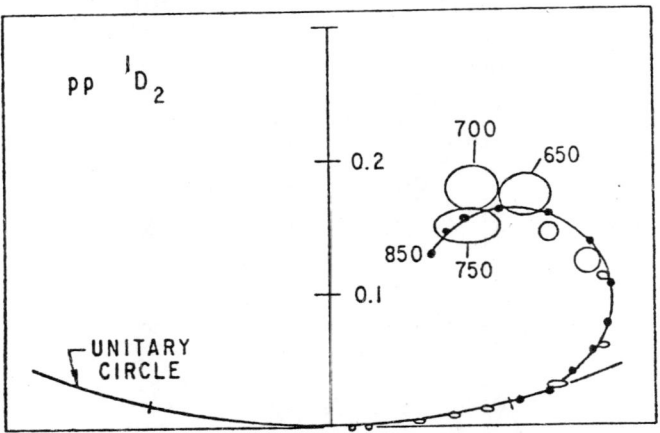

Fig. 14.
Argand diagrams of the $^3F_3$ and $^1D_2$ partial waves based on Arndt's phase shifts. The elipses represent the errors in the real and imaginary parts of the amplitudes for energy-independent solutions. The continuous curves represent the energy-dependent solutions.

Fig. 15a.   Phase shift δ_2 for $N\Delta \to N\Delta$ calculated in coupled channel analysis using δ_1 and η from Arndt's phase shifts (Ref. 20).

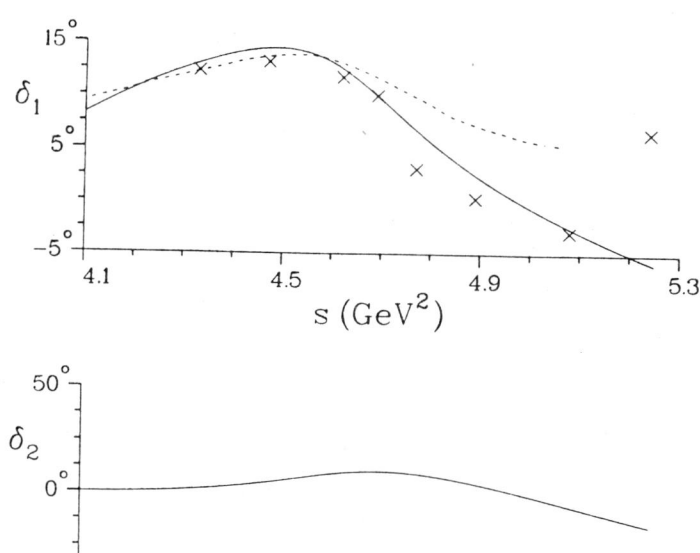
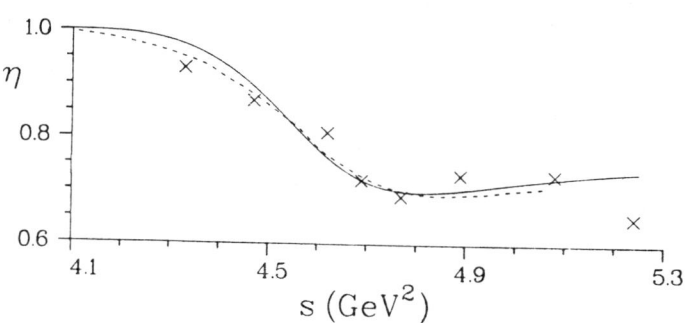

Fig. 15b.   Same as 15a, using Hoshizaki's phase shifts.

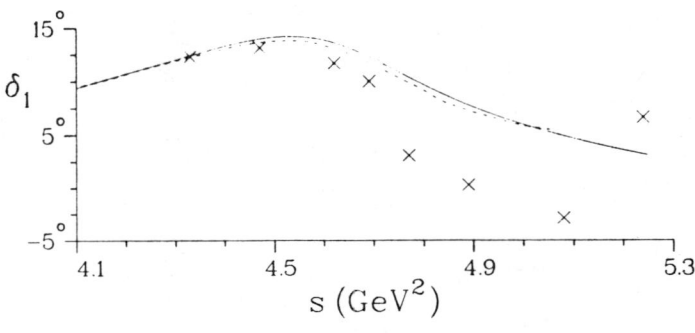
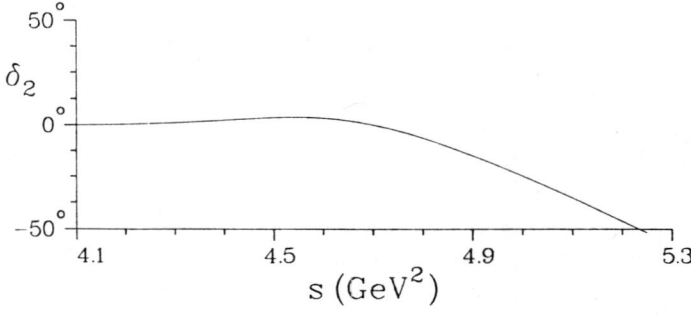
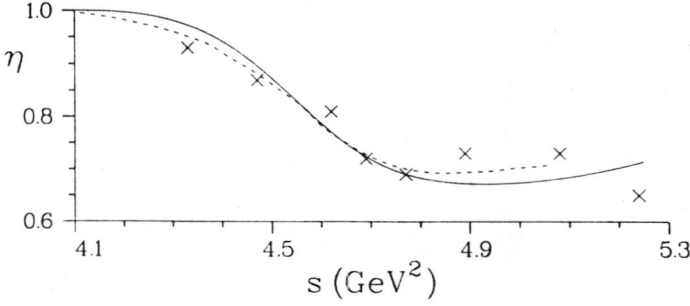

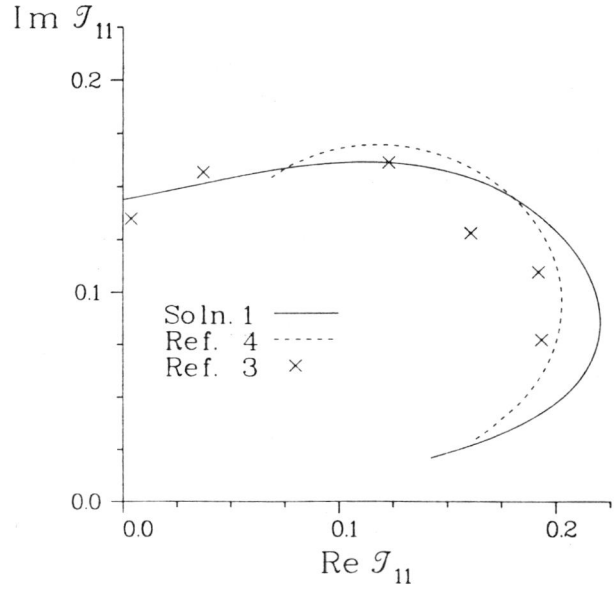

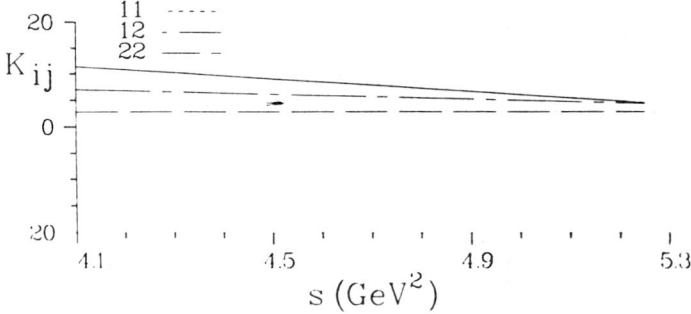

Fig. 15c. Argand plot for $^1D_2$ and K-matrix from calculation of Ref. 20 using phase shifts from 15a.

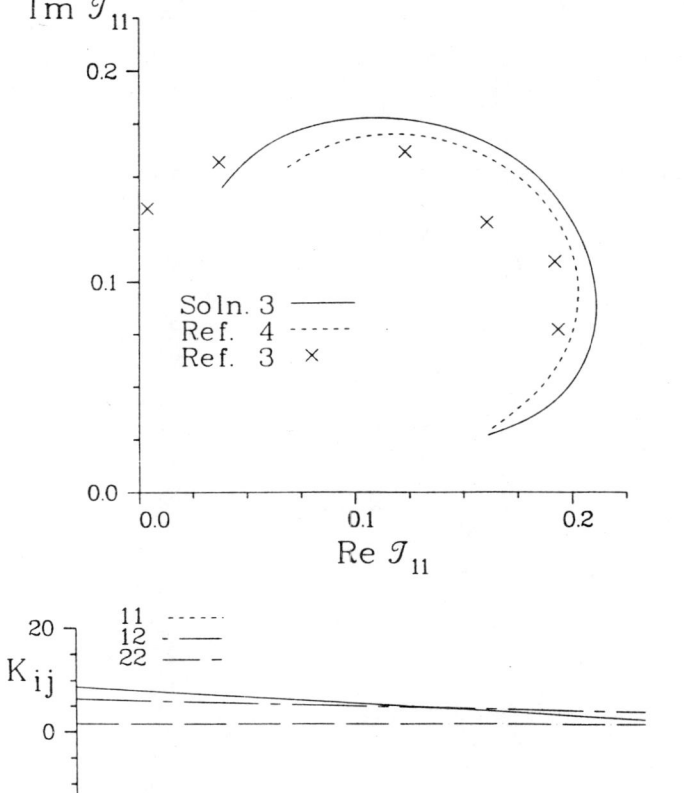

Fig. 15d.   Same as 15c, using phase shifts from 15b.

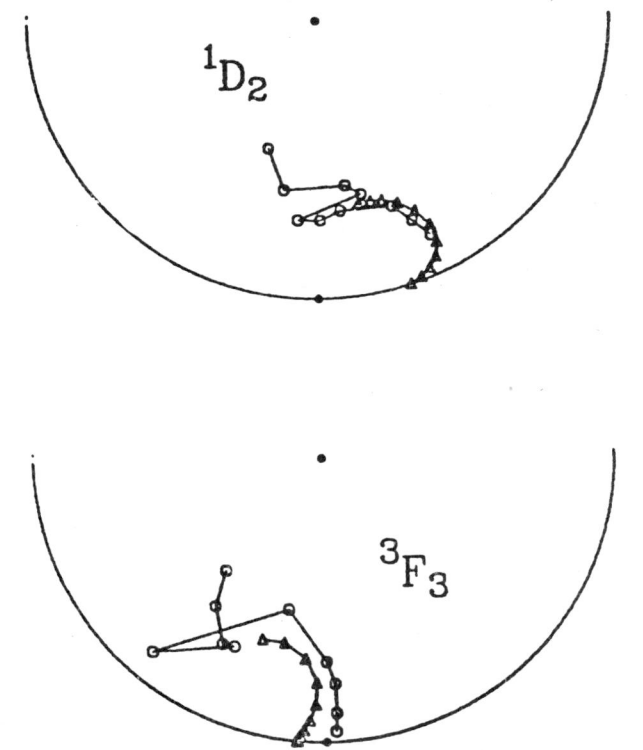

Fig. 16. Sample Argand diagram for $^1D_2$ and $^3F_3$ from unitary dynamical model of Kloet and Silbar.

Fig. 17a. Calculation of cross section $d^5\sigma/dpd\Omega_1 d\Omega_2$ for $pp \to pn\pi^+$ of Umland and Duck using only meson exchange and adding $^1D_2$ and $^3F_3$ dibaryon amplitudes.

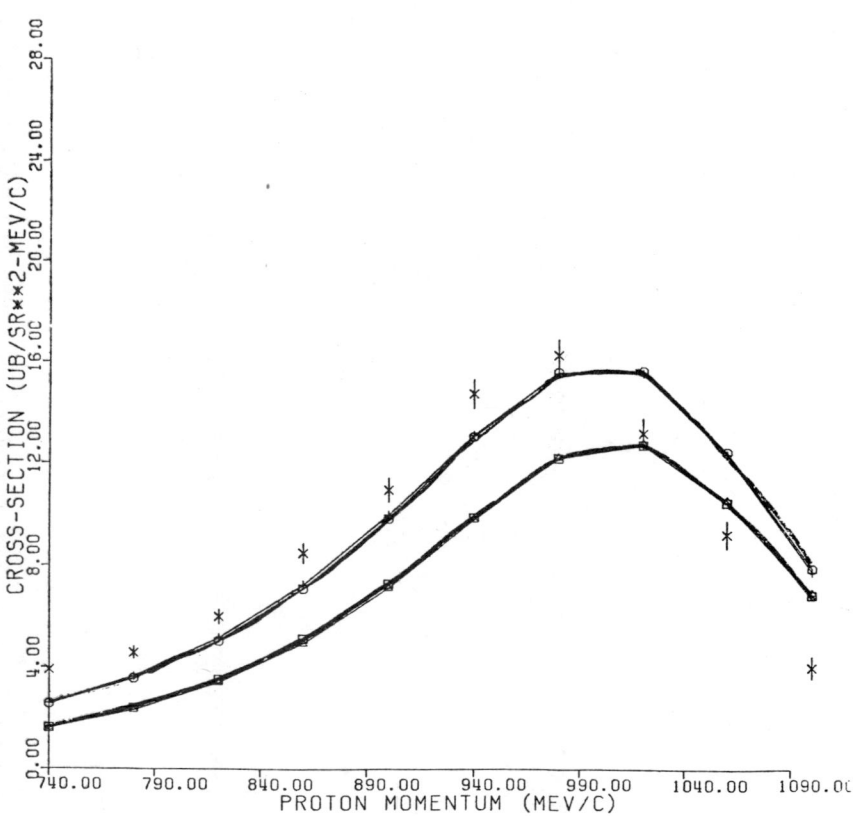

Fig. 17b. Similar calculation to 17a of single spin asymmetry in pp→pnπ⁺.

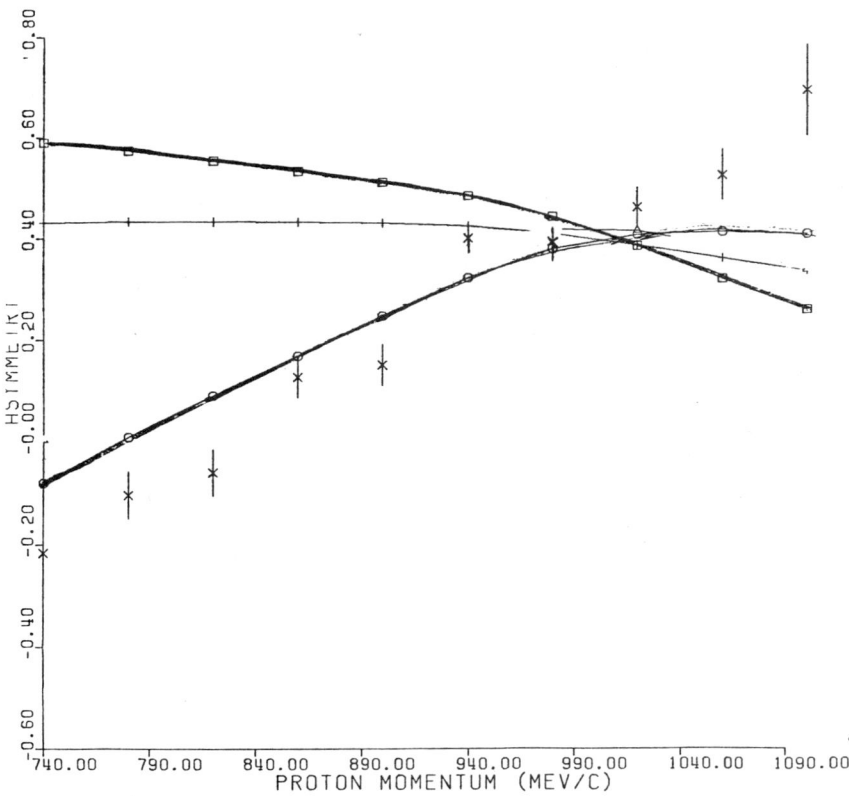

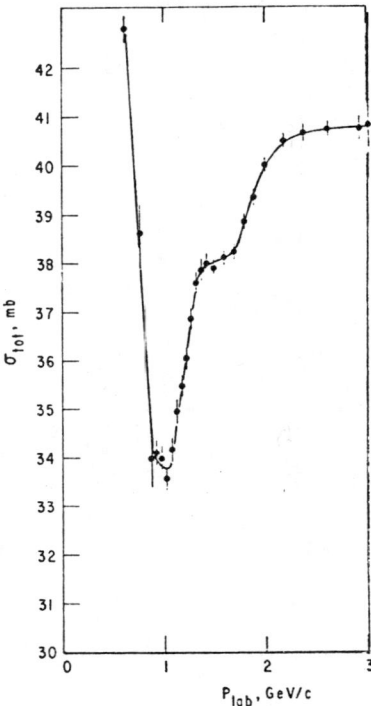

Fig. 18. Neutron-proton total cross section (Ref. 23).

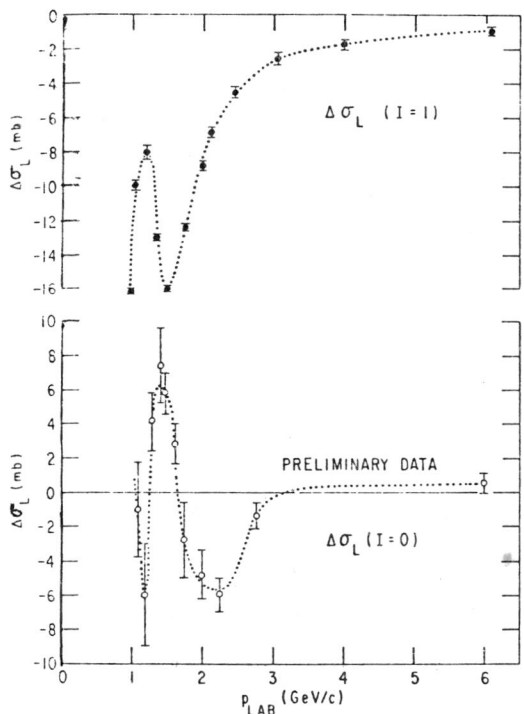

Fig. 19. $\Delta\sigma_L(I=0)$ extracted from $\Delta\sigma_L(pd)$ and $\Delta\sigma_L(I=1)$ plotted along with $\Delta\sigma_L(I=1)$.

Fig. 20a. (Ref. 4).

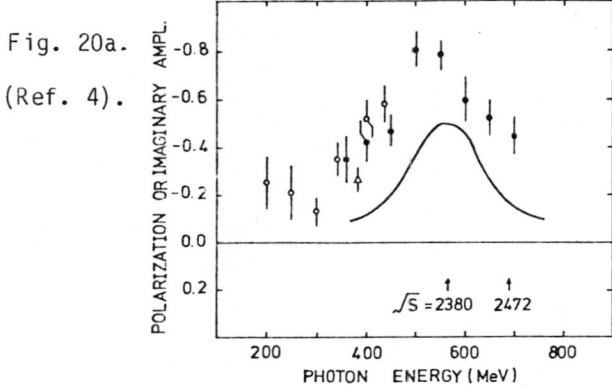

Proton polarization in $\gamma d \to pn$. Filled circles are data from Ref. 1, open circles are data from Ref. 5 of Ref. 1, and triangle is datum from Ref. 6 of Ref. 1. The curve shows the Breit-Wigner-type imaginary and amplitude due to the $\Delta\Delta$ bound state at $\sqrt{s}$ = 2380. Note that the unbound $\Delta\Delta$ phase space opens at $\sqrt{s}$ = 2472.

Fig. 20b. (Ref. 4).

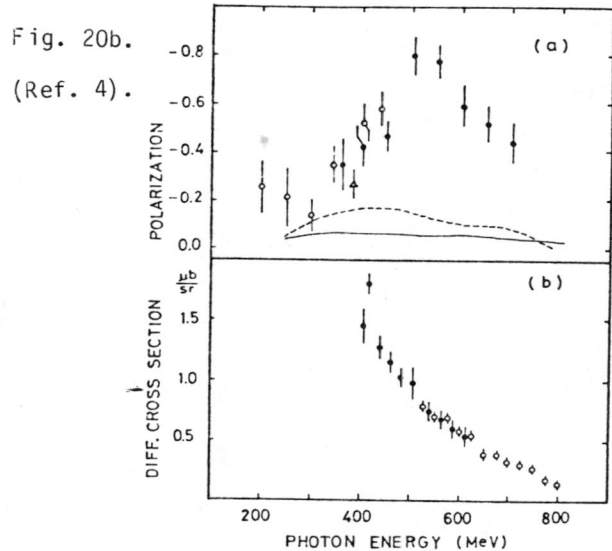

(a) Proton polarization at 90° c.m. system as a function of the photon energy. The solid and the dashed curves are the results of a relativistic-covariant computation and a phenomenological analysis respectively (see Ref. 10). Data points are from Ref. 5 (open circles), Ref. 6 (triangle), and the present experimental (filled circles). (b) Differential cross section at 90° c.m. system as a function of the photon energy. Data points are taken from Ref. 1 (open circles) and Ref. 2 (closed circles).

DISCUSSION

KLOET: The presence of a loop in the Argand plot is not direct evidence for a resonance. I understand that in Thomas' analysis, and also in Arndt's, a pole in the complex S-plane was found. In our dynamical model calculation we also find loops in the $^3F_3$ and $^1D_2$ Argand plots but the analytic structure shows that these loops are not associated with resonance poles.

POLARIZATION PHENOMENA IN (p,π) REACTIONS*

J.A. Niskanen†
Dept. of Physics, SUNY at Stony Brook, N.Y. 11794, USA

and

Research Institute for Theoretical Physics
University of Helsinki, Helsinki, Finland

ABSTRACT

A critical review on the current theoretical understanding of (p,π) reactions in the light of recent experiments is given.

INTRODUCTION

Since the appearance of high intensity "meson factories" high quality experimental data on (p,π) reactions at intermediate energies (from threshold to about 800 MeV) have accumulated with unpolarized [1-3,10] and polarized [4-9,11-13,43] beams and/or targets, and even the measurement of final state polarizations is conceivable.[9] In theory the understanding of $p+p \to d+\pi^+$ below 700 MeV appears to be reasonable, although minor disagreements still exist between theory and experiment as well as between different models. Several calculations are available where rather different methods generally give qualitatively similar results.[14-24] Polarization phenomena in the more general $A(p,\pi)A+1$ reactions still lack calculations with a realistic two nucleon mechanism model (TNM), whereas variations of the more elementary one nucleon model (ONM) have given very varying predictions.[25-27]

In sec. 2 we shall give a short overview of the major models of $p+p \to d+\pi^+$, and a comparison with current experimental data in sec. 3. Pion production on nuclei is briefly discussed in sec. 4.

MODELS FOR THE REACTION $p+p \rightleftarrows d+\pi^+$

The long accepted mechanism for $p+p \to d+\pi^+$ consists of "direct" production with correlated NN wave functions (fig. 1a) and of the dominant rescattering part (fig. 1b), where the pion suffers a rescattering off the second nucleon. The s-wave rescattering dominates near the production threshold, whereas the p-wave

---

*Work supported in part by USDOE Contract No. EY-76-S-02-3001.
†On leave of absence: Dept. of Theoretical Physics, University of Helsinki, Helsinki, Finland
 Address for the academic year 1980-81: TRIUMF, Univ. of British Columbia, Vancouver, B.C., Canada, V6T 1W5

rescattering causes a broad peak around the laboratory energy 600 MeV. In the p-wave rescattering, the rather long range cut-off functions necessary[28] to decrease otherwise far too large cross-sections were later accounted for an intermediate ρ-meson.[14,18]

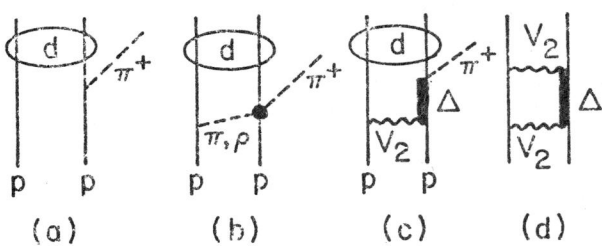

Fig. 1. Contributions to p+p → d+π⁺

As the p-wave pion-nucleon scattering at energies relevant to us is dominated nearly exhaustively by the Δ_{33}-isobar, which can be treated as a short lived elementary particle (an excited state of the nucleon), one can consider the reaction to occur through an intermediate NΔ-state, as shown in fig. 1c. This state is generated in T=1 states by a transition potential V_2 consisting of pion and ρ-meson exchanges, and it decays quite naturally into an NNπ final state. As the "box diagram" with a nucleon and a Δ (fig. 1d) gives a major contribution to the intermediate range attraction of nucleons, probably the most reliable way to treat this intermediate state is by the coupled channels method (CCM)[14], which sums the NΔ box diagrams to all orders. In the coordinate space this model yields coupled Schrödinger equations like

$$[-\frac{d^2}{dr^2} + \frac{L(L+1)}{r^2} - k^2 + \frac{M}{\hbar^2} V_1] u(r) = -\frac{M}{\hbar^2} V_2 \omega(r)$$

$$[-\frac{d^2}{dr^2} + \frac{L'(L'+1)}{r^2} - k^2 + \frac{M}{\hbar^2} ((\Delta-M) - \tfrac{1}{2}i\Gamma + V_3)]\omega(r) = -\frac{M}{\hbar^2}V_2 u(r)$$

(1)

where the second equation, describing the NΔ-state, is modified by the mass difference Δ-M between the Δ and the nucleon and by the width of the Δ. The interaction V_1 is the NN potential modified by removing the box diagram with Δ (fig. 1d), and V_3 the interaction between the nucleon and the Δ. Once the generalized initial NN wavefunction with an admixture of NΔ is thus generated, it is then operated by the (one body) pion production operator

$$H^{prod} = \sum_{i=1,2} \frac{f}{\mu} i \vec{\sigma}_i \cdot \{\vec{\nabla}_\pi \vec{\tau}_i \cdot \vec{\phi}(\vec{x}_i) + \frac{1}{2M} [\vec{p}_i \vec{\tau}_i \cdot \vec{\pi}(\vec{x}_i) + \vec{\tau}_i \cdot \vec{\pi}(\vec{x}_i)\vec{p}_i]\},$$ (2)

where $\vec{\phi}$ and $\vec{\pi}$ are the pion conjugate fields, and the πNN coupling constant f is replaced by the πNΔ coupling constant f^* in case of the Δ. This method was used in refs. [15-17] to calculate the differential cross sections with a polarized and unpolarized beam and target and the polarization of the final deuteron.

A first order perturbation approach has been proposed and used in refs. [18-20]. However, these calculations contain also another even more serious approximation, which is not pointed out by the authors. Namely their implicit use of the NΔ wave function

$$\psi_{N\Delta} = \frac{V_2(r)}{\omega_R - \omega - i\Gamma/2} \psi_{NN} \qquad (3)$$

is an unjustified "closure" approximation, which severely distorts the radial dependence of the exact wave function. At the resonance region the range of $\psi_{N\Delta}(r)$ should be longer than the range of $V_2(r)$, whereas for small r their $\psi_{N\Delta}$ is grossly overestimated. Even more importantly, the L dependence of $\psi_{N\Delta}(r)$ is lost. In fact, of

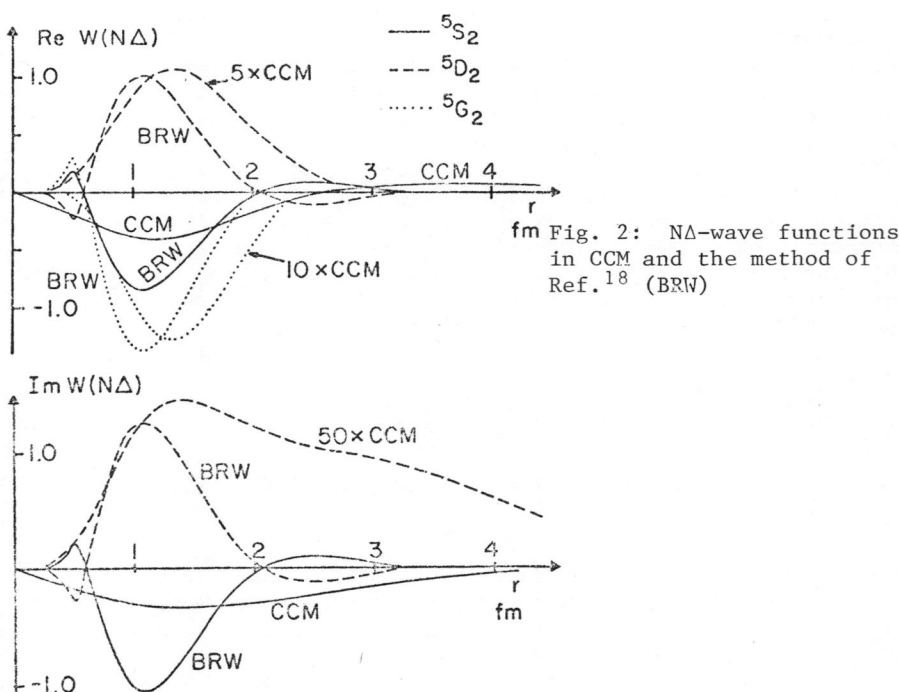

Fig. 2: NΔ-wave functions in CCM and the method of Ref.[18] (BRW)

the NΔ states $^5S_2$, $^5D_2$ and $^5G_2$, coupling to the $^1D_2$ (NN), the $^5G_2$ would be the largest and the $^5S_2$ (which, lacking the centrifugal barrier, should be by far the largest) is the smallest! These wave functions at 600 MeV are shown in fig. 2 together with the exact ones. As seen from table I, only the lucky suppression of the $^5D_2(N\Delta)$ and cancellation of the errors, and the fact that $^5G_2(N\Delta)$ does not give rise to p-wave pions, save this method from a disaster at first sight. But there is no a priori reason for this coincidence and so no reason to trust in this approximation. A reliable first order calculation for pp → d+π^+ still remains to be performed along the lines shown in the pioneering $N\Delta$ works, e.g. ref.[29].

Table I: Accumulation of contributions to $^1D_2$ (p-wave pion) partial wave in CCM and in the approach of ref.[18] (BRW) (amplitudes as defined in ref.[15])

	$^1D_2$(NN)	$^5S_2(N\Delta)$	$^5D_2(N\Delta)$	$p_2=\Sigma$	σ_{abs}(mb)
Re CCM	.105	.270	-.055	.320	8.4
Im CCM	.006	.621	-.030	.597	
Re BRW	.127	.464	-.146	.445	8.9
Im BRW	.023	.813	-.303	.533	

One drawback of the above coordinate space approach is the use of static potentials, which may be unrealistic at high energies. Rinat et al.[22] have attempted to overcome this by performing a coupled channels calculation in momentum space and keeping states with at most one pion. Although, in principle, this might be a step forward, in comparison with experiment their results are worse than those obtained in coordinate space. A simpler re-scattering study in momentum space is by Pong[21]. His results are insensitive on the ρ coupling constant, whereas in coordinate space the total cross section is sensitive on it. According to Pong, the difference could be due to a different treatment of the Green functions.

Recently resonances have been reported in pp scattering[30] at about CM energy corresponding to the $N\Delta$ mass. As these resonances most prominently occur in the same partial waves $^1D_2$ and $^3F_3$, where also the $N\Delta$ intermediate states are the most important, it is difficult to avoid the impression that these diproton or dibaryon resonances could be reflections of this underlying $N\Delta$ mechanism. However, in ref. [23] the extreme standpoint is taken that these resonances are totally distinct from the $N\Delta$ admixture effects and that their contribution should be added to the reaction amplitudes separately. The reason for this is the failure of the authors to reproduce the polarization asymmetry at 591 MeV with only the $N\Delta$ admixture. However, we shall see (fig. 7) that the CCM gives the asymmetry well enough also at this energy, thus eliminating the necessity of introducing ten extra free parameters in the model, as is done in ref.[23].

OBSERVABLES IN $p+p \to d+\pi^+$

Following and expanding the early work of Mandl and Regge[31] the cross sections can be presented as in table II.[16,32] The cross sections not given in the table can be obtained by the symmetries

$$(\sigma_{xo}, \sigma_{ox}, \sigma_{xx}, \sigma_{xz}, \sigma_{zx})_\phi = (\sigma_{yo}, \sigma_{oy}, \sigma_{yy}, \sigma_{yz}, \sigma_{zy})_{\phi+\pi/2}$$

$$\sigma_{oz} = \sigma_{zo} = 0, \quad \sigma_{yx} = \sigma_{xy} \tag{4}$$

$$A(\theta)_{\text{beam polarized}} = -A(\pi-\theta)_{\text{target polarized}}.$$

$d\sigma/d\Omega$	Partial cross section
σ_{00}	$32\pi\sigma_{00} = \gamma_0^{00} + \gamma_2^{00}\cos^2\theta + \gamma_4^{00}\cos^4\theta$
$\sigma_{00} + P_{By}\sigma_{yo}$	$32\pi\sigma_{yo} = (\lambda_0^{yo} + \lambda_1^{yo}\cos\theta + \lambda_2^{yo}\cos^2\theta + \lambda_3^{yo}\cos^3\theta + \lambda_4^{yo}\cos^4\theta)\sin\theta\cos\phi$
$\sigma_{00} + P_{By}\sigma_{yo} + P_{Ty}\sigma_{oy} + P_{By}P_{Ty}\sigma_{yy}$	$32\pi\sigma_{yy} = \gamma_0^{yy} + \gamma_2^{yy}\cos^2\theta + \gamma_4^{yy}\cos^4\theta + (\mu_0^{yy} + \mu_2^{yy}\cos^2\theta)\sin^2\theta\cos 2\phi$
$\sigma_{00} + P_{Bz}P_{Tz}\sigma_{zz}$	$32\pi\sigma_{zz} = \gamma_0^{zz} + \gamma_2^{zz}\cos^2\theta + \gamma_4^{zz}\cos^4\theta$
$\sigma_{00} + P_{By}\sigma_{yo} + P_{By}P_{Tz}\sigma_{yz}$	$32\pi\sigma_{yz} = (\lambda_0^{yz} + \lambda_1^{yz}\cos\theta + \lambda_2^{yz}\cos^2\theta + \lambda_3^{yz}\cos^3\theta + \lambda_4^{yz}\cos^4\theta)\sin\theta\sin\phi$
$\sigma_{00} + P_{By}\sigma_{yo} + P_{Tx}\sigma_{ox} + P_{By}P_{Tx}\sigma_{yx}$	$32\pi\sigma_{yx} = (\mu_0^{yx} + \mu_2^{yx}\cos^2\theta)\sin^2\theta\sin 2\phi$

Table II. Parametrization of the cross section for $pp \to d\pi^+$ when the beam and target are polarized (P_B and $P_T \neq 0$)

Here the notation corresponds to the geometry that the incident beam moves in the z direction (fig. 3), and all angles are in the CM system. Of course, conceptually the parametrization in terms of γ_i's, λ_i's etc. does not give anything new but an easy way to show the cross sections and asymmetries as a function of energy.

First, for the sake of completeness let us shortly review the unpolarized cross sections. The total cross section has a simple structure (fig. 4), which is fairly easy to reproduce by varying the NN and NΔ interactions (the πNΔ and ρNΔ coupling constants, form factors, NΔ diagonal interaction, possible non-static properties, etc.). This

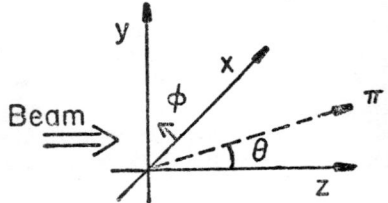

Fig. 3. The geometry of pion production experiments.

has been done early in ref.[28] and later in refs. [15-23]. In the coupled channels calculations of refs. 15-17 a fit has not been attempted, but rather the aim has been to find a characteristic "signature" of some particular interaction parameter in the hope to fix that parameter by data uniquely. Unfortunately the changes in observables were of the order of the variation of the interactions and one could produce a similar effect in various ways. Generally the changes scaled up or down with the total cross section so that the relative quantities, e.g. $A_{yo} = \sigma_{yo}/\sigma_{00}$ or $A = \gamma_o/\gamma_2$, vary

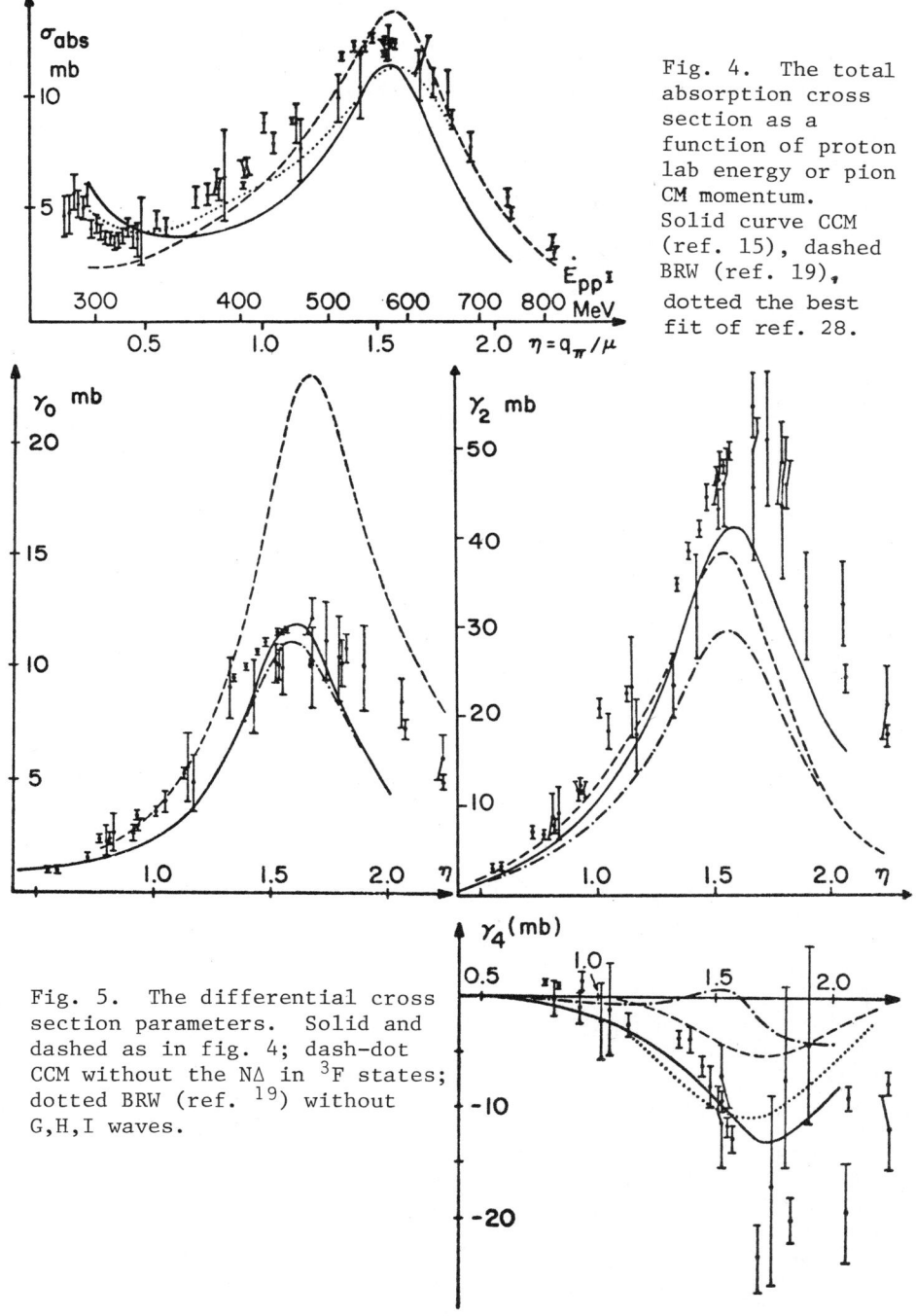

Fig. 4. The total absorption cross section as a function of proton lab energy or pion CM momentum. Solid curve CCM (ref. 15), dashed BRW (ref. 19), dotted the best fit of ref. 28.

Fig. 5. The differential cross section parameters. Solid and dashed as in fig. 4; dash-dot CCM without the $N\Delta$ in ^3F states; dotted BRW (ref. 19) without G,H,I waves.

rather little. One exception to this was adding the NΔ intermediate states to the NN wave $^3F_3$. Singling out some particular wave(s) makes a dramatic effect, in this case improvement, in the differential cross section and asymmetry due to interference effects (see figs. 5 and 6). It is worth noting that this extension of NN space is

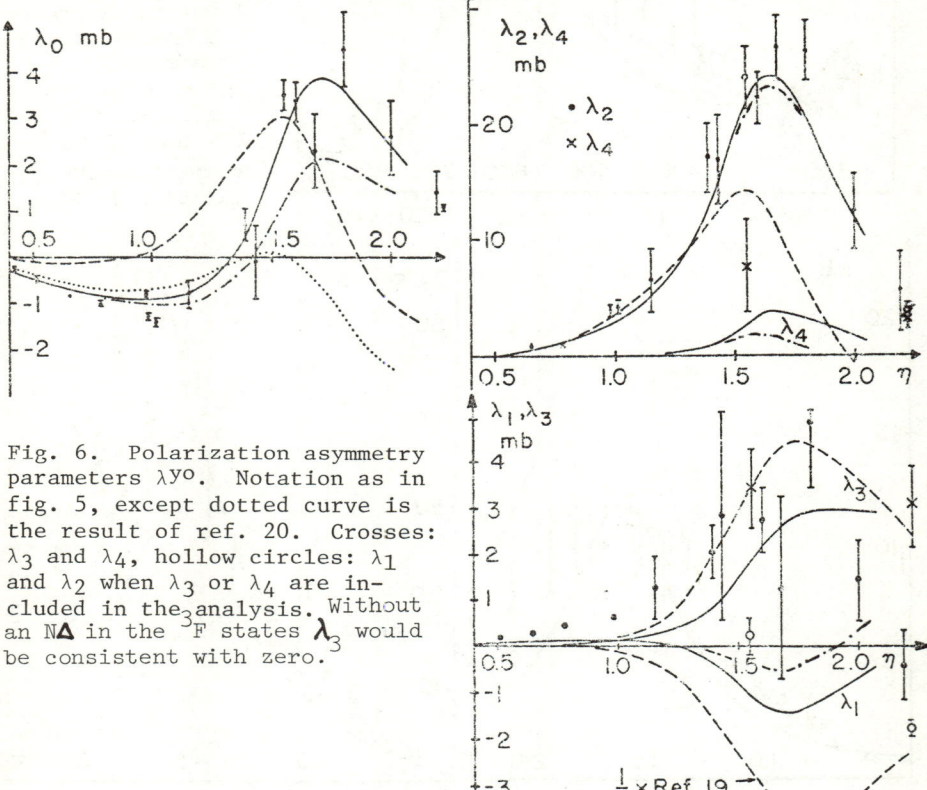

Fig. 6. Polarization asymmetry parameters λ^{yo}. Notation as in fig. 5, except dotted curve is the result of ref. 20. Crosses: λ_3 and λ_4, hollow circles: λ_1 and λ_2 when λ_3 or λ_4 are included in the analysis. Without an NΔ in the $^3F_3$ states λ_3 would be consistent with zero.

particularly important in one of the most celebrated dibaryon waves, and that most of the cross section at energies 400-700 MeV comes from $^5S_2$(NΔ) coupled to $^1D_2$(NN), another dibaryon wave. Unfortunately, F-wave nucleons are the highest included in refs. [15-17]. Possibly $^1G_4$ should be added too. Partial waves up to L=6 have been included in refs. [19-20], but, with the dubious method for calculating the elementary amplitudes the effect of the higher waves may be overestimated. The works performed in momentum space as well as the dibaryon analyses have only given the relative asymmetries directly as a function of scattering angle. In fig. 7 we show a sample of these results together with the dibaryon analysis of Kamo and Watari. The cross section with CCM tends to be too small (the same holds for ref.[22]) especially above the resonance, but the polarization asymmetry is fairly good.

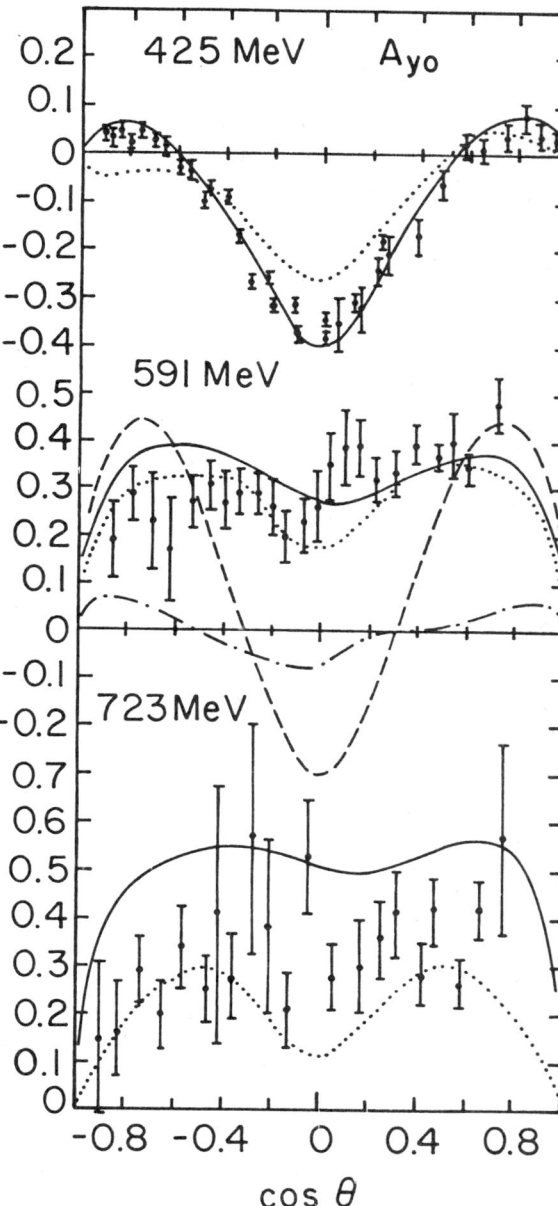

Fig. 7: The relative asymmetry. Solid curve CCM of ref.[15], dotted ref.[23] (dash-dot ref.[23] without the "dibaryon" contribution), dashed ref.[22].

Prior to the new generation of asymmetry measurements the parameter λ_3 was always omitted, and consequently λ_1 was grossly overestimated. The importance of λ_3 was first shown by theory[15] and its inclusion in the analyses considerably lowered λ_1 towards theory[5] and above 500 MeV λ_3 was in fact found to be much larger than λ_1.[6-8] But although theoretical values now are just at the lower end of error limits, theory clearly still needs refinement, which we shall shortly discuss in the last section.

The recent measurements at SIN[6] at 575 and 515 MeV with both spin and target polarized have brought more discrepancy between theory and experiment. Again, the CCM[16] (also in momentum space[22]) gives a qualitative picture (fig. 8), but in details one needs improvement. The dibaryon model[23] falls somewhat behind the $N\Delta$ models. In ref.[16] it was found that again changes in observables were of the order of changes of interactions, but as the accuracy of the data is good, one might finally have a handle on the interactions and on the reaction model.

Next step obviously is to try to measure the polarization of the final state deuteron. So far there is only one old data

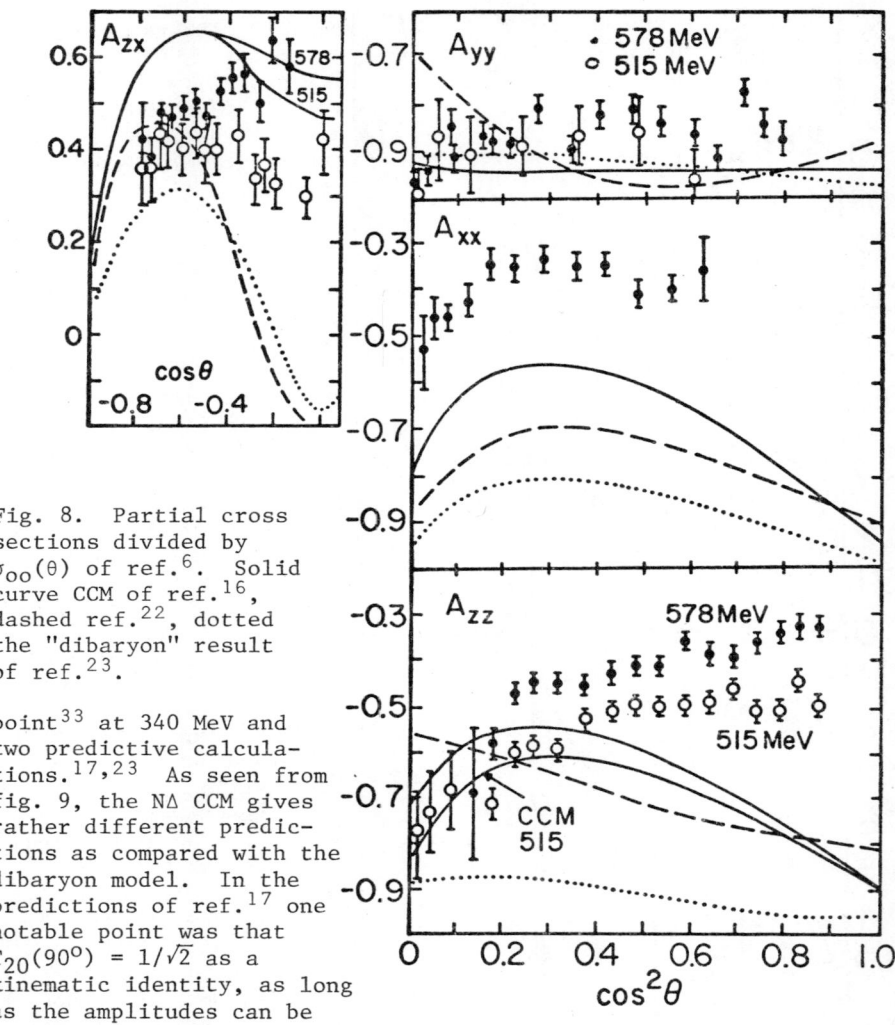

Fig. 8. Partial cross sections divided by $\sigma_{oo}(\theta)$ of ref.[6]. Solid curve CCM of ref.[16], dashed ref.[22], dotted the "dibaryon" result of ref.[23].

point[33] at 340 MeV and two predictive calculations.[17,23] As seen from fig. 9, the NΔ CCM gives rather different predictions as compared with the dibaryon model. In the predictions of ref.[17] one notable point was that $T_{20}(90^\circ) = 1/\sqrt{2}$ as a kinematic identity, as long as the amplitudes can be expressed in terms of matrix elements of an interaction. The prediction of Kamo and Watari does not satisfy this identity. Although at intermediate energies the measurement of the deuteron polarization poses many questions, the measurement of the vector polarization T_{11} is conceivable at present,[9] but the tensor polarization is still elusive. C. Wilkin[34] has, however, shown at forward or backward directions the relation $T_{20}(0^\circ \text{ or } 180^\circ) = 1/2\sqrt{2} \, (3A_{zz}(0^\circ \text{ or } 180^\circ)-1)$, which can be used to obtain T_{20} from the SIN data.[6] He also uses

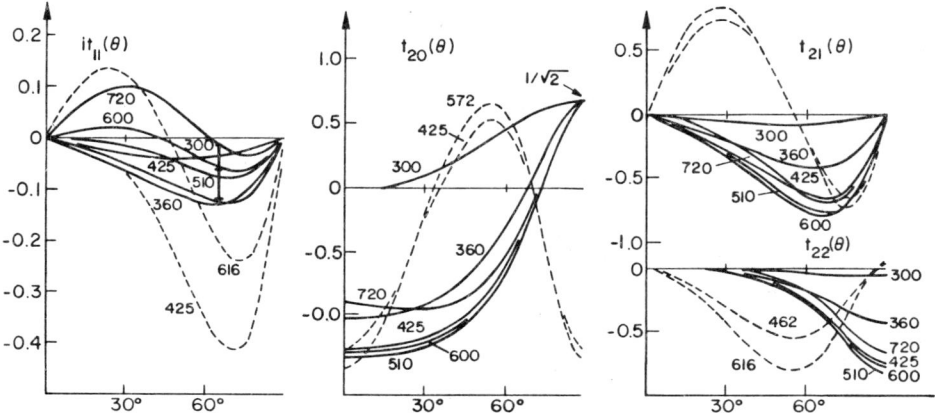

Fig. 9. Polarization of the final state deuteron at various energies as a function of the pion angle. Solid curves CCM (Ref. 17), dotted "dibaryon" calculation of Ref. 23. A numerical error in Ref. 17 is corrected, which effects the high energy t_{11}.

Fig. 10. Contributions in Wilkin's calculation

a simple model to relate graphs shown in fig. 10 to pion-nucleon scattering and calculates $T_{20}(180°)$. In fig. 11 we show his results along with the prediction of ref.[17]. Both models can give a qualitative description of $T_{20}(180°)$, but each needs improvement.

Another interesting finding by Wilkin is a direct proof of the pseudoscalarity of the pion basing on the data of SIN[35]. He shows the relation

$$A_{LL} - 2A_{TT} = 1 \text{ and } A_{LL} + 2A_{TT} = 1 \quad (5)$$

for a pseudoscalar and a scalar pion, respectively, where

$$\begin{aligned}
A_{LL} &= [\sigma^L(++) - \sigma^L(+-)]/2\bar{\sigma} \\
A_{TT} &= [\sigma^T(++) - \sigma^T(+-)]/2\bar{\sigma} \\
\bar{\sigma} &= \frac{1}{2}(\sigma^T(++) + \sigma^T(+-)) = \frac{1}{2}(\sigma^L(++) + \sigma^L(+-)) \\
\sigma^T &= \frac{1}{2}(\sigma^x + \sigma^y) ,
\end{aligned} \quad (6)$$

A_{LL} and A_{TT} are the longitudinal and transverse relative asymmetries. The experiment of SIN at 578 MeV indicates

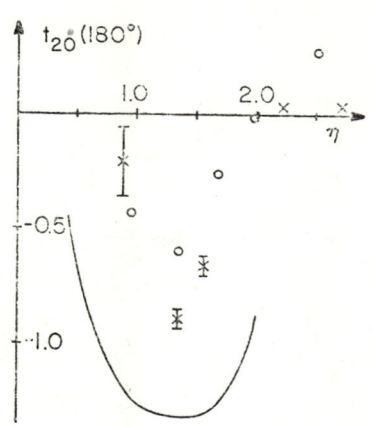

Fig. 11. Tensor polarization T_{20} at $180°$. Crosses experiment, circles Wilkin's calculation,[34] solid curve CCM (ref.[17]).

$$A_{LL} = -0.30 \pm 0.02, \quad A_{TT} = -0.60 \pm 0.02 \qquad (7)$$

which are consistent with the first relation (5), but strongly contradict the scalar pion relation.

THE REACTION $p+A \to (A+1) + \pi$

In pion production from nuclei two principal models appear on the market: various one nucleon models (ONM) and the two nucleon model (TNM). The essence of these models is shown in fig. 12. In ONM the incident proton interacts with the nucleus as a whole, emitting a pion. In different variations the incident proton wavefunction or/and the pion wave function may be distorted (the latter variant approaches TNM by partly taking into account

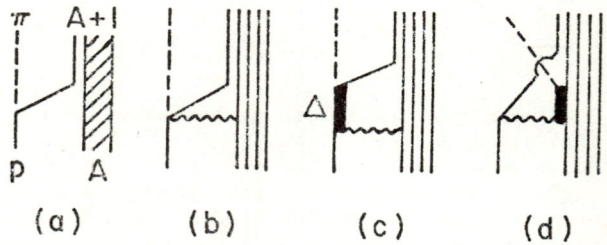

Fig. 12. A diagrammatic presentation of ONM (a) and TNM (b). The diagrams c) and d) present the process, which is dominant in $p+p \to d+\pi^+$, in nuclei.

the process of fig. 12d), but essentially the approach is rather similar to the "direct" part of $p+p \to d+\pi^+$ (fig. 1a). In TNM the incoming proton interacts with one of the nucleons of the target nucleus, emitting a pion. Presumably the excitation of one of the nucleons to a Δ would be the dominant contribution also in this case. In a way, the process could be characterized as a "quasifree $p+p \to d+\pi^+$."

There exist a multitude of calculations in ONM, but only a few in TNM[36-37] and even these few only calculate the cross sections.[44] It is possible to fit also ONM to the cross section data by suitable choice of optical potentials to share the transferred momentum among more particles, and the choice of the model is a question of dispute, which has often been solved to the favour of ONM due to its simplicity and due to a ready access to DWBA computer codes. In a systematic study of the energy dependence on the cross section of $^{40}Ca(p,\pi^+)\,^{41}Ca$ within ONM framework Höistad et al.[38] found first direct experimental evidence in disfavour of ONM. Unfortunately, it appears that there is no corresponding systematic study of the energy dependence within TNM.

The polarization asymmetry has been calculated only within ONM framework.[25-27] Generally these calculations give a large final state dependence, which in ONM seems natural, because the ONM

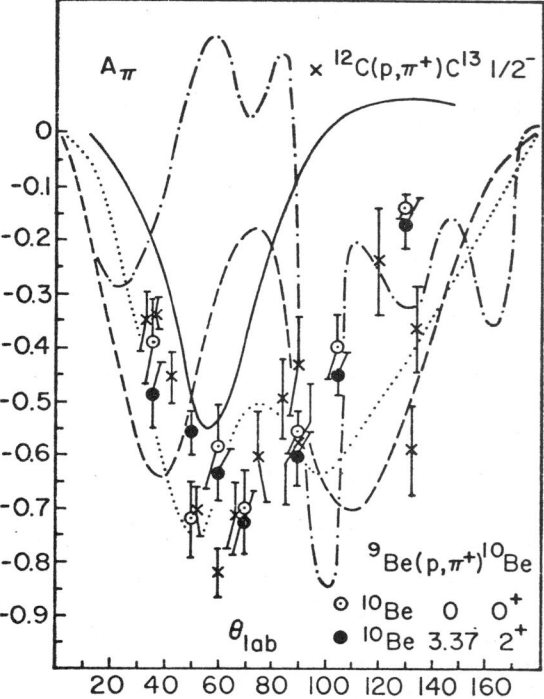

Fig. 13. Asymmetry in π^+ production on the light nuclei ^{9}B and ^{12}C at 200 MeV. The neutron goes into the shell model states $1p_{3/2}$ (g.s.) and $1p_{1/2}$ in ^{10}Be and into $1p_{1/2}$ in ^{13}C. Data for other excited states of ^{10}Be and ^{13}C are very similar.[11] Theoretical results are from ref.[27]: dashed $1p_{1/2}$, dotted $1d_{5/2}$, dash-dot $2s_{1/2}$, and solid Ay ($pp \to d\pi^+$) kinematically transformed.

amplitudes are Fourier like integrals of the bound state wave function (practically the so called form factor at a large transferred momentum), to which the incident nucleon is captured. However, experimentally perhaps the most striking feature has been the insensitivity of the asymmetry on the particular nucleus or the final state in question. All the ONM calculations so far fail to give this behaviour, as seen from Fig. 13, where the most sophisticated ONM calculation of Ref. 27 has been used as example. The similarity to the asymmetry in

$pp \to d\pi^+$, also shown, may be indicative of two nucleon reaction mechanism. A TNM approach by W. R. Gribbs[44] gives a different state dependence than ONM: For the final states published before this conference he finds much more invariant results than ONM, qualitatively similar to the data, but variation in other states. The results of TRIUMF[45] on $(\vec{p},\pi^+)$ and $(\vec{p},\pi^-)$ reactions do indeed show strong state dependence in the newly resolved higher excited states. Many of these states have not been seen earlier by other methods.

WHAT CAN BE LEARNED

The theory of $p+p \to d+\pi$ should be improved aiming at the best possible fit to all measured quantities. Some obvious improvements in CCM can be pointed out immediately[39]: a better treatment of the Δ width taking into account the NNπ final states (so far the width is only given by πd final state kinematics in refs.[14-20]) should improve high energy cross section somewhat and use of NΔ effective mass in the NΔ channel together with ρNΔ coupling adjustment may also give an improvement. Form factors in s-wave rescattering seem to have effects of 10-20% at the threshold, possible ambiguities in the propagator are more important. Taking into account the graph of fig.14 effectively strengthens the transition potential V_2 for large r, thus presumably enhancing $^3F_3$ amplitude and increasing γ_2, γ_4, λ_3 and λ_4 without affecting the already good γ_0 and λ_2. In the light of recent experiments[3,6,8] this appears an improvement. If a good description of $p+p \to d+\pi^+$ (and in another context pp scattering[40]) is obtained by the isobar model, this would strongly support that model as the underlying diproton resonance mechanism. In this search perhaps some uncertain interaction coefficients, e.g. ρNΔ coupling constant, could be fixed. If an acceptable description of all observables in $p+p \to d+\pi^+$ cannot be found, one should first try some more conventional explanation before introducing new phenomenological parameters. At energies above 700 MeV the N'(1470) resonance mixture might be one candidate, relativistic effects another one. Calculationally, no extremely simple and reliable alternative to some kind of CCM is in sight, though a first order perturbation theory should be attempted.

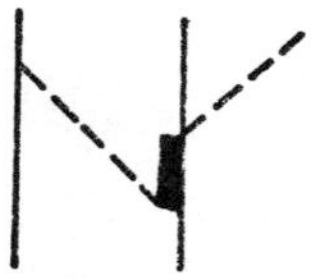

Fig. 14: Process omitted in pure CCM

Experimentally at least above 500 MeV a "complete experiment" to determine all the necessary amplitudes seems still far away due to the great number of amplitudes (perhaps 10-12 complex numbers). However, it appears useful to achieve as complete data as possible with a polarized beam and target, and also attempt to measure the vector and tensor polarizations of the final state deuteron. All of these quantities seem to give new, independent information.

In the reactions $A(p,\pi)$ A+1 currently the open question still appears to be the reaction mechanism. To solve this more experimental data are needed to show the energy dependence of the cross section and of the asymmetry. Asymmetry measurements from 200 MeV up to 400 MeV and a comparison with the energy dependence of $A_y(pp \to d\pi^+)$ would be enough to give hints whether the latter can be taken as a model for the more general reaction. Of course, a systematic theoretical study with TNM is urgently needed for the above quantities and their dependence on energy and nucleus. If a successful description is not obtained, the reason of failure has to be found from new contributions and/or many body corrections. E.g., in nuclei there is nothing against having a single Δ admixture in the final state, and we are back in the study of Δ components in nuclei. An interesting study will also be the asymmetry and the energy dependence of the cross section in the suppressed (p,π^-) channel[41-42], which should be a genuine two nucleon process. At 800 MeV Höistad et al.[41] find a slope in $d\sigma/d\Omega$ (p,π) which is not seen at lower energies and is not understood at present. The recent TRIUMF data[45] on (p,π^+) and (p,π^-) reactions where many high excitations were resolved may revive the hope of using these reactions as probes also in nuclear structure research. In particular, the latter one can be used in getting information on nuclei off the stability region which might be impossible to obtain otherwise.

REFERENCES

1) V.M. Preedom et al., Phys. Lett. 65B, 31(1976).
2) W. Hürster et al., Phys. Lett. 91B, 214 (1980).
3) J. Hoftiezer et al., contr. to this conference 1.19.
4) D. Aebischer et al., Nucl. Phys. B108, 214 (1976), see references here for earlier work on $p+p \rightleftharpoons d+\pi^+$.
5) P. Walden et al., Phys. Lett. 81B, 156 (1979).
6) E. Aprile et al., Proc. 8th Int. Conf. on High Energy Physics and Nuclear Structure, Vancouver 1979, Nucl. Phys. A335, 245 (1980); E. Aprile et al., contr. to this conference no. 1.18; R. Hess, private communication.
7) H. Nann et al., Phys. Lett. 88B, 257 (1979).
8) R. Joseph, private communication, Thesis 1979.
9) B.E. Bonner, private communication.
10) B. Höistad et al., Nucl. Phys. A319, 409 (1979); see references here for earlier work on $A(p,\pi)A+1$.
11) E.G. Auld et al., Phys. Rev. Lett. 41, 462 (1978).
12) P.H. Pile, Bull. Am. Phys. Soc. 24, 614 (1979).
13) G. Jones, Proc. 2nd Int. Topical Conf. on Meson-Nuclear Physics, Houston 1979.
14) A.M. Green and J.A. Niskanen, Nucl. Phys. A271, 503 (1976).
15) J.A. Niskanen, Nucl. Phys. A298, 417 (1978).
16) J.A. Niskanen, Phys. Lett. 79B, 190 (1978).
17) J.A. Niskanen, Phys. Lett. 82B, 187 (1979).

18) M. Brack, D.O. Riska and W. Weise, Nucl. Phys. A287, 425 (1977).
19) J. Chai and D. O. Riska, Nucl. Phys. A338, 349 (1980).
20) O.V. Maxwell, W. Weise and M. Brack, preprint, to be published in Nucl. Phys.; O.V. Maxwell and W. Weise, preprint, to be published in Nucl. Phys.
21) W. S. Pong, preprint, Thesis, Univ. of Colorado 1979.
22) A.S. Rinat, Y. Starkand and E. Hammel, preprint.
23) H. Kamo and W. Watari, Prog. Theor. Phys. 62, 1035 (1979).
    H. Kamo and W. Watari, preprint.
24) M.A. Braun, W. Kallies and V.M. Suslov, Sov. J. Nucl. Phys. 27, 72 (1978).
25) J.V. Noble, Nucl. Phys. A244, 526 (1975).
26) S.K. Young and W. R. Gibbs, Phys. Rev. C17, 837 (1978).
27) H.J. Weber and J.M. Eisenberg, Nucl. Phys. A312, 201 (1978).
28) B. Goplen, W.R. Gibbs and E. Lomon, Phys. Rev. Lett. 32, 1012 (1974).
29) H. Arenhövel, M. Danos and H.T. Williams, Nucl. Phys. A162, 12 (1971).
30) See e.g., J.B. Roberts, invited talk at this conference, 1.B.
31) F. Mandl and T. Regge, Phys. Rev. 99, 1478 (1955).
32) C. Weddigen, Nucl. Phys. A312, 330 (1978).
33) R.D. Tripp, Phys. Rev. 102, 862 (1956).
34) C. Wilkin, private communication.
35) C. Wilkin, J. Phys. G6, L5 (1980).
36) Z. Grossman, F. Lenz and M.P. Locher, Ann. Phys. 84, 348 (1974).
37) M. Dillig and M.G. Huber, Phys. Lett. 69B, 429 (1977).
38) B. Höistad, S. Dahlgren, T. Johansson and O. Jonsson, Nucl. Phys. A319, 409 (1979).
39) J.A. Niskanen, work in progress.
40) J.A. Niskanen and A.W. Thomas, work in progress.
41) B. Höistad et al., Phys. Rev. Lett. 43, 487 (1979).
42) Indiana group contr. Int. Conf. on Nucl. Phys., Berkeley (1980).
43) E.G. Auld et al., Phys. Lett. 93B, 258 (1980).
44) W. R. Gibbs, Proc. of Workshop on Nuclear Structure with Intermediate-Energy Probes, Los Alamos, New Mexico 1980; W.R. Gibbs, private communication.
45) E. G. Auld, invited talk at this conference, 1.E; G. J. Lolos contr. to this conference, 2.55.

An APPENDIX is attached on a proposal
for a notation convention.

(Editor's Note: The references to the conference are given as paper numbers and are cross-indexed to page numbers in the Table of Contents.)

APPENDIX: A PROPOSAL FOR THE NEW NOTATION OF
THE OBSERVABLES IN THE REACTION $p+p \rightarrow d+\pi^+$ *

A proposal for the new notation using orthogonal angular functions is made for differential cross sections and final state polarization in $p+p \rightarrow d+\pi^+$ with polarized or unpolarized initial state. Kinematic forms of observables are shown.

On several occasions, experimentalists (a.o. R. Hess, C. Hollas, K. Seth, Ch. Weddigen) have asked me why we theorists keep on using powers of $\cos \theta$ in expressing the cross sections and asymmetries in $pp \rightarrow d\pi^+$, while orthogonal functions like the Legendre polynomials in $\cos \theta$ have obvious advantages. I have actually been urged to make a proposal for a new notation. With the present accuracy and energies, experiments could and would measure coefficients as high as γ_6, at least in the cross sections, but with strong correlations between these coefficients the inclusion or exclusion of γ_6 give drastic effects on the lower coefficients like γ_4. With orthogonal angular functions, this unpleasant behavior does not arise. So clearly experimental analyses would profit from using orthogonal angular functions, whereas in theory the advantage is not so obvious (except for perhaps in case of spherical harmonics), while the transformation between the two representations would be trivial. On the other side, theorists have not been specially encouraged to use orthogonal functions, because so far, practically all data are given either directly in form of angular distributions or as coefficients of $\cos \theta$ powers! Apparently, the old presentation of Mandl and Regge[1]) has become obsolete, and a new common language is needed. Here I propose a notation which, using orthogonal basis functions, extends the definitions of Mandl and Regge and includes the kinematic restrictions imposed on the angular dependence of the observables.

As to what basis functions to use, theoretically spherical harmonics would have some advantage over the others: all products of spherical harmonics in the amplitudes can be easily reduced to a single spherical harmonic by the Clebsch-Gordan coefficients. The obvious, experimentally decisive, disadvantage is that spherical harmonics are complex and orthogonal in spatial angles, not in a plane. Most often, experiments are performed in one plane only and also usually the dependence on the azimuthal angle ϕ is superfluous and can be determined from the kinematics. So, taking into account that the need for orthogonality is <u>experimentally</u> motivated we should not go to spherical harmonics. The same objection would hold partly for the associated Legendre functions $P_k^m(\cos\theta)$. However, these are real functions of only one angle θ, they do not have (for plane) artifical normalization factors and, luckily, the orthogonality relation

$$\int_{-1}^{1} P_k^m(x) P_{k'}^m(x) dx = \delta_{kk'} \frac{2(k+m)!}{(2k+1)(k-m)!} \tag{1}$$

is enough for the present purpose. These functions introduce the $\sin\theta$ dependence in a natural way and, besides, they have been used earlier to describe polarization asymmetries at low energy.*

My proposal for the representation of the differential cross section $\sigma_{oo}(\theta)$ and of the partial cross sections σ_{ij} (where i and j denote the beam and target polarization, respectively; I assume the same geometry as in my previous talk, beam along the z-axis, which is consistent with the Madison convention[4]; the changeover to the Ann Arbor convention[5] is trivial by assuming $\phi=0$ and replacing $z \to L$, $y \to N$, and $x \to S$) as well as of the final state deuteron polarization (multiplied by σ_{oo}) is:

1. Take out ϕ-dependence as a common multiplicative factor (cos nϕ, sin nϕ, $e^{in\phi}$, n=0,1,2,3);
2. Use a common overall factor $1/4\pi$;
3. Use latin alphabets for the real coefficients of $P_k^m(\cos\theta)$ as follows

a_k: the ϕ independent (m=0, only even k) parts of the partial cross sections (analogous to the γ's of Refs. 1-3

b_k: the parts of the partial cross sections $\alpha \sin\phi$, $\cos\phi$ (m=1; analogous to the λ's of Refs. 1-3).

c_k: the parts of the partial cross sections $\alpha \sin 2\phi$, $\cos 2\phi$ (m=2 and k only even; analogous to the μ's of Refs. 2-3).

d_k-h_k: the deuteron polarization observables $\sigma_{oo} it_{11}$, $\sigma_{oo} t_{10}$, $\sigma_{oo} t_{20}$, $\sigma_{oo} t_{21}$, $\sigma_{oo} t_{22}$, respectively.

Therefore, in general, a partial cross section corresponding to the initial state polarization ij is of the form

$$\sigma_{ij} = 1/4\pi\, f(\phi) \sum_k P_k^{ij} P_k^m(\cos\theta) \tag{2}$$

where now P_k^{ij} stands for a_k^{ij}, b_k^{ij} or c_k^{ij}.
In some cases it is necessary to use two different parts for one partial cross section. Note that here the overall factor is $1/4\pi$ instead of the old $1/32\pi$, which seems preferable, as with orthogonal polynomials a_o^{oo} is directly the total cross section.

* I thank G. Jones for pointing this out to me.

Table I: Parametrization of the differential cross sections as a function of the pion CM angle for various initial state polarizations. The table is superfluous: σ_{yy} and σ_{xx} are related by Eq. (2), σ_{yx} is not so directly obvious. Six observables are enough to exhaust the information obtainable from initial spin correlations.

P_B	P_T	$d\sigma/d\Omega$
0	0	σ_{oo}
y	0	$\sigma_{oo} + P_{By}\sigma_{yo}$
y	z	$\sigma_{oo} + P_{By}\sigma_{yo} + P_{By}P_{Tz}\sigma_{yz}$
z	z	$\sigma_{oo} + P_{Bz}P_{Tz}\sigma_{zz}$
y	y	$\sigma_{oo} + P_{By}\sigma_{yo} + P_{Ty}\sigma_{oy} + P_{By}P_{Ty}\sigma_{yy}$
x	x	$\sigma_{oo} + P_{Bx}\sigma_{yo} + P_{Tx}\sigma_{ox} + P_{Tx}\sigma_{xx}$
y	x	$\sigma_{oo} + P_{By}\sigma_{yo} + P_{Tx}\sigma_{ox} + P_{By}P_{Tx}\sigma_{yx}$

$$4\pi\sigma_{oo} = \sum_{k\ even} a_k^{oo} P_k(\cos\theta)$$

$$4\pi\sigma_{yo} = \cos\phi \sum_k b_k^{yo} P_k^1(\cos\theta)$$

$$4\pi\sigma_{yz} = \sin\phi \sum_k b_k^{yz} P_k^1(\cos\theta)$$

$$4\pi\sigma_{zz} = \sum_{k\ even} a_k^{zz} P_k(\cos\theta)$$

$$4\pi\sigma_{yy} = \sum_{k\ even} a_k^{yy} P_k(\cos\theta) + \cos 2\phi \sum_{k\ even} c_k^{yy} P_k^2(\cos\theta)$$

$$4\pi\sigma_{xx} = \sum_{k\ even} a_k^{yy} P_k(\cos\theta) - \cos 2\phi \sum_{k\ even} c_k^{yy} P_k^2(\cos\theta)$$

$$4\pi\sigma_{yx} = -\sin 2\phi \sum_{k\ even} c_k^{yy} P_k^2(\cos\theta)$$

The definitions of the σ_{ij}'s are summarized in Table I. As in Ref. 2 we omit the unnecessary $\sqrt{2}$'s introduced in Ref. 3. Note also that the set shown in Table 1 is incomplete: only six partial cross sections are independent. One can get his or her favorite set by applying rotations and reflections as

$$(\sigma_{xo}, \sigma_{xx}, \sigma_{xz})_\phi = (\sigma_{yo}, \sigma_{yy}, \sigma_{yz})_{\phi+\pi/2} \quad \text{(rotation)} \tag{3}$$

$$\sigma_{TB}(\theta,\phi) = \sigma_{BT}(\pi-\theta, \phi+\pi) \quad \text{exchange of the beam and target}$$

The deuteron polarization quantities $\sigma_{oo}it_{11}$ and $\sigma_{oo}t_{2q}$ with an unpolarized initial state are all proportional to $e^{iq\phi}$. As before with initial spin correlations, if the initial state is polarized one gets additional terms with different ϕ dependences. Giving each case a new letter would lead to a notation explosion, so it is most practical to use indices: let the superscripts on the right still stand for the initial state polarization and use a superscript on the left to denote the ϕ dependence of the particular term (the multiplier of ϕ). Again, these polarization transfer coefficients t_{kq}^{ij} are the new contributions due to the initial state polarization, to be added to the final state polarization without initial polarization $t_{kq}^{(oo)}$. In Table II we show the kinematic forms of the deuteron polarization along with the simplest polarization transfer coefficients, where the incident beam is polarized. It should be noted that – especially if combined with Table I – all the observables cannot be independent, because the number of spin configuration amplitudes (or t-matrix elements in spin space) is six requiring only eleven observables to be measured at a given angle and energy for complete information. Note, however, that the number of necessary partial wave amplitudes to be determined from the angular dependence is 7-9, perhaps even 12, depending on the energy.

If, as usual, the ϕ dependence is not resolved and the measurement is in the $\phi=0$ plane, the ϕ index can be left out and the coefficient represents the sum of different coefficients. As an example, if the incident beam is polarized in the y-direction, the vector polarization is

$$\sigma_{oo}it_{11/ypol} = [e^{i\phi} \sum_{k\ even} d_k P_k^1(\cos\theta)$$
$$+ P_{By} \sum_k {}^o d^{yo} P_k(\cos\theta)$$
$$+ P_{By} e^{i2\phi} \sum_k {}^2 d^{yo} P_k^2(\cos\theta)] \qquad (4)$$
$$= \sigma_{oo}it_{11}^{oo} + P_{By}\, \sigma_{oo}it_{11}^{yo}$$

and

$$\sigma_{oo}t_{10/ypol} = P_{By}\sin\phi \sum_k {}^1 e_k^{yo} P_k^1(\cos\theta)$$
$$= P_{By}\, \sigma_{oo}\, t_{10}^{yo} \qquad (5)$$

A measurement in the $\phi=0$ plane would give the coefficients $d_k^{yo} = {}^o d^{yo} + {}^2 d^{yo}$ and zero for t_{10}. In some cases, on the other side, it may be advantageous even to exploit the substructure within a given part of a polarization transfer coefficient. If these substructures are resolved (in the cases shown in Table II the two opposite parity parts make it easy and natural), one could

TABLE II: The deuteron polarization quantities t_{kq}.

$$4\pi\sigma_{oo} it_{11}^{oo} = e^{i\phi} \sum_{k \text{ even}} d_k^{oo} P_k^1(\cos\theta)$$

$$4\pi\sigma_{oo} t_{10} = 0$$

$$4\pi\sigma_{oo} t_{20} = \sum_{k \text{ even}} f_k^{oo} P_k(\cos\theta)$$

$$4\pi\sigma_{oo} t_{21} = e^{i\phi} \sum_{k \text{ even}} g_k^{oo} P_k^1(\cos\theta)$$

$$4\pi\sigma_{oo} t_{22} = e^{2i\phi} \sum_{k \text{ even}} h_k^{oo} P_k^2(\cos\theta)$$

$$4\pi\sigma_{oo} it_{11}^{yo} = \sum_k {}^0d_k^{yo} P_k(\cos\theta) + e^{2i\phi} \sum_k {}^2d_k^{yo} P_k^2(\cos\theta)$$

$$= \sum_{k \text{ odd}} {}^0d_k^{,yo} P_k(\cos\theta) + \sum_{k \text{ even}} {}^0d_k^{,,yo} P_k^2(\cos\theta)$$

$$+ e^{2i\phi} \sum_k {}^2d_k^{yo} P_k^2(\cos\theta)$$

$$4\pi\sigma_{oo} it_{11}^{zo} = ie^{i\phi} \sum_k {}^1d_k^{zo} P_k^1(\cos\theta)$$

$$4\pi\sigma_{oo} t_{10}^{yo} = \sin\phi \sum_k {}^1e_k^{yo} P_k^1(\cos\theta)$$

$$4\pi\sigma_{oo} it_{10}^{zo} = \sum_k e_k^{zo} P_k(\cos\theta)$$

$$= \sum_{k \text{ even}} e_k^{,zo} P_k(\cos\theta) + \sum_{k \text{ odd}} e_k^{,,zo} P_k^2(\cos\theta)$$

$$4\pi\sigma_{oo} t_{20}^{yo} = \cos\phi \sum_k {}^1f_k^{yo} P_k^1(\cos\theta)$$

$$t_{20}^{zo} = t_{20}^{oz} = 0$$

$$4\pi\sigma_{oo} t_{21}^{yo} = \sum_k {}^0g_k^{yo} P_k(\cos\theta) + e^{2i\phi} \sum_k {}^2g_k^{yo} P_k^2(\cos\theta)$$

$$= \sum_{k \text{ odd}} {}^0g_k^{,yo} P_k(\cos\theta) + \sum_{k \text{ even}} {}^0g_k^{,,yo} P_k^2(\cos\theta)$$

$$+ e^{2i\phi} \sum_k {}^2g_k^{yo} P_k^2(\cos\theta)$$

$$4\pi\sigma_{oo} t_{21}^{zo} = ie^{i\phi} \sum_k g_k^{zo} P_k^1(\cos\theta)$$

$$4\pi\sigma_{oo} t_{22}^{yo} = e^{i\phi} \sum_k {}^1h_k^{yo} P_k^1(\cos\theta) + e^{3i\phi} \sum_k {}^3h_k^{yo} P_k^3(\cos\theta)$$

$$4\pi\sigma_{oo} t_{22}^{zo} = ie^{2i\phi} \sum_k {}^2h_k^{zo} P_k^2(\cos\theta)$$

use a prime and a double prime on the coefficients d-h. In this case the form given in the examples of Table II is by no means the only one. It might be advantageous to take out e.g., the $\sin^2\theta$ factor explicity and, in particular, one should note that <u>all</u> t_{11} and t_{21} have actually <u>cos θ as a common factor</u> (i.e., the <u>lowest</u> power of cos θ in a power expansion would be one). It seems best to leave the definition of primed quantities free so that in each context there would be some freedom to choose the most practical notation.

I propose that in future all experimental and theoretical results will be given with the orthogonal angular functions as presented above. During the transitional period it appears useful, for comparison with the old works, to give results also in the old way, though this would be multiple presentation of the same thing. In the simplest cases of spin correlations and final state polarizations with an unpolarized initial state the above notation could be readily used, whereas in the more complicated polarization transfer coefficients it may be advantageous for clarity to specify the notation in the context.

REFERENCES FOR APPENDIX

1) F. Mandle and T. Regge, Phys. Rev. <u>99</u>, 1478 (1955).
2) J. A. Niskanen, Phys. Lett. <u>79B</u>, 190 (1978).
3) Ch. Weddigen, Nucl. Phys. A312, 330 (1978).
4) Madison convention, in Proc. 3rd Int'l Symp. on Polarization Phenomena in Nuclear Reactions, eds. H. H. Barschall and W. Haeberli (Madison, University of Wisconsin Press, 1970) p. XXV.
5) Ann Arbor convention, AIP Conf. Proc. No. 42, p. 142 (1977).

DISCUSSION

UEDA: Dr. Niskanen mentioned a strong coupling between the πd channel and the $^3F_3$ di-proton. However, Araki, Koike, and Ueda, myself, obtained a different result. We treated πNN system with I=1, $J^P \simeq 3^-$ in a Faddeev formalism by introducing the P_{11}πN, P_{33}πN, $^3P_2$NN and $^3S_1$NN interactions and could reproduce the $^3F_3$ phase shift and absorption coefficient obtained in Hoshizaki's phase shift analysis, qualitatively. Then we calculated the decay ratio into the πd channel from the $^3F_3$ di-proton and found so small a ratio as o.1%. The reason is simple. The main component of the $^3F_3$ di-proton πNN wave function is the one which has the P states for the sub NN system, while in the πd channel deuteron has S and D states. Therefore, πNN wave functions are othogonal, and then the coupling becomes so small.

NISKANEN: Essentially I would think that in your calculation there is much overlap with mine. But not being very familiar with your work I cannot say why your contribution from $^3F_3$ is so small. In my work the importance of $^3F_3$ comes from a constructive interference of the (rather strongly coupled) $N\Delta$ components and the NN state.

SETH: My question concerns the general lack of success of your calculations at $T_{(p)} \gtrsim 700$ MeV. As you know very well at $T_{(p)} = 800$ MeV, where we have done our experiments at LAMPF, your calculations fail to reproduce σ_T for $p+p \to d+\pi^+$ by a factor ~ 3. Further, even the signs of the asymmetry coeffs. λ_i are predicted incorrectly. So obviously something has to be done. Even though I agree that Chai and Riska may be overestimating contributions of high partial waves, don't you think that you need to include higher partial waves, especially G-wave, in your own work?

NISKANEN: Yes, I agree perfectly. In particular, it seems that $^1G_4$ protons have again a strong coupling to $N\Delta$ and are certainly to be included. My only defense for not including this state is that at the time of my main calculations even ^3F-waves were considered high. Of the higher waves I cannot say, yet, but I think that the effect found by e.g., Maxwell and Weise is an overestimate. I would point out also that the $^1G_4$ state is needed to give a γ_6 at high energies. I also refer to the improvement suggestions in my conclusion which will improve the total cross section at high energies; the effect on the other observables remains to be seen.

SETH: My comment concerns your conclusion about dibaryons. I can not help wondering if we are getting into semantical differences. There are those who call for resonances in $^1D_2$ and $^3F_3$ partial waves in the pp system and they call them "dibaryons." And there is you who finds the same partial waves very important but you like to call it $N\Delta$ channel coupling effects. Since $N\Delta$ is also a "dibaryon" are we not talking about similar, or even the same things, but differing in the semantics?

NISKANEN: For me it is really semantics, but there seem to be people who consider the dibaryons a distinct, even independent, entity from the $N\Delta$ states.

THE INTERPRETATION OF RECENT MEASUREMENTS OF NP AND PP CROSS SECTIONS

J. A. Edgington
Queen Mary College, Mile End Road, London E1 4NS, U.K.

ABSTRACT

I present recent measurements of total and differential np cross sections and compare predictions of the subsequent partial wave analysis with data from LAMPF. A new measurement of $\Delta\sigma_T$(pp) is described and the results of this, and experiments at SIN, interpreted in terms of singlet dominance of the inelastic amplitudes below 600 MeV.

NP CROSS SECTIONS

This spring the BASQUE group reported[1] on np polarisation parameters measured at TRIUMF and the improvements they wrought to Bugg's partial wave analysis. We have now completed a study of np total, and differential elastic, cross sections; some experimental details and preliminary results were reported at Vancouver last year.[2] Figure 1 shows the final values. Our use of calibrated beam monitors ties the two parts of the angular range together, but systematic uncertainties, notably in the variation of neutron detection efficiency with energy, re-impose a relative normalisation error of about 3% between them. The data have been incorporated into the partial wave analysis, to which they contribute a χ^2 of order one per datum. The solutions are greatly stabilised; errors on I = 0 phases are everywhere reduced and the central, tensor and spin-orbit combinations of D and G waves now vary smoothly with energy. Some earlier measurements near 200 MeV from Dubna[3] have been rejected in favour of our new data.

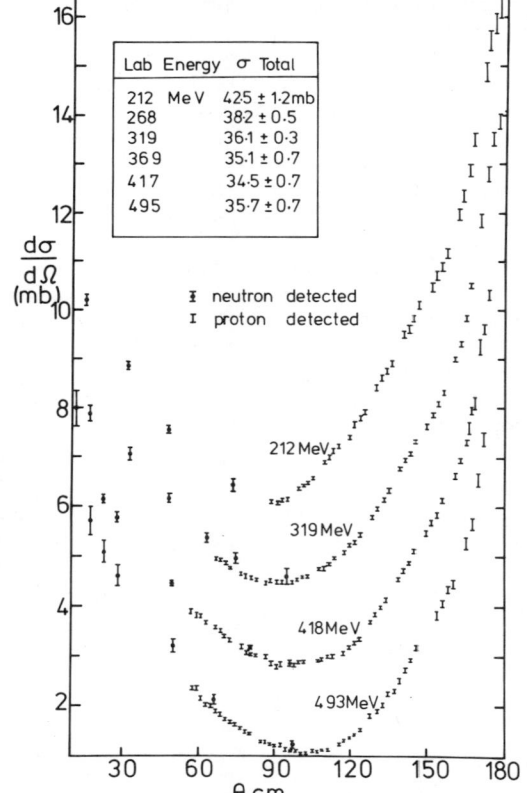

Figure 1. Total, and differential elastic, cross sections in np scattering.

Lab Energy	σ Total
212 MeV	42.5 ± 1.2 mb
268	38.2 ± 0.5
319	36.1 ± 0.3
369	35.1 ± 0.7
417	34.5 ± 0.7
495	35.7 ± 0.7

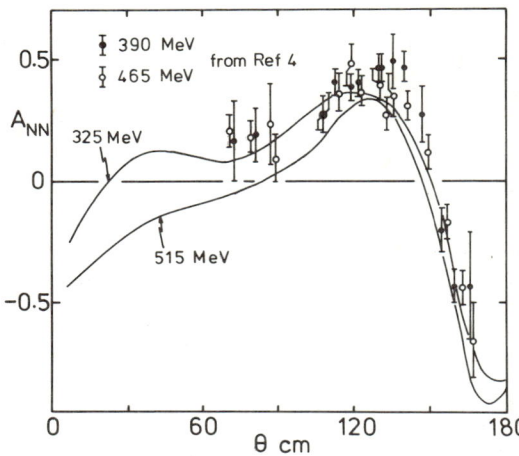

Figure 2. A_{NN} in np scattering; curves are from Bugg's partial wave analysis.

The np observables have been re-calculated using this new analysis, and in Figure 2 the predictions for A_{NN} are compared with the data taken at LAMPF by Bhatia et al, and presented at this Conference[4]. The fit seems to be at least as good as that of the Arndt-VerWest analysis. The extrapolation of $K_{LL}(0^\circ)$ to 800 MeV has also been revised. The new value is 0.62, in agreement with the value of 0.64± 0.03 measured in quasi free scattering and reported here by Riley et al[5].

A significant change in the analysis compared with that reported earlier[1] is that the central combination of G waves is constrained to be positive, in conformity with the very well defined D wave combination. Other changes are insignificant. An important restriction, that all I = 1 inelasticities except $^1D_2$ be fixed at Green and Sainio's values[6], is imposed, although in an analysis published last year[7] and restricted to I = 1 amplitudes this constraint was removed. The fit to existing elastic data is excellent, but we shall see that the reaction channels still give problems.

A NEW MEASUREMENT OF $\Delta\sigma_T(pp)$

The BASQUE group has recently measured $\Delta\sigma_T(pp)$ at TRIUMF. The results are interesting and merit some attention to experimental details. The technique is conventional, the transmission through a polarised target being measured by six circular scintillation counters T1 - T6 (see Figure 3); coincidences are taken between adjacent pairs to eliminate noise. The incident beam passes through S1 and S2, split counters used for horizontal and vertical centering, and S3, a counter of diameter 1 cm mounted in a long air lightguide.

A new TRIUMF beamline, 4C, was used. It is designed to transmit 2×10^5 protons per second, with spin oriented either vertically, as here, or longitudinally. The low intensity is necessary if the fraction of micropulses containing two or more protons is to remain below 1% of the single proton rate. Even with the macroscopically continuous beam at TRIUMF this rate of accidental coincidences can give rise to an effect of the same order as that being measured. However at this level we can with confidence measure the accidental rate electronically, to a precision of 1% or better.

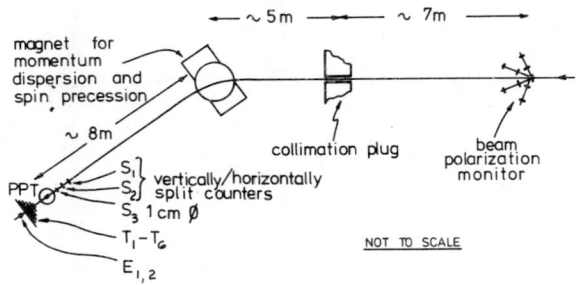

Figure 3. Beamline 4C at TRIUMF: focussing and steering magnets are omitted, as are monitor counter arrays around the target.

The low intensity is achieved in two stages; first, one part in 1000 of the accelerated H⁻ beam is stripped and extracted, and these protons are then focussed to a wide, low-dispersion beam at a copper plug, 20 cm long and bored with a 1 mm diameter axial hole. This collimator effects a further intensity reduction of 1000, and passes a narrow, low divergence ($<$ 0.3 mrad) beam into 4C. A magnet sweeps the beam through 36°, dispersing protons that have suffered energy loss, and providing, in conjunction with an upstream solenoid, a beam polarised longitudinally when needed.

Any material intercepting the beam halo is a potential source of error as the fraction lost may depend on spin orientation. Hence line 4C was set up with great care at each of the seven energies at which measurements were made. Except at 210 MeV no focussing magnets were used in the 20 m distance between the beam polarisation monitor and the target, and the beam was well centered everywhere in the vacuum pipe. Figure 4 shows a typical beam profile; beam movements of 30 μm are readily detected with the split scintillators S1 and S2. Beam profiles were confirmed by Polaroid photographs. Unlike the ANL group[8] we could not see the target outlined on photographs taken downstream of it, since the beam was entirely contained within its cross section. This centering was confirmed by observing the 3% reduction in transmission as the beam was steered across the target. The good beam quality enabled us

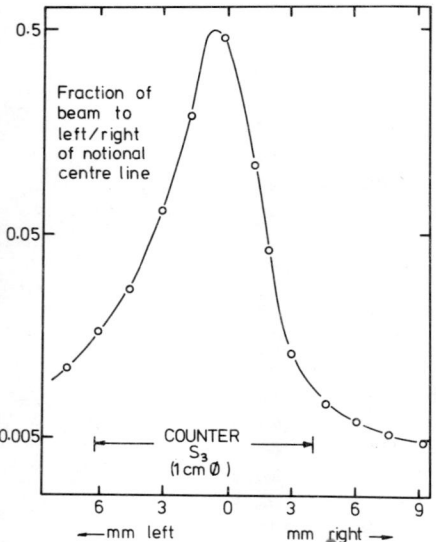

Figure 4. Horizontal beam profile at 425 MeV.

to dispense with halo counters, thus minimising material near the beam.

The target was a 2.4 cm (length) x 1.5 cm (diameter) PTFE cylinder containing butanol beads of order 1 mm diameter, immersed in liquid ^{3}He at 0.9 K. The magnetic field of 2.5 T is provided by superconducting coils. Two independent arrays of scintillators and MWPCs observed elastic pp scatters from the target near a laboratory angle of 24°; these data have not been fully analysed but will provide an absolute measure of the target's polarisation, P_T, using the known pp analysing power at this angle. NMR measurements indicate $P_T \sim 0.65$ during the recent runs. The target monitors will also yield a measure of the beam polarisation, P_B, independent of that measured upstream by the beam polarisation monitor. The polarisation of the beam averaged about 0.63 during these measurements. It was reversed at the ion source at five minute intervals, while the polarisation of the target was reversed every few hours.

The efficiencies of the transmission counters were continuously monitored by small counters E1, E2, and generally exceeded 0.9998.

Let t_{ij} be the transmission ratio

$$t_{ij} = \frac{S1.S2.S3.Ti.Tj}{S1.S2.S3} ,$$

corrected for inefficiencies and for accidental coincidences. Then the difference Δt_{ij} between the two beam polarisation states gives the cross section difference, to a very high precision, as

$$(\Delta\sigma_T)_{ij} = \frac{\Delta t_{ij}}{\langle t \rangle_{ij} N_o \rho P_B P_T} ,$$

where N_o is Avogadro's number and $\rho \approx 0.17$ is the superficial density of free protons in the target, in g cm^{-2}. Results, and linear fits to $(\Delta\sigma_T)_{ij}$ as a function of the solid angle $\Delta\Omega$ subtended by Ti.Tj, are shown in Figure 5. The slopes of the fitted lines are seen to vary fairly smoothly with energy.

Despite our endeavours an instrumental asymmetry remained, which was measured at two energies with the target unpolarised, and was otherwise extracted from the differences between $(\Delta\sigma_T)_{ij}$ in the two states of target polarisation. The asymmetry can be described by an equivalent instrumental cross section, σ_{instr}, and Figure 6 shows the values of σ_{instr} extrapolated to $\Delta\Omega = 0$.

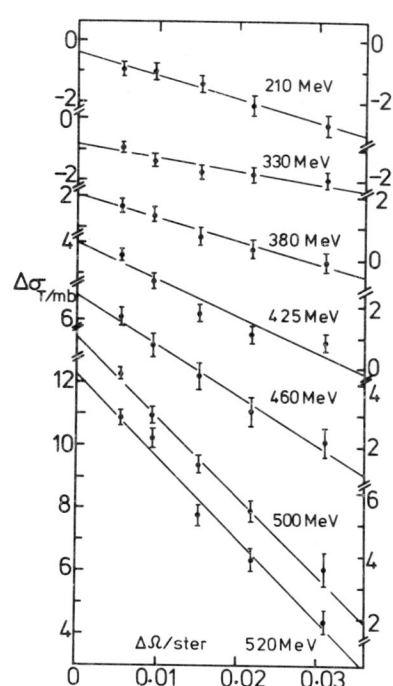

Figure 5. $(\Delta\sigma_T)_{ij}$ versus solid angle subtended by T_{ij}.

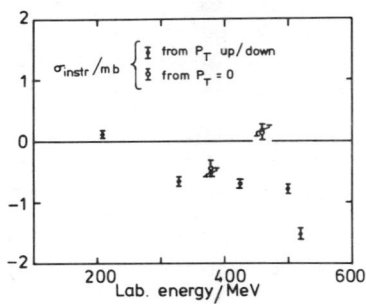

Figure 6. Equivalent cross sections σ_{instr}, extrapolated to $\Delta\Omega = 0$.

The effect is readily explained, with hindsight, by small (and observed) misalignments of the transmission counters which need to be re-positioned for each energy. The sinuous path of the beam through the target's magnetic field makes alignment difficult. The effect is caused by the large polarisation asymmetry in p-nucleus scattering at small angles, which we estimate to give a value of σ_{instr} of about 0.1 mb per mm of offset from the beam centre line. Fortunately our empirical determinations of σ_{instr} by different methods agree well.

Preliminary results for $\Delta\sigma_T$, extrapolated to zero solid angle, are shown in Figure 7. They have not yet been corrected for Coulomb-nuclear interference; this correction is somewhat dependent on the magnitude of $\Delta\sigma_T$ itself, but in any case acts to increase its value by an amount of order 1 mb[10]. Errors are statistical only. Systematic uncertainty in knowledge of the target polarisation and density may be as large as 15% at present, but will be reduced to nearer 5% later. Systematic errors in the values of beam polarisation are believed to be small. A further small correction of order 5% arises from the NMR sampling efficiency being lower in the centre of the target that at its periphery. This effect has been measured[11] but the correction has not yet been applied.

INTERPRETATION OF $\Delta\sigma_T(pp)$

There is clearly some incongruity with the measurements of Biegert et al[12]. I shall examine the consequences of assuming our

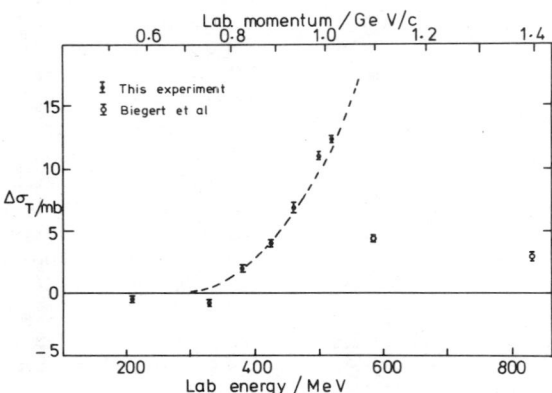

Figure 7. Results of this experiment, compared with data from Biegert et al. Dashed line represents twice the total reaction cross section.

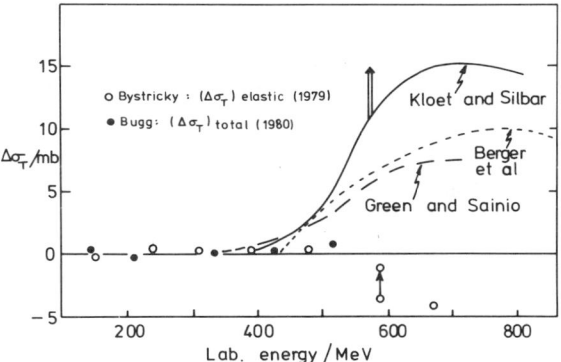

Figure 8. Theoretical estimates of $\Delta\sigma_T$: points are from partial wave analyses, curves from model calculations. The single and double arrows are explained in the text.

data to be correct. It is clearly of interest to separate $\Delta\sigma_T$ into its elastic and inelastic components. I shall first show that the former are small in this energy range. Figure 8 shows values of $(\Delta\sigma_T)_{elastic}$ computed from the partial wave analysis of Bystricky et al[13]. Below 500 MeV these are probably trustworthy - they agree well with the predictions by Bugg of $(\Delta\sigma_T)_{total}$, and his analysis certainly fits other elastic data very well - but above that energy they seem to overestimate the magnitude of $(\Delta\sigma_T)_{elastic}$. To illustrate this I show in Figure 9 the measurements of A_{NN} and A_{SS} at 577 MeV, presented at this Conference by the Geneva group[14]; observe that the phase shift predictions underestimate the magnitude of A_{SS} but get A_{NN} about right. Applying a crude correction to the predicted values of A_{SS} to fit the data better, I find that the prediction for $(\Delta\sigma_T)_{elastic}$, found by integrating A_{NN} plus A_{SS}, is shifted closer to zero, as shown by the single arrow on Figure 8. I conclude that $(\Delta\sigma_T)_{elastic}$ is indeed close to, or below, zero between 150 and 600 MeV.

What of the inelastic contributions? We see that partial wave analyses such as Bugg's grossly underestimate the total $\Delta\sigma_T$, and hence also the inelastic component. Model calculations do much better; even five years ago Berger et al[15] were predicting large effects from $pp \rightarrow pn\pi^+$ alone. I particularly draw attention to the recent coupled channel calculations of Kloet and Silbar[16], whose unitary theory of one pion production is remarkably succesful in reproducing the main features of the partial wave analyses. Their calculation of $\Delta\sigma_T$ and $\Delta\sigma_L$ is described in a recent report[17], and the result for $(\Delta\sigma_T)_{inelastic}$ is shown in Figure 8. Kloet and Silbar express

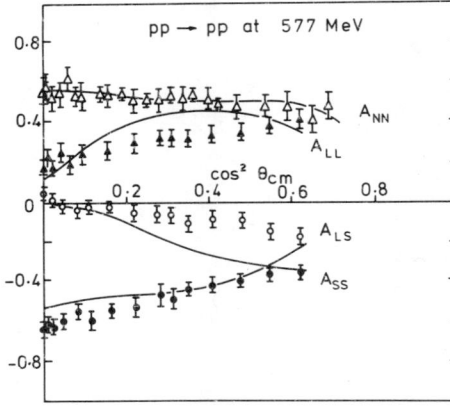
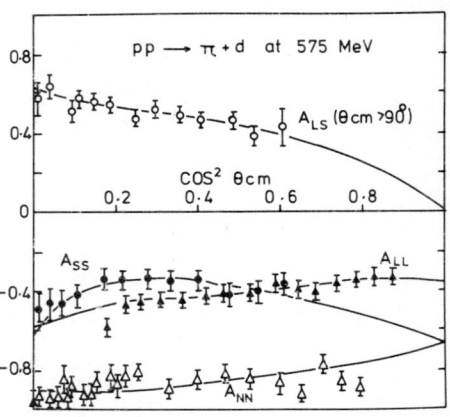

Figure 9. A_{NN}, A_{LL}, A_{SS}, A_{LS} in pp elastic scattering at 577 Mev from Ref. 14; lines are predictions from Ref. 13.

Figure 10. A_{NN}, A_{LL}, A_{SS}, A_{LS} in $pp \rightarrow \pi^+ d$ at 575 MeV from Ref. 18; lines are empirical fits.

reservations about their model calculation of the elastic components of the cross section differences, as they fail to reproduce the spin averaged elastic cross sections. They ascribe this failure to their neglect of short range forces, but the pion production amplitudes, being rather peripheral, should not be badly affected by this neglect.

The two body reaction $pp \rightarrow \pi^+ d$ is specifically excluded in Kloet and Silbar's calculations, and it would be useful to know the size of its contribution from independent sources. Such a source is provided by the recent measurement of polarisation parameters in $pp \rightarrow \pi^+ d$ at 575 MeV, shown in Figure 10 and presented at this Conference by the Geneva group[18]. They have extracted from their data, by integration over the whole angular range, values of $\Delta\sigma_T$ and $\Delta\sigma_L$ due to this reaction. Their value of $\Delta\sigma_T$ is 3.9 ± 0.3 mb. When this is added to Kloet and Silbar's prediction for $pp \rightarrow NN\pi$ one gets a point gratifyingly close to a reasonable extrapolation of our data, as shown by the double arrow on Figure 8.

The clear implication is that our data demand what Kloet and Silbar provide in their dynamical model, namely complete dominance of $(\Delta\sigma_T)_{inelastic}$ below 600 MeV by spin singlet states, in particular the $^1D_2$ state. To appreciate this dominance more clearly recall that the upper bound to $(\Delta\sigma_T)_{inelastic}$ is equal to twice the total reaction cross section, which is shown as a dashed line in Figure 7. This upper bound is attained if $(\sigma^{\uparrow\uparrow})_{inelastic}$ is zero, that is if all spin triplet contributions vanish. Our data are entirely consistent with this being so, as long as one accepts that $(\Delta\sigma_T)_{elastic}$ is less than, or very close to, zero.

Now we see why Bugg's latest phase shifts underestimate $\Delta\sigma_T$. As mentioned earlier, he takes inelasticities other than $^1D_2$ from Green and Sainio, whose model calculation of $(\Delta\sigma_T)_{inelastic}$ is too low because of the large inelasticities they ascribe to P and F waves below 600 MeV. Our data require little or no spin triplet inelasticity here.

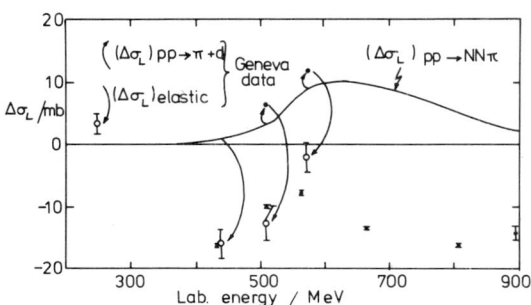

Figure 11. Estimates of $(\Delta\sigma_L)_{pp\to NN\pi}$ from Ref. 17, showing the result of adding other contributions. Data for $\Delta\sigma_L$ are from Ref. 19.

One cannot argue in a similar way from $\Delta\sigma_L$ data because the elastic contribution to this observable is large. By integrating their values of A_{LL} in pp elastic scattering between 440 MeV and 577 MeV, the Geneva group have shown[14] $(\Delta\sigma_L)_{elastic}$ is close to -16 mb in this energy interval. Thus one can find values of $(\Delta\sigma_L)_{total}$, as shown in Figure 11. I take Kloet and Silbar's model calculations of $(\Delta\sigma_L)_{pp\to NN\pi}$ and add successively values of $(\Delta\sigma_L)_{pp\to\pi^+d}$ from Ref. 18 (upward corrections) and $(\Delta\sigma_L)_{elastic}$ from Ref. 14 (downward corrections). The agreement of the final result with the data of Auer et al[19] is reasonably good. On the other hand the uncertainties involved in the summation procedure, and in the use of Kloet and Silbar's calculation, make it impossible at present to make strong statements about the spin structure of $(\Delta\sigma_L)_{inelastic}$.

My interpretation of the data on $\Delta\sigma_T$, and to some extent that on $\Delta\sigma_L$, below 600 MeV thus supports conventional models of the inelasticity in pp scattering. It requires inelasticities different from those used by Green and Sainio, and by Bugg, and is completely at variance with the very large inelasticities in triplet P waves claimed by Hoshizaki[20]. Others must judge its effect on claims of resonant behaviour in the two-nucleon system.

REFERENCES

1. A.S. Clough et al, Phys. Rev. C21, 988 (1980)
   D. Axen et al, Phys. Rev. C21, 998 (1980)
   D.V. Bugg et al, Phys. Rev. C21, 1004 (1980)
2. D.V. Bugg, Nucl. Phys. A335, 171 (1980)
3. Y.M. Kazarinov and Y.N. Simonov, JETP 43, 35 (1962)
4. T.S. Bhatia et al, contribution to this Conference, 1.4.
5. P.J. Riley et al, contribution to this Conference, 3.2.
6. A.M. Green and M.E. Sainio, J. Phys. G5, 503 (1979)
7. D.V. Bugg, J. Phys. G5, 1349 (1979)
8. I.P. Auer, Nucl. Phys. A335, 193 (1980)
9. C. Amsler et al, J. Phys. G4, 1047 (1978)
10. Y. Watanabe, Phys. Rev. D19, 1022 (1979)
11. G.R. Court et al, Daresbury Report DL/P 293E (1980)
12. E.K. Biegert et al, Phys. Letters 73B, 235 (1978)
13. J. Bystricky et al, Saclay Report D. Ph. P. E. 1-79 (1979)
14. E. Aprile et al, contribution to this Conference, 1.12; also, SIN Newsletter 12 (1980)
15. E.L. Berger et al, Argonne Report ANL/HEP-PR-75-72 (1975)
16. W.M. Kloet and R.R. Silbar, Nucl. Phys. A338, 281 (1980)
    R.R. Silbar and W.M. Kloet, Nucl. Phys. A338, 317 (1980)
17. W.M. Kloet and R.R. Silbar, LASL Report LA-UR-79-2139 (1980)
18. E. Aprile et al, contribution to this Conference, 1.18; also, SIN Newsletter 12(1980); and Nucl. Phys. A335, 245 (1980)
19. I.P. Auer et al, Phys. Rev. Letters 41, 354 (1978)
20. N. Hoshizaki, Prog. Theor. Phys. 60, 1790 (1978); and, preprint of talk given at 2nd meeting on Exotic Resonances, Hiroshima, 1980

(Editor's Note: References to this Conference are given as paper numbers and are cross-indexed to page numbers in the Table of Contents.)

DISCUSSION

IGO: Can one understand the drop in $\Delta\sigma_T$ above 600 MeV (where the Rice University data is located) in terms of Silbar's $NN\pi$ model or other models? They appear on the viewgraph to overestimate the measured value of $\Delta\sigma_T$ in this energy region.

EDGINGTON: You are right that Silbar and Kloet's model calculations of $(\Delta\sigma_T)$ (inelastic) remain much higher than the data of Biegert et al.[12] above 600 MeV. However, the phase-shift analyses seem to predict an increasing negative value of $(\Delta\sigma_T)$ (elastic) in this energy region, so there may be a simple cancellation between the two contributions.

POLARIZATION ANALYZING POWER MEASUREMENTS IN COHERENT PION PRODUCTION BY PROTONS

E.G. Auld
University of British Columbia, Vancouver, B.C., Canada, V6T 2A3

INTRODUCTION

The study of the (p,π^+) reaction in nuclei where the residual nucleus is left in a recognizable state dates back approximately ten years,[1,2,3,4] the work being done at CERN(600Mev) and at Uppsala (185 Mev). The first (p,π^-) measurements were also done at these institutions[5,6].

Since these beginning days there has been considerable expansion of the experimental effort; near threshold studies at the Indiana University Cyclotron (I.U.C.F.); wide range of incident proton energies at Saturne and selected nuclei at 800 Mev(LAMPF). There are several good review articles with many references therein[7,8]. The first experiments with polarized protons were done at TRIUMF and were published in 1978[9,10]. Recently the I.U.C.F. has commissioned a polarized proton beam and the TRIUMF group have commissioned a new pion spectrometer. This paper will concentrate on the analyzing power results, but will place them in perspective with the other measurements.

The basic nature of the differential cross-section for the (p,π^+) and the (p,π^-) reaction is shown in Figures 1 and 2[27]. There is a great deal of angular dependence in the (p,π^+) reaction and little

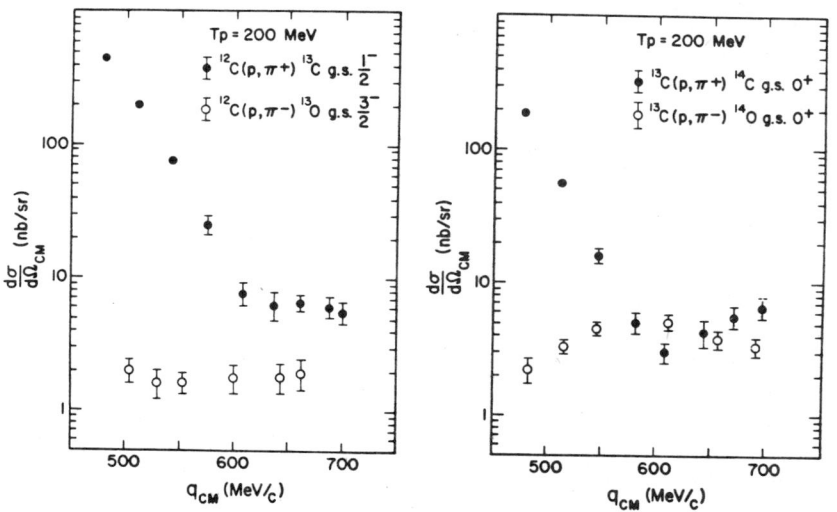

Figure 1. Comparison of (p,π^+) and (p,π^-) differential cross-section measurements for ^{12}C and ^{13}C targets.

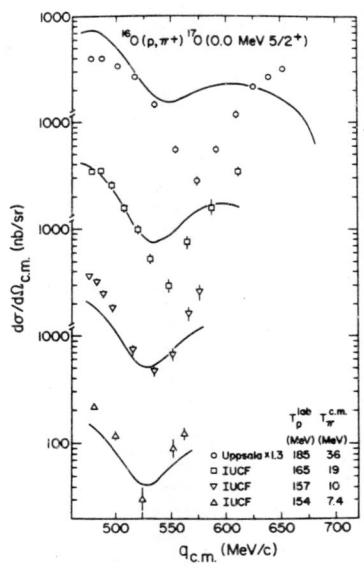

Figure 2: The $^{16}O(p,\pi^+)^{17}O(gs)$ differential cross-section vs momentum transfer. The Uppsala result is reference 28. The curves are the results of a DWBA calculation[29].

in the (p,π^-). Plotting the differential cross-section in this way points out the high value of momentum transfer that occurs in these reactions. The minimum momentum transfer in this $^{16}O(p,\pi^+)^{17}O$ reaction is 480 Mev/c. This is larger than that of most nuclear reactions and is much larger than typical momenta of single nucleons in the nucleus. Much concern has been centred around the nature of the reaction mechanism: whether a "single-nucleon model" (SNM)[7] which is basically a stripping model or whether an impulse approximation type model like a two-nucleon model (TNM)[23] is required. These mechanisms are illustrated in Figure 3.

Although all of the results will not be discussed in the paper, Table 1 represents an attempt to compile all the measured values of A_y for the nuclear $(p,\pi^\pm)$ reaction.

EXPERIMENTAL RESULTS

Both the (p,π^+) and (p,π^-) reaction have now been studied with polarized protons, at the I.U.C.F. and TRIUMF. Much of the work is new and has not as yet been published. Some of the data has been submitted in papers to this conference[11,12] and some has been privately communicated[13].

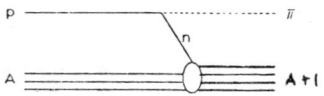

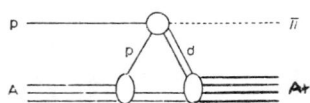

Figure 3. Typical diagrams illustrating pion production.
a) One-nucleon model (ONM)
b) Two-nucleon model (TNM)

(p,π^+) Results: The first measurements of the A_y parameter defined as:

$$A_y = \frac{\frac{d\sigma}{d\Omega}(+) - \frac{d\sigma}{d\Omega}(-)}{p^-\frac{d\sigma}{d\Omega}(+) + p^+\frac{d\sigma}{d\Omega}(-)}$$

for the $(\vec{p},\pi^+)$ reaction in nuclei where transitions to distinctly resolved final states were done on ^{12}C and ^9Be at an incident proton energy of 200 Mev[10]. The results achieved at that time are summarized in Figure 4. The most striking feature is that A_y has approximately the same angular dependence and magnitude for all of the reactions, independent of the target nucleus and the level of excitation in the recoil nucleus[14]. Except for the magnitude and the position of the peak, the shapes of these results are very similar to those for the $pp \rightarrow \pi d$ reaction at similar energies above the pion production threshold[15]. See Figure 5.

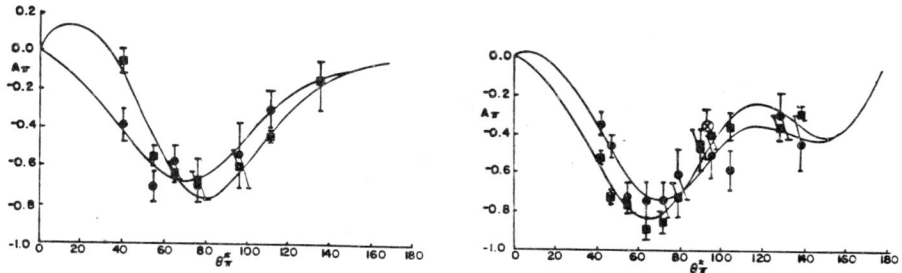

Figure 4. Angular distribution for A_y for a) $^9Be(\vec{p},\pi)^{10}Be$ ground state (circles) and first excited state (3.37 Mev, boxes): b) $^{12}C(\vec{p},\pi)^{13}C$ ground state (circles) and first excited states (~3.5 Mev, boxes). The other point is the only previous data[30]. The lines are Legendre Polynomial fits[16].

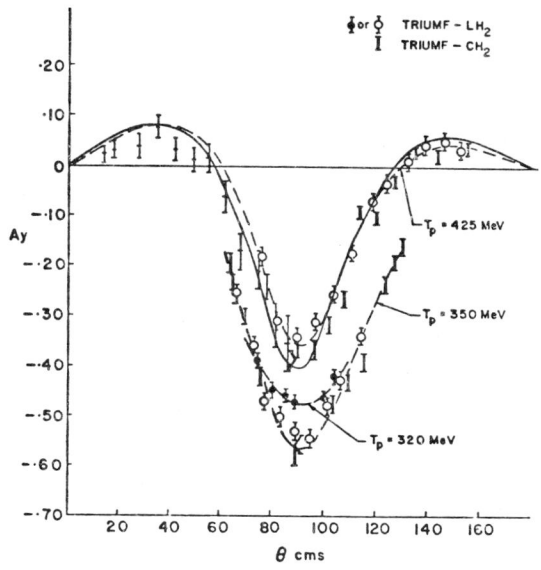

Figure 5. Analyzing powers for the $pp \rightarrow \pi d$ reaction as a fuction of pion cms angle for proton energies of 320, 350, and 425 Mev. The dashed curves are fits to the data[15].

Measurements were also done of the few-body reaction $^2H(\vec{p},\pi)^3H$ at energies of 305, 330[14], and 400 Mev[16] using a magnetic spectrograph and at back angles for 400, 425, 443, and 470 Mev[17] using a counter telescope[17]. The 305 and 330 Mev mesaurements have a sufficient angular range to indicate that A_y for this reaction also is similar to the $pp \to \pi d$ results[14]. Since this initial work, the interest in studying these spin dependent effects has increased. Another TRIUMF experiments has extended the measurements for $^2H(\vec{p},\pi)^3H$ to 500 Mev[12]. Their differential cross-section measurements are consistent with other results in this energy region, however, there is either a very rapid change in A_y with energy between 470 and 500 Mev or an error in one of the experiments (see Figure 6). The 500 Mev results do not extend to sufficiently low angles to determine whether the characteristic "$pp \to \pi d$" A_y shape is still in evidence at these higher energies.

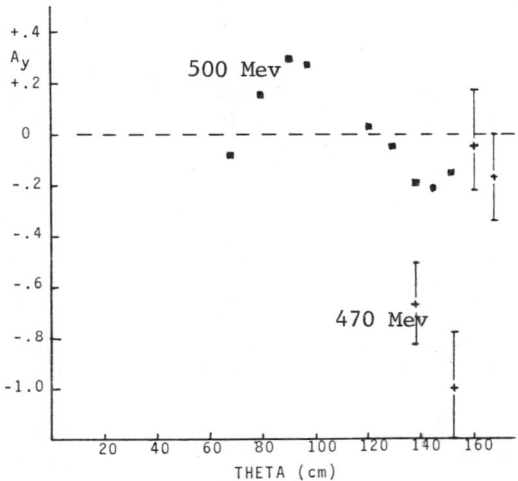

Figure 6. Analyzing power for the reaction $^2H(\vec{p},\pi^+)^3H$ at proton energies of 470[17] and 500[12] Mev.

Several measurements have been done at the I.U.C.F. and they are partly summarized in Figure 7[18,19]. The incident proton energy was chosen to be near threshold so that the outgoing pions would be mainly S and P waves. It limits the number of experimental points needed to determine the angular distribution of A_y to 3 to 4 angles when the corresponding differential cross-section is well established. Several features can be seen in these results: A_y is again negative, and the general shape of A_y as seen at 200 Mev in ^{12}C seems to persist down to near threshold. The $^{16}O(\vec{p},\pi^+)^{17}O(gs)$ transition (not shown, but listed in Table 1.) also has the same characteristic shape. Two reactions that seem to show significant deviation from the general trend are $^{40}Ca(\vec{p},\pi^+)^{41}Ca(gs)$ and $^{16}O(\vec{p},\pi^+)^{17}O(.87$ Mev state) so there are some nuclear state effects afterall.

The energy dependence of A_y for a given reaction is clearly of some importance. With this in mind a measurement has been recently done on the $^9Be(\vec{p},\pi^+)^{10}Be$ reaction for incidents proton energies of 225 and 250 Mev. Figure 8 shows the spectrum produced with the recently commissioned TRIUMF pion spectrometer. The resolution at these energies is 0.8 Mev FWHM, which is largely due to beam energy resolution. The preliminary values for A_y to a selection of the states shown are plotted in Figures 9 and 10 for proton

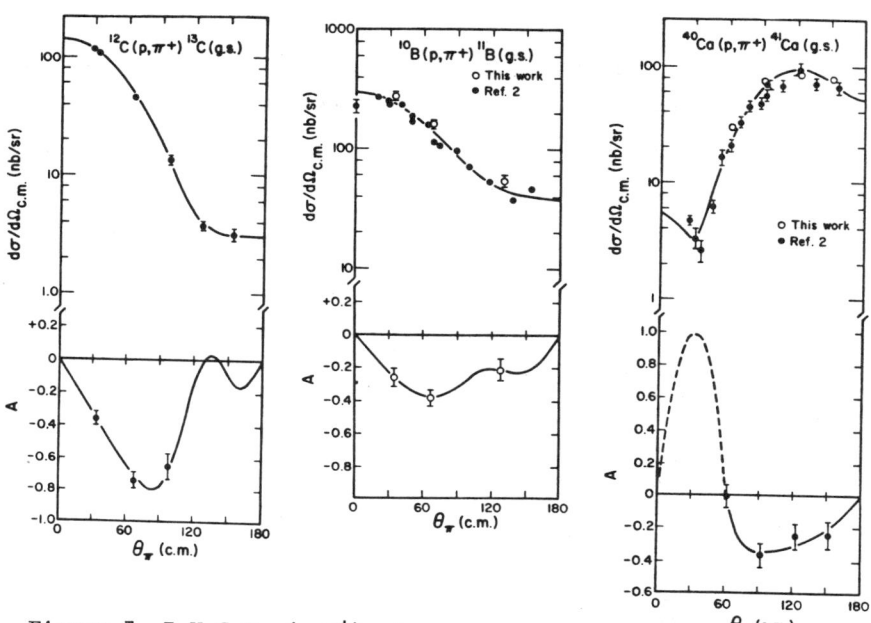

Figure 7. I.U.C.F. (p,π^+) A_y measurements. The reference 2 in the graph is P.H. Pile et al, Phys. Rev. Lett. 42(]979)1461. They have other data which has been tabulated is Table 1.

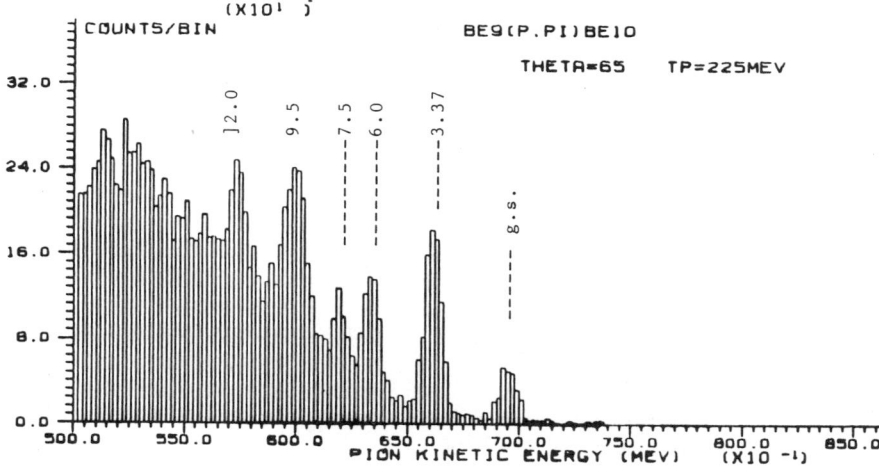

Figure 8. Pion Kinetic Energy spectrum as obtained for the 65 cm Browne-Beuchner TRIUMF spectrometer. $^9Be(\vec{p},\pi^+)^{10}Be$ at theta=65 degrees is the lab. The lines represent the levels of ^{10}Be.

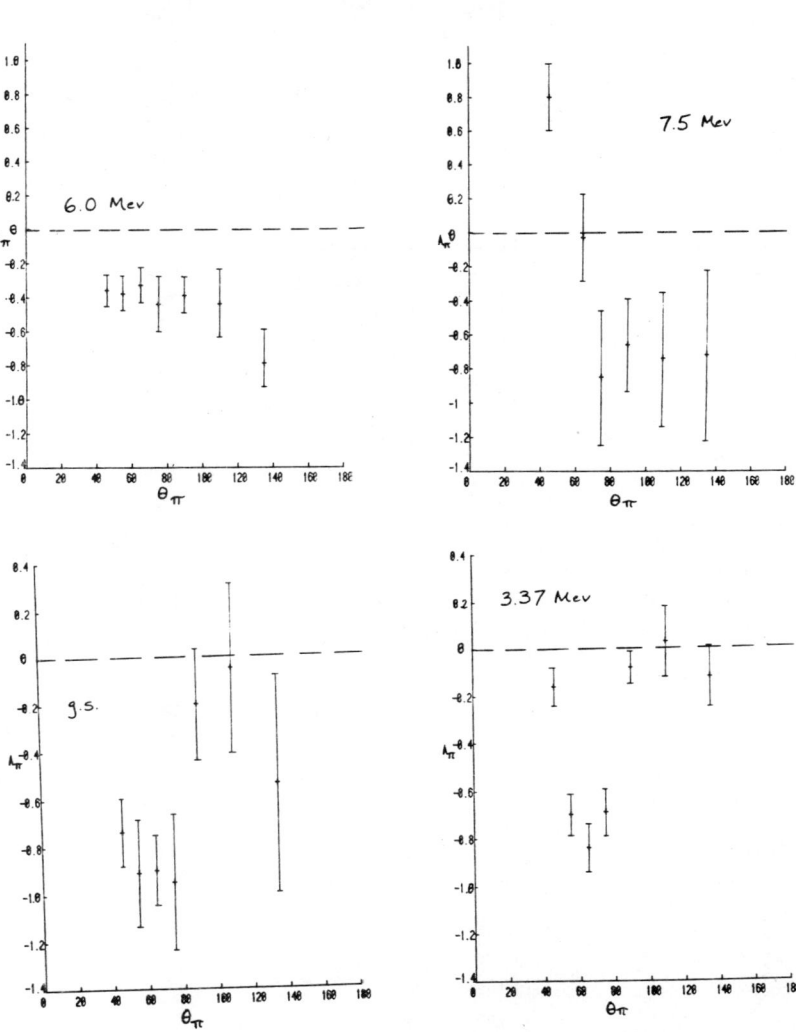

Figure 9. Analyzing power vs laboratory angle for the ^{9}Be($\vec{p},\pi^+$)^{10}Be reaction leaving the Be in the ground state and the excited states of 3.37, 6.0 and 7.5 Mev. Proton energy was 225 Mev. Data for higher states is tabulated in Table 1.

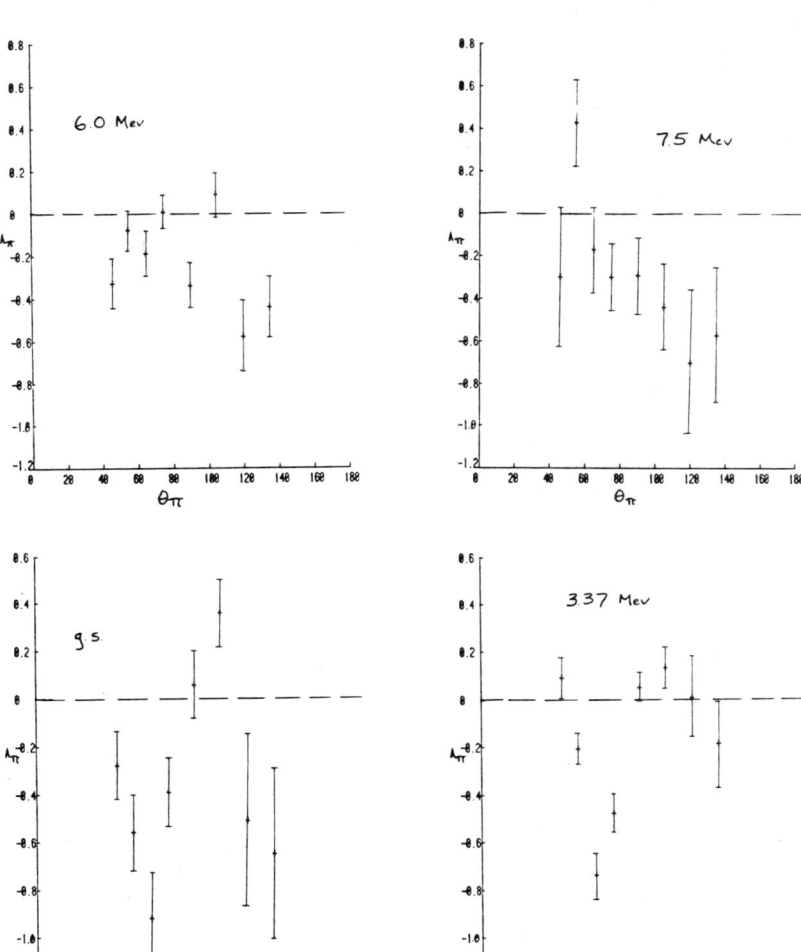

Figure 10. Analyzing power vs laboratory angle for the ^{9}Be$(p,\pi^+)^{10}$Be reaction leaving the Be in the ground state and the excited states of 3.37, 6.0 and 7.5 Mev. Proton energy was 250 Mev. Data for higher states is tabulated in Table 1.

energies of 225 and 250 Mev respectively. The differential cross-sections have as yet not been calculated. The shape of the A_y for the ground state and the 3.37 Mev state of ^{10}Be still show the characteristic negative peak but there is a tendancy for the curve to be shifting to more positive values.

All the other transitions have their own characteristic shape for A_y. The cross-section to the 7.5 Mev level is generally much lower than that for the other transitions and hence the errors on the points are much larger. The errors shown are statistical only with the scale on A_y extended beyond -1 to help the reader assess the errors. The pion continuum under the peaks shows a consistent value of $A_y = -.45 \pm .05$ independent of angle and pion energy over the limited range of the measurements. This is similar to the previously reported results at 200 Mev[10,14].

A limited number of measurements have been taken at higher excitations for both ^{12}C and ^{9}Be targets[11]. Evidence has been seen for transitions to high levels of excitation in ^{13}C(18 Mev) and ^{10}Be(]7.8 and 22.8 Mev) at forward angles of production (46 degrees)

(p,π^-) Results: Although the (p,π^-) reaction to discrete nuclear states is difficult to measure (the total cross-section is typically less than 30 nb) much useful physics is expected to be gleaned from these reactions. The (π^+,π^-) and (π^-,π^+) double charge-exchange reactions have caused considerable interest, but the (p,π^-) reaction

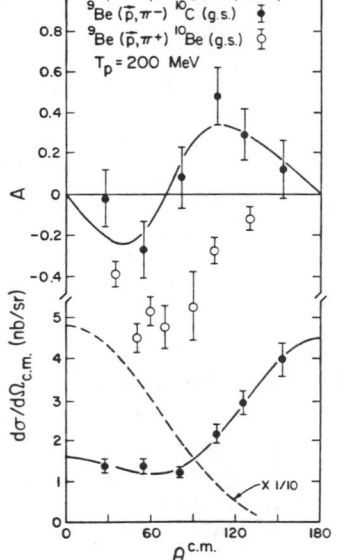

is the only one available that allows spin effects to be measured.
Recent experiments have been performed at TRIUMF and at the I.U.C.F., with ^{9}Be as the target. The 200 Mev results from the I.U.C.F. are shown in Figure 11[13]. The value of A_y is only negative for cm. angles less than 70 and is positive elsewhere is sharp contrast to the values for the (p,π^+) reaction.
A typical pion spectrum for the TRIUMF results is shown in Figure 12. The following states are clearly resolved; gs, 3.37, 5.4, 6.7 and a suggestion of a level at 10 Mev. A sample of the values for A_y in these transitions is shown in Figure 13.

Figure 11. I.U.C.F. ^{9}Be$(\vec{p},\pi^-)^{10}$C data[13]. $A_y, d\sigma/d\Omega$ vs cms angle. The ^{9}Be$(\vec{p},\pi^+)^{10}$Be data is reference 10. The dotted line is the $d\sigma/d\Omega$ for this reaction at 185 Mev (S. Dahlgren et al, Nucl. Phys. A204,(1973)53).

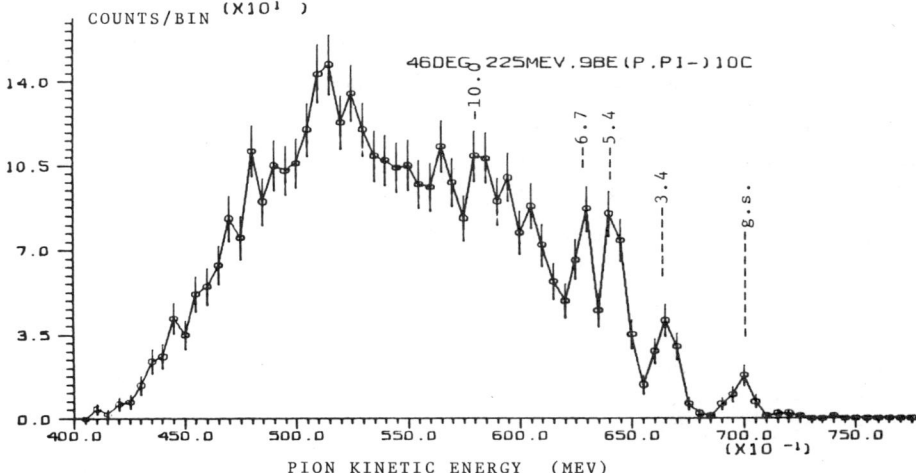

Figure 12. Pion kinetic energy spectrum for $^9Be(\vec{p},\pi^-)^{10}C$ at a laboratory angle of 46 degrees. The lines represent the levels in ^{10}C

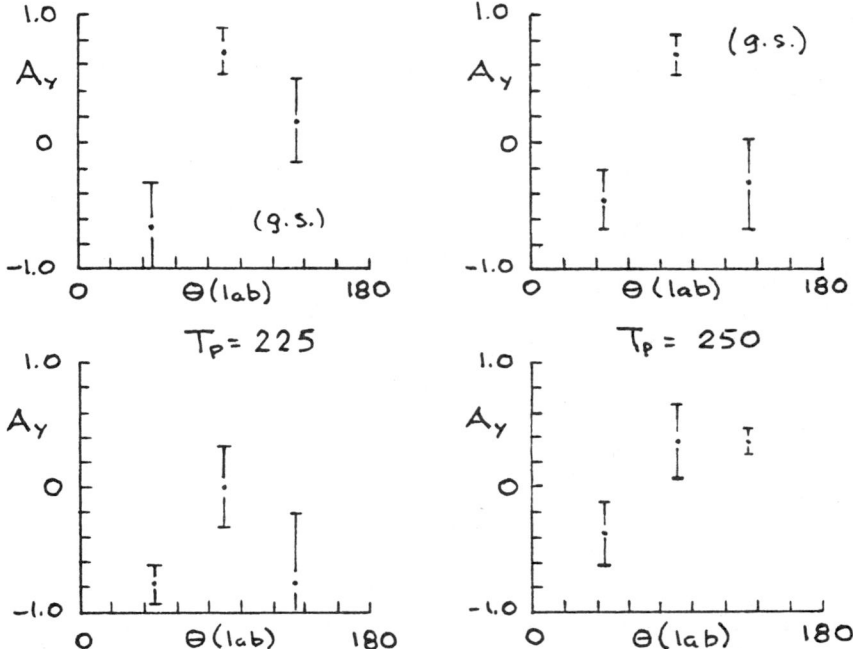

Figure 13. Analyzing power vs laboratory angle for the $^9Be(\vec{p},\pi^-)^{10}C$ reaction where ^{10}C is left in the gs. and the 3.37 mev state. More data is tabulated in Table 1.

DISCUSSION

There have been several attempts at predicting the angular dependence of the (p,π^+) A_y, most of which have used the SNM model with variations. Space does not allow an exhaustive discussion of these results.[20,21,22] The calculations have succeeded in some specific cases, but none predict the general tendency for A_y to be negative as seen in the data presented here. Up until the most recent results of the last year the impression left by the data is that the nuclear $A_y(\theta)$

i) has a shape similar to that for $pp \to \pi d$
ii) is roughly independent of nuclear structure details
iii) is independent of the degree of excitation of the residual nucleus.

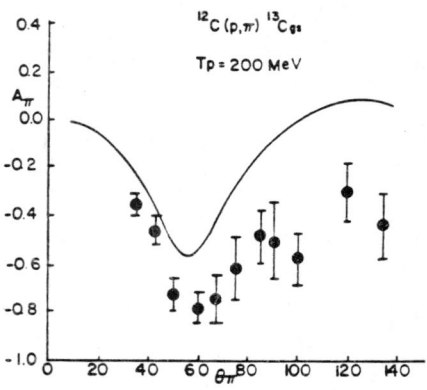

Figure 14. Kinematic transformation of the $pp \to \pi d$ analyzing power applied to $^{12}C(p,\pi^+)^{13}C$.

It is difficult to avoid interpreting the nuclear analysing powers in terms of the basic $\vec{p}p$ reaction (which in turn points to the role of the TNM for such processes). Interpreted this way, the nuclear pion production at 200 MeV would occur through the interaction of the incident proton with a target proton near the top of the Fermi sea (a momentum of about 195 MeV/c is required in order for the pion production to occur via the $pp \to d\pi^+$ reaction). Such a model has previously been applied by a number of authors to pion production by unpolarized protons.[23,24,25,26] Disregarding all nuclear structure effects, we can compare the $\vec{p}$ + nucleus pion production analysing powers with the corresponding results for the $\vec{p}p \to d\pi^+$ reaction by simply performing the appropriate kinematical transformation.

A detailed description of the calculation appears elsewhere[14,16] The results are shown for ^{12}C in Figure 14, where the solid line represents the so-called kinematically transformed $\vec{p}p \to \pi d$ A_y. The qualitative agreement in shape with the experimental measurements, and the quantitative agreement of the positions of the maxima are indeed striking, although it is clear that the magnitude of the maxima in the analysing power results for the nuclear cases are substantially larger than would be expected from the $\vec{p}p \to \pi d$ reaction.

Even the energy dependence of the A_y for the g.s. and 3.37 MeV state in ^{10}Be seems to follow a pattern consistent with the energy dependence for the $pp \to \pi d$ A_y, in that it is tending towards more positive values.

Little theoretical work has been published on the value of A_y for the (p,π^-) reaction. Analysis of the I.C.U.F. data indicates that there is no evidence other than S and P wave pions in the outgoing angular distributions. A recent model proposed by W. Gibbs assuming a direct pion knockout from the nuclear pionic field has predicted maximum values of A_y ranging from +.4 to +.6. Is this predictions unique to this model? Probably not, but there is some encouraging agreement with the data.

The measurement of the spin dependence of the (p,π) reaction has caused a considerable flurry of interest as a possible way to resolve the reaction mechanism. As per usual, A_y and $d\sigma/d\Omega$ will in all probability not be sufficient to fully constrain the models.

One might ask, what other ways can this reaction be studied to further illuminate the situation. Here are two suggestions.

a) Associated particle measurements: To my knowledge no published data exists where a reaction like,

$$A(p,\pi p)$$
$$\pi n)$$
$$\pi d)$$

is measured with sufficient resolution to determine whether the product nucleus is bound or not. Since the pions in the continuum of the present spectra have a definite non-zero value for A_y, a study of this reaction with polarized protons would help unravel the reaction mechanism problem.

b) $(\vec{n},\pi^{\pm})$ Reactions: Despite the lack of resolution, an experiment to measure A_y, for an $(\vec{n},\pi^-)$ reaction at 200 Mev incident proton energy should show maximum values for A_y of around $-.8\pm.2$, for the maximum energy pions. What little data exists for the $(n,\pi)^{31}$ reaction allows an interesting comparison to be made. At 600 Mev protons and a momentum transfer of 600 Mev/c the π^+/π^- ratio to the g.s. transition and to the 3.37 Mev transition in $^9Be(p,\pi^{\pm})^{10}X$ is respectively 85 ± 35 and 125 ± 30^{32}. Whereas the corresponding $\pi^-/\pi^{\pm}$ ratio for the $(n,\pi^{\pm})$ reaction producing a 300 Mev pion, hence leaving the residual nucleus in about 100 Mev of excition, is 10 ± 2.

The $(n,\pi^{\pm})$ reaction is a combination of the following nucleon-nucleon inelastic reactions: $np \to \pi^+ nn$ i)
$$\pi^- pp \quad ii)$$
$$nn \to \pi^- pn \quad iii)$$
where i) and ii) are the same, $\sigma_{np}(\pi^+)=\sigma_{np}(\pi^-)=0.5(\sigma_{11}+\sigma_{01})$, and hence iii) causes the π^- rate to be ten times the π^+ rate. The $(p,\pi^{\pm})$ reaction, of course has similar components except iii) becomes: $pp \to \pi^+ np$ where the np in both these channels also represents d. Nothing, however, too significant should be read into the differences in the $(p,\pi^+/\pi^-)$ ratio compared to the $(n,\pi^-/\pi^+)$ ratio mentioned above as there could be many interpretations. A more thorough study of the inelastic nucleon-nucleon interactions in conjunction with some well chosen nuclear (n,π) experiments might turn up some interesting results.

What of the future for (p,π) physics? There is clearly considerable systematic energy dependence and nuclear dependence measurements to be made to provide information and constraints for the reaction models to be tested. Also, by itself, the (p,π^-) reaction could turn out to be a significant spectroscopy tool for studying neutron deficient isotopes.

ACKNOWLEDGEMENTS

The TRIUMF results referred to is this paper with the exception of the 500 Mev $^2H(p,\pi)^3H$ result were obtained through the joint efforts of a number of colleagues;namely, E.L. Mathie, G.J. Lolos P. Walden, G. Jones and R. Taylor. Considerable assistance was also received from four hard working summer students; K. Dyck, S. Mann D. Frost, and R. Neufeld. D. Sample provided the necessary software knowledge to allow the data analysis to run efficiently.

The author gratefully acknowledges Drs. T. Sjoreen and R. Bent from Indiana University for allowing pre-publication data to be shown is this paper.

Support for this experimental program has been derived from an NSERC grant IEP# 18.

REFERENCES

1. J. J. Domingo, et al. Phys. Lett. 32B (1970) p. 309.
2. K. Gabathuler, et al. Nucl. Phys. B40 (1972) p. 32.
3. S. Dahlgren, B. Hoistad, and P. Grafstrom, Phys. Lett. 35B (1971) p. 219.
4. S. Dahlgren, et al. Proc. Int. Seminar on π-Meson Nucleus Interactions, Strasbourg (1971) p. V-93.
5. J. Rohlin et al. Phys. Lett. 40B (1972) p. 539.
6. S. Dahlgren, et al. Nucl. Phys. A204 (1973) p. 53.
7. B. Hoistad, Advances in Nuclear Physics 11 (1979) p. 135.
8. D. Measday, G. Miller, Ann. Rev. Nucl. Part. Sci. 29 (1979) p. 121.
9. Proceedings of the VIIth Int. Conf. on High Energy Physics and Nuclear Structure, Zurich (1977) B. Hoistad review paper, p. 223.
10. E. G. Auld, et al. Phys. Rev. Lett. 41 (1978) p. 462.
11. G. J. Lolos, et al. Contribution to this conference.
12. R. Abegg, et al. Contribution to this conference.
13. R. Bent and T. Sjoreen, Private communication of a paper submitted for publication.
14. G. Jones, Proceedings Meson-Nuclear Physics Conf., Houston, (1979) AIP Conf. Proc. #54 (1979) p. 116.
15. P. Walden, et al. Phys. Lett. B81 (1979) p. 156.
16. E. L. Mathie, Ph.D. Thesis, Univ. of B.C.
17. E. G. Auld, et al. Phys. Lett. B93 (1980) p. 258.
18. P. H. Pile, IUCF Technical Report 63 (1979).
19. T. Sjoreen, Private communication of preliminary data.
20. J. V. Noble, Nucl. Phys. A244 (1975) p. 256.

21. S. K. Young and W. R. Gibbs, Phys. Rev. C17 (1978) p. 837.
22. H. J. Weber and J. M. Eisenberg, Meson-Nuclear Physics Conf., Houston, 1979, AIP Conf. Proc. 54 (1979) p. 190.
23. M. Dillig and M. Huber, Phys. Lett. B69 (1977) p. 429.
24. M. P. Locker and H. J. Weber, Nucl. Phys. B76 (1974) p. 400.
25. H. W. Fearing, Phys. Rev. C16 (1977) p. 313.
26. C. H. Q. Ingram, et al. Nucl. Phys. B31 (1971) p. 331.
27. B. Hoistad, IUCF Technical Report (1979) p. 62.
28. S. Dahlgren, et al. Nucl. Phys. A227 (1974) p. 245.
29. M. C. Tsangarides, Thesis, Indiana Univ., (1979) IUCF Internal Report #79-4.
30. E. Heer, et al., Phys. Rev. 111 (1958) p. 640.
31. K. O. Oganesyan, J.E.T.P. 27 (1968) p. 679.
32. E. Aslanides, et al. Phys. Rev. Lett. 39 (1977) p. 1654.

Table 1: $(p,\pi^{\pm})$ Analyzing Power A_y

$^2H(p,\pi^+)^3H$

T_p=305 MeV[14]		T_p=330 MeV[14]		T_p=400 MeV[16]	
θ(cm)	A_y	θ(cm)	A_y	θ(cm)	A_y
68	−0.36±0.01	75.3	−0.47±0.02	125.5	−0.27±0.03
89	−0.60±0.01	81.0	−0.53±0.02	138.0	−0.25±0.04
113	−0.40±0.02	90.0	−0.53±0.02	145.5	−0.13±0.03
144	−0.14±0.03	107.0	−0.43±0.02		
		116.5	−0.26±0.03		

T_p=425 MeV[17]		126.0	−0.21±0.02	T_p=400 MeV[17]	
θ(cm)	A_y			θ(cm)	A_y
137.4	−0.33±0.08	T_p=500 MeV[12]		134.8	−0.51±0.07
152.2	−0.64±0.12	θ(cm)	A_y	143.8	−0.44±0.08
167.6	−0.57±0.12	68.0	−0.08±0.02	151.5	−0.31±0.11
		79.0	+0.15±0.02	159.8	−0.41±0.14
T_p=443 MeV[17]		90.0	+0.29±0.02	167.8	−0.31±0.14
θ(cm)	A_y	97.0	+0.27±0.02		
137.1	−0.40±0.06	120.0	+0.03±0.02	T_p=470 MeV[17]	
145.0	−0.40±0.10	129.5	−0.05±0.02	θ(cm)	A_y
152.4	−0.32±0.12	137.4	−0.19±0.02	137.4	−0.67±0.16
159.5	−0.19±0.12	144.6	−0.21±0.02	152.4	−1.00±0.22
167.4	+0.06±0.16	151.2	−0.16±0.02	160.0	−0.05±0.17
				167.5	−0.17±0.17

$^9Be(p,\pi^+)^{10}Be$

T_p=200 MeV[10]

θ(cm)	A_y(g.s.)	A_y(3.37)
39.7	−0.39±0.08	−0.05±0.06
54.9	−0.71±0.08	−0.55±0.05
65.2	−0.58±0.09	−0.63±0.05
75.7	−0.67±0.11	−0.71±0.06
96.0	−0.54±0.17	−0.60±0.07
110.8	−0.30±0.10	−0.44±0.04
134.5	−0.15±0.15	−0.13±0.06

Table 1 (cont.)

T_p=225 MeV (this paper, preliminary data)

θ(cm)	A_y(g.s.)	A_y(3.37)	A_y(6.0)
50.3	-0.74±0.15	-0.16±0.08	-0.36±0.10
59.8	-0.91±0.22	-0.70±0.09	-0.38±0.11
70.3	-0.90±0.17	-0.84±0.10	-0.33±0.11
80.7	-0.94±0.28	-0.69±0.10	-0.44±0.17
95.8	-0.20±0.23	-0.08±0.06	-0.39±0.12
115.4	-0.05±0.37	+0.03±0.16	-0.44±0.20
139.0	-0.54±0.45	-0.12±0.14	-0.79±0.20

θ(cm)	A_y(7.5)	A_y(9.5)	A_y(12.0)
50.3	+0.80±0.20	-0.41±0.12	-0.38±0.20
59.8		-0.53±0.12	-0.70±0.23
70.3	-0.03±0.26	-0.54±0.13	-0.74±0.27
80.7	-0.82±0.40	-0.82±0.24	-0.42±0.15
95.8	-0.66±0.27	-0.33±0.15	-0.12±0.25
115.4	-0.74±0.40	-0.30±0.27	-0.64±0.30
139.0	-0.72±0.50	-0.48±0.27	-0.80±0.46

T_p=250 MeV (this paper, preliminary data)

θ(cm)	A_y(g.s.)	A_y(3.37)	A_y(6.0)
50.1	-0.28±0.15	-0.09±0.08	-0.33±0.12
59.7	-0.56±0.17	-0.21±0.07	-0.08±0.09
70.1	-0.92±0.19	-0.74±0.09	-0.19±0.10
80.4	-0.39±0.15	-0.48±0.08	-0.01±0.07
95.5	+0.06±0.13	+0.05±0.06	-0.34±0.10
110.3	+0.36±0.13	+0.13±0.08	+0.09±0.10
124.7	-0.51±0.35	0.00±0.16	-0.58±0.17
138.9	-0.65±0.35	-0.19±0.17	-0.44±0.14

θ(cm)	A_y(7.5)	A_y(9.5)	A_y(12.0)
50.1	-0.30±0.32	-0.40±0.14	-0.23±0.15
59.7	+0.43±0.21	-0.08±0.11	-0.33±0.16
70.1	-0.17±0.20	-0.44±0.13	-0.66±0.23
80.4	-0.30±0.17	-0.18±0.10	-0.60±0.20
95.5	-0.29±0.19	-0.16±0.13	-0.29±0.21
110.3	-0.44±0.20	-0.06±0.13	-0.33±0.20
124.7	-0.70±0.30		
138.9	-0.57±0.34	-0.90±0.30	-0.60±0.40

Table 1 (cont.)

$^{10}B(p,\pi^+)^{11}B$

T_p=154 MeV[18]

θ(cm)	A_y(g.s.)	A_y(2.12)	A_y(4.44)
34	−0.25±0.05	−0.23±0.10	+0.04±0.10
66.5	−0.38±0.05	−0.42±0.10	−0.28±0.12
126.5	−0.21±0.07	−0.07±0.23	−0.64±0.36

$^{12}C(p,\pi^+)^{13}C$

T_p=159 MeV[18]

θ(cm)	A_y(g.s.)	A_y(3.09)
33	−0.36±0.04	−0.40±0.06
67	−0.74±0.05	−0.82±0.07
92	−0.66±0.08	−0.20±0.13
147		+0.07±0.13

T_p=200 MeV[10]

θ(cm)	A_y(g.s.)	A_y(∼3.5)
38.0	−0.35±0.05	−0.52±0.03
46.4	−0.46±0.06	−0.73±0.03
54.3	−0.73±0.07	−0.76±0.04
63.9	−0.74±0.09	−0.89±0.06
72.0	−0.74±0.10	−0.85±0.06
79.5	−0.61±0.13	−0.73±0.10
89.7	−0.48±0.11	−0.44±0.06
95.1	−0.51±0.12	−0.40±0.06
104.5	−0.59±0.10	−0.36±0.07
123.8	−0.30±0.12	−0.36±0.03
137.0	−0.44±0.14	−0.28±0.03

$^{16}O(p,\pi^+)^{17}O$

T_p=157 MeV[19]

θ(cm)	A_y(g.s.)	A_y(0.87)
32.5	−0.46±0.04	
65	−0.91±0.04	−0.07±0.17
96	−0.64±0.05	−0.21±0.07
124	−0.17±0.06	−0.20±0.08
153	−0.03±0.05	−0.21±0.10

$^{40}Ca(p,\pi^+)^{41}Ca$

T_p=146 MeV[18]

θ(cm)	A_y(g.s.)
61.5	0.00±0.07
92.0	−0.36±0.07
123	−0.25±0.07
152	−0.25±0.08

Table 1 (cont.)

$^9\text{Be}(\vec{p},\pi^-)^{10}\text{C}$

T_p=200 MeV[13)] Scaled from drawing

θ(cm)	A_y(g.s.)
28	−0.02±0.13
55	−0.27±0.14
82	+0.08±0.15
107	+0.48±0.14
126	+0.28±0.13
153	+0.12±0.14

T_p=225 MeV (this paper)

θ(cm)	A_y(g.s.)	A_y(3.4)	A_y(5.3)	A_y(6.7)
50.3	−0.67±0.35	−0.78±0.16	−0.24±0.16	−0.43±0.27
95.8	+0.70±0.18	0.00±0.34	−0.51±0.58	+0.84±0.27
139.0	+0.17±0.32	−0.77±0.67	−0.19±0.30	+0.51±0.44

T_p=250 MeV (this paper)

θ(cm)	A_y(g.s.)	A_y(3.4)	A_y(5.3)	A_y(6.7)
50.1	−0.45±0.23	−0.37±0.25	−0.47±0.19	−0.56±0.37
95.5	+0.69±0.16	+0.37±0.29		
138.9	−0.31±0.37	+0.37±0.08	−0.58±0.42	+1.15±0.14

$^2\text{H}(p,\pi^+)^3\text{H}$

T_p=800 MeV K. K. Seth et al., Northwestern Univ., Private Comm.

θ(cm)	A_y
12.0	+0.25±0.05
16.0	+0.30±0.05
19.0	+0.38±0.04
25.5	+0.44±0.04
27.5	+0.26±0.05
33.5	+0.10±0.05
37.5	−0.15±0.05
46.0	−0.40±0.04
50.5	−0.30±0.04
55.0	−0.22±0.04
59.0	0.00±0.05
64.0	+0.15±0.05
69.0	+0.27±0.05
75.0	+0.40±0.06
81.0	+0.42±0.06

DISCUSSION

HOLT: Do you imply from your simple kinematic model that the neutrons in the nucleus have no role in this reaction process?

AULD: The assumption that the incident proton striking one proton in the target nucleus is the main channel for (p,π^+) is based on the fact that the $pp \to \pi d$ is by far the strongest nucleon-nucleon inelastic reaction.

RAPPORTEUR' REPORT: NUCLEON-NUCLEON AND
HIGH ENERGY POLARIZATION PHENOMENA

L. C. Northcliffe
Cyclotron Inst., Texas A&M Univ.,
College Station, Tx 77843

Of the 32 contributed papers grouped into discussion section 1, twelve are theoretical or phenomenolgical, and of the remaining 20 primarily experimental papers, 19 are concerned with the nucleon-nucleon interaction. It is obviously impractical to display or discuss all of these results here, but a tabular display showing the variety of experimental results reported seems worth presenting. This is done in the Table I.

The proliferation of notations for spin observables is notorious and several different forms appear in the papers. The notation used in Table I is arbitrarily chosen to be the Ann Arbor convention for spin 1/2-spin 1/2 scattering. There is not time to discuss this convention in detail here, but it is helpful to remember that the subscript N means polarization normal to the scattering plane, L means polarization along the momentum vector and S means "sidewise" polarization, perpendicular to both N and L. The double subscripted A's are spin-correlation parameters, the K's are spin-transfer parameters and the D's are spin-alteration parameters.

Several observations seem worth making:
(1) The variety of measurements being reported is considerable;
(2) Few of them are being done at energies below 300 MeV (reasons: few facilities; limited support; experiments difficult at low energy; N-N interaction reasonable well known at low energies); (motivations: persistent discrepancies between model predictions and phase-shift fits for $\delta(^1P_1)$ and the ε_1 coupling parameter);
(3) All of the results from LAMPF (1.4-1.8, 1.13, 1.14) were obtained with 800 MeV incident proton energy. This reflects the fact that, until recently, the LAMPF accelerator was in effect a single-energy machine. Exploration of the 500-800 MeV energy region began at LAMPF only this summer.

The discussion in this session was focused on four topics. The first was a status report on the N-N phase shift analyses of Arndt and collaborators. Arndt noted the flood of new data and discussed the impact of these data on the phase shift sets. Arndt, et al., produce both "single-energy" fits spanning 50 MeV energy bands, and an energy-dependent fit spanning the 0-850 MeV energy region. A few years ago these were poorly determined above 400 MeV, but by now the p-p solutions up to 800 MeV are well-defined, and the n-p solution at 650 MeV is becoming well defined. The improvement at 650 MeV is due most notably to the recent A_{NN} measurements at LAMPF (paper 1.4). Above 650 MeV, the n-p solution remains ill-defined.

He noted further that: (1) the data base now contains over 7400 data; (2) the value of χ_ν^2 for the energy dependent p-p solution is 1.34; (3) that much attention was paid to the evidence for $^1D_2$ and $^3F_3$ dibaryon resonances, and that he is coming to believe in their reality. He also noted again the existence of a telephone dialup system, by which one can access the VPI phase shift code. Interested persons should contact Arndt for details.

An alternative approach discussed was the potential model of the Paris group. Here the contributions due to 1π, 2π, and 3π exchange are calculated for distances greater than ~ 0.7 fm and a phenomenological fit is used for smaller distances. The value of χ_ν^2 obtained is not quite as good (~ 2) but the number of fitting parameters is considerable less.

The remaining two topics were experimental. First, the groups working at SIN presented their results (papers 1.12, 1.18, 1.19). These included measurements of various spin correlation parameters in p-p elastic scattering and in the $pp \to d\pi^+$ reaction. Some of these results are not shown in the contributed paper but have been on display in the poster session.

Finally, in the only work reported from the ZGS accelerator, the Rice roup showed the results of their measurements of $A_{NN}(90°)$ in p-p elastic scattering in the 1-3 GeV/c region (paper 1.15). One motivation for this work was to seek evidence for dibaryon resonances, and indeed they find a relatively sharp peak (0.87) in A_{NN} at P_{lab} = 1.34 GeV/c, the energy of the proposed $^1D_2$ resonance. From this result they conclude that $\sigma_t/\sigma_s \simeq 15$ at this momentum, implying strong triplet dominance in the large angle elastic scattering.

I will conclude by noting that there still appears to be lack of agreement as to the existence or interpretation of these dibaryon resonances.

(Editor's Note: Table follows. References in the text and table are to paper numbers. Paper numbers are cross-indexed to page numbers in the Table of Contents.)

TABLE I.

#	T or P														Notes
		\multicolumn{14}{c}{n-p Elastic Scattering}
1.3	50 MeV	A_N	A_{NN}	A_{LL}	A_{LS}	A_{SL}	A_{SS}	K_{NN}	K_{LL}	K_{SS}	K_{SL}	D_{NN}	D_{SS}	D_{LS}	
1.4	300–665 MeV	A_N													
1.5	300–800 MeV	A_N													
1.6	800 MeV	A_N													
1.7	800 MeV	A_N						K_{NN}	K_{LL}	K_{SS}	K_{SL}				
1.8	800 MeV	A_N													C.E.
1.9	9.88 MeV	A_N													QF
		\multicolumn{14}{c}{p-p Elastic Scattering}
1.10	52 MeV	A_N													
1.12	445–577 MeV	A_N		A_{LL}	A_{LS}		A_{SS}								
1.13	800 MeV	A_N						K_{NN}	K_{LL}	K_{SS}		D_{NN}	D_{SS}	D_{LS}	
1.14	800 MeV														
1.15	1.1–2.75 GeV/c		$A_{NN}(90°)$												
1.18	494–577 MeV	A_N		A_{LL}	A_{LS}	A_{SL}	A_{SS}								$\overline{pp \rightarrow d\pi^+}$
1.19	514–583 MeV	A_N						K_{NN}	K_{LL}						$\overline{pp \rightarrow nx}$ $\overline{pp \rightarrow p\pi^+ n}$
1.20	800 MeV														$\frac{d\sigma^{\pm}}{d\Omega_1 d\Omega_2 dp}$
1.21	800 MeV	$\frac{d\sigma}{d\Omega}$													
1.29	500 MeV	A_N													$\overline{pd \rightarrow p'x}$
1.23															$\overline{\gamma d \rightarrow np}$ assymetry

HIGH-PRECISION np POLARIZATION DATA BETWEEN 13.5 and 16.9 MeV AND THE PARIS-POTENTIAL PREDICTIONS

J. Côté[†], M. Lacombe[*], B. Loiseau[*], P. Pirès, R. de Tourreil and R. Vinh Mau[*]

Division de Physique Théorique[***], IPN, F-91406 Orsay, FRANCE

ABSTRACT

Recent high-precision np polarization data at low energies are compared with the Paris-potential predictions.

New measurements of the polarization $P(\theta)$ for neutron-proton (np) scattering between 13.5 and 16.9 MeV have been recently reported by Tornow et al[1]. They show that their data, especially at 16.9 MeV, differ from results of the Yale-IV[2], Livermore-X[3] and Arndt et al[4] phase-shift analyses. They also found a rather strong influence of the F waves spin-orbit combination Δ^F_{LS}.

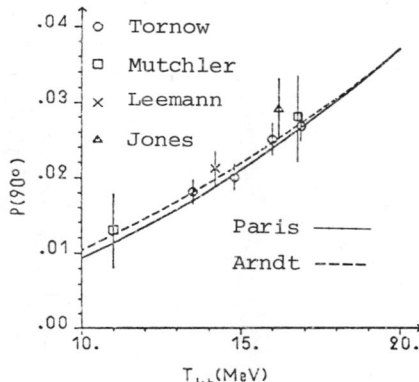

Fig.1. np polarization at 90°c.m. The references for the experimental points can be found in Tornow et al[1].

We present here the predictions of the Paris potential[5] on the polarization and on the phase shifts. We include also for completeness the Polarization measurement of Brock et al[6] at 14.2 MeV. At these energies the P, D and especially the F waves should be mainly given by the intermediate and long range part of the interaction, that part which is well-founded theoretically in the Paris potential.

Our results are summarized in Figs. 1 and 2 and in Tables I, II and III. The agreement with experimental data is excellent for $P(90°)$ and $P(\theta)$ at 14.2 MeV with χ^2/point of .7 and .1 respectively and fairly good at 16.9 MeV with a χ^2/point of 2.6 (Our χ^2/point in np scattering between 11 and 40 MeV is 1.1 for 400 data.). Our T=1 Δ^P_{LS} agrees with that

[*] Also at LPTPE, Université P. et M. Curie, F-75230 Paris, FRANCE.
[***] Laboratoire associé au CNRS.
[†] Post-doctoral fellow of the NSERC of Canada.

of Tornow et al. Our T=0 Δ_{LS}^{D} is very close to that derived from their two-term fit of $\sigma(\theta)P(\theta)$. We however are not in agreement with the Δ_{LS}^{D} and the T=1 Δ_{LS}^{F} of their three-term fit.

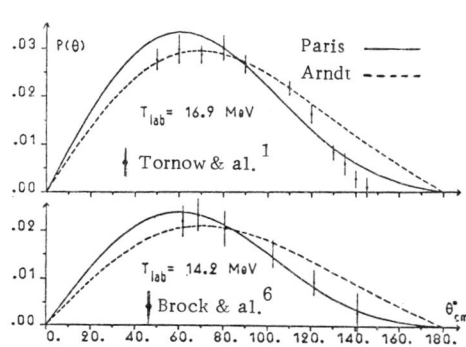

Fig. 2. np Polarization

We believe that, at those low energies, the Δ_{LS}^{F} should be very small as the F phases are mainly given by the one-pion-exchange (OPE) where no L.S force is present. We found indeed the Paris-potential F phases to be very close to those of the Born term of the OPE. This appears also in the aforementioned phase-shift analyses where the long-range OPE constraint was imposed. It is therefore inconsistent to consider Δ_{LS}^{F} in the approximate formula for $\sigma(\theta)P(\theta)$ (cf. eq. 6 of ref. 1) as one then neglects terms of the same magnitude.

In conclusion, the Paris potential gives a very good description of the low-energy NN polarization data for both np scattering as presented here and pp scattering as reported in ref. 7. As far as one is concerned with phase-shifts these new high-precision np data should be used to learn more about the T=0 force, in particular the $^1P_1$, ε_1 and triplet D phases in a locally energy-dependent phase-shift analysis using exact formulae with high partial waves constrained by OPE.

Table I The Paris-potential Δ_{LS}^{P} is compared to that extracted by Tornow[1] from data measured at 90° c.m. and to that of Arndt[4]. The $^3S_1$ phase is also displayed.

E(MeV)	$^3S_1$ (deg)		Δ_{LS}^{P} (deg)		
	Tornow Arndt	Paris Potential	Tornow	Arndt	Paris Potential
13.5	95.8	95.4	.32 ± .03	.33	.30
14.8	93.6	93.2	.35 ± .03	.38	.35
16.0	91.8	91.3	.44 ± .04	.43	.40
16.9	90.5	90.0	.47 ± .03	.47	.44

Table II Paris-potential Phase Shifts at 14.2 and 16.9 MeV.

δ(deg) E(MeV)	$^1S_0$	$^1D_2$	$^3P_0$	$^3P_1$	$^3P_2$	$^1P_1$	$^3S_1$	ε_1	$^3D_1$	$^3D_2$	$^3D_3$
14.2	54.24	.34	5.98	-3.25	1.23	-4.60	94.17	1.38	-1.29	1.66	.012
16.9	52.65	.45	6.95	-3.80	1.56	-5.31	89.99	1.49	-1.69	2.20	.016

Table III Comparison of spin-orbit splittings at 16.9 MeV.

Δ^L_{LS} (deg)	Tornow 2-term fit	Tornow 3-term fit	Arndt	Paris Potential
Δ^P_{LS}	.455 ± .012	.451 ± .012	.469	.443
Δ^D_{LS}	.072 ± .005	.058 ± .005	.037	.073
Δ^F_{LS}		-.015 ± .003	.0007	.0003

REFERENCES

1. W. Tornow, P.W. Lisowski, R.C. Byrd and R.L. Walter, Nucl. Phys. A340, 34 (1980).

2. R.E. Seamon, K.A. Friedman, G. Breit, R.D. Haracz, J.M. Holt and A. Prakash, Phys. Rev. 165, 1579 (1968).

3. M.H. MacGregor, R.A. Arndt and R.M. Wright, Phys. Rev. 182, 1714 (1969).

4. R.A. Arndt, R.H. Hackman and L.D. Roper, Phys. Rev. C15, 1002 (1977) and private communication.

5. M. Lacombe, B. Loiseau, J.M. Richard, R. Vinh Mau, J. Côté, P. Pirès and R. de Tourreil, Phys. Rev. C21, 861 (1980).

6. J.E. Brock, A. Chisholm, J.C. Duder and R. Garrett, Proc. 8th Int. Conf. on few body systems and nuclear forces, Graz, 1978, Lecture Notes in Physics, vol. 82 p. 252.

7. J. Côté, P. Pirès, R. de Tourreil, M. Lacombe, B. Loiseau and R. Vinh Mau, Phys. Rev. Lett. 44, 1031 (1980).

THE INFLUENCE OF MULTIPLE SCATTERING CORRECTIONS ON HIGH ACCURACY
NEUTRON-PROTON ANALYZING POWER DATA MEASURED AT 16.9 MeV

W. Tornow
Physikalisches Institut, Universität Tübingen,
D7400 Tübingen, West Germany

R.L. Walter
Dept. of Physics, Duke University, Durham, N.C. 27706 and
Triangle Universities Nuclear Laboratory*, Duke Station, NC 27706

ABSTRACT

Monte Carlo studies have been made to correct neutron-proton analyzing power data for multiple scattering effects. The presence of carbon in the center (hydrogen) scatterer is taken into account using the new $A_y(\theta)$ data now available for n-^{12}C scattering in the energy region of interest. The asymmetry produced by multiple scattering effects is very close to zero except for scattering angles near 30° and 80° (lab).

In all low-energy (10-20 MeV) neutron-proton analyzing power experiments reported in the literature, organic scintillators were used as the center (hydrogen) scatterer in order to reduce background problems. Usually the time-of-flight technique is employed to detect the scattered neutrons and timing signals are derived from the center scatterer and the neutron side detectors. To further decrease the background, the proton recoil energy information obtained by the scatterer scintillator is usually gated by the neutron time-of-flight information. However, using an organic scintillator as a center scatterer, multiple scattering effects involving carbon are introduced due to the presence of carbon within the scatterer scintillator. It is well known that the analyzing power in n-^{12}C scattering is appreciable at some angles in the energy region of interest. Therefore, even if the ratio of multiple scattering single scattering events is only a few percent, asymmetries due to multiple scattering involving carbon may be a problem if one has to deal with very small analyzing powers as observed in neutron-proton scattering.
Experimental tests concerning such multiple scattering effects done by Leemann et al.[1] at 14.2 MeV and Tornow et al.[2,3] at 13.5, 14.2 and 16.0 MeV neutron energy for a scattering angle θ=45° (lab) gave contradictory results. In the first case[1] a large effect was observed, but in the second case, which used a different technique, strong evidence was obtained that multiple scattering events yield only a very small asymmetry, indicating that the background may be assumed to be unpolarized to a good approximation.
Monte Carlo studies performed by Brock et al.[4] and Tornow et al.[3] shed some light on the magnitude of the effects in question. Unfortunately, due to the previous lack of n-^{12}C analyzing power data

*Supported in part by U.S. Department of Energy

between 7 and 14 MeV, certain assumptions had to be used in these calculations. For instance, in ref. 4 the analyzing power in elastic n-^{12}C scattering was approximated by $A_y(\theta) = -0.6 \sin 2\theta$ for all energies involved. The more realistic assumptions applied in ref. 3 predict results which are in good agreement with the experimental tests also reported in ref. 3.

One of the reasons for performing the ^{12}C($\vec{n}$,$n_{o,1}$)^{12}C analyzing power experiments described in ref. 5 was to have realistic n-^{12}C analyzing power data, especially in the energy range 7 to 14 MeV and around 16.9 MeV, in order to critically assess the very small corrections applied to the high accuracy neutron-proton analyzing power data reported in refs. 3 and 6.

In this paper we report on new Monte Carlo calculations. The calculations simulate the experimental arrangement and techniques used in the neutron-proton analyzing power experiments[3,6] at 16.9 MeV. The same Monte Carlo program as in ref. 3 and the same approximations were used; only the subroutine for calculating the n-^{12}C analyzing power has been updated taking into account the new $A_y(\theta)$ data for elastic and inelastic scattering. Multiple scattering events in the region of the time-of-flight and proton recoil-energy peak due to single n-p scattering are calculated using the cylindrical organic scatterer 2.5 cm in diameter and 3.8 cm in height. Altogether 1.6×10^6 neutron histories were traced per angle in order to get statistical significance in the small effects expected. In fig. 1 the asymmetry produced by a 100% polarized incident neutron beam is given for the two double scattering cases involving elastic scattering on carbon, using broad time-of-flight and recoil-energy windows, i.e., approximately 5% of the peak heights. Here the notation ^{1}H-^{12}C and ^{12}C-^{1}H indicates that a neutron is scattered first from hydrogen and then from carbon and vice versa, respectively. The error bars shown represent the statistical uncertainty of the Monte Carlo calculations. The neutron scattering sequence ^{1}H-^{12}C$_{el}$ yields a negative asymmetry reaching -0.08 at 30° (lab). Near 60° (lab) the asymmetry changes sign reaching +0.05 at 80° (lab). The asymmetry for the neutron scattering sequence ^{12}C$_{el}$-^{1}H is much smaller compared to the opposite sequence. Double scattering processes involving inelastic scattering on carbon account for less than 10% of the whole carbon contribution. Here, in both scattering sequences small asymmetries are observed also.

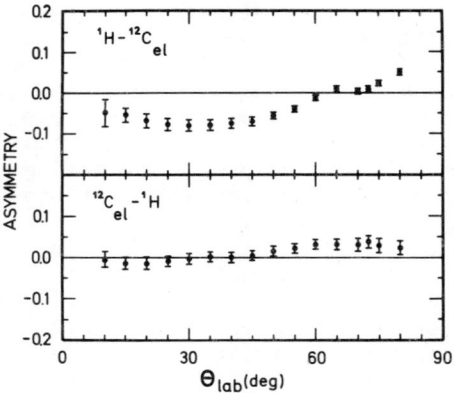

Fig. 1. Calculated asymmetry produced by double scattering processes involving elastic scattering on carbon.

Following refs. 3 and 4, the asymmetry due to multiple scattering in the region beneath the single-scattering neutron-proton peak was calculated from

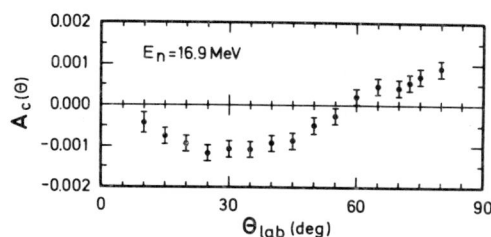

Fig. 2. Calculated asymmetry produced by multiple scattering effects in a neutron-proton analyzing power experiment.

Table I. Calculated $A_c(\theta)$ using experimental conditions

$\theta_{c.m.}$ (deg)	A_c	ΔA_c
50	-0.0012	±0.0002
60	-0.0012	±0.0002
70	-0.0008	±0.0002
80	-0.0004	±0.0002
90	+0.0001	±0.0002
110	+0.0005	±0.0002
120	-0.0001	±0.0002
130	-0.0002	±0.0002
135	-0.0002	±0.0002
140	-0.0001	±0.0002
145	+0.0006	±0.0002

the formula $A_c(\theta)=(M_L-M_R)/(T_L-T_R)$, where M_L and M_R are the multiple scattering events which resulted in neutron scattering into the left and right side detectors, respectively. Equivalently, T_L and T_R are the total number of neutrons (single and multiple scattering events). In fig. 2, A_c is shown as a function of scattering angle. As already observed in ref. 3, scattering angles around 30° and 80° (lab) are influenced most by multiple scattering effects. However, considering the overall uncertainty of ±0.002 achieved in the neutron-proton analyzing power experiment of refs. 3 and 6, the corrections one must apply to the data are not seriously large.

Finally, table I gives the asymmetry $A_c(\theta)$ calculated with the windows actually used on the time-of-flight and proton recoil-energy peaks in the neutron-proton analyzing power experiment of refs. 3 and 6. The present calculations performed with realistic analyzing power data in neutron-carbon scattering yield corrections which are almost identical to the corrections applied in refs. 3 and 6.

REFERENCES

1. B. Leemann et al., Helv. Phys. Acta 47, 479 (1974) and 4th Int. Symp. on Pol. Phenomena in Nuclear Reactions, Zürich 1975, p. 437.
2. W. Tornow, P.W. Lisowski, R.C. Byrd, S.E. Skubic, R.L. Walter, and T.B. Clegg, 4th Int. Symp. on Polarization Phenomena in Nuclear Reactions, Zürich 1975, p. 439.
3. W. Tornow, P.W. Lisowski, R.C. Byrd and R.L. Walter, Nucl. Phys., A340, 34 (1980).
4. J.E. Brock et al., Nucl. Instr. and Meth. 137, 537 (1976).
5. E. Woye et al., contribution to this conference.
6. W. Tornow, P.W. Lisowski, R.C. Byrd and R.L. Walter, Phys. Rev. Lett. 39, 915 (1977).

POLARIZATION TRANSFER IN n-p SCATTERING AT 50 MeV*

H. L. Woolverton, J. C. Hiebert, L. C. Northcliffe, M. J. Marolda,
S. Nath, and W. F. Woodward
Cyclotron Institute, Texas A&M University
College Station, Texas 77843

ABSTRACT

The polarization transfer parameter, $D_t(180°)$ has been measured at 50 MeV for n-p scattering. This measurement is compared with the predictions of 50 MeV phase shift analyses.

A measurement of $D_t(180°)$ provides a new datum which is sensitive to ε_1, the $^3S_1$-$^3D_1$ phase mixing parameter in the 50 MeV phase shift analysis of the n-p scattering data.[1-3] The value predicted by these analyses is ~ -0.15.

The polarized neutrons required for this experiment are produced via the $D(\vec{d},\vec{n})^3He$ reaction using a 50 MeV polarized deuteron beam from the Texas A&M University cyclotron incident on a high pressure gas target. The value of $K_y^{y'}(0°)$ for this reaction has been measured to be 0.63 ± 0.03.[4,5] The proton target (the n-p target) consists of a piece of polyethylene (0.635 cm thick) placed 3.5 m from the gas cell at the exit of the 0° neutron collimator.

A proton polarimeter consisting of a graphite target with left and right side detectors of NE102 plastic scintillator is used to analyze the 0° protons. Carbon was chosen as the analyzer because the p-C analyzing power is reasonably well known in the energy range under consideration.[6]

The carbon analyzer is placed one meter downstream of the n-p target. It is made of two graphite slabs, 0.3175 cm thick, mounted as two sides of an equilateral triangle with its vertex pointing in the beam direction (see Fig. 1). The third side of this equilateral triangle consists of two 20 mil sheets of NE102 (S2). These are used to provide proton time-of-flight (TOF) information and also to help distinguish left scattered events from right scattered events.

The side detectors (S3), also of NE102, are quarter sections of a right circular cylinder, each of which subtends an angular range from 55° to 75° in polar scattering angle and from -45° to +45° (measured from the horizontal plane) in azimuthal scattering angle. To complete the system a 20 mil sheet of NE102 (S1) is placed directly downstream of the n-p target to insure that an event is associated with the ejection of a charged particle from the n-p target.

A triple coincidence between S1, S2, and S3 is required for a good event. TOF and pulse height information are obtained such that seven parameters are acquired for each event and stored on magnetic tape. Each event is tagged so that valid left scattering (S1 + Left S2 + Left S3) and valid right scattering can be identified.

*Supported in part by the National Science Foundation.

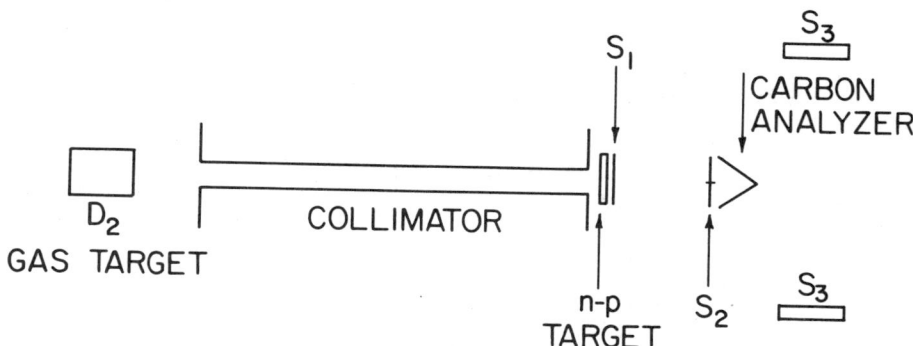

Fig. 1   Schematic diagram of D_t detector system.

Using the $^3\text{He}(\vec{d},\vec{p})^4\text{He}$ reaction as a source of 50 MeV polarized protons, the effective analyzing power for the polarimeter has been measured to be 0.46±0.04. This assumes a value for $K_y^{y'}(0°)$ of ~0.6 in accordance with a simple stripping model which has been verified both at lower energies and at 50 MeV.[5,7] A measured value of $K_y^{y'}(0°)$ for $^3\text{He}(\vec{d},\vec{p})^4\text{He}$ is not yet available.

The analysis makes full use of the multiparameter data. Off-line this consists of first sorting the data into four types: left up (LU), right up (RU), left down (LD), and right down (RD) corresponding to alternate runs with the deuteron beam polarization up and down.

Various one-dimensional spectra are then examined for the purpose of setting gates. The spectrum for the TOF between S1 and S2, T2, is found to have good resolution, and it is easy to distinguish the peak associated with the 50 MeV protons. After gates are set on this peak the same can be done for the other parameters. The remaining counts can be summed for each of the LU, RU, LD, and RD data sets and these sums used to calculate the asymmetry.[8]

The proton asymmetry for the D_t measurement has been found to be -0.05±0.08. With the value given above for the effective analyzing power of the polarimeter, the polarization of the protons is found to be -0.11±0.18.

A propane gas polarimeter is used to monitor the deuteron beam polarization during data acquisition runs by measuring the asymmetry in d-C elastic scattering at 70°. The analyzing power for this reaction at 50.6 MeV is known to be 0.756±0.031.[9] The resulting deuteron beam polarization was 0.46±0.04.

The neutron beam polarization is calculated using the value of $K_y^{y'}(0°)$ for $D(\vec{d},\vec{n})^3\text{He}$ and the deuteron beam polarization and found to be 0.43±0.06. The resulting preliminary value for $D_t(180°)$ is then 0.06±0.17. This is consistent with the prediction given by the phase shift analyses. Additional data will be acquired to improve the precision of the measurement.

References

1. R. A. Arndt, J. Binstock, and R. Bryan, Phys. Rev. D $\underline{8}$, 1397 (1973).
2. J. Binstock and R. Bryan, Phys. Rev. D $\underline{9}$, 2528 (1974).
3. S. W. Johnsen, F. P. Brady, N. S. P. King, M. W. McNaughton, and P. Signell, Phys. Rev. Lett. $\underline{38}$, 1123 (1973).
4. R. L. York, J. C. Hiebert, L. C. Northcliffe, and H. L. Woolverton, Progress in Research 1978-1979, Cyclotron Institute, Texas A&M University, p. 55.
5. R. L. York, Ph.D. dissertation, Texas A&M University, 1979 (unpublished).
6. S. Kato, K. Okada, M. Kondo. A. Shimizu, K. Hosono, T. Saito, N. Matsuoka, S. Nagamachi, K. Nisimura, N. Tamura, K. Imai, K. Egawa, M. Nakamura, T. Noro, H. Shimizu, K. Ogino, and Y. Kadota, Osaka University Research Center for Nuclear Physics, RNCP Report No. R-002, 1979 (submitted to Nuclear Instruments and Methods).
7. R. A. Hardekopf, D. D. Armstrong, W. Grüebler, P. W. Keaton, Jr., and U. Meyer-Berkhout, Phys. Rev. C $\underline{8}$, 1629 (1973).
8. G. G. Ohlsen, Rep. Prog. Phys. $\underline{35}$, 717 (1972).
9. W. D. Cornelius and R. L. York, Progress in Research 1978-1979, Cyclotron Institute, Texas A&M University, p. 57.

THE MEASUREMENT OF A_{nn} FOR FREE N-P SCATTERING
FOR NEUTRON ENERGIES 300-665 MeV*

T. S. Bhatia, G. Glass, J.C. Hiebert, L.C. Northcliffe and W.B. Tippens
Texas A & M University, College Station, Texas 77843

B.E. Bonner and J.E. Simmons
Los Alamos Scientific Laboratory, Los Alamos, N.M. 87545

C.L. Hollas, C.R. Newsom, R.D. Ransome and P.J. Riley
University of Texas, Austin, Texas 78712

ABSTRACT

The spin correlation parameter A_{nn} for free n-p scattering has been measured for 300-665 MeV over an angular range $70°-166°$ cm. This is the first A_{nn} measurement in this energy region. The results are compared with predictions from the existing nucleon-nucleon phase shift solutions.

Measurements of the spin correlation parameter A_{nn} for free np scattering for the 300-665 MeV range of incident neutron energy are reported. These are the first such measurements in the 100-800 MeV energy region. Addition of these results to the data base is expected to be of considerable help in determining a unique phase-parameter representation for n-p scattering in the LAMPF energy region.

The experiment was performed at LAMPF using the polarized neutron beam produced by 800 MeV proton bombardment of a liquid deuterium (LD_2) target. The collimated beam (at $20°$) of polarized neutrons with a broad spectrum of momenta $\vec{p}_i$ was scattered from a polarized proton target (~80% polarization). The incident neutron beam polarization in the vertical direction normal to the scattering plane, of ~20% for almost the entire range of momenta $\vec{p}_i$ was reversed with spin-precession magnets at ~30 minute intervals. The target spin reversal cycling time was roughly 2 hours. Simultaneous measurements of the scattering angle and momentum of the recoiling proton and the direction and velocity of the scattered neutron enabled unambiguous selection of the elastic np events. Such kinematic overdetermination of the two body final state drastically reduced the possible background due to scattering from the walls of the polarized target cryostat etc. Fig. 1 shows a typical incident neutron energy spectrum obtained from the measured direction and momentum of the recoiling proton from the elastic np events. The

*Work performed under the auspices of the U.S. Department of Energy.

The data were analyzed in bins of 100 MeV width. The angular distributions of A_{nn} are shown in Fig. 2. Predictions from the phase shift solutions of Arndt and VerWest (April 1980) are also shown. In a separate contribution to this conference, Arndt and VerWest discuss the impact of the A_{nn} data on their phase shift analyses.[2]

REFERENCES

1) T.S. Bhatia, et al, A polarized Neutron Beam at LAMPF, 5th ISOPPINP, Santa Fe, New Mexico 1980.
2) R.A. Arndt and B.VerWest, Private Communication and Contribution to this conference.

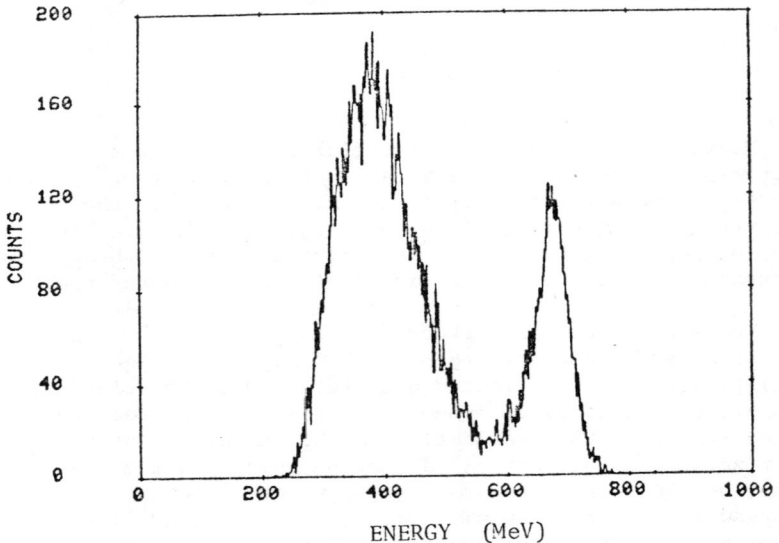

Figure 1. The energy spectrum of the neutron beam obtained from the measurement of the direction and momentum of the recoiling proton from n-p elastic scattering at 126° c.m.

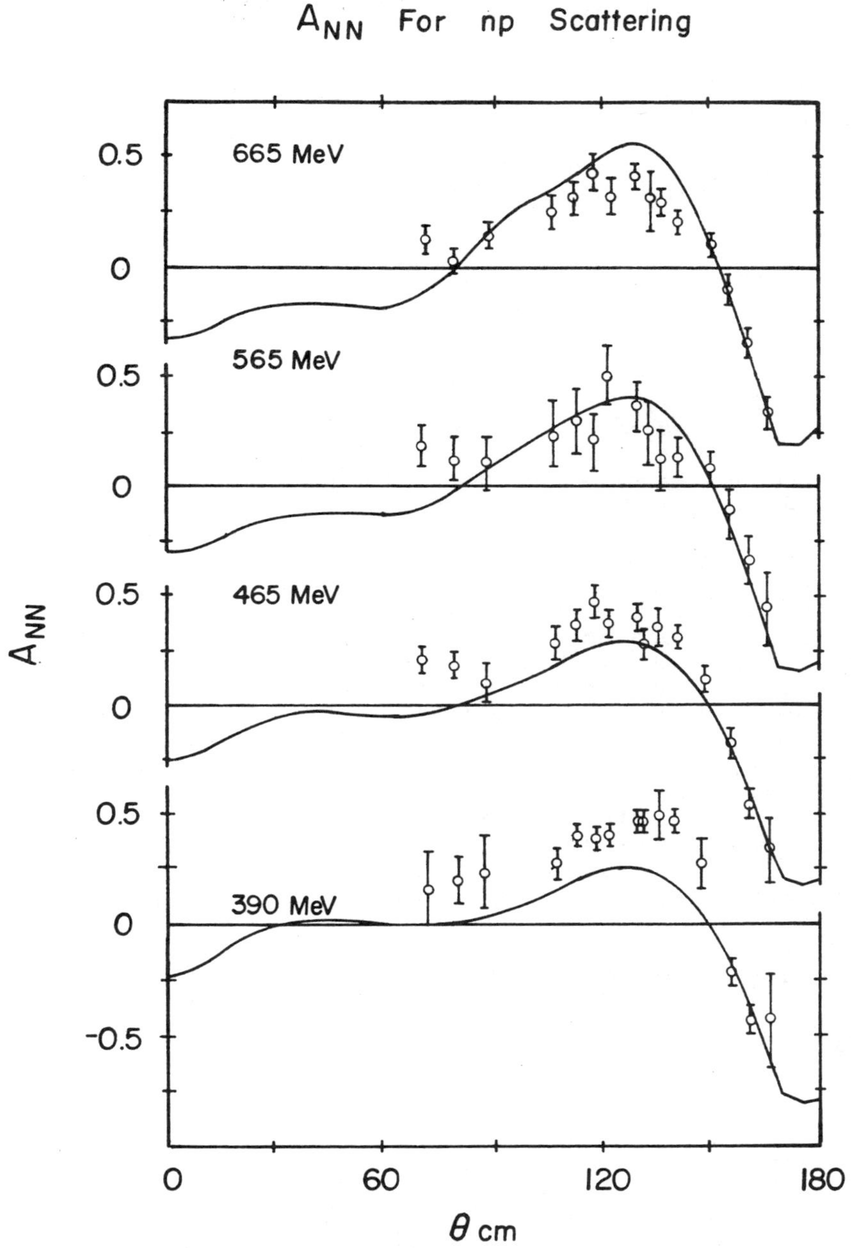

A_{NN} For np Scattering

Figure 2. Spin correlation parameter A_{nn} for np scattering vs. θ_{cm}. The solid curves are predictions from Arndt-VerWest phase shift solution (April 1980).

MEASUREMENT OF THE FREE NEUTRON-PROTON ANALYZING POWER FROM 300 TO 800 MeV*

C. R. Newsom, C. L. Hollas, R. D. Ransome, and P. J. Riley
University of Texas, Austin, Texas, 78712 USA

B. E. Bonner, J. J. Jarmer, M. W. McNaughton, and J. E. Simmons
Los Alamos Scientific Laboratory, Los Alamos, New Mexico 87545 USA

T. S. Bhatia, G. Glass, J. C. Hiebert, L. C. Northcliffe,
and W. B. Tippins

Texas A&M University, College Station, Texas 77843 USA

As a major step in determining the neutron-proton scattering matrix, the free neutron-proton analyzing power has been measured at the Los Alamos Meson Physics Facility in two separate experiments as a function of energy and angle. Additional polarization measurements in this energy range will be presented separately at this conference.

The experiments were performed by scattering a neutron beam from a polarized proton target. The neutron beams were generated by scattering 800 MeV protons from a beryllium or deuterium target. The unpolarized neutrons emerging from the beryllium target at $0°$ provided us with the neutron beam used in the first experiment (Figure 1). Neutrons emerging from the deuterium target at $20°$ (in the laboratory) were used in the second experiment[2] (Figure 2). The analyzing power, A_y, was obtained in the second experiment by averaging over the incident neutron spin. In both experiments, the energy spread of the neutron beam made it possible to measure the analyzing power at different energies simultaneously. Angular distributions were taken from 60 to 170 degrees in the center of mass system (C.M.). The angular distributions obtained for approximately 670 MeV incident neutrons are shown in Figure 3.

The energy dependence of A_y at 110 C.M. is shown in Figure 4. The data exhibit a smooth behavior with a broad minima near 650 MeV. This could be a reflection of the isospin-1 part of the n-p interaction since a similar minimum is seen in the p-p analyzing power.[3]

REERENCES

1. D. Cheng, et al., Phys. Rev. <u>163</u> (1967) 1470.
2. T. S. Bhatia, et al., abstract submitted to this conference.
3. See for example, W. O. Lock, D. F. Measday, "Intermediate Energy Nuclear Physics," Methuen and Co. Ltd., London, 1970, p. 174.
4. A. S. Clough, et al., RL-79-021 (1979).

*Work supported in part by the U.S. Department of Energy.

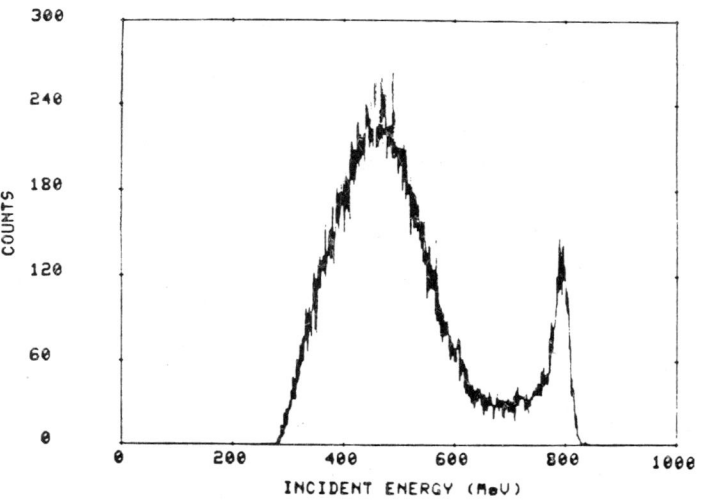

Figure 1. The detected neutron energy spectrum at 0° for 800 MeV protons incident on beryllium.

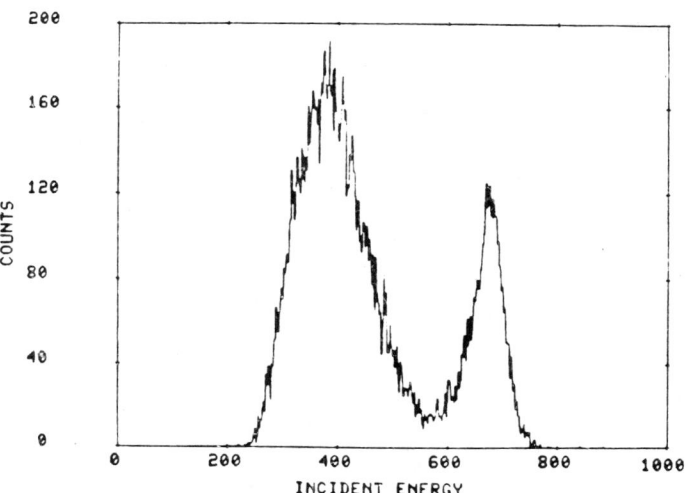

Figure 2. The detected neutron energy spectrum at 20° for 800 MeV protons incident on deuterium.

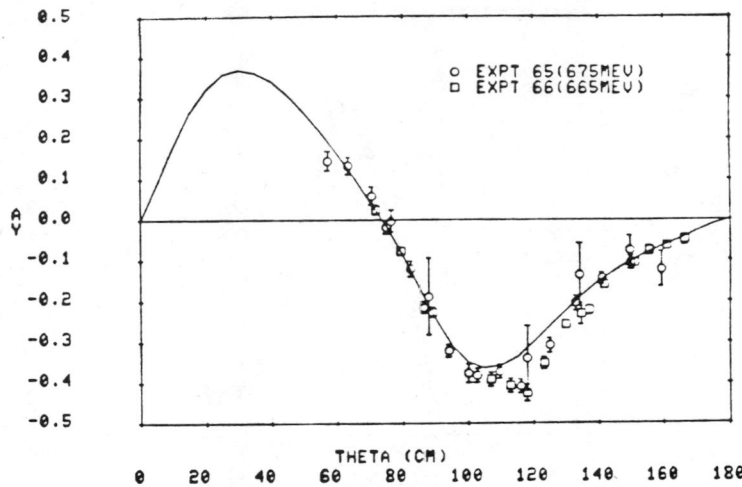

Figure 3. Free n-p analyzing power $A_y(\theta)$ at 675 MeV incident neutron energy. The solid line represents a phase shift calculation by Arndt and VerWest.

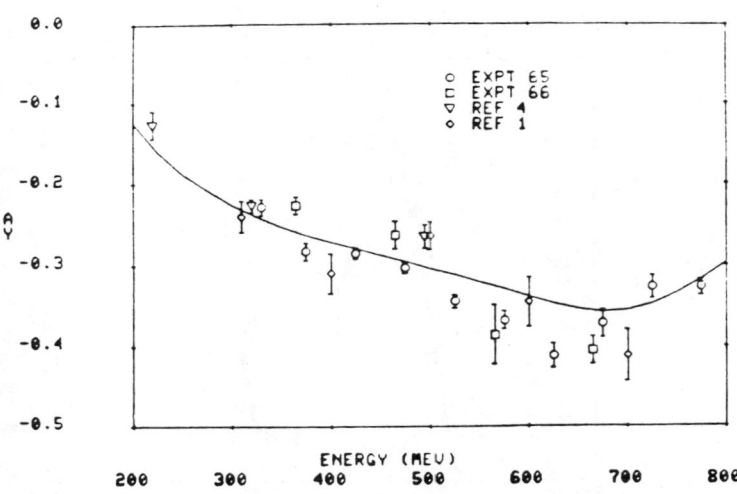

Figure 4. The n-p analyzing power at $110°$ C.M. as a function of incident energy. The solid line represents a a phase shift calculation by Arndt and VerWest.

A Measurement of the Analyzing Power for n-p Elastic Scattering
at 800 MeV for $\theta_{c.m.} > 100°$*

C. L. Hollas, P. J. Riley, C. R. Newsom, R. D. Ransome
University of Texas at Austin, Austin, Texas, 78712

B. E. Bonner, J. E. Simmons
LASL, Los Alamos, New Mexico, 87545

T. S. Bhatia, G. Glass, J. C. Hiebert, L. C. Northcliffe, and
W. B. Tippens
Texas A & M University, College Station, Texas 77843

ABSTRACT

The analyzing power for n-p elastic scattering has been measured using an 800 MeV polarized neutron beam. The measurements extend in the backward scattering region from 100 to 178 degrees.

DISCUSSION

The backward scattering of neutrons from protons at medium energies has been the subject of much experimental and theoretical investigation. The differential cross section in the charge exchange region when plotted against the square of the invariant four-momentum transfer (-u), maintains a simple shape over a large range of incident energies, and suggests that a simple mechanism is responsible.[1]

We have begun an experimental investigation of this backward scattering region using an incident polarized neutron beam at 800 MeV, and report here the results of a relative measurement of the analyzing power $A_N(\theta)$ for center of mass angles from 100 to 178 degrees. Another contribution to this conference reports on measurements of the polarization transfer coefficients K_{NN}, K_{LL}, K_{LS}, K_{SS}, and K_{SL}.[2]

The measurements were carried out using the 800 MeV polarized neutron beam of the nucleon physics laboratory at LAMPF. The neutron beam was produced with longitudinal polarization transfer at 0 degrees from the 800 MeV LAMPF polarized proton beam.[3] Fig. 1 illustrates the neutron momentum spectrum at 0 degrees from protons incident on deuterium. The neutrons within the high momentum peak were ≃40% polarized; the direction of the polarization vector was precessed to normal with a pair of dipole magnets. The neutrons struck a 30 cm long liquid hydrogen target, and recoiling charged particles were momentum analyzed in a magnetic spectrometer. A counting rate asymmetry was produced in the spectrometer by reversing the spin direction of the polarized proton beam at the ion source every three minutes.

0094-243X/81/690129-3$1.50 Copyright 1981 American Institute of Physics

Events were selected to originate only from the high momentum peak of the incident neutron spectrum, and were sorted into laboratory angle bins of one and two degree widths.

The resulting preliminary analyzing powers are illustrated in Fig. 2, and have been normalized absolutely to the 775 MeV data of Newsom et al.[4], which was measured using an unpolarized neutron beam incident on a polarized proton target.

*Work supported by the U. S. Dept. of Energy

REFERENCES

1. B. E. Bonner, et al., Phys. Rev. Lett. $\underline{41}$, 1200 (1978).

2. R. D. Ransome, et al., this conference.

3. P. J. Riley, et al., this conference.

4. C. R. Newsom, et al., this conference.

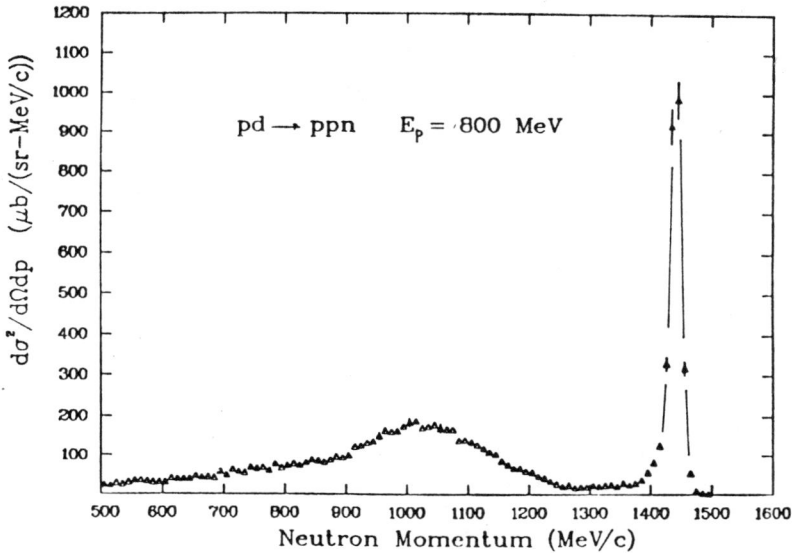

Fig. 1. The neutron momentum spectrum at 0 degrees for 800 MeV protons incident on a liquid deuterium target.

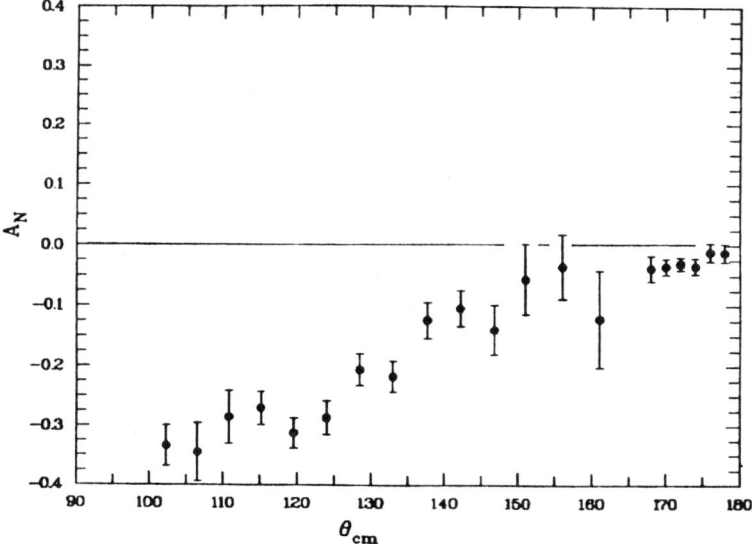

Fig. 2. Preliminary results for the n-p analyzing power at a neutron energy of 800 MeV.

Measurement of K_{NN}, K_{SS}, K_{SL}, and K_{LL} in $\vec{n}p \to \vec{p}n$ at
800 MeV in the CEX Region*

R. D. Ransome, C. L. Hollas, P. J. Riley
University of Texas at Austin, TX 78712

B. E. Bonner, W. R. Gibbs, M. W. McNaughton, J. E. Simmons
LASL, Los Alamos, NM 87545

T. S. Bhatia, G. Glass, J. C. Hiebert, L. C. Northcliffe, W. B. Tip
W. B. Tippens
Texas A&M University, College Station, TX 77843

ABSTRACT

The spin transfer parameters[1] K_{NN}, K_{SS}, K_{SL}, and K_{LL} have been measured for np elastic scattering at 800 MeV between $165°$ and $180°$ c.m. The parameters K_{NN} and K_{LL} are in good agreement with the quasi-free reaction $\vec{p}d \to \vec{n}pp$ at $180°$.[2]

DISCUSSION

The np elastic charge exchange (CEX) region is of interest for several reasons. First, the differential cross section has a sharp peak near $180°$.[3] This peak persists over al large energy range (300 MeV to greater than 60 GeV). The reaction mechanism responsible for this peak is still only poorly understood.[4] A knowledge of the spin transfer parameters will help determine this mechanism. Measurements in this region are important as part of a larger program of determining np elastic scattering amplitudes. Knowledge of these amplitudes will provide a better understanding of the NN force. They are also needed for interpretation of nucleon nucleus scattering data.[5]

The spin transfer parameters K_{NN}, K_{SS}, K_{SL}, and K_{LL} were measured at LAMPF with an incident neutron energy of 800 MeV. The experimental arrangement is shown in Fig. 1. The neutron beam is produced by the reaction $\vec{p}d \to \vec{n}pp$, utilizing the high value of K_{LL} in the quasi-elastic region.[2] Two spin precession magnets allow orientation of the neutron beam in any direction. The recoil protons are momentum analyzed by a magnetic spectrometer. Only elastic scatters from the high energy peak are used. These protons are polarization analyzed by a carbon polarimeter.[6] The incident proton beam polarization is measured by a beam line polarimeter (LBP).[7]

The recoil proton polarization values to be determined are the two transverse components perpendicular to and in the scattering

*Work supported in part by the U.S. Dept. of Energy

plane (σ_n and σ_s) and the component along the momentum vector (σ_ℓ). The use of a magnetic spectrometer causes a precession of the proton spin and a mixing of these parameters. The measured polarizations are of the form:

$$\sigma_{n2} = a\,\sigma_n + b\,\sigma_s + c\,\sigma_\ell$$

$$\sigma_{s2} = d\,\sigma_n + e\,\sigma_s + f\,\sigma_\ell$$

σ_{n2} and σ_{s2} are the two orthogonal, transverse components of the proton polarization after passing through the spectrometer. The unambiguous separation of these components requires two spectrometer magnet settings for both the S and L initial states. Unfortunately, there was only enough time to complete one of the initial state L settings. These preliminaray values for K_{LL} were computed assuming K_{LS} equal to zero (parity conservation requires K_{LS} to be zero at 180° c. m.).

Preliminary results are shown in Figs. 2-5, along with phase shift predictions by Arndt.[8] The quasi-free results from Ref. 2 are also indicated.

REFERENCES

1. The notation follows the Ann Arbor Convention, AIP Conf. Proc. no. 42, p. 142 (1977).

2. P. J. Riley, et al., this conference.

3. B. E. Bonner, et al., Phys. Rev. Lett. 41, p. 1200 (1978).

4. E. Gotsman and U. Maor, Nucl. Phys. B145, p. 459 (1978).
   J. Froyland and G. A. Winbow, Nucl. Phys. B35, p. 351 (1978).
   B. Din and E. Leader, Nuovo Cim. 28A, p. 137 (1975).
   A. Bouquet and B. Din Nuovo Cim. 43A, p. 53 (1978).
   W. R. Gibbs, private communication.

5. L. Ray, Phys. Rev., C19, p. 1855 (1979).

6. R. D. Ransome, et al., this conference.

7. M. W. McNaughton, Los Alamos Scientific Laboratory report LA-8307-MS (1980).

8. R. A. Arndt, et al., private communication, Solution CK 80, (1980).

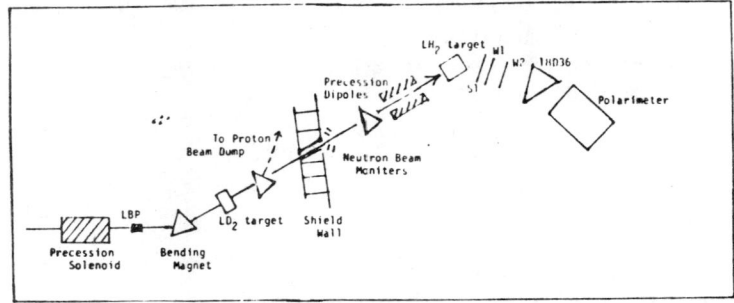

Fig. 1. Experimental arrangement, LAMPF area B.

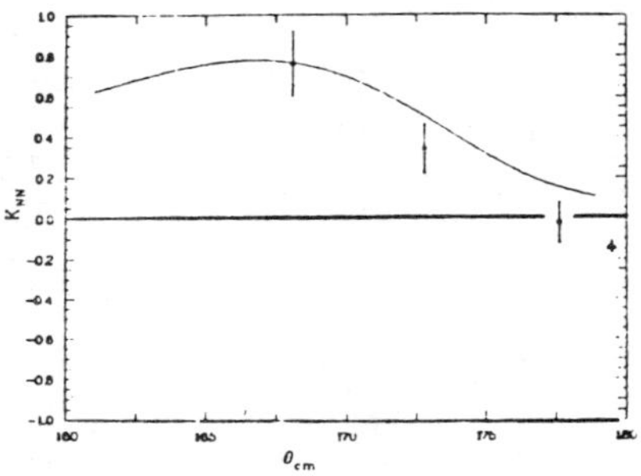

Fig. 2. K_{NN} (▲) indicates quasi-free data of Ref. 2

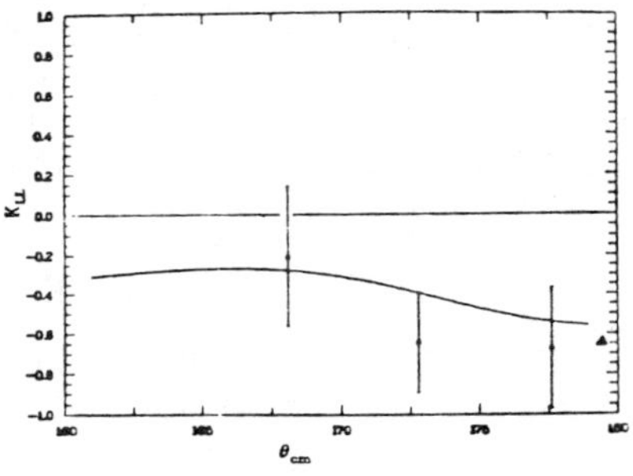

Fig. 3. K_{LL} (▲) indicates quasi-free data of Ref. 2 (→) to (→)= K_{LL} > 0

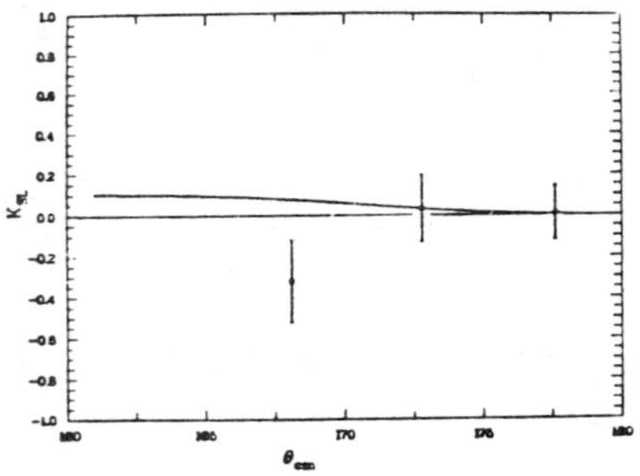

Fig. 4. K_{SL} (↑) to (→)= K_{SL} > 0

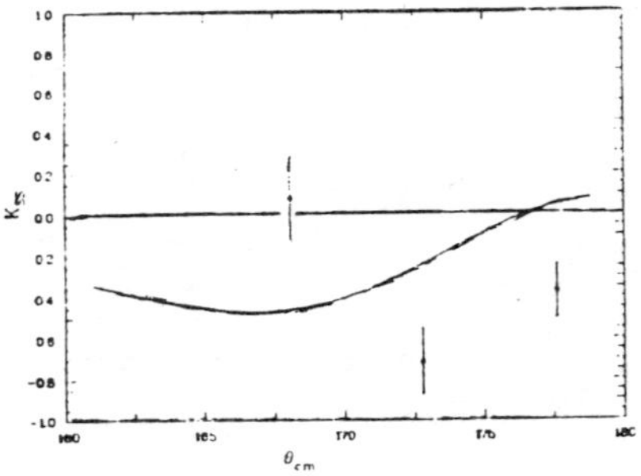

Fig. 5. K_{SS}

Spin transfer parameters for $\vec{n}p \to \vec{p}n$ elastic scattering at 800 MeV between 160° and 180° cm. Phase shift solutions from Arndt.

QUASI-FREE $\vec{p}$ + n ANALYZING POWERS AT 800 MEV*

M. Barlett, G. W. Hoffmann, and J. McGill
University of Texas at Austin, Austin, TX 78712

B. Bonner and B. Hoistad
Los Alamos Scientific Laboratory, Los Alamos, NM 87545

G. S. Blanpied
University of South Carolina, Columbia, SC 29208

ABSTRACT

The $\vec{p}$ + n analyzing power, $A_y(\theta)$, has been measured over the center-of-mass angular range 14 - 75 degrees.

DISCUSSION

Microscopic analyses of medium energy proton-nucleus elastic and inelastic angular distributions are capable of yielding rather unambiguous information about microscopic details of nuclear structure---in particular, ground state neutron densities, neutron transition densities, and the higher multipole moments of the matter densities. However, needed input for the calculations are the empirically determined forward angle fundamental nucleon-nucleon scattering amplitudes, and at 800 MeV these amplitudes are not well determined.

A program to provide some of the forward angle proton-nucleon data required to better determine the 800 MeV nucleon-nucleon amplitudes has been initiated at the High Resolution Spectrometer (HRS) facility at LAMPF.

The initial experiment, a measurement of the $\vec{p}$ + n analyzing powers, has been completed. Quasi-free $\vec{p}$ + n analyzing powers were measured over the center-of-mass angular range from 14 to 75 degrees. A liquid deuterium target was used, and the HRS detected the outgoing high energy protons, while an array of 25 scintillators was used to detect (in coincidence with the HRS) recoil neutrons and protons. A thin veto scintillator in front of the 25-counter array enabled discrimination between recoil neutrons and protons, so that both $\vec{p}$ + p and $\vec{p}$ + n analyzing powers were measured simultaneously.

The quasi-free $\vec{p}$ + p data agrees well with existing analyzing power data[1] at 800 MeV, while the larger angle $\vec{p}$ + n analyzing power data is consistent with n + $\vec{p}$ data[2] obtained independently from another experiment at LAMPF. Preliminary experimental results are shown in Fig. 1.

---

*Work supported by the U.S. Dept. of Energy and the Robert A. Welch Foundation.

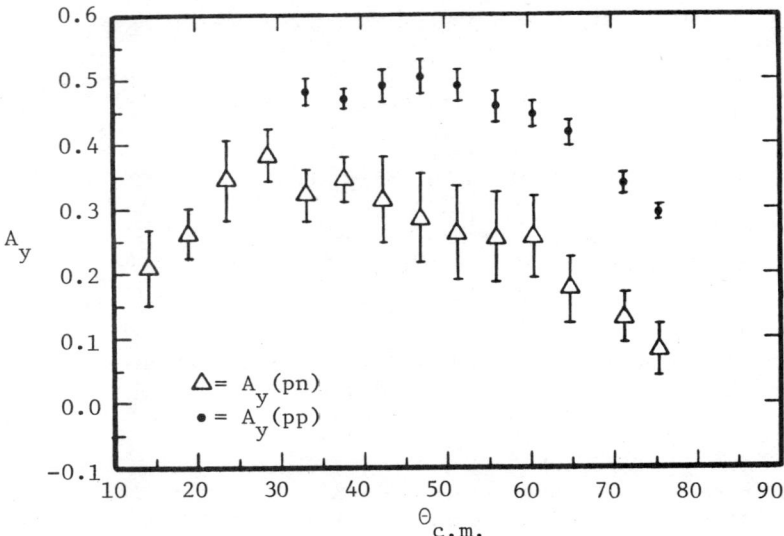

Fig. 1. Preliminary results for $\vec{p} + p$ and $\vec{p} + n$ analyzing powers at 800 MeV.

REFERENCES

1. P. R. Bevington, et al., Phys. Rev. Lett. <u>41</u>, p. 384 (1978).
2. C. Newsome, pri. comm.

ANALYZING POWER OF PROTON-PROTON SCATTERING AT 9.88 MeV[†]

M.D. Barker, P.C. Colby, W. Haeberli
University of Wisconsin, Madison, WI. 53706

J. Ulbricht
Laboratorium für Kernphysik, ETH Hönggerberg
8093 Zürich, Switzerland

The analyzing power of proton-proton scattering at 9.88 MeV has been measured at 14 angles to an uncertainty of $\pm 5 \times 10^{-5}$. This is a factor of four improvement in accuracy over previous measurements.[1] A target of gaseous hydrogen was bombarded with polarized protons and the left-right asymmetry was measured with symmetric detectors.

The polarized beam was produced with a tandem electrostatic accelerator equipped with a colliding-beams ion source. The polarization of the beam was reversed every 0.25 seconds by alternately energizing a weak-field and a strong-field RF transition unit. A spin precessor[2] was used prior to injection to orient the polarization vector perpendicular to the reaction plane. In addition to the fast switching using RF transitions, the overall sign of the polarization vector was reversed with the spin precessor at regular intervals during the measurements. The beam polarization (82-89%) was monitored continuously by observing p-^{4}He scattering in a polarimeter[3] mounted directly behind the main scattering chamber.

The incident beam was defined with a 1.0 mm × 2.0 mm slit located 0.35 m from the target center. A second slit, 2.6 m from the target, defined the direction of the beam to within ±1.1 mrad. The entire scattering chamber was filled with 300 Torr hydrogen gas (purity 99.999%) with the beam entering through a 50 μm Ni foil located 0.36 m from the center of the target. Two antiscattering slits were placed behind the beam defining slit to shield the detectors from slit-edge scattered particles. The two slits were arranged such that a particle scattering from the entrance slit would have to scatter from three more slit edges before it could enter a detector.

Detectors were placed 20 cm from the center of the target, except at $\theta_{lab} = 7.5°$ where the detectors were 40 cm from target center. The extreme angular acceptance of the detector system ranged from ±0.6° to ±2.0°. Both CsI(Th) scintillators and silicon surface barrier detectors were used.

To obtain an accurate measurement of the analyzing power, one must ascertain that the pulse-height spectrum contains no significant background or contaminants under the proton-proton peak. This is especially difficult at forward angles. Figure 1 illustrates that protons scattered from contaminants (C,N,O) could be resolved from p-p protons even at $\theta_{lab} = 7.5°$. In general, the spectra contain background on both the low- and high-energy sides of the p-p peak. It was found that the analyzing power of the low-energy background is equal to that for p-p scattering at the same angle to within $\pm 5 \times 10^{-4}$. The peak-to-background ratio is about 300. The

[†] Work supported in part by the U.S. Department of Energy.

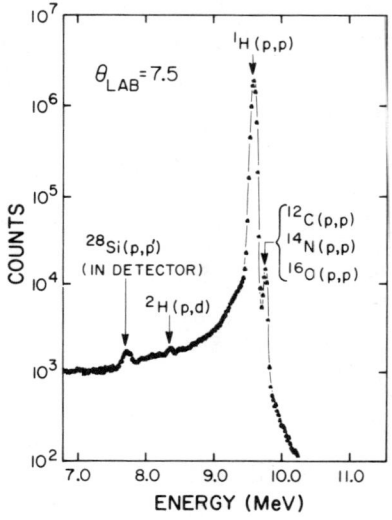

Figure 1. A typical pulse-height spectrum at $\theta_{lab}=7.5°$.

flat background on the high-energy side is negligible. Superimposed on the background are peaks due to protons scattering elastically from deuterium and heavy contaminants, and a peak due to recoil deuterons from p-d elastic scattering. At those angles where the peaks due to p-d scattering are not resolved from the p-p peak, a correction was made to the p-p analyzing power based on known deuterium abundance, p-d cross sections and analyzing powers. The correction is 0.5×10^{-5} or less. The heavy contaminant peak is resolved by at least one detector at all angles. In addition, we observed a peak due to the reaction ^{28}Si(p,p') (Q=-1.78 MeV) in the surface barrier detector itself. This peak is resolved at all angles.

For each scattering angle, the analyzing power was measured in a number of separate runs during which about 8×10^6 counts were collected for each detector in each spin state. The variance of these measurements is consistent with the expected statistical errors. Final analyzing powers are presented in Table I.

TABLE I

Measured values of the analyzing power of proton-proton scattering at 9.88 MeV.

θ_{cm} (deg)	10^4 A	θ_{cm} (deg)	10^4 A
15	-15.28±0.52	50	-5.07±0.36
20	-21.23±0.59	55	-5.04±0.53
25	-18.98±0.44	60	-2.93±0.50
30	-14.48±0.55	65	-2.27±0.53
35	-11.23±0.48	70	-0.78±0.36
40	- 8.47±0.32	80	-0.71±0.51
45	- 5.39±0.52	90	-0.41±0.45

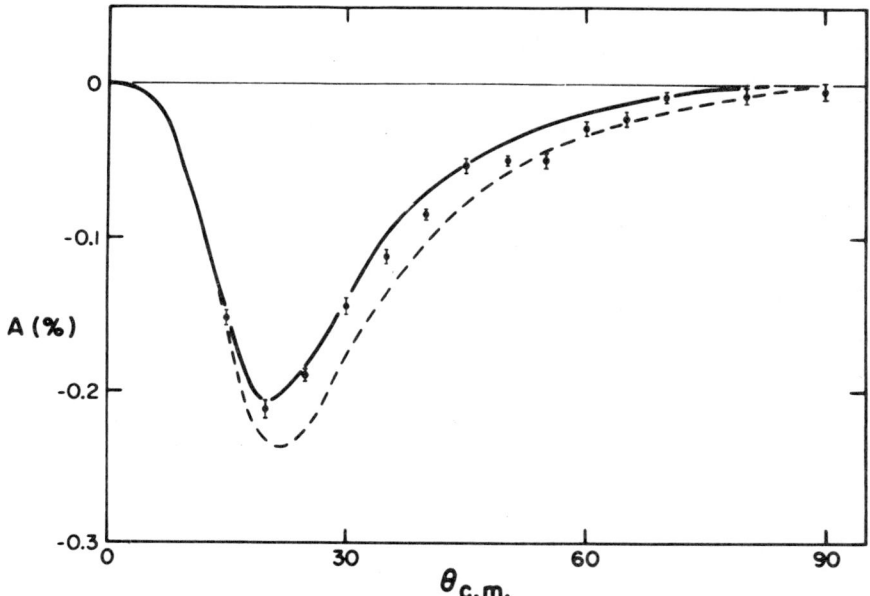

Figure 2. Analyzing power of proton-proton scattering at 9.88 MeV. The dashed curve shows the analyzing power predicted by the Paris potential (Ref. 4) at 10.0 MeV. The solid curve corresponds to the phase shifts of Ref. 1.

A recent paper by Côté, et al.[4] presents the predictions of the Paris potential for the analyzing power at 10.0 MeV. This prediction is shown as a dashed curve in Figure 2, whereas the solid curve is the result of the single-energy phase shift analysis of Hutton et al.[1], also at 10.0 MeV. A phase shift analysis of the present data is forthcoming.

REFERENCES

[1] J.D. Hutton, W. Haeberli, L.D. Knutson, P. Signell, Phys. Rev. Lett. 35 (1975) 429.
[2] The spin precessor is a crossed-field analyzer purchased from ANAC, Inc., Model No. 2171.
[3] The polarimeter measures the asymmetry in p-^{4}He scattering with symmetric detectors at θ_{lab} = 112.3° where the analyzing power is known accurately for 9.66 MeV incident protons from P. Schwandt, T.B. Clegg, W. Haeberli, Nucl. Phys. A163 (1971) 432.
[4] J. Côté, P. Pirès, R. de Tourreil, M. Lacombe, B. Loiseau, R. Vinh Mau, Phys. Rev. Lett. 44 (1980) 1031.

POLARIZATION IN PROTON-PROTON SCATTERING AT 52 MEV
AND EFFECT OF THE ELECTROMAGNETIC LS FORCE

K. Imai, T. Matsusue, H. Shimizu, J. Shirai and K. Nisimura
Department of Physics, Kyoto University, Kyoto, Japan

K. Hatanaka and T. Saito
Research Center for Nuclear Physics, Osaka University, Osaka, Japan

It is wellknown that the polarization effect in nucleon-nuclear scattering can arise from electromagnetically induced spin-orbit force.[1] The large polarization effect was clearly observed in the small angle neutron-nucleus scattering. But in the proton-proton scattering the polarization effect due to the electromagnetic LS force was estimated to be small and no definite indications have been found.

Previously the polarization in proton-proton scattering at 52.3 MeV was measured at angles from θ_{lab}=15° to 50°.[2] At those angles, the effect of the electromagnetic LS force is very small and the correction due to the effect was not necessary to deduce phase shifts. But at this energy, it was expected that the definite indication of the electromagnetic LS effect could be found in an accurate measurement of the polarization at smaller angles.

In order to find the effect and improve the accuracy of phase shifts, the polarization in proton-proton scattering between θ_{lab}=5° and 15° has been measured with accuracies of $1 \sim 2 \times 10^{-3}$ at the same energy.

The measurement has been performed with polarized proton beam from AVF cyclotron at Research Center for Nuclear Physics, Osaka University.

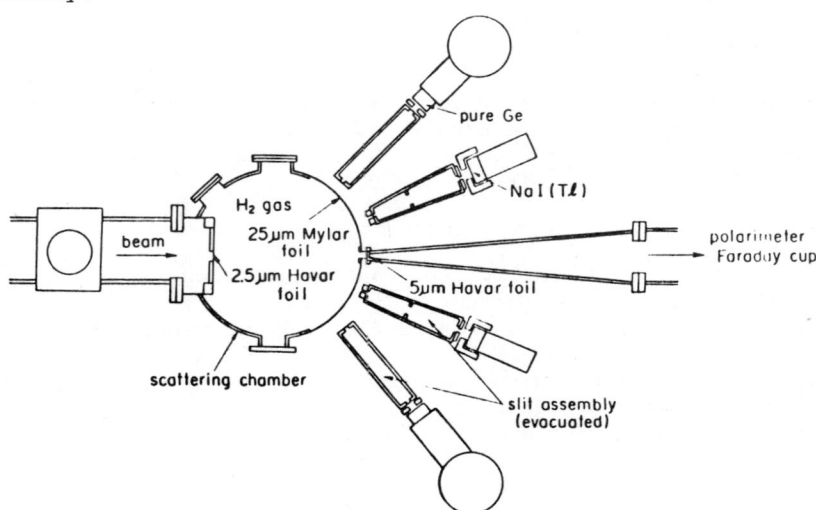

Fig. 1 A schematic view of the experimental set up.

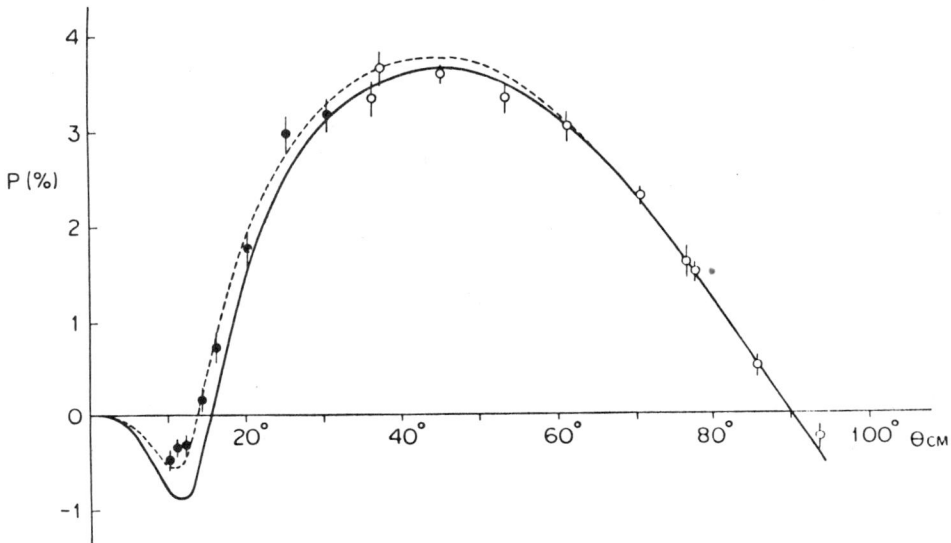

Fig. 2   Polarization in proton-proton scattering at E_p=52.3 MeV. The solid and open circles indicate the present and previous data respectively. The solid curve is calculated from phase shift set obtained previously without the electromagnetic LS correction. The dashed curve is calculated from the same phase shifts set with the electromagnetic LS correction.

A schematic view of the experimental set up is shown in Fig. 1. The target was high-purity (99.999%) hydrogen gas of 1 atm. The level of impurities were monitored with backward detectors. The measured analyzing power was corrected by 7.6×10^{-4} at $\theta_{lab}=5°$ for the effect due to impurities. At angles smaller than $\theta_{lab}=8°$, same amounts of background runs were taken to subtract backgrounds under the proton-proton peak and the resulting correction to the analyzing power was 6×10^{-4} at $\theta_{lab}=5°$.

The asymmetries were calculated using the data of left and right detectors of both spin directions by means of the geometrical mean method.[3] The analyzing powers are shown in Fig. 2 together with previous data at larger angles.

The electromagnetic LS potential between two protons can be written as

$$V_{LS} = -8\mu_0(\mu_T - \tfrac{1}{4}) \frac{1}{r^3} \vec{L}\vec{S}$$
$$\vec{S} = \tfrac{1}{2}(\vec{\sigma}_1 + \vec{\sigma}_2) \quad , \quad \mu_T = 2.79276\mu_0 \quad .$$

The phase shifts resulting from this potential were calculated using a plane wave Born approximation. Then the amplitude due to the electromagnetic LS interaction was obtained by summing up the many partial wave amplitudes (L_{max}=1000) including the Coulomb phase factors. The Coulomb phase factors are very important in the calcu-

TABLE 1   The result of the phase shift analysis. NOF is a number of freedom.

$^1S_0$	38.21±.12	38.16±.06	38.27±.07	38.20±.08
$^3P_0$	13.15±.10	12.99±.16	11.61±.27	12.75±.13
$^3P_1$	-8.30±.09	-8.43±.04	-8.58±.07	-8.37±.04
$^3P_2$	5.80±.06	5.89±.06	6.23±.04	5.91±.03
$^1D_2$	1.62±.02	1.62±.04	1.64±.02	1.61±.02
ε_2	-1.69±.05	-1.66±.06	-1.63±.05	-1.66±.05
χ^2/NOF	0.92	0.90	1.38	0.94
present Data	no	no	yes	yes
El-Mag correction	no	yes	no	yes

lation of polarization parameter. At 10 MeV, the polarization was calculated with this correction and the result showed virtually no effect of the electromagnetic LS force.

The phase shift analysis have been done with and without the electromagnetic LS correction to investigate the effect phenomenologically. The other data used in the present analysis are same as those in the previous analysis.[2]

Results of the phase shift analysis are given in Table 1. Without the present data, the electromagnetic LS correction have almost no effect on phase shifts and χ^2-value. But with the present data, the electromagnetic LS correction clearly reduce the χ^2-value from 1.38 to 0.94. The results show that the present polarization data at small angles cannot be fit without the electromagnetic LS correction. The obtained phase shifts using present data shows that the present data and phase shift analysis confirmed the previous solution of phase shifts.

The present result encourages the idea that the proton-proton polarization due to the electromagnetic LS force could be a polarization analyzer at higher energy as several hundred GeV.

REFERENCES

1. J. Schwinger, Phys. Rev. 73, 407 (1948).
2. N. Tamura et al., J. Phys. Soc. Japan 44, Supplement 289 (1978).
3. G. G. Ohlsen and P. W. Keaton, Jr., Nucl. Instr. & Method 4, 53 (1959).

MEASUREMENT OF THE SPIN CORRELATION PARAMETERS A_{ookk}, A_{ooks} AND A_{ooss} IN THE p-p ELASTIC SCATTERING BETWEEN 400 AND 600 MeV

E. Aprile*, R. Hausammann*, E. Heer*, R. Hess*, S. Jaccard**,
C. Lechanoine-LeLuc*, W. Leo*, Y. Onel*, S. Mango**, D. Rapin*

These measurements are part of a series in progress at SIN with the objective of determining the nucleon-nucleon amplitudes in the p-p elastic scattering between 400-600 MeV from a complete experiment.

We have used a longitudinally polarized beam and a polarized butanol target in the horizontal plane. Due to the restrictive geometric acceptance of the target, the polarization axis of the target had to be oriented at an angle α with respect to the beam direction. Therefore we have measured parameters $A_{kk} + A_{ks}$ as a linear combination at 577, 536, 514, 494 and 445 MeV. These experiments were extended for the measurements of $A_{ks} + A_{ss}$ by using a transversally polarized beam.

Incident beam polarization was 41.65 ± .43% and the polarization in the butanol target was about 60%. Outgoing protons were detected by two telescopes, each containing multiwire proportional chambers. Data were taken with fast on-line event reconstruction technique[1].

In the first case data were taken for two different beam-target angles (α = 78.5°, 32°) and in the second case, beam-target angles were α = 101.5°, -8°, 32°. Our preliminary results of A_{pq} vs θ_{CM} at 577 MeV are shown in fig.2, where the dotted lines are the predictions of the Saclay[2] phase shifts. The predictions of D.V. Bugg are also shown in fig.1 for data sets I and II as a solid line. Only data set V was taken with accelerated polarized beam, where beam polarization was 85 ± 2.0%. Our chosen acceptance for these configurations in ϕ (where ϕ is the angle between y-direction and normal to the scattering plane) was not symmetric, so that non-vanishing terms with factors $\sin^2\phi$, $\cos^2\phi$, $\sin\phi$, $\cos\phi$, $\sin\phi\cos\phi$ contribute to the M-matrix equation. Therefore the measured quantity A_{pq} is not only a function of $A_{ooks} + A_{ookk}$ or $A_{ooks} + A_{ooss}$ but also contains small amounts of A_{oonn} and A_{ooss} in the first combination and $A_{oonn} + A_{ookk}$ in the second. Then A_{pq} is given as $F_1(\phi)A_{oonn} + F_2(\phi)A_{ooss} + F_3(\phi)A_{ooks} + F_4 A_{ookk}$. As a total we have five data sets with four unknowns and A_{nn} has already been measured by this group[1]. Further measurements on A_{nn} made recently were added to this data and are shown in fig.1.

Sets I, II, III, V cover angular region of θ_{CM} = 38° → 60°. For this domain we have extracted the values of A_{kk}, A_{ks} and A_{ss}

*   D.P.N.C., University of Geneva, Switzerland
** S.I.N., Villigen, Switzerland

by linear fit, as shown in fig.1. This fit gives χ^2 = 4.3/pt. indicating that one should take into account the systematic uncertainties due to the fluctuation of target and/or beam polarization (relative error for set I and II 10%, 6% for set III and IV, and 7% for set V). Therefore we have added the systematic errors quadratically to the statistical errors, as shown in fig.2 and the errors on A_{kk}, A_{ks} and A_{ss} have increased from 1-2% to 3-4%. This fit gives χ^2 = 1.6/pt.

Data sets II, IV and V cover the angular domain of θ_{CM} = 62° → 90°. For this domain we have extracted the values of A_{kk}, A_{ks}, A_{ss} by using the exact solution of linear equations. This solution agrees very well with the theoretical predictions at θ_{CM} = 90°, namely $A_{nn} - A_{ss} - A_{kk}$ = 1 and A_{sk} = 0. A study of the possible other sources of systematic errors is in progress.

REFERENCES

1. D. Besset, thesis, University of Geneva (1978)
2. J. Bystricky et al., Saclay Report D Ph P E 1-79
3. D.V. Bugg, Nucl. Phys. A 335 (1980) 171-191.

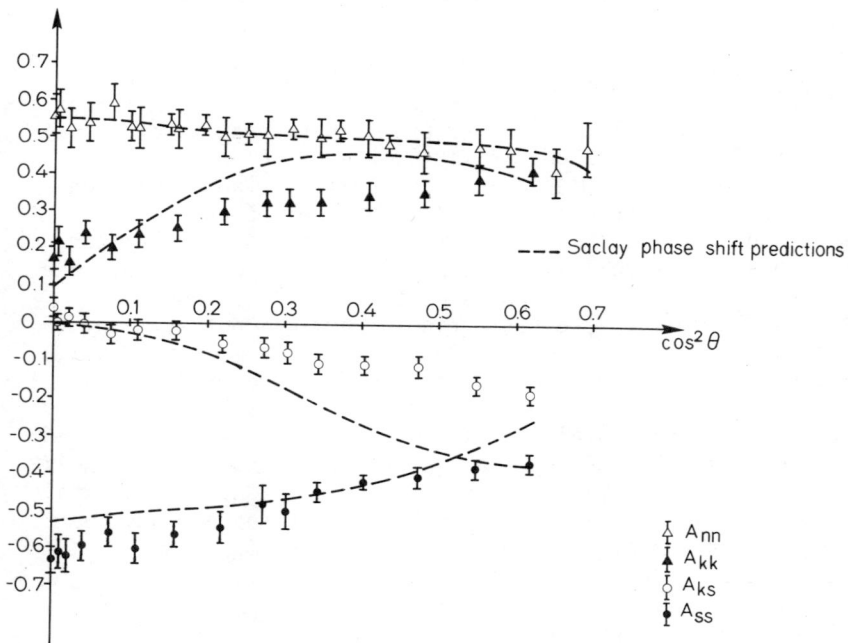

Fig. 1   Preliminary results of A_{kk}, A_{ks} and A_{ss} at 577 MeV

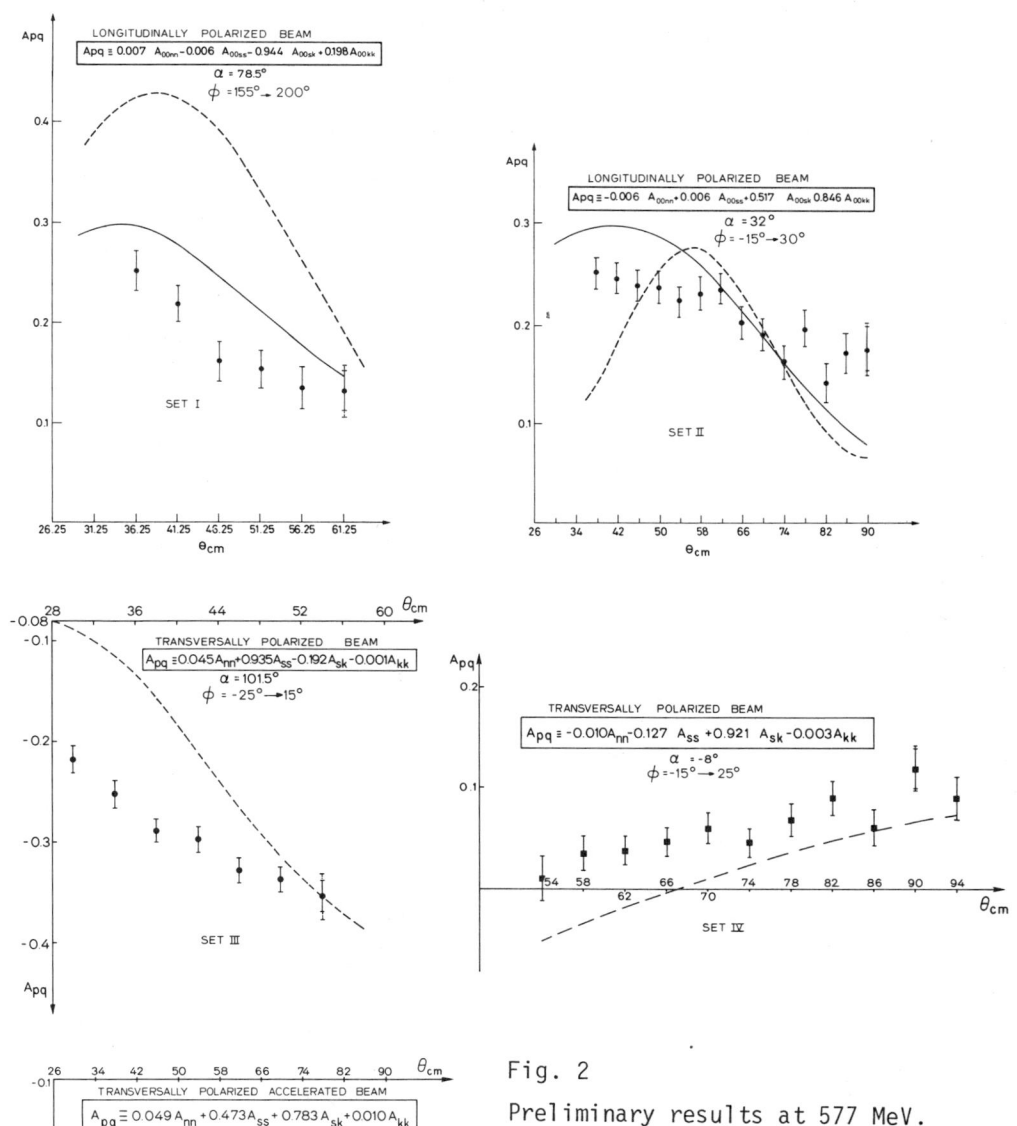

Fig. 2

Preliminary results at 577 MeV. Errors on the data points are statistical (solid line), dotted line on the point which shows the amount has to be added as a systematic error for normalization.

ASYMMETRIES FOR ELASTIC pp SCATTERING IN THE COULOMB INTERFERENCE
REGION AT 800 MeV*

G. Pauletta, G. S. Adams, G. Igo, J. B. McClelland, A. T. M. Wang,
C. A. Whitten, Jr. and A. Wriekat
University of California, Los Angeles, CA 90024

M. Gazzaly
University of Minnesota, Minneapolis, MINN 55455

B. Hoistad
Los Alamos Scientific Laboratoy, Los Alamos, NM 87545

ABSTRACT

Asymmetries for elastic scattering of transversely polarised 800 MeV protons from Hydrogen in the coulomb interference region are presented and compared with the predictions of a recent phase shift analysis.

INTRODUCTION

Small- angle asymmetries for the elastic scattering of transversly polarized protons off hydrogen have been measured at 800 MeV. These measurements are the continuation of the differential elastic cross section measurements reported previously.[1] These measurements are motivated by the need to determine forward p-p scattering amplitudes accurately around 1.4 GeV/C. Forward spin-orbit amplitudes at present are poorly known and needed as input to microscopic p-Nucleus calculations, while the spin-independent and double spin-flip amplitudes are of relevance to the speculations concerning dibaryon resonances in this energy region.

EXPERIMENTAL METHOD

The experiment was performed at LAMPF, using the High Resolution Spectrometer (HRS) to detect protons scattering in a forward direction from a 6-cm LH_2 target with 5-mil Kapton windows. For measurements at the smallest angles, the beam was incident on the focal plane where it passed through the insensitive region of a specially constructed set of multiwire proportional chambers (MWPC's) and was counted by means of a 3-fold scintillator telescope. This permitted the beam related background such as that which would be produced by passing the beam through the bulky MWPC frames to be virtually eliminated. A second set of chambers was sensitive to the beam and could be used to monitor eventual shifts which would produce spurious asymmetries. The spectrometer's good resolution may be relied on to differentiate against events from inelastic scattering at all angles, and against those from elastic scattering off the heavy constituents of the target windows

0094-243X/81/690146-3$1.50 Copyright 1981 American Institute of Physics

(carbon and oxygen) down to 1.2° (lab).

Data below 1.2° can be obtained by subtracting target-empty runs after correcting for multiple-coulomb scattering. At small angles ($\lesssim 2°$ lab) the yields could be normalized both to the measured beam intensity and to a set of 3-fold scintillator telescopes viewing the target at $\pm 20°$ (lab). At larger angles, ($\gtrsim 2°$ lab), the telescopes were used and at yet larger angles ($\gtrsim 4°$ lab), an ion chamber was introduced. In order to avoid deteriorating the beam profile, the beam polarization ($\sim 80\%$) was monitored continuously by means of a polarimeter in another beam line.

RESULTS

Data has been analyzed down to 1.2° (lab) and is shown in Fig. 1. It is seen to agree with previous LAMPF data[2] and with the data from another experiment reported at this conference[3] at the points of overlap. The solid line represents the prediction of the most recent Arndt phase shifts which do not include our data. The data will be extended down to 0.7° (lab) as soon as yields at small angles have been corrected for multiple-coulomb scattering. Both differential cross section and asymmetries will then be analyzed in conjunction to extract zero- degree scattering amplitudes.

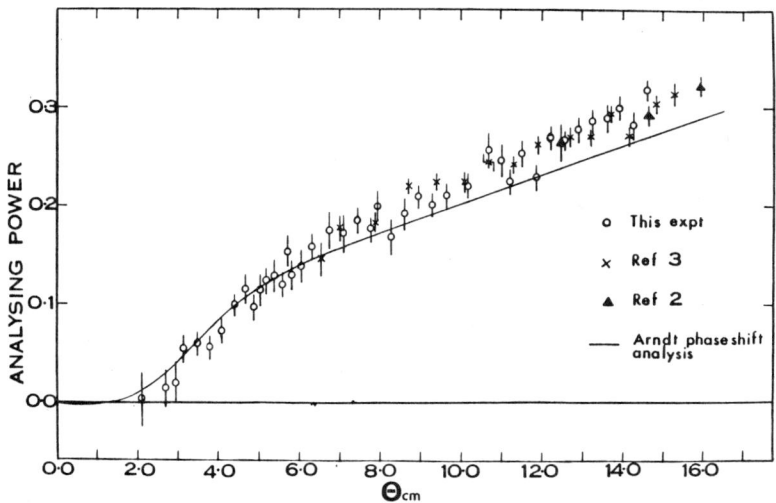

Figure 1.

REFERENCES

*
Work supported in part by the Department of Energy.
1. G. Pauletta et al., "8th Int. Conf. on High Energy Physics and Nuclear Structure," Vancouver (1979),
   A. Wriekat et al., submitted to Phys. Lett. (1980).
2. M. McNaughton, LASL Rep. LA-UR-80-142. Submitted to Phys. Rev. C, (1980).
3. J. B. McClelland et al., contribution to this conference.

MEASUREMENT OF D_{NN}, D_{SS}, D_{LS} IN pp → pp AT 800 MeV

M. W. McNaughton, B. E. Bonner, W. D. Cornelius
E. W. Hoffman, O. B. van Dyck, R. L. York
Los Alamos Scientific Laboratory, Los Alamos, NM 87545

R. D. Ransome, C. L. Hollas, P. J. Riley
University of Texas, Austin, TX 78712

K. Toshioka, H. Spinka
Argonne National Laboratory, Argonne, IL 60439

P. R. Bevington, H. B. Willard
Case Western Reserve University, Cleveland, OH 44106

ABSTRACT

The spin transfer parameters D_{NN}, D_{SS}, D_{LS} have been measured for pp elastic scattering at 800 MeV between 20° and 135° cm. These data bring the number of parameters measured at this energy to a total of 10, which is in general sufficient to determine a set of I = 1 amplitudes.

DISCUSSION

As part of our program to determine the I = 1 nucleon-nucleon amplitudes near 800 MeV, we have measured the spin depolarization parameters D_{NN}, D_{SS}, D_{LS}[1] from about 20° to 135° cm. Because of the identity of particles in pp → pp, the spin transfer parameters K_{NN}, K_{SS}, K_{LS} are related to the above parameters: $K(\theta) = D(\pi-\theta)$. Consequently, these measurements may be thought of as six parameters in total.

It is well known that at least nine independent parameters are required for an unconstrained analysis of the pp → pp amplitudes above pion threshold. These measurements together with previous measurements of cross section,[2] analyzing power,[3] and spin correlation parameters A_{NN}[4] and A_{LL}[5] bring the total to 10. These data should provide a solution near 800 MeV for the I = 1 amplitudes and will clarify the interpretation of the resonance-like structure observed near 800 MeV.[6] The data are also urgently needed for the interpretation of proton-nucleus data.[7]

The experimental method was similar in principle to that used at lower energies at TRIUMF[9] and SIN.[10]

The LAMPF polarized proton beam was oriented in the N, S, or L direction by the combination of a 750 keV Wien filter (ExB fields) and a novel 800 MeV precessor.[11] Three components of spin direction were monitored by the combination of two beam line polarimeters[12] with 46° of precession between them.

The first scatter was from liquid hydrogen (25 mm thick). Both primary and conjugate protons were detected in multiwire drift chambers (MWDC) selecting elastic scattering by the precise angular correlation that is characteristic of the two-body final state. Inelastic background was always less than 1% and was ignored.

The spin after the first scatter was measured by a second scatter from a carbon target ranging in thickness from 32 to 254 mm. Three MWDC x-y pairs defined the proton trajectory incident on the carbon, and three pairs defined the charged particle trajectory after the carbon. Only minimal restrictions were placed on the final state charged particle. Further details are contained in Ref. 8.

The methods of extracting the spin depolarization parameters from the azimuthal distribution are well described in Refs. 9 and 10. We are grateful to these groups for freely sharing their experience with us.

Preliminary results are shown in comparison with phase shift predictions by Arndt[13] and Hoshizaki[14] in Fig. 1. Corrections have been considered for instrumental asymmetries, finite angular acceptance and imprecise alignment of the beam spin. These corrections are small and have been applied where significant.

REFERENCES

[1] For notation see AIP Conf. Proc. No. 42, p. 142 (1977).
[2] H. B. Willard et al., Phys. Rev. C14, p. 1545 (1976).
[3] P. R. Bevington et al., Phys. Rev. Lett. 41, p. 384 (1978).
   M. W. McNaughton et al., preprint LA-UR-80-1412, submitted to Phys. Rev. C.
[4] M. W. McNaughton et al., to be submitted to Phys. Rev. C.
[5] I. P. Auer et al., Phys. Rev. Lett. 41, p. 1436 (1978).
[6] I. P. Auer et al., Phys. Rev. Lett. 41, p. 354 (1978); E. K. Biegert et al., Phys. Lett. 73B, p. 235 (1978), and ref. therein.
[7] L. Ray, Phys. Rev. C19, p. 1855 (1979).
[8] R. Ransome et al., this conference.
[9] G. Waters et al., Nucl. Instr. 153, p. 401 (1978); D. Bugg et al., J. Phys. G4, p. 1025 (1978).
[10] D. Besset et al., Nucl. Instr. 166, p. 379 and p. 515 (1979).
[11] E. W. Hoffman, IEEE Trans. on Nucl. Sci. NS-26, No. 3, p. 3995 (1979).
[12] M. W. McNaughton, LASL report LA-8307-MS (1980).
[13] R. A. Arndt et al., Solutions CD79, CK80.
[14] N. Hoshizaki, AIP Conf. Proc. No. 51, p. 399 (1978).

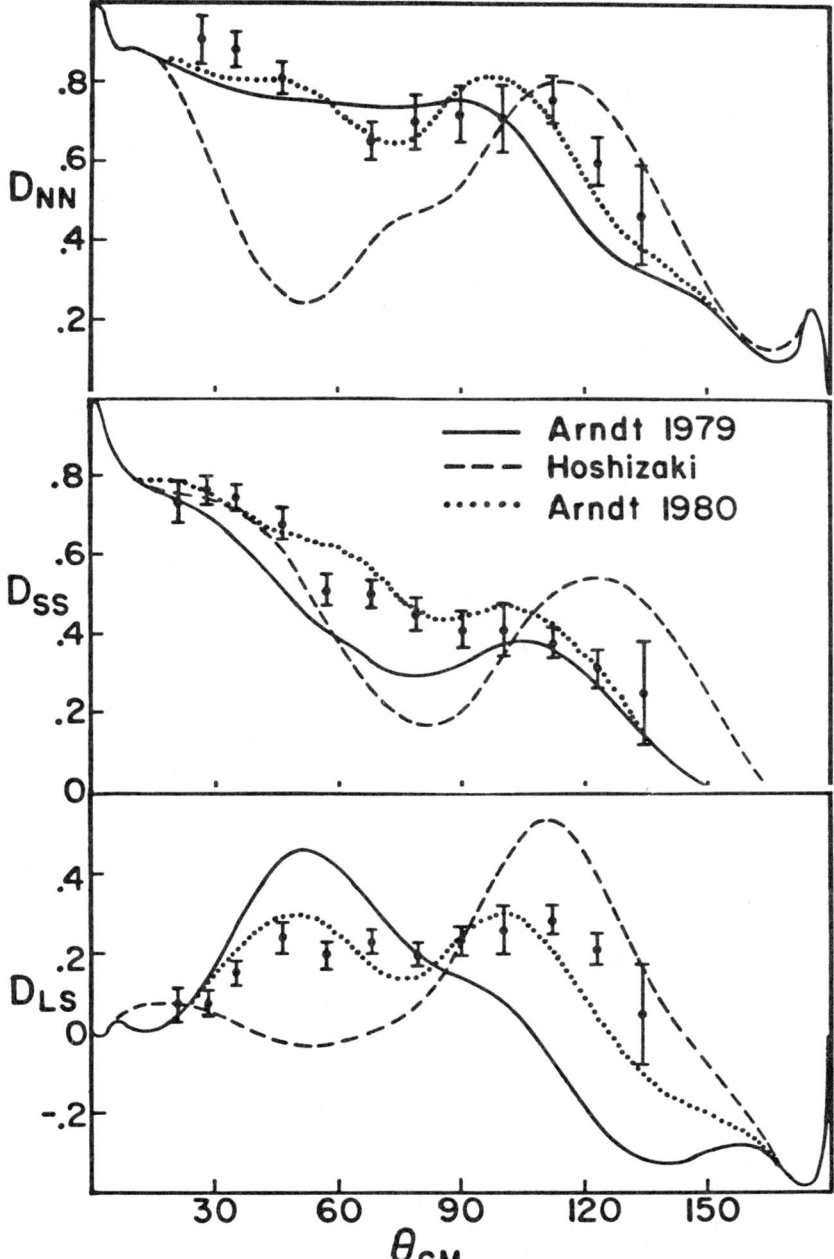

Fig. 1. D_{NN}, D_{SS}, D_{LS} for pp → pp at 800 MeV in comparison with phase shift predictions by Arndt[13] and Hoshizaki.[14]

ENERGY DEPENDENCE OF $A_{nn}(90°)$ IN pp ELASTIC SCATTERING FROM 1.10 to 2.75 GeV/c*

H.E.Miettinen, D.A.Bell, J.A.Buchanan, M.M.Calkin, J.M. Clement, W.H.Dragoset[1], M. Furić[2], K.A.Johns, J.D.Lesikar, T.A.Mulera[3], G.S.Mutchler, G.C. Phillips, J.B.Roberts, and S.E.Turpin
T.W.Bonner Nuclear Laboratories and Physics Department
Rice University, Houston, TX 77005

ABSTRACT

We have measured the spin correlation parameter A_{nn} in pp elastic scattering near $\Theta_{CM}=90°$ from 1.10 to 2.75 GeV/c. We find that $A_{nn}(90°)$ has a pronounced peak at $p_{lab}\approx 1.34$ GeV/c, reaching a value of about 0.8-0.9.

INTRODUCTION

Recent experiments on the spin dependence of pp scattering have revealed striking energy dependent structures both in the pure spin total cross-sections[1,2] and in large angle elastic scattering[3] in the 1-3 GeV/c region. The interpretation of these structures still remains unclear: attempts have been made to explain them in terms of dibaryon resonances,[4] or as the effects of crossing of inelastic thresholds.[5]

In this report we present measurements of the spin correlation parameter A_{nn} in pp elastic scattering near $\Theta_{CM}=90°$ at eight incident momenta between 1.10 and 2.75 GeV/c. At $\Theta_{CM}=90°$ the pure spin differential cross-sections are related to A_{nn} through

$$\sigma(\uparrow\uparrow) = \sigma(\downarrow\downarrow) = \sigma(1+A_{nn})$$
$$\sigma(\uparrow\downarrow) = \sigma(\downarrow\uparrow) = \sigma(1-A_{nn}), \qquad (1)$$

where σ is the spin-averaged cross-section. The initial spins are polarized normal to the scattering plane.

EXPERIMENTAL METHOD

The experiment utilized the Argonne ZGS polarized proton beam focused onto the Argonne-Rice polarized proton target (PPT-VI). The beam polarization, measured with the 50 MeV polarimeter[6] just before

---

*Supported by the U.S. Department of Energy
[1] Now at Western Geophysical Co., Houston, TX.
[2] Visiting Scientist from Institut "Ruđer Bošković," Zagreb, Yugoslavia.
[3] Now at Lawrence Berkeley Laboratory, Berkeley, CA.

injection into the ZGS, was reversed each spill and averaged about 70% over the run.

The polarized target was a conventional He^4-He^3 cryostat operating at 0.5°K in a magnetic field of 25 kG. The target cavity, 7 cm long by 2 cm in diameter, was filled with ethanediol ($C_2H_6O_2$) beads doped with chromium paramagnetic complexes. The free hydrogen protons were dynamically polarized by a 70 GHz microwave system and the target polarization ($P_T \simeq 75\%$) was measured by NMR techniques.

Elastic scattering events were detected in a two-arm spectrometer consisting of an analyzing magnet, scintillation counters, and multiwire proportional chambers. Relative beam intensities were monitored by scintillator telescopes viewing polyethylene targets upstream of the PPT. Measurement of the momentum and angles of the "forward" particle and the angles of the "recoil" particle, together with pulse height and time-of-flight information, allowed a clean separation of elastic events from free hydrogen. The quasi-elastic background, due to scattering from bound protons in carbon and oxygen, was typically 8-10%.

With the beam and the target only partically polarized ($P_B, P_T < 1$), the event rates $N(ij)$ in each of the four initial spin states (i,j= beam, target), normalized to the incident beam intensity, are given by

$$N(ij) = \lambda(j)[1+\{P_B(ij)+P_T(j)\}A+P_B(ij)P_T(j)A_{nn}], \quad (2)$$

where A is the asymmetry parameter (A=0 at $\Theta_{CM}=90°$) and λ is a normalization factor. For a given set of runs (ij=↑↑,↑↓,↓↑,↓↓) P_T and λ are independent of the beam polarization due to fast reversal of the beam spin. A_{nn} and A are obtained by inverting Eqs. (2).

RESULTS

Our results on $A_{nn}(90°)$ are shown in Fig.1, along with previous measurements[7-14] in the 1-3 GeV/c region. The errors shown for this experiment are purely statistical. We estimate an additional relative systematic error of ±8-10% in A_{nn}, due primarily to uncertainties in P_B, P_T and the background subtraction.

The most distinct feature of these data is the pronounced peak in A_{nn} centered around $p_{lab} \simeq 1.34$ GeV/c. We note that $A_{nn}(90°)$ is related in a simple way to the ratio of the triplet and singlet cross sections:

$$\frac{1+A_{nn}}{1-A_{nn}} = \frac{\sigma_t}{\sigma_s} \qquad (\Theta_{CM}=90°). \quad (3)$$

At 1.34 GeV/c our value $A_{nn}=0.872\pm0.035$ implies $\sigma_t/\sigma_s \simeq 15$, thus indicating very strong triplet dominance in large angle elastic scattering at this momentum.

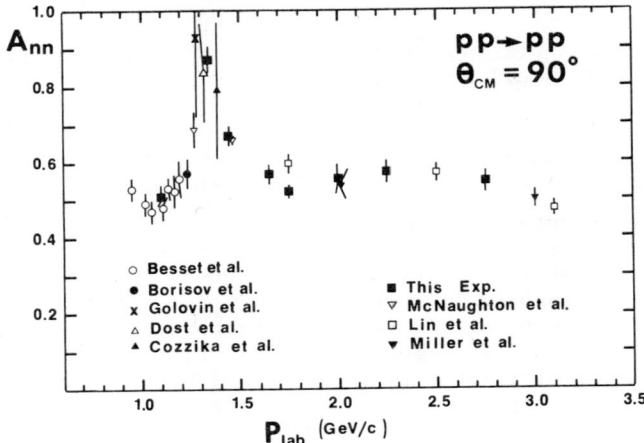

Fig.1

A_{nn} at $\Theta_{CM}=90°$ as a function of P_{lab}. Previous data are from Refs. 7-14.

It is interesting to note that the observed structure occurs in the vicinity of the reported $^1D_2(2.17$ GeV$)$ and $^3F_3(2.22$ GeV$)$ di-proton resonances[4] (p_{lab}=1.26 and 1.40 GeV/c, respectively), and it may therefore have a bearing on the existence and properties of these states. An alternative interpretation, without any resonant behavior, might arise as a result of different momentum dependences of the singlet and triplet contributions, reflecting the production properties of the S- and P-wave $N\Delta$ final states and their coupling to the pp system.[5]

REFERENCES

1. I.P.Auer et al., Phys.Rev.Lett. 41, 354 (1978).
2. Ed.K.Biegert et al., Phys.Lett. 73B, 235 (1978).
3. I.P.Auer et al., Phys.Rev.Lett. 41, 1436 (1978).
4. N.Hoshizaki, Progr.Theor.Phys. 60, 1796 (1978); 61, 129 (1979); A.Yokosawa, Argonne Preprint ANL-HEP-CP-80-07 (1980).
5. S.Minami, Phys.Rev. D18, 3273 (1978); C.L.Hollas, Phys.Rev.Lett. 44, 1186 (1980).
6. H. Spinka, Argonne Preprint ANL-HEP-PR-80-02 (1980).
7. D.Besset et al., SIN Preprint PR-80-003 (1980).
8. N.S. Borisov et al., AIP Conf.Proc. 35, 59 (1976).
9. B.M.Golovin et al., results quoted by Yu.M. Kazarinov, Rev.Mod.Phys. 39, 509 (1967).
10. H.E.Dost et al., Phys.Rev. 153, 1394 (1967).
11. G.Cozzika et al., Phys.Rev. 164, 1672 (1967).
12. M.W.McNaughton et al., to be published.
13. A.Lin et al., Phys.Lett. 74B, 273 (1978).
14. D.Miller et al., Phys.Rev. D16, 2016 (1977).

SPIN-DEPENDENT FORCES IN pp→pp AT 6 GeV/c

S. Wakaizumi and M. Sawamoto
Department of Physics, Hiroshima University, Hiroshima 730, Japan

ABSTRACT

We present a fit to all the available kinds of spin-observables of pp→pp at 6 GeV/c with the eikonal model involving the five independent spin-dependent eikonals. The eikonals have proved to require three components; long-range(2-2.5fm), intermediate-range(∼1.5fm) and short-range(0.3-0.5fm) ones. The long-range eikonals resemble quite well in range, sign and magnitude the "one-pion-exchange" established in the low-energy NN scattering. Some discussions are given on the scattering amplitudes obtained.

Measurements of spin-correlation parameters for pp→pp at 6 GeV/c at ANL have suggested that spin-dependent forces are still strong at 6GeV/c than expected before when Regge-pole exchange view was promoted in 1960's.[1] Moreover, if combined with the data on the other spin-observables, it seems to have become possible to determine the five independent scattering amplitudes in the wide $|t|$ range.[2,3]
We presented before a method of analysis for pp→pp by taking the five independent spin-dependent eikonals in the eikonal model,[4] being developed from Durand and Halzen's model.[5] Here, we report the results of an improved fit to all the available kinds of observables in the same model with the addition of a short-range(0.3-0.5fm) eikonal to the intermediate-range(∼1.5fm) and long-range(2-2.5fm) eikonals. For the latter two eikonals are taken the modified Fermi function which has the Yukawa-type tail at large impact-parameter b, being tended to zero at b=0 by multiplying a factor $(1-\exp(-(b/d_0)^2))^n$. For the short-range one is taken the Gaussian function.
The scattering matrix is written with the eikonal in the spin-space as

$$M = \frac{ip}{2\pi}\int d^2\vec{b}(1-\exp(-\chi(\vec{b})))\cdot\exp(-i\vec{q}\cdot\vec{b}) \ , \qquad (1)$$

where $\vec{q}$ is momentum transfer. The eikonal function $\chi(\vec{b})$ is expressed as

$$\chi(\vec{b}) = \chi_C(b) - i(\vec{\sigma}_1+\vec{\sigma}_2)\cdot(\vec{b}\times\vec{1})\chi_{LS}(b) - i\vec{\sigma}_1\cdot\vec{\sigma}_2\chi_{SS}(b)$$
$$-i\{\vec{\sigma}_1\cdot(\vec{b}\times\vec{1})\}\{\vec{\sigma}_2\cdot(\vec{b}\times\vec{1})\}\chi_Q(b) - iS'_{12}\chi_T(b) \ , \qquad (2)$$

where χ_C is the spin-independent central eikonal, the second to the fifth terms are spin-orbit, spin-spin, quadratic spin-orbit and tensor coupling eikonals, respectively; $\vec{\sigma}_1$ and $\vec{\sigma}_2$ are the Pauli spin matrices of the two incident protons, $\vec{1}=(\vec{p}+\vec{p}')/|\vec{p}+\vec{p}'|$, S'_{12} is the "tensor" operator symbolically written in the b-plane. The five scattering amplitudes are obtained by introducing the eikonal into eq.(1) and see ref.3 for the explicit expressions.
The fit has been done to $d\sigma/dt$, P, C_{NN}, C_{LL}, C_{SS}, C_{SL}, D_{NN}, D_{SS}, D_{LS}, K_{NN}, 'K_{SS}', 'H_{SNS}', σ_T, ρ, $\Delta\sigma_T$ and $\Delta\sigma_L$ with 225 data points for

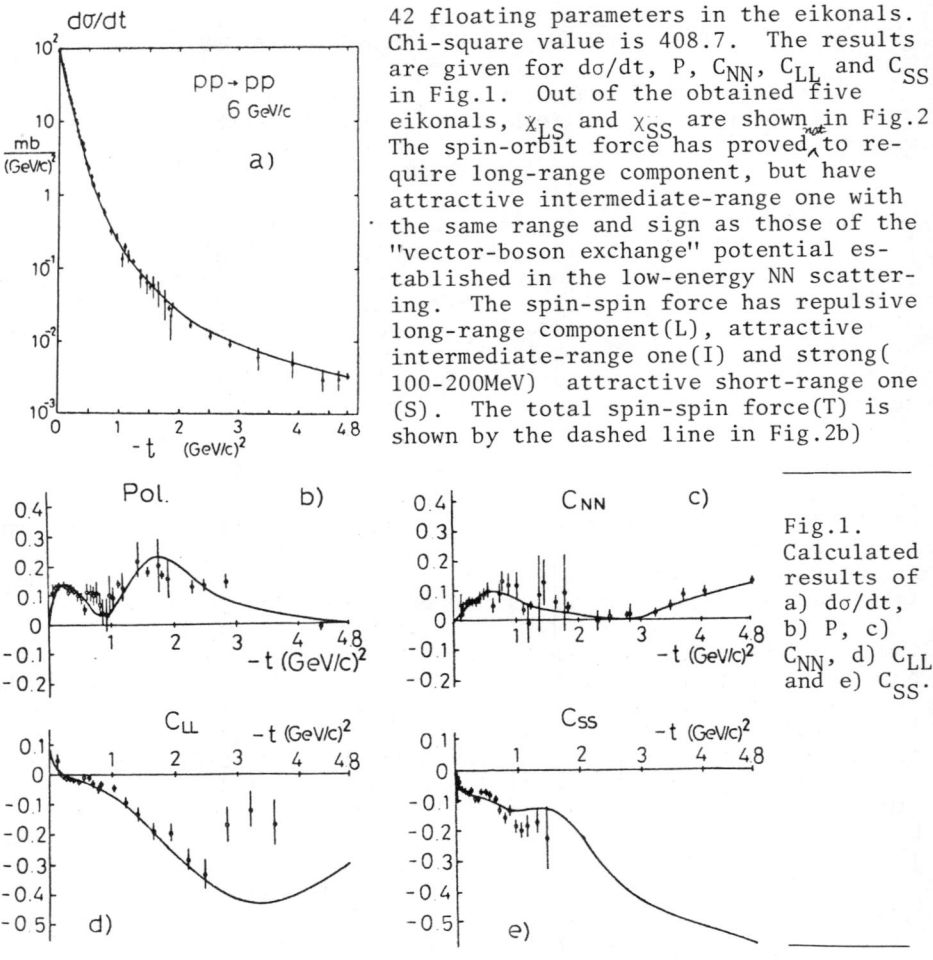

42 floating parameters in the eikonals. Chi-square value is 408.7. The results are given for dσ/dt, P, C_{NN}, C_{LL} and C_{SS} in Fig.1. Out of the obtained five eikonals, χ_{LS} and χ_{SS} are shown in Fig.2. The spin-orbit force has proved not to require long-range component, but have attractive intermediate-range one with the same range and sign as those of the "vector-boson exchange" potential established in the low-energy NN scattering. The spin-spin force has repulsive long-range component(L), attractive intermediate-range one(I) and strong(100-200MeV) attractive short-range one (S). The total spin-spin force(T) is shown by the dashed line in Fig.2b)

Fig.1. Calculated results of a) dσ/dt, b) P, c) C_{NN}, d) C_{LL} and e) C_{SS}.

and, for the spin-singlet state, it happens to be attractive in the wide range 0.4≤b≤2.5 fm and repulsive for b≤0.3fm just like the "soft core" for $^1S_0$-state. The wide-range attractive force might be responsible for the formation of the diproton resonances of singlet state like $^1D_2$(2.14-2.17GeV) and $^1G_4$(2.43-2.50GeV). The tensor force is quite similar to the spin-spin force. The "repulsive" feature of the long-range component of both the spin-spin and the tensor force reconciles well with that of the one-pion-exchange potential established in the low-energy NN scattering and the strength remains nearly the same as that of the potential with $G_\pi^2/4\pi=14.4$ if calculated at b=3 and 4fm for the spin-spin force.

The scattering amplitudes obtained are shown in Fig.3 for N_0, N_1 and U_2. The natural-parity exchange, helicity non-flip amplitude N_0 has no zero up to $|t|=3(GeV/c)^2$ for both the real and the imaginary part. The real part of the unnatural-parity exchange, helicity

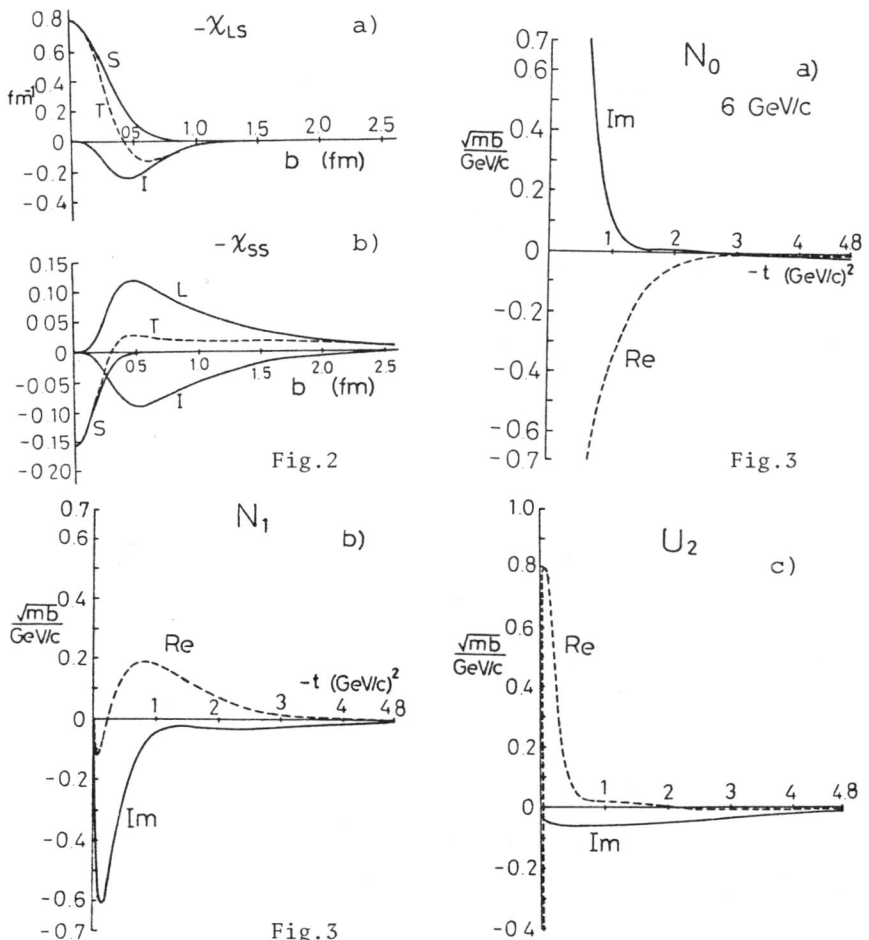

Fig.2

Fig.3

double-flip amplitude U_2 has a very sharp negative peak around t=0 $(GeV/c)^2$, which reproduces the forward "spike" in np→pn, being consistent with the result of phase-shift analysis.[6] The amplitude N_1 has a negative imaginary part fairly larger than its real part at $|t|$ =0.1-0.4$(GeV/c)^2$, which is different from N_1 obtained from the Regge pole model[2] and also from the result from the phase-shift analysis.[6] Precise measurement of D_{SS} for $0.1 \leq |t| \leq 0.4 (GeV/c)^2$ with much smaller errors than now available is needed to resolve this problem of N_1.

REFERENCES

1. A. Yokosawa, Proc. of 19th Int. Conf. on High Energy Physics,p39.
2. E. L. Berger et al., Phys. Rev. D17, 2971 (1978).
3. M. Matsuda et al., Prog. Theor. Phys. 62, 1436 (1979).
4. M. Sawamoto and S. Wakaizumi, Prog. Theor. Phys. 62, 1293 (1979).
5. L. Durand and F. Halzen, Nucl. Phys. B104, 317 (1976).
6. M. Matsuda, H. Suemitsu, W. Watari and M. Yonezawa, Hiroshima University preprint HUPD-8013(June,1980).

INCLUSIVE SCATTERING OF PROTONS ON HELIUM AND NICKEL
AT 500 MeV

C. Roy, L.G. Greeniaus, G.A. Moss, D.A. Hutcheon, R. Liljestrand
University of Alberta, Edmonton, Alberta T6G 2N5

R.M. Woloshyn
TRIUMF, Vancouver, B.C. V6T 2A3

D. Boal
Simon Fraser University, Burnaby, B.C. V5A 1S6

A.W. Stetz
University of Oregon, Corvallis, Oregon 97331

K. Aniol
University of Washington, Seattle, Washington 98195

A. Willis, N. Willis
Institut de Physique Nucleaire, 91406 Orsay, France

R. McCamis
University of Manitoba, Winnipeg, Manitoba R3T 2N2

ABSTRACT

Inclusive scattering on ^{4}He and Ni has been measured at 500 MeV proton energy with polarized beam at laboratory angles of 65°, 90°, 120°, and 160°. The cross section data are well fitted by the direct knockout model. The analyzing powers show major differences in going from ^{4}He to Ni. This result may have strong implications for inclusive scattering theories.

INTRODUCTION

We have measured inclusive scattering analyzing powers and cross sections for proton induced inclusive scattering (p+A → p+X) on ^{4}He and Ni. A liquid helium cryostat was used as a target; the target windows yielded the nickel data. Reaction protons were detected in a telescope which consisted of 2 ΔE detectors for time-of-flight information, a copper absorber and a 3" x 5" NaI detector for energy information. Several thicknesses of copper absorber were used to yield energy spectra extending out to the p-^{4}He elastic peak. Data analysis included identification of the reaction protons, correction for energy losses in the proton path and correction for reaction losses.

Figures 1 and 2 show the cross-section data for ^{4}He and Ni. The curves are fits to the data using the direct knockout model[1]. This model assumes that the incoming proton interacts with a nucleon inside the nucleus with sufficient internal momentum that the internal nucleon ends up as the detected proton. The residual A-1 nucleus is basically a spectator. The only freely adjustable parameter in the calculation is related to the momentum distribution of the nucleons inside the nucleus. Our value for this parameter is consistent with

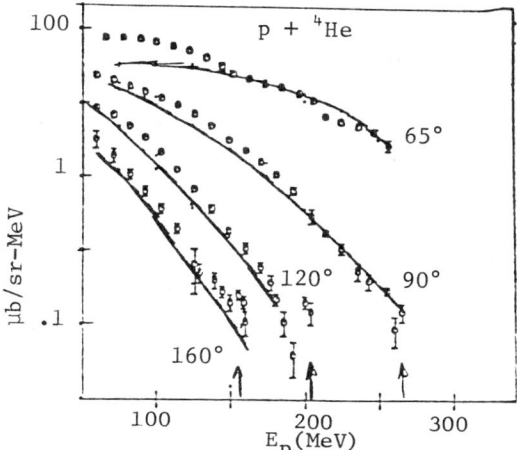

Fig. 1. p+⁴He inclusive cross section. The arrows point to the elastic peak at the appropriate angle. The curves are theoretical calculations.

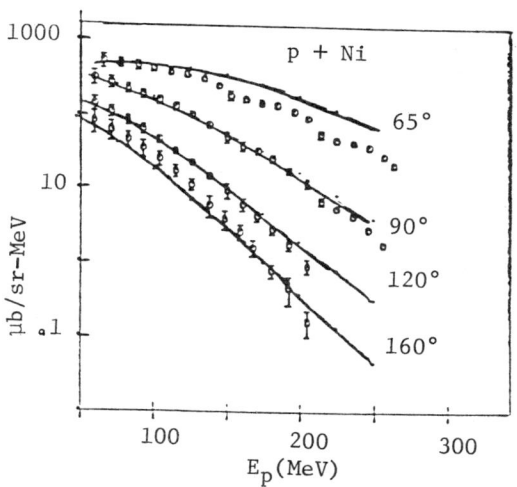

Fig. 2. p+Ni inclusive scattering. The curves are theoretical calculations.

fits to a wide range of data obtained at 800 MeV bombarding energy.

Figures 3 and 4 show the analyzing power data. The analyzing powers at 120° and 160° are small and consistent with zero, until the elastic scattering peak for ⁴He is reached. At 65° and 90°, the analyzing powers for ⁴He and Ni start off small but negative, and increase with increasing detected proton energy. However, while the ⁴He analyzing powers never become positive, the Ni results do reach substantial positive values. This latter result is consistent with the results of Frankel *et al.*[2] on various nuclei from ^{6}Li to ^{181}Ta at 800 MeV bombarding energy. This demonstrates a fundamental difference in the inclusive reaction process between proton and ⁴He and nuclei heavier than ⁴He. A possible explanation is that alpha-clusters inside the nucleus lead to the positive analyzing powers seen in nuclei heavier than ⁴He. The polarization map for p + ⁴He elastic scattering[3] shows widespread regions where the polarization has large positive values, consistent with this interpretation of our results.

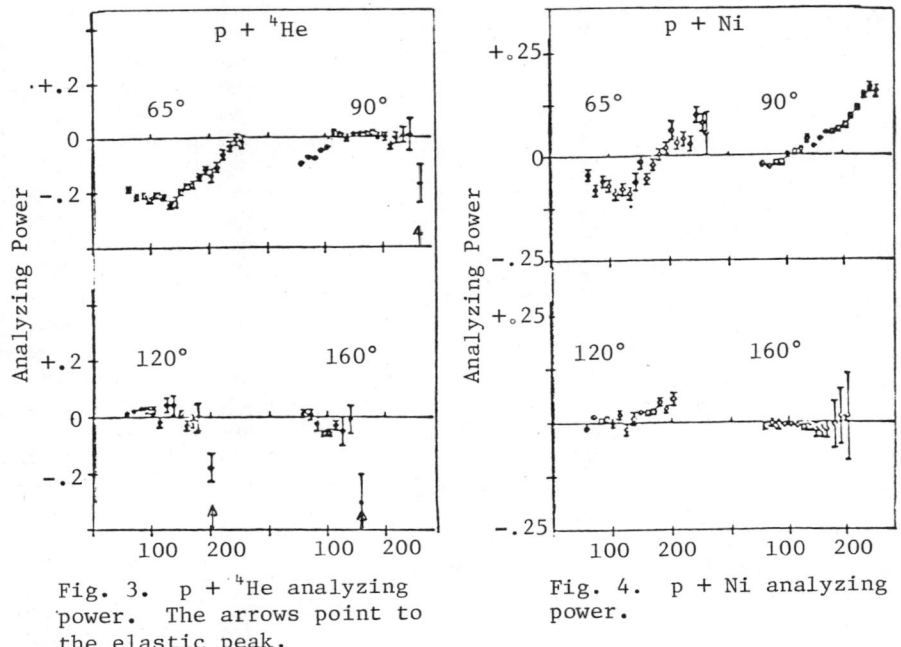

Fig. 3. p + ⁴He analyzing power. The arrows point to the elastic peak.

Fig. 4. p + Ni analyzing power.

REFERENCES

1. D. Boal, Phys. Rev. C21, 1913 (1980).
2. S. Frankel, W. Frati, M. Gazzaly, G.W. Hoffman, O. van Dyck, and R.M. Woloshyn, Phys. Rev. Lett. 41, 148 (1978).
3. G.A. Moss, L.G. Greeniaus, D.A. Hutcheon, R.L. Liljestrand, C.A. Miller, G. Roy, B.K.S. Koene, W.T.H. van Oers, A.W. Stetz, A. Willis, and N. Willis, Phys. Rev. C21, 1932 (1980).

MEASUREMENT OF POLARIZATION PARAMETERS IN pp → π^+d AT INTERMEDIATE ENERGIES

E. Aprile, R. Hausammann, E. Heer, R. Hess,
C. Lechanoine-Leluc, W. Leo, Y. Onel, D. Rapin

DPNC, University of Geneva, Geneva, Switzerland

J.M. Cameron, S. Mango, S. Jaccard

SIN - Villigen, Switzerland

As part of our measurements of the pp elastic scattering process using a polarized beam and target at SIN[1]), we have also measured polarization parameters for the inelastic reaction pp → π^+d. These parameters include the asymmetries A_{oy}, A_{yo}, and the spin correlation parameters A_{xx}, A_{yy}, A_{zz}, A_{zx}, and A_{xz}, where the two subscripts indicate the orientation of the beam and target polarizations respectively. In these experiments, the z-axis was defined as being along the incident beam direction and y along the vertical. Some of the data, it should be noted, are of linear combinations of the spin correlation parameters rather than of the individual parameters themselves. The data were taken mainly at 577 and 514 MeV with some measurements at 536 and 494 MeV also being made.

The experimental apparatus employed was similar to that used in our pp measurements. Detection of the outgoing particles was made by two telescopes each consisting of three MWPC's and plastic scintillation counters. The telescopes were mounted on movable platforms which could be rotated about the target axis so that several angular regions could be accessed. A time-of flight method was also used to help identify pion deuteron events.

In these experiments, two types of polarized beam were used; one produced by scattering the main unpolarized proton beam off a thin Be target and one via a Stern-Gerlach type polarized ion source. In both cases, a beam polarization along the y axis is obtained; however, orientations along the x and z axes could be produced by using an appropriate arrangement of a superconducting solenoid and deflecting magnet to rotate the spin vector. The polarization of the scattered beam was 41.65%, while that obtained from the ion source was about 85%. The bulk of the data were taken with the scattered beam.

The polarized target consisted of tiny beads of frozen butyl alcohol immersed in a bath of liquid helium-3. This mixture was kept at a temperature of 2K by pumping on the helium and was placed in a homogeneous magnetic field of 25 kG. Polarization of the free protons was produced by the method of dynamic nuclear orientation which yielded average polarization values of about 60%. Due to the flexible design of the target, polarizations along all three axes, x, y and z, could be achieved by simply changing the orientation of the magnet coils. No change of the cryostat structure was required.

The pion-deuteron events were taken simultaneously with the pp measurements and were reconstructed on-line using a new technique[2] based on a linearization of the reconstruction equations. The data were then written onto magnetic tape in the form of histograms which could be further analyzed off-line. To correct for background events from the carbon and helium atoms in the target, measurements were also made with a dummy target consisting of carbon and helium alone.

Preliminary results of much of our measurements at 577 MeV have already been presented at the Nucleon-Nucleon conference[3] at Vancouver last year. Final analysis of these data plus some data on A_{xz} at this energy is still in progress along with a similar set of measurements at 514 MeV. This latter set is still in the preliminary stages of evaluation, however. At 536 and 494 MeV, a smaller set of measurements consisting mainly of A_{xx}, and A_{zz}, A_{zx} in a somewhat smaller angular range, is also currently under analysis. Available results mainly at 514 MeV will be presented.

REFERENCES

1. E. Aprile et al., Contribution to this meeting
2. D. Besset, Ph.D thesis, University of Geneva, 1978.
3. E. Aprile et al., Proc. of Eighth Int'l Conf on High Energy Physics and Nuclear Structure, Vancouver, 1979, Nucl Phys A335(1980)245

DIFFERENTIAL CROSS SECTION FOR PION PRODUCTION IN THE REACTION
pp→πd BETWEEN 500 AND 600 MeV

J. Hoftiezer, Ch. Weddigen
Institut für Kernphysik, Universität und Kernforschungszentrum,
Karlsruhe, Federal Republic of Germany

B. Favier, S. Jaccard
Suiss Institute for Nuclear Research (SIN)

P. Chatelain, F. Foroughi, C. Nussbaum
Institut de Physique, Université de Neuchâtel, Switzerland

The unpolarized differential cross section $d\sigma/d\Omega$ has been measured at SIN for seven energies with a relative precision of about 1%. The experimental procedure has been described elsewhere [1]. Preliminary data have been normalized from measured pp elastic scattering rates and analysed to obtain the usual anisotropy parameters γ_i (Table I) defined by

$$\frac{d\sigma}{d\Omega} = \frac{1}{32\pi} (\gamma_0 + \gamma_2 \cos^2\Theta_{CM} + \gamma_4 \cos^4\Theta_{CM}).$$

Table I Anisotropy parameters γ_i [mb/sr] for the reaction pp→πd.
E_p = proton lab. kinetic energy, W_p = kinetic energy of one proton in the C.M. system.

E_p	W_p [MeV]	γ_0	γ_2	γ_4
514	120.7	9.47 ± .11	34.81 ± .67	− 3.67 ± .80
527	123.6	9.96 ± .14	38.40 ± .99	− 3.73 ± 1.23
540	126.5	10.68 ± .12	40.91 ± .79	− 6.06 ± .94
554	129.6	11.11 ± .18	44.42 ± 1.43	− 8.03 ± 1.84
569	132.8	11.57 ± .14	46.56 ± .91	− 9.39 ± 1.08
576	134.4	11.57 ± .13	47.87 ± .79	−11.88 ± .89
583	135.9	11.64 ± .14	49.43 ± 1.00	−12.99 ± 1.19

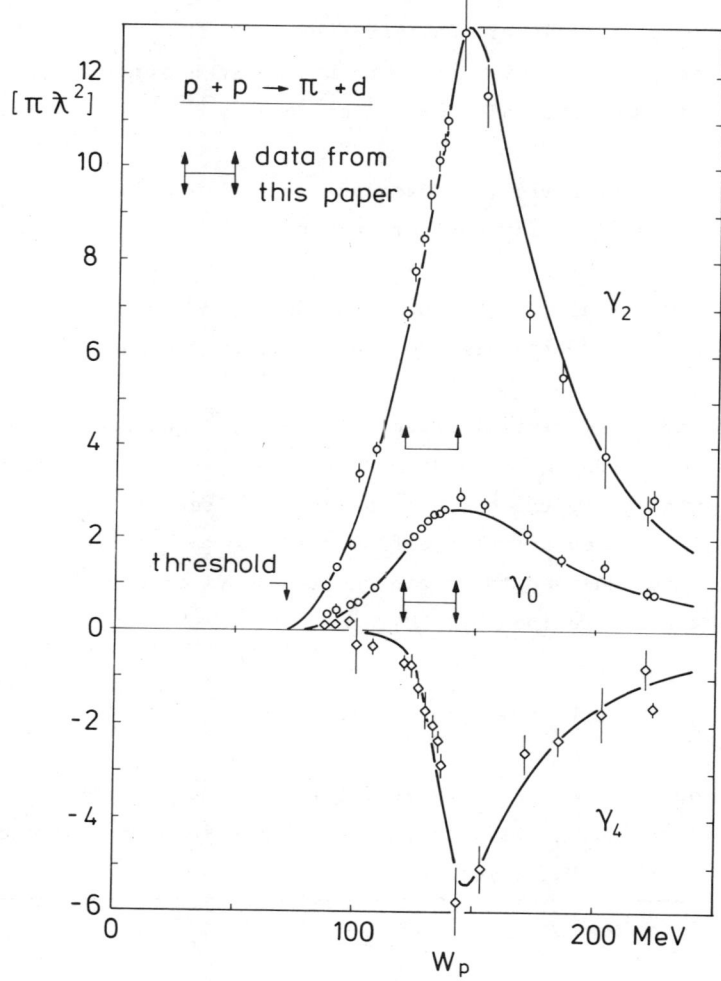

Fig. 1. Anisotropy parameters γ_i $[\pi\lambdabar^2]$ as a function of W_p. The curves are calculated [2] using a Breit-Wigner formalism. $\lambdabar$ = wave length of one proton in the C.M. system.

Fig. 1 shows the results together with a selection [2] of data from other experiments. The γ parameters, and especially γ_4 show a much stronger energy dependence than predicted theoretically [3,4].

In order to clarify whether such a strong energy dependence calls for a possible $^1D_2$ dibaryon resonance in addition to the well known $N\Delta$ resonant production mechanism a qualitative interpretation [2] was tried by means of a Breit-Wigner energy dependent width approximation. Only initial singlet state amplitudes were taken into account: a_2 and a_7 which describe the decay of an initial resonant $^1D_2$ configuration with angular momenta $l_\pi = 1$ and 3 respectively, and a_o a non resonant ^1S_o background with $l_\pi = 1$. The a's are defined as in ref. 5. In this simplified model, γ_4 is given by the expression

$$\gamma_4 = \frac{15}{2} |a_7|^2 - 15\sqrt{6} \operatorname{Re} a_2 a_7^*.$$

The sudden increase of γ_4 as a function of energy can be explained as an interference between one predominant amplitude a_2 and a small contribution a_7, which differ in threshold behaviour from a_2. At resonant energy one finds $|a_7|^2/|a_2|^2 = 3\times10^{-3}$. (The importance of a_7 was first pointed out by NISKANEN [3].)

1. B. Favier et al., SIN Newsletter 12, 56 (1979)
2. Ch. Weddigen, Technical Report KFK 2996 B, Karlsruhe (1980)
3. J.A. Niskanen, Nucl. Phys. A298, 417 (1978)
4. O.V. Maxwell, W. Weise, and M. Brack, University of Regensburg, Preprint (1980)
5. F. Mandl and T. Regge, Phys. Rev. 99, 1478 (1955)

The Measurement of K_{NN} and K_{LL} in $\vec{p}p \to \vec{n}X$ at 800 MeV*

T. S. Bhatia, G. Glass, J. C. Hiebert
L. C. Northcliffe, W. B. Tippens
Texas A&M University, College Station, TX 77843

C. L. Hollas, C. R. Newsom, R. D. Ransome, P. J. Riley
University of Texas at Austin, Austin, TX 78712

G. P. Pepin
Rice University, Houston, TX 77001

B. E. Bonner, J. E. Simmons
Los Alamos Scientific Laboratory, Los Alamos, NM 87545

ABSTRACT

The spin transfer parameters, K_{NN} and K_{LL} have been measured in $\vec{p}p \to \vec{n}X$ at 0° and 800 MeV for neutron momenta between 700 and 1200 MeV/c. Peak values of K_{NN} and K_{LL} are -.3 ± .05 and -.5 ± .1 respectively.

Text

The spin transfer parameters K_{LL} and K_{NN} for the reaction $\vec{p}p \to \vec{n}p\pi^+$ with $\vec{n}$ at 0° have been measured on the neutron beam line at LAMPF. The 800 MeV proton beam was polarized vertically for the K_{NN} part and almost longitudinally for the K_{LL} part. A small horizontal component existed for the latter. In both cases the beam struck a liquid hydrogen (LH$_2$) target and then deflected through 60° into a beam dump. The 0° neutrons traversed a ~ 3.7m (12 ft.) steel collimator which terminated with a ~ 5cm (2 in.) diameter aperture. The neutron flux was cleared of charged particles with a sweep magnet. In the K_{LL} case the sweep magnet was also used to compensate for the precession caused by the beam line bending magnets which the neutrons had to traverse. Use was then made of the recently measured[1] analyzing power in the charge exchange scattering, $\vec{n}p \to pn$. An LH$_2$ radiator, ~ 30 cm thick, scattered the incoming vertically polarized neutrons. The longitudinal case required 90° spin precession prior to the scattering at the LH$_2$ radiator in order to put the neutron polarization into a vertical orientation. Since the neutrons in this experiment were polarized and the protons (LH$_2$ radiator) were not, the assumption of charge symmetry invariance ($A(\vec{n}p) = A(n\vec{p})$) was invoked in order to use the previously measured analyzing power to get the neutron polarization. The proton beam polarization was measured with beam line polarimeters[2] and was typically about 75%.

The major source of background in these measurements was the inelastic reactions in the LH$_2$ radiator. This background as well as that caused by target walls was removed through an analysis procedure that made use of the 5 nsec micro structure of the proton beam. Time of flight measurements with respect to this structure yielded another measurement of the neutron momentum which when compared to the spectrometer momentum measurement provided a means of distinguishing between elastic and inelastic reactions in the LH_2 radiator.

The results for the K_{NN} experiment are shown in Fig. 1, and the K_{LL} results are shown in Fig. 2.

The only calculations[3] of these parameters available to us are in considerable disagreement with our present K_{NN} and K_{LL} results. The model used in reference 3 does not comply with unitarity and it is felt that when unitarity is brought into the picture the calculation should be more in line with our measurements. There exists another approach[4] based on Aaron, Amado and Young's[5] three body final state solution which complies completely with unitarity. We await these results to see whether in fact unitarity is as important in these spin transfer processes as it is felt to be or whether still other exchange mechanisms, not taken into account, play unexpected roles here.

We would like to thank Professor G. C. Phillips for the use of the Rice/Houston equipment for the K_{LL} measurement. We also appreciate the valuable support provided by LAMPF personnel.

* Work supported by the U.S. Department of Energy.

1. C. R. Newsom, et al., 8th ICOHEPANS, Vancouver, B.C., 1979.
2. M. W. McNaughton LA-8307-MS (unpublished) 1980.
3. B. J. Ver West Phys. Lett. B 83 p 161, 1979.
4. W. M. Kloet and R. R. Silbar, Nuc. Phys. A338 281, 1980.
5. R. Aaron, R. D. Amado, J. E. Young, Phys. Rev. 174 2022, 1968.

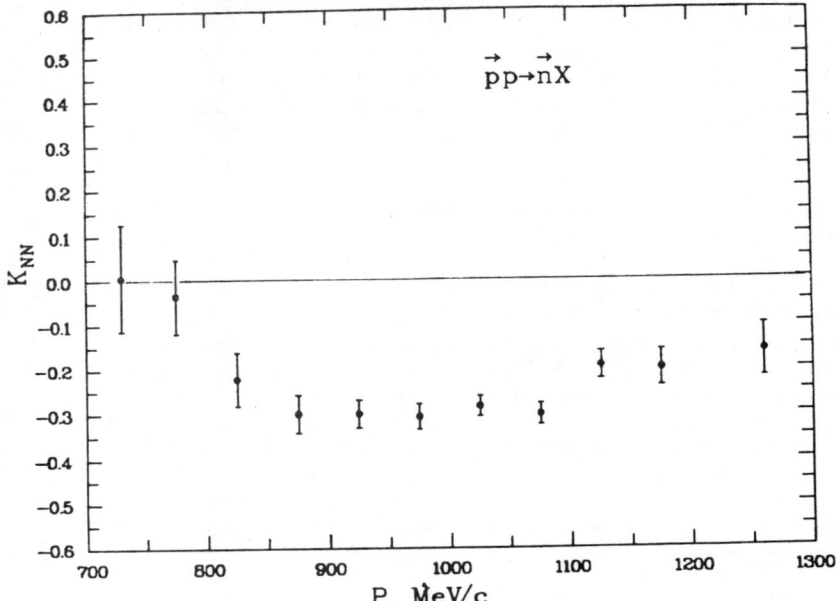

Fig. 1. Transverse spin transfer results.

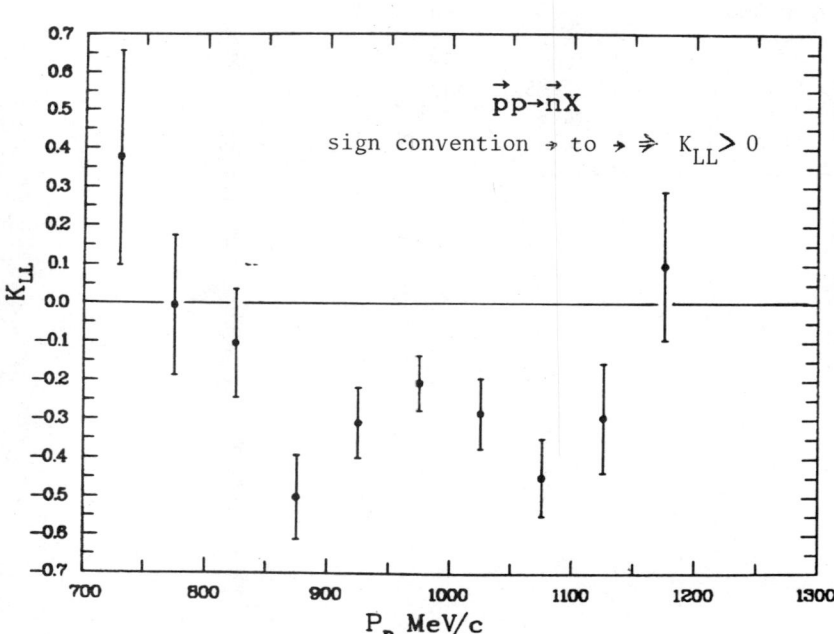

Fig. 2. Longitudinal spin transfer results.

POLARIZED CROSS SECTION FOR THE REACTION p↑+p→pπ⁺n*

A.D. Hancock, R.W. Hackenburg, E.V. Hungerford, B.W.
Mayes, L.S. Pinsky, J.C. Allred, T.M. Williams
University of Houston, Houston, Texas 77004

S.D. Baker, J.A. Buchanan, J.M. Clement, M. Copel, D.M. Judd,
G.S. Mutchler, G.P. Pepin, E.A. Umland, and G.C. Phillips
Rice University, Houston, Texas 77001

M. McNaughton and C. Hwang
Los Alamos Scientific Laboratory, Los Alamos, New Mexico 87545

ABSTRACT

We have measured the spin dependent fifth differential cross section for the reaction p↑+p→pπ⁺n through the 3/2, 3/2 resonances in a kinematically complete experiment at 800 MeV.

The recent discovery of dramatic structure in the spin dependent pp observables has aroused interest in the spin effects in pion production mechanisms. The spin dependence of the reaction pp→pπ⁺n is essential for the understanding of the pp interaction near incident laboratory energies of 800 MeV. Since the inelastic channels are dominated by this reaction at that energy, and since the total inelastic cross section is such a significant part of the total cross section, any attempt to understand the spin orbit part of the total interaction without understanding its role in the inelastic channels is of doubtful value. In addition, energy dependent measurements of the transverse, $\Delta\sigma_T$, and longitudinal $\Delta\sigma_L$ total cross section differences in pure spin states, as well as other nucleon-nucleon observables at laboratory momenta between 1 and 3 GeV/c have led to the suggestion of possible direct channel dibaryon resonances in the pp system or the $N\Delta^{++}$ system. Phase shift analyses, such as those of Hoshizaki, indicate possible dibaryon resonances at ($^1D_2$) 2.175 GeV and at ($^3F_3$) 2.22 GeV. If such resonances exist, they will have large inelastic decay widths and therefore strong effects should appear in the channel.

We have measured the fifth order differential polarized cross section for the reaction pp→pπ⁺n in a kinematically complete experiment using the Los Alamos Meson Physics Facility 800 MeV polarized proton beam. The target was a 5.1 cm diameter by 6.9 cm long kapton cylinder filled with liquid hydrogen. A magnetic spectrometer arm consisting of two scintillation counters, a magnet, and four multiwire proportional counters (MWPC) measured the momentum, angles, and time of flight of the final state proton. The angular

---

*Supported by U.S.DOE Contracts No. DE-AS05-76ER03948 and No. DE-Ac-5-76ER0-1316

acceptance of the spectrometer was ±3.5° in theta and ±3.0° in phi. The momentum resolution was $\Delta p/p = 2\%$. The time of flight (TOF) arm measured the pion TOF and angles. The angular acceptances of the TOF arm were $\Delta\Theta = 2.8°$ and $\Delta\phi = 2.0°$. The incident beam intensity was measured with a Faraday cup and two ion chambers. The MWPC's in both arms were triggered by a fourfold coincidence of scintillators. Time of flights, pulse heights, wire plane coordinates and beam monitoring information were streamed to magnetic tape by a PDP 11/45 computer.

The beam polarization was typically 0.7. The beam polarization was measured using the LAMPF EPB beam line polarimeter consisting of four pair of scintillation counters viewing a CH_2 target near the maximum of the pp analyzing power (17° lab). This polarimeter has been calibrated to have an analyzing power of 0.481±.002. The beam polarization was flipped from spin up to spin down every three minutes to minimize false asymmetries.

The polarized cross sections are defined as:

$$\frac{d\sigma^{\pm}}{d\Omega_1 \, d\Omega_2 \, dp} = N^{\pm} - U/Live^{\pm} \; EFF^{\pm} \; N_O \; N_T \; SA \; \Delta p$$

where:
- $N^{\pm}$ = number of good events passing all TOF, pulse height and geometrical cuts for spin up (down)
- $Live^{\pm}$ = system Live time for spin up (down)
- $EFF^{\pm}$ = MWPC system efficiencies
- N_O = number of incident polarized protons
- N_T = number of target protons
- SA = solid angle for particular momentum bin
- Δp = final state proton momentum bin width
- u = number of good events due to unpolarized component of the beam

The data shown in figure 1 is for Δ^{++} production angles of 0°, 30°, 60°, 90° (top to bottom). All errors are stastistical only. These results are preliminary. While the data cannot be fit perfectly with the dibaryon resonances, the model with dibaryon resonances gives much better results than the peripheral model. Theoretical results will be discussed in a paper to this symposium by Eric Umland.

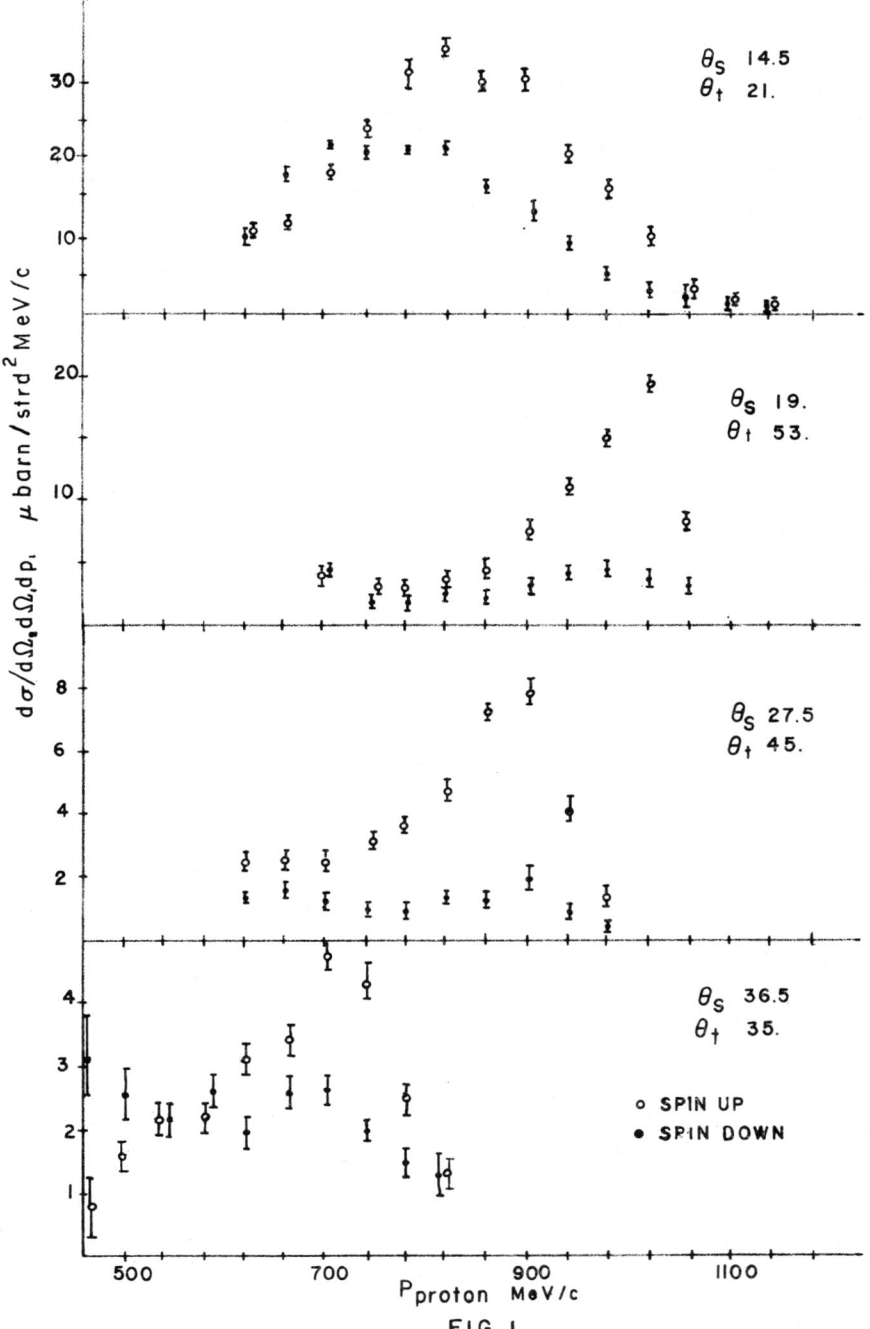

FIG 1

CALCULATION OF ASYMMETRIES AND CROSS SECTIONS
FOR THE REACTION p↑p→pπ⁺n*

E.A. Umland, I.M. Duck, G.S. Mutchler
T.W. Bonner Nuclear Laboratories
Rice University, Houston, TX 77001

The recent kinematically-complete, polarized beam experiment of Hancock et al.[1] performed at LAMPF, provides high quality data on spin up and spin down differential cross sections ($d\sigma/d\Omega_p d\Omega_\pi dP_p$) for the reaction p↑p→pπ⁺n and is a stringent test of single pion production (SPP) theories at medium energies.

Our model employs one pion exchange with the full pion-nucleon (πN) scattering amplitude included between the two πN vertices; one rho exchange coupled to a Δ-resonance intermediate state; and $^3F_3$ and $^1D_2$ dibaryon resonant contributions (see Fig.1a-f).

The πN scattering amplitude $f + i\vec{\sigma}\cdot\hat{n}g$,[2] is taken off shell, following Rinat and Thomas,[3] with the continuation

$$e^{i\delta_{LJ}} \sin\delta_{LJ} \rightarrow e^{i\delta_{LJ}} \sin\delta_{LJ} \left[\begin{matrix} q \\ q_i \end{matrix}\right]^L \left[\begin{matrix} q^2 + \beta^2 \\ q_i^2 + \beta^2 \end{matrix}\right]^L ; \qquad (1)$$

$\beta = 355$ MeV/c.

The πNN, ρNN, and ρNΔ vertices are modified using standard, off-shell momentum dependent, monopole form factors:

$$F_\pi(t) = \frac{\Lambda_\pi^2 - m_\pi^2}{\Lambda_\pi^2 + t_\pi} \quad \text{and} \quad F_\rho(t) = \frac{\Lambda_\rho^2 - m_\rho^2}{\Lambda_\rho^2 + t_\rho} . \qquad (2)$$

The value of Λ_π has been fixed at 1000 MeV as determined by Dominquez and Clark,[4] from pion photoproduction and charge exchange data. The value of Λ_ρ was set equal to 1800 MeV to fit the small angle unpolarized pp→pπ⁺n data of J. Hudomalj-Gabitzsch et al.[5]

Recent phase shift analyses of NN elastic scattering admit the possibility of highly inelastic resonances in the $^3F_3$ and $^1D_2$ partial waves at medium energies. Since the only significant inelasticity at these energies is SPP, we expect a strong coupling to an NΔ out channel (NN→R→NΔ). We model these resonant amplitudes with a Breit-Wigner form and complex coupling constants at the RNN and RNΔ from the partial decay widths of the R→NN and R→NΔ channels; the

---

*Supported by the U. S. Department of Energy.

overall phases of the $^3F_3$ and $^1D_2$ amplitudes are not determined and are adjusted to fit the data.

Preliminary results are presented in Fig.2. We present the spin averaged cross section and the asymmetry, defined as $\sigma^+-\sigma^-/\sigma^++\sigma^-$, for several different (proton, pion) angle pairs. It is clear that the addition of the dibaryon amplitudes (solid curves) improves upon the peripheral model (dashed curves), which fails completely to reproduce the asymmetries and large angle cross sections. However, more work is needed to achieve detailed fits.

Fig.1: Feynmann Graphs

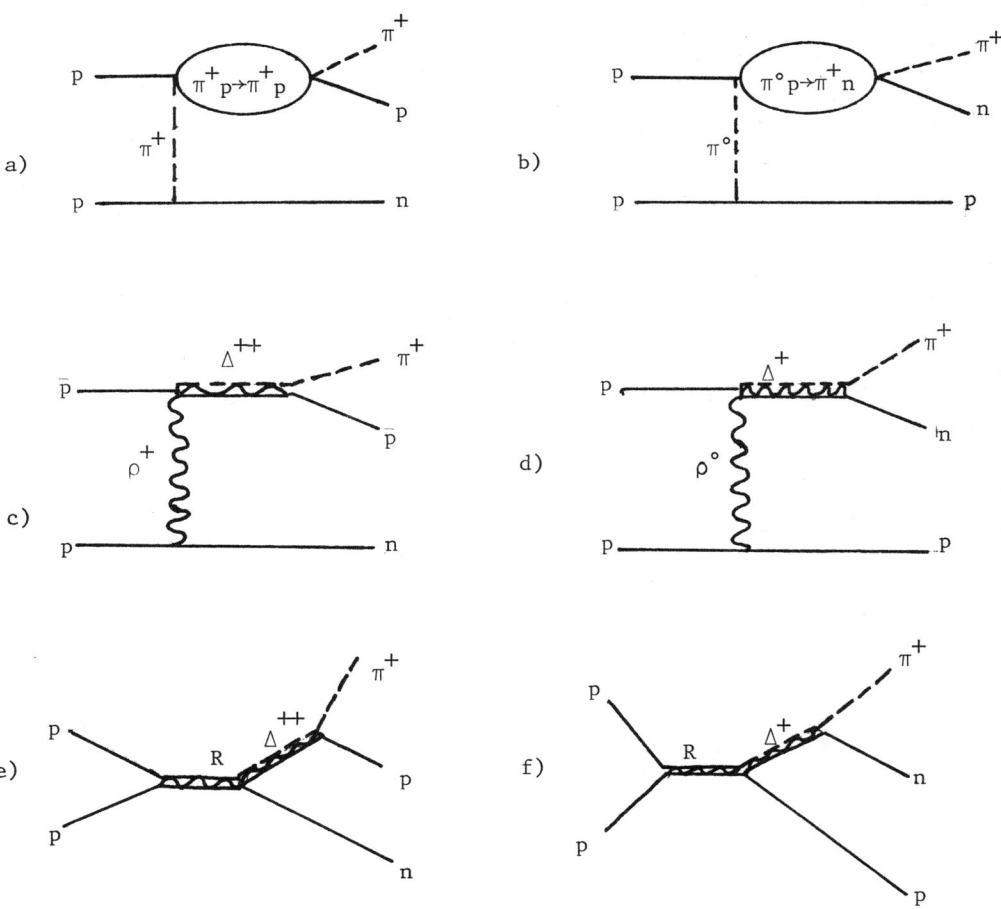

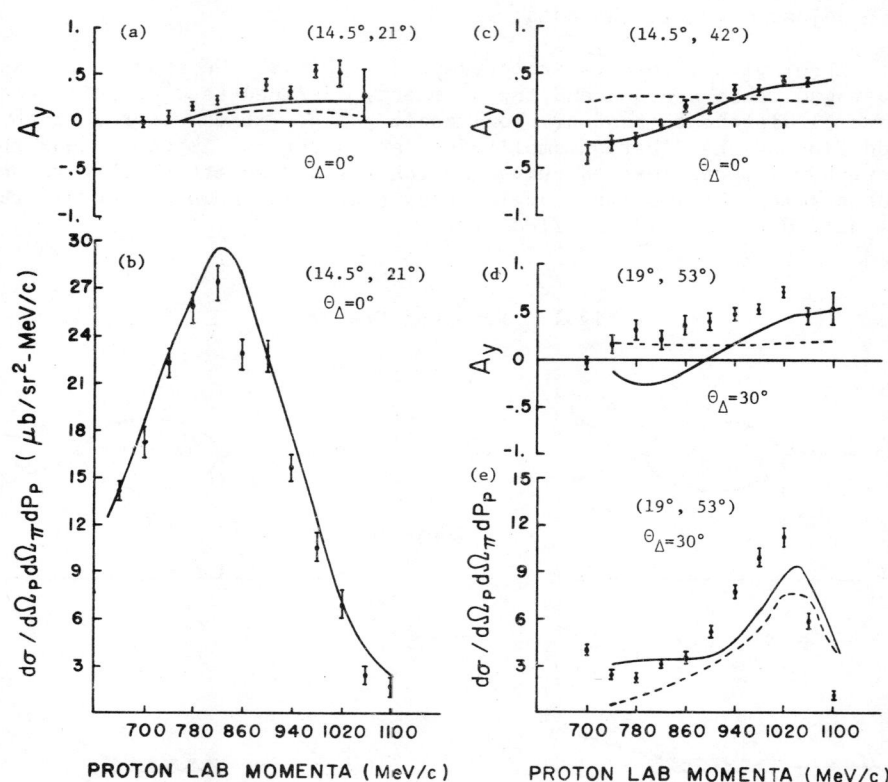

Θ_Δ=center of mass Δ-production angle
Figs. 2a, 2c, 2d are asymmetries
FIGURE 2: Fits to Data

REFERENCES

[1] A. D. Hancock et al. [unpublished].
[2] V. S. Zidell, R. A. Arndt, and L. D. Roper, Phys.Rev. D21, 1255 (1980).
[3] A. S. Rinat, A. W. Thomas, Nucl.Phys. A282, 365 (1977).
[4] C. A. Dominquez, R. B. Clark, Phys.Rev. C21, 1944 (1980).
[5] J. Hudomalj-Gabitzsch et al., Phys.Rev. C18, 2666 (1978).

DEUTERON PHOTODISINTEGRATION INDUCED BY
MONOCHROMATIC, LINEARLY POLARIZED GAMMA-RAYS

W. Del Bianco[(x)], L. Federici, G. Giordano, G. Matone,
G. Pasquariello, P. Picozza
INFN - Laboratori Nazionali di Frascati

R. Caloi, L. Casano, L. Ingrosso, M. P. De Pascale,
M. Mattioli, E. Poldi, C. Schaerf
Istituto di Fisica dell'Università di Roma, and
INFN - Sezione di Roma

P. Pelfer, D. Prosperi
Istituto di Fisica dell'Università di Napoli, and
INFN - Sezione di Napoli

S. Frullani, B. Girolami
Istituto Superiore di Sanità, and INFN - Sezione Sanità

H. Jeremie
Montreal University

ABSTRACT

Measurement of the asymmetry factor in D(γ, np) is reported and the result is discussed in the framework of Partovi's theory.

In the present measurement, we have used the monochromatic and polarized γ-rays of the newly developed Ladon facility[1] at the Frascati National Laboratory (LNF-INFN).
The differential cross section for the deuteron photodisintegration induced by linearly polarized γ-rays can be written in the form:

$$\frac{d\sigma}{d\Omega} = I_0(\theta) + P I_1(\theta) = I_0(\theta)\left[1 + P \Sigma(\theta) \cos 2\phi\right]$$

where θ and ϕ are the angles that one of the ejected nucleons makes with the direction of the momentum and with the electric vector of the incident photon beam, respectively, P represents the degree of linear polarization of the photon beam and $\Sigma(\theta)$ is equal to the ratio of $I_1(\theta)$ and $I_0(\theta)$. Including EM multipoles of order $L \leq 2$, the functions $I_0(\theta)$ and $I_1(\theta)$ can be expressed as follows:

$$I_0(\theta) = a + b \sin^2\theta + c \cos\theta + d \sin^2\theta \cos\theta + e \sin^4\theta,$$

$$I_1(\theta) = f \sin^2\theta + g \sin^2\theta \cos\theta + h \sin^4\theta.$$

By measuring the photoneutron yield at $\theta_n = 90°$ and at the azimuthal angles $\phi_n = 90°$ and $\phi_n = 0°$, we have determined in this experiment the asymmetry factor $\Sigma(\theta_n = 90°)$ in the energy interval between 10 MeV and 40 MeV.

A 6.3 cm diam. x 6.3 cm long NE-213 deuterated scintillator, was used both as a deuterium target and as a proton detector. A 30.5 cm diam. x 15.2 cm long NE-213 scintillator, placed at a distance of 100 cm from the deuterium target was employed to detect the photoneutrons. The energy of the protons and the time of flight of the neutrons from the $^2H(\gamma,n)H$ reaction were measured in coincidence together with the electron bunch in the storage ring and the events recorded in a bidimensional spectrum. In figure, $\Sigma(90°)$ has been plotted as a function of γ-ray energy. In the same figure is also reported the result of Del Bianco et al.[2] at 20.3 MeV and the points of Liu[3] at γ-ray energies above 80 MeV. In addition, the solid curve has been determined from Partovi[4] coefficients. If should be noted the good agreement between our result for Σ and that obtained by Del Bianco et al. Furthermore at all energies the values for Σ obtained in this experiment are, within the errors, consistent with Partovi's theory.

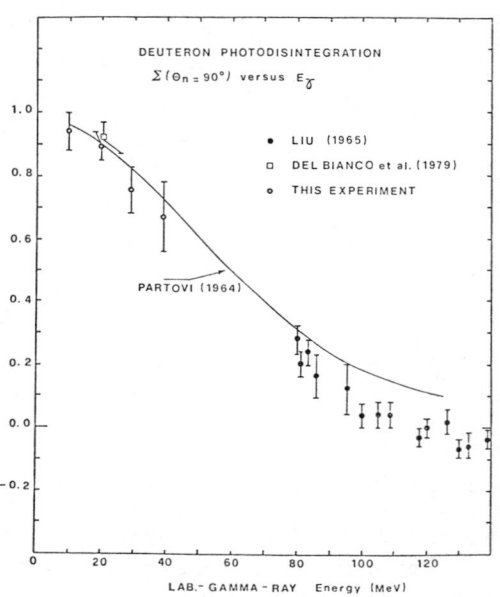

(x) On leave of absence from the Montreal University.
1. L. Federici, G. Giordano, G. Matone, P. Picozza, R. Caloi, L. Casano, M. P. De Pascale, M. Mattioli, E. Poldi, C. Schaerf, P. Pelfer, D. Prosperi, S. Frullani, B. Girolami, in: Nucler Physics with Electromagnetic Interactions, Ed. by H. Arenhövel and D. Drechsel, Lecture Notes in Physics 108 (Springer, 1979), pagg. 234-239; and this Conference.
2. W. Del Bianco, H. Jeremie, M. Irshad and G. Kayrus, Nuclear Phys., to be published.
3. F. F. Liu, Phys. Rev. B138, 1443 (1965).
4. F. Partovi, Ann. of Phys. 27, 79 (1964).

PHASE SHIFT ANALYSES OF np AND pn ELASTIC SCATTERING BETWEEN 10-750 MeV

J. BYSTRICKY*, C. LECHANOINE-LE LUC**, F. LEHAR*

*   D.Ph.P.E., CEN-Saclay, France

**   D.P.N.C., University of Geneva, Switzerland

To determine the $T = 0$ phase shifts, the $T = 1$ phase shifts obtained from a locally energy dependent pp analysis were used as fixed parameters. The difference between the p and n masses, as well as the difference in the virtual pion exchange for $T = 1$ and $T = 0$ are taken into account. Study of total inelastic cross-sections have shown that the isotopic state $T = 0$ contributes to the inelastic total cross-section for an amount of 60% with respect to $T = 1$ contribution. Therefore we have introduced imaginary parts of phase shifts also in $T = 0$ phase shifts. Contributions for $^3D_2$ is dominant and starts from the π production threshold. At higher energy important contribution for $^3D_1$ was found.

PHASE SHIFT ANALYSES OF pp ELASTIC SCATTERING

BETWEEN 10-750 MeV

J. BYSTRICKY*, C. LECHANOINE-LE LUC** and F. LEHAR*

*   D.Ph.P.E., CEN-Saclay, France

**   D.P.N.C., University of Geneva, Switzerland

A locally energy dependent phase shift analysis has been carried out for pp scattering for five overlapping intervals in the energy region 10-750 MeV. The pp phase shifts including inelastic contributions in the imaginary parts of seven phase shifts at most ($^1S_0$, $^3P_0$, $^3P_1$, $^3P_2$, $^1D_2$, $^3F_2$ and $^3F_3$) were calculated. The threshold of each imaginary part was always calculated as a free parameter. Up to 500 MeV a good compatibility with existing predictions is found, but above 550 MeV two slightly different solutions exist which cannot be distinguished using statistical criteria

EFFECT OF RECENT NP AY, AYY MEASUREMENTS ON SCATTERING ANALYSES BELOW 850 MEV

R. A. Arndt, Physics Department
Virginia Polytechnic Institute and State University
Blacksburg, VA

and

B. J. VerWest, Physics Department
Texas A&M University
College Station, TX

Recent NP scattering measurements of LAMPF have begun to define stable I=0 solutions for scattering analyses below 800 MeV. To illustrate the effect of this newer data, we have performed analyses in overlapping energy bins from 425 MeV to 700 MeV with, and without, recently measured AY and AYY data from LAMPF.[1,2] Table 1 indicates the range (Tlab in MeV) and quality of fit for 6 such solutions; they indicate an average increase in X^2 of about 1 per newly included data point. The analyses were performed by varying both I=0 and I=1 phases against a combined data set (PP+NP), but the numbers are for NP only since the I-1 waves did not tend to vary much and were essentially determined by PP data.

Figure 1 shows the systematic changes in some I=0 phase created by including the new AY, AYY points. Perhaps the most significant changes are in the errors which are reduced (e.g. $^3D_2$ at 700 MeV) by as much as a factor of 3. The energy dependent fit (solid curve shown in figure 1, had not been adjusted to the newer data.

It is apparent that the recent NP and PP data flowing from LAMPF is contributing heavily to our ongoing determination of Nucleon-Nucleon scattering amplitudes below 850 MeV. "Current" solutions and all predictions and data pertaining thereto are available through an interactive dialin computer program at VPI&SU. Information on the use of the system is available upon request from either VPI&SU or TAMU.

This analysis effort is supported by the Department of Energy and is part of a collaboration of the analysis efforts at VPI&SU, TAMU and P. Signell at Michigan State University.

REFERENCES

1. T. Bhatia, Paper Submitted to this conference.
2. C. Newsom, Paper submitted to this conference.

Table 1 Combined PP+NP Phase Shift Analyses, May 1980
Numbers are for NP only

Range (MeV)	X^2/data (without)	X^2/Data (with)
375–450	476/347	512/362
450–550	587/490	609/505
490–612	709/585	727/600
540–665	963/787	997/817
590–709	825/702	849/726
640–759	625/550	656/574

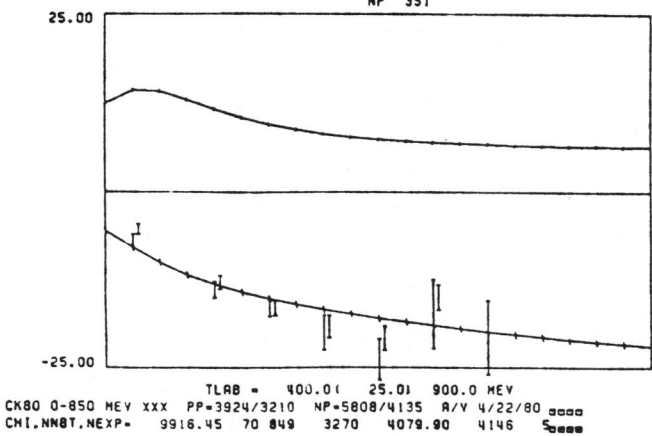

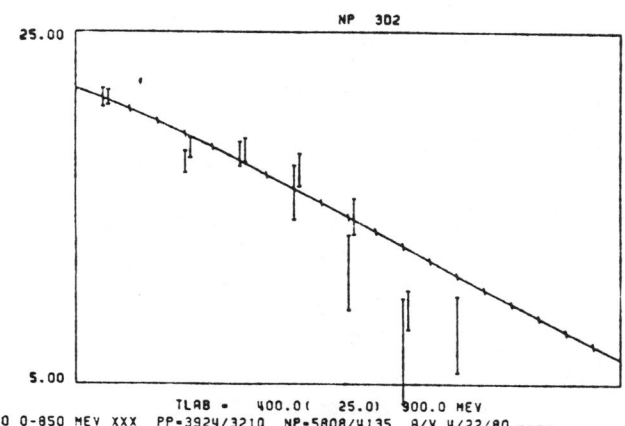

Figure 1 Some I=0 phases from analyses with (and without) new NP AY, AYY data. Smaller error bars are from analyses with MeV data. The single point at 750 MeV did not contain newer data.

COULOMB CORRECTIONS IN PROTON-PROTON OBSERVABLES

W. Plessas
Inst. f. Theor. Physik, Universität Graz, A-8010 Graz

The attempt of extracting from available experimental data the purely nuclear scattering amplitude for the proton-proton (p-p) system and likewise for other charged hadrons brings about several severe difficulties. The problem of separating strong and electromagnetic interactions is still not solved rigorously and much less in a model-independent way. Rather the situation remains unsatisfactory not only from the theoretical point of view but also with regard to practical needs. So e.g. the question of charge independence or charge symmetry of the nuclear (strong) interaction cannot be answered with confidence; in phenomenological phase shift analyses of the neutron-proton (n-p) system one needs to supply - because of the lack of sufficient and accurate enough data - the isospin T = 1 part from the p-p system in order to reduce ambiguities in the solution. For that one is eager to know the purely nuclear p-p scattering amplitude perhaps via a model-independent way. In this respect one was so far mostly content to use an amplitude, that was derived from the total scattering amplitude by subtracting all direct electromagnetic contributions via the amplitudes of the pure Coulomb interaction, of the vacuum polarization, etc., and disregarded the fact that the remaining rest contained further electromagnetic effects [1]. Of the latter the most important one (at least for the p-p system) is certainly the Coulomb distortion (CD) of the strong interaction; it manifests the fact that the strong interaction is embedded in the long-ranged Coulomb field, wherefore its forces act in a modified manner.

The CD effect has been studied in many respects mainly on the basis of scattering phase shifts and especially for the p-p and π^+-p systems [2,3,4]. The most important results being that the CD effect in phase shifts can be calculated exactly (within potential theory to all orders in e^2) - for some particular interaction models even in closed form [5,6,7] - and further that the phase-shift difference

$$\Delta = \delta^{sc}(p\text{-}p) - \delta^{s}(p\text{-}p \text{ purely nuclear}) \qquad (1)$$

with δ^{sc} defined by

$$\delta^{sc}(p\text{-}p) = \delta^{tot}(p\text{-}p) - \sigma^{c}(p\text{-}p \text{ pure Coulomb}) \qquad (2)$$

can be approximated satisfactorily well in a model-independent way by means of a formula [2-4] containing the derivative of δ^S; for some partial wave L the latest version of this prescription reads

$$\Delta_L = - \frac{\mu e^2}{\hbar} a_L \left[\frac{d}{dk} \delta_L^S(k) + \frac{1}{2k} \sin 2\delta_L^S(k) \right] \qquad (3)$$

with a_L being a numerical constant [4], μ the reduced mass.
In view of that it seems to be worthwhile to investigate the role of the CD effect also on the level of scattering observables. For the p-p case this can be done both within a specific interaction model (nucleon-nucleon potential) by e.g. solving the two-potential formalism [8] in a Lippmann-Schwinger approach or in a model-independent way by employing prescription (3). In particular it is interesting to study observables obtained according to the following cases, where each time a different phase shift δ is used as input:

1. <u>Full p-p observables</u> corresponding to

$$\delta = \delta^{tot} = \sigma^C + \delta^{SC} \qquad (4)$$

2. <u>p-p observables with the CD effect subtracted</u> corresponding to

$$\delta = \sigma^C + \delta^S \qquad (5)$$

3. <u>Coulomb-distorted nuclear observables</u> (Coulomb subtracted p-p observables) corresponding to

$$\delta = \delta^{SC} \qquad (6)$$

4. <u>Purely nuclear observables</u> (n-n observables) corresponding to

$$\delta = \delta^S. \qquad (7)$$

By comparing cases 1 and 2 one finds the influence of the CD effect on p-p observables (with the pure Coulomb contribution included). Contrasting cases 1 and 3 (equivalently cases 2 and 4) shows the effect of the pure Coulomb part of the scattering amplitude, while a comparison of cases 3 and 4 reveals the difference in the "nuclear" part of the amplitude caused by the CD effect. The latter comparison thus allows an estimate of the error made in the above-mentioned n-p phase-shift analyses by disregarding the CD effect in the (Coulomb-subtracted) "nuclear" part of the p-p interaction. Furthermore case 4 can be considered to predict neutron-neutron observables (under the assumption that further - certainly less important - electromagnetic effects may be neglected).

A study of that kind was carried out for the nucleon-nucleon system using the meson-theoretic PARIS potential, the non-local separable GRAZ potential, and the Arndt-Hackman-Roper phenomenological phase shifts [9]. For the latter case the approximation formula (3) was used as prescription for the CD effect in the phase shifts. Corresponding results are presented in the following contribution [10].

1. Cf. e.g.: E.M. Henley, in Isospin in Nuclear Physics ed. by D.H. Wilkinson (North-Holland Publishing Company, Amsterdam, 1969), p. 14,
   M.S. Sher, P. Signell, and L. Heller, Ann. Phys. 58, 1 (1970),
   A. Gersten, Nucl. Phys. A290, 445 (1977) and Phys. Rev. C18, 2252 (1978).
2. See e.g. J. Hamilton, Fortschr. Phys. 23, 211 (1975).
3. W. Plessas, L. Streit, and H. Zingl, Acta Phys. Austr. 40, 272 (1974).
4. J. Fröhlich, L. Streit, H. Zankel, and H. Zingl, J. Phys. G, to appear.
5. L. Crepinsek, C.B. Lang, H. Oberhummer, W. Plessas, and H.F.K. Zingl, Acta Phys. Austr. 42, 139 (1975).
6. G. Cattapan, G. Pisent, and V. Vanzani, Nucl. Phys. A241, 204 (1975).
7. H. van Haeringen and R. van Wageningen, J. Math. Phys. 16, 1441 (1975), H. van Haeringen, Nucl. Phys. A253, 355 (1975) and J. Math. Phys. 18, 927 (1977).
8. C.J. Joachain, Quantum Collision Theory (North-Holland Publishing Company, Amsterdam, 1975), p. 442.
9. R.A. Arndt, R.H. Hackman, and L.D. Roper, Phys. Rev. C9, 555 (1974).
10. W. Plessas and L. Mathelitsch, Contribution to this conference.

THE EFFECT OF THE COULOMB DISTORTION ON PROTON-PROTON OBSERVABLES

W. Plessas and L. Mathelitsch
Inst. f. Theor. Physik, Universität Graz, A-8010 Graz

We present results of proton-proton scattering observables obtained from a study of the Coulomb-distortion effect as it was outlined in the preceding contribution[1]. The calculations were performed for the PARIS[2] and GRAZ[3] potentials as well as for the phenomenological p-p phases of Arndt et al. (AHR)[4]; for the latter case an approximation formula for the Coulomb-distortion effect in the phase shifts[5] was tested (cf. eq. (3) of ref.[1]).

We examined the influence of the Coulomb-distortion effect on various p-p polarization observables in the low- and medium-energy domain. The results presented here concern the cases where we found the Coulomb-distortion effect to be of considerable importance, i.e. for the differential cross section I_0 and the polarization P; we give these quantities for laboratory kinetic energies of E_{Lab} = 10 and 20 MeV in Figs. 1 and 2. Further (also more complicated) observables, for which the Coulomb-distortion effect is certainly less important, are presented at even more energy points in a forthcoming paper[10]. The various curves in Figs. 1 and 2 refer to the case studies explained in ref.[1]; their description is as follows:

```
———— AHR  ⎱ case 1 for p-p    ······ AHR   ⎱ case 3 for p-p
— — GRAZ  ⎰ case 2 for n-n    ------ GRAZ  ⎰ case 4 for n-n
—·— PARIS   (i.e. with the    —··— PARIS    (i.e. without
            CD effect)                      the CD effect)
```

By comparing the curves to each other (cf. also the discussion of their relevance in refs.[1,10]) one finds the contribution of the pure Coulomb interaction as well as the effect of the Coulomb distortion. The cases where both of them are switched off can be regarded as predictions for neutron-neutron observables. From the result for AHR the usefullness of an approximation like in refs.[5,11] is evident.

1. W. Plessas, Contribution to this conference.
2. R. Vinh Mau, in <u>Mesons in Nuclei</u>, Vol. I ed. by M. Rho et al. (North-Holland Publ. Co.,Amsterdam,1979),p.151.
3. L. Crepinsek et al., Acta Phys. Austr. <u>42</u>, 139 (1975).
4. R.A. Arndt et al., Phys. Rev. <u>C9</u>, 555 (1974).
5. J. Fröhlich et al., J. Phys. G, to appear.
6. N. Jarmie et al., Phys. Rev. Lett. <u>25</u>, 34 (1970).
7. J.W. Burkig et al., Phys. Rev. <u>113</u>, 290 (1959).
8. J.D. Hutton et al., Phys. Rev. Lett. <u>35</u>, 429 (1975).
9. P. Catillon et al., Phys. Rev. Lett. <u>20</u>, 602 (1968).
10. W. Plessas et al., Univ. Graz Preprint UTP04/80 (1980).
11. W. Plessas et al., Acta Phys. Austr. <u>40</u>, 272 (1974).

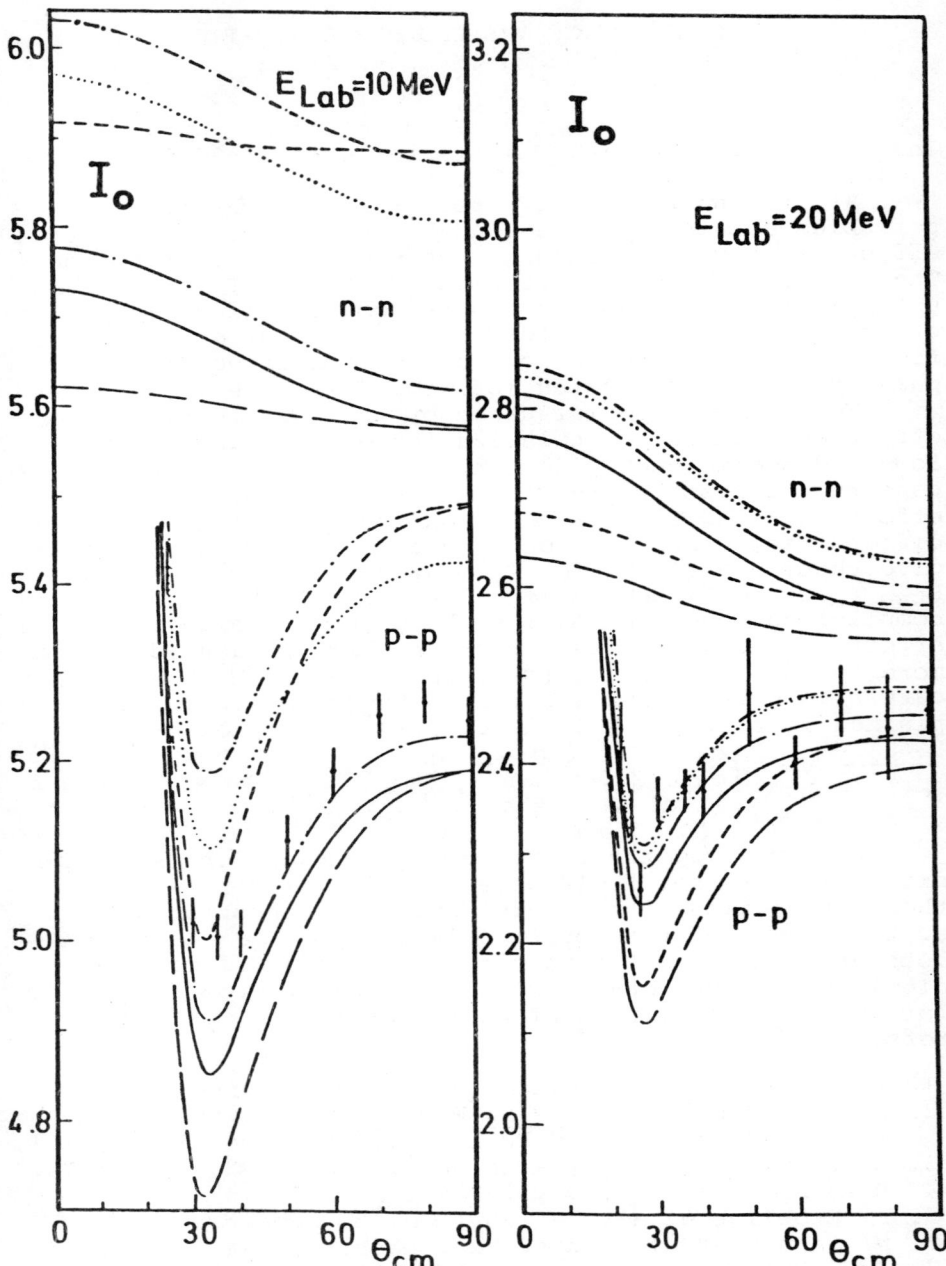

Fig. 1. Differential cross sections for N-N scattering. The curves are described in the text. Experimental data were taken from refs. 6,7.

187

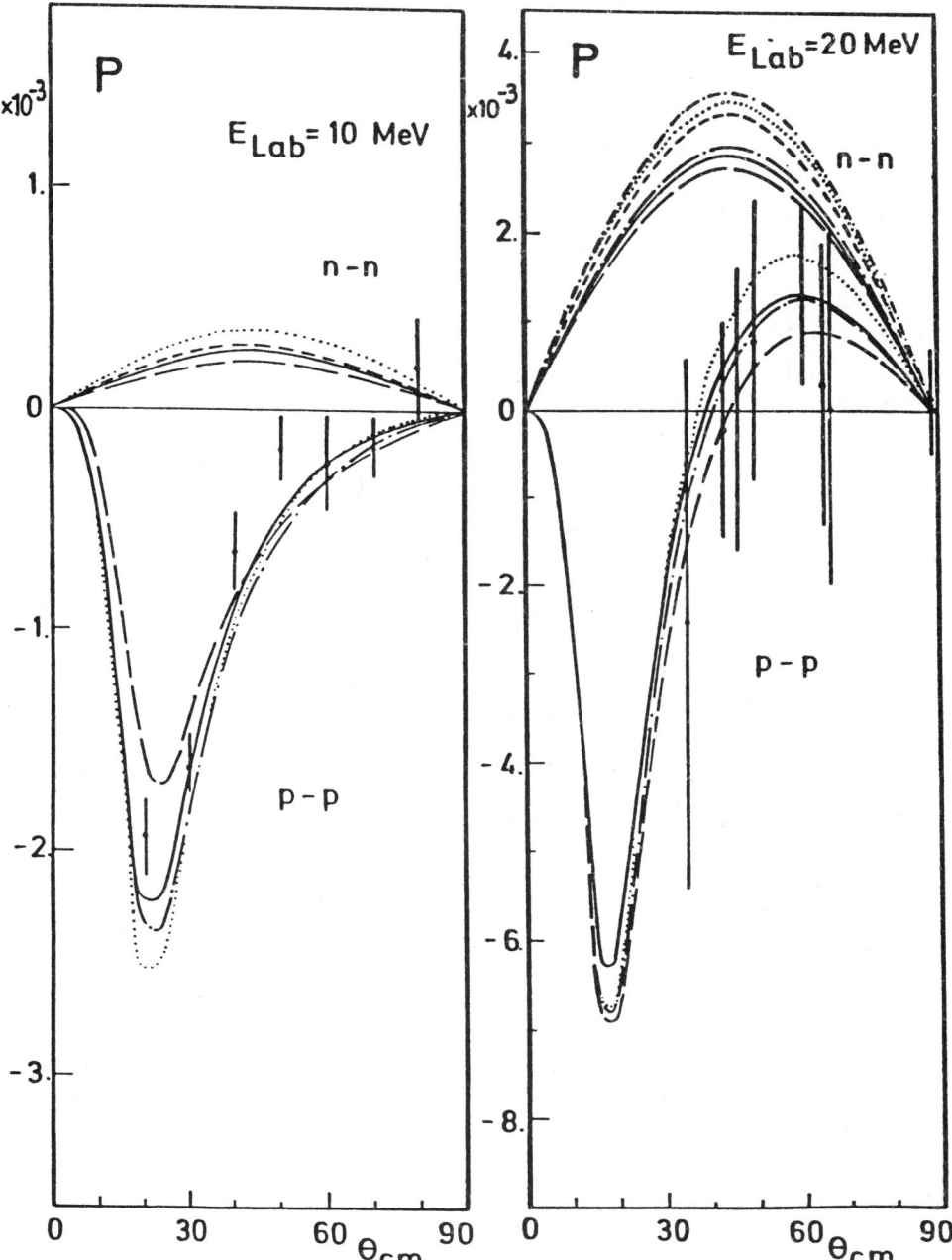

Fig. 2. Polarizations for N-N scattering. The curves are described in the text. Experimental data were taken from refs. 8,9.

SEARCH FOR THE DIBARYON BANDHEAD

R. Abegg
University of Manitoba, Winnipeg, Manitoba R3T 2N2 Canada

J. M. Cameron, D. A. Hutcheon, R. P. Liljestrand, W. J. McDonald,
C. A. Miller and L. E. Antonuk
University of Alberta, Edmonton, Alberta T6J 2J1 Canada

C. E. Stronach
Virginia State University, Petersburg, Virginia 23803 USA

J. R. Tinsley
University of Oregon, Eugene, Oregon 97403 USA

ABSTRACT

Analyzing power measurements were obtained for the reaction $^2H(\vec{p},p')$ at T_p=500 MeV using the 1.6 GeV/c TRIUMF magnetic spectrometer in an attempt to find evidence for the possible existence of the ppπ bandhead of the dibaryon system. Coincidence measurements of A_y at θ_{lab}=25°, 30° and 35° did not reveal any evidence for the existence of the ppπ bandhead in this initial experiment.

INTRODUCTION

The existence of proton-proton dibaryon resonances is suggested by several recent reports. Auer et al.[1] associate observed dips in the energy dependence of the spin-spin correlation parameter C_{LL} in proton-proton elstic scattering around θ_{cm}=90° with the possible existence of proton-proton dibaryon resonances $^1D_2$, $^3F_3$ and $^1G_4$ ($^{2s+1}\ell_j$) with respective energies of approximately 2.14, 2.26 and 2.43 GeV. Further evidence was presented[2] in the analysis of C_{LL} and of cross section differences between parallel and antiparallel longitudinal ($\Delta\sigma_\ell$) and transverse ($\Delta\sigma_T$) total cross section measurements in proton-proton elastic scattering. Additional effects that could be due to dibaryon resonances were reported in a total cross section measurement[3], the polarization[4] at 90° in the reaction $\gamma + ^2H \to p + n$ and in the excitation function measurement at 180° of $\pi^+ - ^2H$ elastic scattering[5].

MacGregor[6] interprets the proton-proton dibaryon resonances at 2.14, 2.26 and 2.43 GeV as members of a nuclear-physics-type rotational band with a virtual ppπ bandhead. This approach leads to the prediction of the energies of the bandhead and first excited state at

2.020 and 2.060 GeV, respectively. This is within the energy region that can be explored in a ^{2}H($\vec{p}$,p') missing mass experiment using the TRIUMF polarized ion source, cyclotron and 1.6 GeV/c magnetic spectrometer. We report on the initial experiment ^{2}H($\vec{p}$,p') that was undertaken to find evidence for the existence of the bandhead with a total energy of 2.02 GeV.

EXPERIMENTAL ARRANGEMENT AND RESULTS

A beam of 500 MeV polarized protons from TRIUMF was incident on a ^{2}H gas cell at 101 kPa. The beam current and polarization were monitored with a previously calibrated in-beam polarimeter and were about 5 nA and 0.69, respectively. Inelastically scattered protons from the reaction ^{2}H($\vec{p}$,p') were detected at Θ_{lab}=25°, 30° and 35° using the TRIUMF 1.6 GeV/c magnetic spectrometer ($\Delta p/p \simeq \pm 10\%$) in coincidence with a decay proton from the ppπ bandhead detected at Θ_{lab}=50° on the same side of the beam in a 0.64 cm thick NE 102 scintillator followed by a veto counter. This somewhat arbitrary choice of dibaryon decay product angle was severely limited by target cell geometry and was also based on the need to avoid kinematic regions where proton quasi-free scattering is dominant. Tests were carried out by detecting elastically scattered protons from ^{2}H at Θ_{lab}=25°.

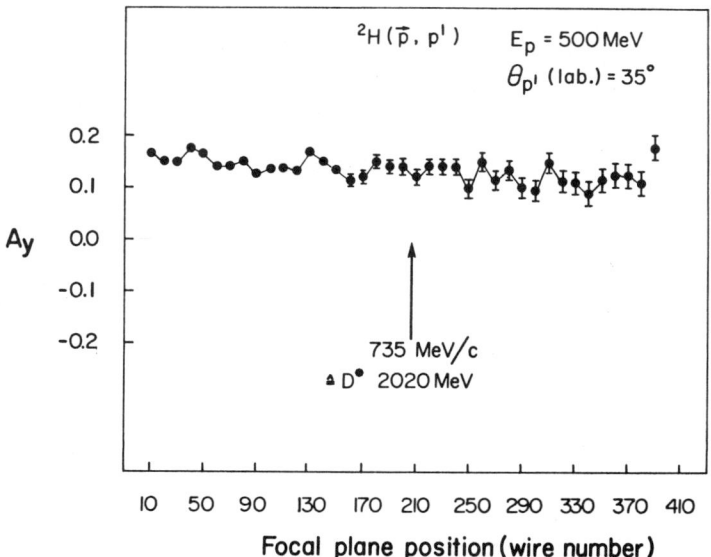

Fig. 1. Analyzing power A_y as a function of focal plane position. A_y was calculated from singles spectra.

The second test observed quasi-elastic scattered protons from ^{2}H at $\Theta_{lab}=25°$ and $130°$, respectively. The spectrometer was adjusted such that the momentum corresponding to a dibaryon total energy of 2.02 GeV was centered in the focal plane. Figure 1 shows the analyzing power A_y as obtained from singles spectra at $\Theta_{lab}=35°$. A_y is displayed as a function of the focal plane position which was binned in channels of 10 ($\hat{=}$ 2.33 MeV at the midpoint of the focal plane). The errors shown are only the statistical ones. A_y does not indicate the existence of the bandhead in this initial experiment. No evidence was found in the coincidence spectra either. It is hoped to further extend the experiment to include an excitation function measurement spanning the energy range of interest.

REFERENCES

1. I. P. Auer, A. Beretvas, E. Colton, H. Halpern, D. Hill, K. Nield, B. Sandler, H. Spinka, G. Theodosiou, D. Underwood, Y. Watanabe and A. Yokosawa, Phys. Rev. Lett. 41,1436(1978).
2. A. Yokosawa, ANL Report No. ANL-HEP-CP-78-52 and H. Spinka, ANL Report No. ANL-HEP-CP-78-56.
3. J. C. Keck and A. V. Tollestrup, Phys. Rev. 101,360(1956).
4. T. Kamae and I. Arai, T. Fujii, H. Ikeda, N. Kajiura, S. Kawabata, K. Nakamura, K. Ogawa, H. Takeda and Y. Watase, Phys. Rev. Lett. 38,468(1977) and H. Ikeda, I. Arai, H. Fujii, H. Iwasaki, N. Kajiura, T. Kamae, K. Nakamura, T. Sumiyoshi, H. Takeda, K. Ogawa and M. Kanazawa, Phys. Rev. Lett. 42,1321(1979).
5. R. Frascaria, I. Brissaud, J. P. Didelez, C. Perrin, J. L. Beveridge, J. P. Egger, F. Goetz, P. Gretillat, R. R. Johnson, C. Lunke, E. Schwarz and B. M. Preedom, Phys. Lett. 91B,345(1980).
6. M. H. MacGregor, Phys. Rev. Lett. 42,1724(1979).

SOME EVIDENCE FOR THE EXISTENCE OF THE $^1S_0$-
DIBARYON RESONANCES

V.V.Komarov, A.M.Popova, Yu.V.Popov
Institute of Nuclear Physics, Moscow State University,
Moscow 117234, USSR

In some recent experiments there was indicated the existence of the dibarion resonances in the two-nucleon system. The evidence includes, on the one hand, the experimental detection of such a resonance in the F-state[1] and, on the other hand, the calculations of the six-quark bound states in the "bag"-model [2,3]. In its turn, the existence of such states is the indication of a sufficiently specific behaviour of the nucleon-nucleon potentials at short distances and definitely requires that the models of hard core potentials should be out of consideration. Direct experimental evidence for the existence of $^1S_0$-dibarion resonances are still absent, probably because of their high instability. However, interesting indirect evidence may be obtained by comparing between the dependence of the $^1S_0$-phases of the n-p and p-p scatterings at the incident-proton energies of about 400 MeV in laboratory system[4]. The data show that $\Delta \delta_0 = \delta_{np} - \delta_{pp} \simeq 0.22 \pm 0.07$ at $E_{lab} \simeq 450$ MeV. This phase difference is due to the interference of the nuclear and Coulomb interactions so that the scattering phase of two charged nuclear particles may be presented as

$$\delta_0 \equiv \delta_{pp} = \delta_0^{Coul.} + \delta_0^{Nucl.} - \Delta \delta_0$$

The analytical expression of the function $\Delta \delta_0$ can conveniently be obtained on the basis of phase approach[5]. For this we first have to find the phase δ_0 and after that, to decompose this phase in the set using the parameter $\nu = \alpha/\kappa$, where κ is the relative momentum of nucleons; $\alpha = Me^2/2\hbar^2 = 1.74 \times 10^{-2} \text{fm}^{-1}$ is the Coulomb constant of nucleon. The input phase equation is of the form

$$\frac{d\delta_o(z,\kappa;\alpha)}{dz} = -\frac{1}{\kappa}\left[V(z) + \frac{2\alpha}{z}\right] \sin^2[\kappa z + \delta_o(z,\kappa;\alpha)]$$

$$\delta_o(0,\kappa;\alpha) = 0$$

The phase difference $\Delta\delta_o$ in the first order of the parameter ν has been calculated[6]. The eventual result is

$$\Delta\delta_o \approx 0.22 \approx 2\nu \int_0^R \frac{dz}{z} \varphi_o^2(z,\kappa) \qquad (1)$$

where $\varphi_o(z,\kappa)$ is the wave function of the S-wave interaction of nucleons with potential $V(z)$, R is the radius of the potential which is about 2-3 fm. It can be shown that the main contribution to the observable phase difference is given by the formula (1), because at the energy $E_{lab} = 450$ MeV ($\kappa = 2.35$ fm^{-1}), the relativistic corrections (i.e. the effects of the electromagnetic nucleon form factors, vacum polarization and indirect electromagnetic processes) can be neglected since their contribution is an order as small as the observable value. Because of the smallness of the parameter ν in (1) one can obtain the value $\Delta\delta_o$ close to the experimental data if the absolute value of the integral in (1) is big. Using, for example, the hard-core potential models one can not obtain the proper value of $\Delta\delta_o$[6]. It is only when a two-nucleon system includes a quasi-bound state, we have a strong increase in the value of the integral in (1). Indeed, if the energy of the interacting particles is close to the value of the real part of the energy of the possible quasi-bound state the wave function $\psi_0(z,\kappa)$ of the two-nucleon system increases very fast, and hence the total integral in (1) may have a big value. Thus, the observable increase of the phase difference in the elastic n-p and p-p scattering at large energies may be the indication of the existence of the dibarion resonance in the $^1S_0$-state.

References

1. H.Hidaka et al. Phys.Lett., 70B (1977) 479.
2. A.Th.M. Aerts, P.J.G. Mulders, J.J. de Swart. Phys. Rev., D17 (1978) 260.
3. C. De Tar. Phys. Rev., D17 (1978) 323.
4. M.H.Mac Gregor, R.A.Arndt, R.W.Wright. Phys. Rev., 169 (1969) 1122.
5. F.Calogero. Variable phase Approach to Potential Scattering. Acad. Press, New-York - London, 1967.
6. V.V.Komarov, A.M.Popova, Yu.V.Popov. Elementary Particles and Atomic Nuclei, 9 (1978) 1213.

THE 3F3 DIPROTON IN πNN DYNAMICS

M. Araki and T. Ueda
Faculty of Engineering Science, Osaka University, Toyonaka 560

Y. Koike
Department of Nuclear Engineering, Kyoto University, Kyoto 606

ABSTRACT

A Faddeev calculation is made for the πNN system of IJP = 13-. A resonance pattern is generated in the 3F3 NN phase parameter in qualitative agreement with experiment. The decay rate of the resonance into the πd channel is so small as 0.1%.

Polarization experiments for N-N systems have presented remarkable evidences for many dibaryons to exist[1,2,3]. About interpretation of these new phenomena the model by one of us (T.U.) where the dibaryons are quasibound states of πNN and ππNN systems, generated by the nuclear forces and the force between the two nucleons keeping one or two pions in the Δ state in common, has made extensively qualitative agreement with the experimental observations[4]. As a development a Faddeev calculation of the 3F3 diproton in the πNN system has been presented by the present authors[5]. We summarize here a remarkable result of this work and a subsequent study of the πd decay of the 3F3 diptoton[6].

At laboratory energies E_L = 0 - 1200 MeV we consider pp elastic scattering and one-pion-production processes in the I = 1 and JP = 3- state. In the Faddeev formalism the processes are represented by transition amplitudes between various πNN channels[7]. The pp channel is represented by the πNN channel in which a πN subsystem is in the bound state equivalent to proton. This bound state and the spectator proton are correctly antisymmetrized. In the form of the partial wave expansion the Faddeev equation of AGS type[8] is written for the transition amplitudes[7] and solved by the Pade approximant method.

In the I = 1 and JP = 3- system we select the 9 channels from the 1st to the 9th in Table 1 as important for numerical calculation, since their two-body interactions are strong. The two-body interactions are the ones in the πN P11, πN P33 and NN 3P2 states. We find by calcultion that the NN 3S1 interaction is negligible.

The separable potentials for the two πN states and the NN 3P2 states are taken from ref.6 and 8 respectively. The pions and nucleons are treated as non-relativistic particles. We checked our calculation by comparing it with similar non-relativistic calculations of other processes[7,10].

First, we consider channels 1, 3 and 7. This model involves very important channels and is similar with the ones in ref.4. In the 3-channel model the pattern of the inelastic resonance is clearly generated in phase shift δ and absorption coefficients η as is shown in Fig.1. Namely the phase shifts cross the non-resonant background contribution, which is the OBEP[11] or the Born term contribution of the AGS equation, from upside to downside with energy increasing and the energy of the crossing the absorption coefficients show the dip structure. However the 3-channel model gives the resonance energy in c.m.s. too high approximately by 50 MeV, compared with the experimental one.

Second, we consider all the 9 channels. Then additional binding energy of approximately 100 MeV is provided by the contribution from the additional 6 channels. Among these channels, channels 2 and 5 are important. The 9-channel model gives the resonance energy too low approximately by 50 MeV, compared with the experimental one.

The deviation from the experimental values in the 9-channel model would be remedied by the following elements. First, if we treat the pion as a relativistic particle the attractive πN potentials are weakened. Second, the possible ρ-exchange contribution, which is ignored in this calculation, works repulsively. Furthermore several channels which are not listed in Table I make weak, but repulsive contribution.

Next we discuss the πd decay of the 3F3 diproton. We consider channels 1, 2, 3, 5, 9 and 10. This 6-channel model is approximately the same with the 9-channel model in describing the pp phase-parameters. To introduce channel 10 allows us to discuss the πd decay of the 3F3 diproton. For the 3S1-NN interaction we take Tabakin potential[12].

Table II shows that the branching ratio of the 3F3 diproton to the πd channel is of the order of 0.1% around the resonance energy of $E_L \sim 600$ MeV. The 3F3 diproton in our model is composed primarily by the NN subsystem, while in the πd channel the NN subsystem is in the S state and the pion is in the d state with respect to the NN subsystem. Therefore the wave function of the 3F3 diproton and the πd channel are approximately orthogonal. This is the main reason why the branching ratio to the πd channel is so small.

In conclusion we remark that the Faddeev approach to the 3F3 diproton as a πNN resonance is promising and that an experimental check of the prediction of the small πd decay rate from the 3F3 diproton would make a useful test of our model, since the test investigates the composition of the diproton.

Table I The πNN channels. S represents the sum of the spins of the spectator and the interacting pair and L does their relative angular momentum.

Channel	1	2	3	4	5	6	7	8	9	10
Pair states	P11	P11	P33	P33	P33	P33	3P2	3P2	3P2	3S1
S	1	0	2	1	2	2	2	2	2	1
L	3	3	1	3	3	5	1	3	5	2

Table II The total and partial cross-sections in the IJP =13- state in units of mb.

E_L(MeV)	Total	pp-pp	pp-πd	pp-πNN
400	1.43	0.041	0.0012	1.39
500	6.59	0.95	0.00140	5.63
550	12.14	2.85	0.027	9.55
600	16.8	4.98	0.028	11.8
650	17.5	6.07	0.0205	11.4
700	16.2	6.33	0.0131	9.91
800	13.0	5.99	0.0055	7.03

Fig.1. The 3F3 phase-shifts and absorption coefficients.

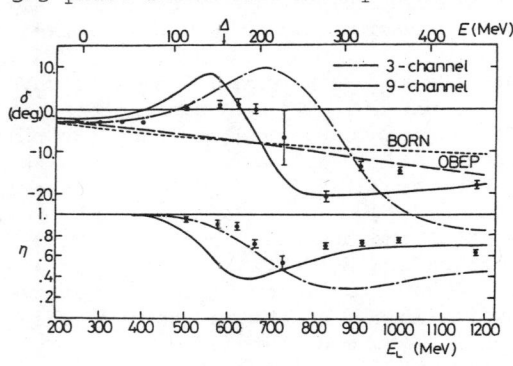

REFERENCES

1. N. Hoshizaki, Prog. Theor. Phys. 60, 1796 (1978); 61, 129 (1979).
2. K. Hidaka et al., Phys. Lett. 70B, 479 (1977).
3. H. Ikeda et al., Phys. Rev. Lett. 42, 1321 (1979).
4. T. Ueda, Phys. Lett. 74B, 123 (1978); ib. 79B, 487 (1978).
5. M. Araki, Y. Koike and T. Ueda, Prog. Theor. Phys. 63, 335 (1980).
6. M. Araki, Y. Koike and T. Ueda, Prog. Theor. Phys. 63, No.6 (1980), to appear.
7. I.R. Afnan and A.W. Thomas, Phys. Rev. C10, 109 (1974).
8. E.O. Alt, P. Grassberger and W. Sandhas, Nucl. Phys. B2, 167 (1967).
9. T.R. Mongan, Phys. Rev. 178, 1597 (1968).
10. Y. Koike, Prog. Theor. Phys. 59, 87 (1978).
11. T. Ueda and A.E.S. Green, Phys. Rev. C18, 337 (1978).
12. F. Tabakin, Phys. Rev. 174 1208 (1968).

ELASTIC WIDTHS OF NN̄-RESONANCES FROM A MOMENTUM-SPACE ONE-BOSON-EXCHANGE POTENTIAL[+)]

L. Heins, K. Holinde and D. Schütte
Institut für Theoretische Kernphysik der Universität Bonn
Nußallee 14-16, D-5300 Bonn, W.-Germany

ABSTRACT

Widths and positions of NN̄-resonances are calculated, starting from a momentum-space one-boson-exchange potential. Compared to former results based on r-space potentials, the resulting widths are comparatively narrow.

INTRODUCTION

One theoretical approach to describe the low-energy nucleon-antinucleon (NN̄) scattering problem is in terms of an NN̄-potential. It is often believed[1] that such potential models cannot provide sufficiently narrow resonances with quantum numbers suggested by experiment. However, up to now, calculations have only been done with essentially local (r-space) NN̄-potentials like e. g. the Bryan-Phillips potential[2] or the Paris potential[3], in which nonlocalities and off-shell effects generated by meson field theory have necessarily been neglected almost completely.

Therefore, we present in this letter the results for the NN̄-resonance energies and their elastic widths starting from a momentum-space one-boson-exchange NN-potential[4] including $\pi, \rho, \omega, \sigma, \delta$-exchange, which is non-local and, moreover, energy-dependent. It is based on the use of a field-theoretic Hamiltonian and old-fashioned perturbation theory, see ref.[5], and describes the nucleon-nucleon scattering data and the deuteron properties in a satisfactory way, see ref.[4]. The corresponding NN̄-potential is, as usual, obtained by changing the sign of the contributions of the mesons with negative G-parity.

RESULTS

The resulting resonances (in the interesting region $E_{CM} < 2$ GeV) together with quantum numbers and elastic widths are given in table I. Since the D-wave resonances lie roughly 100 MeV above threshold, their width is relatively large. Nevertheless, the relation between width and position has considerably improved compared to the (r-space) Paris-potential[3]. This is demonstrated in fig. 1, where the width is shown as function of the energy above threshold. The

[+)] Supported in part by Deutsche Forschungsgemeinschaft

solid curves are obtained by varying form-factor parameters (cut-off masses) in our $V(z)^4$ (in the range of 20%), which modifies mainly the short-range part of the NN--interaction. Consequently, the NN-phases in the corresponding channel are only slightly changed by such a modification; we have checked that they still lie in the range of the experimental uncertainties.

We see from the figure that, in the case of the D-waves, the resonance position can be shifted to that of the S-meson ($\simeq$ 60 MeV) with a reasonably small width ($\simeq$ 10 MeV for $^{33}D_2$, only $\simeq$ 1 MeV for $^{11}D_2$).

For $^{33}D_2$, the corresponding curve for the Paris--potential is given by the dashed line taken from ref.[1]. With our (momentum-space) quasi-potential $V(z)$ the situation has greatly improved: at the S-position, our potential yields a width nearly 3 times smaller than in the case of the Paris-potential.

How can we understand such a big discrepancy? (Both potentials predict roughly the same NN-phases in the $^3D_2$--channel). We believe that, apart from nonlocality effects in $V(z)$ (which are possibly small in D-waves), the main reason lies in the different treatment of the short-range part of the NN-interaction in both potentials. The situation is shown schematically in fig. 2. Whereas our NN-potential (solid line) is used throughout the whole range (in momentum space), the Paris-potential is sharply cut off at $r<r_0$ setting arbitrarily $V(r) = V(r_0)$ for $r<r_0$. Thus, our potential is more attractive in the inner part and, consequently, must be more repulsive in the outer-range part ($r \simeq$ 1fm) if both potentials predict the same resonance energy. Obviously, this feature leads to a comparatively smaller width. Thus, it is of considerable advantage to keep the full attraction in the inner part of the NN--potential.

Table I   Quantum numbers, energies and widths of resonances predicted by $V(z)^4$

$^{2I+1,2S+1}L_J$	$J^{PC}(I^G)$	E_{CM} (MeV)	Γ (MeV)
$^{11}P_1$	$1^{+-}(0^-)$	1916	40
$^{11}D_2$	$2^{-+}(0^+)$	2000	40
$^{33}D_2$	$2^{--}(1^+)$	1988	50

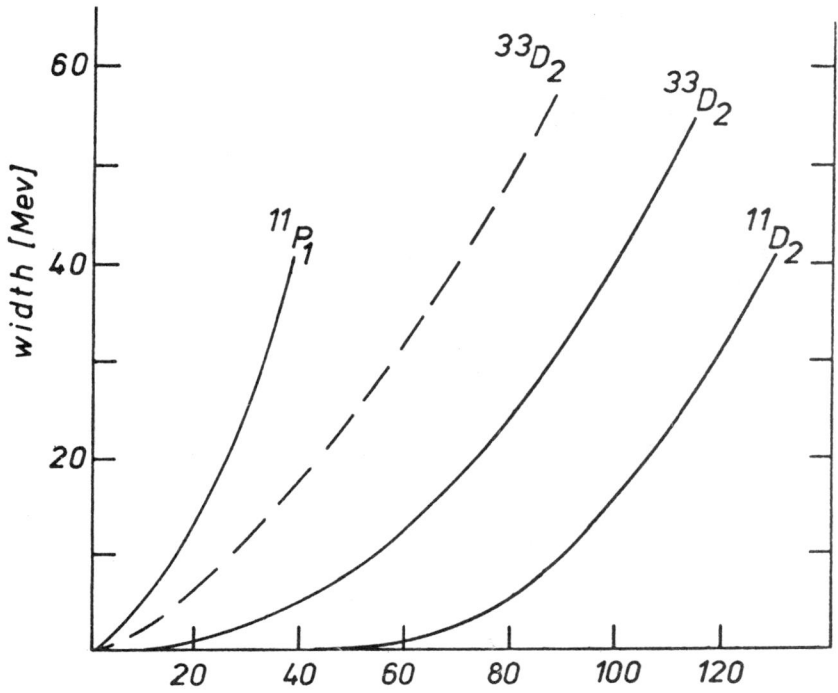

Fig.1. Elastic widths as function of energy (MeV) above threshold for $V(z)$[4] (solid line) and the Paris-potential[5] (dashed line).

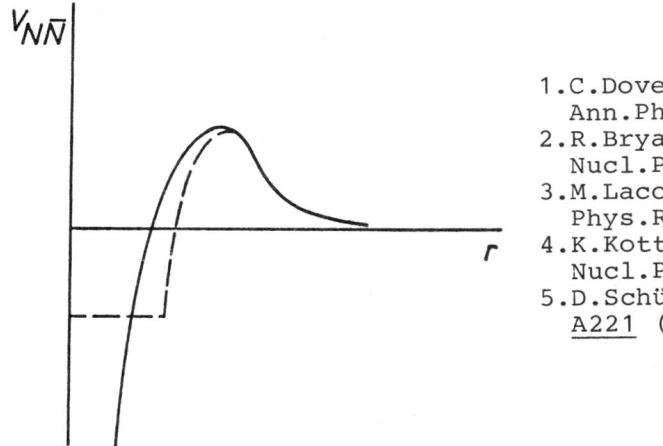

Fig.2. Behaviour of $V(z)$[4] (solid line) and the Paris-potential[5] (dashed line).

REFERENCES

1. C.Dover and J.Richard, Ann.Phys. 121 (1979) 70
2. R.Bryan and R.Phillips, Nucl.Phys. B5 (1968) 201
3. M.Lacombe et al. Phys.Rev. D12 (1975) 1495
4. K.Kotthoff et al. Nucl.Phys. A264 (1976) 484
5. D.Schütte, Nucl.Phys. A221 (1974) 450

EFFECT OF NON-ITERATIVE ISOBAR DIAGRAMS ON NN-SCATTERING DATA

K. Holinde and R. Machleidt
Institut für Theoretische Kernphysik der Universität Bonn
Nussallee 14-16, D-53oo Bonn, W.-Germany

A. Faessler and H. Müther
Institut für Theoretische Physik der Universität Tübingen
Auf der Morgenstelle 14, D-74oo Tübingen, W.-Germany

M. R. Anastasio
Brooklyn College of CUNY
Brookly, N. Y. 1121o, USA

ABSTRACT

The fourth-order non-iterative diagrams involving nucleon-isobar intermediate states and including π-exchange are calculated in momentum space, in the framework of non-covariant perturbation theory. The effect of these diagrams on NN-scattering phase shifts is studied.

INTRODUCTION

An explicit description of the 2π-exchange contribution to the NN-interaction is essential for a realistic treatment of the modifications of the NN-interaction in the medium (because of Pauli and dispersive effects), which play a decisive role in obtaining a consistent description of light and heavy nuclei.

Our general scheme is to start from a field-theoretic Hamiltonian $H = H_0 + W$, containing as interaction part W meson-nucleon-nucleon and meson-nucleon-isobar vertices[1]. Note that antiparticles are neglected from the beginning.

H is treated in old-fashioned perturbation theory in order to have a handable framework for the two- and the many-body problem. This leads to a series expansion for the nucleon-nucleon transition matrix T, which can be partially summed by solving an integral equation of Lippmann-Schwinger type, containing as driving term an energy-dependent quasi-potential $V(z)$. In principle, $V(z)$ is the (infinite) sum of all diagrams with at least one meson or one Δ-isobar present in each intermediate state.

Here, we want to study the non-iterative diagrams involving $N\Delta$ intermediate states and including π-exchange. (Corresponding diagrams involving two-nucleon intermediate states have been studied in ref. 2, whereas the effect of iterative isobar diagrams has been demonstrated in ref.3). They are shown in fig. 1:

0094-243X/81/690200-3$1.50 Copyright 1981 American Institute of Physics

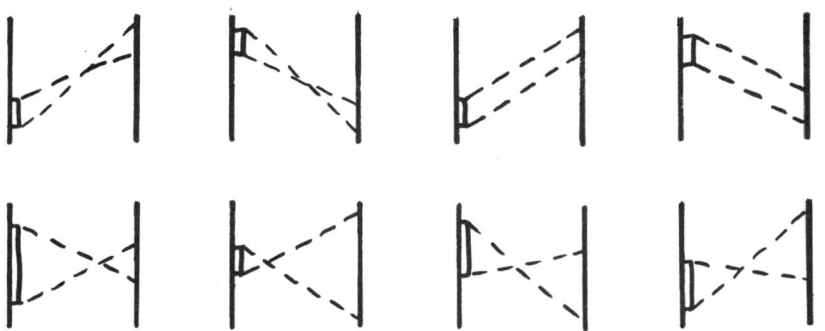

Fig. 1. Non-iterative diagrams involving NΔ-intermediate states.

Note that the corresponding diagrams in which the Δ appears on the right-hand side, are also included in the calculation.

RESULTS

Within the approximation of putting all particles at rest (i.e. energies are replaced by corresponding masses) and neglecting isospin, the sum of all time orderings for crossed and uncrossed exchanges looks like the iteration of a Yukawa-potential of pion range, usually used in the transition potential concept. Our explicit calculations show that this approximation is roughly true in the $^1S_0$-channel, but considerably overestimates the true result in the $^3S_1$-channel.

In order to show the effect of these diagrams on NN-scattering phase shifts, we use as effective potential

$$V(z) = V_{OBE}(z) + M_{N\Delta}(z) + M_{\Delta\Delta}(z) + M_{NN}^{'\pi}(z) + M_{N\Delta}^{'\pi}(z) \qquad (1)$$

Here, V_{OBE} is the OBE-potential of ref. 4. $M_{N\Delta}$ ($M_{\Delta\Delta}$) denotes the sum of iterative diagrams with $N\Delta$ ($\Delta\Delta$) intermediate states including π- and ρ-exchange. $M_{NN}^{'\pi}$ ($M_{N\Delta}^{'\pi}$) denotes the sum of non-iterative diagrams with NN (NΔ) intermediate states including only π-exchange.

The parameters (coupling constants and masses) are partly adjusted in order to obtain a reasonable description of the NN-scattering phase shifts. The results for $^1S_0$ and $^3S_1$ are shown in fig. 2. It is clearly demonstrated that the non-iterative diagrams ($M_{N\Delta}^{'\pi}$) are as important as isobar box-diagrams ($M_{N\Delta} + M_{\Delta\Delta}$). (We believe that ρ-exchange will not drastically reduce the importance of $M_{N\Delta}^{'\pi}$). Especially in isospin-zero states (where $M_{NN}^{'\pi}$ does not contribute), the isobar contribution is drastically enlarged.

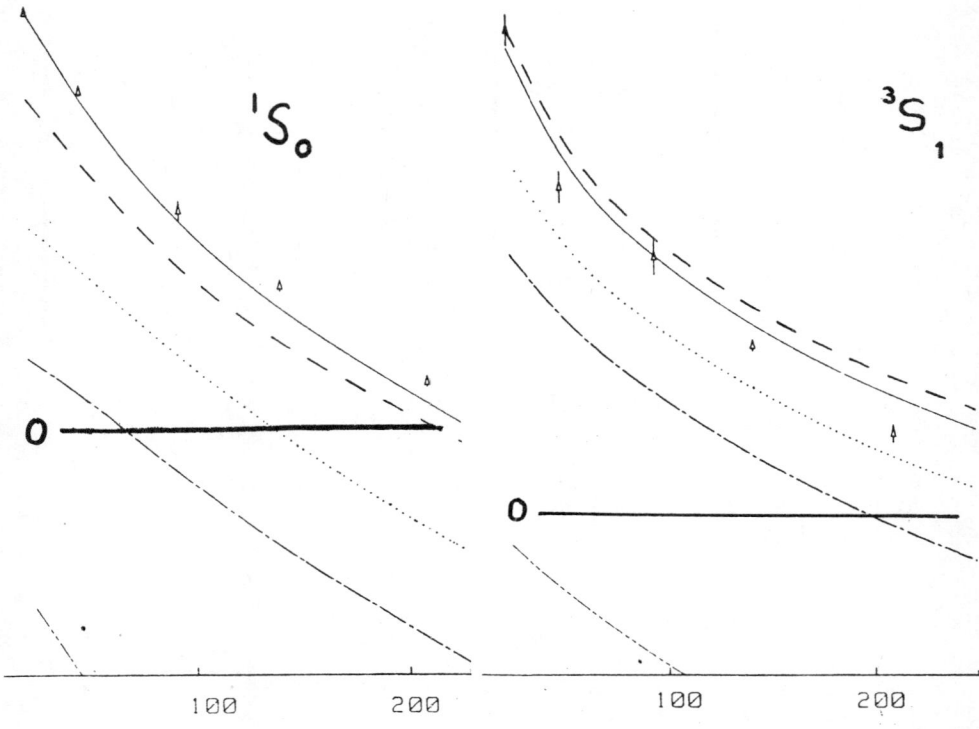

Fig. 2. NN S-wave scattering phase shifts as function of E_{lab} (Mev). The solid lines are obtained when V (eq.(1)) is used in the scattering equation. The dashed lines result when M'^{π}_{NN} is omitted; for the dotted line, $(M'^{\pi}_{NN} + M'^{\pi}_{N\Delta})$ is omitted; the dash-dot curve is based on the use of V_{OBE} only, whereas the dash-double-dot curve is obtained when, in addition, the σ-contribution in V_{OBE} is omitted describing effectively the part of the whole 2π-exchange which is not described explicitly, (mainly rescattering contributions in which the two exchanged pions interact). The error bars are taken from the Livermore analysis[5].

REFERENCES

1. K.Holinde, Nucl.Phys. A328 (1979) 439
2. K.Holinde, R.Machleidt, M.R.Anastasio, A.Faessler and H.Müther, Phys.Rev. C19 (1979) 948
3. K.Holinde, R.Machleidt, M.R.Anastasio, A.Faessler and H.Müther, Phys.Rev. C18 (1978) 870
4. K.Kotthoff, K.Holinde, R.Machleidt and D.Schütte Nucl.Phys. A242 (1975) 429
5. M.Mc Gregor, R.Arndt and R.Wright, Phys.Rev. 182 (1969) 1714

SECTION 2

POLARIZATION IN NUCLEAR STRUCTURE PHYSICS

ELASTIC SCATTERING POTENTIALS FOR NUCLEONS
AND COMPOSITE PARTICLES

W. J. Thompson *
Department of Physics and Astronomy, University of North Carolina,
Chapel Hill, NC 27514, USA, and
Triangle Universities Nuclear Laboratory, Durham, NC

ABSTRACT

A review is made of elastic-scattering optical-model potentials for n, p, d, t, ^{3}He, and ^{6,7}Li nuclear scattering from targets with A>6 in the bombarding energy range 4 to 1000 MeV. Summaries of cross-section and polarization angular-distribution data for n through ^{3}He are presented. Emphasis is placed on polarization phenomena leading to better understanding of the spin-dependent terms in optical potentials. Recent advances in microscopic theories of nucleon scattering and folding models of composite-particle scattering are surveyed.

1. INTRODUCTION

The goals of an optical-potential parameterization of elastic-scattering data are to produce, starting with the nucleon-nucleon interaction, a global description as a simple function of projectile, bombarding energy E, target charge Z and mass number A. In this review I emphasize the relation between polarization phenomena and the spin-dependent terms in the optical potential. Isospin terms can be elucidated because of the extra constraints provided by polarization data, so these are also discussed. Increased availability of polarized beams at intermediate energies now makes determination of the energy dependence of parameters a realistic goal. Theoretical and empirical progress in the past five years has been rapid, so that, especially for protons, the spectrum -20<E<1000 MeV can be viewed fairly clearly.

Specific nuclear-structure effects on the optical potential are superimposed on global effects for any given E and A. Their elucidation is necessary if the underlying global potential is to be extracted, and their interpretation produces insight into the connection between nuclear structure and nuclear reactions.

For uniformity of notation I write the local optical potential in parameterized form as

$$V(\vec{r}) = -V_0 f(r,r_0,a_0) - iW_s f(r,r_s,a_s) + iW_D g(r,r_D,a_D)$$
$$+ (V_{so} + iW_{so}) g(r,r_{so},a_{so})/r \; \vec{\ell}\cdot\vec{s}/s + V_c(r,r_c). \qquad (1)$$

---

* Research supported in part by the U.S. Department of Energy.

This 14-parameter potential has, in most representations, a Woods-Saxon radial form factor for each f

$$f(r, r_i, a_i) = 1/(1+\exp\{(r-R_i)/a_i\}), \quad R_i = r_i A^{1/3} \quad (2)$$

and $g = df/dr$. In Eq.(1) V_c is the Coulomb potential from a sphere of uniform charge density and radius $r_c A^{1/3}$. For particles with spin greater than 1/2, tensor potentials (Sec.4.2) are added. Each parameter in Eqs.(1) and (2) is implicitly dependent on projectile, E, and A.

In Sec.2 I summarize the data for optical-potential studies; in Sec.3 progress in nucleon-scattering formalism and its comparison with data to investigate central, spin-orbit, spin-spin and isospin components of the nucleon optical potential are reviewed. Developments in composite-particle scattering are presented in Sec.4, in which I also review the formalism of folding models, spin dependence and target-structure effects in (d,d), spin-dependent potentials for triton and helion scattering and heavy ions, as well as the relations between potential integrals for elastic scattering of composite particles and nucleons.

2. DATA RESOURCES

Data on elastic scattering are summarized in Fig.1, plotted against log(E) (appropriate for the energy dependence of potentials at E>50MeV) for E>4 MeV, and against $A^{1/3}$ (indicating the predominance of nuclear sizes in diffractive scattering) for A>6. As a function of E, the large gaps above the tandem accelerator limit and below about 100 MeV/amu are undesirable, because in this energy range scattering is usually less sensitive to target-structure effects than at lower E, but the analysis is not limited by using non-relativistic quantum mechanics.

The gaps in A arise from limitations of target preparation, for example, noble-gas targets (whose nuclear structure is often intriguing), by the difficulty of resolving elastic scattering (as in the rare earths and actinides), and the limited availability of large isotopically-enriched samples for (n,n). The predominance of data for closed-shell nuclei, ^{16}O, ^{40}Ca, ^{90}Zr, Sn isotopes, and ^{208}Pb, prejudices global potential parameterizations. The lack of data and analyses for A>208 hinders extrapolation of nucleon potentials to predict the structure of super-heavy nuclei, especially since this is sensitive to spin-orbit potentials[1]

Polarization observables are involved in fewer than 30% of the more than 1000 angular distributions indicated in Fig.1. The polarization data are concentrated in (p,p) and (d,d) experiments. The number of (n,n) polarization data has been doubled by contributions to this conference. The polarization data are mostly for vector analyzing powers, and very few polarization-transfer or charge-exchange (p,n) or (^{3}He,t) data are available. For heavy ions only a few ^{6,7}Li analyzing power angular distributions at low energies are available, as reviewed in Sec.4.5.

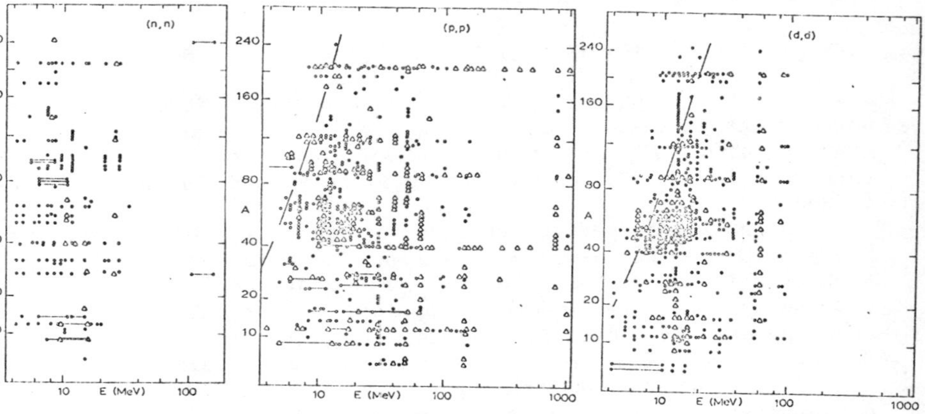

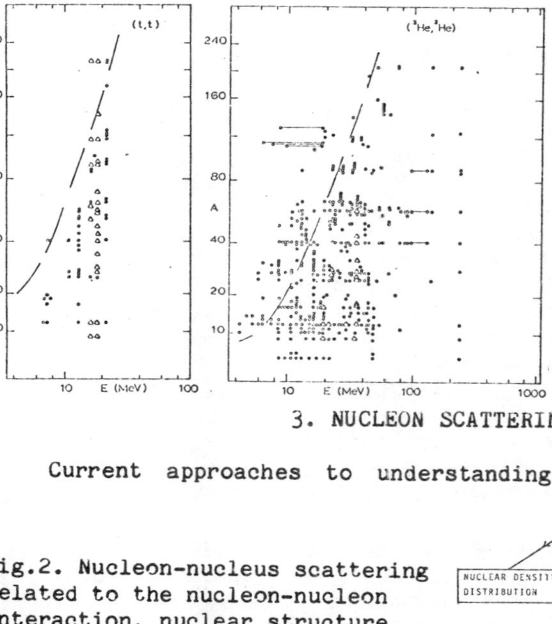

Fig.1. Data resources for light-ion scattering. Angular distributions marked by circles for cross sections and triangles for polarization. Horizontal lines join points with close energy steps. Dashed curves show empirical Coulomb barriers below which elastic polarization effects are small.

3. NUCLEON SCATTERING

Current approaches to understanding nucleon-nucleus optical

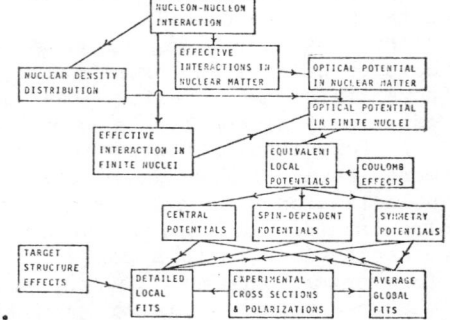

Fig.2. Nucleon-nucleus scattering related to the nucleon-nucleon interaction, nuclear structure, the optical potential, and the analysis of elastic-scattering data.

potentials are summarized in Fig.2.

The past five years have seen much progress in the formalism and its relation to empirical optical potentials.

3.1 Progress in formalism

The theory of microscopic optical potentials was reviewed in topical conferences at Pavia,[2] 1976, and Hamburg,[3] 1978. There are

two major approaches (Fig.2) to constructing this potential:
(1) The nuclear matter approach of Jeukenne, Lejeune and Mahaux (JLM)[4] and of Brieva and Rook (BR)[5] first calculates the complex optical potential in nuclear matter of constant density at energy E; the potential is approximated for finite nuclei by evaluating at each radial point r the potential at the corresponding matter density $\rho(r)$, which is usually that determined from electron scattering. This is called the local-density approximation (LDA), and the details of its use are the main distinction between JLM and BR.
(2) The nuclear-structure approach of Vinh Mau, Bouyssy and Ngo (VBN)[6] in which the optical potential in a finite nucleus is calculated directly, but approximately, and nuclear-surface effects (including absorption) are included explicitly.

Each of the methods, JLM, BR and VBN, allows both real and imaginary central and spin-orbit potentials to be calculated. In JLM the ρ and E dependence of the potential is tabulated, so that empiricists can construct their own potentials for a chosen $\rho(r)$. Each method produces a non-local potential, especially because of exchange effects at low energies, and further approximations are needed to produce the equivalent local potentials favored by phenomenologists. This step introduces an approximately-linear energy dependence of potential depths for E<30 MeV, and predominates over the dynamic energy dependence of the potential depths, which

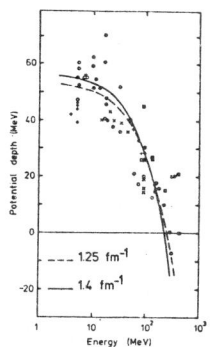

Fig.3. Energy dependence of V_o predicted by the JLM model for two values of the Fermi momentum 1.25 fm^{-1} and 1.4 fm^{-1}, compared with empirical values. (From Ref.4.)

are nearly logarithmic for E>100 MeV, as shown in Fig.3.
Predictions of elastic-scattering observables with no free

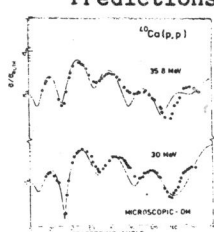 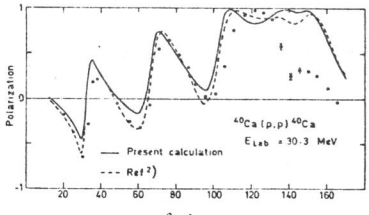

Fig.4. Cross sections and polarizations in the BR theory. For the left curve, variations of 10% in the potentials have been allowed. (From Refs.7, 8.)

parameters are very satisfactory, as indicated in Fig.4. The central absorptive potential is predicted to change from surface to volume absorption in the region $E \sim 50$ MeV. A Coulomb correction predicted for the absorption has been verified by Rapaport[9] in the analysis of nucleon scattering by ^{40}Ca.

Comparison of the three microscopic methods with data has not yet been made systematically. However, the theoretical and empirical volume integrals and rms radii are in fair accord.

3.2 Central potentials

A detailed understanding of nucleon central absorptive and refractive potentials is important for polarization phenomena because of the well-established relations $r_p > r_0 > r_{so}$. A new result in understanding central potentials is that allowing them to be angular-momentum (ℓ) dependent significantly improves cross-section and analyzing-power fits, as shown by Kobos and Mackintosh[10], and by van Hall[11] in a contribution. Much of this dependence can be attributed to coupled reaction channels, such as (p,d)(d,p) in (p,p), and is therefore quite A and E dependent. A means of constructing the optical potential in a fairly model-independent way by combining phase-shift searches with an inversion construction of the optical potential has been suggested recently[12].

Proton absorptive potentials at sub-Coulomb energies are not well understood, as recent data and analyses of very small (p,p) A_y values by the Erlangen group[13], and of (p,n) total cross sections by the Kentucky and Livermore groups[14] have shown. A small value $a_p \sim 0.4$ fm is suggested. Near the Coulomb barrier a non-uniform dV_0/dE arising from non-locality, first discussed by Eck and Thompson[15], has been confirmed by recent investigations[16].

3.3 Nucleon spin-orbit coupling

Recent nucleon A_y data (Fig.1) are of improved quality and greater E and A ranges than previously, which has allowed refined determination of spin-orbit potentials. Absorptive spin-orbit terms are predicted from the BR formalism to have $W_{so}/V_{so} \sim -0.05$ at $E \sim 20$ MeV, which is not inconsistent with the empirical analyses by Jackson and Abdul-Jalil, de Swiniarski and Pham[17].

Small values of a_{so}, $0.2 < a_{so} < 0.4$ fm, are suggested by analyses of sub-Coulomb $(\vec{p},p)$ data[13], by improved fits to $(\vec{n},n)$ data[18], and by the small a_{so} values in $(^3\vec{He},^3He)$ described in Sec.4.3. The origins of this empirical result are unknown.

Nuclear-structure dependence of the strength V_{so} is expected for spin-unsaturated subshells, for which $j=\ell+1/2$ is full but $j=\ell-1/2$ is not. Exchange contributions from two-body spin-dependent interactions modify V_{so} in bound[19] and scattering[20] states, by as much as 20%. The contribution by Sakaguchi et al.[21] is evidence of this effect. Bound-state calculations[22] indicate that r_{so} should decrease slightly between closed shells.

The energy dependence of V_{so} is not well understood. For bound states and E<50 MeV $dV_{so}/dE = 0.023 \pm 0.012$ (Ref.23), attributed to non-locality effects[24], but at higher energies a logarithmic dependence of V_{so} on E is suggested by the IUCF group[25]. At E>100 MeV the spin-orbit potential can produce spin-flip cross sections of the same magnitude but out of phase with non-spin-flip cross sections ,

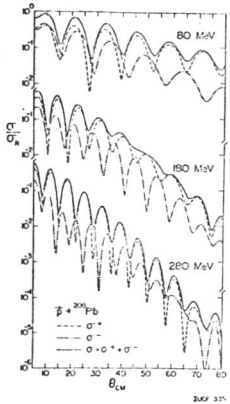

Fig.5. Intermediate-energy cross sections resolved into spin-up and spin-down components, showing the origin of the damped oscillations observed near 180 MeV. (From Ref.26.)

as shown in Fig.5. At E>200 MeV, results from LAMPF and TRIUMF indicate that V_o changes sign at about 600 MeV, in agreement with a microscopic relativistic theory , and that V_{so} changes sign in the same energy region.[28] For 800<E<1000 MeV, the data have been used primarily to determine nuclear-matter distributions and n-p radius differences, as reviewed recently by Whitten[29].

3.4 Spin-spin interaction

This interaction has been generally ignored, because of the difficulty of experiments and interpretation. In nuclear matter the scalar term $V_{ss} f_{ss}^{(o)}(r) \vec{I}.\vec{s}$ is predicted to have V_{ss} between 0 and 2 MeV in magnitude[3,0] but this will be quenched in finite nuclei by core polarization[31]. Experimentally, two techniques have been used: (1) Depolarization, which has been shown[32] to be unambiguously attributed (in the direct-reaction energy regime) to spin-spin potentials only if I=1/2, since for I>1/2 coupling with the nuclear quadrupole moment can produce depolarization. Preliminary data from Osaka[33] on $^{31}P(\vec{p},\vec{p})^{31}P$ indicate $2<V_{ss}<5$ MeV. (2) Spin-spin total cross sections from the transmission of polarized neutron beams through polarized targets have been made so far only for $^{59}\vec{Co}$ and $^{165}\vec{Ho}$, which have large hyperfine magnetic fields. At low energies the data can be interpreted as due to compound-elastic scattering only[3]. At T~10 mK and B~7 Tesla there are 20 nuclei 6≤A≤209 which can be brute-force polarized to >20%. The necessary facilities, described in the review by Heeringa, are under development at Karlsruhe, and at TUNL for the 10- to 20-MeV neutron energy range[35].

3.5 Isospin potentials for nucleons

The isovector interaction in the Lane model, $(V_1+iW_1)f_1(r)\vec{t}.\vec{T}$, produces splitting of the potential between (p,p) and (n,n), and a mechanism for direct (p,n) reactions between isobaric-analog states.

Microscopic derivations of the Lane potential in the nuclear-matter formalism[30,4,36] underestimate V_1 and W_1 by about a factor of 2 compared with empirical potentials for nucleon elastic scattering, (p,n) differential cross sections[3,7] and (n,γ) total cross sections[38]. However, the radial dependence may be appropriate since the (p,n) angular distributions are well fitted[37]. The isovector non-locality range may be 60% larger than that for isoscalar interactions[39], making the transition from non-local to equivalent local potentials (Fig.2) ambiguous. Coulomb effects and energy dependence in (p,p) also obscure genuine isospin dependences.

For light nuclei, a detailed study of the Lane potential for ^{9}Be (Ref.40) was constrained by simultaneous analysis of $d\sigma/d\Omega$ data for (n,n), (p,p) and (p,n) as well as $(\vec{p},p)$ and $(\vec{p},n)$ analyzing powers A_y, and is supplemented by $(\vec{n},n)$ results in a contributed paper[41]. The symmetry potential is primarily of volume form for the the real part and surface form for the imaginary part. Empirically, there is no evidence for isospin dependence of the nucleon spin-orbit potential[23,40]. Also in light nuclei, differences between P_y and A_y in (p,n) reactions of up to 0.1 have been predicted by shell-model calculations[42] and observed in ^{15}N(p,n)^{15}O (Ref.43.) Since $P_y \neq A_y$ requires both isospin-symmetry breaking and transverse spin-flip interactions, it is a sensitive probe of effective interactions in nuclei. Angle-integrated polarization-transfer reactions for cases from ^{19}F$(\vec{p},n)^{19}$Ne to ^{39}K$(\vec{p},n)^{39}$Ca can be fitted by a spin-independent Lane potential, as described in a contribution from Stanford[44].

Elastic (n,n) data for the even isotopes of Sn for 11≤E≤26 MeV have been reported by the Ohio University group[3,7]. The values of V_1 and W_1 are in general agreement with the nuclear-matter calculations. Ambiguities in the A dependence of potential radii did not allow determination of unique isospin potentials, even with the constraints of fitting low-energy (p,p) data and matter radii from 800-MeV proton scattering. A similar ambiguity is indicated for the 1f-2p shell in $(\vec{p},p)$ by a contributed paper[45].

4. COMPOSITE-PARTICLE SCATTERING

4.1 Formalism of folding models

Elastic-scattering potentials for the loosely-bound composite projectiles d, t and ^{3}He are usually constructed directly from nucleon-nucleus optical potentials by averaging over the internal wave function of the projectile ground state. This is called the single-folding model, and relates models and data as indicated in Fig.6. An alternative prescription, the double-folding model, favored for alpha particles and heavy ions,[46] averages nucleon-nucleon effective interactions (Fig.2) over both target and projectile ground-state wave functions.

Projectile-structure effects are important in polarization studies of spin-dependent effects, especially for d and ^{7}Li because of their non-spherical ground states. This aspect of folding models is considered for (d,d) in Sec.4.2, as are those modifications from global potentials which are attributable to target structure. The

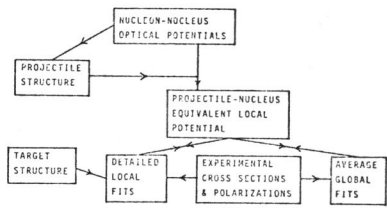

Fig.6. Composite-particle scattering related to nucleon elastic scattering, projectile and target structure, and the analysis of elastic-scattering data. Compare with Fig.2 for nucleon scattering.

puzzling disagreements between single-folding models and the spin-orbit potentials for trion scattering are described in Secs.4.3 and 4.4.

Folding-model potentials which are linear in both projectile and target densities have an interesting property: Integrals of the potentials can be found which are independent of both the radial distributions of the densities and of the particular collision partners. These integrals gauge the overall strength of the interaction and are insensitive to those continuous ambiguities in optical-model analyses which occur for low energies. An empirical potential can thus be compared for consistency with folding models and with potentials for other targets and projectiles. The appropriate integrals for spherically-symmetric central potentials are volume integrals, and for spin-orbit and the usual T_r and T_p tensor potentials r^4-weighted radial integrals are appropriate. Model potentials of Woods-Saxon type have simple analytic expressions[47] for these integrals. They are used in Sec.4.4 to discuss the consistency of phenomenological potentials with nuclear-matter and folding-model estimates of central and spin-orbit potentials.

4.2 Spin-dependent deuteron potentials

The increase in accurate deuteron polarization data (Fig.1) during the past five years, especially over extended E and A ranges, has motivated extensive parameterizations and comparisons with models, especially for the central and spin-orbit potentials, and (to a lesser extent) the tensor potentials. Among such data and analyses are those from Madison[48] ($10 \leqslant E \leqslant 15$ MeV, $46 \leqslant A \leqslant 122$), Zürich[49] (E=12 MeV, $46 \leqslant A \leqslant 90$), Osaka[50] (E=56 MeV, $16 \leqslant A \leqslant 208$). A global parameterization using about 4000 data points for $12 \leqslant E \leqslant 90$ MeV, $27 \leqslant A \leqslant 238$, including recent Indiana data[51] near 80 MeV, has just been published by Daehnick and coworkers[52].

Although folding-model deuteron potentials are in qualitative agreement with empirical potentials, there are four topics which are not well understood; (1) the effects of deuteron breakup, (2) the nature of the imaginary spin-orbit potential, (3) the mechanisms producing tensor analyzing powers, and (4) the effects of target structure on deuteron potentials.

(1) Deuteron breakup affects the central potential at least in the absorptive term[53]; the spin-dependent potentials through the strong ℓ-dependence of breakup (see the contribution by Rawitscher and Mukherjee[54]), and the tensor potentials mainly through the inclusion

of spin dependence (including the deuteron D state) in breakup. The approximations required to solve the bound-continuum coupled-channels problem usually restrict the reliability of the predictions to E>20 MeV, a range where there are relatively few data. As yet there is no prescription for parameterizing breakup potentials suitably for inclusion in search codes. Experiments and analyses of the breakup process ($\vec{d}$,np) would probably give valuable insight into its mechanism.

(2) Imaginary spin-orbit potentials, W_{so}, are expected from three sources: (i) Nucleon-nucleus equivalent local potentials have a complex spin-orbit term arising from complex spin-dependent interactions in nuclei, as discussed in Sec.3.3. Folding with the deuteron wave function will produce a complex (d,d) spin-orbit potential, which should not be strongly ℓ-dependent. This effect does not seem to have been investigated. (ii) Stripping to weakly-bound states coupled to the (d,d) channel can produce significant contributions to vector analyzing powers, as shown by Wong and Quin in a contribution[55]. This source of an effective W_{so} will be largest for Coulomb-barrier energies and will be strongly target-dependent through Q values and available stripping orbitals. (iii) Breakup produces strongly ℓ-dependent W_{so} values which are predominantly

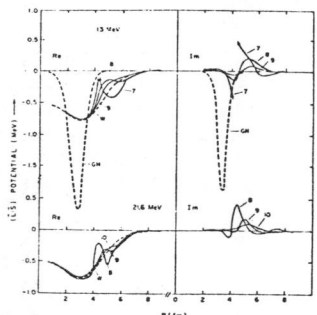

Fig.7. Spin-orbit potentials for ^{64}Ni(d,d)^{64}Ni at E=13 and 21.6 MeV corrected for breakup, labelled by partial waves. The dashed curve is the empirical potential . (From Ref.56.)

positive (Fig.7), of opposite sign, and are an order of magnitude smaller than the empirical potential[57]. However, all of the extensive parameterizations mentioned above, and a contributed paper[58], are in accord with the empirical result.

(3) Deuteron tensor potentials and their relation to tensor analyzing powers T_{2q} are not well understood, partly because there are relatively few data available for systematic analysis, and because second-order effects of V_{so} produce T_{2q} values comparable in magnitude to the data. In a contribution Ohnishi et al[59] suggest new linear combinations of the T_{2q} which may be especially sensitive to tensor potentials. There are three types of tensor potentials for spin-1 particles:

$$T_r(\vec{r}) = F_r(r)\{(\vec{s}.\hat{r})^2 - 2/3\},$$

$$T_p(r) = F_p(r)\{(\vec{s}.\hat{p})^2 - 2/3\},$$

$$T_L(r) = F_L(r)\{(\vec{s}.\vec{\ell})^2 + \vec{s}.\vec{\ell}/2 - 2\ell(\ell+1)/3\}. \qquad (3)$$

The T_r potential arises from folding by including the deuteron D state, and it is the major potential studied so far. Breakup also produces significant T_r potentials[5]. Empirical studies of the T_p form were motivated by the work of Ioannides and Johnson[60] on the effects of the Pauli principle on the S and D states for different deuteron spin alignments relative to its direction of propagation through nuclear matter. Optical-model studies by Goddard[61] have shown that, at low energies and within the approximations made to render the problem tractable in finite nuclei, T_p potentials from this source are difficult to distinguish from T_r potentials. The T_L potential is second-order in the nucleon-nucleus (complex) spin-orbit potential, and is generally believed to have small values of $F_L(r)$ (Ref.62.) However, it is diagonal in j and ℓ, and the splitting between the j=ℓ+1 and j=ℓ-1 states is roughly $3f_L(r)\ell^2/4 \dot{\infty} E$, so that its effects could be straightforwardly investigated and may be significant as E increases.

Empirically, the T_r potential is predominantly imaginary, in disagreement with the folding model and breakup effects. . Since $V_{so}\vec{\ell}.\vec{s}$ increases as E, whereas T_r and T_p are essentially energy-independent, by 80 MeV the spin-orbit interaction dominates in tensor analyzing powers[63].

Target-structure effects in (d,d) have received recent attention. For the Se isotopes enhancement in back-angle iT_{11} at 12 MeV as A increased was attributed[64] to decreasing absorption because of decreasing inelastic scattering to the 2^+ collective states. This interpretation is questioned in a contribution[65] which includes measurement and analysis of this inelastic scattering. The effect is not so clear in the A~90 region, where other neutron-excess effects may be important[66]. Coupled-channel analyses for collective 2^+ and 3^- states with E=20 MeV and 16≤A≤208, and other studies by the Munich group[67], produce a smooth A dependence of the imaginary central potentials. A neutron-shell-closure effect has been found by Daehnick et al.[52] to be best described as an effect on the absorption diffusivity.

4.3 Trion-scattering interactions

The isospin doublet t and ^{3}He (the trion) are considered together. In the past five years $(\vec{t},t)$ data from LASL at 15 and 17 MeV for 18 targets in the mass range 9≤A≤208 and a wide angular range of 20° to 160° have become available[68]. Examples for light nuclei are given in contributions by Hardekopf et al. A comprehensive conventional optical-potential analysis for the 17-MeV data on 40≤A≤208 finds[68] that the central potentials are consistent with folding models, but that the spin-orbit parameters V_{so}=6.2 ± 1.5 MeV, r_{so}=1.10 fm, a_{so}=0.60 ± 0.20 fm, result in spin-orbit integrals about three times larger than a simple folding-model prediction. Coupled reaction channels, such as (t,d)(d,t), can produce large analyzing powers[69], but no systematic analyses have been performed. The evidence for $W_{so} \neq 0$ is weak[68,70].

Analyzing powers in ($^3\vec{He}$,^{3}He) at E=33 MeV have been measured at Birmingham for 11 targets with $9 \leq A \leq 58$ at angles up to 110°, and analyzed by the Birmingham and London groups[71,72]. Conventional optical potentials, guided by the folding model, produce anomalously small $a_{so} \sim 0.2$ fm, as well-defined chi-squared minima in analyzing

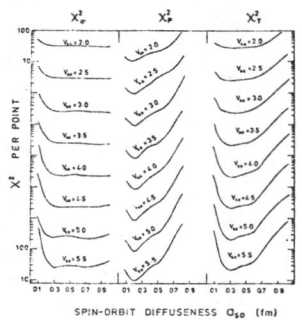

Fig.8. Sensitivity of chi-squared in 33-MeV ^{3}He scattering from ^{58}Ni to spin-orbit diffuseness, shown for fits to cross sections and analyzing powers separately, then together. (From Ref.73.)

powers, shown in Fig.8. However, as shown by Barnwell et al.[72], discreet couples of (r_{so}, a_{so}) can produce similar fits. The small a_{so} occur for both ($\vec{n}$,n) (Sec.3.3) and ($^3\vec{He}$,^{3}He) scattering (^{3}He having an unpaired neutron), but for neither ($\vec{p}$,p) nor ($\vec{t}$,t) scattering (t having an unpaired proton). Coupled reaction channel effects[71], as discussed above for ($\vec{t}$,t), reduce the sensitivity to the value of a_{so}. Nuclear-structure effects on the light nuclei investigated in the ^{3}He studies might also be important. Spline-fitting methods, as used for ^{3}He scattering analyses at higher energies[75] might provide model-independent determinations of trion potentials.

4.4 Volume integrals for elastic scattering

The folding model in Sec.4.1 predicts at a given E/amu, for central potentials invariant volume integrals per interacting nucleon pair, while for the spin-orbit potentials r^4-weighted radial integrals J_4 divided by the fraction of the projectile nucleons with unpaired spins, f_{so}, should be invariant. Comparisons for $E \sim 10 \pm 4$ MeV/amu are given in Table I, where fair agreement for central potentials is indicated, but the spin-orbit integrals are anomalously large (about a factor 3) for both t and ^{3}He scattering, even though they have quite different radial form factors. This result is as intriguing as the small values of a_{so} for ^{3}He.

4.5 Heavy-ion spin-orbit coupling

I consider here only the spin-orbit terms in heavy-ion potentials; surveys of shape (tensor) effects and other aspects of heavy-ion reactions are presented by Steffens, Fick and Graw. The first predictions of heavy-ion spin-orbit potentials and the possibilities of inelastic spin-flip measurements and large elastic analyzing powers were made by Thompson[78] for ^{13}C and ^{6}Li scattering

Table I. Potential integrals for elastic scattering.

Projectile	Central isoscalar J_0	Central isovector J_1	Central absorptive, J_W	Unpaired fraction f_{so}	Spin-orbit J_4/f_{so}	Refs.
n & p	435±35	180				4,47
			100±30			4,76
				1	17±2	47
d	400±50					52
	480±20		115±20			52,76
				1	18±2	47
t	490±10	220±60				47
			140±40			76
	460±15	260±120		1/3	3(20±5)	47
^{3}He	470±30	-210±220				47
	355±55					77
			160±25			76
				1/3	3(16±6)	47

using ^{12}Cαn and ααd single-folding models. Analyzing-power data for ^{6}Li scattering on light nuclei[79] were successfully fitted by such models, and double-folding models for ^{6}Li and ^{13}C were soon developed[80]. In a contribution Moroz et al.[81] describe ^{7}Li scattering data and analyses which indicate a strong influence of $p_{1/2}$ orbitals on the ^{7}Li spin-orbit potential.

The inelastic-spin-flip gamma-ray technique has recently been applied to ^{13}C scattering to the 2^+ states in ^{58}Ni (Ref.82), ^{24}Mg (Ref.83) and ^{12}C (contribution by Tanaka et al.[84]) For inelastic scattering the analysis is very sensitive to the absorptive potential; using conventional heavy-ion absorptive potentials, V_{so} is found to be an order of magnitude larger than folding-model predictions.

In transfer reactions, such as (^{13}C, ^{14}N) and (^{19}F, ^{16}O), the DWBA predictions can be matched to oscillations in the data if entrance-channel spin-orbit coupling is included[85] with a strength two orders of magnitude larger than from folded potentials; for ^{13}C, V_{so} is four times larger than from the spin-flip analysis, while for ^{19}F a large enhancement can be produced if an ^{16}Oαt cluster state is used for ^{19}F and triton-scattering V_{so} values (Sec.3.4) are used[85]; but V_{so} is still a factor of seven smaller than that needed in the transfer reaction.

5. CONCLUSIONS

Progress in understanding the fundamental basis of the optical potential for nucleon scattering is rapid. The relation between nucleon scattering and composite-particle scattering is also becoming clearer, especially for deuteron elastic scattering.

In nucleon scattering the energy region 30<E<1000 MeV is being reconnoitered, allowing tests of the predicted energy dependence of the optical potential, and new insights into scattering phenomena. However, the potential for protons in the sub-Coulomb energy region for medium to heavy nuclei is still unclear. It is also apparent that new polarized-beam data for neutron scattering will help determine the neutron spin-orbit diffuseness parameter a_{so}, and the effects of spin-unsaturated subshells on V_{so}. These three problematic topics also bear on the interpretation of alpha-radioactivity in heavy nuclei and the properties of superheavy nuclei.

In composite-particle scattering, data on (d,d) tensor analyzing powers obtained systematically as a function of E and A would be very helpful in unravelling the contributions from the several reaction processes. Analyzing powers for trion scattering are anomalously large compared with folding-model predictions, as are the spin-dependent effects observed in heavy-ion elastic scattering. A solution to both these puzzles may lie in the effects of absorption potentials, if these are being overestimated.

I thank M. Suskin for help in preparing Fig.1 and G. Amdahl for help with typing.

REFERENCES

1. J. R. Nix, in Proc. Int. Conf. on Dynamical Properties of Heavy-Ion Reactions, Johannesburg, 1978, LA-UR-78-1689.
2. 'Nuclear Optical Model Potential', ed. S. Boffi & G. Passatore, Springer Lecture Notes in Physics, 55 (1976).
3. 'Microscopic Optical Potentials', ed. H. V. von Geramb, Springer Lecture Notes in Physics, 89 (1979).
4. C. Mahaux in Ref.3, p.1.
5. F. A. Brieva in Ref.3, p.84.
6. N. Vinh Mau in Ref.3, p.40.
7. F. A. Brieva, H. V. von Geramb & J. R. Rook, Phys. Lett. 79B, 177 (1978).
8. F. A. Brieva & J. R. Rook, Nucl. Phys. A297, 206 (1978).
9. J. Rapaport, Phys. Lett. 92B, 233 (1980).
10. A. M. Kobos and R. S. Mackintosh, J. Phys. G5, 97 (1979).
11. P. J. van Hall et al., contribution to this conference.
12. R. S. Mackintosh, J. Phys. G5, 1587 (1979).
13. W. Kretschmer et al., Phys. Lett. 87B, 343 (1979); W. Drenckhahn et al., Nucl. Phys. A339, 13 (1980); A. Feigel et al., contribution to this conference.
14. R. Schrils et al., Phys. Rev. C20, 1706 (1979); S. M. Grimes, Phys. Rev. C, August 1980.
15. J. S. Eck & W. J. Thompson, Nucl. Phys. A237, 83 (1975).
16. B. Gyarmati et al., J. Phys. G5, 1225 (1979); S. Kailas et al., Phys. Rev. C20, 1272 (1979).

17. D. F. Jackson and l. Abdul-Jalil, J. Phys. G5, 1699 (1979); R. de Swiniarski & D-L. Pham, contribution to this conference.
18. W. Tornow et al., contribution to this conference.
19. R. R. Scheerbaum, Phys. Lett. 63B, 381 (1976), & refs. therein.
20. W. G. Love, Phys. Rev. C20, 1638 (1979).
21. H. Sakaguchi et al., contribution to this conference.
22. J. Dudek, W. Nazarewicz & T. Werner, Nucl. Phys. A341, 253 (1980).
23. S. G. Cooper & P. E. Hodgson, J. Phys. G6, L21 (1980).
24. J. A. Ramirez & W. J. Thompson, J. Phys. G6, L113 (1980).
25. P. Schwandt et al., contribution to this conference.
26. A. Nadasen et al., to be published.
27. D. A. Hutcheon et al., contribution to this conference.
28. M. Jaminon, C. Mahaux & P. Rochus, Phys. Rev. Lett. 43, 1097 (1979).
29. C. A. Whitten, Jr., Nucl. Phys. A335, 419 (1980).
30. J. Dabrowski & P. Haensel, Can. J. Phys. 52, 1768 (1974).
31. G. R. Satchler, Particles & Nuclei, 1, 397 (1971).
32. J. S. Blair, M. P. Baker & H. S. Sherif, Phys. Lett. 60B, 25 (1975).
33. H. Nakamura et al., in Proc. of the Tsukuba Symposium on Polarization Phenomena in Nuclear Reactions (1979), ed. K. Yagi, p.50.
34. W. J. Thompson, Phys. Lett. 65B, 309 (1976).
35. C. R. Gould et al., contribution to this conference.
36. F. A. Brieva & R. G. Lovas, Nucl. Phys. A341, 377 (1980).
37. J. Rapaport et al., Nucl. Phys. A341, 56 (1980).
38. D. R. Chakrabarty & S. K. Gupta, Nucl. Phys. 83B, 271 (1979).
39. M. M. Giannini, G. Ricco & A. Zucchiatti, Ann. Phys. (N.Y.) 124, 208 (1980).
40. R. C. Byrd, R. L. Walter & S. R. Cotanch, Phys. Rev. Lett. 43, 260 (1979).
41. C. E. Floyd et al., contribution to this conference.
42. R. J. Philpott & D. Halderson, Phys. Rev. Lett. 43, 1785 (1979).
43. R. C. Byrd & R. L. Walter, contribution to this conference.
44. J. W. Hugg & S. S. Hanna, contribution to this conference.
45. T. Noro et al., contribution to this conference.
46. G. R. Satchler & W. G. Love, Phys. Rep. 55, 183 (1979).
47. W. J. Thompson, Phys. Lett. 85B, 180 (1979).
48. R. P. Goddard & W. Haeberli, Nucl. Phys. A316, 116 (1979).
49. H. R. Buergi et al., Nucl. Phys. A334, 413 (1980).
50. K. Hatanaka et al., Nucl. Phys. A340, 93 (1980), and contribution to this conference.
51. E. J. Stephenson et al., contributions to this conference.
52. W. W. Daehnick, J. D. Childs & Z. Vrcelj, Phys. Rev. C21, 2253 (1980).
53. G. Baur et al., Phys. Rev. C21, 2668 (1980).
54. G. H. Rawitscher & S. N. Mukherjee, contribution to this conference.
55. W. H. Wong & P. A. Quin, contribution to this conference.

56. G. H. Rawitscher & S. N. Mukherjee, Ann. Phys. (N.Y.) 123, 330 (1979); Nucl. Phys. A342, 90 (1980).
57. R. P. Goddard & W. Haeberli, Phys. Rev. Lett. 40, 701 (1978).
58. W. W. Daehnick, contribution to this conference.
59. H. Ohnishi, M. Tanifuji & H. Noya, contribution to this conference.
60. A. A. Ioannides & R. C. Johnson, in Ref.3, p.244; R. C. Johnson, in Proc. of the European Symp. on Few Body Problems in Nuclear and Particle Physics, Sesimbra, Portugal, June 1980.
61. R. P. Goddard, Nucl. Phys. A291, 13 (1977).
62. A. P. Stamp, Nucl. Phys. A159, 399 (1970).
63. E. J. Stephenson et al., IUCF preprint No. 108, July 1980.
64. J. Nurzynski et al., Phys. Lett. 67B, 23 (1977); H. Ohnishi, H. Noya & M. Tanifuji, Phys. Lett. 76B, 256 (1978).
65. Y. Tagishi et al., contribution to this conference.
66. J. A. Bieszk & J. Ulbricht, Madison preprint, May 1980.
67. H. Clement, contribution to this conference; R. Frick et al., contribution to this conference.
68. R. A. Hardekopf et al., Phys. Rev. 21C, 906 (1980); R. A. Hardekopf et al., contribution to this conference.
69. J. R. Shepard & J. J. Kraushaar, U. Colorado NPL Progress Report (1976) p.70.
70. J. Meyer & E. Elbaz, contribution to this conference.
71. Y-W. Lui et al., Nucl. Phys. A333, 205 (1980), and contributions to this conference on He scattering.
72. J. Barnwell et al., J. Phys. G5, L69 (1979).
73. S. Roman et al., Nucl. Phys. A284, 365 (1977).
74. N. M. Clarke, M. D. Cohler & R. J. Griffiths, J. Phys. G5, 1233 (1979).
75. A. Djaloeis and S. Gopal, submitted to Nucl. Phys. A, May 1980.
76. S. Kailas, preprint, April 1980.
77. H-J. Trost et al., Nucl. Phys. A337, 377 (1980).
78. W. J. Thompson, in Proc. of the Symp. on Heavy Ion Elastic Scattering (Rochester, New York, 1977) ed. R. M. DeVries, p.321, and refs. therein.
79. W. Weiss et al., Phys. Lett. 61B, 237 (1976).
80. H. Amakawa & K-I. Kubo, Nucl. Phys. A266, 521 (1976); P. J. Moffa, Phys. Rev. C16, 1431 (1977); F. Petrovich et al., Phys. Rev. C17, 1642 (1978).
81. Z. Moroz et al., contribution to this conference.
82. C. Chasman, P. D. Bond & K. W.Jones, Bull. Am. Phys. Soc. 20, 55 (1975).
83. R. Albrecht et al., MPI Annual Report, Heidelberg, 1976; W. Duennweber et al., Phys. Rev. Lett. 43, 1642 (1979).
84. M. Tanaka et al., contribution to this conference.
85. S. Kubono et al., Phys. Rev. Lett. 38, 817 (1977).
86. M. B. Golin & S. Kubono, Phys. Rev. C20, 1347 (1979).

DISCUSSION

DAEHNICK: I was very intrigued by your comment that new and better optical potentials may differ from the traditional Woods-Saxon shape. Do data of higher energies point to specific changes? What evidence is available?

THOMPSON: From alpha-particle scattering at energies above 100 MeV it has been shown that a Woods-Saxon-squared form factor is appropriate for alpha particles, which is in agreement with folding models. Similar investigations should be made for light ions.

CRAMER: I wonder if you would comment on the possible non-locality of the spin orbit potential (particularly in the proton + nucleus potential), on its effects on the energy dependence of the spin orbit potential as compared to the central potential, and on possible spin-dependent effects in the reaction channels arising from different amounts of Perey damping in the different spin channels.

THOMPSON: Non-locality of the nucleon spin-orbit potential has been studied only in local-energy approximation as it influences the energy dependence of spin-orbit well depths (Refs. 23 and 24). At low energies the spin-orbit non-locality is smaller than that of the central potential, which would likely reduce its spin-dependent effects in reaction channels.

SLOBODRIAN: Is there evidence of improvement in reaction theory, like DWBA, using their more refined potentials in the calculation of wave functions for the initial and final states?

THOMPSON: The refinements in the nucleon optical potential in Refs. 4, 5 and 6 concern mostly the dynamic origin and strength of real and imaginary potential depths. The geometric parameters of the potentials are determined primarily by empirical nuclear density distributions, as indicated in Fig. 2. Since DWBA fits, at a given energy, are often dominated by geometric effects, improvements might be difficult to detect. However, over energy ranges $\sim$ 100 MeV potential depth changes as indicated in Fig. 3, would be evident in reactions.

POLARIZATION EFFECTS IN INELASTIC PROTON SCATTERING
AT INTERMEDIATE ENERGIES

A. D. Bacher
Indiana University, Bloomington, IN   47401

ABSTRACT

A review is presented of recent measurements of proton inelastic scattering for bombarding energies between 80 and 800 MeV. In this energy region, data of sufficient precision are now available to allow a test of some features of the impulse approximation description of the inelastic excitation of both natural and unnatural parity transitions. Attention is focused on specific examples where polarization measurements are available. These measurements provide a more stringent test of the local t-matrix interaction than do cross-section measurements alone. The sensitivity of the proton probe to nuclear structure effects is demonstrated and several promising new directions are identified which, when considered along with information from electron and pion probes, should enable new types of nuclear structure information to be extracted.

INTRODUCTION

Since the pioneering work by Watson[1] and by Kerman, McManus, and Thaler[2], there has been the hope of understanding the elastic and inelastic scattering of fast nucleons from nuclei in terms of free nucleon-nucleon scattering and functions describing the distribution of nucleons in the nucleus. Such a microscopic understanding would enable one to use the nucleon probe to examine new features of nuclear structure and to investigate effects due to the modification of the nucleon-nucleon interaction in the nuclear medium.

Attempts at obtaining a microscopic interpretation of measurements at low energies (<80 MeV) have been frustrated by the difficulty in establishing a direct connection between the effective nucleon-nucleon interaction and free nucleon-nucleon scattering. The intermediate energy region (80-800 MeV) has attracted considerable attention for studies of this nature for several reasons. First of all, theoretical considerations suggest that the impulse approximation, in which the effective interaction is simply the free nucleon-nucleon t-matrix, may provide a suitable framework for initiating the microscopic description of nucleon-nucleus scattering in this energy region. In addition, experimental measurements of sufficiently high quality to provide tests of the microscopic descriptions have recently become available from new accelerators such as IUCF (60-200 MeV), TRIUMF (200-500 MeV), and LAMPF (500-800 MeV). The availability of polarized beams at these facilities has in many cases enabled measurements of spin-dependent observables which have not previously been accessible at intermediate energies. Measurements of these spin-dependent observables (e.g. analyzing powers and spin-flip probabilities) can be expected to provide more stringent tests of the microscopic descriptions for both elastic and inelastic scattering.

The present paper will focus on attempts to understand the mechanism for inelastic proton scattering at intermediate energies. A microscopic description which employs the distorted wave impulse approximation and a local representation of the free nucleon-nucleon t-matrix will be briefly reviewed. This form of the interaction, which includes central, spin-orbit, and tensor components, will be examined for a variety of nuclear transitions. These transitions serve to isolate particular components of the effective interaction and thereby allow a comparison with the predicted energy dependence to be made.

The incident energy range from 100-200 MeV will be emphasized since the nucleon-nucleon parameters are well known and there are indications that the impulse approximation is beginning to provide a reasonable description. Few inelastic transitions have been examined at TRIUMF and those from LAMPF occur at an energy where the spin-dependent parts of the nucleon-nucleon interaction are not as well determined. Emphasis will also be placed on those states of well-known character for which existing electron scattering results can be used to constrain the nuclear structure parameters in the microscopic calculations.

Examples of the types of transitions to be discussed are shown in Fig. 1 for the inelastic scattering of protons from ^{208}Pb. These include the low-lying collective states (3^- and 5_1^-), natural parity states of both collective and single particle character ($2^+ \rightarrow 12^+$), and high-spin unnatural parity states (12^- and 14^-) of the "stretched" configuration. One of the most exciting features of inelastic proton scattering at these energies is the way in which high-spin particle-hole states of rather pure configuration are selectively excited at large momentum transfer. As will become apparent in subsequent sections, the observation of these states presents an excellent opportunity to test features of the DWIA description of inelastic scattering.

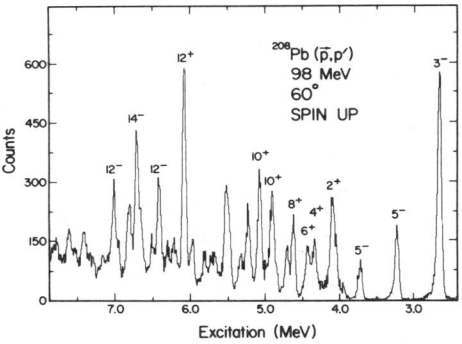

Fig. 1. The inelastic proton spectrum for the scattering of 98 MeV polarized protons from ^{208}Pb.

EFFECTIVE INTERACTION FORMALISM

The description of nucleon-nucleus scattering in terms of free nucleon-nucleon scattering is complicated and a discussion of the various prescriptions which make use of the impulse approximation is beyond the scope of the present paper. Those prescriptions which provide this connection through a multiple scattering formalism[2,3] are more appropriate at higher energies (e.g. 400-800 MeV) where the multiple scattering expansion is expected to converge and various corrections have been shown to be small.[4] In the energy range of primary interest here (100-200 MeV), where the multiple scattering expansion is less appropriate, an alternative formulation in which exchange effects are treated explicitly has been developed. This formulation has been extensively refined by Love and Petrovich and is discussed in detail elsewhere.[5] Only a few essential features of the formalism will be presented here in order to establish the dependence of nucleon-nucleus inelastic scattering on specific parts of the nucleon-nucleon interaction.

In the impulse approximation employed here[5,6], the effective interaction (v) is assumed to have a local, complex form with parameters adjusted to reproduce each spin and isospin state of the on-shell, free nucleon-nucleon t-matrix. As such, v must explicitly include central (C), spin-orbit (LS), and tensor (T) terms

$$v = v_0^C(r) + v_\sigma^C(r)\vec{\sigma}_1\cdot\vec{\sigma}_2 + v_\tau^C(r)\vec{\tau}_1\cdot\vec{\tau}_2 + v_{\sigma\tau}^C(r)\vec{\sigma}_1\cdot\vec{\sigma}_2\vec{\tau}_1\cdot\vec{\tau}_2$$
$$+ \{v_0^{LS}(r) + v_\tau^{LS}(r)\vec{\tau}_1\cdot\vec{\tau}_2\}\vec{L}\cdot\vec{S} + \{v_0^T(r) + v_\tau^T(r)\vec{\tau}_1\cdot\vec{\tau}_2\}S_{12} \quad (1)$$

with the analytic form of the radial dependence taken to be a sum of Yukawa terms (C and LS) or r^2 times a sum of Yukawa terms (T). The long range portion is constrained to have the form of OPEP while the shorter range terms are adjusted to provide a fit to the nucleon-nucleon scattering. Such a representation is especially reliable at energies below about 500 MeV where new nucleon-nucleon measurements are now available.[7]

This effective interaction is then combined with anti-symmetrized nuclear wave functions and scattering wave functions to determine the transition amplitude and hence obtain expressions for the differential cross section, $\sigma(\theta)$, the analyzing power, $A(\theta)$, and other spin-dependent observables. Following Petrovich,[5] we reproduce here Born approximation expressions for the cross sections for $0\to J$ natural parity transitions (no spin flip)

$$\sigma(\theta) = 4\pi\left(\frac{\mu}{2\pi\hbar^2}\right)^2 \frac{k_f}{k_i}(2J+1)\left\{\left|\sum_\alpha v_\alpha^C(q)\rho_\alpha^J(q)\right|^2 + \left|\sum_\alpha v_\alpha^{LS}(q)\rho_\alpha^J(q)\right|^2\right\} \quad (2)$$

and for $0\to J$ unnatural parity transitions

$$\sigma(\theta) = 4\pi\left(\frac{\mu}{2\pi\hbar^2}\right)^2 \frac{k_f}{k_i}(2J+1)\left\{\left|\sum_{\lambda,\alpha} v_\alpha^{LS}(q)B_\lambda \rho_{J\lambda}^{s\alpha}(q)\right|^2 + \sum_\lambda\left|\sum_\alpha [v_\alpha^C(q)\rho_{J\lambda}^{s\alpha}(q)\right.\right.$$
$$\left.\left. + v_\alpha^T(q)\sum_{\lambda'} Z_{\lambda\lambda'}^J \rho_{J\lambda'}^{s\alpha}(q) + v_\alpha^{LS}(q)A_\lambda \rho_J^{\ell\alpha}(q)]\right|^2\right\} \quad (3)$$

since they serve to illustrate the dependence of these inelastic transitions on the components of the effective interaction and on the nuclear structure. In these expressions, α = p (proton) or n (neutron), λ is the orbital angular momentum transfer, and the v(q) are the appropriate Bessel transforms of the interaction components defined in Eq. (1).

The dependence of inelastic proton scattering on nuclear structure is contained in the transition density functions $\rho(q)$. For natural parity transitions, $\rho_p^J(q)$ and $\rho_n^J(q)$ correspond to the proton and neutron transition densities. Measurements of the longitudinal electron scattering form factor can be used to deduce $\rho_p^J(q)$. While it is usually assumed that $\rho_n^J(q) \simeq \rho_p^J(q)$, one can in principle use the electron and proton probes to examine possible differences in the two transition densities. Strictly speaking, these natural parity transitions also contain a term involving a transverse form factor that couples to spin and current distributions in the target, but these effects are small and have been neglected.

For the unnatural parity transitions the longitudinal form factor vanishes. However, the dependence of the transverse form factor on the orbital current ($\rho_{J\lambda}^{\ell\alpha}$) and spin ($\rho_{J\lambda}^{s\alpha}$) transition densities is complicated and does not allow a direct connection to be made between electron and proton inelastic scattering. For transitions to particle-hole excited states of "stretched" configuration (i.e., $0 \to (j_p j_h^{-1})J_{max}$), only the λ = J-1 spin transition density is non-zero. For this case it is again possible to obtain a direct connection between the electron and proton inelastic scattering.

The calculations of proton inelastic scattering presented here have been performed in the framework of the distorted wave impulse approximation (DWIA) primarily using a modified version of the code DWBA-70.[8] Several effects that have been necessarily neglected in the previous Born approximation expressions have been properly treated in the DWIA calculations. These include: (1) distortion effects, which are incorporated in the form of phenomenological optical potentials derived from elastic scattering measurements,[9] and (2) exchange effects, which are explicitly included by an exact treatment of the knockon exchange amplitudes.

The primary motivation of the present work is to investigate the validity of the DWIA description of inelastic scattering by examining transitions to a variety of well-known nuclear states. As input, one necessarily requires an accurate knowledge of: (1) the nucleon-nucleon interaction components, (2) the effects of distortion on the scattered waves, and (3) the transition densities from electron scattering. It should be pointed out that tests of this detailed a nature have only recently become possible and that effects that have not explicitly been incorporated here may take on a new significance. The influence of the nuclear medium on the effective nucleon-nucleon interaction can arise from effects such as Pauli blocking, meson exchange processes, and fluctuations in the nuclear pion field. These effects may be enhanced by studying inelastic transitions that depend on specific components of the nucleon-nucleon interaction. Detailed comparisons of electron and proton inelastic scattering can also be expected to improve our knowledge of nuclear wave functions. For transitions where the DWIA provides an accurate description, use may be made of the complementary sensitivity of the electron and proton probes to determine neutron and proton transition densities.

The manner in which the individual components in the effective nucleon-nucleon interaction contribute to inelastic transitions can be understood by examining the momentum transfer dependence of the magnitudes of these components at a fixed proton energy. This is illustrated in Fig. 2 for an energy of 140 MeV. Central (C), spin-orbit (LS), and tensor (T) components are grouped according to spin and isospin transfer to the target. The effects of knockon exchange can be approximately represented in a plot of this type only for central and spin-orbit components. The excitation of natural parity states (usually dominated by $\Delta S=0$) occurs primarily through the central component at small q; however, for both isoscalar and isovector transitions, the spin-orbit becomes more important at larger q. For unnatural parity transitions (only $\Delta S = 1$), the central component is again dominant at small q. At larger values of q, the spin-orbit component is seen to be dominant for isoscalar transitions, while for isovector transitions both spin-orbit and tensor components can be expected to contribute.

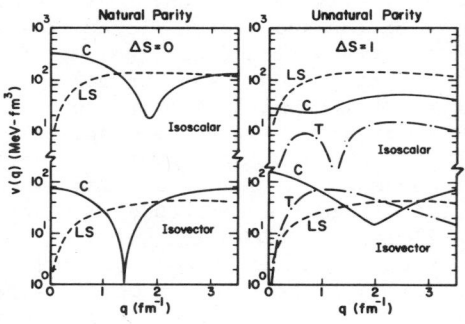

Fig. 2. Momentum transfer dependence of the magnitudes of the effective interaction components at 140 MeV. (From ref. 5.)

The relative importance of the central (C) and non-central (LS and T) components of the effective interaction in exciting both natural and unnatural parity transitions is expected to exhibit an energy dependence which reflects that of free nucleon-nucleon scattering. A schematic representation of this energy dependence is shown in Fig. 3 as a function of proton bombarding energy. In order to simplify the discussion of the momentum and energy dependence of the individual components of the t-matrix, $|t(q,E)|$, the central term has been evaluated at its maximum value

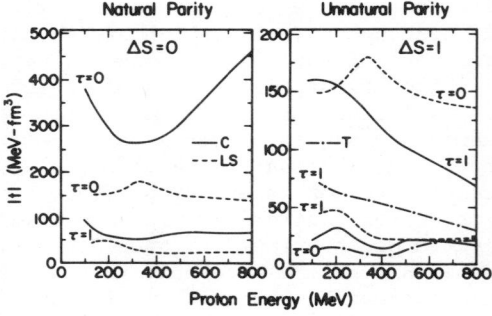

Fig. 3. Schematic energy dependence of the effective interaction components. (From ref. 6.)

(q = 0) while the non-central terms have been evaluated at their maximum value for the range $0 \lesssim q \lesssim 2 \mathrm{fm}^{-1}$. The most significant feature of this energy dependence occurs in the central components of the interaction. For isoscalar natural parity transitions, this contribution is observed to decrease at low energies and to increase dramatically at energies above 400 MeV. The resulting minimum in the energy range 200-400 MeV may provide an energy window in which unnatural parity transitions are enhanced relative to the background of natural parity transitions. This plot should be used with caution, however, since the relative excitation of particular states may involve comparisons of the force components at fixed q rather than comparisons of the maximum values depicted here.

Changes in the relative magnitudes of the components in the effective interaction as a function of energy can be expected to significantly alter the relative population of states in the inelastic spectra observed at different bombarding energies. Recent measurements of proton inelastic scattering to the unnatural parity 6^-, T = 0 and T = 1 levels in ^{28}Si have been obtained for 180 MeV protons at IUCF[10] and for 800 MeV protons at LAMPF.[11] Spectra measured at the maxima of the angular distributions are shown in Fig. 4. These results, which demonstrate the selective excitation of unnatural parity states at the lower bombarding energies, tend to confirm the qualitative predictions of the energy dependence presented in Fig. 3.

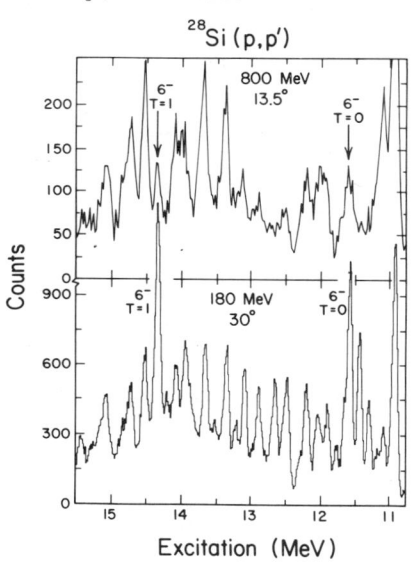

Fig. 4. Inelastic proton spectra for the scattering of 180 MeV and 800 MeV protons from Silicon.

The next two sections discuss specific applications of the effective interaction formalism to the excitation of natural and unnatural parity states. Some additional information on the energy dependence of the excitation mechanism is presented in the final section on future areas of interest.

NATURAL PARITY TRANSITIONS

In order to directly test the DWIA formalism for inelastic scattering reviewed in the previous section, it is necessary to separate effects due to nuclear structure from those related to the description of the effective interaction. This test is most readily accomplished by examining natural parity transitions for which a direct connection can be established between electron and proton inelastic scattering.

Inelastic scattering to the natural parity states in ^{208}Pb provides an excellent testing ground for examining the momentum dependence of the spin-independent central and spin-orbit components of the effective nucleon-nucleon interaction. The simultaneous consideration of transitions to levels with spins covering the range J = 2 to 10 allows a wide range of momentum transfer to be sampled. (See Fig. 1 in the previous section.) Proton inelastic scattering to these states has been measured[12] at 135 MeV and a microscopic description[13] of these excitations has been carried out using transition densities derived from inelastic electron scattering along with the assumption that $\rho_n^J(q) = \rho_p^J(q)$. The proton transition densities are shown in Fig. 5 along with the magnitudes of the components of the effective interaction for $\Delta S = 0$. It is clear from an inspection of Fig. 5 that a measurement of these transitions will be most sensitive to the interaction components over the range $q \approx 0.5 - 2.0$ fm^{-1}. Over this range of q, the scattering cross sections should exhibit a transition from a

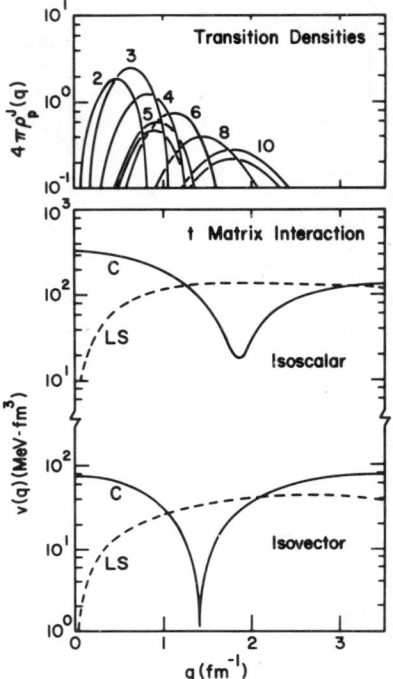

Fig. 5. Proton transition densities for the J = 2 to 10 transitions in ^{208}Pb from electron scattering and the magnitudes of the $\Delta S = 0$ components of the effective interaction.

central interaction shape to a spin-orbit dominated shape since the isoscalar central interaction changes sign near q = 1.8 fm^{-1} due to the short-range repulsion in the nucleon-nucleon interaction. This effect is apparent in the DWIA calculations for the 3$^-$ and 10$^+$ transitions displayed in Fig. 6. The solid curves are the result of using (C + LS) components; the dashed curves use (C) only. The generally good agreement is especially exciting because it provides an explanation in terms of a microscopic formalism for the dramatic change in shape of the high-spin transitions that has been noted in earlier collective model analyses.[12]

The calculations described above reproduce the essential features of the measurements for all of the transitions. However, there are some indications, especially evident in the details of the calculations for the low-spin transitions (J≤6), that corrections to the formalism may be required. Modifications to the low-q portion of the central components of the effective interaction, such as might arise from Pauli blocking effects, would improve the overall agreement. The need for corrections of this type has also been noted in impulse approximation calculations of the elastic scattering.[9,14] These effects may also be related to the difficulty in describing recent analyzing power measurements[15] for the low-spin transitions. Further discussion of the need for these corrections will be presented in a later section.

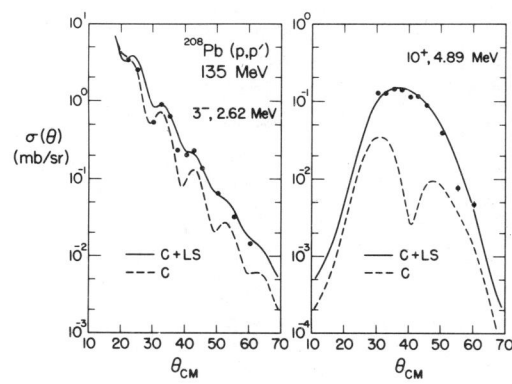

Fig. 6. DWIA calculations for the 135 MeV proton inelastic excitation of 3^- and 10^+ levels in ^{208}Pb.

An additional explanation for some of the small differences between the DWIA calculations and the experimental results may arise from nuclear structure effects. These effects are evident in differences in the shapes of the cross sections for the 5_1^- and 5_2^- excitations at 3.198 and 3.709 MeV. As is indicated

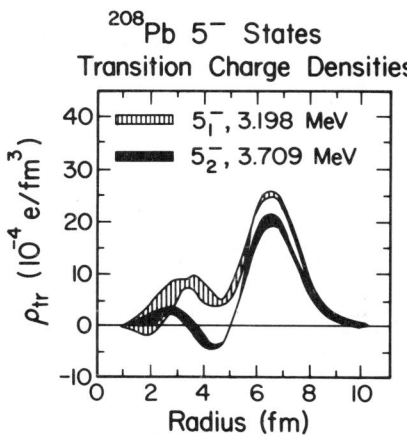

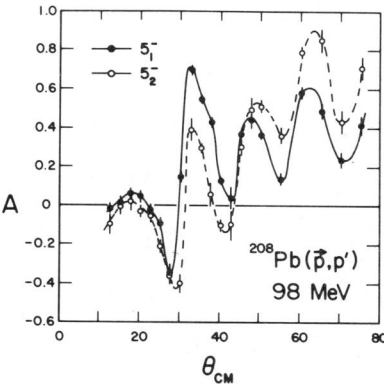

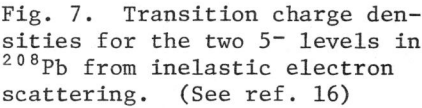

Fig. 7. Transition charge densities for the two 5^- levels in ^{208}Pb from inelastic electron scattering. (See ref. 16)

Fig. 8. Analyzing powers for the inelastic excitation of the two 5^- levels in ^{208}Pb.

in Fig. 7, the transition charge densities deduced from electron scattering[16] indicate substantial differences between these two states. Recent analyzing power measurements for these two transitions are shown in Fig. 8. Striking differences are observed which indicate the sensitivity of analyzing power measurements to nuclear structure effects. However, it has not yet been possible to understand the observed differences by using the electron scattering measurements alone. This may indicate additional differences between the assumed forms of the proton and neutron transition densities.

More stringent tests of the DWIA formalism can be made by studying both cross section and analyzing power measurements for the excitation of high-spin particle-hole states of known configuration in an N = Z nucleus where the assumption $\rho_n^J(q) = \rho_p^J(q)$ will be valid. An example of such an excitation is the 5^-, T = 0 state at 9.70 MeV in ^{28}Si which has recently been examined by both electron and proton inelastic scattering. The longitudinal form factor for this natural parity transition has been measured in inelastic electron scattering[17] and is well described by an open-shell RPA calculation which contains a dominant $(f_{7/2}, d_{3/2}^{-1})$ particle-hole component with small additional components of $(f_{7/2}, d_{5/2}^{-1})$ and $(f_{5/2}, d_{5/2}^{-1})$. The transverse form factor is small, indicating that $\Delta S = 1$ contributions to the excitation are not significant.

The results of the 134 MeV proton inelastic scattering[18] to this level are displayed in Fig. 9. The solid curves represent DWIA calculations assuming a $(f_{7/2}, d_{3/2}^{-1})$ configuration. The individual contributions of the central (C), spin-orbit (LS), and tensor (T) components to the cross section are indicated. Both the central and spin-orbit components

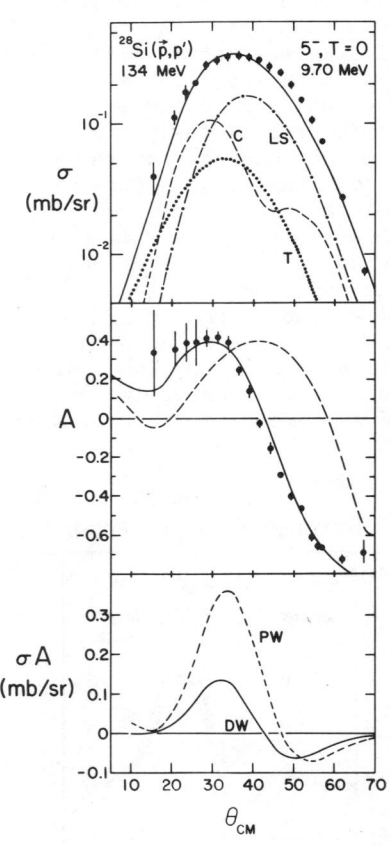

Fig. 9. Cross section and analyzing power measurements for the 5^-, T = 0 level in ^{28}Si. The curves are discussed in the text.

are seen to be important. The DWIA calculation reproduces the shape but overestimates the magnitude of the cross section. A renormalization factor of 0.576 is required relative to the (e,e') result, indicating that some corrections to the effective interaction may be required for this natural parity transition.

For a natural parity transition in which transverse contributions can be neglected, the analyzing power in the plane-wave impulse approximation (PWIA) is given by[2,5]

$$A(\theta) = \frac{2\left(t_R^{LS} t_I^C - t_I^{LS} t_R^C\right)}{|t^C|^2 + |t^{LS}|^2}$$

This expression shows that the analyzing power contains information about the interference between the spin-orbit and central terms in the effective interaction. For this case one expects large analyzing powers since the real part of the spin-orbit term (t_R^{LS}) and the imaginary part of the central term (t_I^C) are substantial.

This expectation is borne out by the measurements for $A(\theta)$ presented in Fig. 9. Excellent agreement is obtained using the full DWIA calculation that is indicated by the solid curve. The most significant feature of these results is the zero crossing observed near 43°. The origin of this effect is associated with the short-range repulsion in the nucleon-nucleon force which produces a change in sign of the imaginary part of the spin-independent central component of the effective interaction (see Fig. 2) at a momentum transfer of 1.8 fm^{-1}. The dashed curve, which was obtained using an imaginary central interaction of zero range that has no sign change, shows the sensitivity of the analyzing power to this feature of the central component. Since the two forms of the central interaction provide an equally good description of the cross section (recall that the spin-orbit component dominates here), analyzing power measurements can provide new information about the form of the effective interaction.

The effect of distortion is best illustrated by referring to plots of σA for both PWIA and DWIA calculations. These are shown at the bottom of Fig. 9. For the 5$^-$ transition, the main effect of distortion is to decrease σA by about a factor of 2, primarily through the reduction of σ. Such a reduction leaves intact the shapes of the σ and A angular distributions.

The examples discussed so far have referred to inelastic scattering measurements from IUCF in the energy range 100-200 MeV. At higher energies, near 800 MeV, a number of impulse approximation analyses of elastic and inelastic proton-nucleus scattering have been reported[4] and measurements of the elastic scattering have been used extensively for the purpose of obtaining ground-state neutron density distributions.[4,19] For a few cases, analyzing power measurements have been obtained[20] for inelastic proton scattering to low-lying collective states and these have been interpreted using collective model form factors. Progress in obtaining a microscopic description of these transitions has been hampered by an inadequate knowledge of the spin-dependent terms in the nucleon-nucleon interaction near 800 MeV. For the case of natural parity transitions in ^{16}O, DWIA calculations[21] have demonstrated the importance of spin-orbit components in the nucleon-nucleon interaction for explaining forward angle analyzing powers.

A recent analysis[22] of proton inelastic scattering at 800 MeV suggests that for collective excitations a simple relationship exists between analyzing power and cross section angular distributions. This model implies that there is little independent nuclear structure information to be obtained from asymmetry measurements that is not already contained in measurements of the cross section. An example of this data-to-data prescription is illustrated in Fig. 10 for low-lying states in ^{54}Fe. For the 2^+ and 3^- transitions the agreement is comparable to that obtained with a collective model analysis.[20] For the 4^+ transition, which is less collective, the quality has deteriorated.

Data-to-data prescriptions of this type have a general utility in providing a simple explanation for some analyzing power measurements at 800 MeV. However, care should be taken in extending such considerations to transitions which are non-collective and to energies where the scattering is no longer dominated by strong absorption effects. In particular, at lower energies (100-200 MeV) it appears that analyzing power measurements may provide important constraints in determining differences in the microscopic structure of nuclear levels.

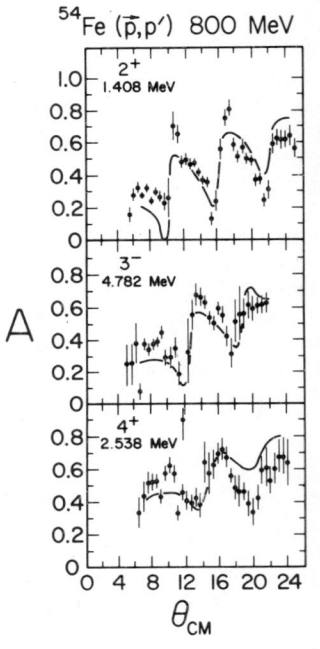

Fig. 10. Analyzing powers for 800 MeV proton inelastic excitation of low-lying levels in ^{54}Fe. (After ref. 22).

An example of the sensitivity of analyzing power measurements to nuclear structure effects has been demonstrated by comparative measurements of the low-lying states in ^{90}Zr and ^{92}Zr. (See Fig. 11). While some differences in the cross sections and analyzing powers have been seen at 800 MeV,[23] far more dramatic results have been observed at 160 MeV.[24] At this energy, the elastic scattering and inelastic scattering to the collective 3^- states are nearly identical for the two targets, whereas the results shown in Fig. 11 indicate significant differences in the analyzing power measurements for the excitation of the lowest 4^+ states. These differences are indicated by the smooth curves drawn through the data for the two transitions. In ^{90}Zr the lowest 4^+ state is predominantly a proton state with the configuration $\pi(1g_{9/2})^2$, whereas the lowest 4^+ state in ^{92}Zr is believed to be a neutron state with the dominant configuration $\nu(2d_{5/2})^2$. The differences in the analyzing powers for these two transitions can arise from two sources: 1) differences between the p-p and p-n parts of the nucleon-nucleon effective interaction, and 2) differences in the shell-model configurations for these two states. Calculations for the 4^+ transition in ^{90}Zr are indicated, along with the data, at the top of Fig. 11. These include: 1) microscopic DWIA valence calculations (dotted curve), 2) DWIA valence calculations with collective core polarization (solid curve), and 3) collective calculations with deformed spin-orbit (dashed curve). While the sensitivity to nuclear structure differences is clearly demonstrated by the data, the attempts at a semi-microscopic DWIA fail to reproduce the observed features. This perhaps indicates the need for more complete wave functions in the microscopic analysis.

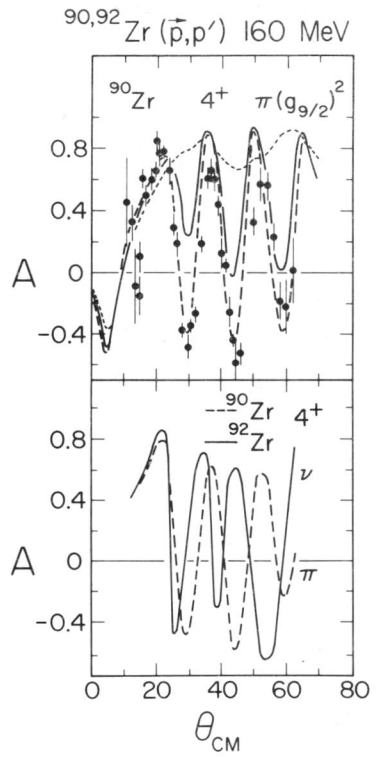

Fig. 11. Analyzing powers for 160 MeV proton inelastic excitation of the 4^+ proton state in ^{90}Zr (top) and a comparison of 4^+ proton and neutron states in ^{90}Zr and ^{92}Zr (bottom). (Adapted from ref. 24)

UNNATURAL PARITY TRANSITIONS

Studies of the excitation of nuclear states which can be described by only a few shell-model configurations are especially attractive because their simple structure provides a favorable opportunity to examine basic features of the microscopic description of inelastic scattering. Transitions to unnatural parity states of "stretched" configuration are particularly noteworthy because of their unique particle-hole description in a 1-ℏω basis and their preferential excitation at high momentum transfer by 100-200 MeV protons.[5,25] Transitions of this type have also been studied extensively in high-resolution inelastic electron scattering.[26] As discussed earlier, the special character of these transitions enables a direct connection to be made between proton inelastic scattering and nuclear structure information which can be determined from electron scattering.[5,27] The excitation of "stretched" configuration states has been observed in nuclei ranging in mass from ^{16}O to ^{208}Pb. The present discussion will focus of those cases where both cross section and analyzing power measurements have been obtained since they provide a more stringent test of the DWIA formalism.

The proton inelastic scattering spectrum presented in Fig. 12 demonstrates the selective excitation of the 6^-, $T = 0$ and $T = 1$ states at 11.58 MeV and 14.35 MeV in ^{28}Si. These "stretched" states are described by the $(f_{7/2}, d_{5/2}^{-1})6^-$ particle-hole configuration. The transverse form factor for the 6^-, $T = 1$ state has been determined from inelastic electron scattering[28] and the resulting parameters have been used as input for microscopic DWIA calculations of the proton inelastic scattering.

The results of the 134 MeV proton inelastic scattering to these states are displayed in Fig. 13. For the cross section measurements (at the top), the solid curve represents the full DWIA calculation. The contributions arising from the individual terms in the effective interaction are also displayed and it is apparent that the transition to the 6^-, $T = 1$ state is dominated by the isovector tensor (T) component. The renormalization factor of 1.17, which is required to match the absolute magnitude of the cross section, indicates that the (p,p') and (e,e') descriptions of this transition are in good agreement.

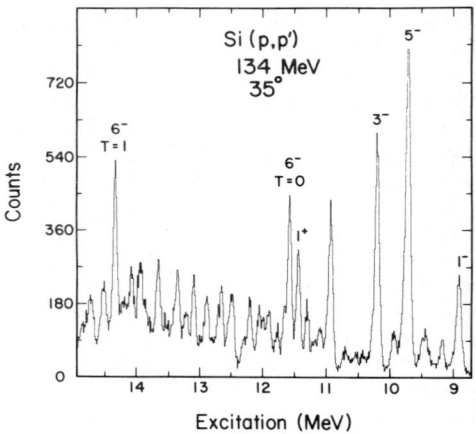

Fig. 12. Inelastic proton spectrum for the scattering of 135 MeV protons from Silicon.

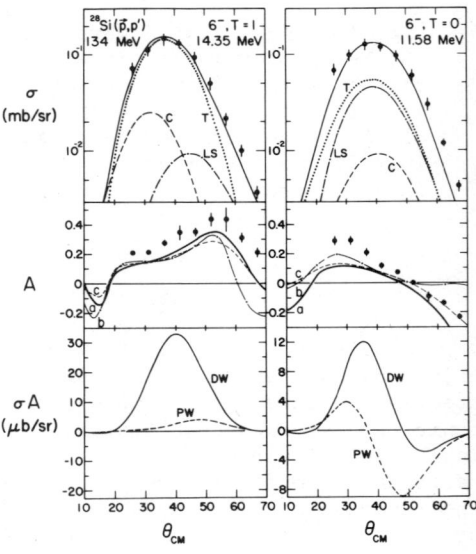

Fig. 13. Cross section and analyzing power measurements for the 6^-, $T = 0$ and $T = 1$ states in ^{28}Si.

The transition to the 6^-, $T = 0$ state contains substantial contributions from both the tensor (T) and spin-orbit (LS) components. In this case, the isovector tensor component contributes to the isoscalar transition through the knockon exchange term since direct excitation through the isoscalar tensor component is small (see Fig. 2). For the $T = 0$ transition, electron scattering results are not available. DWIA calculations have been performed using the same geometry as the 5^-, $T = 0$ transition and, assuming that the microscopic description is valid, squared spectroscopic amplitudes have been extracted. The resulting value for the $T = 0$ transition is about a factor of 3 smaller than that determined for the $T = 1$ transition. Differences of this magnitude in the strengths of the isoscalar and isovector transitions are surprising and suggest that the 6^-, $T = 0$ wave function is considerably more complicated than the simple form assumed for the 6^-, $T = 1$ state.[29] If the origin of these differences lies in the nuclear structure of these states, it can not be understood in the framework of existing shell-model calculations. Further discussion of this effect will be presented in the next section.

The analyzing powers for these unnatural parity transitions have also been determined and the results are given in the middle of Fig. 13. For an unnatural parity transition to a "stretched" configuration state, the analyzing power in the plane-wave impulse approximation (PWIA) is given by[2,5]

$$A(\theta) = \frac{2\left[t_R^{LS}\left(t_I^C + t_I^{T\beta}\right) - t_I^{LS}\left(t_R^C + t_R^{T\beta}\right)\right]}{|t^C + t^{T\alpha}|^2 + |t^{LS}|^2 + |t^C + t^{T\beta}|^2 + \frac{2J}{J+1}|t^C + t^{T\gamma}|^2}$$

where $t^{T\alpha}$, $t^{T\beta}$, and $t^{T\gamma}$ are three different linear combinations of the direct and exchange tensor terms and t^C now refers to the spin-dependent ($\Delta S = 1$) central force. The analyzing power can be seen to arise from interference between the spin-orbit term and a linear combination of central and tensor terms. In this case, in contrast to the situation for natural parity transitions, one expects small analyzing powers since all of the imaginary terms t_I^C, t_I^{LS}, and t_I^T are small. The latter two terms have little effect on the resulting values for σ and A, and have been neglected in the calculations presented here.

In fact, these analyzing powers are observed experimentally to be rather large. The good agreement of the full DWIA calculations that are represented by the heavy solid curves in Fig. 13 suggests that the PWIA expectations are not sufficient as a guide for understanding these transitions. An indication of the importance of distortion for these transitions is apparent from an examination of the plots of σA shown in Fig. 13 for both PWIA and DWIA calculations. In spite of the fact that σ is reduced by including distortion, the product σA is substantially greater, indicating that A has undergone a marked change. The large analyzing powers arise from the presence of spin-orbit distortion which introduces a correction term[30] to the PWIA expression for $A(\theta)$ that is proportional to $(t_R^T)^2$ and is substantial because t_R^T is large compared to the other components in the PWIA expression. This explanation has been verified[18] by observing a sign change in the DWIA result for A(θ) when the sign of the spin-orbit distortion is changed. Analyzing powers with the opposite sign to that observed experimentally are also predicted if the tensor term in the effective interaction is taken to be purely imaginary.

The differences in the shapes of the analyzing power angular distribution for the two transitions are well reproduced by the DWIA calculations presented here, although there appear to be some small differences in the magnitude of the predicted analyzing powers. The good agreement for these transitions, which are strongly influenced by the tensor component of the effective interaction, indicates that both the magnitude and phase of the tensor component are reasonably well understood in the region of high-momentum transfer. It is worthwhile to note that the predominantly real nature of the tensor component observed in the present work is consistent with the free nucleon-nucleon t-matrix and is expected from considerations of π- and ρ- exchange.[31] These general conclusions regarding the form of the tensor interaction are not significantly affected by changes in the parameters describing either the optical potential (dashed curve c in the middle of Fig. 13) or the tensor component (dash-dot curve b in Fig. 13). Further discussion of these transitions and a preliminary examination of their energy dependence will be presented in the next section.

Among the examples of inelastic scattering considered here, we have focused on the 5^-, T = 0; 6^-, T = 0; and 6^-, T = 1 transitions in ^{28}Si which are excited by quite different combinations of components in the effective interaction. The general success in reproducing the observed cross sections and analyzing powers for these transitions is particularly gratifying. For the isovector transition, there is good agreement with the absolute magnitude of the cross section as predicted from inelastic electron scattering. For the isoscalar transitions, the analyzing powers and the shapes of the differential cross sections are well described, but the necessity for renormalizing the magnitude of the predicted cross sections is not understood at present. The examination of high-spin, unnatural parity levels in ^{208}Pb enables an extension of these tests to a nucleus where the "stretched" states allow a separate examination of the neutron and proton transition strengths.

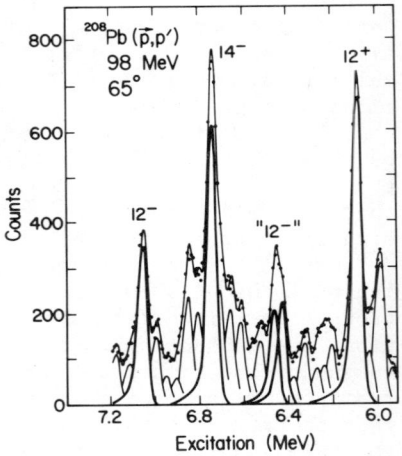

Fig. 14. The inelastic proton spectrum for ^{208}Pb in the region of the high-spin states. The contribution of individual levels to the overall fitted spectrum are shown.

The high-spin unnatural parity states in ^{208}Pb, which are apparent in the inelastic proton spectrum of Fig. 1, are displayed on an expanded scale in Fig. 14. These include the "stretched" configuration states at 6.74 and 7.06 MeV, which are predominantly due to the configurations $(\nu j_{15/2}, \nu i_{13/2}^{-1})14^-$ and $(\pi i_{13/2}, \pi h_{11/2}^{-1})12^-$, respectively, as well as the state near 6.43 MeV which is believed to be the 12^- member of the neutron configuration $(\nu j_{15/2}, \nu i_{13/2}^{-1})12^-$.

These states have been observed in high-momentum-transfer inelastic proton[32] and inelastic electron[33] scattering measurements in which the spin, parity, and configuration assignments are deduced from both the q-dependence of the angular distributions and the closeness of the observed excitation energies to the corresponding particle-hole energies. In addition, the electron scattering data also indicate that these excitations are transverse. Transition densities derived from the electron scattering measurements have been used in DWIA calculations of the proton inelastic scattering.

In the initial studies of these levels by inelastic proton scattering at 135 MeV, only differential cross-section measurements were obtained. While these transitions are excited predominantly by the tensor component of the effective interaction, the influence of the remaining spin-orbit and central components is quite different for the neutron and proton configurations. In each of the three cases, good agreement was found between the experimental and theoretical (DWIA) shapes of the angular distributions. The calculated magnitude of the cross section for the 14^- "stretched" neutron transition was also in good agreement with the prediction based on the electron scattering results. For both (e,e') and (p,p') a renormalization factor of 0.50 was required if a pure particle-hole configuration is assumed for the 14^- level. On the other hand, an intriguing result was observed in similar analyses of the 12^- states. For the (p,p') measurements, the upper (proton) 12^- state at 7.06 MeV required a renormalization of 0.20 while the lower (neutron) 12^- state at 6.43 MeV required a renormalization of 0.80. This is in contrast to the electron scattering results which required a renormalization of about 0.5 for each of the 12^- states.

The (e,e') results are thus consistent with a simple pure-configuration description of these states while the (p,p') results, if they are attributed to mixing of the two 12^- configurations, require an unusually large value for the mixing parameter.[32] These results are particularly puzzling since theoretical considerations of these states indicate a small, but non-negligible, mixing between the two 12^- configurations that should be apparent in both (e,e') and (p,p') studies.

In an attempt to understand the origin of this discrepancy, proton inelastic scattering measurements of both cross sections and analyzing powers were undertaken since calculations[35] indicated that $A(\theta)$ was substantially different for the neutron and proton configurations. These measurements were performed at 100 MeV where an improved energy resolution ($\simeq$50 keV) could be obtained.

The resulting cross section and analyzing power angular distributions for the upper (proton) 12^- state at 7.06 MeV are shown in Fig. 15. DWIA calculations assuming proton (solid curve labeled π)

and neutron (dashed curve labeled υ) configurations for this state are also displayed. A renormalization factor, which is consistent with that obtained for this state in the earlier (p,p') measurements at 135 MeV, has been applied to the predicted cross sections shown here. The results indicate that the (p,p') measurements at the two energies are in good agreement. Further evidence for the predominantly proton nature of this state is apparent from a comparison of the predicted and experimental analyzing powers. The sensitivity of the analyzing power measurements to the basic configuration of the state is clearly established and offers some new hope for understanding the character of these high-spin states. However, any conclusions regarding configuration mixing would be premature. A similar analysis of measurements for the levels with the predominantly neutron configuration (both 14^- and 12^-) is essential for a proper examination of the effects of configuration mixing. This work is still in progress.

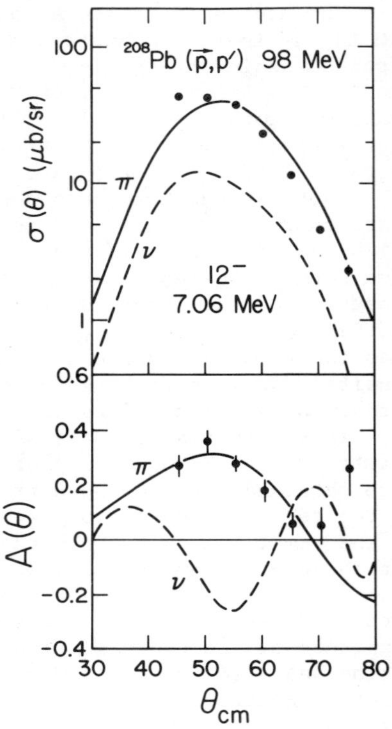

Fig. 15. Cross section and analyzing power measurements for the 12^- "stretched" configuration proton state at 7.06 MeV in ^{208}Pb.

It is also clear from an examination of Fig. 14 that the peak near 6.43 MeV consists of a doublet which was unresolved in the (p,p') measurements at 135 MeV. In the measurements with improved resolution at 100 MeV, the structure is barely visible. The two members both appear to be of high-spin, to peak at a momentum transfer similar to that observed for the proton 12^- state at 7.06 MeV, and to be excited with nearly equal cross sections. The existence of several other high-spin states with the same particle-hole structure is expected[34] in this region of excitation, but this is the first observation reported. The determination that the 6.43 MeV level is a closely spaced doublet will have important implications on the deduced mixing parameter for the 12^- configurations. However, preliminary indications are that the existence of the doublet will not alter the mixing parameter enough to bring the (p,p') and (e,e') results into agreement.

It is clear that further experimental information with better resolution will be required to understand the nature of these high-spin states in ^{208}Pb. There is also a need for additional theoretical work. In particular, it would be useful to have a consistent treatment of core polarization, the effects of mixing with higher-order configurations, and the proton-neutron configuration mixing considered here.

The previous examples have focused on transitions to high-spin unnatural parity states which have been used to test the DWIA formalism at high momentum transfer where the tensor component of the effective interaction plays an important role. Low-spin, unnatural parity transitions are also of interest since they tend to emphasize the spin-flip components of the effective interaction at low momentum transfer.

In particular, the 1^+, T = 1 level at 15.11 MeV in ^{12}C has attracted considerable attention recently. The transverse form factor for this transition has been accurately determined[36] by inelastic electron scattering and the large second maximum near q = 300 MeV/c has been interpreted[37] as possible evidence for precritical fluctuations in the pion field. In addition to using this transition to test the DWIA formalism at low momentum transfer, it has also been suggested[38] that manifestations of precritical phenomena ought to be more evident in inelastic proton scattering since the proton can couple more directly to the pionic degrees of freedom in the nucleus.

Cross section and analyzing power angular distributions have recently been obtained for inelastic proton excitation of the 15.11 MeV level in ^{12}C at proton bombarding energies of 65, 120, 155, 200, and 800 MeV.[39-44] These results are displayed as a function of the momentum transfer in Fig. 16 using smooth curves drawn through the data to emphasize the energy dependence. The cross section measurements decrease exponentially at 65 and 800 MeV. In the energy region between 120 and 200 MeV the exponential behavior is modified by the appearance of a minimum near q = 250 MeV/c. At values of the momentum transfer away from the minimum, the cross sections are remarkably similar over a wide range of bombarding energy.

The analyzing power measurements, displayed at the bottom of Fig. 16, exhibit a similar pattern for the measurements up to 200 MeV. At 800 MeV, where the measurements cover only the low momentum transfer region, the analyzing power remains close to zero, indicating that the spin-orbit term in the nucleon-nucleon interaction is relatively small.[44]

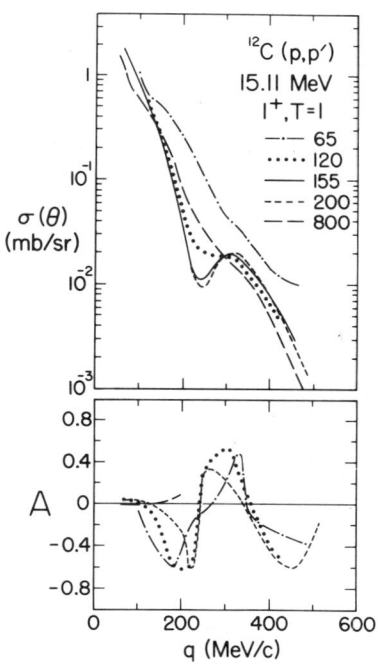

Fig. 16. Momentum transfer dependence of σ and A for proton excitation of the 1^+, T = 1 level in ^{12}C.

Attempts have been made to relate the enhancements seen in inelastic electron scattering to the structures observed in inelastic proton scattering, particularly in the energy region between 100 and 200 MeV. DWIA calculations at 120 MeV[40] reproduce the cross section measurements out to a momentum transfer of about 275 MeV/c. In this region of momentum transfer, the transition is dominated by the isovector central ($\Delta S = 1$) component of the effective interaction and the good agreement is obtained using a renormalization factor of 1.3. Additional calculations at 155 MeV[45] correctly predict the formation of the minimum near 225 MeV/c. At momentum transfers q > 300 MeV/c, the DWIA calculations fall rapidly below the observed cross sections. Calculations[38,45] which incorporate precritical effects in the (p,p') description of the transition significantly overestimate the magnitude of the cross section at momentum transfers in the region q > 300 MeV/c.

The analyzing power angular distribution for 120 MeV proton excitation[41] of the 15.11 MeV state is shown in Fig. 17. The result of a DWIA calculation, which used the 1p-shell wave function of Cohen and Kurath,[46] is seen to provide a reasonable description of the measurements out to a center-of-mass angle of 30°. This is the same angular range where agreement was obtained for the differential cross sections. Dubach and Haxton have shown[47] that modifications of this wave function can be made within the 1p-shell basis that provide a better description of the (e,e') form factor. However, when this new form of the wave function is employed in the DWIA calculations of the (p,p') excitation, the re-

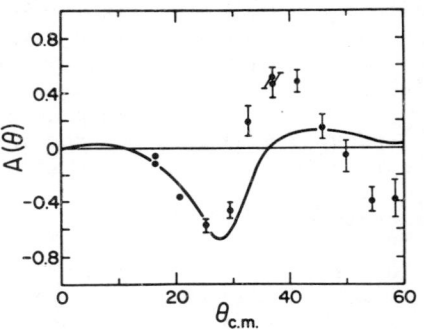

Fig. 17. Analyzing power angular distribution for the 120 MeV proton excitation of the 1^+, T = 1 level at 15.11 MeV in ^{12}C.

sulting cross section and analyzing power predictions are only slightly modified. Additional calculations also indicate that the analyzing power angular distribution is rather insensitive to the inclusion of precritical effects.

By examining the energy dependence of the (p,p') excitation of this level, it appears that the dramatic change that occurs in the energy range between 100 and 200 MeV is the appearance of a minimum in the differential cross section near q = 225 MeV/c rather than an enhancement for q > 300 MeV/c. The search for an enhancement that might be indicative of precritical effects was suggested by an interpretation of the (e,e') form factor at similar values of the momentum transfer. At present there is no conclusive evidence for such an enhancement in proton inelastic scattering.

Regardless of whether or not one believes in the existence of precritical phenomena, it is important to search for alternative explanations for the effects that have been observed in (p,p') as well as in (e,e'). For example, there is a recent suggestion[48] that the inclusion of ρ exchange may provide an alternative explanation for much of the observed enhancement in (e,e'). Non-resonant multiple pion exchange may also be important. This is not unrelated to (p,p') studies since the effective interaction employed has uncertainties in the high momentum transfer region that also arise from the exchange of heavier mesons.

It should be noted that the effects observed for these two probes may arise from quite different origins. An examination of similar transitions in other nuclei may provide a useful perspective in these studies.

FUTURE AREAS OF INTEREST

In the previous two sections, we have examined specific examples of natural and unnatural parity transitions and used them to test the applicability of the effective interaction formalism. In those cases where it is possible to compare the microscopic description with both cross section and analyzing power measurements, the results are rather encouraging. In the present section we consider some of the future areas in which inelastic proton scattering measurements can be used to further test predictions of the microscopic DWIA description, to examine features of the reaction mechanism, and to extract new information about nuclear structure.

<u>Energy dependence of inelastic proton scattering</u>. Measurements of the energy dependence of inelastic proton scattering enable one to further test predictions that are inherent in the microscopic DWIA formalism. In this description the overall energy dependence for a particular transition arises from the individual energy dependences of 1) the underlying free nucleon-nucleon scattering, 2) the knockon exchange amplitude, and 3) the effects of distortion. The final state can be chosen to emphasize particular regions of momentum transfer (e.g., low-spin versus high-spin states) or to focus on selected components of the effective interaction (e.g., high-spin natural parity versus unnatural parity states).

The importance of studying the energy dependence of inelastic scattering was illustrated in Fig. 4 for the region of high excitation in ^{28}Si which includes the 6^-, T = 0 and T = 1 states. These states are observed to be selectively populated for energies between 100 and 200 MeV, but are only barely visible at 800 MeV. While these results can be understood in a qualitative manner from an examination of the energy dependence shown in Fig. 3, the interpretation is limited by a rather poor knowledge of the spin-dependent terms in the nucleon-nucleon interaction at 800 MeV. Quantitative comparisons of the energy dependence require additional measurements in the region below 500 MeV where our knowledge of the nucleon-nucleon interaction is considerably more complete. Once measurements have been obtained in the energy range between 100 and 500 MeV, it will be instructive to compare these results with calculations based on the "fixed-energy" effective interactions of Love and Franey[49] and a similar, local effective interaction by Picklesimer and Walker[50] which is somewhat more general in that it is fitted to nucleon-nucleon amplitudes for bombarding energies in the range between 50 and 400 MeV. New measurements[51] from LAMPF in the energy range between 400 and 500 MeV will be of great value in these comparisons.

Measurements of the energy dependence for high-spin states have been obtained at IUCF in the energy range between 80 and 180 MeV for a few selected transitions. These are illustrated in Fig. 18 where the peak cross sections[10,18,52] are plotted (on a logarithmic scale) as a function of bombarding energy for the 5^-, $T = 0$; 6^-, $T = 0$; and 6^-, $T = 1$ transitions in ^{28}Si. For the two 6^- states, the peak cross sections are observed to increase by a factor of about 1.5 over the energy range between 80 and 180 MeV. The peak cross section for the 5^-, $T = 0$ transition exhibits a decrease from 80 to 135 MeV and then a modest increase from 135 MeV to 180 MeV.

Over this energy range the components of the effective interaction that are primarily responsible for the excitation of these states do not change dramatically. DWIA calculations have been performed for these transitions using the effective interaction[5] derived at 140 MeV and optical model parameters[53] that properly account for the energy dependence of the elastic scattering. The results of these calculations, which correctly reproduce the general features of the measurements, indicate that the primary origin of the energy dependence in this energy range lies in the effects of distortion and, to a lesser extent, the energy dependence of the knockon exchange amplitudes.

The ratio of the peak cross section for the excitation of the 6^-, $T = 1$ state to that for the 5^-, $T = 0$ state is indicated (on a linear scale) at the bottom of Fig. 18. The significance of this ratio is that it effectively removes the energy dependence of distortion. It thereby allows an empirical observation of the enhanced excitation of high-spin, unnatural parity states relative to high-spin, natural parity states in the range of bombarding energies between 100 and 200 MeV.

The 6^-, $T = 0$ and $T = 1$ states in ^{28}Si have also been studied[54] recently at 65 MeV. The shapes of both the cross section and analyzing power angular distributions are substantially different from those observed at 135 MeV. DWIA calculations,[54] which employed a variety of different forms for the effective

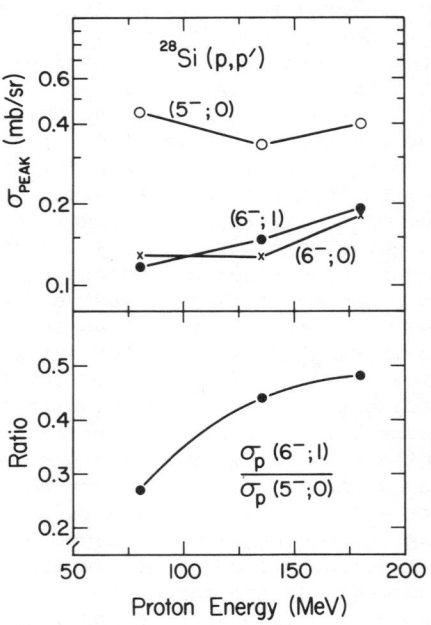

Fig. 18 Energy dependence of the peak cross sections for the inelastic excitation of high-spin states in ^{28}Si.

interaction, were unable to simultaneously reproduce both the T = 0 and T = 1 cross section and analyzing power angular distributions. This is not too surprising since, at these low energies, the effects of distortion and exchange processes are expected to have a substantial influence on the shapes of the angular distributions. There is also the more fundamental concern regarding the validity of the impulse approximation at such low energies.

Studies of the energy dependence of inelastic proton scattering can also be expected to shed some light on the importance of alternative reaction mechanisms which may compete with the one-step direct excitation implied by the DWIA formalism. While there are some calculations[55] which indicate that two-step processes become less important as the bombarding energy is increased, there are only a few measurements to support these expectations.

Influence of the nuclear medium. The influence of the nuclear medium on the form of the effective nucleon-nucleon interaction is expected, on theoretical grounds,[56,57] to become more significant at the low energy end of the intermediate energy range. The impulse approximation, in which the effective interaction is just the free nucleon-nucleon t-matrix, ignores effects such as Pauli blocking, Fermi motion of the target nucleons, and the off-shell nature of the nucleon-nucleon interaction. These effects have been shown to be less important at higher energies (near 800 MeV) and to increase in importance as the bombarding energy is reduced.[4,58]

At energies near 200 MeV, the most important correction to the impulse approximation involves the inclusion of Pauli blocking[14] which causes a truncation of the momentum space available to the nucleons in the target. This effect is particularly apparent in a phenomenological analysis of recent measurements of the cross section and analyzing power for proton elastic scattering from ^{12}C at 200 MeV. A pronounced depression in the real central potential, which is located inside the nuclear surface region, is required in order to reproduce the observed large oscillations in the analyzing power angular distribution. The effects of Pauli blocking on inelastic proton scattering are expected to primarily influence low-spin, natural parity transitions for which the transition densities peak in the high-density region of the nuclear interior where the effects on the elastic scattering are most pronounced.

The modifications of the nucleon-nucleon interaction which occur in the nuclear medium have been determined in a microscopic treatment using Brueckner-Hartree-Fock techniques and a local density approximation (LDA) to establish the connection with finite nuclei.[56] Kelly and Petrovich[60] have recently incorporated a form of this density-dependent interaction in folding model calculations of elastic and inelastic proton scattering. In this manner a direct comparison can be made between predictions based on the LDA and predictions obtained from the impulse approximation (IA) treatment.

Comparisons of the LDA and IA predictions[60] are presented in Fig. 19 together with analyzing power measurements for the 135 MeV inelastic proton excitation of the 2^+, 4^+, and 5^- natural parity states in ^{28}Si. Since both the LDA and the IA do not provide a particularly good account of the elastic scattering data in this energy range, predictions are shown in which the distorted waves are determined either by the density-dependent effective interaction (consistent OM) or by a phenomenological optical model parametrization of the elastic scattering (phenom. OM).

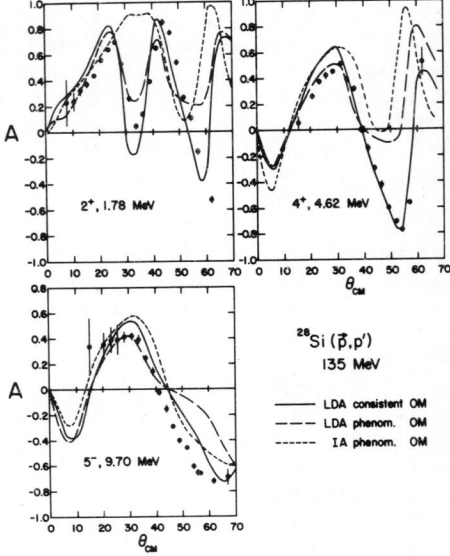

Fig. 19. Analyzing powers for the inelastic proton excitation of natural parity states in ^{28}Si.

For the 2^+ and 4^+ transitions, the IA provides a rather poor representation of the analyzing power data, particularly in the region of the first minimum. On the other hand, the LDA predictions give an especially good account of the structures observed in these data. While the consistent treatment of the distorted waves produces a somewhat better agreement for these transitions, it should be noted that this prescription results in a poorer account of the elastic scattering. In any case, the LDA provides, for the first time, a microscopic description of the inelastic excitation of low-spin natural parity states that previously could be understood only in terms of a macroscopic collective model treatment.

For the 5^- transition, the LDA and IA prescriptions provide comparable representations for the analyzing power data, although somewhat different renormalizations of the differential cross sections are required. Predictions for the excitation of high-spin states are less sensitive to the inclusion of the density-dependent interaction since their excitation depends primarily on the spin-dependent components of the nucleon-nucleon interaction which are not significantly altered[61] by these effects in the nuclear medium. It should be noted that many of the transitions discussed in the earlier sections, which were used to examine the validity of the DWIA, correspond to the excitation of high-spin, natural and unnatural parity states which are not substantially affected by the nuclear medium corrections considered above.

In addition to these nuclear medium corrections, there are other processes which can modify the overall strength or the momentum transfer dependence of an inelastic transition. These processes affect the nuclear structure portions of the transition density and are mentioned here only because they may be confused with nuclear medium corrections to the basic nucleon-nucleon interaction. These include: core polarization, configuration mixing, meson exchange and precritical phenomena. As a general comment, what one needs is a more consistent treatment of these effects. Our present knowledge of the mechanism for inelastic scattering is not sufficiently well developed to allow a separation of these effects. However, while they may appear to be hopelessly intertwined, there are several ways in which their individual contributions might be isolated. These include differences in their dependence on the projectile type and energy, the nature of the transition and the momentum transfer. Attempts at unraveling these effects should provide both experimentalists and theorists with an intriguing puzzle for the next few years.

Comparative studies with different probes. One of the most exciting features of current investigations at intermediate energies is the variety of projectiles available to probe[62] the nucleus. The different sensitivities of electrons, protons and pions to nuclear structure effects present a promising opportunity. Before the full potential of these probes can be exploited, however, a more detailed microscopic description of their interaction with the nucleus is required. While the present discussion is focused on attempts directed at obtaining an understanding of the proton-nucleus interaction, considerable effort has also been devoted to understanding the pion-nucleus interaction.[63,64]

Many of the examples cited in the earlier sections involved the inelastic excitation of states for which nuclear structure information that is available from electron scattering can be used directly as input to DWIA calculations of proton scattering. In self-conjugate nuclei, where the neutron and proton transition densities can be assumed to be equal, one can examine the form of the effective interaction.

An example[65] of such a comparison is illustrated in Fig. 20 for the electron and proton inelastic excitation of the "stretched" configuration, 6^-, $T = 1$ level in ^{24}Mg. The transverse form factor

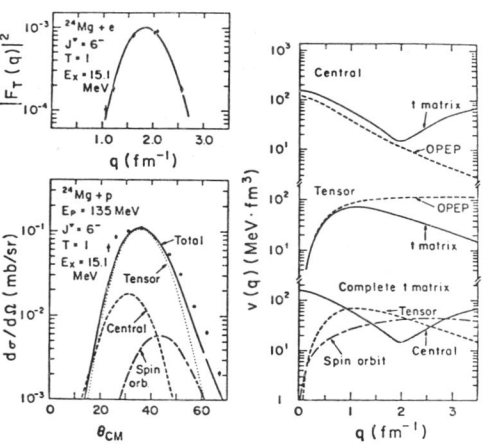

Fig. 20. Inelastic excitation of the 6^-, $T = 1$ state in ^{24}Mg by electrons (top) and protons (bottom) and the magnitudes of the isovector $\Delta S = 1$ components of the effective interaction. (From ref. 62.)

from (e,e') measurements has been used as input to the DWIA calculations of the (p,p') excitation. The dependence of the proton induced excitation on the individual components of the nucleon-nucleon t-matrix is indicated.

The excitation of these "stretched" configuration, unnatural parity states is dominated by the isovector, tensor component of the effective interaction. Comparative studies of this nature have been made for a number of "stretched" configuration states. The quantitative agreement obtained for the two probes suggests that the isovector tensor interaction is well described for momentum transfers near 2 fm^{-1}. In both the electron and proton measurements, only a fraction of the pure particle-hole strength is observed for these transitions. Current theoretical efforts are directed toward understanding the reduction of these transition strengths in terms of core-polarization effects[66] or other models for the mixing of these states with higher-lying configurations.[67]

The T = 0 counterparts of the T = 1 "stretched" configuration states have also been studied in several self-conjugate nuclei using inelastic proton scattering. A discussion of these results for ^{28}Si has been presented earlier. (See Fig. 13 and the accompanying discussion.) The factor of 3 difference in the transition strengths for the excitation of these two states suggests that the T = 0 wave function is considerably more complicated than that for the T = 1. While such comparisons are not possible using electron scattering, these two transitions have been investigated using π^+ and π^- inelastic scattering. The pion probe is especially attractive in this regard since the isospin character of the pion-nucleon interaction can be used to examine the individual proton and neutron components of these states.

The π^+ and π^- inelastic scattering measurements[68] have been carried out at 162 MeV, near the 3,3 resonance, where the pion-nucleon interaction is reasonably well understood. The observed ratio of the isoscalar to isovector transition strengths is 1.6 ± 0.4. This is to be compared with the value 3.7 that is predicted from the known isoscalar and isovector components of the pion-nucleon interaction under the assumption that the wave functions for the two levels are of similar character. This difference is similar to that observed in the (p,p') reaction.[5,25] Several alternative explanations have been offered which ascribe these effects either to nuclear structure differences[29] or to differences in the tail region of the radial wave functions.[68]

In attempting to compare measurements with different probes, it is clearly important to understand and properly include the ways in which these probes are sensitive to different regions of the nuclear wave function. Such effects can be investigated by exploring, for example, the energy dependence of the excitation mechanism for each probe.

The electron, pion and proton probes can also be used in a comparative fashion to study the isospin mixing of nuclear states. While this type of study may be carried out in a more controlled manner using electromagnetic transitions, there are cases where this information may be obtained more easily through comparative studies of inelastic scattering using several projectiles. For example, the 4⁻ "stretched" states in ^{16}O have recently been examined using inelastic proton,[69] pion,[70] and electron[71] scattering. A consistent picture[72] of the isospin mixing of the T = 0 states at 17.74 and 19.80 MeV and the T = 1 state at 18.98 MeV can be obtained using these measurements. It is not particularly useful, at this stage, to make absolute statements as to the value of the different probes in these measurements. Attention was initially focused on the "stretched" states in ^{16}O by their observation in inelastic proton scattering. However, the proton probe is not very sensitive to isospin mixing in this case. While comparisons of π^+ and π^- inelastic scattering are especially sensitive to isospin mixing, the pion-nucleus interaction is not yet well understood. The observation of two of these states in inelastic electron scattering provides a more definitive test of the isospin mixing since, in this case, the interaction is well known.

There are other cases where measurements with the three probes may prove illuminating because of their quite different sensitivities to the nuclear structure. A direct comparison of electron and proton elastic and inelastic scattering from ^{16}O, ^{17}O and ^{18}O has been undertaken[61] at Bates and IUCF. For the measurements with ^{16}O, where the neutron and proton transition densities can be assumed to be equal, one can minimize the nuclear structure uncertainties and use inelastic transitions to examine the effective interaction. Studies of low-spin transitions have already led to an appreciation of the role of density-dependent corrections to the nucleon-nucleon interaction. Once the form of the effective interaction is established, one can proceed to study nuclear structure effects in the other oxygen isotopes by incorporating proton transition densities from electron scattering and using the proton probe to study the role of neutron transition densities in the inelastic excitations. For those cases where the resolution is not a significant factor, it will also be valuable to make comparisons with π^+ and π^- inelastic scattering.

An additional area of interest involves an examination of "stretched" configuration states in medium-mass nuclei where the particle-hole strength may be distributed among a number of discrete states. Splitting of the $(g_{9/2}, f_{7/2}^{-1})8^-$ strength has been observed in ^{54}Fe and ^{58}Ni by inelastic electron[73] and proton[74] scattering. Shell model calculations[75] predict the observed fragmentation of the 8⁻ strength in these nuclei. These calculations suggest that important constraints on the wave functions of the fragmented states can be obtained from comparative studies of their excitation with electrons, protons and pions.

<u>Studies of the nuclear continuum region</u>. The examples which have been presented in the earlier sections have focused on inelastic transitions to low-lying bound states or to narrow, discrete states at high excitation that can easily be isolated from the underlying nuclear continuum. At high excitation, it is also of interest to examine transitions to giant resonances as well to the nuclear continuum itself.

The study of giant resonances[76] in the nuclear continuum has attracted considerable interest in recent years, as evidence for the existence of multipole excitations other than the well-known giant dipole resonance has been accumulated. Since the identification of these resonances can provide new information about the residual interaction between nucleons in nuclei, such studies are of fundamental importance. Inelastic scattering measurements with protons and alpha particles have been useful in this effort because of selection rules, sensitivity to multipolarity in the transition, and favorable count rates which have enabled the collection of extensive systematics, especially for the isoscalar quadrupole and monopole giant resonances. The search for higher-order multipole excitations continues to be of vital importance, especially since recent theoretical calculations[77] have predicted their presence both in the region of the giant quadrupole resonance and at higher excitation energies.

The energy range above 100 MeV is especially well suited to inelastic proton scattering studies of the giant resonance region since cross section angular distributions show a pronounced sensitivity to higher multipole excitations. Analyzing power measurements exhibit a similar sensitivity and are expected to aid in unraveling the multipolarity content of the observed structures.

An example of such cross section and analyzing power measurements[78] is shown in Fig. 21 for the giant quadrupole resonance (GQR) region near 14 MeV excitation in ^{92}Zr. The cross section data (at the top) can be better understood in terms of a small addition of $L = 4$ multipole content in the $L = 2$ GQR peak. The curves represent collective model distorted wave Born approximation (DWBA) calculations assuming 50% of the $L = 2$ energy-weighted sum rule (EWSR) strength (solid curve) and the addition to this of 2% of the $L = 4$ EWSR (dashed curve).

The analyzing power measurements for this giant resonance structure are shown at the bottom of Fig. 21. Here the curves represent DWBA calculations assuming: 1) 50% GQR, $\beta_{SO} = \beta_C$ (dashed-dot curve); 2) 50% GQR, $\beta_{SO} = 1.5 \beta_C$ (solid curve); and 3) 50% GQR + 2% $L = 4$, $\beta_{SO} = \beta_C$ (dashed curve).

The worsening of the agreement for the analyzing power data when one includes the small additional $L = 4$ multipole contribution does not support the explanation based on cross section data alone. Eventually, data of

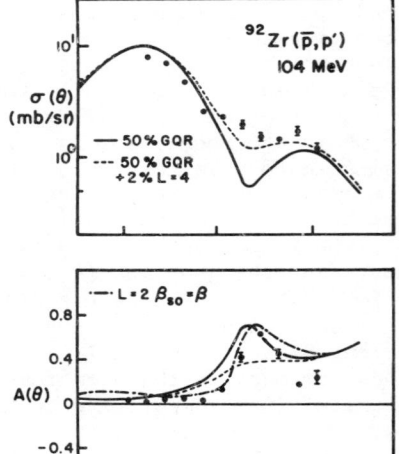

Fig. 21. Cross section and analyzing power measurements for the 104 MeV proton excitation of the giant quadrupole resonance (GQR) region in ^{92}Zr.

this nature can be expected to put stringent limits on the multipole content of giant resonance structures.

Several contributions to this conference deal with the measurement[79] and interpretation[80] of analyzing powers of the nuclear continuum at high excitation. For 65 MeV proton excitation of the continuum, these analyzing powers are small at forward angles but increase to reach surprising large positive values at backward angles. This can be explained by including two-step processes in the calculations.[80]

Analyzing power information of this nature can provide important tests of the mechanism by which protons dissipate energy in their interaction with the nucleus. Tracking the energy dependence of this effect toward higher energies may aid in our understanding of proton-nucleus interactions.

Measurements of other spin-dependent observables. One of the indisputable advantages of the proton probe is its sensitivity to spin-dependent effects in the inelastic excitation of nuclear states. In the preceding discussions only cross section and analyzing power measurements have been considered. At intermediate energies measurements of spin-flip probabilities are especially relevant because of their direct connection to spin transfer ($\Delta S = 1$) processes which are enhanced in this energy region. While measurements of this type are generally considered to be more difficult experimentally, in fact they become more amenable to direct measurement as one goes to higher energy. This results from the fact that at these higher energies the polarimeters that are used to determine the outgoing polarization have substantially higher efficiencies than those previously used at lower energies.[81]

Measurements of spin-flip probabilities are already underway at IUCF and at LAMPF. The initial measurements will focus on inelastic transitions to states that are considered to proceed primarily by a spin-flip mechanism and hence should exhibit large transverse spin-flip probabilities. Examples of the transitions to be measured include the 1^+, $T = 0$ and $T = 1$ states in ^{12}C and the "stretched" configuration, unnatural parity states in ^{28}Si. Calculations for these cases indicate that measurements of the spin-flip probability are especially sensitive to details of the wave functions for these excited states. In addition, measurements of this type can be used to test the assumption that these excitations proceed primarily by a one-step reaction mechanism. For the transitions to the 1^+ states in ^{12}C, there is also the suggestion[82] that additional information on both the nuclear form factors and the nucleon-nucleon spin-flip amplitudes can be determined from detailed measurements of the Wolfenstein parameters, D and R.

A more detailed description of spin-transfer measurements is included in the contribution by Moss.[81]

Selectivity of the (p,n) reaction. In a general sense, the (p,n) reaction is a form of nucleon inelastic scattering. Some mention of this reaction must be included here because of its exciting potential for determining new nuclear structure information at intermediate energies. The (p,n) reaction is more selective than inelastic proton scattering since only isovector excitations are allowed. By making measurements at $0°$, one can also focus on low-spin transitions near q = 0 which depend primarily on the central components of the nucleon-nucleon interaction. Considerations[6] of the energy dependence of these central components indicate that spin transfer ($\Delta S = 1$) is strongly enhanced over non-spin transfer ($\Delta S = 0$) in the energy range between 120 and 350 MeV. This suggests that the (p,n) reaction at intermediate energies provides a unique tool for identifying isovector spin-flip giant resonances.

This expectation is born out by recent measurements at IUCF[83] at energies above 100 MeV. At these energies the (p,n) reaction has been established as a quantitative probe of Gamow-Teller strength. It is being employed to map the mass dependence of this strength throughout the periodic table, but especially for heavier nuclei where its location had previously eluded experimenters working at lower energies. Other isovector spin-flip giant resonances are expected[84] and there is recent evidence[85] from the $^{89}Y(p,n)^{89}Zr$ reaction at 200 MeV of an L = 1 transition with possible $J^\pi = 1^-, 2^-, 3^-$.

CONCLUSION

This review of inelastic proton scattering has focused on the intermediate energy region where recent experimental measurements of high quality are allowing detailed tests to be made of our microscopic understanding of the interaction of protons with nuclei. Polarization measurements are playing an important role in these tests since the spin-dependent components of the underlying nucleon-nucleon interaction are relatively stronger here, especially for energies between 100 and 500 MeV.

By examining inelastic proton transitions to both natural and unnatural parity states of special character, it is possible to isolate the dependence of the transition on specific components of the nucleon-nucleon interaction. For those cases where the nuclear structure information is available from inelastic electron scattering, the validity of the microscopic distorted wave impulse approximation has been directly tested. The result is the establishment of a new level of reliability for the microscopic description of these transitions for energies above about 100 MeV. When nuclear medium corrections are properly incorporated, the variety of nuclear states that can be described with this formalism can be expected to increase dramatically.

The work reviewed here represents an important step in establishing inelastic proton scattering as a useful probe for nuclear structure. In several instances, new phenomena have been identified which have not previously been accessible at other energies. The selective excitation of high-spin particle-hole states of rather pure configuration at energies near 200 MeV are worthy of mention since they provide important benchmarks for the tests of the microscopic description in this energy range.

One of the most encouraging results of the comparative studies of inelastic scattering that have been made with the electron and proton probes is the common meeting ground that has been established for those who study nuclear structure with the weakly and strongly interacting probes. As a result, we are in a better position to view realistically the advantages and shortcomings of the individual probes. This can be expected to continue in its importance as attempts, of a more demanding nature, are made to include meson degrees of freedom in our microscopic treatment. Those developing the utility of the pion probe can also be expected to benefit from comparative studies of this nature.

In the course of trying to understand the proton probe, we are also gaining additional insight into features of the nucleon-nucleon interaction. This, in turn, can be expected to have a direct influence on our ability to understand features of the nuclear many-body problem in terms of the underlying nucleon-nucleon variables.

The tone of this report and the selection of topics have been guided by a feeling of optimism for the field. Another individual might have chosen a more cautious and deliberate stance. Reality, of course, lies somewhere in between, but the present approach more adequately represents the enthusiasm and exuberance of the participants.

It is important to realize that this area of intermediate energy physics is undergoing a rapid change in outlook. While we can point with some pride to the progress we have made in understanding the proton probe, there remains an additional gap to be bridged before one can use this probe to extract nuclear structure information with complete confidence. That the continuing efforts in this direction are in a healthy condition can be determined from the large number of new topics that are considered in the section on future areas of interest.

ACKNOWLEDGEMENTS

Numerous experimental colleagues have made important contributions to the results presented here both through their direct collaborations and through their participation in spirited discussions. The author is especially indebted to G. T. Emery and C. Olmer for their continued involvement in this enterprise. S. Yen, T. E. Drake and J. Kelly have contributed extensively to this work. Much of the theoretical effort which has been of crucial importance in the understanding of these results has been enthusiastically provided by W. G. Love and F. Petrovich. This work was supported in part by the National Science Foundation.

REFERENCES

1. K. M. Watson, Phys. Rev. $\underline{89}$, 575 (1953).
2. A. K. Kerman, H. McManus and R. M. Thaler, Ann. Phys. $\underline{8}$, 551 (1959).
3. R. J. Glauber, in Lectures in Theoretical Physics, eds. W. E. Brittin and L. G. Dunham (Interscience, 1959), p. 315.
4. L. Ray, Phys. Rev. C20, 1857 (1979) and invited talk at this symposium.
5. W. G. Love, in The (p,n) Reaction and the Nucleon-Nucleon Force, C. Goodman et al., eds. (Plenum, New York, 1980) p. 23; F. Petrovich, in The (p,n) Reaction and the Nucleon-Nucleon Force, C. Goodman et al., eds. (Plenum, New York, 1980), p. 115; F. Petrovich and W. G. Love, Proc. LAMPF Workshop on Pion Single Charge Exchange, Los Alamos, New Mexico, 1979, document LA-7892-C addendum.
6. W. G. Love, Proc. LAMPF Workshop on Nuclear Structure with Intermediate-Energy Probes, Los Alamos, New Mexico, 1980, document LA-8303C, p. 26.
7. See, for example, the invited talk at this symposium by J. A. Edgington.
8. J. Raynal and R. Schaeffer, computer code DWBA-70, as modified by W. G. Love, unpublished.
9. P. Schwandt, A. D. Bacher, W. W. Jacobs, H.-O. Meyer and S. E. Vigdor, contribution to this symposium and IUCF Scient. and Techn. Report 1979, p. 47.
10. A. D. Bacher and C. Olmer, private communication.
11. N. M. Hintz, private communication.
12. G. S. Adams, A. D. Bacher, G. T. Emery, W. P. Jones, D. W. Miller, W. G. Love and F. Petrovich, Phys. Lett. $\underline{91B}$, 23 (1980); and references therein.
13. F. Petrovich, W. G. Love, G. S. Adams, A. D. Bacher, G. T. Emery, W. P. Jones, and D. W. Miller, Phys. Lett. $\underline{91B}$, 27 (1980); and references therein.
14. P. Schwandt and F. Petrovich, IUCF Scient. and Techn. Report 1978, p. 27.
15. A. D. Bacher, G. T. Emery, W. P. Jones, S. Kailas, D. W. Miller, H. Nann, C. Olmer, W. G. Love, and F. Petrovich, Bull. Am. Phys. Soc. $\underline{24}$, 522 (1980).
16. J. Lichtenstadt, Ph.D. Thesis, MIT, 1979.
17. S. Yen, R. J. Sobie, H. Zarek, B. O. Pich, T. E. Drake, C. F. Williamson, S. Kowalski and C. P. Sargent, to be published.
18. A. D. Bacher, G. T. Emery, W. P. Jones, D. W. Miller, P. Schwandt, S. Yen, R. J. Sobie, T. E. Drake, W. G. Love and F. Petrovich, to be published.
19. G. W. Hoffmann, Proc. LAMPF Workshop on Nuclear Structure With Intermediate-Energy Probes, Los Alamos, New Mexico, 1980, document LA-8303C, p. 99; and references therein.

20. G. S. Adams, Th. S. Bauer, G. Igo, G. Pauletta, C. A. Whitten, Jr., A. Wriekat, G. W. Hoffmann, G. R. Smith and M. Gazzaly, Phys. Rev. C21, 2485 (1980).
21. G. S. Adams, Th. S. Bauer, G. Igo, G. Pauletta, C. A. Whitten, Jr., A. Wriekat, G. W. Hoffmann, G. R. Smith, M. Gazzaly, L. Ray, W. G. Love and F. Petrovich, Phys. Rev. Lett. 43, 421 (1979).
22. J. A. McNeil and D. A. Sparrow, preprint; R. D. Amado, F. Lenz, J. A. McNeil and D. A. Sparrow, submitted to Phys. Rev. C.
23. F. T. Baker, C. Glashausser, A. Scott, G. Adams, M. Grimm, G. Hoffmann, G. Igo, W. G. Love, J. Moss, V. Penumetcha, W. Swenson and B. E. Wood, contribution to this symposium.
24. A. Scott, F. T. Baker, M. A. Grimm, Jr., J. H. Johnson, V. Penumetcha, R. C. Styles, W. G. Love, W. P. Jones and J. D. Wiggins, Jr., submitted for publication.
25. G. S. Adams, A. D. Bacher, G. T. Emery, W. P. Jones, R. T. Kouzes, D. W. Miller, A. Picklesimer and G. E. Walker, Phys. Rev. Lett. 38, 1387 (1977).
26. See, for example, R. A. Lindgren, in Giant Multipole Resonances, F. E. Bertrand, ed. (Harwood Academic Publishers, New York, 1980), p. 399; J. Heisenberg, Proc. LAMPF Workshop on Nuclear Structure With Intermediate-Energy Probes, Los Alamos, New Mexico, 1980, document LA-8303C, p. 13; and references therein.
27. P. J. Moffa and G. E. Walker, Nucl. Phys. A222, 140 (1974).
28. S. Yen, R. Sobie, H. Zarek, B. O. Pich, T. E. Drake, C. F. Williamson, S. Kowalski and C. P. Sargent, Phys. Lett. 93B, 250 (1980).
29. F. Petrovich, W. G. Love, A. Picklesimer, G. Walker and E. Siciliano, accepted for publication in Phys. Lett.
30. G. P. McCauley and G. E. Brown, Proc. Phys. Soc. 71, 893 (1958).
31. M. R. Anastasio and G. E. Brown, Nucl. Phys. A285, 516 (1977).
32. A. D. Bacher, G. T. Emery, W. P. Jones, D. W. Miller, G. S. Adams, F. Petrovich and W. G. Love, submitted to Phys. Lett.
33. J. Lichtenstadt, J. Heisenberg, C. N. Papanicolas, C. P. Sargent, A. N. Courtemanche and J. S. McCarthy, Phys. Rev. C20, 497 (1979).
34. W. W. True, C. W. Ma and W. T. Pinkston, Phys. Rev. C3, 2421 (1971); W. W. True, private communication; D. Haldersen, private communication.
35. W. G. Love, private communication.
36. J. B. Flanz, R. S. Hicks, R. A. Lindgren, G. A. Peterson, J. Dubach and W. C. Haxton, Phys. Rev. Lett. 43, 1922 (1979).
37. J. Delorme, M. Ericson, A. Figureau and N. Giraud, Phys. Lett. 89B, 327 (1980).
38. H. Toki and W. Weise, Phys. Rev. Lett. 42, 1034 (1979) and preprint, 1980 and Z. Physik A292, 389 (1979).
39. K. Hosono, M. Kondo, T. Saito, N. Matsuoka, S. Nagamachi, S. Kato, K. Ogino, Y. Kadota and T. Noro, Phys. Rev. Lett. 41, 621 (1978).
40. J. R. Comfort, S. M. Austin, P. T. Debevec, G. L. Moake, R. W. Finlay and W. G. Love, Phys. Rev. C21, 2147 (1980).

41. J. R. Comfort, C. C. Foster, C. D. Goodman, D. W. Miller, G. L. Moake, P. Schwandt, J. R. Rapaport and R. E. Segel, contribution to this symposium; J. R. Comfort et al., IUCF Scient. and Techn. Report 1979, p. 3.
42. R. E. Segel, J. R. Comfort, D. W. Miller and G. L. Moake, IUCF Scient. and Techn. Report 1979, p. 4.
43. J. R. Comfort, private communication.
44. J. M. Moss, C. Glashausser, F. T. Baker, R. Boudrie, W. D. Cornelius, N. Hintz, G. Hoffmann, G. Kyle, W. G. Love, A. Scott and H. A. Thiessen, Phys. Rev. Lett. $\underline{44}$, 1189 (1980).
45. J. R. Comfort and W. G. Love, Phys. Rev. Lett. $\underline{44}$, 1656 (1980).
46. S. Cohen and D. Kurath, Nucl. Phys. $\underline{73}$, 1 (1965).
47. J. Dubach and W. C. Haxton, Phys. Rev. Lett. $\underline{41}$, 1453 (1978).
48. J. Delorme, A. Figureau and N. Giraud, Phys. Lett. $\underline{91B}$, 328 (1980).
49. W. G. Love and M. A. Franey, private communication.
50. A. Picklesimer and G. Walker, Phys. Rev. $\underline{C17}$, 237 (1978).
51. N. M. Hintz, spokesman, HRS experiment #451, private communication.
52. G. S. Adams, Ph.D. Thesis, Indiana University (1977), unpublished.
53. P. Schwandt, private communication.
54. K. Hosono, N. Matsuoka, T. Saito, K. Hatanaka, M. Kondo, T. Noro, H. Shimizu, S. Kato, K. Okada, K. Ogino and Y. Kadota, contribution to this symposium.
55. J. R. Comfort, in Proc. Int. Symp. on Nuclear Direct Reaction Mechanisms, Fukuoka, Japan, Oct. 25-28, 1978, eds. M. Tanifuji and K. Yazaki.
56. H. V. von Geramb, F. A. Brieva and J. R. Rook, in Proc. Hamburg Topical Workshop on Nucl. Phys., Hamburg 1978, ed. W. Beiglböck, Lecture Notes in Physics No. 89, Springer, Berlin 1979, p. 104.
57. M. Jaminon, C. Mahaux and P. Rochus, Phys. Rev. Lett. $\underline{43}$, 1097 (1979).
58. L. Ray, Phys. Rev. $\underline{C19}$, 1855 (1979); D. R. Harrington and G. K. Varma, Nucl. Phys. $\underline{A306}$, 477 (1978).
59. H.-O. Meyer, P. Schwandt, G. L. Moake and P. P. Singh, submitted for publication.
60. J. Kelly and F. Petrovich, private communication.
61. J. Kelly et al., Proc. LAMPF Workshop on Nuclear Structure With Intermediate-Energy Probes, Los Alamos, New Mexico, 1980, document LA-8303C, p. 461; J. Kelly et al., IUCF Scient. and Techn. Report 1979, p. 6.
62. S. Harris, What's So Funny About Science?, (William Kaufmann, Inc., Los Altos, Calif., 1977).
63. See, for example, contributions by T.-S. H. Lee and J. F. Dubach, in Proc. of LAMPF Workshop on Nuclear Structure With Intermediate-Energy Probes, Los Alamos, New Mexico, 1980, document LA-8303C, p. 2 and p. 72.
64. F. Petrovich, in Proc. International Conference on Nuclear Physics, Berkeley, California, August, 1980, to be published.

65. R. A. Lindgren, W. J. Gerace, A. D. Bacher, W. G. Love and F. Petrovich, Phys. Rev. Lett. $\underline{42}$, 1524 (1979).
66. I. Hamamoto, J. Lichtenstadt and G. F. Bertsch, Phys. Lett. $\underline{93B}$, 213 (1980).
67. J. Speth, E. Werner and W. Wild, Phys. Rep. $\underline{33C}$, 127 (1977); J. Speth, private communication.
68. C. Olmer, B. Zeidman, D. F. Geesaman, T.-S. H. Lee, R. E. Segel, L. W. Swenson, R. L. Boudrie, G. S. Blanpied, H. A. Thiessen, C. L. Morris and R. E. Anderson, Phys. Rev. Lett. $\underline{43}$, 612 (1979).
69. R. S. Henderson, B. M. Spicer, I. D. Svalbe, V. C. Officer, G. G. Shute, D. W. Devins, D. L. Friesel, W. P. Jones and A. C. Attard, Aust. J. Phys. $\underline{32}$, 411 (1979).
70. D. B. Holtcamp et al., Phys. Rev. Lett. $\underline{45}$, 420 (1980).
71. W. Bertozzi, private communication.
72. F. Petrovich, private communication.
73. R. A. Lindgren, C. F. Williamson and S. Kowalski, Phys. Rev. Lett. $\underline{40}$, 504 (1978); R. A. Lindgren, private communication.
74. A. D. Bacher and C. Olmer, private communication.
75. R. D. Lawson, private communication.
76. F. E. Bertrand, Ann. Rev. Nucl. Sci. $\underline{26}$, 457 (1976).
77. S. Krewald and J. Speth, Phys. Lett. $\underline{52B}$, 295 (1974).
78. S. Kailas, P. P. Singh, A. D. Bacher, C. C. Foster, D. L. Friesel, P. Schwandt and J. Wiggins, IUCF Scient. and Techn. Report 1979, p. 21.
79. H. Sakai, K. Hosono, N. Matsuoka, S. Nagamachi, K. Okada, K. Maeda and H. Shimizu, contribution to this symposium and Phys. Rev. Lett. $\underline{44}$, 1193 (1980).
80. H. Lenske, T. Tamura and T. Udagawa, contribution to this symposium.
81. J. M. Moss, invited talk at this symposium.
82. E. Bleszynski, M. Bleszynski and C. A. Whitten, Jr., contribution to this symposium.
83. See contributions to IUCF Scient. and Techn. Report 1979, pp. 27-46.
84. A. Bohr and B. Mottelson, Nuclear Structure, Vol. 2 (Benjamin, Reading, Mass., 1975).
85. C. D. Goodman, private communication.

REACTION MECHANISM STUDIES WITH POLARIZED IONS

Kohsuke Yagi
Institute of Physics and Tandem Accelerator Center,
University of Tsukuba, Ibaraki 305, Japan

ABSTRACT

Measurements of analyzing powers for nuclear reactions are regarded as the powerful probe for the sensitive detection of interference effects between various reaction transition amplitudes. Based on such a point of views, studies of nuclear reaction mechanisms and effective interactions as well as microscopic structures of nuclei are made.

INTRODUCTION

After introducing a completely new type of experimental probes, the understanding of a phenomenon which has been assumed to be already complete can sometimes turn out to be by no means complete and even short of essential points. Contrary to differential cross sections $\sigma(\theta)$ for reactions with unpolarized beams, analyzing powers $A(\theta)$ for reactions with polarized beams can be regarded as such probes. Realizing that analyzing powers are quantities which are very sensitive to interferences between various competing reaction processes, it will be shown in this talk that $A(\theta)$ is indeed a new and powerful probe not only for the investigation of reaction mechanisms but also for the study of microscopic structures of nuclei.

The study of reaction mechanism is to elucidate what reaction processes or reaction paths compete with one another as shown in Fig. 1 for a reaction X(a,b)Y, and to analyze relative magnitudes and signs of transition amplitudes T_1, T_2, T_3, etc. of the competing processes. The cross section $\sigma(\theta)$ is expressed schematically in terms of the total transition amplitude $T \equiv T_1 + T_2 + T_3 + \ldots$ as

$$\sigma(\theta) \propto |T|^2, \qquad (1)$$

while the corresponding analyzing power $A(\theta)$ is

$$A(\theta) \propto \mathrm{Im}[TT'^*] / |T|^2, \qquad (2)$$

where T' is the transition amplitude whose value of transferred magnetic quantum number m is different by unity from that of T. Here we assume that the incident polarized particles have spin of 1/2. From these difinitions, it is obvious that $A(\theta)$ is much more sensitive to interference effects between various reaction transition amplitudes than $\sigma(\theta)$. By utilizing this characteristic of analyzing powers, extensive studies of

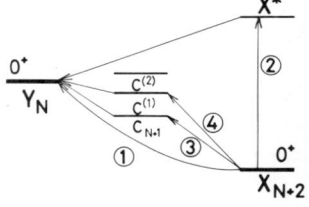

Fig. 1

nuclear reaction mechanisms as well as microscopic structures of
nuclei are being performed with a 22.0 MeV polarized proton beam
accelerated with the University of Tsukuba 12 UD Pelletron[1,2]. The
beam intensity was about 100 na on target with a typical beam polarization of 85 %. Emitted particles were analyzed with a magnetic
spectrograph and detected with a single-wire proportional counter.

SEQUENTIAL TRANSFER PROCESS IN (p,t) REACTIONS

(i) <u>Anomalous Analyzing Powers for (p,t) Ground-State Transitions:
N = 84 → 82</u> I want to start my talk by discussing the result of (p,t)
ground-state (g.s.0^+) transitions from two isotones of N = 84 to 82:
^{144}Nd(p,t) and ^{142}Ce(p,t). The two transitions show quite similar
$A(\theta)$ as well as $\sigma(\theta)$ with each other as shown in Fig. 2. The g.s.
of both ^{144}Nd and ^{142}Ce can be simply assumed to be neutron $(f_{7/2})^2$
configuration.

The conventional one-step DWBA calculation can reproduce well
the experimental cross section $\sigma(\theta)$ (Fig. 3(a)). But it cannot
reproduce the experimental analyzing power $A(\theta)$ around $\theta \simeq 20°$ at all
where the $\sigma(\theta)$ has a deep minimum. Can we recover the failure in
reproducing the observed $A(\theta)$ by modifying some optical potential
parameters in the proton and/or triton channels? The answer is no
because a derivative relation[1]

$$A(\theta) \simeq [d\sigma(\theta)/d\theta]/\sigma(\theta) \qquad (3)$$

is valid in good approximation for the present $0^+ \to 0^+$(p,t) transition
within the frame of the one-step DWBA. According to this relation,
a sharp oscillation with a negative dip and a positive peak in the
$A(\theta)$ must appear around $\theta \simeq 20°$ as far as the one-step DWBA theory
has once reproduced the first minimum in the cross section at $\theta \simeq 20°$
(Figs. 3(a) and (a')).

In consequence it can be concluded that other reaction mechanisms than the direct one-step mechanism are essential to interpret
the observed 'anomalous' $A(\theta)$ around $\theta \simeq 20°$. Then the contribution
of inelastic multistep processes via the first $2^+(2_1^+)$ states of the
initial (see the path 2 in Fig. 1) and final nuclei is taken into
account by use of random-phase-approximation (RPA) wave functions[3]
for the 2_1^+ states. The calculated cross section is, however, much
smaller than the experimental one by a factor of less than 1/20.
Therefore no improvement in fitting the experimental $A(\theta)$ can be
obtained by including the inelastic multistep processes.

In the next step two-step sequential transfer processes (the
path 3 and/or 4 in Fig. 1) are taken into account in terms of second-order DWBA theory . As is shown in Fig. 3(b'), the calculated $A(\theta)$
for the sequential transfer process is quite different not only from
the calculated one-step $A(\theta)$ but also from the experimental $A(\theta)$.
However, inclusion of both one- and two-step processes results in a
marked improvement of the analyzing power and the experimental $A(\theta)$
is reproduced quite well (Fig. 3(c')). It should be emphasized that
the <u>interference</u> between one- and two-step processes is essential
to reproduce the 'anomalous' $A(\theta)$.

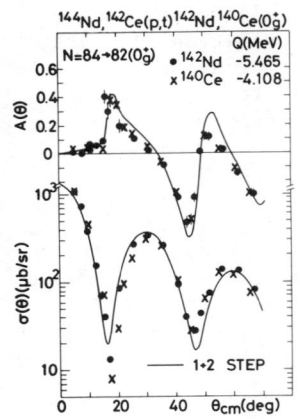

Fig. 2

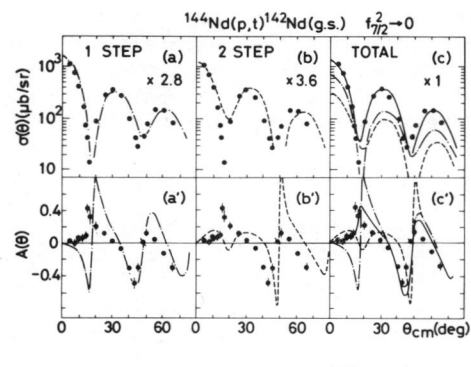

Fig. 3

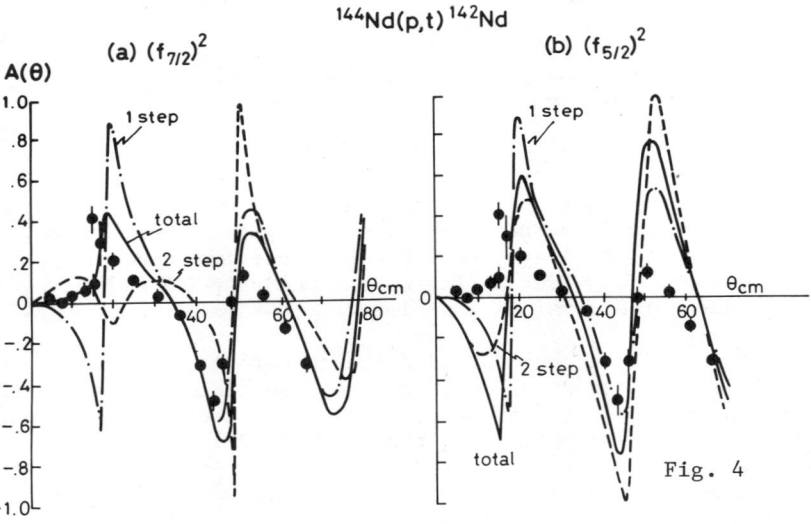

Fig. 4

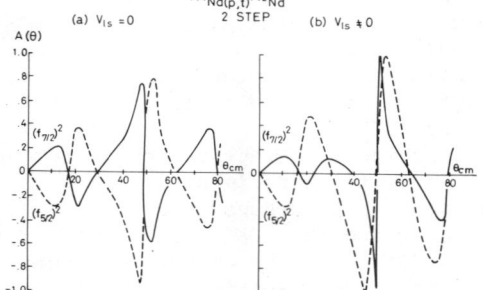

Fig. 5

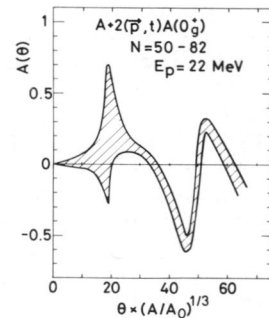

Fig. 6

In addition to $A(\theta)$, the observed $\sigma(\theta)$ is also reproduced quite well in its shape as well as in magnitude by including the sequential transfer process (Fig. 3(c)). The contribution of the two-step (p,d)(d,t) process is as much as that of the direct one-step process in the strong (p,t) g.s. transition. This important conclusion is achieved through the measurement of the analyzing power and not through the measurement of the cross section. This is because theoretical prediction for the absolute magnitudes of (p,t) cross sections has an uncertainty of a factor of at least 2 even if exact-finite range DWBA calculations are carried out; see subsection (iii).

If two-step (p,d)(d,t) calculation is done by assuming $(f_{5/2})^2$ shell configuration in stead of the $(f_{7/2})^2$, the calculated two-step analyzing power (Fig. 4(b)) shows a very different angular distribution around $\theta \approx 20°$ from that obtained from the $(f_{7/2})^2$ configuration (Fig. 4(a)). As a consequence, the sign of calculated total analyzing power at forward angles becomes negative, as shown by the solid line in Fig. 4(b), and then fails in reproducing the measurements. Fig. 4 also indicates that one-step calculation is rather insensitive to the choice of shell configurations. Consequently, this change of sign in the total analyzing power is caused by inclusion of two-step sequential transfer contribution and is quite sensitive to the choice of nuclear shell orbit, from which two neutrons are picked-up sequentially. Such nuclear shell effects appeared in two-step calculation are more clearly compared for the cases of $(f_{7/2})^2$ and $(f_{5/2})^2$, with and without spin-orbit distortions, in Fig. 5. When the spin-orbit distortions are neglected in calculation, the oscillations of two results are completely out of phase. Indeed it can be easily proved[5] by using geometrical properties of the transition amplitude that the analyzing power for the two-step sequential transfer in the case of $0^+ \to 0^+$ transition changes its sign with respect to picking-up from the shell $j_> = l + 1/2$ or $j_< = l - 1/2$, and the ratio of these analyzing powers is expressed as

$$A(\theta)_{j_>}/A(\theta)_{j_<} = -l/(l+1), \qquad (4)$$

where l and j denote the orbital and the total angular momentum, respectively, from which two neutrons are picked-up. The above-mentioned j dependence of analyzing powers is similar to that appeared in one-nucleon transfer reactions[6]. When spin-orbit distortions are included, such simple sign change described in eq. (4) is violated, as can be seen in Fig. 5(b). However, the two analyzing powers show still opposite phase oscillations at forward angles where the distortion effects are negligiblly small.

(ii) <u>Anomalous Analyzing Powers: Superconducting Nuclei</u> As an extension of a previous work[1], analyzing powers for (p,t) g.s. transitions to 21 nuclei of ^{92,94,96}Mo, ^{98,100,102}Ru, ^{102,104,106,108}Pd, ^{110,112,114}Cd, ^{116}Sn, ^{120,126,128}Te, ^{136}Ba, ^{138}Ce, ^{140}Nd, ^{142}Sm having neutron number of $N \approx 50 - 82$ have been measured. All the nuclei investigated show 'anomalous analyzing powers' around $\theta \approx 20°$, which cannot be reproduced by one-step DWBA calculations. In addition the $A(\theta)$ display distinguishable change in going one nucleus to the other in the angular distribution only around $\theta \approx 20°$ where the corresponding $\sigma(\theta)$ have the minimum. This feature is shown in Fig. 6: a hatched area covers all the data points after correcting the effect due to the nuclear radius. The observed anomalies in $A(\theta)$ and their variation with mass numbers can be explained only by including (p,d)(d,t) processes. The detailed results have been given elsewhere[1,7].

(iii) <u>Enhancement of Cross Section for ^{208}Pb(p,t)^{206}Pb Reaction due to Sequential Transfer Process</u> The neutron-orbit effect on the (p,t) analyzing powers suggests an interesting phenomenon. According to Fig. 4(b), the calculated total analyzing power becomes very similar to the one-step analyzing power when the two-step process involved is assumed to proceed via an $f_{5/2}$ orbit in stead of the $f_{7/2}$. This fact suggests that the analyzing power for a $0^+ \to 0^+$(p,t) reaction in which two neutrons are picked-up from $j_<(= l - 1/2)$ orbits like $f_{5/2}$ and/or $p_{1/2}$ must resemble the one-step analyzing power even if two-step sequential transfer processes are dominant. In this case a so-called 'anomalous' analyzing power is not observed regardless of existing intense sequential transfer processes. This is indeed the case of ^{208}Pb(p,t)^{206}Pb(g.s.0^+) reaction.

Neutron-single hole structure of ^{207}Pb is shown in Fig. 7 and the shell-model wave function[8,9] of ^{206}Pb(g.s.0^+) is given in Table I. The spectroscopic amplitude for $p_{1/2}$ orbit is largest. By using this wave function, first- and second-order DWBA calculations are carried out. Optical potential parameters for protons, deuterons, and tritons are obtained from the works of Becchetti and Greenlees[10] (BG), Perrey and Perrey[11] (PP), and Becchetti and Greenlees[12] (BG), respectively (Set I). Then experimental $A(\theta)$ and $\sigma(\theta)$ are reproduced very well (Fig. 8). As suggested, each of the three analyzing powers, which mean the total, one-step and two-step analyzing powers are similar to one another.

Table I Wave functions of the lowest two 0^+ states of ^{206}Pb

States	$p_{1/2}^2$	$f_{5/2}^2$	$p_{3/2}^2$	$i_{13/2}^2$	$f_{7/2}^2$	$h_{9/2}^2$	Ref.
g.s.0^+	0.822	0.401	0.369	-0.109	0.130	0.059	True, 8
	0.795	0.390	0.393	0.165	0.167	0.076	Orihara, 9
0_2^+(1.17 MeV)	0.495	-0.834	-0.090	0.152	-0.145	-0.083	True, 8
	0.534	-0.786	-0.122	-0.204	-0.183	-0.082	Orihara, 9

It should be emphasized that the two-step cross section is much larger than the one-step cross section. A remarkable enhancement in the two-step cross section is clearly shown in Fig. 9, where contributions of each orbit are indicated. Indeed the two-step transition amplitudes from the individual orbits add up so coherently that the following full in-phase relation is almost valid:

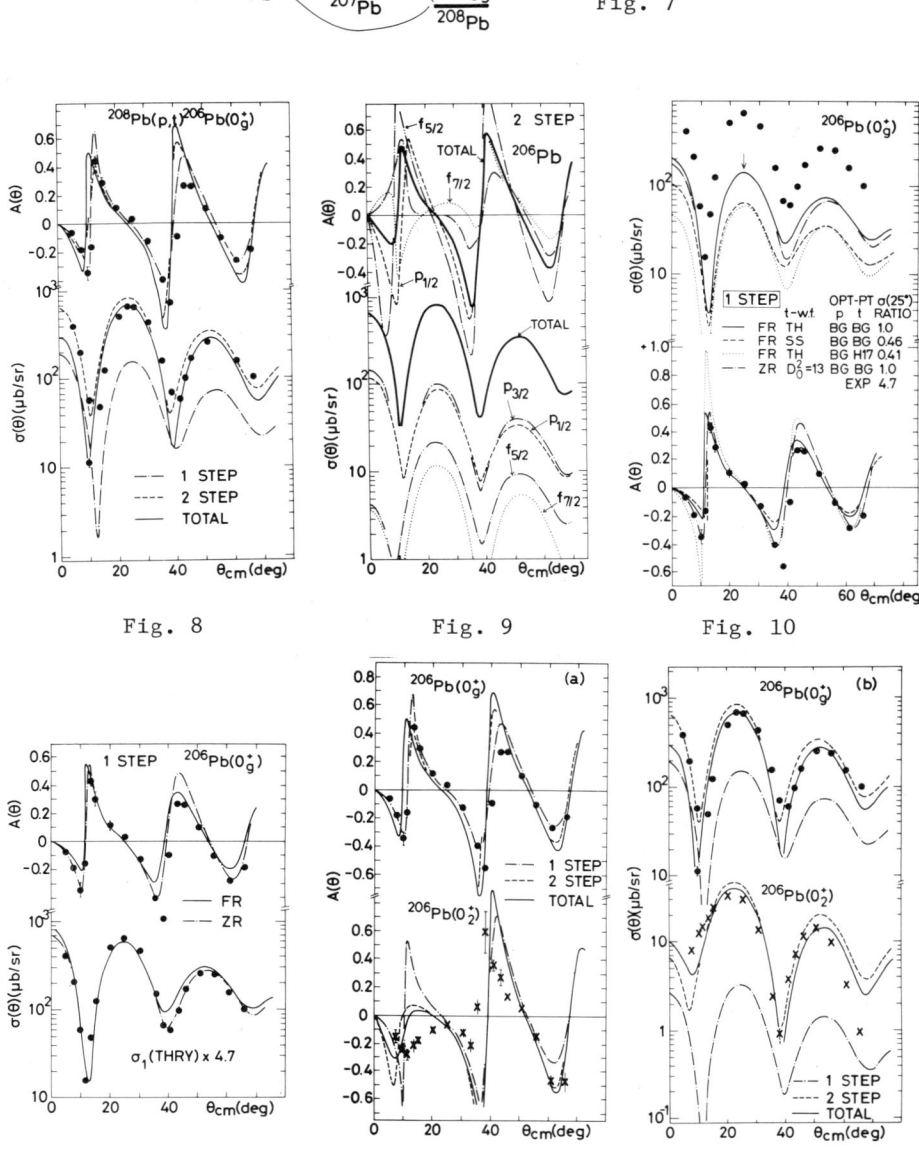

Fig. 7

Fig. 8

Fig. 9

Fig. 10

Fig. 11

Fig. 12

$$\sigma_2(2 \text{ step total}) \simeq [\sqrt{\sigma_2(p_{1/2})} + \sqrt{\sigma_2(p_{3/2})} + \sqrt{\sigma_2(f_{5/2})} + \ldots]^2. \quad (5)$$

What we have to explain next is the following point: is the one-step cross section really so small as shown in Fig. 8? In order to obtain the absolute magnitude of the theoretical one-step cross sections, exact-finite range (EFR) DWBA calculations have been performed[13] by employing realistic triton wave functions as well as realistic nucleon-nucleon interactions which cause the neutron transfer. The Tang and Herndon (TH) triton wave function[14] obtained from a potential with a hard core by a variational calculation is employed. Then EFR one-step calculations are carried out by using two sets of the optical potential parameters in order to study the effect of different optical potential parameters. The one-step calculation with Set I is shown by solid lines in Fig. 10 and that with Set II, which replaces the triton parameters in Set I by those obtained from the LASL polarization data at $E_t = 17$ MeV[15] (H17), is shown by dotted lines. An especially large amplitude oscillation found in the $A(\theta)$ near $\theta \simeq 20°$ is due to a large spin-orbit potential $V_{ls} = 6$ MeV for tritons[15]. In addition EFR calculations by employing the Sasakawa and Sawada (SS) triton wave function[16] which is obtained as a solution of the three-body Faddeev equation with Reid soft core potentials are carried out and the results are shown by broken lines. The zero range (ZR) calculation is also performed by using the potential Set I and the resultant cross section (dot-dashed line) is found to be very similar to the corresponding FR cross sections. A normalization constant $D_0^2 = 13.3 \times 10^4$ MeV^{2}fm^3 is obtained by normalizing the ZR cross section to the FR one calculated with the TH wave function using the potential Set I (solid line). All the results thus obtained are summarized in Fig. 10. Now we are ready to answer the question. The predicted one-step cross section is really much smaller than the experimental cross section by a factor from 5 to 10. Therefore we need other reaction process than the one-step process. And that is the sequential transfer process!

But for recognition of important contribution of the two-step sequential transfer process in (p,t) reactions, we should have tried to explain the ^{208}Pb(p,t) data only in terms of the direct one-step process because the one-step calculation could reproduce not only the angular shape of the observed $\sigma(\theta)$ but also the observed $A(\theta)$ so well as shown in Fig. 11. And we should have tried to justify a normalization constant 4.7 by saying that the (p,t) reaction theory has such an ambiguity in the absolute magnitude of the cross section. But the present study denies such interpretation. As a consequence, we now arrive at a quite reasonable conclusion that strong two-step sequential transfer processes in (p,t) reactions, which have been first recognized in the nuclear region of $N = 50 - 82$ as an origin of the 'anomalous' analyzing powers do exist also in the Pb region where the observed analyzing powers are 'normal'.

A strong support for the existence of the sequential transfer processes in the ^{208}Pb(p,t) reaction comes from the data of the (p,t) transition to the excited $0^+(0_2^+)$ state at 1.17 MeV in ^{206}Pb (Fig. 7). Observed $A(\theta)$ and $\sigma(\theta)$ for the 0_2^+ state are very different from those for the g.s. 0^+ in forward angles $\theta < 30°$ as shown in Fig. 12. The difference is more remarkable in the analyzing powers than in the cross sections. One- and two-step DWBA calculations for the 0_2^+ transitions are carried out by using the shell-model wave function of the ^{206}Pb(0_2^+) state of Table I in a similar manner as in the case of the g.s.0^+ transition. The one-step calculation predicts very similar $A(\theta)$ and $\sigma(\theta)$ for both of the 0^+ transitions and cannot explain the observed difference at all. However, if we include the contributions of the sequential transfer processes in the both transitions, marked difference appears in the total analyzing powers as well as in the corresponding cross sections, so that the experimental $A(\theta)$ and $\sigma(\theta)$ for the both transitions are reproduced well.

There is room for improvement in fitting the $A(\theta,0_2^+)$. This can be made easily by using a refined wave function of the 0_2^+ because the $A(\theta,0_2^+)$ is quite sensitive to the 0_2^+ wave function employed as shown in Fig. 13. Indeed two kinds of the 0_2^+ wave functions which are only slightly different in shell configurations from each other (Table I) result in a large difference in $A(\theta)$ at forward angles. In other words, we can determine the 0_2^+ wave function preceisely by the measurement of the $A(\theta,0_2^+)$. In contrast to the 0_2^+, the g.s. $A(\theta)$ is quite stable for the choice of the wave functions. It should be emphasized again that analyzing powers are more sensitive in detecting contributions of two-step processes than the cross sections.

In addition to the ^{208}Pb(p,t), (p,t) experiments on other lead isotopes ^{206,204}Pb have been done. The result of the ^{206}Pb(p,t) ^{204}Pb(g.s.0^+) together with an analysis based on the one- and two-step calculations employing the same optical potential parameters[10,11,12] as before and the similar shell-model wave functions[9] is shown in Fig. 14. A good fit to the experimental data is obtained both in $A(\theta)$ and $\sigma(\theta)$. We can find again that two-step sequential processes play an essential role in reproducing the observations. The two-step cross section decreases from that of the ^{208}Pb(p,t) because of a decrease of the neutron occupation probability of the $p_{1/2}$ orbit in ^{206}Pb. In contrast, the one-step cross section increases due to stronger pairing correlation in non-closed shell nuclei so that we have an increase of the total cross section. A decrease of the positive peak in the observed $A(\theta)$ around $\theta \approx 20°$ can be understood by taking into account of the decrease of the contribution of the $p_{1/2}$ sequential transfer process in the ^{206}Pb(p,t).

(iv) Effect of Deuteron-Distorted Wave in Sequential Transfer Process In the previous paper[1], an important effect of the wave propagation in the intermediate channel in (p,d)(d,t) sequential processes on the anomalous analyzing powers was found. Let us discuss this problem lucidly. In Fig. 15, $A(\theta)$ and $\sigma(\theta)$ for (p,t) g.s. transitions from two isotones of N = 52 to 50 are shown. A marked difference is observed in $A(\theta)$ around $\theta \approx 20°$: a sharp negative dip in ^{90}Zr and a sharp positive peak in ^{92}Mo. One- and two-step calculations are performed by assuming a simple neutron configuration of $d_{5/2}^2$ ($d_{5/2}$) for the targets (the intermediate nuclei). Exactly the same optical potential parameters are used for the both reactions[10,17,18]. Then the calculations (solid lines) can reproduce the experimental results beautifully. We can find that the inclusion of the two-step sequential processes is essential. A delicate

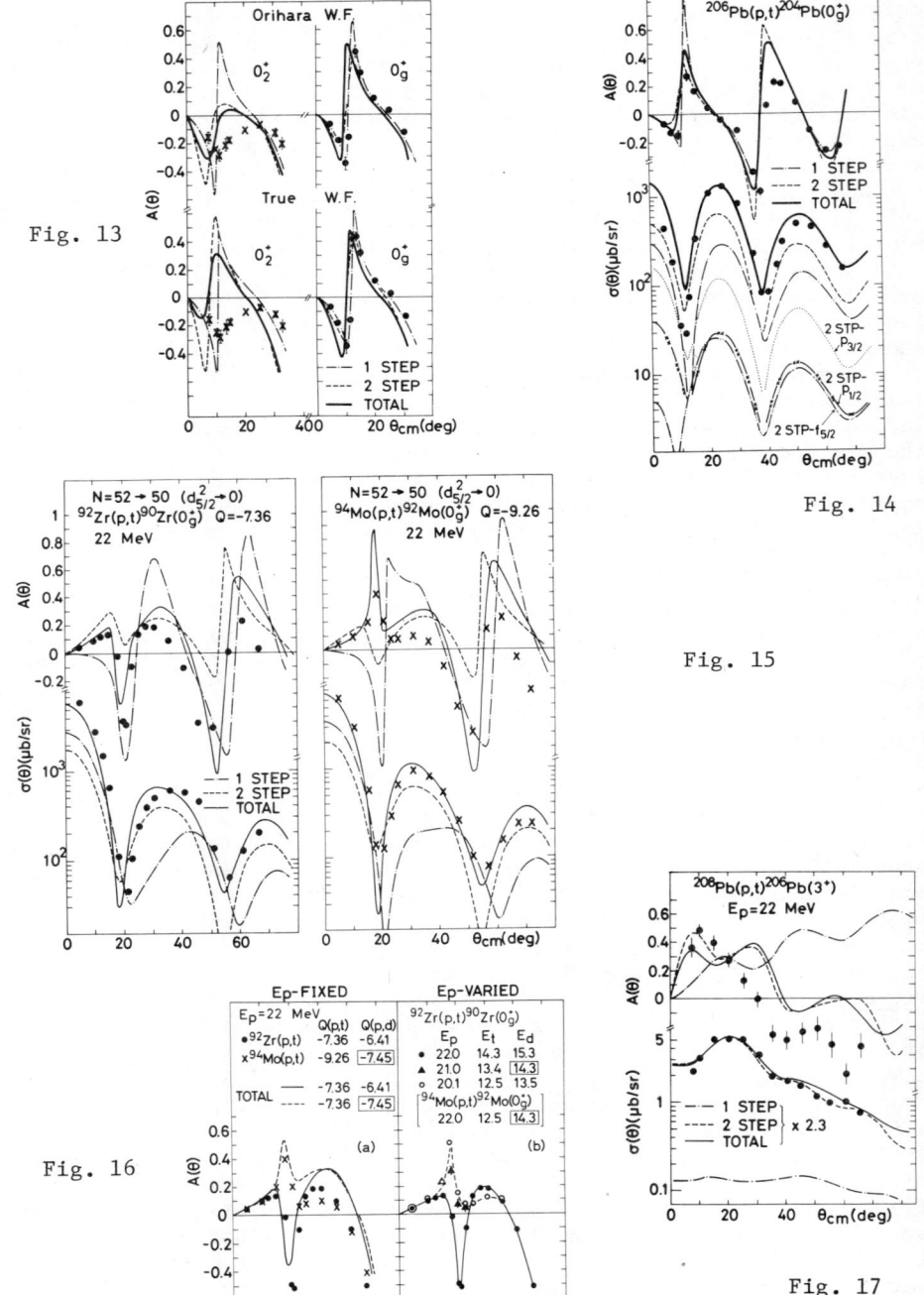

Fig. 13

Fig. 14

Fig. 15

Fig. 16

Fig. 17

interference effect between the one- and two-step processes produces a sharp negative dip in $A(\theta)$ in the case of ^{90}Zr on one hand, and a sharp positive peak in the case of ^{92}Mo on the other hand. It should be noticed that the $\sigma(\theta)$ data are also reproduced well only by including the two-step processes: we can get a flat top of the second peak in the ^{90}Zr case while a sharp peak is obtained in the ^{92}Mo case.

Now let go to the next problem, namely, to find out an origin which makes the drastic change in $A(\theta)$ for the two transitions. The original one- and two-step calculations of the ^{92}Zr(p,t)^{90}Zr are modified artificially by only replacing the Q-value of the ^{92}Zr(p,d) process by that of the ^{94}Mo(p,d). The resultant $A(\theta)$ is given by a broken line in Fig. 16(a). You can obtain an abrupt change in $A(\theta)$ from the negative dip to a positive peak. Therefore we can conclude that a striking difference in the $A(\theta)$ is caused by the difference in the propagation of the deuteron waves in the intermediate channel of the two two-step processes. It should be worth to say that the propagation of the deuteron waves is expressed by a Green's operator $G(d) = [E(A+1) - T(d) - U(d) + i\varepsilon]^{-1}$ in the transition amplitude of the two-step $(A+2)(p,d)(A+1)(d,t)(A)$ process.

In order to confirm the dependence of the deuteron energy, an incident energy of protons for the ^{92}Zr(p,t)^{90}Zr is decreased from 22 to 21 MeV so that the deuteron energy of the ^{92}Zr(p,d)^{91}Zr(g.s.) process just equals to that of the ^{94}Mo(p,d)^{93}Mo(g.s.): $E(d) = 14.3$ MeV. As denoted by triangles in Fig. 16(b), the resultant $A(\theta)$ shows a sudden change from the negative dip to a positive peak. The reason why the deuteron energy is so critical in this (p,d)(d,t) process is still an open problem.

UNNATURAL-PARITY (p,t) TRANSITION: ^{208}Pb(p,t)^{206}Pb(3^+)

There is a long story for the interpretation of the reaction mechanism for an unnatural-parity (p,t) transition to the 3^+(1.34 MeV) state in ^{206}Pb, i.e., a one-step process[19] in terms of the finite-range DWBA employing a realistic triton wave function or two-step sequential transfer process[20]. In order to settle this conflict between the interpretations, one-step vs. two-step mechanisms, the analyzing power $A(\theta, 3^+)$ for this transition has been recently measured at Tsukuba[21]. A large difference in the $A(\theta, 3^+)$ can be expected between the two reaction mechanisms because mechanism of spin transfers in the one- and two-step processes is quite different from each other.

In a contribution by Igarashi and Kubo[22], exact evaluations of the both processes have been carried out employing realistic triton and deuteron wave functions and realistic interaction which causes the transfers of each step. The Sasakawa-Sawada wave function[16] for triton and the Reid soft core potentials are used. Although the calculations are not the final ones, the following characteristic features can be found in Fig. 17. The calculated one-step analyzing power is so different from the observed $A(\theta, 3^+)$ that dominance of the one-step process in the 3^+ transition is impossible. On the contrary, the two-step calculations are reasonably successful in reproducing the principal features of the observed $A(\theta,3^+)$. Therefore the dominant process of this transition is the two-step one. Actually the calculated cross sections show that the one-step process is much weaker than the two-step one by about one order in magnitudes of the cross sections. Lack of the magnitude of the theoretical cross sections by a factor of 2.3 is traced back to that of the theoretical (d,t) cross section in the sequential transfer process. Possible other processes in the excitation of the 3^+ state are an inelastic multistep process via the 3^- state in ^{208}Pb and a deuteron-breakup process in the (p,d)(d,t) sequential transfers. Both contributions have been estimated by Toba et al.[23] and found to be negligible.

IMPORTANCE OF TWO-STEP PROCESS IN ^{14}N(p,p')^{14}N(2.31 MeV) REACTION

An important contribution of two-step sequential transfer processes recognized so far in (p,t) reactions suggests that an effect due to such processes should be carefully investigated in other reactions such as an inelastic scattering of nucleon by nuclei. A nice example is presented by Aoki et al.[24] who have studied an inelastic scattering ^{14}N(p,p')^{14}N(2.31 MeV) and succeeded in extracting an effective interaction responsible for the inelastic scattering by taking into account the effect of a strongly competing sequential transfer process. In order to get reliable information for the spin- and isospin-dependent part of the effective interaction from this spin- and isospin-flip scattering $(1^+,T=1) \rightarrow (0^+,T=0)$ (Fig. 18), a contribution of competing (p,d)(d,p') processes through ^{13}N should be assessed. Spin and isospin are transferred by a nucleon in the transfer reactions so that second-order DWBA term of spin- and isospin <u>independent</u> interaction can compete with the first-order DWBA term of the weak spin- and isospin-dependent interaction. Indeed the inelastic cross section is two orders in magnitude smaller compared to the ^{14}N(p,d)^{13}N.

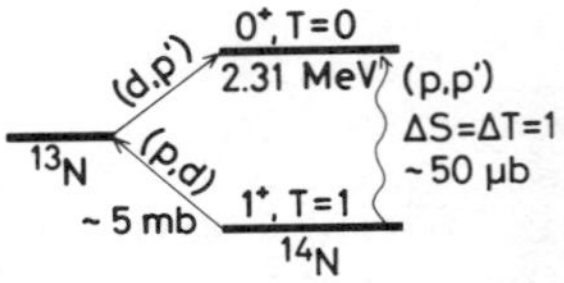

Figure 2 by Aoki et al.[24] compares the experimental data with the one-step predictions of interaction strengths around 3.5 MeV for both central (V_C) and tensor (V_T) terms. It is found that the analyzing power data are not reproduced within the frame of the one-step calculations. However, inclusion of the (p,d)(d,p') process can reproduce the observed $A(\theta)$ as well as the $\sigma(\theta)$ as shown in Fig. 4 of the contribution. The inclusion reduces the strength of the effective interaction by a factor of 2; $V_C = -1.74$ and $V_T = -1.31$ MeV. It should be emphasized that the analyzing power data play an essential role in obtaining the effective interaction.

Fig. 18

ANALYZING POWERS FOR (p,t) REACTIONS AT E_p = 52 MeV

It is interesting to investigate the degree of contribution of two-step processes in (p,t) reaction at higher incident energies by utilizing analyzing powers as a probe. Measurements of $A(\theta)$ for ^{104,110}Pd(p,t)^{102,108}Pd(g.s.0^+ and 2_1^+) have been done by using 52 MeV

polarized protons accelerated by the RCNP
Osaka cyclotron. It should be noted first
that absolute magnitudes of the observed $A(\theta)$
(Fig. 19 and 20) are generally larger than
those obtained with the 22 MeV protons. This
is due to the fact that the strength of spin-
orbit potentials in the incident and/or exit
channels becomes larger because of an increase
of effective orbital angular momenta at higher
energies. Fig. 19 shows that the observed
$A(\theta)$ for the g.s.0^+ transition can be reproduced at forward angles only by including the
(p,d)(d,t) processes. The two-step processes
are as strong as the one-step process.

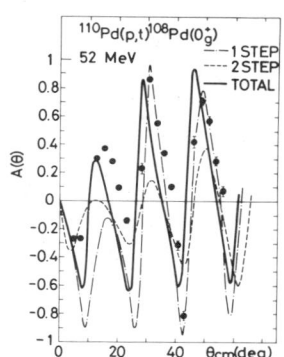

Fig. 19

In the excitation of the 2_1^+ state,
inelastic multistep processes are dominant
as shown by a dashed line in Fig. 20. A
small cross reaction of the one-step process
(dash-dotted line) is due to a large cancellation between the spectroscopic amplitude
arising from the forward-scattering amplitude
in RPA and that arising from the backward-
scattering amplitude[3,25]. The nuclear wave
functions for the g.s.0^+ and 2_1^+ states are
constructed under the BCS and quasiparticle-
RPA model by using both the pairing and Q-Q
interactions. The $A(\theta,2_1^+)$ results clearly
demonstrate first that the interference between the one- and inelastic two-step processes is essential to reproduce the observed
analyzing power and secondly that a destructive interference predicted by the BCS-RPA
theory is definitely confirmed. If a constructive interference is artificially assumed
by replacing the direct form factor F_2 with -
F_2, the resultant analyzing power (dotted
line) cannot reproduce the observations at
all. The cross section data also confirm the
destructive interference.

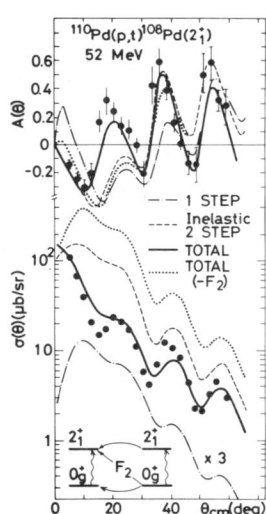

Fig. 20

It should be noted that the original BCS-RPA wave functions of
2_1^+ states are valid enough to predict the observed destructive interference but are insufficient to obtain good fitting to the experimental data. Actually a modified form factor which is obtained by
multiplying the original form factor by a factor 4 ($4F_2$) is used to
get a good fit as shown in Fig. 20. This modification, which is
exactly consistent with the previous calculations[3,25], arises from
the inadequacy of the RPA wave functions of the 2_1^+ states because
of large anhamonicity in the quadrupole oscillation of Pd isotopes[25,26]. In any way, the results thus obtained at $E(p) = 52$ MeV completely
agree with those previously obtained at $E(p) = 22$ MeV[3].

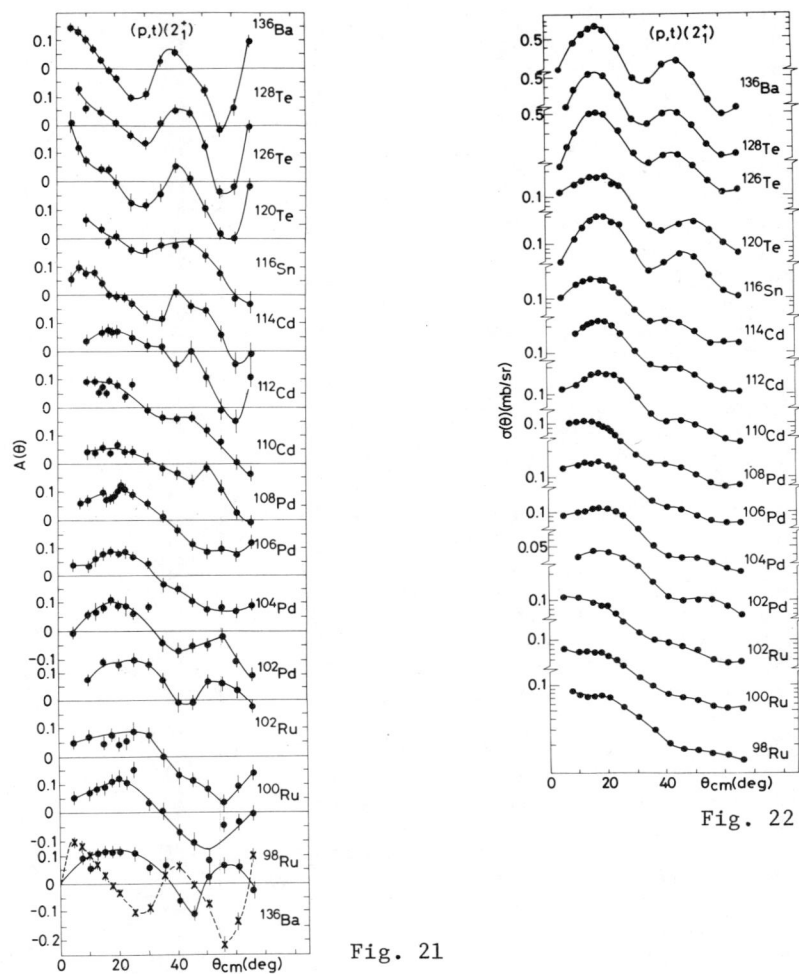

Fig. 21

Fig. 22

FURTHER STUDY OF $A(\theta, 2_1^+)$ FOR (p,t) TRANSITIONS IN MEDIUM-MASS VIBRATIONAL NUCLEI

As an extension of the previous works[2,3] in which microscopic features of collective quadrupole oscillations of nuclei have been studied, both $A(\theta, 2_1^+)$ and $\sigma(\theta, 2_1^+)$ for (p,t) reactions with 22 MeV protons have been measured systematically for 21 nuclei of $N \approx 50 - 82$ (Figs. 21 and 22). An analysis within the frame of the RPA method is found to be insufficient for the nuclei having large quadrupole deformation parameters. A further analysis by taking into account anharmonicity effect is needed.

In summary, measurements of polarization quantities can be regarded as the useful probe for the sensitive detection of interference effects between various reaction transition amplitudes. Based on such a new-point of views, the understanding of nuclear reaction mechanisms as well as microscopic structures of nuclei is being achieved. This is indeed a qualitatively new application of polarization experiments in nuclear physics.

I would like to thank Y. Aoki, M. Igarashi, H. Iida, T. Kishimoto, K.-I. Kubo, S. Kunori, K. Nagano and Y. Toba for their collaborations. This work is supported in part by Nuclear and Solid State Research Project, University of Tsukuba.

REFERENCES

1. K. Yagi, S. Kunori, Y. Aoki, K. Nagano, Y. Toba, and K.-I. Kubo, Phys. Rev. Lett. $\underline{43}$, 1087 (1979).
2. K. Yagi, S. Kunori, Y. Aoki, K. Nagano, Y. Tagishi, and Y. Toba, Phys. Rev. $\underline{C19}$, 285 (1979).
3. K. Yagi, S. Kunori, Y. Aoki, Y. Higashi, J. Sanada, and Y. Tagishi, Phys. Rev. Lett. $\underline{40}$, 161 (1978).
4. N. Hashimoto and K. Kawai, Prog. Theor. Phys. $\underline{59}$, 1245 (1978).
5. K.-I. Kubo, Phys. Rev. $\underline{C}$, to be published.
6. T. Yule and W. Haeberli, Phys. Rev. Lett. $\underline{19}$, 756 (1967).
7. S. Kunori, in Proc. Tsukuba Symp. Polarization Phenomena in Nuclear Reactions, Tsukuba, 1979, ed. K. Yagi (Univ. of Tsukuba, 1980) p.160; S. Kunori et al., to be published.
8. W. True, Phys. Rev. $\underline{168}$, 1388 (1968).
9. H. Orihara, private communication (1980); T. T. S. Kuo and G. H. Herling, Naval Research Laboratory Report $\underline{2258}$ (1971).
10. F. D. Becchetti, Jr., and G. W. Greenlees, Phys. Rev. $\underline{182}$, 1190 (1969).
11. C. M. Perey and F. G. Perrey, Phys. Rev. $\underline{132}$, 775 (1963).
12. F. D. Becchetti, Jr., and G. W. Greenlees, in Polarization Phenomena in Nuclear Reactions, ed. H. H. Barschall and W. Haeberli (Univ. of Wisconsin, 1971) p.682.
13. M. Igarashi, computer code FNR-TWOSTEP, unpublished.
14. Y. C. Tang and R. C. Herndon, Phys. Lett. $\underline{18}$, 42 (1965).
15. R. A. Hardekopt, R. F. Haglund, Jr., G. G. Ohlsen, W. J. Thompson, and L. R. Vesser, Phys. Rev. $\underline{C21}$, 906 (1980).
16. T. Sasakawa and T. Sawada, Phys. Rev. $\underline{C19}$, 2035 (1979).
17. S. A. Hjorth, E. K. Lin, and A. Johnson, Nucl. Phys. $\underline{A116}$, 1 (1968).
18. E. R. Flynn, D. D. Armstrong, J. G. Beery, and A. G. Blair, Phys. Rev. $\underline{182}$, 1113 (1969).
19. M. A. Nagarajan, M. R. Strayer, and M. F. Werby, Phys. Lett. $\underline{68B}$, 421 (1977).
20. N. B. de Takacsy, Phys. Rev. Lett. $\underline{31}$, 1007 (1973); L. A. Charlton, Phys. Rev. $\underline{C14}$, 506 (1976).
21. Y. Toba, Y. Aoki, S. Kunori. K. Nagano, and K. Yagi, Phys. Rev. $\underline{C20}$, 1204 (1979).
22. M. Igarashi and K.-I. Kubo, Contribution to this Conference, 2. 112.
23. Y. Toba, Y. Aoki, H. Iida, S. Kunori, K. Nagano, and K. Yagi, Contribution to this Conference, 2.110.

24. Y. Aoki, K. Nagano, Y. Toba, S. Kunori, and K. Yagi, Contribution to this Conference, 2.48.
25. K. Yagi, Y. Aoki, M. Matoba, and M. Hyakutake, Phys. Rev. $\underline{C15}$, 1178 (1977).
26. K. Yagi, S. Kunori, Y. Aoki, K. Nagano, Y. Tagishi, and Y. Toba, in Proc. Intern. Symp. Direct Reaction Mechanism, Fukuoka, 1978, ed. M. Tanifuji and K. Yazaki (INS, Univ. of Tokyo, 1978) p.137.

(Editor's Note: References to this Conference are given as paper numbers abd are cross-indexed to page numbers in the Table of Contents.)

DISCUSSION

SLOBODRIAN: Could you explain, how did you calculate the final cross section in the palladium case, is this done with some fudge factor or is it parameter-free calculation? Are you adding coherently the amplitudes? Are two- and one-step processes assumed coherent?

YAGI: The amplitudes are added coherently. Solid lines in the figure for the ^{110}Pd(p,t)^{108}Pd(2^+_1) transition (Fig. 20) are the final results of the cross section and the analyzing power arising from the destructive interference between the direct one-step process (dash-dotted line) and the inelastic multistep processes (dashed line). There are no free adjusting parameters in this calculation.

HANNA: Has the very striking effect that you observed in the A=92 nuclei been observed in any other cases?

YAGI: The (p,t) reaction on the pair of isotones ^{92}Zr and ^{94}Mo has been so far the only example of the isotones which shows such a drastic change in the analyzing powers.

DAEHNICK: You have shown that the inclusion of sequential stripping amplitudes in (p,t) results in dramatic changes in the predicted analyzing powers. You also pointed out that in $0^+ \to 0^+$ transitions the sequential amplitudes add constructively and become larger than the one-step term. Given that 0^+ ground-state wave functions tend to have admixtures of many j^2 terms what criteria do you use to limit the configurations (and intermediate states) which are considered in the computation of the sequential stripping amplitudes?

YAGI: The configurations arising from the one-major shell are considered. Thus, in the calculations of the ^{208,206,204}Pb (p,d)(d,t) ^{206,204,202}Pb transitions, the six states $p_{1/2}$, $p_{3/2}$, $i_{13/2}$, $f_{5/2}$, $f_{7/2}$, and $h_{9/2}$ were taken into account; see Fig. 7 and Table I. In the case of Pd isotopes, the five quasiparticle states $d_{5/2}$, $s_{1/2}$, $g_{7/2}$, $h_{11/2}$ and $d_{3/2}$ were considered (Ref. 1).

ROST: Do you include unbound (n-p) states in your 2-step sequential transfer calculations?

YAGI: The contribution of deuteron breakup effect in the (p,d)(d,t) processes is not taken into account in the calculations of my talk. However, this effect has been estimated in another case (Contribution 2.110 by Toba et al.) and found to be small at E_p = 22 MeV.

ROST: How do you treat non-orthogonality corrections?

YAGI: The non-orthogonal terms in the (p,d)(d,t) processes are included in the present calculations by using the method described by Kawai and Hashimoto in Ref. 4.

SURVEY OF DIRECT REACTION STUDIES WITH POLARIZED TRITONS

E. R. FLYNN

Los Alamos Scientific Laboratory
Los Alamos, NM USA 87545

ABSTRACT

There now exists a large amount of direct reaction data utilizing the LASL polarized triton beam. Systematic studies of elastic scattering $(\vec{t},\alpha)$ and $(\vec{t},p)$ reactions have been performed on a variety of medium and heavy nuclei and the results satisfactorily interpreted in terms of DW calculations. Limited (t,d) and (t,t') results are also available. The sensitivity of the measured analyzing powers to reaction theory is also being explored in a variety of tests involving both forbidden and allowed transitions. Comparison to coupled channel calculations does show improvement in a number of cases.

INTRODUCTION

The polarized triton source at the LASL Van de Graaff facility has been in operation now for several years and has produced a large amount of interesting and exciting results. The source, which may be used to inject triton ions into a FN Van de Graaff for acceleration up to 17 MeV, produces on-target intensities of up to 100 na with 50-60 na being an average value.[1] Except for the elastic scattering and few nucleon experiments, the majority of the work has been done using the Q3D magnetic spectrometer with helical focal plane detector system.[2] This apparatus with its large solid angle has permitted the measurement of very small cross sections for polarized work, <1 μb, while at the same time allowing excellent energy resolution, usually ∿ 15 keV. These attributes thus have permitted a very diversified program ranging from detailed spectroscopy of complex nuclei to investigating the complex reaction mechanism responsible for the excitation of weakly excited forbidden states.

The triton has several unique properties which may be utilized in a variety of direct reaction programs. These have previously been summarized in a paper which details earlier efforts with the polarized triton beam.[3] However, it is worth commenting on some of these again as they illustrate the usefulness of such a beam. Figure 1 lists these properties. The present paper emphasizes the $(\vec{t},\alpha)$ and $(\vec{t},p)$ reactions with some discussion of $(\vec{t},d)$ and other results.

The $(\vec{t},\alpha)$ is chosen because of the enormous success of using a polarized beam instead of unpolarized. Because of a large momentum mismatch in ths reaction, the differential cross sections tend to be somewhat structureless in the heavier masses and may yield ambiguous angular momentum transfer, ℓ. The very large, well structured

0094-243X/81/690270-12$1.50 Copyright 1981 American Institute of Physics

analyzing powers, A_y, on the other hand, give distinctive spin information and have made this a most valuable tool for proton hole state information throughout the periodic table. The two nucleon transfer reactions have long been a source of interest both in their spectroscopic application and in the reaction mechanism. The application of two nucleon transfer reactions to the study of elementary excitations has led to the distinctive role of the pairing excitation (or vibration) in this theory. The (t,p) data have played a major role in this program being the principle two neutron stripping reaction employed. The addition of the measurement of A_y to the cross section has permitted a more definitive spin measurement in the case where the shape of the differential cross section was not sufficiently distinctive to differentiate between adjacent L-transfers. Another important aspect of the (t,p) reaction is the sensitivity of A_y to the reaction mechanism.

It has been considered an important aspect of reaction theory for some time to understand the mechanism involved in multinucleon transfer reactions. The addition of A_y data to cross section information is a further test of whether particles are transferred as a pair or one by one or how other reaction channels are involved in the process. Some of the new polarized triton data directly address this question.

THE ($\vec{t},\alpha$) REACTION

There have been extensive ($\vec{t},\alpha$) measurements already published in the lead region and throughout the rare earths. As mentioned earlier, these results have been extraordinarily successful and

TRITON PROPERTIES
$S = \frac{1}{2}, t = \frac{1}{2}, Z = 1, (n)_o^2$

Principal Reactions Studied

(t,α)-proton pickup, (t,d)-neutron strip
(t,t')-$t_z=\frac{1}{2}$ inel. scat., (t,p)-two neut.
transfer, (t,^{3}He)-charge exch. t=1

Beam and Experimental Properties

i_a = 50-60na, P_a = 0.75 spot = .75x2 mm
$\delta E(Q3D)$= 15-20 KeV Solid Ang. = 14.3 msr

Fig. 1. A summary of properties of a polarized triton beam

provided encouragement for further use of this spectroscopic tool. Discussed here are three recent measurements: (1) a measurement of masses and spins of neutron rich nuclei near the transition region at A=100, (2) a measurement on Ir isotopes to examine aspects of a supersymmetry theory recently proposed, and (3) measurements on a nonzero spin target.

In the first of these studies a beam of 17 MeV polarized tritons of ~ 50 na intensity was used to bombard targets of ^{110}Pd, ^{104}Ru, ^{96,98,100}Mo and ^{96}Zr. A spectrum of ^{110}Pd(t,α)^{109}Rh is shown in Fig. 2. Large A_y values for this reaction are seen in these Q3D results, and examples of a number of A_y values for different spins are shown in Fig. 3 where they are compared to distorted wave (DW) results as calculated with standard t and α potentials[8,9] using the measured spin-orbit strength for the triton. This study represents the first examination of the nucleus ^{109}Rh and yields a nuclear mass excess of $-84.800 \pm .04$ MeV as compared to the systematic estimate of -85.11 MeV.[10] The results of Fig. 3 indicate a ground state spin of $9/2^+$ with eight other spin assignments made among the 18 levels observed. These data indicate the power of this technique for measuring ground state mass, excited state energies, total spins and spectroscopic factors of such neutron rich nuclei. There also exists in this region an interesting group of states whose spins cannot be assigned because of their anomalous A_y values. These apparently are particle-vibration states formed by coupling of the type $[3^- \times 1/2^-]_{5/2^+, 7/2^+}$ and are excited by two-step processes involving inelastic excitation. Other possible multiplets in this region are the $[2^+ \times 9/2^+]_{5/2^+ \ldots 13/2^+}$ and $[2^+ \times 1/2^-]_{3/2^-, 5/2^-}$ although the latter will mix with allowed $p_{3/2}$ and $f_{5/2}$ components

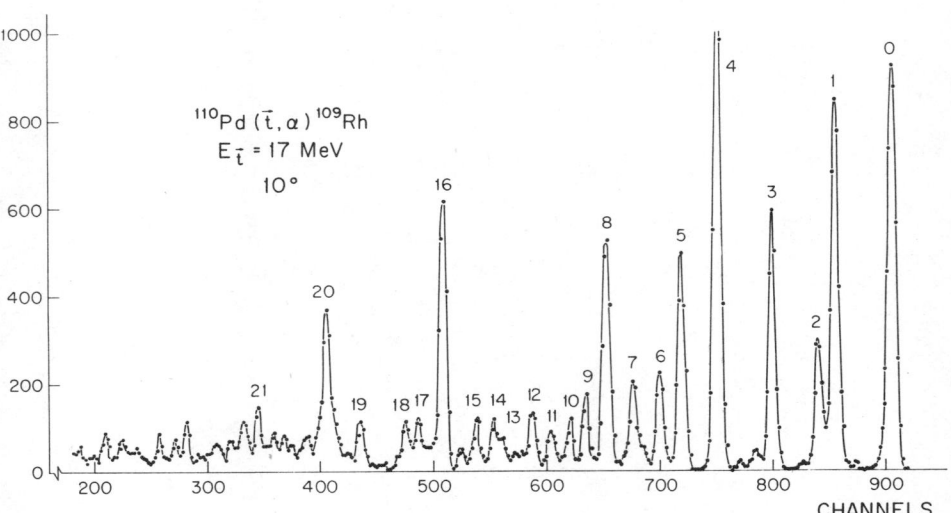

Fig. 2. Spectra of the ^{110}Pd($\vec{t}$,α)^{109}Rh Reaction.

and probably explain some
of the frgmentation of
these states. These re-
sults will soon be pub-
lished in total.[11]

In this same study, we examined the ^{96,98,100}Mo $(\vec{t},\alpha)$ reaction to explore the possible role of the proton $g_{9/2}$ orbital in the onset of a slope deformation with increasing A as suggested by Federman.[12] The observed $g_{9/2}$ strength in these three nuclei as determined by the A_y and $d\sigma/d\Omega$ values is <u>1.5</u>,<u>1.1</u>, and <u>1.4</u> for 98, 98 and 100 targets, respectively. These results are being combined with a (t,p) study of the Mo nuclei to compare to the IBA model[13] and the Federman predictions.[12]

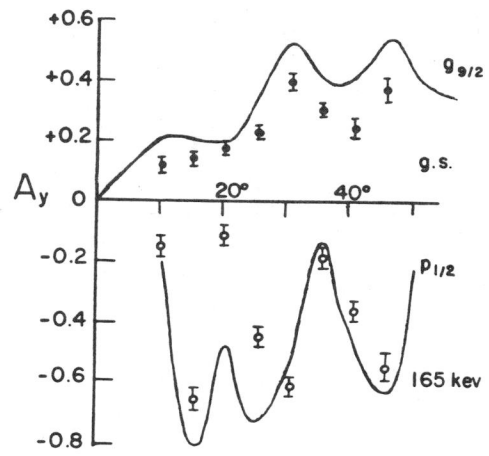

Fig. 3. Examples of A_y for various spins for the ^{110}Pd$(\vec{t},\alpha)^{109}$Rh reaction.

The application of the $(\vec{t},\alpha)$ reaction in the rare earth region has produced such marked success that it was logical to use it to test the new theory of supersymmetry suggested to be applicable to the Ir nuclei.[14] This was done through the Pt$(\vec{t},\alpha)$ reaction using targets of A=194, 196 and 198 to study ^{193,195,197}Ir. In Fig. 4 we show spectra from the ^{198}Pt target for the spin up and down cases at θ=30°. One can see from this figure the large A_y values encountered again in this region. The three ground state transitions, which are all identified here as $3/2^+$, are shown in Fig. 5 where they are compared to DW calculations assuming a $3/2^+$ transfer.

A preliminary analysis of these data is shown in Fig. 6 in a comparison to supersymmetry predictions. Here N is the total number of neutron and proton bosons outside of the lead closed shell. A normalization factor of 23 for the DW calculations is assumed (see Ref. 7 for optical model parameters) which considerably overestimates the lead $d_{3/2}$ state but gives the correct value for the $s_{1/2}$. The results indicate that the Ir ground states see a rapidly increasing fraction of the $d_{3/2}$ strength reaching 50% of the total available at ^{197}Ir. ^{193}Ir agrees with the supersymmetry prediction of $3/2^+$ splitting, although no limit could be placed on the $1/2^+$ strength which should be small. The heavier Ir isotopes disagree substantially with this model both in the $3/2^+$ ratio as well as exciting significant $1/2^+$ strength. Thus ^{193}Ir is the only case which qualifies from these data as a possibility for supersymmetry.

As a final example of the application of the $(\vec{t},\alpha)$ reaction, we consider here the case of mixed ℓ-values as possible in using a non-zero spin target. The reaction discussed is the ^{159}Tb$(\vec{t},\alpha)^{158}$Gd case[16] and we shall only consider the ground state band of ^{158}Gd.

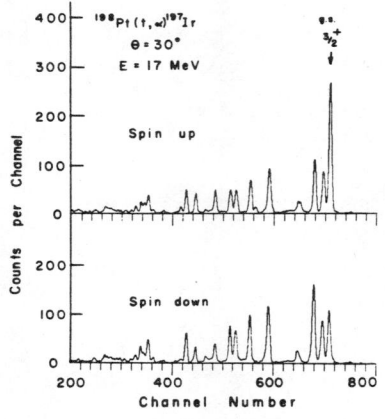

Fig. 4. Spectra of the $^{198}(t,\alpha)^{197}$Ir reaction for both spin up and spin down.

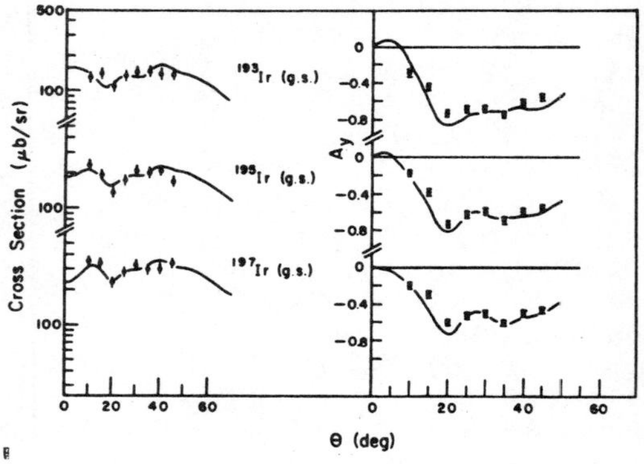

Fig. 5. A_y and cross section results for the ground state transitions in the ^{194,196,198}Pt$(\vec{t},\alpha)$ reaction.

Search for Supersymmetry in Ir Nuclei by Pt(t,α) Reaction

Quantum numbers give the relation:

$$\frac{\sigma(N-\tfrac{1}{2},\tfrac{1}{2})}{\sigma(N+\tfrac{1}{2},\tfrac{1}{2})} \approx \frac{N}{N+4} \sim .70 \quad \text{for N bosons out of closed shell}$$

Predicts two $3/2^+$ states will be seen with above ratio. Predicts no excitation of $1/2^+$ state.

Results

Final Nucleus	E_x (KeV)	J	S_{rel}^a	S_{ratio}
^{193}Ir	0	$3/2^+$	0.19	1.0
	180	$3/2^+$	0.01	0.07
	460	$3/2^+$	0.14	0.72
	73	$1/2^+$	(unr)	-
^{195}Ir	0	$3/2^+$	0.26	1.39
	288	$3/2^+$	0.06	0.31
	69	$1/2^+$	0.48	-
^{197}Ir	0	$3/2^+$	0.39	2.07
		$1/2^+$	0.78	-

a) relative to ^{207}Tl

Fig. 6. Preliminary interpretation of the Pt$(\vec{t},\alpha)$IR results in terms of a supersymmetry model.

The ground state band is populated by picking up the $3/2^+[411]$ orbital which originates in the $2d_{5/2}$ shell model orbital and is thus dominated by transitions involving it. The ground state involves pickup of a $3/2^+$ particle and is thus very weak as seen in Fig. 7, although it is well described by the DW calculations. The 2^+ is dominated by the $5/2^+$ orbital as the data indicates. The 4^+ contains strong admixtures of $7/2^+$ and $9/2^+$ which should reduce A_y as the solid curve indicates but the data are fit better by assuming again only $5/2^+$ as indicated by the dashed line. The 6^+ is only populated by a $9/2^+$ direct transition but the A_y data disagree substantially with the DW predictions. Thus the 4^+ and 6^+ indicate serious problems with normal DW calculations and weaker $(\vec{t},\alpha)$ transitions. We shall return to this point later.

THE $(\vec{t},d)$ REACTION

The $(\vec{t},d)$ reaction is discussed only briefly here but the results are encouraging. To date, two experiments have been performed utilizing this reaction, a study of the Ni isotopes[17] and a study of ^{227}Ra.[18] The latter reaction has been used to limit spin values and give spectroscopic values in a large collaboration using a variety of techniques to study ^{227}Ra. The former study is being used to examine the trend toward deformation of the heavier Ni isotopes. An example of A_y values and typical DW calculations are shown in Fig. 8. The A_y values ae not as large as in the $(\vec{t},\alpha)$ case but are still significant and very well described by DW theory. The advantage of the $(\vec{t},d)$ over $(\vec{d},p)$ lies in the larger cross section to higher spin states and the excellent DW results due to the strong

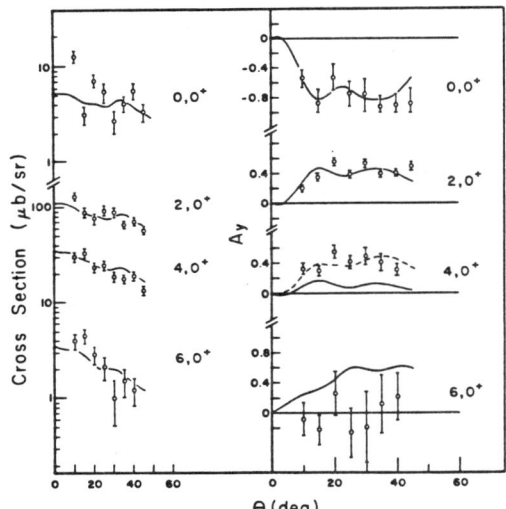

Fig. 7 Cross section and A_y results for the ground state band excited in the ^{159}Tb$(\vec{t},\alpha)^{158}$Gd reaction.

absorption of particles in ingoing and outgoing channels. The $(\vec{t},d)$ reaction thus appears to be an excellent spectroscopic tool for single neutron stripping.

THE $(\vec{t},p)$ REACTION

The use of a polarized triton beam in the two nucleon transfer reaction $(\vec{t},p)$ has led to surprisingly little new spectroscopic information. In large part this is due to the fact that, for even-even target nuclei, the final spin is given by the transferred ℓ-value. Thus, the only additional information gained is from the ability to distinguish different spins based on A_y characteristics when the differences in the shapes of the angular distributions are too small.

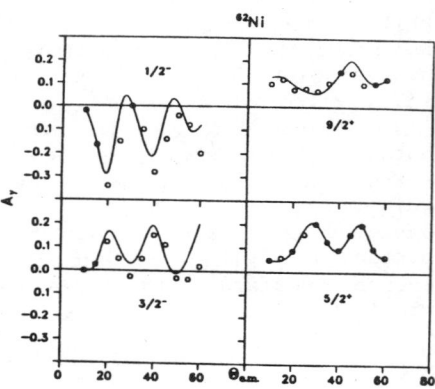

Fig. 8. Typical values of A_y for the ^{62}Ni$(\vec{t},d)$ reaction.

Such a situation does exist in a variety of cases such as the Zr(t,p) reaction and this is illustrated in Fig. 9 in work from Ref. 5.

Another attempt to utilize the A_y information for spectroscopic information has been in the examination of the nuclear shape-transition region which exists near A=100. Here the ^{96,100}Mo$(\vec{t},p)$ reaction was employed to see if changes in the characteristic L=0 or 2 A_y values could be seen depending upon whether the final state was deformed or spherical.[19] Unfortunately, the data as shown in Fig. 10 indicate that the reaction picks out only that piece of the wavefunction which has a strong overlap with the target ground state. This is indicated by a conservation of pairing boson strength which goes from initially being all in the ground state transition for the 96 → 98 case to being split between ground state and the 0.696 MeV state in the 100 → 102 case. The observed L=0 A_y values are thus essentially identical since they represent pieces of the same photon and no interference with the deformed state is noted. There are some differences in the L=2 A_y values.

In general, the A_y values obtained from the $(\vec{t},p)$ reaction are rather well described by normal DW techniques and coupled channel analysis are not required. Examples of this behavior are shown in Figs. 11 and 12 where the differential cross sections and A_y values

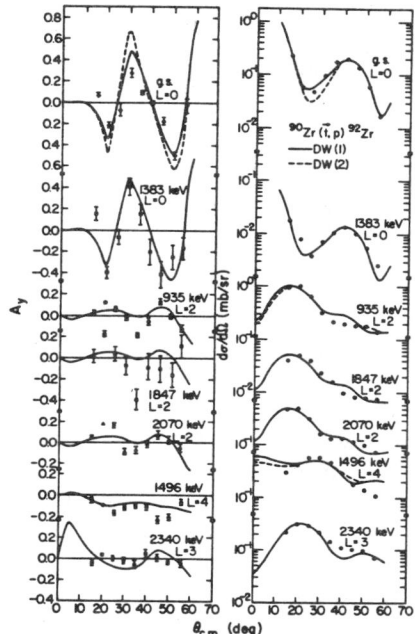

Fig. 9. A_y results from the ^{90}Zr$(\vec{t},p)^{92}$Zr reaction.

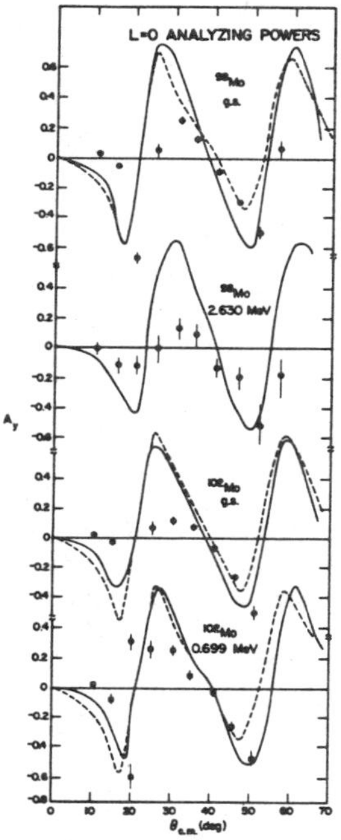

Fig. 10. A_y results from the (Mo$(\vec{t},p)$) reaction to 0^+

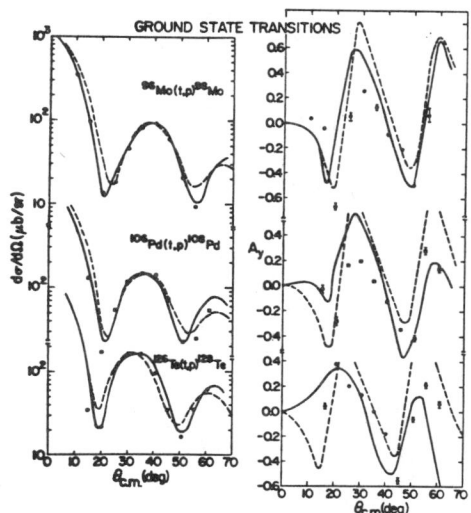

Fig. 11. A_y results for L=0 transitions on targets of ^{96}Mo, ^{106}Pd and ^{126}Te.

are shown for the $(\vec{t},p)$ reaction on targets of ^{96}Mo, ^{105}Pd and ^{129}Te using two different optical model potentials. There are certain exceptions, however, and these are described in the next section.

COMPLEX REACTION MECHANISM PROCESSES

The distinctive A_y characteristics and their sensitivity to the nuclear reaction process make measurements of A_y a useful test of various reaction theories. These properties show up in anomalies observed in angular distributions of weak states or in the process which permits the excitation of states forbidden in first order. Such transitions have been observed in all the reactions studied and are always more obvious in the measured A_y values than the cross section.

The $(\vec{t},\alpha)$ reaction in the rare earths has produced a number of examples of anomalous A_y cases and some of these are pointed out in Fig. 13. It is seen that the $7/2^+$ spin from the $3/2^+$[411] configuration gives consistent A_y values for several targets but vastly different from the $7/2^+$ spin of the $5/2^+$[413]. Such effects on (t,α) cross sections have recently been described in terms of the CCBA using inelastic scattering through the quadrupole state.[20] Calculations which also describe the A_y are now in progress.

In a contribution to this conference, Alford et al.[21] have measured the ^{206}Pb$(\vec{t},p)$ cross section and A_y to the 4^- and 5^- states of ^{208}Pb. These results are shown in Fig. 14 where they are compared to either DW or CCBA calculations. In the case of the 5^- state, this is an allowed transition in first order and a DW calculation is shown assuming a $(p_{1/2}\ g_{9/2})_{5^-}$ configuration for the final state. The figure indicates that the cross section is well fit in shape but A_y is not. The 4^- state being of unnatural parity is not allowed in a direct transition but must be excited through a more complex process. If one assumes the reaction follows the (t,d) (d,p) consecutive particle transfer process, then a CCBA calculation gives the curves shown for the 4^- state. Here A_y is well fit but the cross section is not well described. The overall quality of the fits are encouraging but the remaining discrepancies are a puzzle. More complex calculations appear necessary, perhaps containing more complete wave functions of the ^{206}Pb ground state and 4^-, 5^- states or finite range effects including the d-state of the triton.

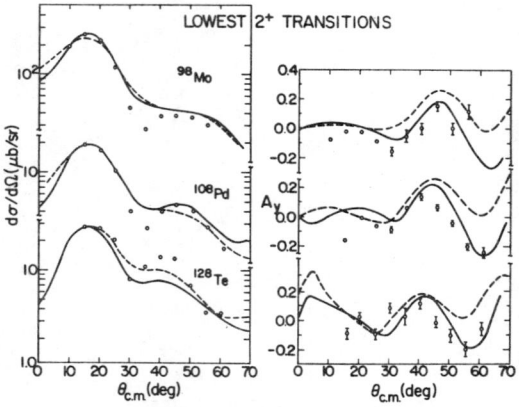

Fig. 12. A_y results for L=2 transitions on targets of ^{96}Mo, ^{106}Pd and ^{126}Te.

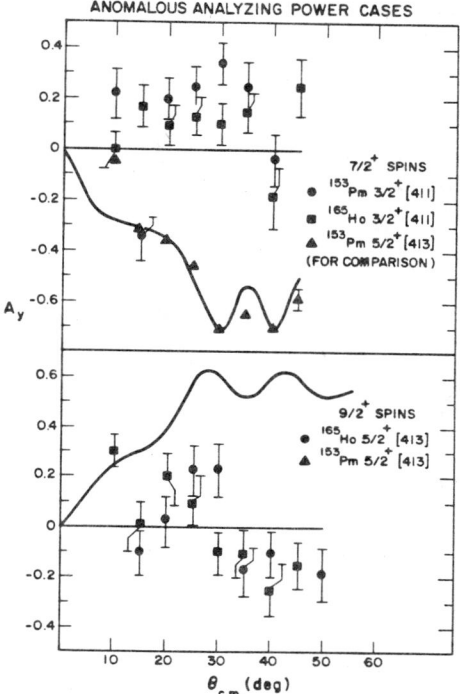

Fig. 13. Anomalous A_y results in the $(\vec{t},\alpha)$ reaction on rare earth isotopes.

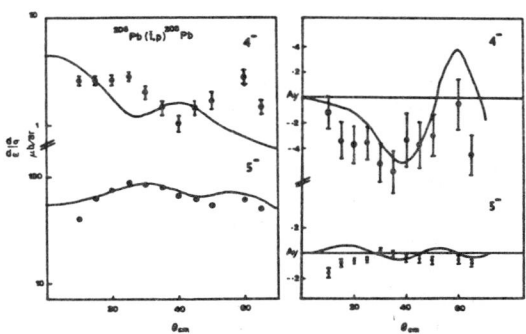

Fig. 14. Cross section and A_y results from the ^{206}Pb$(\vec{t},p)^{208}$Pb reaction to the 4^- and 5^- states.

CONCLUSIONS

Experiments involving the use of the polarized triton source have been both prolific and exciting. The combination of this source with the high efficiency and good resolution of the Q3D spectrometer have produced a unique facility for direct reaction studies. Recently the emphasis on these experiments has turned from only spectroscopic information to a combination of spectroscopy and reaction mechanism studies. It remains to be seen which of these aspects will prove to be the most significant but the renewed interest over the entire nuclear physics community in polarization phenomena would seem to quarantee that LASL's unique polarized triton facility will continue to play an important role.

REFERENCES

1. R. A. Hardekopf, R. F. Haglund, Jr., G. G. Ohlsen, W. J. Thompson and L. R. Veeser, Phys. Rev. C$\underline{21}$, 906 (1980).
2. E. R. Flynn, S. Orbesen, J. D. Sherman, J. W. Sunier and R. Woods, Nucl. Instr. and Methods $\underline{128}$, 35 (1975).
3. E. R. Flynn, "Proceedings of 1978 INS International Symposium on Nuclear Direct Reaction Mechanisms," 255 (1978).
4. A Bohr and B. Mottelson, "Nuclear Structure" Vol. II, W. A. Benjamin Publ. (1975).
5. E. R. Flynn, R. A. Hardekopf, J. D. Sherman and J. W. Sunier, Phys. Letts $\underline{61B}$, 433 (1976).
6. E. R. Flynn, R. A. Hardekopf, J. D. Sherman, J. W. Sunier and J. P. Coffin, Nucl. Phys. $\underline{A279}$, 394 (1977).
7. D. G. Burke, G. Lovhoiden, E. R. Flynn and J. W. Sunier, Nucl. Phys. $\underline{A315}$, 90 (1979).
8. E. R. Flynn, D. D. Armstrong, J. G. Beery and A. G. Blair, Phys. Rev. $\underline{182}$, 1113 (1969).
9. L. McFadden and G. R. Satchler, Nucl. Phys. $\underline{84}$, 177 (1966).
10. A. H. Wapstra and K. Bos, Nucl. Data Sheets $\underline{15}$, 177 (1977).
11. F. Ajzenberg-Selove, E. R. Flynn, R. A. Brown,, J. A. Cizewski and J. W. Sunier, contribution to this conference and to be published.
12. P. Federman and S. Pittel, Phys. Letts. $\underline{77B}$, 29 (1978).
13. A. Arima and F. Iachello, Ann. Phys. $\underline{99}$, 253 (1976).
14. F. Iachello, Phys. Rev. Lett. $\underline{44}$, 772 (1980).
15. J. A. Cizewski, E. R. Flynn, J. W. Sunier, R. E. Brown and D. G. Burke, contribution to this conference.
16. D. G. Burke, E. Hammaren, C. L. Swift, J. A. Cizewski, E. R. Flynn and J. W. Sunier, to be published.
17. E. R. Flynn, J. A. Cizewski, R. E. Brown and J. W. Sunier, to be published.

18. T. von Egidy et al., to be published.
19. E. R. Flynn, R. A. Brown, J. A. Cizewski, J. W. Sunier, W. P. Alford, E. Sugarbaker, and D. Ardouin, Phys. Rev., to be published.
20. T. F. Thorsteinsen, J. S. Vaagen, G. Lovhoiden, D. G. Burke, and E. R. Flynn, Phys. Letts., **93B**, 223 (1980).
21. W. P. Alford, R. N. Boyd, E. Sugarbaker, E. deBoer, R. E. Brown, and E. R. Flynn, contribution to this conference.

DISCUSSION

OHNUMA: If I remember correctly, Tsukuba results on forbidden two-neutron transfer are just the opposite to yours. Namely they obtained reasonable fits to the cross section angular distributions, while the fits to the analyzing powers were very poor. I wonder why there is such a difference between the (p,t) and (t,p) results.

FLYNN: The 4^- CCBA calculations were carried out with consecutive particle transfer only and inelastic channels were not included. Inclusion of the 3^- inelastic state may indeed affect the calculations.

POLARIZATION EFFECTS IN ^{3}He INDUCED REACTIONS

S. Roman

Department of Physics, The University of Birmingham, England

ABSTRACT

Studies of direct nuclear reactions with polarized ^{3}He are reviewed. Analysing powers of the one-nucleon transfer (^{3}He,d) and (^{3}He,α) reactions on a range of nuclei exhibit a strong sensitivity to the value of the transferred angular momentum j and useful regular features of the j-dependence have been established. Whereas certain aspects and gross features of the reactions can be explained in terms of a simple semiclassical model, a full description is often sensitive to the details of the assumed reaction mechanism and in particular to contributions of multistep processes.

The polarized ^{3}He beam has been available for experiments at Birmingham since mid-1974 and preliminary experimental results were reported to the Zurich Symposium[1]. In spite of some difficulties with the interpretation and certain unexpected features of the early results, the large analysing powers with a strong j-dependent behaviour observed prompted further development of the source and the injection system, doubling the beam polarization and increasing the available beam current[2]. To date analysing power data have been obtained for some 160 (^{3}He,d) and (^{3}He,α) transitions on the following targets: ^{6,7}Li, ^{9}Be, ^{11}B, ^{12,13}C, ^{16,17,18}O, ^{24,26}Mg, ^{27}Al, ^{28}Si, ^{32}S, ^{40,44,48}Ca, ^{54,56}Fe and ^{58}Ni. Some of the measurements have been carried out in collaboration with the experimental group from King's College, London. Simultaneously with the reactions, elastic scattering data have also been acquired for all the targets and findings of this work are discussed by Thompson[3]. Also not included here are the important (^{3}He,t) charge exchange reactions, which are subject of a separate contribution[4].

Polarized ^{3}He beams are being developed at the Texas A. & M. and Laval universities[5,6], but no experimental results have so far been reported. The scope of the polarized ^{3}He reaction studies is at this point limited by the specifications of the Birmingham facility, which are briefly: a) ^{3}He beam energy 33 MeV, fixed energy cyclotron; b) beam polarization $p = \pm 0.60$, switchable; c) on-target polarized ^{3}He beam current $i \leqslant 2$ nA. Since the cross sections of the ^{3}He reactions fall very rapidly with angle at this energy, typically by more than a factor of 100 between 5° and 60°, polarization measurements at large angles are prohibitively time consuming.

The interest in nuclear reaction studies with a polarized ^{3}He beam is first of all in the expected sensitivity of the spin-dependent pattern of the analysing power to the nature of the reaction mechanism. At 33 MeV the one-nucleon transfer reactions are essentially direct reactions and, in analogy with the well known deuteron reactions, the analysing power is expected to be sensitive to the total angular momentum transfer j, determined by the spin of the bound state of the transferred nucleon. Secondly, the possibility to apply the established sensitivity of the analysing power is of interest to spectroscopy. The method of using the analysing power of (d, p) reactions to determine the j-value of the transferred neutron has been used for a long time and is probably the most convenient and unambiguous technique available today [7]. To extend this method to include the wider class of states accessible to ^{3}He reactions would be very useful. For a direct reaction on a spin-0 target the transferred ℓ and j values uniquely specify the angular momenta of the final state of the reaction. It is convenient therefore, to examine the j-dependence for reactions on spin-0 targets leading to states whose spin assignments are well known.

Firstly we shall consider the j-dependence of the (^{3}He, d) reactions, expected to proceed by direct stripping, examining a selection of analysing powers for different ℓ and j transfers including transitions with ℓ = 0, 1, 2 and 3. Typical of (^{3}He, d) transitions with even ℓ-values are the data for ^{24}Mg and ^{16}O target nuclei [8,9] shown in fig. 1. For ℓ = 2, the 5/2$^+$ transitions are characterized by a deep broad minimum in the angular range between 20° and 35°, whereas the analysing powers of the 3/2$^+$ transfers have an extra oscillation in this range. This characteristic pattern changes very little from ^{24}Mg to ^{16}O and thus spin assignments can be made simply by visual inspection of the data, without need for a good theoretical "fit". The ℓ = 0, j = $\frac{1}{2}^+$ analysing powers are also clearly identified by their characteristic sharp oscillations of large amplitude.

Typical of the odd ℓ-values are examples of the ℓ = 1 transfers on ^{58}Ni [10] and ^{54}Fe [11], shown in fig. 2. Equally, there is no difficulty with recognizing the 3/2$^-$ from $\frac{1}{2}^-$ transfer: the sign of the analysing power of the first oscillation is positive or negative respectively for the $\ell + \frac{1}{2}$ and $\ell - \frac{1}{2}$ transfers. Apart from the sign change, there is also a pronounced difference in the amplitude of the analysing power oscillations: the magnitude of the 3/2$^-$ analysing power being much smaller than that of the $\frac{1}{2}^-$ transfers.

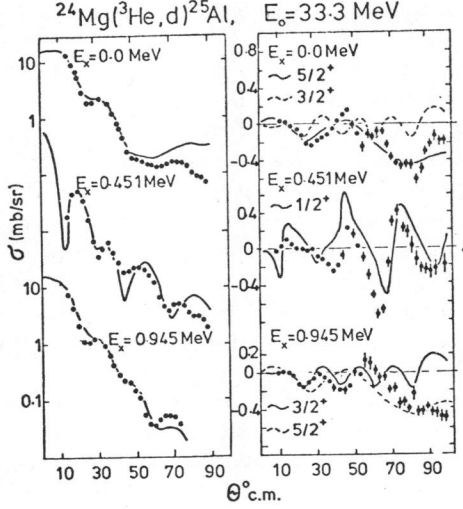

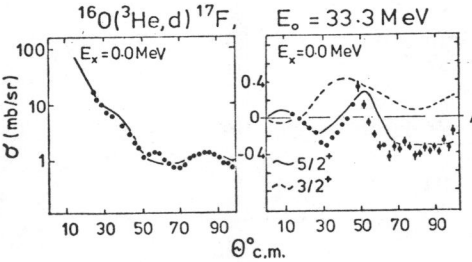

Fig. 1. Examples of $\ell = 0$ and 2 (^{3}He,d) transitions. The curves are DWBA predictions for the j-values indicated.

Similar patterns have been observed for (^{3}He,d) reactions on the light targets ^{9}Be and ^{12}C [12]. It is interesting to note how little do the patterns of the analysing power change with change of the target mass and excitation energy.

Fig. 3 shows an example of the j-dependence in the $\ell = 3$ transfer for the $7/2^-$ and $5/2^-$ transitions on ^{54}Fe. The $5/2^-$ pattern is characterized by a distinct minimum below 30°, whereas the analysing power for the $7/2^-$ transfer is positive in this region [11].

The characteristic j-dependent patterns, whereby it is possible to distinguish the analysing power for $\ell + \frac{1}{2}$ from $j = \ell - \frac{1}{2}$, are clearly identified for all angular momentum values. This is in spite of the fact that, unlike typical (d,p) analysing powers (especially at sub-coulomb energy), the $\ell + \frac{1}{2}$ and $\ell - \frac{1}{2}$ patterns do not appear usually as mirror reflections of each other, not even in the angular region of the stripping peak. There is, therefore, no simple sign rule available.

DWBA calculations have been carried out using the best-fit ^{3}He optical-model potentials, obtained from analyses of the elastic cross-section and polarization data. Potentials for the outgoing deuteron distortion were taken from the literature. Apart from the very light nuclei [12], where the DWBA description is unsatisfactory, the calculations reproduced the cross-sections and gave a reasonable overall account of most of the analysing power data. In detail, however, the analysing power distributions are not very well reproduced, especially at large angles.

Investigations of the sensitivity of the DWBA predictions to different parameters in the incident and outgoing channels can be

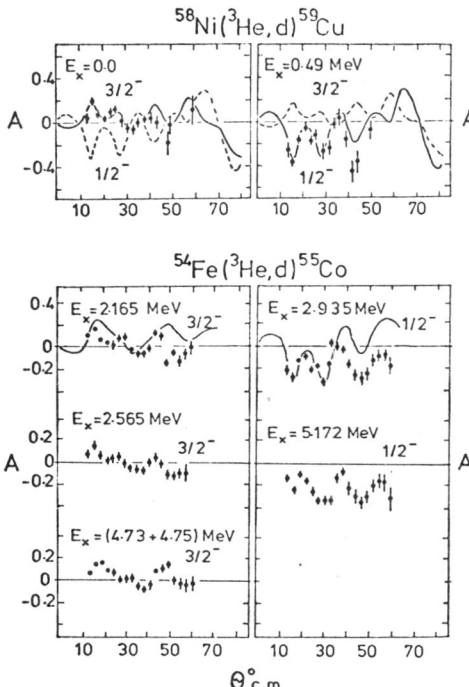

Fig. 2. Examples of $\ell = 1$ (^{3}He, d) transitions. The curves are DWBA predictions for the j-values indicated.

summarized as follows:
1) Both channels are, in general equally important, the predictions being strongly affected by changes of the ^{3}He and deuteron potential. In particular instances, however, one channel may be dominant. For example, the analysing power predictions at large angles are usually more strongly affected by changes to the ^{3}He potential [10], especially the imaginary part.
2) Although the predicted analysing powers of (^{3}He, d) reactions are usually more sensitive to the choice of the ^{3}He potential than are the ^{3}He elastic polarizations given by the optical-model calculations - attempts to use the reaction data to resolve the discrete ambiguity of the ^{3}He central potential have failed to give a conclusive answer. An overall preference for potentials characterized by volume integral between 450 and 700 MeV·fm^3 is indicated, but within this range there are differences which can be explained by an argument whereby the difference between the incident and the outgoing central potential strength is related to the strength of the captured particle bound state potential [9].
3) The DWBA predictions are not very sensitive to the choice of the spin-orbit potential geometry parameters. Improvements

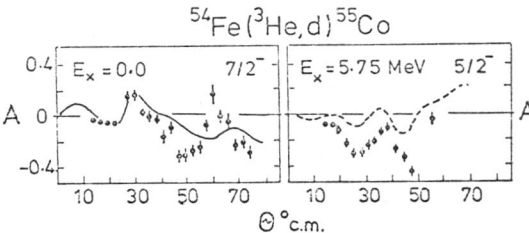

Fig. 3. Examples of $\ell = 3$ (^{3}He, d) transitions. The curves are DWBA predictions for the j-values indicated.

have been achieved by using folding models to obtain the ^{3}He and deuteron spin-orbit potential [13, 14, 15].

4) Of all the parameters, the most important was found to be the choice of the imaginary potentials. Adjustments of the imaginary term can be justified on the basis that the changes are required to compensate for the effects of other open channels, which are not explicitly taken into account by the conventional DWBA theory.

5) The finite range corrections were found to have an insignificant effect on the predictions of the (^{3}He,d) reactions cross sections and analysing powers.

Based on the established j-dependence of the analysing power, spectroscopic configurations can be assigned in reactions leading to states of the final nucleus where more than one transferred j-value is possible [16]. Where good DWBA fits are expected, the spectroscopic factors for the mixed transitions are obtained by means of a procedure whereby the analysing power data are compared with DWBA curves obtained by mixing the two $j = \ell + \frac{1}{2}$ and $\ell - \frac{1}{2}$ contributions and optimizing the mixing ratio for best fit to the experimental data. For (^{3}He,d) reactions, where details of the analysing power distributions are not very well reproduced by DWBA, it is more effective to use "reference" curves measured for known pure transitions on the same or a neighbour nucleus [17].

Next the (^{3}He,α) reactions are considered, expected to proceed by direct neutron pick-up at this energy. To establish the j-dependence of the analysing powers, a representative selection of transitions with different ℓ and j values is examined. Analysing powers of (^{3}He,α) reactions on the oxygen isotopes are shown in fig. 4, as an example of $\ell = 1$ transfers including both $\frac{1}{2}^-$ and $3/2^-$ cases. The striking feature of these reactions is the very large positive analysing power throughout the angular range for the $\frac{1}{2}^-$ ($\ell - \frac{1}{2}$) transfers, reaching A = 0.8 for many angles measured. The $3/2^-$ ($\ell + \frac{1}{2}$) transfers have negative analysing power, whose magnitude is about one half of that for the $\frac{1}{2}^-$

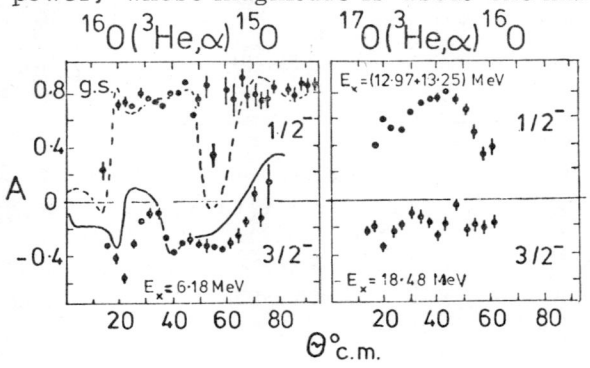

Fig. 4. Examples of $\ell = 1$ (^{3}He,α) transitions. The curves for the reaction on ^{16}O are DWBA predictions for the j-values indicated

transfers. Figs. 5 and 6 give examples of $\ell = 2$ transfers on ^{24}Mg, ^{28}Si, ^{32}S and ^{40}Ca targets. Here again the $3/2^+$ ($\ell - \frac{1}{2}$) analysing powers are very large and positive, reaching A = 1.0 within the

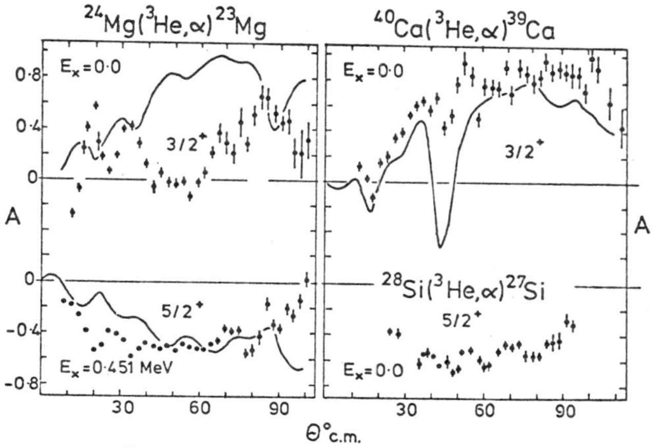

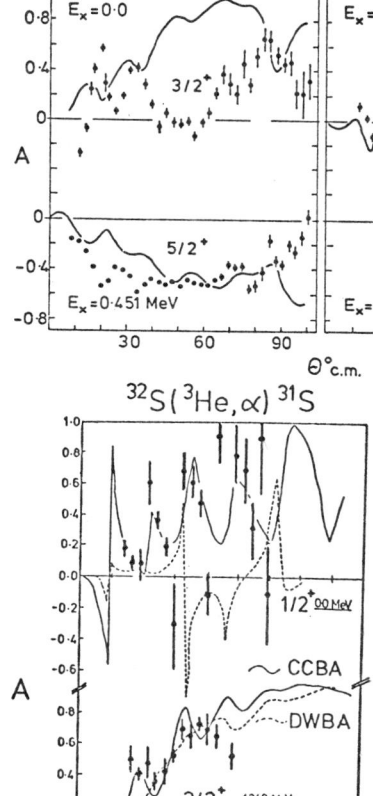

Fig. 5. Examples of $\ell = 2$ (^{3}He,α) transitions on ^{24}Mg, ^{28}Si and ^{40}Ca. The curves are DWBA predictions for the j-values indicated.

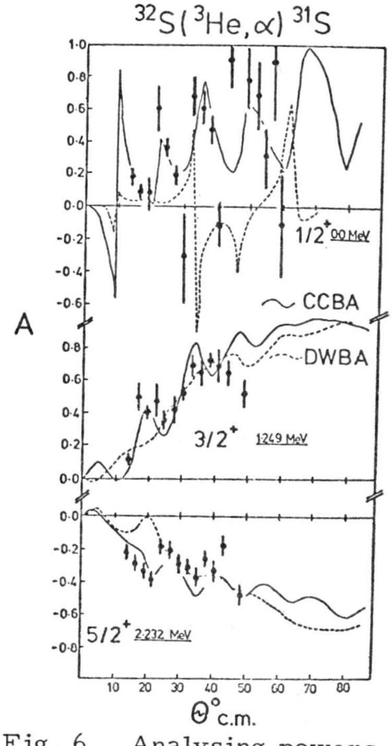

experimental errors, whereas the $5/2^+$ ($\ell + \frac{1}{2}$) transfers have negative analysing power throughout, somewhat smaller in magnitude.

With this unusual and at the same time simple character of the analysing power following a simple sign rule reminiscent of the sub-coulomb (d, p) reactions, it is tempting to invoke a simple semiclassical model analogous to that of Newns [18] or Vigdor et al. [19]. Considering the semiclassical model [9] which is explained in fig. 7, it is immediately obvious that the sign rule established for the (^{3}He,α) reactions is opposite to that expected at low energy, when the particle trajectories are due to the repulsive Coulomb force. At an energy well above the Coulomb barrier the observed sign of the analysing power is consistent with an attractive nuclear force being dominant.

Fig. 6. Analysing powers for the $\ell = 0$ and 2 (^{3}He,α) transitions on ^{32}S (Ref. 20).

It is interesting to note the striking contrast between the analysing power of the $\ell = 0$ ground state transition on ^{32}S (fig. 6), which has sharp oscillations about zero and a number of sign changes, and the $\ell > 0$ analysing powers, which are featureless and of a constant sign throughout the angular range. Whereas for $\ell = 0$ the non-zero analysing power must arise entirely from a spin-dependent interaction, for $\ell > 0$ the interaction need not be spin-dependent; the different character of the analysing power and calculations discussed below suggest that for $\ell > 0$ the spin-dependent distortions are of secondary importance. An independent

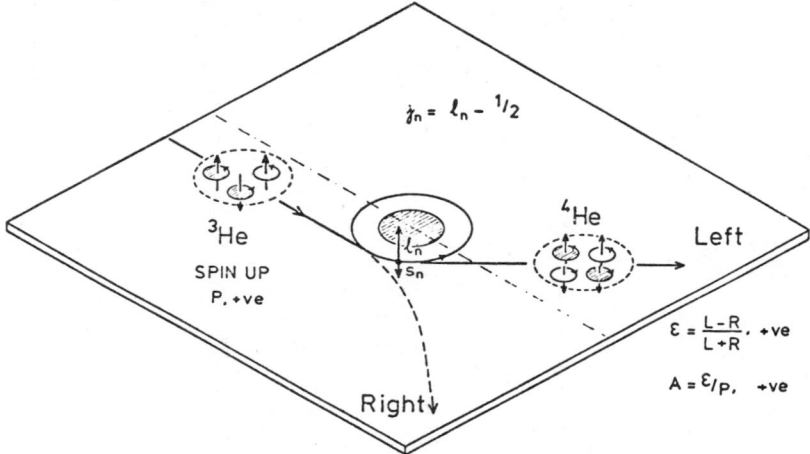

Fig. 7. Explanation of the analysing power of (^{3}He,α) reactions above the Coulomb barrier, showing the pick-up of a $j = \ell - s$ neutron, such as in the $\ell = 1$, $j = \frac{1}{2}$ reaction on ^{16}O (Ref. 9). The incident ^{3}He beam has positive polarization (spin-up). The positive analysing power observed implies that an attractive force is required to produce α-particles emerging to the left. A repulsive force (sub-Coulomb model of Vigdor et al.[19]) would produce α-particles to the right and hence a negative analysing power, contrary to the experiment.

verification of the absence of the spin-orbit distortion is possible in terms of the simple relation given by Satchler[21]

$$A_{\ell+\frac{1}{2}}(\Theta) = -\ell/(\ell+1) \cdot A_{\ell-\frac{1}{2}}(\Theta),$$

obtained from angular momentum coupling considerations assuming zero spin-orbit distortion. This relation is in an excellent agreement with the analysing power measurements for the $3/2^+$

and $5/2^+$ transitions in the $^{32}S(^3He,\alpha)^{31}S$ reaction, confirming that the spin-dependent distortions are insignificant in this case.

The simple picture of the $(^3He,\alpha)$ reactions discussed here is expected to work best well above the Coulomb barrier, that is it should be better at a high energy and for light nuclei than at lower energies and (or) for heavier targets. It is interesting to look for evidence of such change taking place. For this purpose let us examine first transitions on heavier nuclei at the same energy of 33 MeV. As an example analysing powers of a pair of $3/2^-$ and $\frac{1}{2}^-$ ($\ell = 1 \pm \frac{1}{2}$) transitions on ^{56}Fe is shown in fig. 8. The analysing

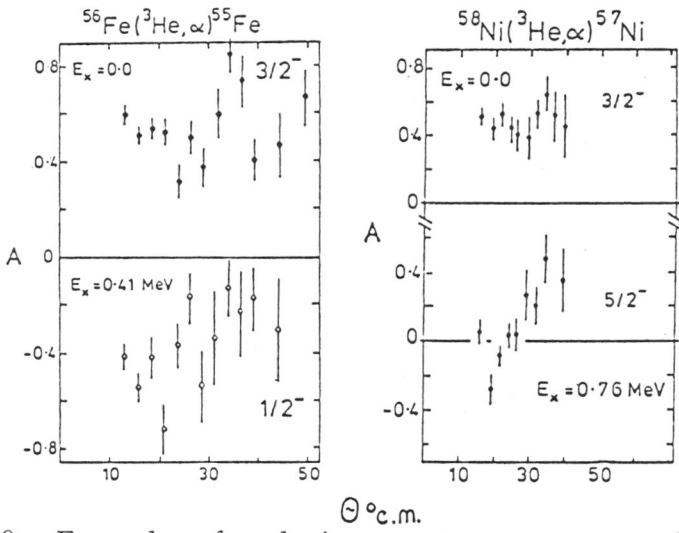

Fig. 8. Examples of analysing power measurements for $(^3He,\alpha)$ reactions on ^{56}Fe and ^{58}Ni at $E_0 = 33.3$ MeV. The sign of the analysing power is opposite to that observed for the lighter nuclei.

power for the $3/2^-$, ($\ell + \frac{1}{2}$) neutron pick-up is positive throughout the angular range whereas the $\frac{1}{2}^-$, ($\ell - \frac{1}{2}$) analysing power is negative, in agreement with the sub-Coulomb sign rule. This is to be compared with the analysing powers measured at the same energy of 33 MeV for the same agular momentum transfers on the oxygen isotopes, shown in fig. 4, where the sign rule is opposite. The pattern for the $3/2^-$ transition on ^{58}Ni is nearly identical to that for ^{56}Fe, while the analysing power for the $\ell = 3$, $5/2^-$ transfer on ^{58}Ni has an oscillation about zero, so that no simple sign rule can be applied in this case.

Next the energy dependence of the $(^3He,\alpha)$ reactions is examined, considering the measurements of the $^{12}C(^3He,\alpha)$ reaction analysing power for the $E_x = 1.995$ MeV, $\frac{1}{2}^-$ state in ^{11}C, carried

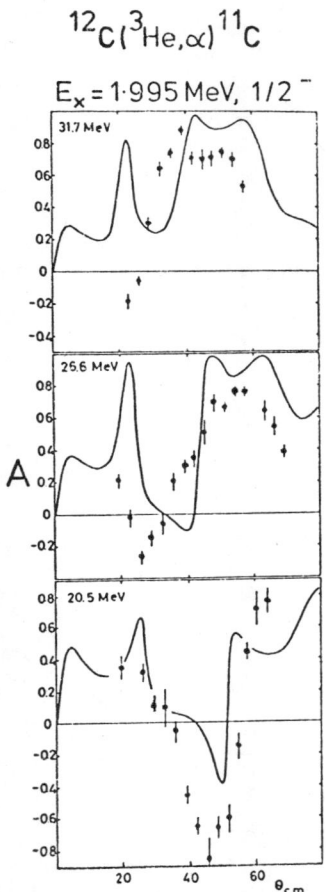

Fig. 9. Analysing power of the $^{12}C(^{3}He,\alpha)^{11}C^{*}$, $\frac{1}{2}^{-}$, reaction at 31.7, 26.6 and 20.5 MeV and DWBA predictions.

out at three ^{3}He beam energies[22]. At 32 MeV the analysing power is predominantly positive and very large, consistent with the simple picture of the $j = \ell - \frac{1}{2}$ transfer in the pick-up reactions above the Coulomb barrier. This pattern changes with energy: at 26 MeV a minimum appears around 30°; at 20 MeV the minimum is broader and deeper, so that the analysing power has practically the opposite sign to that at 32 MeV (fig. 9).

The DWBA calculations carried out for the (^{3}He,α) reactions in general give a reasonable account of the shape of the cross sections and analysing powers, but the angular distributions are not reproduced in detail; for some transitions the theory clearly fails altogether. For example, the analysing power of the $^{24}Mg(^{3}He,\alpha)^{23}Mg$ g.s. reaction, shown in fig. 5, could not be reproduced by the calculations. Thus, in spite of the simplification implied by the absence of spin in the outgoing channel, the DW description of the (^{3}He,α) reactions including cross section and analysing power data, compares unfavourably with that for the (^{3}He,d) reactions.

Attempts to improve the description fall into two categories: firstly to overcome the inadequacy of the DW approximation and the optical-model potentials used and secondly to account for the reaction mechanisms other than the direct neutron pick-up from a single-particle orbit. In the first category, the (^{3}He,α) reaction is particularly sensitive to: a) The momentum mismatch effects, when the semi-classical momentum $|K_f - K_i| R$ differs from the transferred angular momentum ℓ. It has been shown[23] that this effect can be compensated for by a suitable choice of the distorting

potentials in relation to the depth of the bound state potential.
b) The finite-range effects; it has been demonstrated using the $^{40}Ca(^{3}He,\alpha)^{39}Ca$ g.s. reaction as an example [9], that corrections using the local-energy approximation may involve singularities and should be used with extreme caution. Calculations either in zero-range or full finite-range are preferred. c) The choice of the optical model potentials is important as it often compensates for inadequacies of DW description and the presence of non-direct processes. It has been found that with the best ^{3}He potentials determined from optical model analyses of the elastic scattering data, the ($^{3}He,\alpha$) reaction data were not reproduced by DWBA. For example, to describe the analysing power for the $\ell = 0$, $j = \frac{1}{2}^{+}$ $^{32}S(^{3}He,\alpha)^{31}S$ g.s. transition (fig. 6)[20] which is particularly sensitive to the spin-orbit term of the ^{3}He potential, required a change from the sharply surface-peaked to a more diffuse form, although the latter is unable to fit the elastic polarization. Equally, changes of the central part are often required. Further difficulties are encountered with finding suitable alpha-particle potentials for the outgoing channel of the reaction, since average expressions for the alpha-particle potential are not available.

To account for other reaction mechanisms, attempts have been made to include in the calculations contributions of multi-step processes. Preliminary results of two such attempts reported to this Symposium look promising [17, 20]. The reaction paths assumed in the multi-step calculations for the $^{32}S(^{3}He,\alpha)^{31}S$ and $^{24}Mg(^{3}He,\alpha)^{23}Mg$ reactions are indicated in fig. 10. In the reaction on ^{32}S the contribution to the pick-up process of the first three levels in ^{31}S $\frac{1}{2}^{+}$, $3/2^{+}$ and $5/2^{+}$ is included. Because the $\ell = 0$ transfer can produce no "feedback" into the elastic channel and it can be shown that the effects of the two $3/2^{+}$ and $5/2^{+}$ transfers cancel each other — a coupled channels Born approximation (CCBA) calculation should be sufficient (one way coupling). Results of these calculations

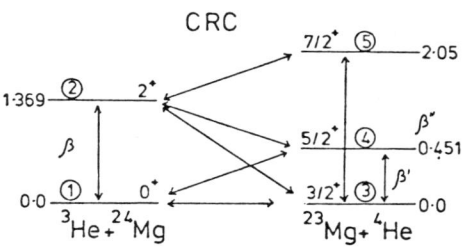

Fig. 10. Coupling diagrams for the $^{32}S(^{3}He,\alpha)^{31}S$ and $^{24}Mg(^{3}He,\alpha)^{23}Mg$ reactions.

are shown in fig. 6, obtained in successive steps where the magnitudes and phases of the transition amplitudes were adjusted to give best fit to all the data sets. The results of these calculations show a marked improvement over the DWBA predictions.

Full coupled reaction channels (CRC) calculations (two way coupling) according to the diagram in fig. 10 have been necessary for the ^{24}Mg(^{3}He,α)^{23}Mg reaction ^{24}, where no simpifying assumptions could be justified. This reaction is interesting in that

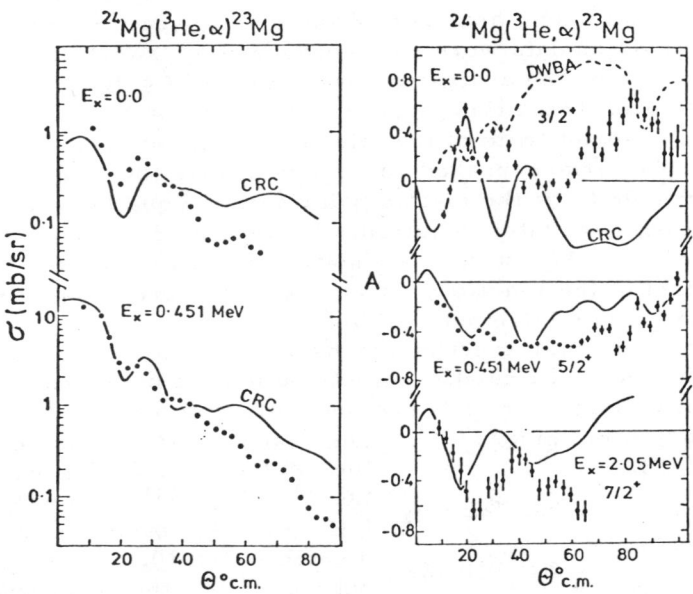

Fig. 11. Cross section and analysing power measurements for the ^{24}Mg(^{3}He,α)^{23}Mg reaction^{24} compared with results of multi-step calculations.

the 7/2$^+$ transition to the 2.05 MeV state in ^{23}Mg is forbidden by shell-model considerations^{25} and, therefore, can only be reached by a multistep process. The shell-model spectroscopic amplitudes were taken from ref. 25, assuming rotational model deformation of the ^{24}Mg nucleus with a deformation parameter $\beta = 0.30$. Preliminary results of the calculations are compared with experimental data in fig. 11. It is encouraging to note that, although the details of the angular distributions are not reproduced, the main features of the measurements are well described by the coupled reaction channels calculations, especially at small angles. An improvement of the description of the reaction data is expected by adjustment of the optical model parameters used in the calculations.

REFERENCES

1. S. Roman, Proc. 4th Int. Polarization Symposium (Zurich, 1975) p. 255
2. W. C. Hardy, R. G. Green, G. W. Guest, O. Karban, W. B. Powell and S. Roman, This Symposium, 3.16.
3. W. J. Thompson, This Symposium, Invited paper, 2.A.
4. A. K. Basak, O. Karban, P. M. Lewis, G. C. Morrison and S. Roman, This Symposium, 2.70.
5. Y.-W. Lui, D. P. May, S. D. Baker, S. Nath, E. Takada, D. Tanner and Y. Yamaya, Texas A.&M. Univ. Progr. Report 1980, p. 93
6. R. J. Slobodrian, This Symposium, Invited paper, 3.C.
7. S. E. Darden and W. Haeberli, Proc. 4th Int. Polarization Symposium (Zurich, 1975), p. 229
8. F. Entezami, A. K. Basak, O. Karban, P. M. Lewis and S. Roman, This Symposium, 2.68.
9. Y.-W. Lui, O. Karban, S. Roman, R. K. Bhowmik, J. M. Nelson and E. C. Pollacco, Nucl. Phys. $\underline{A333}$, 221 (1980)
10. S. Roman, A. K. Basak, J. B. A. England, O. Karban and G. C. Morrison, Nucl. Phys. $\underline{A284}$, 365 (1977)
11. O. Karban, A. K. Basak, C. O. Blyth, F. Entezami, P. M. Lewis and S. Roman, Univ. of Birmingham, Nucl. Progr. Report 1980, unpublished
12. O. Karban, A. K. Basak, J. B. A. England, G. C. Morrison, J. M. Nelson, S. Roman and G. G. Shute, Nucl. Phys. $\underline{A269}$, 312 (1976)
13. P. V. Drumm, A. K. Basak, O. Karban, P. M. Lewis, S. Roman and G. C. Morrison, Univ. of Birmingham, Nucl. Progr. Report 1980, unpublished
14. J. A. Ramirez and W. J. Thompson, Proc. 4th Int. Polarization Symposium (Zurich, 1975) p. 618
15. P. W. Keaton Jr., E. Aufdembrink and L. R. Veeser, Los Alamos S. L. Report No LA-4379-MS, 1970
16. N. E. Sanderson, A. K. Basak, J. B. A. England, J. M. Nelson and S. Roman, Phys. Rev. Lett. $\underline{37}$, 1672 (1976)
17. O. Karban, A. K. Basak, G. C. Morrison, J. M. Nelson and S. Roman, This Symposium, 2.108.
18. H. C. Newns, Proc. Phys. Soc. (London) $\underline{A66}$, 477 (1953)
19. S. Vigdor, R. D. Rathmell and W. Haeberli, Nucl. Phys. $\underline{A210}$, 93 (1973)
20. J. M. Barnwell, N. M. Clarke and R. J. Griffiths, This Symposium, 2.39.
21. G. R. Satchler, Nucl. Phys. $\underline{55}$, 1 (1964)
22. O. Karban, W. E. Burcham, J. B. A. England, R. G. Harris and S. Roman, Univ. of Birmingham, Nucl. Progr. Report 1976, p. 71, unpublished
23. R. Stock, B. Bock, P. David, H. H. Duhm and T. Tamura, Nucl. Phys. $\underline{A104}$, 136 (1967)
24. O. Karban, A. K. Basak and S. Roman, This Symposium, 2.109.
25. R. O. Nelson and N. R. Roberson, Phys. Rev. $\underline{C6}$, 2153 (1972)

(Editor's Note: References to This Symposium are given as paper numbers and are cross-indexed to page numbers in the Table of Contents.)

DISCUSSION

SLOBODRIAN: These (^{3}He,α) results are most interesting. Have you tried an approach based on the direct knock-out of an alpha particle by the ^{3}He projectile? Knock-out processes were quite easy to deal with in the PWBA for cross sections. Perhaps they are an alternative to multistep processes as an explanation, as ^{3}He is already fairly well bound, and knock-out yields a sizable contribution to the transition amplitude whenever α-particle clustering exists and/or is enhanced.

ROMAN: We have not done direct α-particle knock-out calculations for the (^{3}He,α) reactions.

RAWITSCHER: When you do the multistep calculations, do you include coupling back to the ground state? Similarly, does Dr. Flynn also include such couplings for the triton? The reason for asking is that it is possible that the difference in the optical potentials for the t and ^{3}He projectiles could be connected with the difference between the intermediate stripping channels which participate for either projectile. Hence comparison between coupled channel calculations for the two projectiles would be very useful.

ROMAN: The answer to the first part of the question is that we do include the backward and forward coupling to the ground state, as indicated in the C.R.C. scheme for the ^{24}Mg(^{3}He,α)^{23}Mg reaction shown.

NEUTRON DENSITY DISTRIBUTION STUDIES USING INTERMEDIATE ENERGY POLARIZED PROTONS

L. Ray
Department of Physics, The University of Texas at Austin
Austin, Texas 78712

ABSTRACT

The role of spin-dependence in the intermediate energy proton-nucleus interaction is investigated. When elastic differential cross section data are analyzed to obtain nuclear size information, uncertainties in the deduced neutron densities arise as a result of ambiguities in the proton-nucleus spin orbit potential. Generally these errors are small; however, they are enhanced for light nuclei especially when isotopic neutron density differences are constructed. The improvements which result when analyzing power and spin rotation data are considered are quantitatively discussed. Results for p + ^{16}O, ^{40}Ca at 0.8 GeV are given. Absolute analyzing power and spin rotation predictions are also calculated from recent nucleon-nucleon phase shifts using the impulse approximation.

INTRODUCTION

The study of the distributions of nucleons in nuclei has occupied a prominent position in nuclear physics research over the past two decades.[1,2] The emphasis of this effort is twofold. We seek to obtain precise information about the nuclear density distributions with which to test nuclear structure models of the ground state[3] and at the same time attempt to fundamentally understand the projectile-nucleus reaction process. A great variety of probes have been employed in this effort including the electron and muon[1], protons[4-10], alphas[2] and pions[11], the latter three having been utilized in attempts to probe the elusive matter distributions in nuclei.

In this work attention will be focused on the study of nuclear matter densities with ∿1 GeV polarized protons. Rather than deal with the aggregate empirical matter density results which are well documented in the literature[4-10], the concern here will be with the role played by the proton-nucleus spin-dependent interaction in the extraction of nuclear matter densities. Information about the ∿1 GeV proton-nucleus spin dependence has accumulated rapidly during the last decade. The original 1.04 GeV proton-nucleus elastic scattering survey from Saclay[12] lacked any polarization measurements and analyses of these data were always plagued by uncertainties resulting from the assumed spin-dependent interaction. However, a few years later analyzing power [$A_y(\theta)$] measurements on many nuclei at 800 MeV were provided by the high resolution spectrometer (HRS) at the Clinton P. Anderson Meson Physics Facility (LAMPF).[8,13] Polarization measurements on a few nuclei were also obtained at Gatchina.[14] These new $A_y(\theta)$ and polarization data enabled some constraints to be imposed on the empirical spin-dependent proton-nucleus interaction, thus increasing our confidence in the deduced matter densities.[5,7] Only in the last year or so have nucleon-nucleon (N-N) amplitudes become available above 500 MeV which provide absolute microscopic predictions of $A_y(\theta)$ which are in reasonable agreement with the data. Even so, some phenomenological adjustment is needed still. Since proton-nucleus phenomenology has been introduced into what should strictly be a microscopic analysis, one must determine the impact of this empiricism on the deduced neutron densities.

Thus, of the many problems associated with proton-nucleus scattering near 1 GeV the specific questions to be dealt with in this article are: (1) how knowledge of the proton-nucleus spin-dependence affects the extraction of nuclear matter densities and (2) how well the spin-dependence of the proton-nucleus interaction is understood microscopically. To this end a brief outline of the multiple scattering calculations employed in the analyses will be presented followed by representative results for p + ^{16}O and ^{40}Ca at 800 MeV. Spin orbit deformation effects in ^{24}Mg will also be examined.

METHOD OF ANALYSIS

The calculations presented here have employed local, spin-dependent forms for both the first and second order optical potentials as derived by Kerman, McManus and Thaler (KMT).[15] The first order optical potential for spin zero target nuclei is given by[16]

$$U^{(1)}(r) = 4\pi\eta(A-1)/A \sum_{i=p,n} \int q^2 dq A_{pj}(q) F_j(q) j_0(qr)$$
$$+ 4\pi\eta(A-1)/A \frac{1}{r}\frac{d}{dr} \sum_{i=p,n} \int q^2 dq \tilde{C}_{pj}(q) F_j(q) j_0(qr) \vec{\sigma}\cdot\vec{\ell} \quad (1)$$

where q is the momentum transfer, η converts the N-N amplitude in the two-nucleon center-of-momentum (c.m.) system to a t-matrix in the proton-nucleus c.m. system[16], $F_j(q)$ are the target nucleon form factors where $F_p(0) = Z$ and $F_n(0) = N$, $\tilde{C}_{pj}(q)$ is approximately $C_{pj}(q)/(k_N q)$ where k_N is the momentum in the proton-nucleus c.m. system and the N-N amplitudes A_{pj} and C_{pj} are defined as [16]

$$f_{pj}(q) = A_{pj}(q) + C_{pj}(q)\vec{\sigma}\cdot\hat{n} \quad . \quad (2)$$

The optical potential in Eq. (1) makes use of the "impulse" approximation[15] since free N-N amplitudes are assumed. The point proton densities have been obtained from empirical charge distributions.[5,17,18] The second order KMT optical potential contributions arising from Pauli, short range dynamical and center-of-mass correlations amongst the target nucleons have been approximated by local, density squared terms as in Ref. 5. The Pauli correlation correction to the spin orbit optical potential has also been incorporated.[5] The point neutron densities have been assumed to be of the three parameter Fermi form,

$$\rho_n(r) = \rho_n^o(1+w_n r^2/c_n^2)/(1+\exp((r-c_n)/z_n)), \quad (3)$$

where the parameters w_n, c_n, and z_n are adjusted to optimize the fit to the angular distribution data. The Coulomb interaction has been included according to the prescription in Ref. 6.

The N-N amplitudes of Eq. (2) are parametrized as

$$A_{pj}(q) = (k_o \sigma_{pj}^T/4\pi)(i+\alpha_{pj})\exp(-B_{pj}q^2),$$

$$C_{pj}(q) = (ik_o \theta_{pj}/4\pi)[i+\alpha_{spj}(1+\bar{\alpha}_{spj}q^2)](q^2/4M^2)^{\frac{1}{2}}\exp(-B_{spj}q^2), \quad (4)$$

where j = proton or neutron, k_o is the nucleon momentum in the two-nucleon c.m. system, σ_{pj}^T is the spin averaged total cross section, and M is the nucleon mass. The recent phase shift solution CL80 of Arndt[19] has also provided N-N amplitudes where the Coulomb distortions of A_{pp} and C_{pp} are properly included. The parameters of A_{pj} are determined by the N-N cross section and polarization data[5] while the parameters of C_{pj} are adjusted to provide the best fit to the proton-nucleus analyzing power data. The neutron density is varied to recover the fit to the proton-nucleus angular distribution. Of course the phenomenological parameter values for C_{pj} affect the resulting ρ_n and vice versa so that ambiguities exist. Such uncertainties are examined in the next section.

RESULTS FOR ^{16}O AND ^{40}Ca

The best phenomenological fits to the 800 MeV p + ^{16}O, ^{40}Ca analyzing power data[8,13] are indicated by the dashed curves in Fig. 1. Absolute predictions for these data based on the recent N-N phase shift analysis of Arndt[19] are exhibited by the solid curves in this same figure. Although the absolute predictions are quite good, some phenomenological "fine tuning" is required until further optical potential corrections are investigated. From the point of view of determining the neutron densities the most important angular region of the differential cross section is from about 10° - 20° for ^{16}O and 7° - 20° for ^{40}Ca. The phenomenological fits in Fig. 1 have therefore been optimized over these angular regions. The best fits to the differential cross sections for p + ^{16}O, ^{40}Ca obtained by adjusting the neutron densities are shown by the solid curves in Fig. 2. The N-N parameters assumed in these calculations are given in Table I.

To begin with it is instructive to examine the total effect of the spin orbit potential. The dashed curves in Fig. 2 result when the spin orbit potentials used in the optimized fits are set to zero. The differences in the cross sections are about 15-50% in ^{16}O and 10-30% in ^{40}Ca. The dash-dot curves in Fig. 2 result when the second order correlation terms are additionally eliminated.

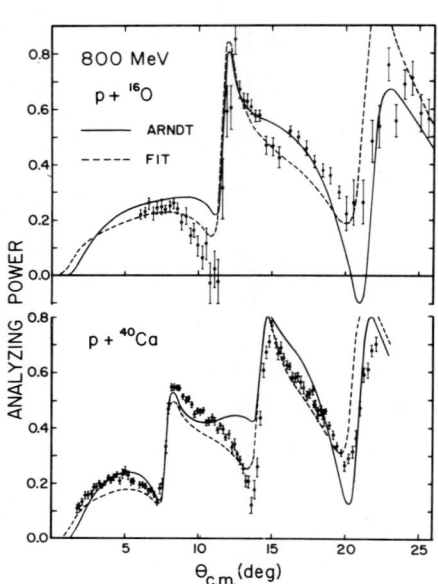
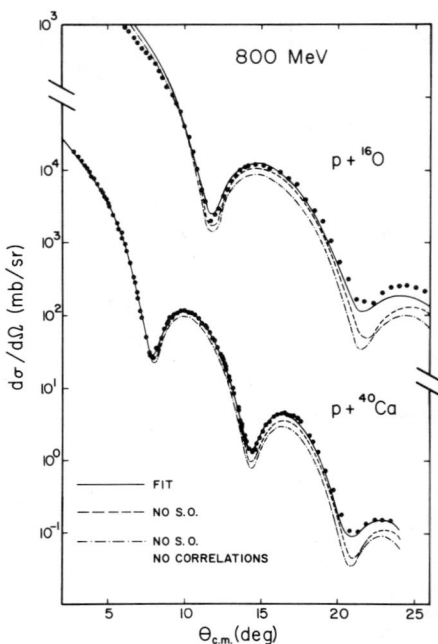

Fig. 1. Analyzing power calculations and data at 800 MeV.

Fig. 2. Differential cross section calculations and data at 800 MeV. (^{40}Ca data, Ref. 20).

Thus the spin orbit potential has a perturbative effect on the angular distribution comparable to the correlation terms. If the fits to the data in Fig.2 are recovered by adjusting the neutron densities with (without) the spin orbit potential, then the model densities given in Fig. 3 by the solid (dashed) curves result. The differences are rather small, both in the radial distributions and in the root mean square (rms) radii r_n (see Table II) which differ by 0.044 (0.073) fm for ^{40}Ca(^{16}O). For comparison the empirical neutron densities in ^{40,48}Ca extracted from model-independent analyses are indicated by the shaded bands in the top half of Fig. 4 (Ref.5). The dashed curves are Hartree-Fock predictions.[3] Other deduced neutron densities are given in Refs. 4-10 and 13. It is customary to compare the extracted neutron-proton rms radius difference, Δr_{np}, with theory. Typical values of Δr_{np} range from 0.0 to 0.2 fm (Refs. 4-10 and 13) which are comparable in magnitude to the above spin related errors.

Three independent measurements are needed to uniquely determine the proton- (spin 0) nucleus amplitudes. Since only two types of ∼1 GeV proton-nucleus data currently exist, $d\sigma/d\Omega$ and A_y, ambiguities must occur in the phenomenological determination of the optical potential. The strength of the real part of the spin orbit optical potential is proportional to $\theta_{pj}\alpha_{spj}$ and is well determined

TABLE I

800 MeV proton-nucleon
amplitude parameters

Spin-independent

	p-p	p-n
σ^T_{pj} (mb)	47.3	37.9
α_{pj}	0.06	-0.2
B_{pj} (fm^2)	0.09	0.12

Effective spin orbit
parameters[a]

	^{16}O	^{40}Ca
θ_{pj} (mb)	107.	89.8
α_{spj}	0.63	0.76
B_{spj} (fm^2)	0.2	0.2
$\bar{\alpha}_{spj}$ (fm^2)	-0.03	-0.03

[a]The spin parameters for p-p and p-n are assumed to be equal.

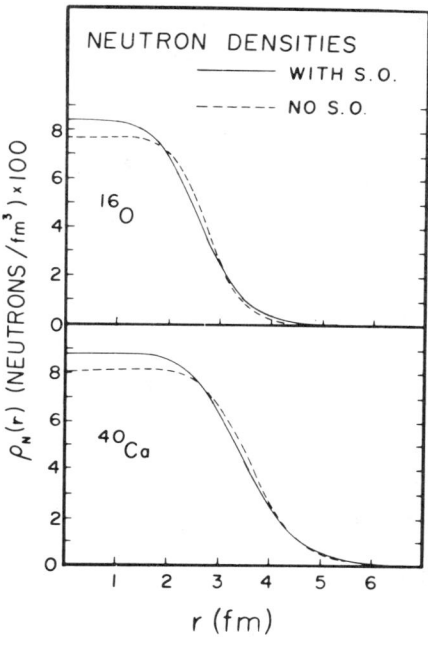

Fig. 3. Model-dependent neutron densities in ^{16}O and ^{40}Ca computed with and without a spin orbit interaction.

to about ±10% by the $A_y(\theta)$ data. However, the imaginary part of the spin orbit potential is very poorly determined and can vary by ±20% without the fit to the $A_y(\theta)$ data deteriorating. In addition the range parameter B_{spj} can vary by ±0.1 fm^2 without unduly affecting the fit to the analyzing powers. Several sets of C_{pj} are thus obtained which result in different differential cross section predictions and hence different deduced neutron densities. Calculations have been performed in which θ_{pj} and B_{spj} are separately varied by the above amounts with the neutron distribution being adjusted to regain an optimum fit to the differential cross section data of Fig. 2. The $A_y(\theta)$ predictions are all very similar to the dashed curves of Fig. 1 and lie within the shaded bands given in the tops of Figs. 5 and 6. The effect of this ambiguity on the deduced neutron rms radii is given in Table II. An error of ±0.044 (±0.055) fm in r_n for ^{40}Ca(^{16}O) is seen to result from this remaining spin orbit ambiguity. It is interesting to note that this remaining error in r_n is of the same magnitude as that which one would have to assign in the complete absence of $A_y(\theta)$ data. These spin orbit potential uncertainties can be compared to the estimated total

TABLE II

Effect of $A_y(\theta)$ and Q data on the error in r_n[a]

	^{16}O	^{40}Ca
Error due to S.O. only:		
1) $d\sigma/d\Omega$ only	0.073[b]	0.044[b]
2) with A_y	0.055	0.044
3) A_y and Q	0.017	0.014
Total error:		
1) $d\sigma/d\Omega$ only	0.22	0.09
2) with A_y	0.16	0.07
3) A_y and Q	0.15	0.05

[a] All errors are ± values in fm.
[b] These are combined linearly with the other errors in r_n (see Ref. 7).

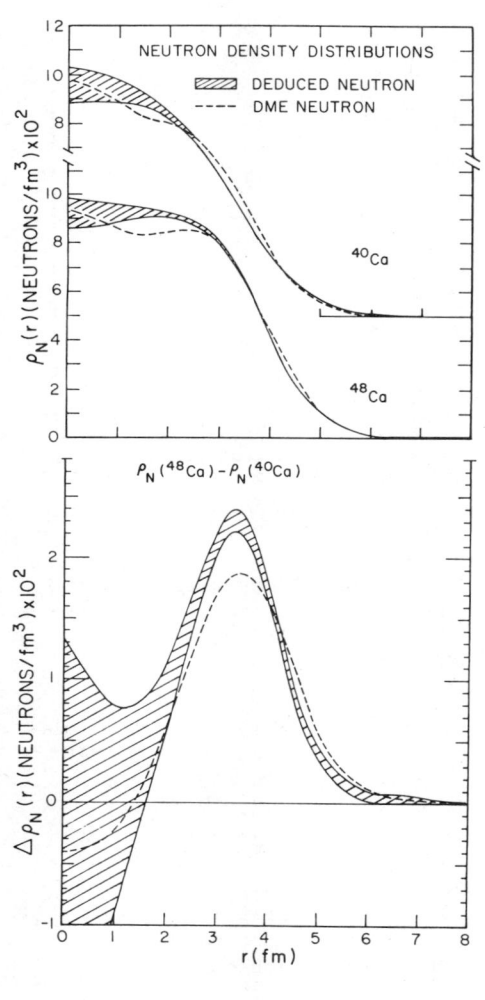

Fig. 4. Deduced and theoretical neutron densities in ^{40,48}Ca and their isotopic difference (from Ref. 5).

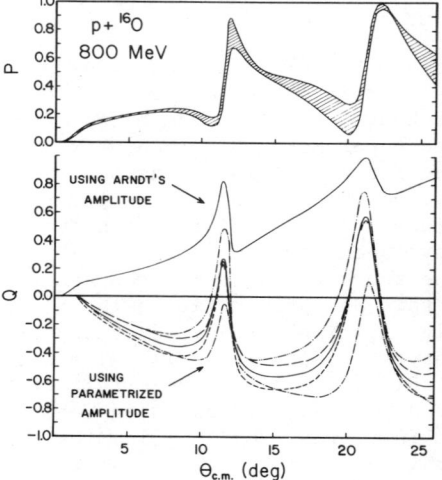

Fig. 5. Various polarization and spin rotation calculations for p + ^{16}O at 800 MeV.

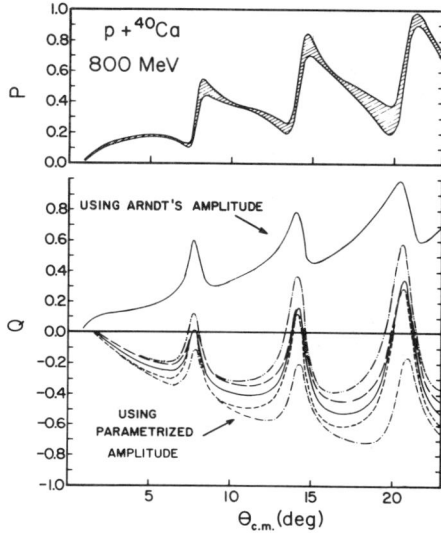

Fig. 6. Same as Fig. 5, except for ^{40}Ca.

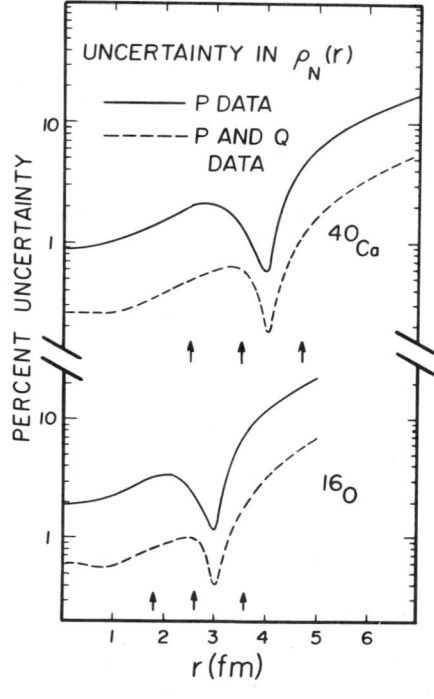

Fig. 7. Percent error in the neutron density distributions of ^{16}O and ^{40}Ca arising from spin orbit ambiguities.

error in r_n for ^{40}Ca and ^{16}O of ±0.067 fm and ±0.16 fm, respectively.[20,8] The inclusion of $A_y(\theta)$ data permits some reduction in the total error in r_n (see Table II).

The percent uncertainties in the deduced neutron radial distributions due to these spin orbit ambiguities are indicated by the solid curves in Fig. 7 for ^{16}O and ^{40}Ca. The three arrows indicate the value of the radius at which the density is 90%, 50% and 10% of the central value. Comparison of the errors in Figs. 3 and 7 indicates that the inclusion of $A_y(\theta)$ data in the analysis has brought about a moderate decrease in the uncertainty in $\rho_n(r)$ due to spin effects.

The above discussion suggests that in order to significantly reduce the uncertainty in r_n and $\rho_n(r)$ due to spin orbit ambiguities, additional experimental constraints are required. This further constraint could be obtained by measurement of the spin rotation quantity Q proposed by Glauber and Osland.[21] If the proton-nucleus scattering amplitude is written as

$$f_{p-A}(q) = F(q) + G(q)\vec{\sigma}\cdot\hat{n} \qquad (5)$$

then Q is defined by

$$Q = 2\text{Im}\{F(q)G^*(q)\}/(d\sigma/d\Omega) \quad . \qquad (6)$$

The predictions of Q resulting from the various C_{pj} amplitudes discussed above are shown in Figs. 5 and 6 for ^{16}O and ^{40}Ca, respectively. The Q predictions corresponding to θ_{pj} increased and decreased by 20% and to B_{spj} raised and lowered by 0.1 fm^2 relative to the values in Table I are indicated by the short dashed, long dashed, dash-double-dot and dash-dot curves, respectively, in both figures. The solid curves in each group of curves in the lower halves of Figs. 5 and 6 correspond to the parameter values of Table I. Thus the above spin orbit ambiguities could be reduced significantly by the additional measurement of Q. If the experimental errors in Q were for example twice that of the $A_y(\theta)$ data, the contribution to the uncertainty in r_n due to spin-dependence could be reduced to about ±0.014 (±0.017) fm in ^{40}Ca (^{16}O). The total error in r_n would correspondingly decrease to ±0.051 (±0.15) fm in ^{40}Ca (^{16}O) (see Table II). The error in $\rho_n(r)$ due to the spin orbit potential uncertainty could probably be reduced to the values indicated by the dashed curves in Fig. 7.

Because of various experimental[7,13], theoretical[4-7], and N-N amplitude uncertainties[5] the empirical isotopic neutron density differences are more reliable than the individual neutron densities.[5] Typical isotopic density differences are given in Ref. 5 and the result for $\rho_n(^{48}Ca) - \rho_n(^{40}Ca)$ from this reference is displayed in Fig. 4. For these large nuclei definite trends in the isovector dependence of the effective spin orbit potential are known[7,22] so that it can be reasonably argued that the above spin orbit ambiguities cancel when studying relative density differences. This might not be a safe assumption in the light p and s-d shell nuclei where large deformation and shell structure effects could be quite different between various isotopes. For illustrative purposes the uncertainty in the isotopic neutron density difference, $\Delta\rho_n(r) = \rho_n(^{18}O) - \rho_n(^{16}O)$, due to the above spin orbit ambiguities is given in Fig. 8. If Q data were available this uncertainty in $\Delta\rho_n(r)$ could be reduced as indicated by the central cross hatched region of Fig. 8. Similarly the isotopic neutron rms radius difference, $\Delta r_{nn'}$, could be better determined in light nuclei by simultaneously fitting $d\sigma/d\Omega$, $A_y(\theta)$, and Q data for both isotopes. The uncertainty in $\Delta r_{nn'}$ due to spin orbit ambiguities could be reduced from about ±0.08 fm to ±0.024 fm for $^{18}O - ^{16}O$. Typical values of $\Delta r_{nn'}$ vary from 0.1 to 0.2 fm (Refs. 5 and 20). These calculations are illustrative only since no actual p + ^{18}O data at 800 MeV exist. Pseudo-data generated with the KMT potential, the amplitudes of Table I and Hartree-Fock densities[3] were used instead.

From the above discussion it is seen that the uncertainties in ρ_n and r_n due to spin orbit ambiguities diminish with increasing target mass. This trend is easily understood by considering the ratio of the volumes of the spin orbit and the central potentials. Evaluating the spin orbit term at an angular momentum corresponding to $k_N R_N$ (R_N is the nuclear radius) results in a ratio of volumes which is proportional to $A^{-0.24}$ for typical central and spin orbit geometries. This trend is observed explicitly in the errors in r_n for ^{16}O and ^{40}Ca of Table II and continues for larger nuclei.

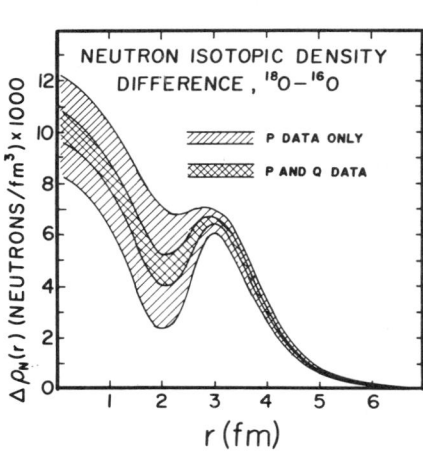

Fig. 8. Uncertainty in the neutron isotopic density difference for $^{18}O-^{16}O$ due to spin effects.

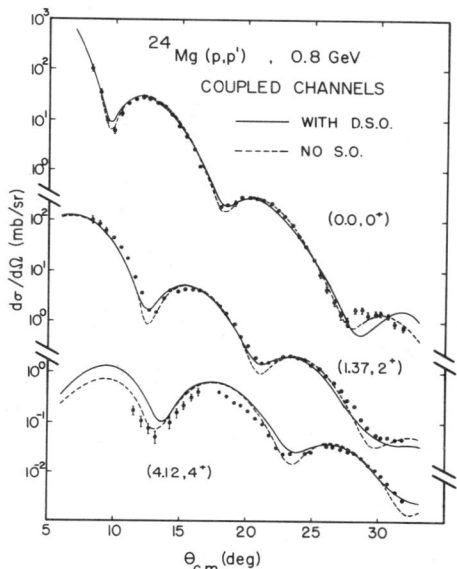

Fig. 9. Experimental and computed angular distributions for p + ^{24}Mg at 800 MeV.

Absolute tests of the predicted proton-nucleus spin orbit optical potential have been partially completed at 800 MeV as indicated by the solid curves in Fig. 1. Predictions of Q for p + ^{16}O, ^{40}Ca at 800 MeV are given by the positive definite solid curves in the lower halves of Figs. 5 and 6. These calculations employ the second order KMT optical potential, Arndt's CL80 800 MeV Coulomb distorted N-N amplitudes[19] and the impulse approximation. The $A_y(\theta)$ predictions are very good even compared with the phenomenological fits. The Q predictions using Arndt's amplitudes and those of Eq. (4) are very different as seen in Figs. 5 and 6. This large difference in Q is due to the fact that the real-to-imaginary ratio of the spin orbit potential is positive (negative) for the Gaussian parametrization (Arndt's amplitude).[21] The positive values for α_{spj} are needed in order to reproduce the sharp "sawtooth" structure of the analyzing power data which rises very rapidly and falls more slowly with increasing angle. The calculations using Arndt's amplitudes reproduce this "sawtooth" pattern even though the real-to-imaginary ratio of the spin orbit term is negative because Arndt's p-p amplitudes include Coulomb distortions. These distortions drastically alter the low momentum behavior of the Im$\{C_{pp}\}$, causing it to depart from the $q \times \exp(-B_{spj}q^2)$ form. Predictions of $A_y(\theta)$ using Arndt's amplitudes without Coulomb distortions are very poor at forward angles. Thus, in proton-nucleus calculations the Coulomb distortion effects in the p-p amplitudes must always be included.

The above discussion and calculations pertain to spherical nuclei for which coupling to low lying discrete inelastic channels is not expected to be important. For deformed nuclei coupled-channels (CC) KMT calculations should be performed. Though such is not formidable theoretically, the problem does present some computational challenges and as yet no such work has been carried out in the KMT approach. However, useful nuclear structure information can be obtained from conventional phenomenological CC analyses[23] by taking advantage of the well proven applicability of the folding model, Eq. (1), for ∿1 GeV proton scattering and by invoking Satchler's theorem.[24] These allow the multipole moments of the deformed ground state matter density and the optical potential to be trivially related.[25]

Regarding the effect of spin-dependence on extracted matter density information, CC calculations with full deformed spin orbit potentials and alternately with no spin orbit interaction were carried out for $p + {}^{24}Mg$ at 800 MeV for the 0.0 MeV 0^+, 1.37 MeV 2^+, and 4.12 MeV 4^+ members of the ground state rotational band.[26] The calculations were performed with the coupled channels code ECIS[27] assuming Woods-Saxon optical model forms and relativistic kinematics.[7] The solid and dashed curves in Fig. 9 correspond to the best fits obtained with and without the spin orbit interaction included, respectively. The two sets of calculations are similar but those which include the spin orbit potential are in somewhat better agreement with the data near the diffractive minima. The standard optical model parameters[23,28] and multipole moments $M(E\lambda)$[29] are given in Table III for the calculations with and without the spin orbit terms. From Table III it is seen that the extracted matter density multipole moments, M(E2) and M(E4), are not greatly affected by the spin orbit interaction. This result had been previously observed in single step inelastic transitions[30] and in this case is shown to be true for the 4^+ which is predominantly excited by multistep processes.[26]

TABLE III

Deformed optical model parameters for $p + {}^{24}Mg$ at 800 MeV[a]

V	r	a	W	r_I	a_I	W_{SF}	r_{SF}	a_{SF}	V_{so}	W_{so}	r_{so}	a_{so}	r_c	β_2	β_4
4.7	0.93	0.45	92.5	0.93	0.55	16.0	0.45	0.40	0.7	1.5	1.0	0.78	1.0	0.6	0.01
-5.3	0.93	0.45	100.	0.93	0.54	16.0	0.45	0.40	0.0	0.0	-	-	1.0	0.6	-.02

Multipole moments in $eb^{\lambda/2}$ (b)

	M(E2)	M(E4)
With D. S. O.	0.189	0.0099
No D. S. O.	0.184	0.0076

[a] Well depths are in MeV, radii and diffusenesses are in fermis.
[b] Moments of the imaginary potential are given.

CONCLUSIONS

In this work the impact of the spin-dependence of the ∼1 GeV proton-nucleus interaction upon empirically deduced matter densities has been assessed. Generally these effects are small but are enhanced in light nuclei and can be important in isotopic density differences. Ambiguities in fitting the available proton-nucleus analyzing power data result in uncertainties in the deduced neutron rms radii which continue to be of the same magnitude as that which would result in the complete absence of $A_y(\theta)$ data. The uncertainty in the deduced neutron radial distributions due to spin-dependence ambiguities is significantly reduced when $A_y(\theta)$ data are available. The additional measurement of the spin rotation quantity Q could further reduce the error in the deduced neutron densities. The reduction in the <u>total</u> error in r_n would, however, only be about 0.01 to 0.02 fm with Q data since numerous other sizeable contributions to the total error exist.[7]

The microscopic description of ∼1 GeV proton-nucleus $A_y(\theta)$ and Q data is fundamentally important as this provides a test for the reaction theory and subsequent approximations. It has been demonstrated here that local, second order KMT calculations together with the impulse approximation and Arndt's CL80 Coulomb distorted N-N amplitudes provide fairly good absolute predictions of the 800 MeV $A_y(\theta)$ data for ^{16}O and ^{40}Ca. The Q predictions provided here will make possible further interesting tests in the future.

REFERENCES

1. R. C. Barrett and D. F. Jackson, <u>Nuclear Sizes and Structure</u> (Clarendon, Oxford, 1977).
2. H. Rebel, in <u>Radial Shape of Nuclei</u>, edited by A. Budzanowski and A. Kapuscik (Jagellonian University, Cracow, 1976), p.164.
3. J. W. Negele and D. Vautherin, Phys. Rev. C<u>5</u>, 1472 (1972); The density matrix expansion code of Negele provided the numerical results given here.
4. L. Ray, Nucl. Phys. A<u>335</u>, 443 (1980).
5. L. Ray, Phys. Rev. C<u>19</u>, 1855 (1979).
6. L. Ray, G. W. Hoffmann, and R. M. Thaler, Phys. Rev. C (in press).
7. L. Ray, W. R. Coker, and G. W. Hoffmann, Phys. Rev. C<u>18</u>, 2641 (1978).
8. G. S. Adams, et al., Phys. Rev. Lett. <u>43</u>, 421 (1979).
9. I. Brissaud and X. Campi, Phys. Lett. <u>86B</u>, 141 (1979).
10. A. Chaumeaux, V. Layly, and R. Schaeffer, Ann. Phys. (N. Y.) <u>116</u>, 247 (1978).
11. R. R. Johnson, et al., Phys. Rev. Lett. <u>43</u>, 844 (1979).
12. G. Bruge, Report No. DPh-N/ME/78-1, Saclay, 1978 (unpublished).
13. G. W. Hoffmann, et al., Phys. Rev. Lett. <u>40</u>, 1256 (1978); and G. Igo, et al., Phys. Lett. <u>81B</u>, 151 (1979).
14. G. D. Alkhazov, et al., Sov. Acad. Sci. <u>26</u>, 715 (1977).

15. A. K. Kerman, H. McManus, and R. M. Thaler, Ann. Phys. (N. Y.) $\underline{8}$, 551 (1959).
16. E. Kujawski and J. P. Vary, Phys. Rev. C$\underline{12}$, 1271 (1975).
17. I. Sick and J. S. McCarthy, Nucl. Phys. A$\underline{150}$, 631 (1970).
18. I. Sick, et al., Phys. Lett. $\underline{88B}$, 245 (1979).
19. R. A. Arndt, private communication.
20. L. Ray, et al., preprint (unpublished).
21. R. J. Glauber and P. Osland, Phys. Lett. $\underline{80B}$, 401 (1979).
22. G. W. Hoffmann, et al., Phys. Lett. $\underline{76B}$, 383 (1978).
23. T. Tamura, Rev. Mod. Phys. $\underline{37}$, 679 (1965); and T. Tamura, Oak Ridge National Laboratory Report No. ORNL-4152 (1967), (unpublished).
24. G. R. Satchler, J. Math. Phys. $\underline{13}$, 1118 (1972).
25. M. L. Barlett, et al., Phys. Rev. C (in press).
26. G. Blanpied, et al., Phys. Rev. C$\underline{20}$, 1490 (1979).
27. J. Raynal, private communication.
28. C. M. Perey and F. G. Perey, At. Data Nucl. Data Tables $\underline{13}$, 293 (1974).
29. C. H. King, J. E. Finck, G. M. Crawley, J. A. Nolen, Jr., and R. M. Ronningen, Phys. Rev. C$\underline{20}$, 2084 (1979).
30. L. Ray and W. R. Coker, Phys. Lett. $\underline{79B}$, 182 (1978).

DISCUSSION

FICK: Can you comment on a comparison of α and $\vec{p}$ experiments to determine details of nuclear shapes?

RAY: Alpha particle probes are useful for isotopic density difference measurements since they avoid the spin ambiguities of the proton probes. However, at energies for which the folding model for alpha scattering works, above 100 MeV, the strong absorption prevents the alpha probe from penetrating as far into the nucleus as does the 800 MeV proton, so the protons are more useful than alphas in this respect.

IGO: Does the new analysis of elastic proton scattering data using Arndt's phase shifts affect the rms radius of neutron distribution in nuclei?

RAY: Using Arndt's CL80 N-N amplitude results in neutron radii about 0.2 fm smaller than in previous analyses with parametrized amplitudes. It is possible that the proton-neutron amplitudes are still not well known or there may be some interesting proton-nucleus physics left out of the calculations.

IGO: What is the physics reason that Q narrows the error corridor in the ^{16}O - ^{18}O difference plot, especially at small radius?

RAY: The spin related error in the neutron difference for $^{18}O - ^{16}O$ has assumed the worst possible situation, in which no assumption about the isovector dependence of the spin-orbit potential is assumed, so that the spin-orbit terms in $^{16,18}O$ are completely independent. Knowledge of Q minimizes this spin related error and produces the largest reduction where the error is greatest, namely at small r.

HANNA: How do the charge radii obtained in electron scattering fit in to your analysis of neutron densities?

RAY: The empirical charge densities from electron scattering are used to determine the point proton distributions in the analysis of 800 MeV proton elastic scattering data.

THOMPSON: How important are relativistic effects, especially for spin-orbit coupling?

RAY: The Schrödinger equation with the exact relativistic wave number and the reduced relativistic total energy has been used. The effects are large compared to using non-relativistic kinematics. The relativistic effects on the polarization are kinematic only in these calculations.

JOHNSON: Did I understand you to say that the correlation corrections are of the same order as the $\underline{l \cdot s}$ contributions.

RAY: Yes.

JOHNSON: Could you please explain why the expression you gave for the correlation correction only involved the N-N amplitude A?

RAY: The largest correlation correction is due to the spin-independent amplitude squared term and Pauli correlations. The actual calculations do include the Pauli correlation corrections to the spin-orbit potential. Of course all five N-N amplitudes enter in the second order optical potential, but the remaining terms are very small at 1000 MeV.

POLARIZED NEUTRON CAPTURE STUDIES IN THE GIANT RESONANCE REGION

H. R. Weller
Duke University and Triangle Universities Nuclear Laboratory
Durham, North Carolina 27706

ABSTRACT

The technique of fast polarized neutron capture is briefly described. The results of three $(\vec{n},\gamma)$ experiments (on ^{40}Ca, ^{13}C and ^{3}He) are presented. Results are, where appropriate, compared to the predictions of direct-semi-direct model calculations. Effects which can be attributed to the giant quadrupole resonance and to direct M1 radiation are observed in the data. The ^{3}He$(\vec{n},\gamma)^{4}$He data indicate that the ^{4}He(γ,p)-to-(γ,n) total cross-section ratio is about 1.8 near 28 MeV, while the percentage of E2 and spin-flip E1 amplitudes are similar in the n and p channels.

The study of fast neutron capture reactions has received considerable attention in the past few years. This has in part been a result of renewed interest in capture experiments generated by proton capture studies using polarized beams. These experiments were motivated by the realization that, by measuring the analyzing powers for a $(\vec{p},\gamma)$ reaction, it would be possible, with a minimum number of assumptions, to determine the amplitudes and phases of the contributing transition matrix elements.[1] If one assumes that the capture reaction proceeds only by E1 and E2 radiation, these results provide an almost unambiguous determination of the E2 cross section in the capture reaction channel. Of course it was the discovery of the giant quadrupole resonances which made the interest in this E2 cross section so keen.[2]

The utility of <u>polarized</u> capture measurements can be illustrated with a simple example. If the unpolarized cross section is measured as a function of θ it can be expanded in terms of Legendre Polynomials as

$$\sigma(\theta) = A_0 [1 + \sum_{k=1}^{n} a_k P_k(\cos\theta)] \quad (1)$$

to obtain the a_k coefficients. If the reaction proceeded by pure E1 radiation there would only be an a_2 coefficient. If we have a spin zero target, and we end up in a final state of $J^\pi = 1/2^-$, for example, we could have two contributing transition matrix elements corresponding to formation of a $1/2^+$ or formation of a $3/2^+$ state which then E1 decays to the $1/2^-$ final state. Taking these matrix elements as $s_{1/2} e^{i\phi_s}$ and $d_{3/2} e^{i\phi_d}$ we can write

$$1.0 = (s_{1/2})^2 + 2.0(d_{3/2})^2$$

$$a_2 = -(d_{3/2})^2 - 2.0(s_{1/2})(d_{3/2}) \cos(\phi_s - \phi_d).$$

With only two equations and three unknowns ($s_{1/2}$, $d_{3/2}$ and $\phi_s - \phi_d$) these equations cannot be solved. The use of polarized nucleons, however, allows for the determination of another relationship among the three unknowns. If the analyzing power $A(\theta) = (N_+ - N_-)/(N_+ + N_-)P$ is measured, it can be expanded according to

$$\frac{\sigma(\theta) A(\theta)}{A_0} = \sum_{k=1}^{n} b_k P_k^1(\cos\theta) . \qquad (2)$$

For our pure E1 assumption this results in a b_2 coefficient which can be expressed in terms of the transition matrix elements in this case as

$$b_2 = (s_{1/2})(d_{3/2}) \sin(\phi_s - \phi_d) .$$

We now have three equations and three unknowns so that a solution can be found. Although two solutions result, it has been shown that the physical solution can be chosen by using a model calculation[3] and, in a few cases, by measuring the interference of the "giant resonance" background[4] with a "special state", if such occurs in the energy region of interest.

This is the basic idea behind polarized capture studies. When E2 is considered along with E1, the equations get considerably more complex, and the experiments must be done to a rather high degree of precision to extract the necessary coefficients (n=4 in Eqs. 1 and 2). In the best cases there will be two E1 and two E2 amplitudes and three relative phases. Since one can measure 9 independent coefficients, the problem is overdetermined. If M1 radiation also contributes, there will in general be 11 unknowns and a solution cannot be determined.

Polarized proton capture studies have been performed on a number of targets.[5] An E1+E2 analysis of the $^{13}C(\vec{p},\gamma_1)^{14}N^*$ reaction[6], for example, leads to the preferred E2 solutions shown in Fig. 1. The solid line in this figure represents the results of a calculation which assumes a pure direct-E2 capture mechanism. A spectroscopic factor of $C^2S=0.85$ was assumed in this calculation. This direct cross section can account for much of the observed E2 strength, and clearly represents an annoying "background" if we are trying to see the effects of the GQR in this channel.

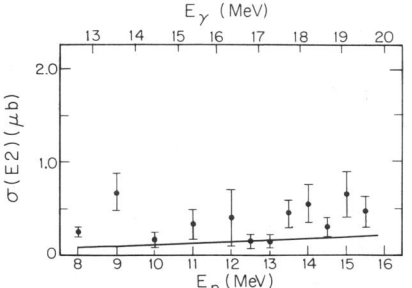

Fig. 1. Small solutions for the E2 cross section in $^{13}C(p,\gamma_1)^{14}N$. Solid line is the direct E2 calculation.

This result for $^{13}C(\vec{p},\gamma_1)^{14}N$, along with similar results obtained in several other studies,[5] was the principle reason for our decision to perform $(\vec{n},\gamma)$ measurements. This can be understood by noting that the direct E2 cross section is scaled by the square of the kinematic effective charge. Since the ratio of the effective charge for neutrons to protons for a target (Z,A) is $Z/(A^2+Z)$, the direct E2 cross section can be virtually eliminated by using neutrons in place of protons.

The ^{2}H(d,n)^{3}He reaction is used to provide our neutron beam in the energy range of 5 to 18 MeV. The experimental arrangement is shown in Fig. 2. The 1" gas cell is typically operated at about 45 psia which gives us a neutron energy spread of about 160 keV for 10 MeV neutrons. The ^{2}H(d,n) reaction has several attractive features as a source reaction: the high and rather well known cross section[7] and the strong forward peaking of the angular distribution of the neutrons are two of these. We typically obtain a neutron flux of 2×10^8 n/μC/sr. In order to perform these experiments it is important to carefully shield the NaI-plastic spectrometer system and to process the signals so as to minimize pulse pile-up effects. A pulsed beam, used to establish a time of flight criterion, is important especially for angles less than 60° and for neutron energies below 7.0 MeV. The ^{2}H(d,n) reaction is also an excellent source of polarized neutrons at 0° when a polarized deuteron beam is used.[8] Our typical polarized deuteron beam has $p_3=0.7$ and varies between 0.1 and 0.2 μA on target. In some cases we have been able to compensate for the reduced beam intensity by increasing the deuterium gas cell pressure. Of course the increased neutron energy spread may sometimes be undesirable. The neutron polarization which would be obtained at 0° for an incident polarized deuterium beam having $p_3=p_{33}=1.0$ is shown in Fig. 3.[8] We typically run with a neutron polarization of ~60%. Our recently developed three stage bunching system[9] allows us to transport ~80% of the normally transported D.C. beam to the target and to bunch ~90% of this beam into pulses having a time width of about 2.0 nsec. This development permits us to use a TOF criterion in our polarized neutron capture experiments.

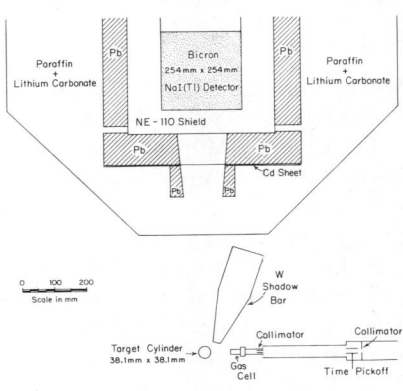

Fig. 2. Experimental arrangement

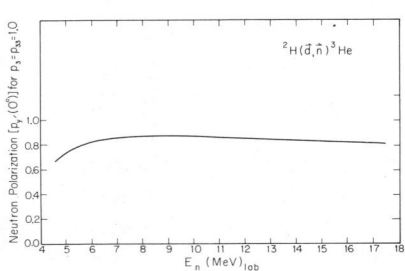

Fig. 3. Neutron polarization at 0° vs neutron energy

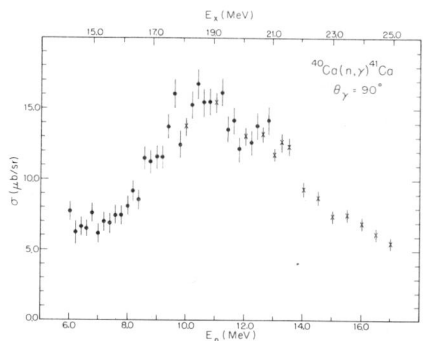

Fig. 4. ^{40}Ca(n,γ)^{41}Ca yield. Data are from Ref. 21 (dots) recently extended to higher energies (x's).

The first polarized neutron capture experiment which we performed[10] was for ^{40}Ca($\vec{n}$,γ)^{41}Ca. The unpolarized yield curve is shown in Fig. 4. Our absolute cross sections are in good agreement with previous values.[11] These data have been corrected for finite geometry, neutron attenuation and multiple scattering, γ-ray attenuation, and the energy and angular dependence of the ^{2}H(d,n) reaction. The variation of the neutron angular distribution with energy is primarily responsible for the energy dependent correction factor (see Fig. 5) which is applied to the data after normalizing to the ^{2}H(d,n) 0° yield. Angular distributions of cross sections and analyzing powers were measured at four energies in the region of the GDR. The cross section data were corrected for the effects mentioned above. The spin dependent effects have not yet been included in our Monte-Carlo codes, but preliminary estimates indicate that these effects are small in comparison with our statistical uncertainty. The angular distribution data were fitted using Eqs. (1) and (2) to obtain the a_k and b_k coefficients of Table 1. For E1 capture to the $7/2^-$ final state of ^{41}Ca we can have $g_{9/2}$, $g_{7/2}$ and $d_{5/2}$ T-matrix elements. Since most calculations suggest that the $g_{7/2}$ (~spin flip) strength is rather small, we neglect it and use A_0, a_2 and b_2 to obtain the $g_{9/2}$ and $d_{5/2}$ amplitudes and phases. These are shown in Fig. 6. The curves are the result of DSD calculations where a form factor proportional to r as well as a complex coupling form factor[12] have been assumed.[13] These results indicate that ~70% of the E1 strength in this reaction is due to capture of $g_{9/2}$ neutrons while ~30% is due to $d_{5/2}$ capture.

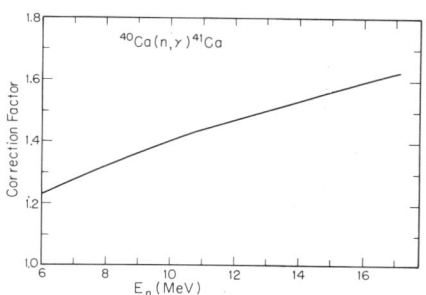

Fig. 5. Correction factor for ^{40}Ca yield curve.

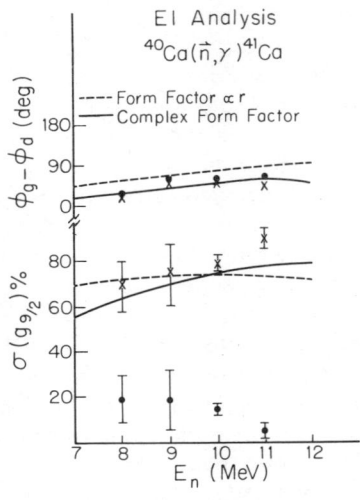

Fig. 6. E1 solutions for ^{40}Ca($\vec{n},\gamma$)^{41}Ca. Remaining cross section is due to $d_{5/2}$ n capture.

The a_k and b_k coefficients of Table 1 are shown in Fig. 7. Higher order coefficients were not statistically significant. The solid curves are the result of a direct-semidirect (DSD) calculation which includes the isoscalar GQR at 18 MeV with Γ=4.0 MeV.[13] The energy dependence of the a_1 coefficient is reasonably well accounted for by this calculation. The absence of direct E2 capture appears to make the effect of the GQR more observable than similar p capture results. The observation of coherent E2 strength which can be attributed to the IS-GQR suggests that the GQR decays in a non-statistical manner-at least in the neutron channel leading to the ground state of ^{40}Ca. Although the calculations for a_2 and b_2 are in good agreement with experiment, the DSD calculations for b_1 disagree, especially at the lower energies where b_1 is seen to increase. The dotted line of this figure represents an extended calculation which includes direct M1 strength due to $f_{5/2}$ neutron capture.[14] No adjustable parameters were involved in adding this strength. I emphasize the sensitivity of these coefficients to small amounts of non-E1 radiation by noting that at E_n=10 MeV the E2 strength in the DSD calculation amounts to only ~0.15% of the E1 cross section. The direct M1 strength is slightly less than 0.3(2)% of the E1 cross section at E_n=10(6) MeV. Our ability to observe the effects of small non-E1 amplitudes in the capture channels clearly represents a powerful tool for studying nuclear structure.

Table 1 ^{40}Ca($\vec{n},\gamma$)^{41}Ca

E_n(MeV)	a_1	a_2	b_1	b_2
6.0	−0.14±0.07	0.26±0.10		
7.0	0.02±0.06	0.05±0.10		
8.0	−0.04±0.04	0.38±0.07	0.16±0.06	−0.05±0.04
9.0	−0.08±0.03	0.12±0.05	0.09±0.06	−0.17±0.04
10.0	−0.12±0.17	0.10±0.03	0.13±0.02	−0.15±0.02
11.0	−0.13±0.03	−0.01±0.06	0.06±0.04	−0.10±0.03
12.0	−0.08±0.04	−0.20±0.08		
13.0	−0.04±0.06	−0.29±0.12		

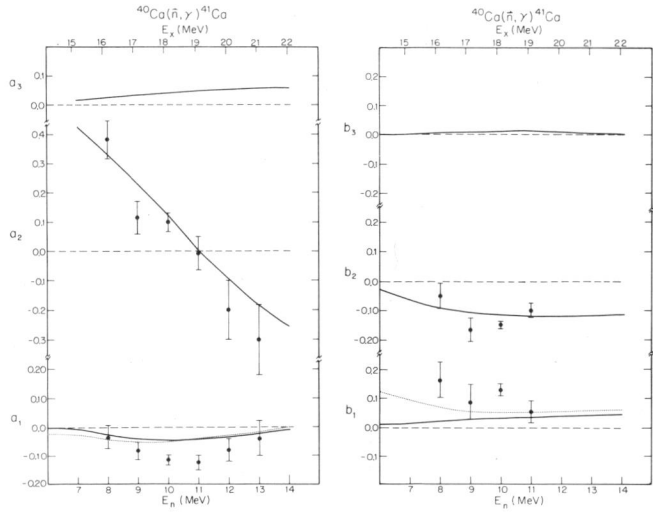

Fig. 7. The a and b coefficients for ^{40}Ca$(\vec{n},\gamma)^{41}$Ca and DSD calculations (see text).

Since there are five E2 and three E1 amplitudes in the ^{40}Ca(n,γ) case, a model independent E1-E2 analysis is impossible. We can, however, attempt an analysis guided by the DSD model calculation. For the E1 amplitudes we choose the predominant $g_{9/2}$ solution. The choice of E2 amplitudes depends upon the E2 form factor of the DSD calculation. If a Goldhaber-Teller form factor $(-rdU/dr)$ is used,[15] one finds that the $h_{11/2}$ and the $p_{3/2}$ terms together account for ~85% of the E2 strength, while if the form factor is taken to be proportional to r^2 then $h_{11/2}$ and $f_{7/2}$ together account for around 90% of the E2 strength. The solutions found under these two choices indicate that the E2 cross section accounts for 6.0±4.2 or 3.3±2.7% of the total. Using the position and width of the GQR of Youngblood et al.[16], this E2 strength would amount to 6 or 3% of the IS-E2-EWSR, respectively.

The large number of E1 and E2 terms involved in the ^{40}Ca$(n,\gamma)^{41}$Ca experiment makes it difficult to extract unambiguous information on the contributing amplitudes and phases of this reaction. This is one of the reasons that we chose the ^{13}C$(\vec{n},\gamma)^{14}$C reaction as our second case for study. Our interest in this reaction was further stimulated by the fact that we had previously studied the closely-related ^{13}C$(\vec{p},\gamma_1)^{14}$N$(0^+,T=1)$ reaction.[6]

Figure 8 shows the yield curve (after corrections) which we measured at 90° for the ^{13}C$(n,\gamma)^{14}$C reaction. The ^{13}C$(p,\gamma_1)^{14}$N* yield curve is also shown here. These two reactions are examining the GDR built on the ground state of ^{14}C$(T=1, T_z=1)$ and its analogue, the first excited state of ^{14}N$(T=1, T_z=0)$. The isospin selection rules imply that we should see the T=0 strength of the GDR in ^{14}N via its proton decay to ^{13}C, and the T=1 strength of this same GDR in ^{14}C via its neutron decay to ^{13}C. One might expect the relative strengths to be determined by the ratio of the isospin Clebsch-Gordon coefficients $C^2(\frac{1}{2}\frac{1}{2}, \frac{1}{2}\frac{1}{2}; 11)/C^2(\frac{1}{2}\frac{1}{2}, \frac{1}{2}-\frac{1}{2}; 00) = 2$. When

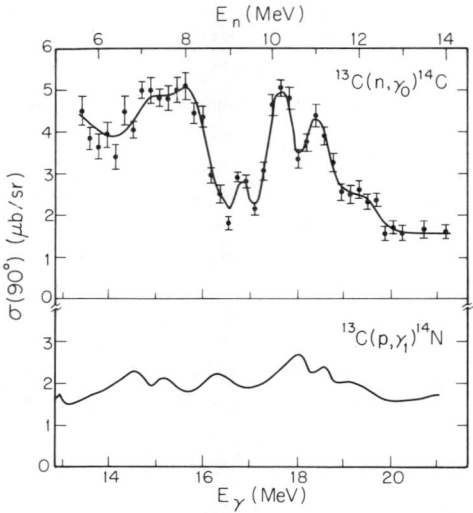

Fig. 8. The $^{13}C(n,\gamma)^{14}C$ and $^{13}C(p,\gamma_1)^{14}N$ yield curves. Smooth curves were drawn through the data points.

we integrate the detailed balanced experimental cross sections over energy and angle, we find that the neutron data exhaust 16% of the classical dipole sum while the proton data exhaust 9% in the same energy region giving a ratio of 1.8.

Angular distributions of cross section and analyzing powers have been measured at several energies in the GDR region of this experiment. The results at three energies are shown in Fig. 9 along with the curves generated from polynomial fits to the data. In this case the D_2 gas cell pressure was run at 90 psia, so that the energy spread of the neutron beam ranged from 280 to 400 keV.

For this reaction there are only two E1 T-matrix elements: $s_{1/2} e^{i\phi_s}$ and $d_{3/2} e^{i\phi_d}$, and two E2 terms: $p_{3/2} e^{i\phi_p}$ and $f_{5/2} e^{i\phi_f}$. If an E1-E2 analysis of our data is performed, the E1 amplitudes and phases of Fig. 10 are found. Again we have two solutions. The DSD calculation clearly favors the predominantly $d_{3/2}(E1)$ solution. We conclude that the E1 strength in this reaction is due primarily to capture of $d_{3/2}$ neutrons.

The a and b coefficients which we have obtained to date in this experiment are listed in Table 2. These data indicate the presence of E2 radiation which is almost certainly not direct. The a_1 and a_3 coefficients obtained here tend to have signs which are opposite to those in the $^{13}C(p,\gamma_1)^{14}N$ experiment suggesting that the E2 strength present is mostly isoscalar.

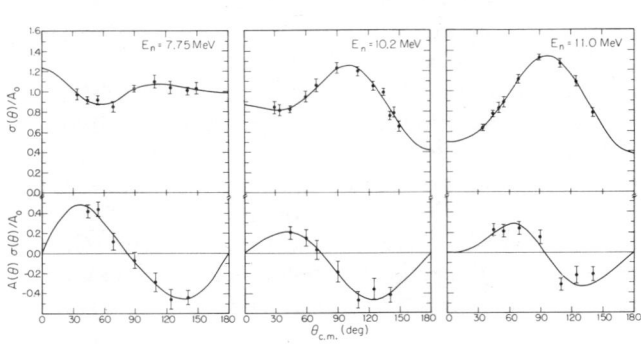

Fig. 9. $^{13}C(n,\gamma)$ angular distributions.

315

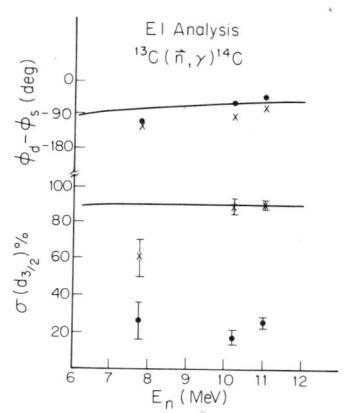

Fig. 10 (left). E1 solutions for $^{13}C(\vec{n},\gamma)^{14}C$. Remaining cross section is $s_{1/2}(E1)$. Curves are DSD calculation.

Fig. 11 (below). Angle integrated cross section for $^{4}He(\gamma,n)^{3}He$ reaction. Penn, Ref. 22; U of S, Ref. 23; Toronto, Ref. 24; TUNL, present work; LLL, Ref. 18.

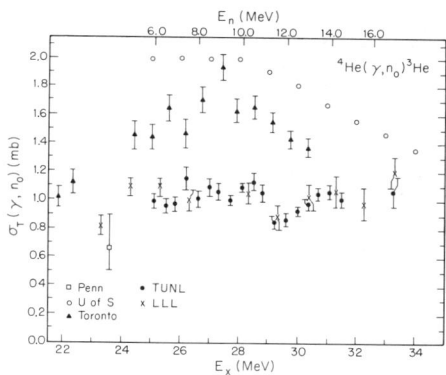

Table 2 $^{13}C(\vec{n},\gamma)^{14}C$

E_n(MeV)	a_1/b_1	a_2/b_2	a_3/b_3	a_4/b_4
7.75	-0.04±0.03 / -0.05±0.04	-0.03±0.05 / 0.30±0.02	0.17±0.05 / 0.02±0.02	0.11±0.06 / 0.02±0.03
10.2	0.01±0.02 / -0.19±0.04	-0.43±0.04 / 0.22±0.02	0.19±0.05 / 0.01±0.03	0.05±0.06 / -0.01±0.03
11.0	-0.06±0.01 / 0.00±0.03	-0.65±0.01 / 0.19±0.02	0.12±0.01 / -0.05±0.02	0.05±0.02 / -0.03±0.02

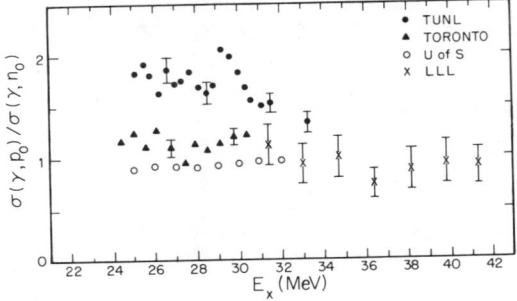

Fig. 12. The ^{4}He(γ,p) -to-(γ,n) total cross-section ratios. TUNL, present work with (γ,p) of Ref. 19; Toronto, Ref. 24 and Ref. 19; U of S, Ref. 23 and Ref. 19; LLL, Ref. 25. Typical errors are shown.

The final experiment which I will discuss is our recent work on the ^{3}He$(n,\gamma)^4$He reaction. The target for this experiment consisted of a high pressure gas cell filled to 140 atm. We began by measuring the 90° differential cross section as a function of energy. Our results, after detail balancing and multiplying by $8\pi/3$, are shown in Fig. 11. These data have been corrected for the factors mentioned earlier as well as the energy dependence of our detection efficiency. These results were somewhat surprising to us since they indicate that the $(\gamma,p)/(\gamma,n)$ cross section ratio near $E\gamma = 28$ MeV is about 1.8! Our absolute cross section was determined using the neutron flux, and our carefully measured detection efficiency.[17] Several checks, including a recoil counter flux measurement and a direct normalization to the ^{40}Ca(n,γ) cross section all support our results. We associate an error of about 15% with the absolute cross section reported here. The most recent (γ,n) data of Berman et al.[18] are shown in Fig. 11; there is excellent agreement. The ratios of the $\overline{(\gamma,p)}$ to the (γ,n) cross sections implied by our (and Berman's new) results are shown in Fig. 12. The (γ,p) cross sections used here were taken from Ref. 19.

Fig. 13. The angular distributions of cross section and analyzing powers. $(\vec{n},\gamma)$, present work; $(\vec{p},\gamma)$, Ref. 20. The curves are the result of polynomial fits which give coefficients shown.

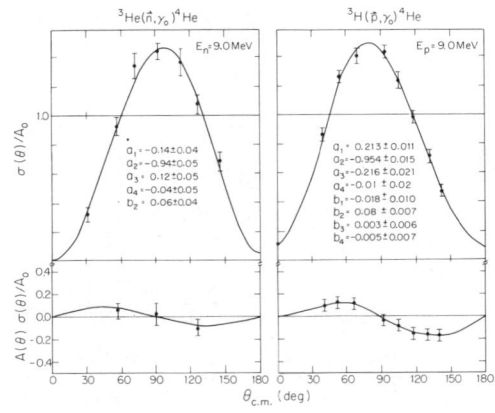

We have made a preliminary measurement of the ^{3}H(p,γ)^{4}He cross section in our laboratory which is in good (~10%) agreement with that of Ref. 19. So, in conclusion, our ^{3}He(n,γ) cross section measurements support the result (which has come and gone during the past ten years) that the (γ,p) cross section for ^{4}He is about twice the (γ,n) cross section, at least near 28 MeV. As can be seen in Fig. 12, the available data suggest that this ratio approaches 1.0 near $E_\gamma = 35$ MeV.

The preliminary results of our measurements of the angular distribution of cross section and analyzing powers for the ^{3}He($\vec{n}$,γ) reaction are shown in Fig. 13 for $E_n = 9.0$ MeV. The cross section data shown here have been corrected for finite geometry effects. This angular distribution is consistent with the relation $\sigma_T = \frac{8\pi}{3}\sigma(90°)$. The data from the ^{3}H($\vec{p}$,γ)^{4}He reaction[20] are shown here for comparison. The a and b coefficients implied by these results are listed on Fig. 13.

In these reactions we can have two E1 and two E2 amplitudes corresponding to S=0 and S=1 in the entrance channels. The a_1 and a_3 coefficients are dominated by terms representing interference between S=0(E1) and S=0(E2) amplitudes. However b_2 arises from the interference between S=0 and S=1 terms. It would be impossible to obtain any information on the S=1 (spin-flip) strength without the use of polarized beams. As can be seen in Fig. 13, the b_2 coefficients appear to be quite similar, albeit with considerable statistical uncertainty, in the n and p channels. This suggests that the relative S=0 to S=1 E1 strength is about the same in both n and p channels.

If the equations for the a and b coefficients are expressed in terms of the S=0 and the S=1 E1 amplitudes and the S=0(E2) amplitude, a fit to the data indicates that the S=0(E2) amplitude accounts for ~2.5% of the total cross section while the S=1(E1) amplitude accounts for ~1% of the total. The ($\vec{p}$,γ) experiment gave similar results for these two numbers.[20] The relative phases between the three amplitudes obtained from the ($\vec{n}$,γ) data are also in good agreement with those found from the ($\vec{p}$,γ) data. A detailed comparison of these two experiments must await a reduction in the statistical uncertainties of our ($\vec{n}$,γ) data.

We have seen several results of polarized neutron capture measurements, a field which has only begun to be exploited. In the giant resonance region such experiments can now be performed with a precision comparable to that obtained in ($\vec{p}$,γ) studies. These experiments can help us to refine our reaction theories as we have seen in ^{40}Ca where the effects of direct M1 radiation seem apparent. They also provide information on the decay properties of the E1 and E2 giant resonances. And this is done without the complication of a direct E2 "background". In the four-body problem polarized neutron and proton capture measurements provide information on small amplitudes which are sensitive to spin dependent components of the effective nuclear force.
I am certain that we can look forward to more data having greater precision than that presently available, and that these data will continue to surprise and enlighten us.

This work was performed at the Triangle Universities Nuclear Laboratory (TUNL) in collaboration with Drs. N. R. Roberson, S. A. Wender and D. R. Tilley. We acknowledge the significant contributions of M. Jensen and L. Ward. Theoretical support from D. R. Seyler, H. Hasan, H. Kitazawa, F. Dietrich and S. Cotanch is gratefully acknowledged. TUNL is partially supported by the U. S. D. O. E.

REFERENCES

1. H.F. Glavish, S.S. Hanna, R. Avida, R.N. Boyd, C.C. Chang, and E. Diener, Phys. Rev. Letts. 28, 766 (1972).
2. F.E. Bertrand, Ann. Rev. Nucl. Sci. 26, 457 (1976), and references therein.
3. H.R. Weller, N.R. Roberson, S.R. Cotanch, Phys. Rev. C18, 65 (1978).
4. K.A. Snover, et al., Phys. Rev. Letts. 44, 927 (1980).
5. H.R. Weller and N.R. Roberson, Rev. Mod. Phys. (to be published) Oct. 1980; and references therein.
6. J.D. Turner, N.R. Roberson, S.A. Wender, H.R. Weller and D. R. Tilley, Phys. Rev. C21, 525 (1980).
7. M. Drosg, LA-UR 77-2097. To be published in "Nucl. Science and Engineering".
8. P.W. Lisowski, R.L. Walter, C.E. Busch and T.B. Clegg, Nucl. Phys. A242, 298 (1975).
9. S. A. Wender et al., Nucl. Instr. and Methods (to be published).
10. M. Jensen, D.R. Tilley, H.R. Weller, N.R. Roberson, S.A. Wender and T.B. Clegg, Phys. Rev. Letts. 43, 609 (1979).
11. I. Bergqvist, D.M. Drake, and D.K. McDaniels, Nucl. Phys. A231, 29 (1974).
12. M. Potokar, Phys. Lett. 46B, 346 (1973).
13. F. Dietrich, private communication.
14. Hashima Hasan, private communication.
15. W. Uberall, Electron Scattering from Complex Nuclei (Academic, New York, 1977), Pt. B.
16. D.H. Youngblood, J.M. Moss, C.M. Rozsa, J.D. Bronson, A.D. Bacher, and D.R. Brown, Phys. Rev. C13, 994 (1976).
17. R.E. Marrs et al., Phys. Rev. Letts. 35, 202 (1975).
18. B.L. Berman, D.D. Faul, P. Meyer and D.L. Olson, to be published and B.L. Berman, private communication.
19. W.E. Meyerhof and S. Fiarman, Proc. Int. Conf. Photonuclear Reactions and Applications (ed. B.L. Berman, Lawrence Livermore Laboratory, Livermore, 1973), p. 385.
20. G. King III, Ph.D. Dissertation, Stanford University (1978).
21. S.A. Wender, N.R. Roberson, M. Potokar, H.R. Weller and D.R. Tilley, Phys. Rev. Letts. 41, 1217 (1978).

22. R. W. Zurmühle, W. E. Stephens, and H.H. Staub, Phys. Rev. 132, 751 (1963).
23. C.K. Malcom, D.V. Webb, Y.M. Shin and D.M. Skopik, Phys. Letts. 47B, 433 (1973).
24. J.D. Irish, R.G. Johnson, B.L. Berman, B.J. Thomas, K.G. McNeill, and J.W. Jury, Can. J. Phys. 53, 802 (1975).
25. T.W. Phillips, B.L. Berman, D.D. Faul, J.R. Calarco and J.R. Hall, Phys. Rev. C19, 2091 (1979).

DISCUSSION

HANNA: What is the strength of the form factor in the semi-direct term that gives you 10% of the experimental E2 strength?

WELLER: Our DSD calculation was performed using a Goldhaber-Teller type form factor (-rdU/dr) and the assumption that the GQR of ^{14}C exhausts 100% of the E2-EWSR. The calculation implies an E2 strength in the n capture channel which is an order of magnitude smaller than the E2 strength which we extract from the data when we perform an E1-E2 analysis.

SNOVER: If you really have direct M1 contributions in ^{40}Ca(n,γ) wouldn't you expect large effects in (p,γ) and (n,γ) at lower energies?

WELLER: I don't know of any polarized neutron capture data other than ours, nor of any a_1's for ^{40}Ca(n,γ) besides ours. I have only calculated a_1 down to 6 MeV. The effects in a_1 are very small. I would worry about compound nucleus contributions at lower energies. As for (p,γ), this may be complicated by the presence of direct E2 strength (which is negligible in the (n,γ) case). But I have not yet looked at this in any detail.

SNOVER: In the ^{4}He (γ,n)/(γ,p) ratio that you showed, it appears that two sets of data give a ratio of about unity at low energies, and two other sets give a different ratio. So, the situation would appear as uncertain as ever.

WELLER: If one takes all previously reported measurements at face value, then it appears as though there are a number of experiments which give a ratio near 1.0 even below $E_\gamma \sim 35$ MeV. Several of these, however, seem quite suspect. The most recent data of Berman, obtained in a (γ,n) experiment, are in excellent (point for point) agreement with our detail balanced capture data. This, I feel, constitutes strong evidence for the factor of ~ 1.8 in the region near 30 MeV. After all, the technologies are so completely different. However, perhaps a total absorption cross section measurement should be performed to convince anyone who still doubts these results.

ADELBERGER: Do I understand correctly that you have a $\pm 18°$ spread of neutron angles in your measurements? Are all effects of this included in your analysis?

WELLER: The detector subtends an angle of about 13°. Although the incident neutrons have an angular range of $\pm 17°$ on the target, most of the target and most of the neutrons are in a much smaller range. Our Monte Carlo codes correct the data for the finite geometry of both the target and the detector. However, due to the slow variation of the angular distribution with θ, the effects are rather small (a measured a_1 of -0.13 ± 0.03 in ^{40}Ca(n,γ) changes by +0.01 due to these corrections). Although spin dependent effects have not yet been included in these codes, experimental results suggest that these corrections will be small in comparison to our statistical errors.

POLARIZED PROTON RADIATIVE CAPTURE STUDIES OF GIANT RESONANCES

K.A. Snover
Department of Physics, University of Washington
Seattle, WA 98195, USA

ABSTRACT

Several interesting E1, M1 and E2 resonance studies in $(\vec{p},\gamma)$ reactions are discussed. These include a unique determination of E1 amplitudes in the $^{12}C(\vec{p},\gamma_0)^{13}N$ reaction, E2 strength in the $^{15}N(\vec{p},\gamma_0)^{16}O$ reaction, M1 decays to the ground states and to the excited 0^+ states of the doubly magic ^{16}O and ^{40}Ca nuclei, and the M1 γ-decay of the stretched 4^-, T=1 particle-hole state in ^{16}O.

INTRODUCTION

Radiative proton capture is a powerful tool for investigations of electromagnetic decays from nuclear resonances in the continuum. Measurements of polarized proton capture can in many cases provide the necessary information to determine resonance multipolarities E1, E2 or M1. For isolated resonances, it is well-known that experiments which measure only the angular dependence of the γ-ray intensity cannot determine the parity of the radiation (ML or EL). However, continuum resonances are never completely isolated, and a sensitivity to the multipole character, including the parity of the radiation, comes about through resonance-background interference in the angular distribution of the capture radiation, along with some knowledge of the background character (usually predominantly E1).

For most (p,γ) reactions, on the average, E1 radiation is dominant and E2 radiation is present with an intensity 1-2 orders of magnitude weaker than E1. M1 radiation appears to be weakest, although strong M1 resonances may occur at low ($E_p \lesssim 10$ MeV) energies. The pioneering $(\vec{p},\gamma)$ studies of E1 and E2 radiation were performed at Stanford University[1,2] and were described by H.F. Glavish at the previous polarization conference.[3] Since then we have learned much more about E2 strength, which I will discuss briefly. The sensitivity of $(\vec{p},\gamma)$ to the identification of M1 resonances, discovered at Seattle, has led to several interesting studies of M1 resonances. I will mainly discuss these studies, since both the techniques and the physics of these measurements are interesting and different from the earlier $(\vec{p},\gamma)$ studies.

TECHNIQUE

The cross section $\sigma(E,\theta)$ and analyzing power $A(E,\theta)$ for the capture of polarized particles may be defined in the usual manner as

$$\sigma(E,\theta) = [\sigma\uparrow(E,\theta) + \sigma\downarrow(E,\theta)]/2 \qquad (1)$$

and

$$\sigma(E,\theta)A(E,\theta) = [\sigma\uparrow(E,\theta) - \sigma\downarrow(E,\theta)]/2P \qquad (2)$$

where $\sigma\uparrow$ and $\sigma\downarrow$ are the cross sections for an incident beam of energy E and vector polarization of magnitude P oriented along ($\uparrow$) or against ($\downarrow$) the normal $\hat{n} = \vec{K}_{in} \times \vec{K}_{out}$ to the reaction plane.

The dependence on γ-ray emission angle θ may be expanded as

$$\sigma(E,\theta) = \sum_{K=0}^{2L_{max}} A_K(E) Q_K P_K(\cos\theta) \qquad (3)$$

and

$$\sigma(E,\theta)A(E,\theta) = \sum_{K=1}^{2L_{max}} B_K(E) Q_K P_K^1(\cos\theta) \qquad (4)$$

Here $\sigma_{total} = 4\pi A_0$, L_{max} is the maximum multipole which contributes ($L_{max} = 2$ for dipole + quadrupole) and the Q_K are the usual angular attenuation factors. It is often convenient to define fractional Legendre coefficients $a_K = A_K/A_0$, $b_K = B_K/A_0$. For the capture of polarized spin-1/2 particles on unpolarized targets, with only the γ-ray intensity (at a given energy and angle) observed in the outgoing channel, the above equations completely specify the (parity-allowed) capture process.

The usual angular momentum coupling rules tell us that interference between opposite (same) parity radiations contributes to the odd (even) coefficients so that E1-M1 interference contributes to A_1 and B_1, E1-E2 to A_1, B_1, A_3, B_3, etc. The exact relations may be written down as

$$A_k = \sum_{tt'} D_{tt'k} \, \text{Re} \, T_t T_{t'}^*, \qquad (5)$$

and

$$B_k = \sum_{tt'} f_k(tt') D_{tt'k} \, \text{Im} \, T_t T_{t'}^*, \qquad (6)$$

where T_t, $T_{t'}$ are the reaction amplitudes for different channels t and t', the $D_{tt'k}$ are angular momentum coupling factors and

$$f_k(tt') = [j'(j'+1) + \ell(\ell+1) - j(j+1) - \ell'(\ell'+1)]/2k(k+1) \qquad (7)$$

where ℓ, j, ℓ', j' are the interfering orbital and total angular momenta for the incident nucleon.

For the simplest spin sequence ($J_{target} = 1/2$, $J_{residual} = 0$ or vice versa), only 2 complex reaction amplitudes contribute for each multipole. For cases of this sort involving $1p_{1/2}$-shell targets, the amplitudes are

$$E1 : s_{1/2}e^{i\phi_{s_{1/2}}}, d_{3/2}e^{i\phi_{d_{3/2}}}$$

$$E2 : p_{3/2}e^{i\phi_{p_{3/2}}}, f_{5/2}e^{i\phi_{f_{5/2}}}$$

$$M1 : p_{1/2}e^{i\phi_{p_{1/2}}}, p_{3/2}e^{i\phi'_{p_{3/2}}}$$

If only electric multipoles contribute at a given energy, the problem is overdetermined (e.g., 9 independent A_K, B_K versus 7 amplitude parameters for E1 + E2) whereas if magnetic multipoles contribute, the problem is underdetermined (E1 + E2 + M1 requires 11 amplitude parameters). However, as we show below, one may use $(\vec{p},\gamma)$ to uniquely determine E1 amplitudes, identify the multipolarity E1, E2 or M1 of resonances, and provide E2 cross sections for broadly distributed strength in the continuum.

E1 CAPTURE AMPLITUDES

For the simple spin sequences described above, only 2 complex E1 amplitudes contribute, and $(\vec{p},\gamma)$ angular distribution measurements restrict the E1 amplitudes to 2 possible solutions. This 2-fold ambiguity is inherent, resulting from the quadratic nature of the equations relating the amplitudes to the data. A typical example in light nuclei is the $^{12}C(p,\gamma_o)^{13}N$ reaction,[4,5] illustrated in Fig. 1. The GDR region extends from $E_p \sim 8$-30 MeV with $(\vec{p},\gamma_o)$ angular distribution results[4] for E_p = 10-17 MeV. Only incoming s- and d-wave amplitudes (with j = 1/2 and 3/2, respectively) may contribute to E1 capture, and Fig. 1 shows that one of the 2 solutions is predominantly d-wave ($d_>$) and the other predominantly s-wave ($s_>$).

Similar $d_>$ and $s_>$ solutions are obtained for other capture reactions such as $^{14}C(\vec{p},\gamma_o)$[6] and $^{15}N(\vec{p},\gamma_o)$[2,7,8] (Fig. 2), indicating that one is observing a general feature of the GDR build on $1p_{1/2}$-shell nuclei. The $d_>$ solution is expected on theoretical grounds—virtually all models of radiative capture through the GDR, such as the doorway-state[10] or the direct-semidirect (DSD)[9,4] models, predict that d-waves should dominate, with results in reasonable agreement with the experimental data if the $d_>$ solution is the correct (physical) one.

However, one does not need a detailed calculation to understand why d-wave capture is expected to dominate. Relative to the ^{13}N ground-state, the configurations in the GDR which contribute to pro-

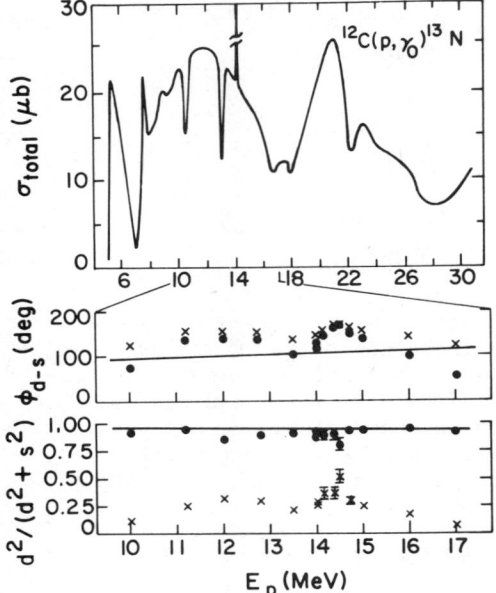
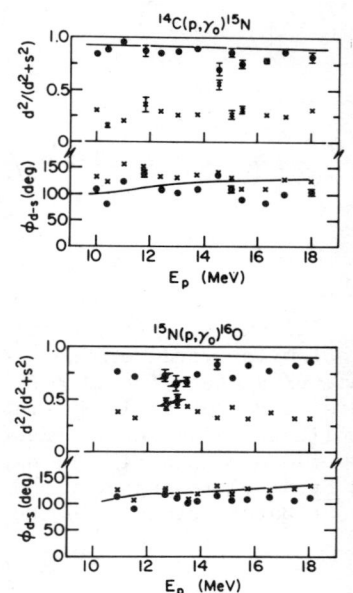

Fig. 1. Upper part: σ_{total} for $^{12}C(\vec{p},\gamma_0)$ (refs. 4-5). Lower part: The d-s phase difference and the relative d-wave intensity for E_p = 10-17 MeV (ref. 4 plus ref. 12 for 14 < E_p < 15 MeV). The points and crosses correspond to the $d_>$ and $s_>$ solutions, respectively. The solid lines are DSD calculations described in ref. 4.

Fig. 2. The relative d-wave intensity and the d-s phase difference for $^{14}C(\vec{p},\gamma_0)$ (ref. 6) and $^{15}N(\vec{p},\gamma_0)$ (refs. 7,9). The solid curves are DSD model predictions (see ref. 9).

ton capture are $(1d)^1 (1p_{1/2})^{-1}$ and $(2s)^1 (1p_{1/2})^{-1}$, and the schematic model predicts the amplitudes for these configurations should be in the ratio of the E1 matrix elements connecting these configurations to the ground state. Since $\langle 1d|E1|1p_{1/2}\rangle \gg \langle 2s|E1|1p_{1/2}\rangle$ (the first matrix element has a good radial overlap while the second involves a node change resulting in radial cancellations), this predicts d-wave capture should dominate. This is a special application of a more general rule discussed many years ago by Wilkinson[11] that E1 photoabsorption in the GDR should be dominated by nucleon excitations of the form $n\ell \rightarrow n'\ell'$ where $n\ell$ is an occupied shell model orbital and $n'\ell'$ an unoccupied orbital with $n' = n$, $\ell' = \ell+1$.

At Seattle we recently determined[12] that the $d_>$ solution is the physically correct one by making the first unique E1 amplitude determination in radiative capture. We did this by studying the

interference between the lowest T = 3/2 M1(E2) resonance at
E_p = 14.23 MeV and the E1 background in the $^{12}C(\vec{p},\gamma_0)$ reaction. The
basic idea is to use interference with a known resonance to determine
unknown properties of the background.

The dominant M1-E1 interference effects should appear in the
A_1 and B_1 coefficients; hence we measured excitation curves at 90°
with a polarized beam and 55° and 125° with an unpolarized beam.
The results are shown in Fig. 3 for

$$[\sigma(55°) + \sigma(125°)]/2 = A_0 - 0.39 A_4 \simeq \sigma_{total}/4\pi$$

$$\sigma A(90°) = B_1 - 1.53 B_3$$

$$[\sigma(55°) - \sigma(125°)]/2 = 0.57 A_1 - 0.39 A_3$$

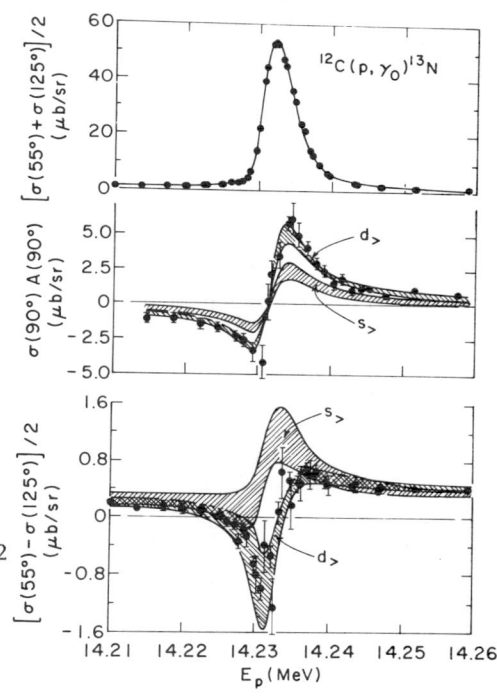

The experimental data for the
latter 2 quantities clearly
show pronounced interference
effects.

We calculated resonance
curves using the known T =
3/2 resonance parameters of
ref. 13 and background E1 and
E2 amplitudes determined from
off-resonance angular distributions. For the interference
shapes in $\sigma(90°)A(90°)$ and
$\sigma(55°) - \sigma(125°)$, the only free
parameter was the phase of the
T = 3/2 resonance relative to
the E1 background. The results
clearly select the $d_>$ solution
as the physically correct one.

These measurements also
provide restrictions on the E2
contributions to this reaction.[12]
In principle, more extensive
measurements of this sort could
uniquely determine the E2 amplitudes. Also this technique
could be applied to other
nuclei where known multipolarity
resonances occur at sufficiently
high excitation energy to be
used for unique determinations
of giant resonance amplitudes
and phases.

Fig. 3. Excitation curves taken
near the lowest T = 3/2 resonance
in $^{12}C(p,\gamma_0)^{13}N$ (ref. 12). The
solid curve in the top part is a
calculated fit. The bands in the
lower 2 parts represent the spread
of calculated curves for the $d_>$
and $s_>$ solutions consistent with
off-resonance angular distributions.

E2 STRENGTH

Effects of E2 radiation are apparent in most (p,γ) studies in and above the GDR region. Direct E2 capture contributes strongly in the GDR region, and in many cases E1 capture plus direct E2 capture accounts for the observed cross sections and angular distributions; in many other cases it is difficult to discern whether or not additional (collective) E2 strength is present (see refs. 4, 6, 8, 14, 15). $(\vec{n},\gamma)$ studies, as discussed by H.R. Weller at this conference, should be free of direct E2 effects, but are quite difficult experimentally. Decay studies of the isoscalar giant quadrupole resonance (GQR) show[16] that in most cases p_0 is a weak decay channel, in agreement with direct-semidirect calculations.[6] This is due to the large spreading width and, in light nuclei, non-statistical decay to the α-channel. Nevertheless, (p,γ) E2 strength remains interesting because of the possibility of collective isovector contributions, about which little is known. There remain a few cases where $(\vec{p},\gamma)$ studies seem to indicate a significant excess of E2 strength over direct capture.

One such case is $^{15}N(\vec{p},\gamma_0)^{16}O$. The E2 cross sections deduced from $^{15}N(\vec{p},\gamma_0)^{16}O$ measurements at Seattle are shown in Fig. 4 for E_p = 1.4 to 18.0 MeV (E_x = 13.4 to 29 MeV). The data include a reanalysis of the work of Bussoletti et al.,[7] plus new results[17] mainly at the lower energies. Here we show results only at energies where the data are consistent with no M1 radiation (see the following section). The region below $E_x \sim$ 20 MeV is made up of a number of small resonances, where one needs to perform a resonance analysis of the

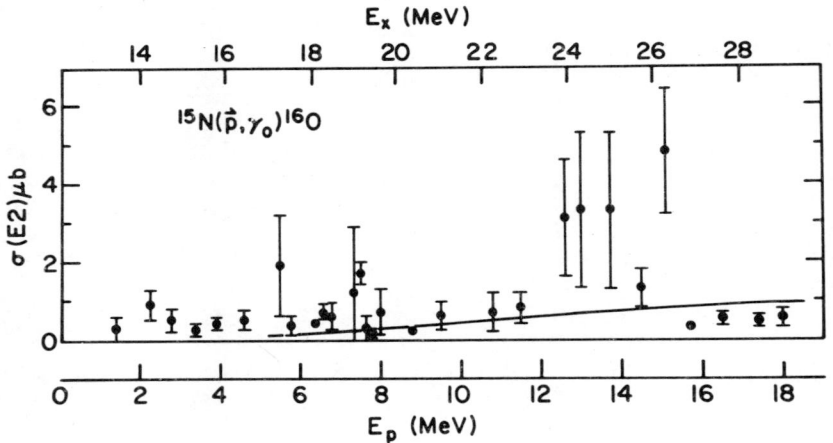

Fig. 4. E2 cross sections for the $^{15}N(\vec{p},\gamma_0)^{16}O$ reactions (refs. 7, 17). The solid line represents calculated direct E2 capture.

angular distribution coefficients in order to understand the strength (this is in progress[17]).

Above E_x = 20 MeV, the only region of possibly significant structure is for E_x = 23-27 MeV, where the present data indicate a strength of roughly 5-10% of the isoscalar E2 energy weighted sum rule (EWSR)[18] in excess of a smooth "background" estimated from the lower points in this region. These data show less evidence for a "GQR" than do previous Stanford data.[2,19] For E_x = 17.9-27.3 MeV where $(\alpha,\alpha'p_0)$ coincidence decay studies[20] show 9% of the EWSR we find in (p,γ_0) 12-22% of the EWSR (calculated direct E2 capture accounts for ~6% of the EWSR). Thus the <u>integrated</u> E2 strength seen in this region in $(\alpha,\alpha'p_0)$ and (p,γ_0) may be compatible when one accounts for coherent direct capture, without the need to invoke the presence of significant isovector (IV) E2 strength. For E_x = 13.4 to 29 MeV (p,γ_0) shows 20-30% of the EWSR, with direct capture accounting for ~11% of the EWSR.

M1 DECAYS TO THE GROUND AND THE FIRST EXCITED 0^+ STATES OF ^{16}O

Until recently very little was known about ground-state M1 decays in doubly magic light nuclei. Such decays were generally expected to be weak, since the doubly magic closed shell component of the ground state wavefunction cannot contribute.

Recently we explored this phenomenon at Seattle, with the discovery that M1 excitations can be uniquely identified in radiative capture.[9,21] This was done in the region of semi-isolated resonances below the GDR in the $^{15}N(\vec{p},\gamma_0)^{16}O$ reaction, as illustrated in Fig. 5. Pronounced structure in $A(90°)$ or a_1, which can be non-zero only due to interfering radiations of opposite parity, indicate possible M1 or E2 resonances interfering with the E1 background. The M1 assignments come from fits to detailed angular distributions assuming a model-independent parameterization in terms of E1 and E2 reaction amplitudes. The χ^2 for these angular distribution fits is also shown in Fig. 5. Strong deviations from acceptable values ($\chi^2 \lesssim 2$) indicate areas of concentrated M1 strength. Analysis of the a_i and b_i near these energies shows that the prominent <u>resonances</u> at 16.22 and 17.14 MeV are M1, with a third M1 resonance near 18.8 MeV which in the cross section is unresolved from a neighboring E1 resonance.

These M1 resonances in ^{16}O correspond to a total ground-state M1 strength $B(M1)\downarrow \gtrsim 0.24\ \mu_0^2$. This is quite sizable compared to a non-closed shell A = 4n nucleus such as $^{12}C (B(M1)\downarrow = 0.93\ \mu_0^2)$. The observed M1 decays stem from the ground-state correlations (primarily 2 particle-2 hole) and are in reasonable accord with recent shell model calculations by Arima and Strottman.[22] Between 16 and 20 MeV several M1 states are predicted, with a total strength of 0.27 μ_0^2, which is quite comparable to experiment. At higher energies up to 29 MeV an additional strength of 0.6 μ_0^2 is predicted to be fragmented over a number of levels. The ground-state wavefunction generated in this calculation has a 2p-2h intensity of 17%.

The experimental search for ground state M1 strength predicted at higher energies in ^{16}O represents an intriguing experimental challenge. We have measured A(90°) in fine (100 keV) energy steps from E_p = 9-16 MeV in the $^{15}N(p,\gamma_0)^{16}O$ reaction, and we find no pronounced structure.[17] Additional angular distributions in this region show no significant evidence for M1 strength. However, M1 resonances in this energy region may be either too broad or have too weak a proton formation probability Γ_p/Γ to be observed in $(\vec{p},\gamma_0)$. At lower energies, we find no evidence (Γ_{γ_0} < 1 ev) for a previously reported[23] M1 (p,γ) resonance, in agreement with an earlier electron scattering experiment.[24]

The utility of $(\vec{p},\gamma)$ for discovering M1 transitions has so far been demonstrated only in the one case discussed above. In the future it will be interesting to extend this technique to other nuclei, and to see if M1 resonances can be identified in reactions which do not have the simplest spin sequences (see the discussion below).

More recently we have measured γ-decay branches from the 16.22 and 17.14 MeV 1^+, T = 1 states to the 0_2^+ (6.05 MeV) final state (see Fig. 6) for which we find[25] preliminary values of B(M1, $1^+ \to 0_2^+$)/B1 M1, $1^+ \to 0_1^+$) = 0.45 ± 0.03 and 0.55 ± 0.04, respectively (the strength of these decays rules out a significant contribution from an unresolved M2/E3 branch to the 3^- (6.13 MeV) level). These relatively strong M1 decays should provide

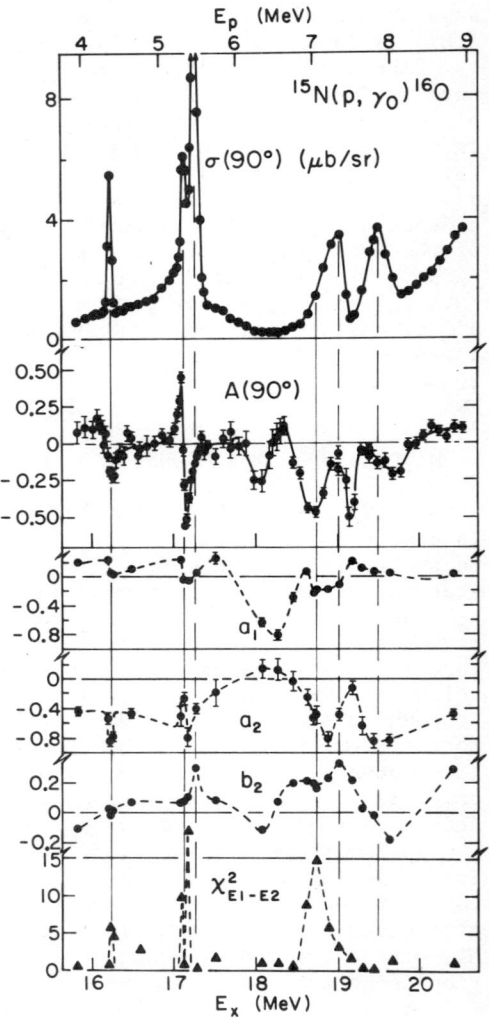

Fig. 5. Excitation curves for $^{15}N(\vec{p},\gamma_0)^{16}O$: $\sigma(90°)$, A(90°) and the a_1, a_2 and b_2 coefficients (3rd and 4th order coefficients not shown) and the reduced χ^2 for angular distribution fits assuming only E1 and E2 radiation. The curves are to guide the eye. Vertical solid and dashed line lines indicate M1 and E1 resonances, respectively (ref. 21).

important restrictions on the character of the 0_2^+ state, notably its 2p-2h composition which is not known very well. Unfortunately no theoretical calculations of these decays are currently available. Hopefully the future will bring experimental investigation of higher energy resonances which undergo M1 decays to the 0_2^+ state. Particularly interesting is the question of whether a "normal" giant M1 resonance exists built on the 0_2^+ state. Weak coupling arguments along with the known ground-state M1 strength in

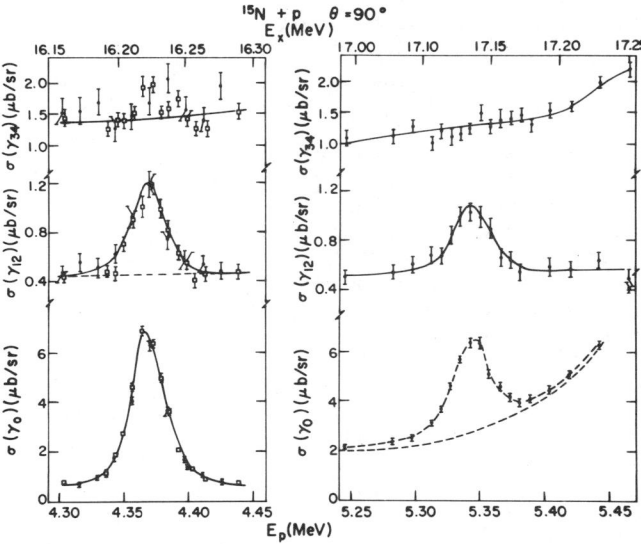

Fig. 6. ^{15}N(p,γ) yields to the 0_1^+ state (γ_0), the 0_2^+-3^- doublet (γ_{12}) at 6.1 MeV, and the 2^+-1^- doublet (γ_{34}) at 6.9-7.1 MeV, in the vicinity of the 16.22 and 17.14 MeV 1^+, T = 1 resonances (ref. 25).

^{12}C and ^{20}Ne suggest such strength in ^{16}O would lie in the $E_x \sim$ 17-21 MeV region.

M1 DECAYS TO THE GROUND AND FIRST EXCITED 0^+ STATES OF ^{40}Ca

A recent electron scattering experiment[26] has resulted in a definitive M1 assignment for a strong (Γ_{γ_0} = 4.74 ± 0.30 eV, B(M1)↓ = 0.37 μ_0^2) ground state transition from a level at 10.32 MeV in ^{40}Ca. It is interesting to note that this one state in ^{40}Ca carries more M1 strength than the total of all the known ground state M1 strength in ^{16}O. This is consistent with the belief that ^{40}Ca is not as good a closed shell nucleus as ^{16}O.

The 10.32 MeV state is a well-known resonance in the ^{39}K(p,γ)^{40}Ca reaction at E_p = 2.043 MeV, with a capture strength $\Gamma_p\Gamma_\gamma/\Gamma$ = 10.3 ± 1.7 eV (ref. 27). This resonance strength has been used as a standard upon which other strength measurements in this mass region are based,[27] and is clearly incompatible with the radiative width quoted above. We have remeasured this capture strength with the result[28] $\Gamma_p\Gamma_{\gamma_0}/\Gamma$ = 4.33 ± 0.35 eV, compatible with the radiative width derived from electron scattering. We also observed decay branches to the excited 0_2^+ (3.35 MeV) and 2^+ (3.90 MeV) final states, with $B(M1, 1^+ \to 0_2^+)/B(M1, 1^+ \to 0_1^+) = 0.43 \pm 0.03$ and $B(M1, 1^+ \to 2^+)/B(M1, 1^+ \to 0_1^+) = 0.16 \pm 0.02$ (assuming E2/M1 = 0 for the $1^+ \to 2^+$ transition). The observed decays in ^{16}O and ^{40}Ca are shown in Fig. 7. It is truly remarkable that the reduced B(M1) branching ratios for decays to the excited 0_2^+ state relative to the 0_1^+ ground state are equal within errors for the three 1^+ states studied (the 16.22 and 17.14 1^+, 1 states in ^{16}O and the 10.32 1^+, 1 state in ^{40}Ca).

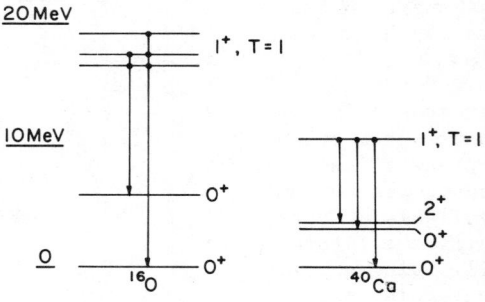

Fig. 7. Observed decays of 1^+, T = 1 states in ^{16}O and ^{40}Ca.

At Seattle, we looked for interference in $(\vec{p},\gamma)$ between this M1 resonance and the non-resonant E1 background. We did not see such effects, probably because the background is very weak and the resonance is very narrow ($\Gamma \ll 1$ keV). However, it is important to search for M1 strength in ^{40}Ca at higher energies, where such interference effects may be much easier to observe.

GAMMA DECAY OF THE 4^-, T = 1 STRETCHED PARTICLE-HOLE STATE IN ^{16}O

Narrow "stretched" high spin particle-hole states are found in nuclei from ^{12}C to ^{208}Pb and have been studied in high energy electron, proton, and pion scattering and in some cases in direct transfer reactions. The lowest T = 1 levels of this sort such as $[d_{5/2}, p_{3/2}^{-1}](4^-1)$ in ^{12}C and ^{16}O (refs. 29, 30) and $[f_{7/2}, d_{5/2}^{-1}](6^-1)$ in ^{24}Mg and ^{28}Si (see ref. 31) are believed to be predominantly 1 particle-1 hole states, which is part of the reason why they are so interesting. In ^{16}O, for example, the 4^-,1 state at 18.98 MeV has nearly all of the expected (d,t) pickup strength,[32] and has $\sim 1/2$ of the M4 "single particle" inelastic electron scattering strength.[31] However, very little is known about the decays of these levels.

Recent pickup measurements[32] and decay coincidence measurements[33] along with previous $^{15}N(p,\gamma_{12})$ measurements[34] strongly suggest assignments of 3^-, T = 1, and 4^-, T = 1 for resonances at 18.03 and 18.98 MeV in ^{16}O, respectively. We recently remeasured the capture reaction over these 2 resonances.[25] We also detected γ-rays from p_{12} and α_1 decay channels (Fig. 8) with strengths which confirm the identification of these resonances with the states seen in the pickup/decay studies. We find $\Gamma_p \Gamma_\gamma / \Gamma = 1.96 \pm 0.27$ eV and 0.85 ± 0.10 eV for the 3^-,1 and 4^-,1 resonance decays to the 3^-, 0 (6.13 MeV) final state.

Using $\Gamma_{p_0}/\Gamma = 0.46 \pm 0.15$ and 0.12 ± 0.05 (ref. 33) leads to $\Gamma_\gamma = 4.8 \pm 1.9$ eV and 7.1 ± 3.1 eV, corresponding to $B(M1) = 0.24 \pm 0.10$ μ_0^2 and 0.29 ± 0.13 μ_0^2 for the $3^-,1$ and $4^-,1$ decays to the $(3^-,0)$ level. We also obtain total widths $\Gamma = 23 \pm 12$ and 8 ± 4 keV for the $3^-,1$ and $4^-,1$ resonances, respectively, by comparing our resonance strengths for the p_{12} and α_1 exit channels with the coincidence decay results of ref. 33.

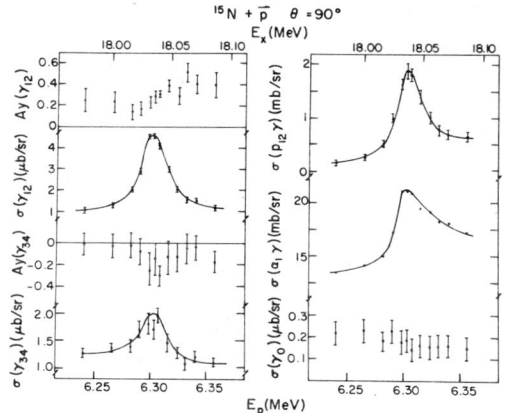

The $(4^-,1) \to (3^-,0)$ decay strength is in accord with the shell model value of 0.41 μ_0^2 calculated[35] by J. Millener (using a $1\hbar\omega$ basis for the $4^-,1$ state). A real test of the 1p-1h purity of the $(4^-,1)$ level must await an improved value for Γ_γ, which depends mainly on an improved measurement of Γ_{p_0}/Γ. The reasonably strong $(3^-,1) \to (3^-,0)$ decay strength is also interesting since this level is not particularly strong in pickup,[32] implying it should be mostly 3p-3h.

Also shown in Fig. 8 is $A(90°)$ for the capture γ-rays. Now the non-resonant background is almost certainly E1, so that one would expect M1-E1 resonance-background interference effects in $A(90°)$, whereas the striking aspect

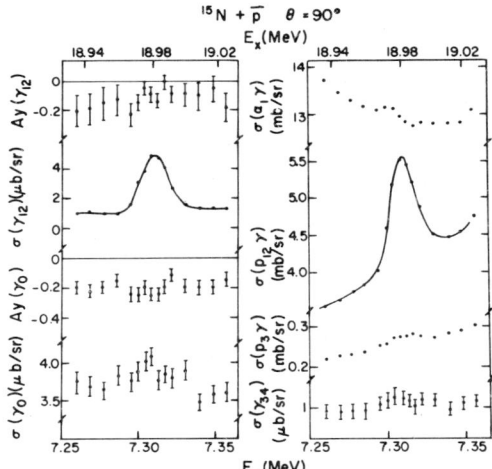

Fig. 8. ^{15}N + p yields near the 18.03 3^-, T = 1 and 18.98 4^-, T = 1 resonances (ref. 25).

of these data is the absence of a significant resonance in the analyzing power. The most likely explanation for this may be that the many reaction amplitudes present in the E1 background tend to wash out interference effects.

ACKNOWLEDGEMENTS

I am indebted to my colleagues who have shared in experiments performed in Seattle in the past few years, particularly P.G. Ikossi, without whose close collaboration much of the work described here would not be completed, and E.G. Adelberger, with whom discussions and collaborations are always valuable. Special thanks are due also to J.E. Bussoletti, K. Ebisawa, K.T. Lesko, and T.A. Trainor.

This work has been supported by the United States Department of Energy.

REFERENCES

1. H.F. Glavish, S.S. Hanna, R. Avida, R.N. Boyd, C.C. Chang, and E. Diener, Phys. Rev. Lett. $\underline{28}$, 766 (1972).
2. S.S. Hanna, H.F. Glavish, R. Avida, J.R. Calarco, E. Kuhlman, and R. LaCanna, Phys. Rev. Lett. $\underline{32}$, 114 (1974).
3. H.F. Glavish, Proc. of the Fourth Internat. Symp. on Polarization Phenomena in Nuclear Reactions, Zurich, 1975 (Birkhäuser Press, Basel, 1976), p. 317.
4. R. Helmer, M.D. Hasinoff, J.E. Bussoletti, K.A. Snover, and T.A. Trainor, Nucl. Phys. $\underline{A336}$, 219 (1980).
5. D. Berghofer, M.D. Hasinoff, R. Helmer, S.T. Lim, D.F. Measday, and K. Ebisawa, Nucl. Phys. $\underline{A263}$, 109 (1976).
6. K.A. Snover, J.E. Bussoletti, K. Ebisawa, T.A. Trainor, and A.B. McDonald, Phys. Rev. Lett. $\underline{37}$, 273 (1976).
7. J.E. Bussoletti, Ph.D. Thesis, University of Washington, 1978; K.A. Snover, J.E. Bussoletti et al., to be published.
8. K.A. Snover, Proc. of the Giant Multipole Resonance Topical Conference, Oak Ridge, Tenn., October 1979, F.E. Bertrand, Ed., to be published.
9. K.A. Snover, Proc. of the 3rd Internat. Symp. on Neutron Capture Gamma Ray Spectroscopy and Related Topics, Brookhaven, 1978 (Plenum Press, NY, 1979), p. 319.
10. D.G. Mavis, H.F. Glavish, and D.C. Slater, Proc. of the 4th Internat. Symp. on Polarization Phenomena in Nuclear Reactions, Zurich, 1975 (Birkhäuser Press, Basel, 1976), p. 749.
11. D.H. Wilkinson, Physica $\underline{22}$, 1039 (1956).
12. K.A. Snover, P.G. Ikossi, E.G. Adelberger, and K.T. Lesko, Phys. Rev. Lett. $\underline{44}$, 927 (1980)
13. R.E. Marrs, E.G. Adelberger, and K.A. Snover, Phys. Rev. C $\underline{16}$, 61 (1977).
14. J.D. Turner, N.R. Robertson, S.A. Wender, H.R. Weller, and D.R. Tilley, Phys. Rev. C $\underline{21}$, 525 (1980).
15. C. Fitzpatrick et al., and R.D. Ledford et al., contributions #2.74 and 2.75 to this conference. See the Table of Contents for the page numbers.

16. K.T. Knöpfle, Proc. of the Internat. Conf. on Nuclear Physics with Electromagnetic Interactions, Mainz, Germany, June 1979; Lecture Notes in Physics 108, 311 (1979); G.J. Wagner, Proc. of the Giant Multipole Resonance Topical Conference, Oak Ridge, Tenn., October 1979, F.E. Bertrand, Ed., to be published.
17. K.A. Snover, P.G. Ikossi, and K.T. Lesko, to be pub. (see also ref. 8).
18. M. Gell-Mann and V.L. Telegdi, Phys. Rev. 91, 169 (1953). We use values for $<r^2>$ taken from electron scattering: C.W. de Jager et al., Atomic and Nuclear Data Tables 14, 479 (1974).
19. S.S. Hanna, Proc. of the Internat. Conf. on Nuclear Physics with Electromagnetic Interactions, Mainz, Germany, June 1979; Lecture Notes in Physics 108, 288 (1979).
20. K.T. Knöpfle, G.J. Wagner, P. Paul, H. Breuer, C. Mayer-Böricke, M. Rogge, and P. Turek, Phys. Lett. 74B, 191 (1978).
21. K.A. Snover, P.G. Ikossi, and T.A. Trainor, Phys. Rev. Lett. 43, 117 (1979).
22. A. Arima and D. Strottman, Los Alamos Preprint LA-UR-78-2969.
23. C. Rolfs and W.S. Rodney, Nucl. Phys. A235, 450 (1974).
24. M. Stroetzel, Z. Phys. 214, 357 (1968).
25. K.A. Snover, P.G. Ikossi, E.G. Adelberger, and K.T. Lesko, to be published.
26. W. Gross, D. Meuer, A. Richter, E. Spamer, O. Titze, and W. Knüpfer, Phys. Lett. 84B, 296 (1979); see also P. Burt et al., contributed paper to Internat. Conf. of Nuclear Physics with Electromagnetic Interactions, Mainz, Germany, June 1979.
27. See for example, P.M. Endt, Atomic and Nuclear Data Tables 23, 3 (1979) and references therein.
28. K.A. Snover, W. Rösch, E.G. Adelberger, and P.G. Ikossi, to be published.
29. T.W. Donnelly, J.D. Walecka, I. Sick, and E.B. Hughes, Phys. Rev. Lett. 21, 1196 (1968).
30. I. Sick, E.B. Hughes, T.W. Donnelly, J.D. Walecka, and G.E. Walker, Phys. Rev. Lett. 23, 1117 (1969).
31. R.A. Lindgren, W.J. Gerace, A.D. Bacher, W.G. Love, and F. Petrovitch, Phys. Rev. Lett. 42, 1524 (1979) and references therein.
32. G. Mairle, C.J. Wagner, P. Doll, K.T. Knöpfle, and H. Breuer, Nucl. Phys. A299, 39 (1978).
33. H. Breuer, P. Doll, K.T. Knöpfle, G. Mairle, and C.J. Wagner, preprint.
34. A.R. Barnett and N.W. Tanner, Nucl. Phys. A152, 257 (1970). S.H. Chew, J. Lowe, J.M. Nelson, and A.R. Barnett, Nucl. Phys. A229, 241 (1974); Nucl. Phys. A286, 451 (1977).
35. J. Millener, private communication.

POLARIZATION TRANSFER IN INELASTIC SCATTERING

J. M. Moss
Los Alamos Scientific Laboratory
Los Alamos, NM 87545

ABSTRACT

Polarization transfer experiments are now feasible for inelastic scattering experiments on complex nuclei. Experiments thus far have dealt with the spin-flip probability; this observable is sensitive to the action of spin-spin and tensor forces in inelastic scattering. Spin-flip probabilities at $E_p \sim 40$ MeV in isoscalar transitions in $^{12}C(12.71$ MeV$)$ and $^{16}O(8.88$ MeV$)$ show considerable deviation from DWBA-shell model predictions; this indicates evidence for more complex reaction mechanisms. Experiments at intermediate energies will soon be possible and will yield data of much higher precision than is possible at lower ($E < 100$ MeV) energies. These experiments hold exciting promise in such areas as nuclear critical opalescence.

INTRODUCTION

A glance through the contributed papers in nuclear structure and reactions at this conference reveals that a majority of them deal with measurements of analyzing powers. These observables are relatively easy to obtain with present day polarized beams since they require only measurement of single intensities for various beam polarization states. Polarization transfer observables (triple scattering parameters), on the other hand, usually require not only a polarized beam, but analysis of the final polarization state as well; until recently such experiments were limited to light nuclear systems where poor resolutions and (or) low efficiencies could be tolerated.

What I hope to convince you of in the course of this talk is that polarization transfer can lead to a new and interesting class of physics, quite different from that obtained from analyzing powers, and that these experiments can now be performed with sufficiently high efficiency and resolution for nuclear structure studies. The particular observable which I will focus on is the transverse spin-flip (SF) probability, S, in inelastic scattering. In terms of the Wolfenstein parameter, D, the SF probability is $S = 1/2(1 - D)$.

Measurement of S can be made in certain restricted cases without resorting to polarization analysis[1,2], but in general, one needs a polarimeter with both good resolution and high efficiency. These requirements can be met by coupling a polarimeter to a magnetic spectrograph as is shown schematically in Fig. 1. The properties of three such polarimeters constructed for different proton energy ranges are given in Table 1. It is obviously a great

TABLE I - POLARIMETER CHARACTERISTICS

Polarimeter	Energy Range (MeV)	Efficiency	Analyzing Power
Texas A&M	30	4×10^{-5}	0.42
Indiana	150	2×10^{-3}	0.4
LAMPF-HRS	500-800	7×10^{-2}	0.25

advantage to use medium-energy protons where the longer range can be exploited to produce a very high efficiency double-scattering target. The Texas A&M and Indiana polarimeters, which resemble the schematic of Fig. 1, are designed for measuring S for one final momentum at a time. The LAMPF-HRS polarimeter, on the other hand, utilizes an array of multi-wire drift chambers together with a software reconstruction of double-scattered particle trajectories to give polarization analysis for a broad range of momenta in the focal plane of the high-resolution spectrometer (HRS).

The unique features of S as an observable arise from its close connection to spin transfer, s, and to the action of spin-dependent forces in inelastic scattering. One can show in the Eikonal approximation[3] for central spin-dependent forces that the SF probability associated with a given ℓ, s, and j transfer, $S_{\ell sj}$, is:

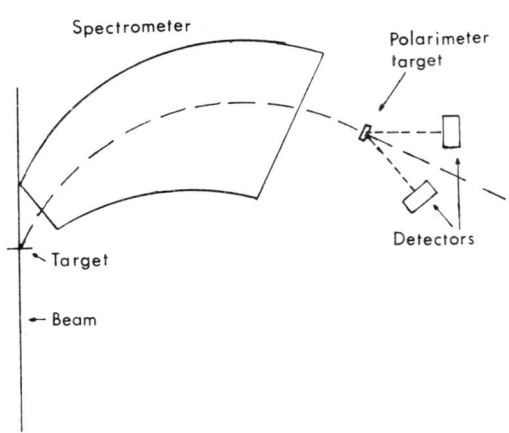

Fig. 1. Schematic focal-plane polarimeter.

$$S_{\ell s j} = \frac{3j+1}{2(2j+1)} \qquad j = \ell + 1$$

$$S_{\ell s j} = \frac{3j+2}{2(2j+1)} \qquad j = \ell - 1 \qquad\qquad s = 1$$

$$S_{\ell s j} = 1/2 \qquad j = \ell$$

$$S_{\ell s j} = 0 \qquad\qquad\qquad\qquad\qquad s = 0$$

More relatistic calculations of S over a wide range of energies show that S is large (S $\geq$ 0.5) only when s = 1 and that S ~ 0 in the absence of spin transfer. As examples one might expect S ~ 0 for a collective $0^+ \to 2^+$ excitation in the (p,p') reaction, while for a $0^+ \to 1^+$ transition, one should see S $\geq$ 0.5. As a final note DWBA calculations of S show a remarkable insensitivity to spin-orbit distortion from the proton-nucleus optical potential; this is in contrast to the large sensitivity exhibited by vector analyzing powers at energies below 100 MeV.

SPIN FLIP AT LOWER ENERGIES

Rather than review all of the cases studied thus far (see Refs. 3 and 4) I want to concentrate on a few which illustrate the nature of S as an observable and its potential at intermediate energies where much more precise data should be easily obtainable.

The $1^+ \to 0^+$ (T=1, 3.56 MeV) transition in ^{6}Li is a well-studied isovector "spin flip" transition. In the (p,p') reaction excitation may proceed by $\ell s j$ = 011 and 211 transfer. Figure 2 shows measurements of S at two angles at E_p = 32 MeV. In accordance with our simple arguments, S is large in this transition where, due to angular momentum selection rules, spin transfer must occur in a single-step reaction. A DWBA calculation employing the Cohen and Kurath[5] (CK) wave function in conjunction with a N-N interaction derived from the Reid potential[6] gives a good account of the data. One might be tempted to infer deep meaning from the success of the calculation, however over the entire angular range shown in Fig. 2 the predicted value of S is very close to the value S = 2/3 expected from pure $\ell s j$ = 011 transfer. Thus any combination of interactions and wave functions which do not suppress this matrix element will give similar results. In this case then, at the level of precision of the present data, S only confirms what was known before: this reaction proceeds by single step spin transfer.

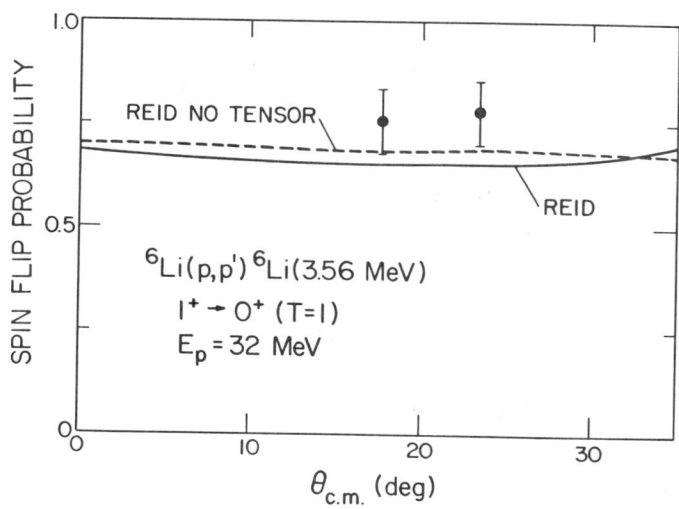

Fig. 2. Spin-flip probability for ^{6}Li(3.56 MeV)

All is not so simple though even when the data have fairly large error bars. Figures 3 and 4 show the SF probability and cross section data for the 12.71 MeV, 1^+, T=0 state of ^{12}C at E_p = 42 MeV. The best calculation utilizing the CK wave functions together with the Reid interaction substantially overestimates S. This is not a simple problem of optical potential ambiguities, since a plane wave calculation yields results scarcely distinguishable from the DWBA result. Removing the tensor part of the effective interaction seems to help; but now the magnituide of $d\sigma/d\Omega$ is no longer well described.

A key to where the solution to this discrepancy may lie is suggested by the spectacular failure of conventional theory in the case of the 8.89 MeV, 2^-, T=0 state of $^{16}_8$O. Here DWBA calculations (Fig. 5) with either the Gillet-Vinh Mau or single-particle wave functions (not shown) yield large values of S (as they must since s=1. Experimental SF probabilities in contrast hover around 20-25%. Our interpretation of this dilemma is that the main excitation strength for this state at E_p = 40 MeV cannot be single step s = 1 transfer. Other possibilities include multiple s = 0 excitation proceeding through the 6.13 MeV, 3^-, T=0 state and

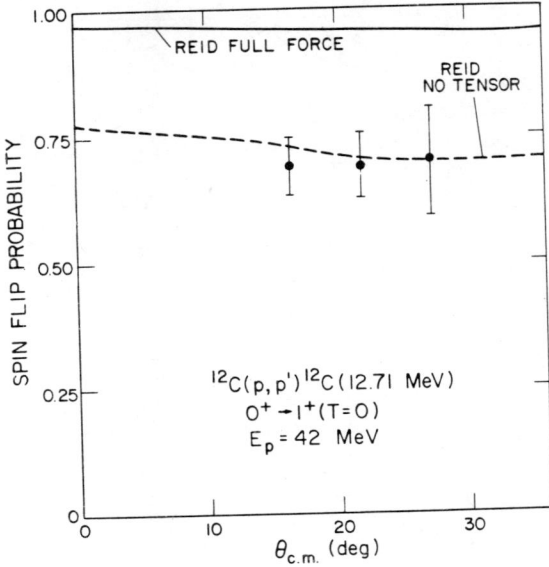

Fig. 3. Spin-flip probability for ^{12}C (12.71 MeV)

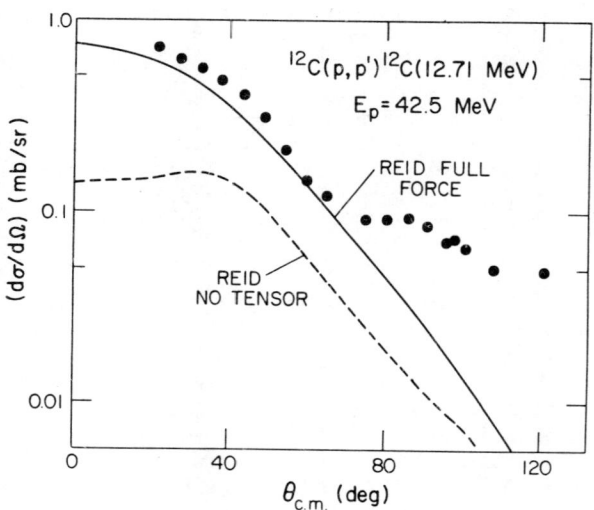

Fig. 4. Differential cross section for ^{12}C (12.71 MeV).

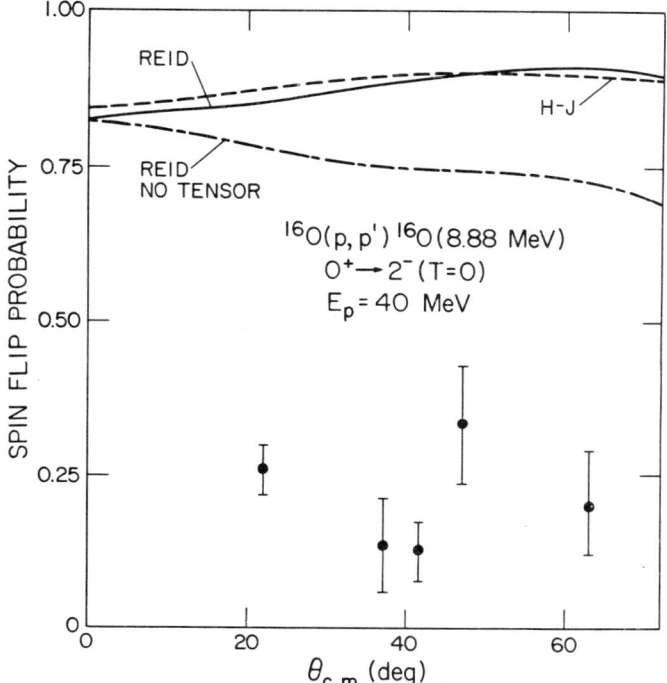

Fig. 5. Spin-flip probability for ^{16}O (8.88 MeV).

perhaps other states, sequential transfer, i.e. (p,d)(d,p), and the excitation of giant resonances.[9] Now it is plausible that, whatever the more complex mechanism, (there may of course be several) it does not produce much spin flip. This follows from the general observation that spin dependent effective forces are substantially weaker than spin independent ones. In the limit that the SF probability, associated with the more complex mechanism is zero, one can write for the experimentally observed value S_{exp} as:

$$S_{exp} = S \frac{(d\sigma/d\Omega)}{(d\sigma/d\Omega)_{exp}}$$

where $(d\sigma/d\Omega)$ and S are the quantities associated with pure spin transfer. It is still possible to calculate the product $(d\sigma/d\Omega)_{SF} \equiv S_{exp}(d\sigma/d\Omega)_{exp} = S(d\sigma/d\Omega)$. The comparison of experiment and calculation based on the $(d\sigma/d\Omega)_{SF}$ for the 2^- state is shown in Fig. 6 for both the Reid and Hamada-Johnston (HJ) effective interactions.

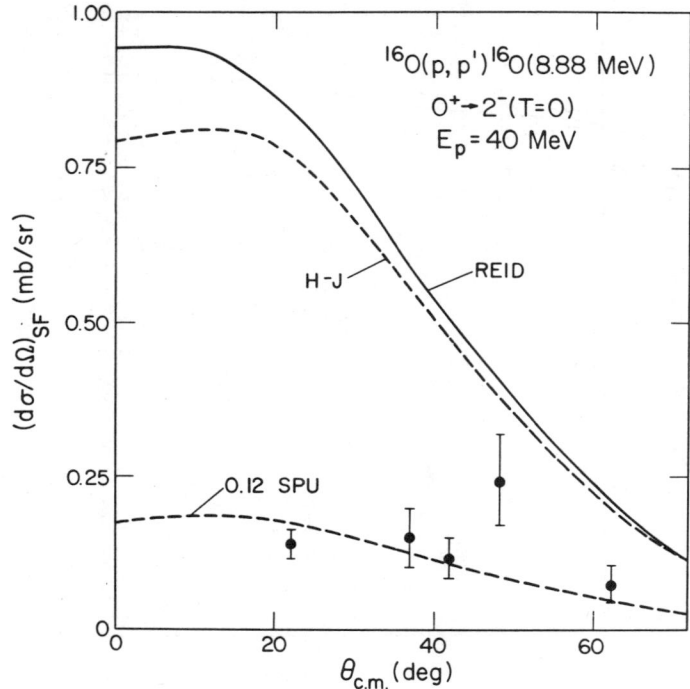

Fig. 6. Spin-flip cross section for ^{16}O (8.88 MeV)

There is obviously still a large discrepancy. Now, however, the culprit would seem to be the Gillet-Vinh Mau wave function. This is perhaps not surprising since isoscalar "magnetic" transition densities have not previously been subjected to much scrutiny. The isoscalar coupling in transverse magnetic electron scattering, for example, is very weak and is therefore sensitive to completely different coherences in wave functions than is isoscalar spin flip in inelastic proton scattering.

SPIN FLIP AT INTERMEDIATE ENERGIES

Polarization transfer at intermediate energies is a very exciting prospect for two reasons: First the theoretical tools which may be applied are more powerful than at lower energies, and second, polarimeters with orders of magnitude higher scattering efficiencies can be constructed thereby yielding data of much higher quality than is possible at lower energies. Work has just begun at Indiana, and only a small amount of preliminary data is available. This fall extensive programs of polarization transfer are scheduled both at LAMPF and Indiana. At LAMPF experiments will measure not only S but other triple scattering parameters as well.

As an illustration of the type of investigation currently envisaged, consider the question of nuclear critical opalescence. This phenomenon, first suggested by Ericson and Delorme[10], is a type of precritical enhancement of the nuclear pion field which is expected to occur if nuclei are sufficiently close to the critical density for pion condensation. Toki and Weise[11] have made explicit calculations for critical opalescence is the $^{12}C(p,p')^{12}C$ (15.11 MeV, 1^+, T=1) reaction at 800 MeV. The precritical enhancement is seen as a substantial increase in the cross section for momentum transfers in the range $q = 2$ to 3 m_π. A discrepancy between impulse approximation calculations and experimental differntial cross section data for the 15.11 MeV state in this range of momentum transfer has been seen at E_p = 122,185 and 800 MeV. The lower energy cross section from Ref. 12 are shown in Fig. 7. The obvious question to ask is, "Does this discrepancy imply an enhancement of the spin dependent N-N effective interaction, i.e. critical opalescence, or is the reaction mechanism poorly described in this region of q?" To see why a measurement of S may provide the answer, assume that the SF probability is measured in the region of critical opalescence and that the result is $S \sim 0$. This would be a strong indication of the dominance of a more complex reaction mechanism having nothing to due with enhanced spin transfer (this argument was

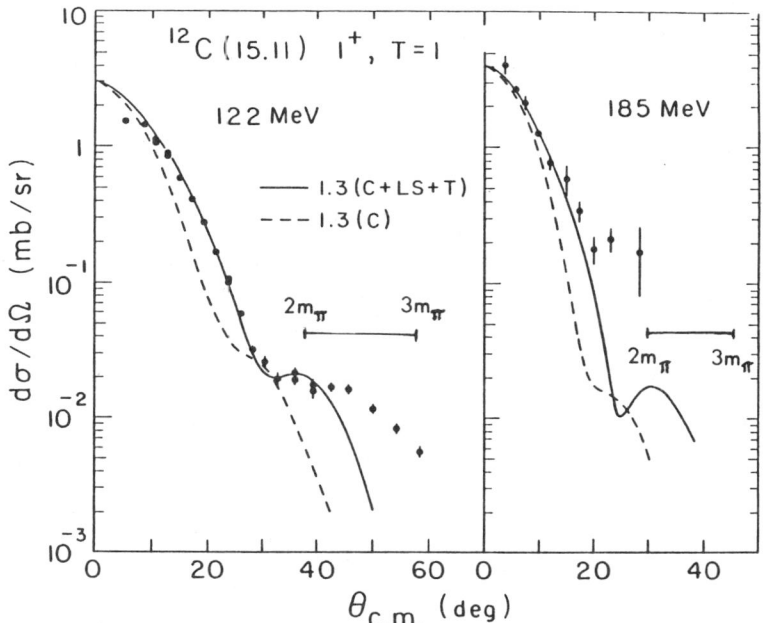

Fig. 7. ^{12}C (15.11 MeV) data from Indiana. The solid curves are distorted-wave impulse approximation calculations.

used in the case of the 2^- in ^{16}O). A large SF probability would not immediately verify the idea of critical opalescence, but at least it would strengthen the evidence in favor of it.

The coming year will see a vast increase in both the quality and quantity of polarization transfer data on complex nuclei. There are already many important questions to be addressed by these experiments, and undoubtedly new surprises and questions will arise when the experimental results are known. At the next International Polarization Conference it would not be surprising to see a much larger number of contributions in the area of polarization transfer experiments.

References

1. F. H. Schmidt et al., Nucl. Phys. 52, 353 (1964).

2. R. H. Howell et al., Phys. Rev. C21, 1153 (1980).

3. J. M. Moss, Invited Talk, International Symposium on Nuclear Reaction mechanisms; Fukuoka, Japan, Nov. 1978; W. D. Cornelius and J. M. Moss, to be published.

4. J. M. Moss, W. D. Cornelius and D. R. Brown, Phys. Rev. Lett. 41, 930 (1978); Phys. Lett. 71B, 87 (1977); Phys. Lett. 69B, 154 (1977).

5. S. Cohen and D. Kuruth, Nucl. Phys. A101, 1 (1967).

6. G. F. Bertsch et al., Nucl. Phys. A284, 399 (1977).

7. T. Hamada and I. P. Johnston, Nucl. Phys. 34, 382 (1962).

8. V. Gillet and N. Vinh Mau, Nucl. Phys. 54, 321 (1964).

9. H. V. Geramb et al., Phys. Rev. C12, 1697 (1970).

10. J. Delorme et al., Phys. Lett. 89B, 327 (1980).

11. H. Toki and W. Weise, Phys. Lett. 92B, 265 (1980).

DISCUSSION

LENSKE: Did you include exchange effects into your DWBA calculations?

MOSS: Yes.

LENSKE: You mentioned the possible importance of higher order processes for the excitation of unnatural parity states. Did you check this point in your analysis by, e.g., the inclusion of an imaginary part into the form factor? I believe that your results for the spin flip probability would be reduced and would come closer to the experimental results.

MOSS: No, we haven't.

COMFORT: I have two brief comments on some items that you raised. First, I certainly agree that multistep contributions could be very important for S, especially for light energy with energies less than about 60 MeV. But it is not true that they always affect S. An example is the 15.11 MeV transition in (p,p') reactions near 150 MeV. My calculation of certain types of multistep processes were found to affect $d\sigma/d\Omega$ to some degree, but were negligible for S. Second, for the same transition, my calculations of nuclear critical opalescence also showed very small affects for the spin-flip probability.

MOSS: It is premature to say critical opalescence has a small effect on the SF probability since the theoretical formulation of these effects can take different forms.

DEUTSCH: It would be, of course, interesting to extend the polarization measurements to the 12.7 MeV ($J^\pi = 1^+$, T = 0) level where no precursor effects would be expected in first order.

MOSS: It is not obvious that this state offers a good null test of critical opalescence effects. There are, in fact, effects on the cross section for this state from the two-body tensor force.

WALTER: Would you please comment about the desirability of obtaining additional data for the polarization transfer in (p,n) reactions to study the spin-flip mechanism.

MOSS: These studies are very interesting. However, the number of interesting cases which can be looked at is limited by the intrinsic poor resolution of neutron polarimeters.

ADVANCES IN NEUTRON POLARIZATION STUDIES

Richard L. Walter

Department of Physics, Duke University, Durham, NC 27706 and
Triangle Universities Nuclear Laboratory,* Duke Station, NC 27706

ABSTRACT

Recent investigations of neutron polarization phenomena are reviewed with an emphasis on the energy region between 4 and 17 MeV. Technological advances on beam pulsing, polarization sources and analyzers are mentioned. New results for $(\vec{p},n)$ and $(\vec{d},n)$ reactions are briefly discussed. A lengthy description of new measurements for the elastic scattering of polarized neutrons and the interpretation in terms of the optical model parameterization are presented.

I. INTRODUCTION

Although the number of laboratories actively involved in neutron polarization measurements has apparently decreased in the five years since the Fourth Polarization Symposium, some impressive work still is being carried out. A number of major advances which have occurred in the last few years have been made at our tandem laboratory, and the primary emphasis of this paper will be on the research program at the Triangle Universities Nuclear Laboratory (TUNL). In describing related work elsewhere, we will tend to concentrate on the energy range available with conventional tandem electrostatic accelerators, i.e., 4 to 17 MeV.

In his closing address at the Fourth Polarization Symposium, Professor Barschall[1] reviewed some of the problems and questions existing in neutron physics and expressed his disappointment that greater progress had not been made along specific research directions. We now are able to report that several of the problems which he noted have been resolved now and that new and accurate data are being obtained at a rapid pace. Furthermore, we note that classical analyses are proceeding even more rapidly because of the availability of a wide range of elegant computer codes and also because powerful computers are on hand at most nuclear research laboratories.

Although the accuracy of neutron measurements in the 4 to 25 MeV range have been lagging behind that of similar proton experiments by a decade or more, one should recognize that neutron data nevertheless still play a unique role in the development of nuclear models. Most of the neutron analyses or interpretations obviously benefit from the previous proton studies--in fact, the uniqueness of the neutron data frequently can be exploited in analyses that use the immense proton data bank as complementary or supplementary information. Such studies will be apparent below.

The present paper will be divided into four main topics: 1) Technological advances, 2) Deuteron-induced reactions, 3) Proton-in-

---

*Work supported in part by the U.S. Department of Energy

duced reactions, and 4) Scattering of polarized neutrons. We will omit work on (n,γ) reactions and polarized targets because these subjects are the topics of other invited talks at this conference. We will concentrate heavily on topic 4) mainly because of the considerable progress made just during the past twelve months.

II. TECHNOLOGICAL ADVANCES

Four recent developments in neutron work will have a significant role in measurements in the future. These deal with sources of polarized neutrons, calibrated analyzers, and neutron detection methods, and are discussed in this section on technological advances.

In 1973 Holt et al.[2] showed that an intense beam of polarized neutrons can be created by scattering from ^{12}C photoneutrons which were produced with the electron LINAC. In fact, they double scattered the neutrons from ^{12}C and unfolded the polarization produced in $^{12}C(n,n)$ scattering. In this way, they were in effect calibrating a new source of polarized neutrons that would be useful for future scattering studies. Over much of the energy ranges between 2.1 and 3.4 MeV and between 4.5 and 5.0 MeV, the source of ^{12}C-scattered neutrons is over 40% polarized. One advantage of this LINAC source is that it is pulsed in narrow bursts; one disadvantage is that it, like all LINAC neutron sources, has a continuous spectrum of neutron energies. The latter property restricts the usefulness to second-scattering samples which have a negligible contribution of inelastic scattering for the energies or angles being studied. In a contribution to this conference, Ahmed and Firk[3] describe small-angle scattering measurements from ^{209}Bi between 2 and 5 MeV, using this source of polarized neutrons. The data agree with optical model calculations that include the Mott-Schwinger term. The results were also compared to several calculations which included the polarizability of the neutron. Although the data shown for 2.57 MeV are accurate to about ±0.01, the systematic behavior was not sufficiently well determined to distinguish between zero polarizability and the currently accepted upper limit. Other small-angle measurements for which the polarizability is less sensitive to nuclear model uncertainties are in progress.

At our laboratory, a neutron polarization program has existed for nearly twenty years. Probably 99% of the experiments have been conducted with d.c. beams, both unpolarized and polarized. An attempt in 1975 to pulse and bunch the polarized beam obtained from the TUNL Lamb-shift ion source gave such feeble beams on target, (I < 2nA) that only a few nuclear measurements were conducted. Last year, however, a major breakthrough at TUNL occurred when Wender et al.[4] carried to completion a proposal first studied by Lawrence et al.[5] at Los Alamos. By applying a 100 V ramp potential to the anode of the duoplasmatron, the polarized proton or deuteron beam can be prebunched into 70 ns wide bursts which occur at 4 MHz (or 2 MHz) repetition rate. This technique is successful with the TUNL Lamb-shift source because the beam drifts for about 1.5 meters at 550 or 1100 eV before it is accelerated to the 50 keV injection energy.

The beam bursts are further compressed using a two-stage double-drift buncher system located at the entrance of the tandem accelerator. A schematic of the system is shown in fig. 1. It is possible to transport to the target about 75% of the d.c. polarized beam that one normally obtains on the target. Furthermore, 90% of this pulsed beam appears in a burst of width 2 ns (FWHM). A measure of this is

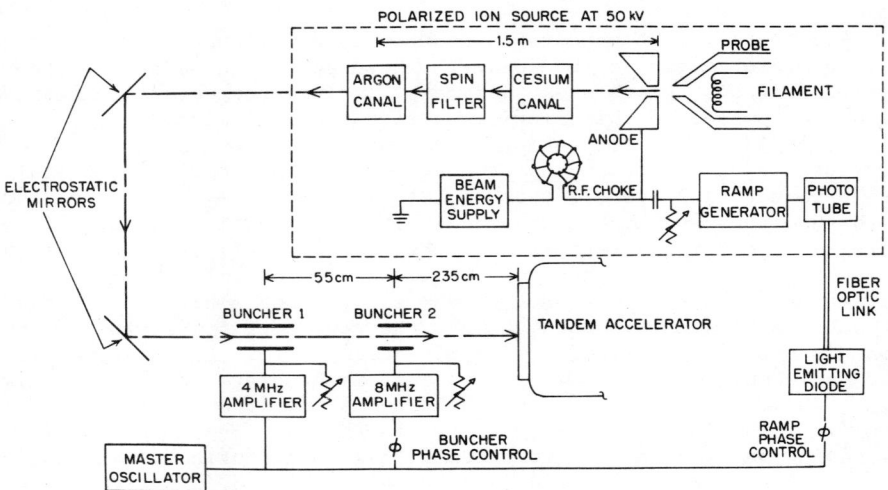

Fig. 1. Diagram of the bunching system for polarized beams at TUNL.

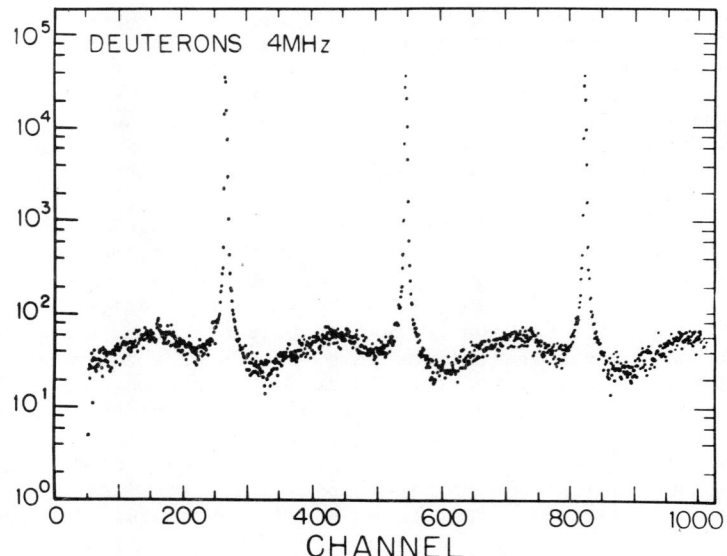

Fig. 2. Number of deuterons vs. time of arrival at target. (FWHM≈2ns)

illustrated in fig. 2. The additional 10% of the beam on the target
is spread across the remaining 250 ns (or 500 ns) interval between
bursts. Cyclotrons have been accelerating polarized beams for many
years so this achievement does not represent the first time narrow
pulses of polarized beam have been produced. However, to our knowledge it is the first time that currents of up to 150 nA of polarized
protons and deuterons have been available in narrow pulses at a 4 MHz
repetition rate, a rate low enough to make low energy neutron time-
of-flight spectroscopy feasible. Additionally, with pulsed tandem
beams one has access to the region between 3 and 17 MeV, a range not
readily accessible with cyclotron beams. This new TUNL pulsed-beam
facility has already been employed in about fifteen neutron polarization experiments during the past year, and most of them are reported
in these proceedings.

The third advance regards the use of the polarization transfer
reaction ^{2}H($\vec{d},\vec{n}$)^{3}He at 0° as a source of polarized neutrons for
elastic and inelastic scattering studies. At the Third Symposium
Ohlsen and Simmons[6] and Walter[7] suggested that this method of producing
polarized neutron beams would provide a relatively clean, prolific
source of monoenergetic neutrons. At the Fourth Symposium new
measurements were reported which provided a calibration for the neutron polarization and which showed that about 90% of the deuteron polarization is transferred to the neutrons emitted at 0° for deuteron
energies from 3 to 15 MeV. The newest advance is that at the present
symposium Tornow et al.[8] reported a measurement of the "effective
polarization" for the ^{2}H($\vec{d},\vec{n}$)^{3}He reaction for angles between 0° and
20° (lab). Their conclusions are summarized in fig. 3 where
the "effective polarization" is shown to be large and nearly constant from 0° out to 15° (lab). For scattering
experiments this feature is important because it is often
necessary to subtend a large solid angle to obtain both
good counting rates and good signal-to-noise ratios. In
addition, the corrections to the observed analyzing powers
for multiple scattering and finite sizes are simpler to
handle if the source's effect-
ive polarization is independent of θ. The TUNL setup for
using the ^{2}H($\vec{d},\vec{n}$)^{3}He reaction
as a source of polarized neu-
trons is described in a later
section.

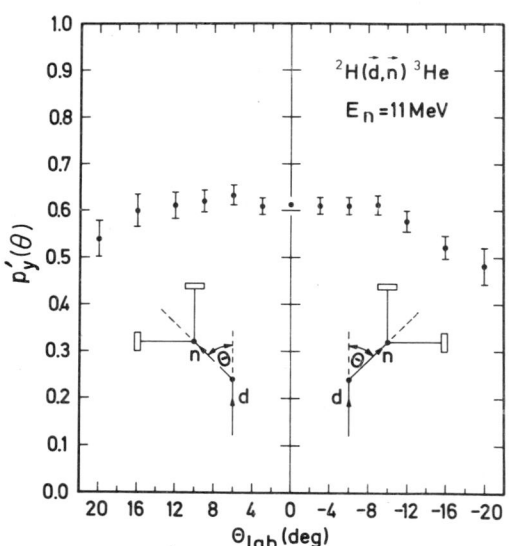

Fig.3. Effective neutron polarization
induced with m=1 state deuterons.

The fourth item relates
to analyzers for beams of polarized neutrons. During the

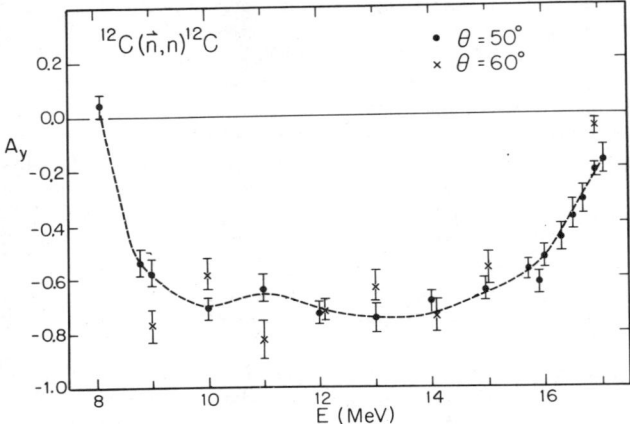

Fig. 4.
Analyzing power for $^{12}C(n,n)$ scattering at 50° and 60°.

past fifteen years scattering from ^{4}He has been the primary analyzer for neutrons mainly because of three important features: 1) the analyzing power is large, 2) it is well known, and 3) ^{4}He can be used as a scintillation medium. The primary disadvantage is that unless one uses liquid helium the density is low. For pulsed-beam time-of-flight neutron experiments it is possible to circumvent the need for an analyzer which can also scintillate and to use a dense scatterer like graphite. At the present conference Woye et al.[9] report measurements of $A_y(\theta)$ for $^{12}C(n,n)^{12}C$ scattering from 9 to 17 MeV. Data from their work at $\theta=50°$ and 60° have been plotted in fig. 4 in order to demonstrate the apparent smoothness and the large magnitude of A_y over the region between 8 to 16 MeV. Because graphite has a very high nuclear density, it is an attractive scatterer, and now that the analyzing power has been shown to be large, ^{12}C is a suitable substitute for ^{4}He as an analyzer in some cases. Already, two polarization transfer experiments at TUNL[8,10] have utilized a ^{12}C analyzer for high energy neutron beams. However, before high accuracy measurements can be made, the error bars on A_y for ^{12}C need to be reduced and data must be obtained in smaller steps.

III. DEUTERON INDUCED REACTIONS

Most of the data that exist for (d,n) polarization studies is for nuclei with $Z < 8$. The reason for this has been the difficulty to resolve different neutron groups emitted from close-lying excited states. About the time of the Zurich conference, the pulsed polarized beam at Wisconsin was being used by Quinn and coworkers[11] in neutron time-of-flight measurements of the j- dependence of A_y for low-lying states in a few medium weight nuclei. Since Zurich, almost all of the A_y data reported has been for very light nuclei. At this conference Brooks et al.[12] have reported a new system which utilizes a deuterated scintillator for neutron detection. Their method can be used for reactions in which the low-lying states are separated in energy down to about 5% of the outgoing neutron energy. They report here measurements and some DWBA results for A_y for (d,n) reactions on

^{9}Be, ^{12}C and ^{28}Si.

At Ohio State University[13] the vector and several tensor analyzing powers for ^{2}H(d,n) and ^{3}H(d,n) reactions have been obtained between 2 and 5 MeV for comparison to the charge symmetric reactions ^{2}H(d,n) and ^{3}He(d,p) and to few-nucleon calculations. At this conference the Erlangen group has reported[14] an eight-detector arrangement for obtaining tensor analyzing powers for (d,n) reactions. Their first results are a comparison of the four tensor analyzing powers for the reactions ^{12}C(d,n) and ^{12}(d,p) at 10 MeV. Conventional DWBA calculations were not too successful in predicting the observed patterns.

The first measurement made at our laboratory with the pulsed polarized beam was a determination of A_y for ^{2}H(d,n)^{3}He. It was done primarily to check the system and to investigate the ultimate accuracy that can be achieved with the method. The results are reported in this conference by Guss et al.[15] for energies between 5.5 and 11.5 MeV. In the experiment nanosecond pulses of polarized deuterons were incident on a ^{2}H gas cell located at the pivot point for two massively shielded neutron detectors located at 4m and 6m from the target. In addition to the normal time-of-flight separation between neutrons and gamma rays, an electronic discrimination circuit was employed with the liquid NE 213 scintillators to preferentially select neutron events over gamma-ray events in the detectors. High accuracy data were obtained over the energy range studied. An example of the data is shown in fig. 5. Here the error bars are typically less than ±0.005. Guss et al. compared their data to that obtained in the experiment of Hilscher and Liers[15] (H-L) at 10 MeV. The comparison to the data of Hilscher and Liers is also shown in fig. 5. (The H-L data were renormalized by 1.05 which is within their quoted scale error.) The data of König et al. differ systematically from the data of Guss et al. within each angular distribution, but not the same way at every energy. (See ref. 15.) Significantly, the two earlier measurements detected the outgoing charged particle ^{3}He in the ^{2}H(d,n)^{3}He reaction, and our neutron experiment was intended to complement the ^{3}He data by filling in angular regions that were impossible to obtain because the ^{3}He particles were too low in energy. In fact, from the polynomial fits of the data for the product $A_y(\theta) \cdot \sigma(\theta)$, it appears that the new neutron data is superior to the charged particle data. The neutron data will be added to the existing file of four-nucleon data to provide additional information for studying excited states in the mass-4 system. Additional $(\vec{d},n)$ measurements on other targets will be pursued with the TUNL time-of-flight system in the near future.

IV. PROTON-INDUCED REACTIONS

The main interest in (p,n) polarization experiments in the past few years has centered on the question of differences (or similarities) in the observables P^y and A_y, the first being the polarization in the outgoing beam in the experiment X(p,$\vec{n}$)Y, the second being the analyzing power for the reaction X($\vec{p}$,n)Y. Recall that for elastic scattering of spin 1/2 particles, time reversal requires that

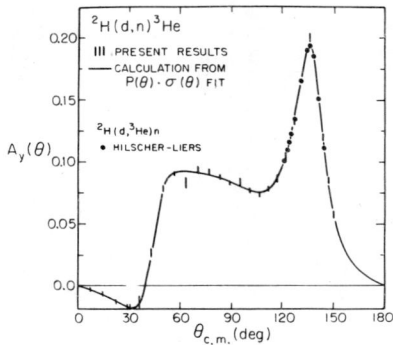

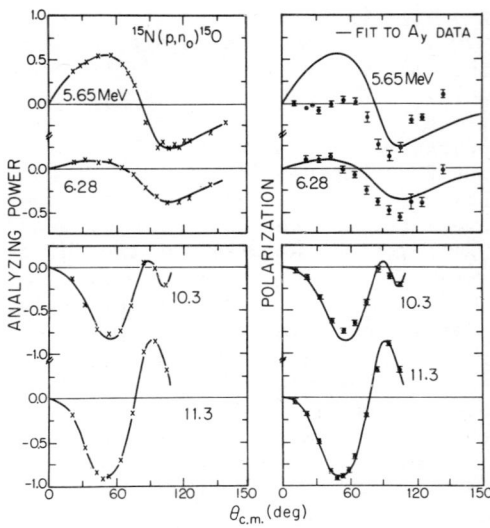

Fig.5. Analyzing power for ^{2}H(d,n)^{3}He. Data of Guss et al. are compared to polynomial fit and to data of Hilscher and Liers.

Fig.6. Comparison of P^y and A_y.
Left side: A_y data and fit. Right side: P^y data and fit to A_y.

P$^y(\theta)$=$A_y(\theta)$. One can argue that if charge symmetry is not broken by some mechanism, the same identity must hold for (p,n) reactions to isobaric analog states. (See ref. 16.) At the Zurich conference, TUNL reported that for the energies studied, P^y=A_y for ^{9}Be(p,n)^{9}B and ^{15}N(p,n)^{15}O, but P^y≠A_y for ^{3}H(p,n)^{3}He between 2 and 4 MeV. The current status for (p,n) reactions is reviewed at this conference by Byrd et al.[16] We have now shown that P^y=A_y for ^{9}Be(p,n) over an extended region, i.e., from 2.7 to 10 MeV. We have also shown that although to within experimental uncertainties, P^y=A_y for ^{15}N(p,n) at 10.3 and 11.3 MeV, sizeable differences exist from 5 to 8 MeV. The magnitude of the effect is illustrated in fig. 6.

Philpott[17] and Byrd et al.[16] at this conference discuss the source and the significance of the difference for ^{15}N(p,n). At first it was believed that differences in P^y and A_y would signal the need to introduce a charge-symmetry breaking term into the nuclear force. It was not clear that the Coulomb interaction, a term which breaks charge symmetry, could account for the sizeable differences in ^{15}N(p,n) since in all other known cases, P^y and A_y are at least approximately equal. However, Philpott and Halderson[17], without employing any unusual term in the nuclear two-body interaction successfully reproduced the qualitative nature of the differences. Their calculations are of a continuum shell-model type which use a microscopic model and which employ a one-particle one-hole approximation for the states in ^{16}O. Their model implicitly contains the possibility for asymmetric spin-flip, a term required for P^y to be different from A_y in (p,n) reactions.

In collaboration with Donoghue, we have remeasured[18] P^y for the ^{3}H(p,n) reaction and conclude that when the backgrounds are properly taken into account that P^y and A_y are equal between 2 and 4 MeV, the energy range in which values for P^y had been observed previously to be a relative 20% lower than A_y data. For ^{3}H(p,n) there are two recent calculations in the region around 3 MeV. Philpott and Halderson's continuum shell model calculation[17] for the four-nucleon system predicts that A_y exceeds P^y by about 0.02 around 45° (c.m.). Hale and Dodder's R-matrix fit[19] to the four-nucleon system predicts that A_y is about 0.05 higher than P^y. Neither of these calculations required any special interaction to produce the differences in P^y and A_y. The new P^y measurements are accurate to about ±0.015 and systematically favor the observation that $P^y = A_y$.

In summary, the observation of the existence of differences in P^y and A_y requires accurate and careful measurements. We have found that $P^y \approx A_y$ in most of the (p,n) reactions studied to date. In only one instance, ^{15}N(p,n)^{15}O, have large differences been observed. However, this difference and a possible slight difference in the ^{3}H(p,n) reaction have been explained without requiring the introduction of some special charge-symmetry breaking term into the nuclear force.

Analyzing power measurements for (p,n) reactions have also been used in Lane optical model calculations, most recently by Byrd et al.[16] In this model, one can connect (p,p), (n,n) and (p,n) observables for the same targets and for mirror targets using a single potential containing explicit isospin terms. In fact, (p,n) reactions are probably the most sensitive method for determining details of the nuclear isospin interaction because this term is responsible for the (p,n) channel. Byrd converted the computer code T-wave into a search code for the purpose of fitting data from all three (N,N) channels. He also modified the code to include the isospin spin-orbit interaction. An example of the "global" description for the ^{9}Be + nucleon system is illustrated in fig. 7. At the time of the search no suitable P^y or A_y data were available for ^{9}Be(n,n). Some data have just recently been obtained by Floyd et al.[20] for the purpose of completing the ^{9}Be + nucleon data set. This new A_y measurement is discussed below, but the data between 11 to 15 MeV are compared in fig. 8 to predictions using the optical model obtained by Byrd et al.[16] in the above analysis. The predictions are encouraging, but not perfect. The calculation will be repeated with this new data set to observe what effect it has, and to attempt to parameterize the elusive isospin spin-orbit interaction.

V. SCATTERING OF POLARIZED NEUTRONS

At the Third symposium the use of the ^{2}H($\vec{d}$,$\vec{n}$)^{3}He reaction at 0° was proposed as an excellent source of polarized neutrons in the 5- to 18-MeV range, a region where very little accurate data existed. By the Fourth symposium a few experiments at TUNL had utilized this source but the method had limited applicability for a programmatic series of measurements with a wide range of nuclei because the polarized deuteron beam could not be efficiently pulsed for time-of-flight neutron detection to eliminate background. Therefore it could only be used successfully in a d.c. mode for scattering samples that could also serve as scintillators, e.g., ^{1}H, ^{2}H, ^{3}He, and ^{4}He.

Since Zurich, Tornow et al.[21] used the ^{2}H($\vec{d}$,$\vec{n}$)^{3}He reaction as a

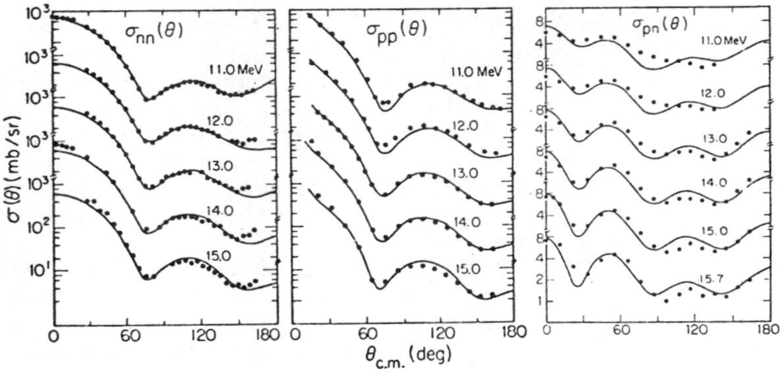

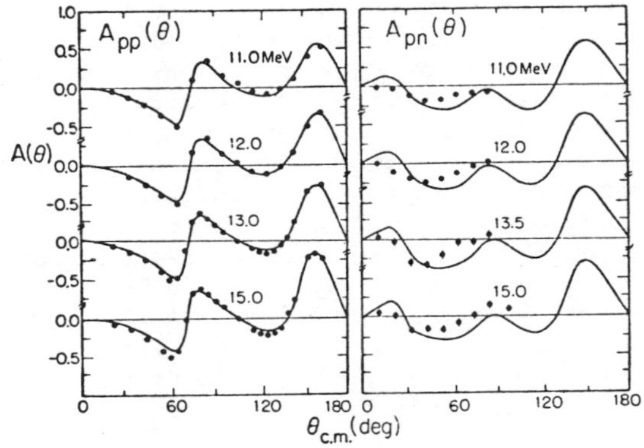

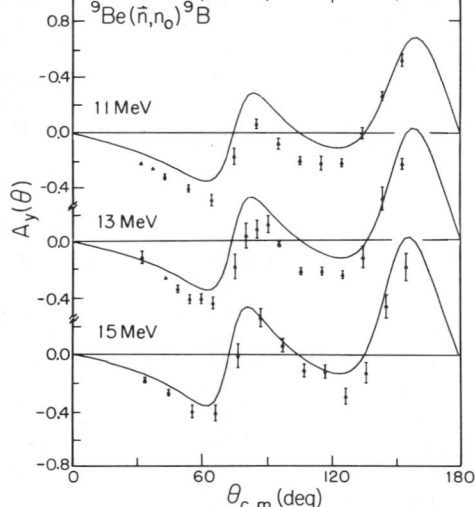

Fig. 7. Global fits to ^{9}Be(N,N) reactions using Lane model description.

Fig. 8. Lane model prediction for ^{9}Be(n,n).

source to complete their ^{1}H(n,n) scattering experiments first reported at Zurich. The experiments eliminated one of the controversies in n-p scattering at 14 MeV that Barschall had singled out. Tornow et al. claim an accuracy of about ±0.002. To our knowledge, this is the best accuracy ever reported in a neutron scattering experiment. The data at 16.9 MeV are compared in fig. 9 to phase shift predictions using three global potential models and a single-energy prediction of the former Hamburg group

(HH-75). None of the global models are successful in predicting the observed $A_y(\theta)$ function. In the lower half of fig. 9 the data are compared to polynomial fits containing two and three terms. The new data, along with the earlier data of Morris et al., clearly require the existence of a third term. This fact permits the determination of the spin-orbit interaction for the F-state, as well as for the P- and D-states for nucleon-nucleon scattering. Although the analyzing power for p-p scattering has been measured to an accuracy an order of magnitude better, the size of A_y is so small in p-p scattering that the n-p data provide better information on the spin-orbit interaction for nucleon-nucleon scattering in the 10 to 20 MeV region. Significantly, we note that the newest potential for nucleon-nucleon scattering, the "Paris potential", does not predict the structure[22]

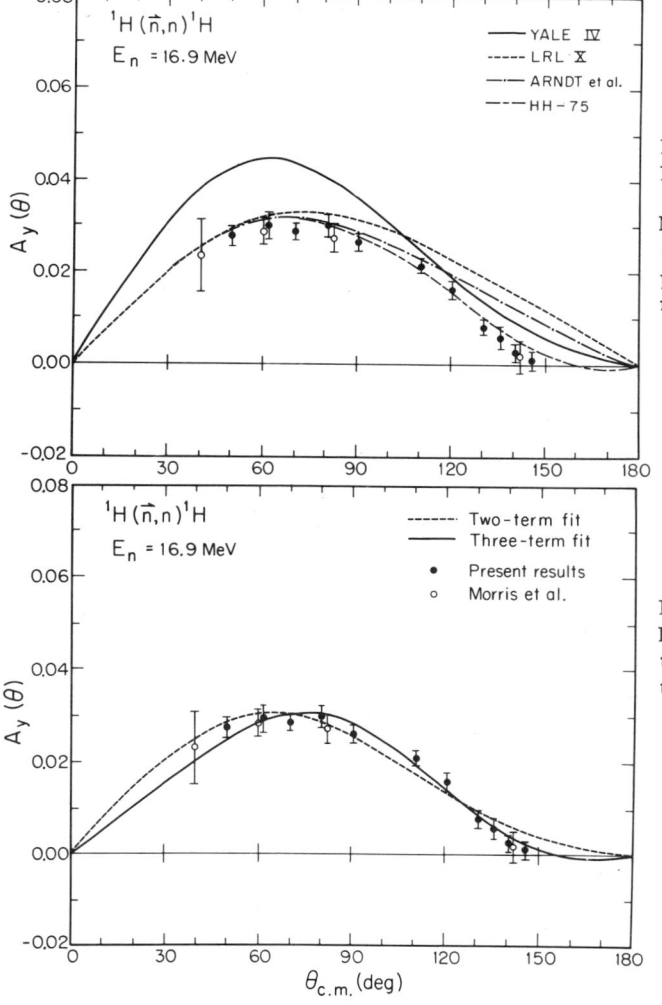

Fig. 9a. Data of Tornow et al. (solid circles) and Morris et al. (open circles) compared to predictions for n-p scattering.

Fig. 9b. Data for n-p compared to two- and three-term polynomial fits.

seen in the data of Tornow et al. Although more work will be done to investigate some of the approximations in the Paris model, it would be satisfying to have an independent check on the behavior of the n-p data, in particular, for angles greater than 120°.

Polarization data for neutron scattering have also been reported for n-^{2}H, n-^{3}He, and n-^{4}He scattering using the ^{2}H($\vec{d}$,$\vec{n}$) reaction as a source of neutrons.[23] These data are included in extensive calculations for the three-nucleon to five-nucleon systems. The only controversy in these systems seems to be the size of the differences in $A_y(\theta)$ for p-^{2}H and n-^{2}H scattering. At energies of 12 MeV and below, no differences exist within experimental accuracies. On the other hand, at this conference Chisholm et al.[24] conclude that at 14 MeV, differences of 0.04 at the maximum in A_y, where A_y=0.21, have been seen if the data from three neutron experiments are carefully combined. One normally expects the Coulomb effects to be more manifest at lower energies than higher energies, so this new analysis suggest an inconsistency in the data at 14 MeV with that at lower energies. Perhaps there is some unusual sensitivity to Coulomb effects around 14 MeV due to the particular dependencies of $\sigma(\theta)$ and $A_y(\theta)$, but it is important to check the data, both at 12 and 14 MeV, to verify for the few-nucleon theorists who are working on the Coulomb problem whether the difference is too small to be measured or as large as Chisholm et al. indicate that it seems to be.

For heavier nuclei, the parameterization of polarization effects is usually tied to the optical model. Two features have a significant bearing on the success of the program of ($\vec{n}$,n) experiments discussed below. First, as can be seen in the early review by Rosen et al.[25], the polarization produced in (p,p) elastic scattering is small at 10 MeV and below for nuclei having Z >30. Coulomb scattering is responsible for this small magnitude. Second, very little reliable neutron polarization data exist for optical model studies for energies above 4 MeV. Although Zijp and Jonker[26] obtained a systematic set of data at 3.2 MeV, the interpretation of the data is complicated by compound nucleus effects which can markedly dilute the structure in $A_y(\theta)$.

At TUNL, a program is underway to fill the void in neutron polarization data in the region above 4 MeV for targets with A >4. Initially we are concentrating the polarization studies on 10 and 14 MeV. The polarization data complement cross-section data either taken earlier at TUNL, or taken alternately with polarization runs. Nine targets have been investigated so far. The technique employs the time-of-flight method for separating the groups of scattered neutrons. The pulsed polarized neutron beam is produced by bombarding a deuterium gas target with the pulsed polarized deuteron beam mentioned in section II. The arrangement in the target area is shown in fig. 10. The apparatus is also used for cross-section measurements in almost the same way, the main difference being that the incident deuteron beam is unpolarized and 20 times more intense. The two neutron detector shields are massive, weighing about 2 x 10^4 N each. The detectors can be positioned at flight paths up to 4 and 6 m respectively.

The first experiment undertaken was a study[20] of ^{9}Be($\vec{n}$,n) to

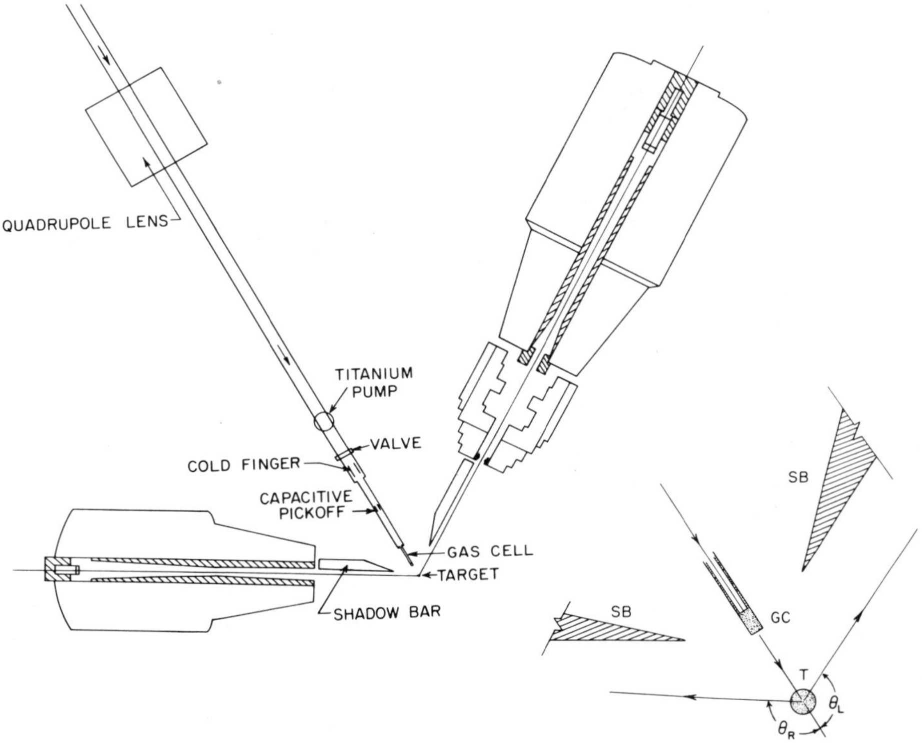

355

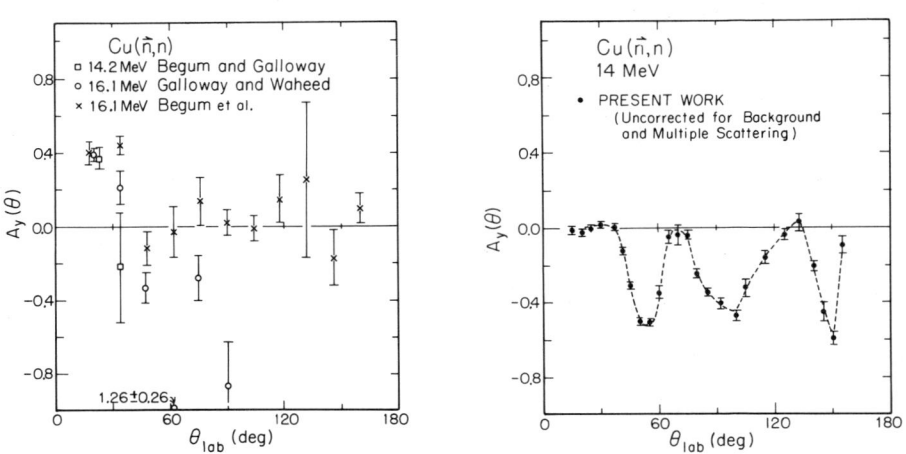

Fig. 10. Schematic of the target and detector arrangement at TUNL.

Fig. 11. Analyzing power for Cu(n,n) for energies near 14 Mev.
Left side: Previous data. Right side: Present TUNL results.

complement the Lane analysis mentioned in section IV. Some of the data were illustrated in fig. 8. Inelastic scattering to the 2.4-MeV state of ^{9}Be was also investigated in the experiment and the results compare favorably to preliminary coupled-channel predictions.[20]

In the precision n-p measurement of Tornow et al.[21] corrections for ^{12}C(n,n) scattering were intimately involved, because the scatterer was an organic compound. Some of the corrections were based on estimates of $A_y(\theta)$ for ^{12}C. In order to be able to check the accuracy of the corrections, we measured $A_y(\theta)$ for ^{12}C from 6 to 17 MeV in at least 1 MeV steps. Some of the data were reported at this conference by Woye et al.[9] Along with available cross-section data, the results are also being used in coupled-channel optical model calculations.

The optical model study on medium-to-heavy weight nuclei started with ^{40}Ca, ^{54}Fe and ^{65}Cu and then was extended to ^{58}Ni, ^{116,120}Sn, and ^{208}Pb. To emphasize the significant improvement in the quality of data that the new method provides, in fig. 11 we compare some 14-MeV data obtained one month ago in our laboratory to data that Galloway and coworkers[27] have reported over the past two years; this is essentially all the data available for Cu above 4 MeV. Other data comparable to those of Galloway for Cu also exist at about 16 MeV for five other nuclei; except for several measurements near 24 MeV, very little additional neutron data has been reported for energies above 4 MeV.

In order to have an acceptable counting rate in neutron scattering experiments, one must compromise on the size of the scatterer: a large scatterer subtends a large angular window, so the finite size must be carefully considered in the experimental design and in the eventual corrections. Secondly, multiple scattering events increase approximately in proportion to the diameter and height of the scatterer. A balance must be chosen between the ultimate statistical accuracy of the experimental data and the accuracy that Monte Carlo codes can estimate the corrections for multiple scattering. With more experience, we should be able to determine the optimum size. An example of the magnitude of the calculated multiple scattering

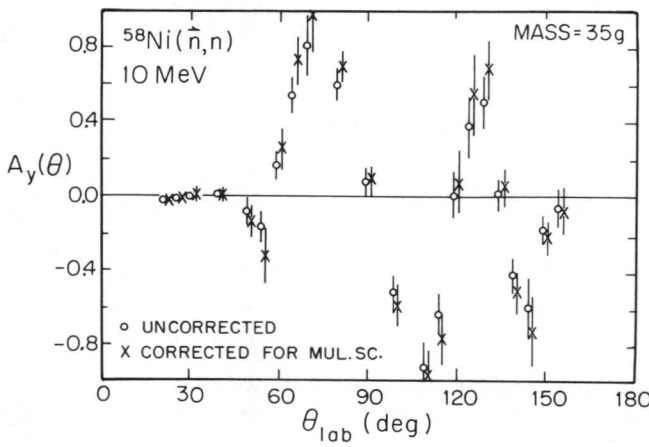

Fig. 12. Multiple scattering corrections to ^{58}Ni(n,n) data.

corrections are shown in fig. 12 for our ^{58}Ni case. The code that
was used in calculating the corrections was developed at Tübingen
and TUNL by E. Woye.

A composite of our neutron polarization data is shown in fig. 13.
The energies are between 10 and 14 MeV, and all the measurements
used the ^{2}H($\vec{d}$,n)^3 source reaction. Data were obtained only between
20° and 150°, but the curves are drawn from 0° to 180° to aid the
eye. (The Mott-Schwinger polarization was ignored in constructing
the drawing at extreme forward angles, and the extra oscillation
near 170° which causes $A_y(\theta)$ to approach 180° from positive values
was also omitted in the sketch.) The data for Sn and Pb in fig. 14
have not been corrected for multiple scattering yet.

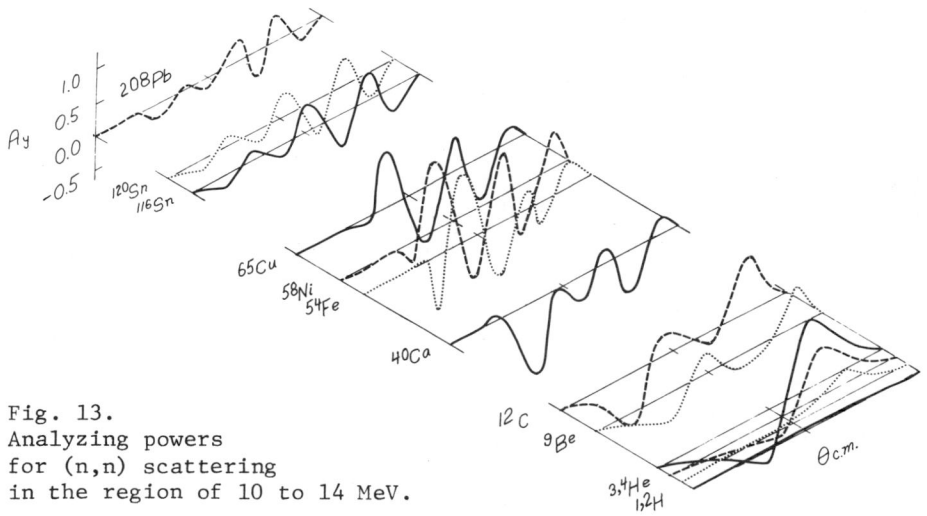

Fig. 13.
Analyzing powers
for (n,n) scattering
in the region of 10 to 14 MeV.

Until now, our optical model analyses have concentrated on the
medium weight data. The global search code GENOA obtained from F.
Perey has been used for this purpose. Some preliminary results of
the optical model searches are reported by Floyd et al. in a contribution to this conference and will be discussed below. Most of the
searching to date has been performed with the original version of
GENOA which does not include Mott-Schwinger scattering, i.e.,
scattering produced by the interaction of the magnetic moment of the
neutron with the electromagnetic field of the nuclear charge Z. (In
the scattering of charged particles, this effect is negligible at
forward angles in comparison to Rutherford scattering, particularly
below 20 MeV.)

Just two weeks prior to this conference, Floyd introduced Mott-
Schwinger scattering into GENOA. This was done using the Born
approximation, the method having been proposed and tested by Hogan
and Seyler.[28] The results of an optical model search on $\sigma(\theta)$ and
$A_y(\theta)$ for ^{54}Fe is illustrated by the solid curve in fig. 14. The
data are quite well represented by the curve. The effect of just

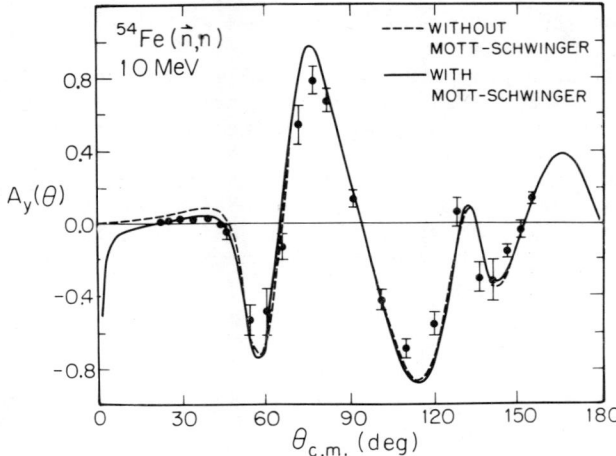

Fig. 14 Data and optical model calculations for ^{54}Fe(n,n).

turning off the Mott-Schwinger scattering is shown by comparing the dashed and solid curves. The effect is not insignificant, even out to 150°! More importantly, at forward angles our A_y data are obtained to about ±0.005. Our early searches that used the original GENOA gave poor fits in this region. This forced us to reduce the relative weighting of these data in order to be able to get suitable fits elsewhere. Our preliminary findings are that our optical model conclusions, based on this "reduced-weighting" analyses, are closely borne out by the recent calculations with Mott-Schwinger scattering included.

One of the most interesting features in the searches on the ^{12}C and ^{40}Ca data is exhibited in fig. 15. We have been able to describe the global properties of the cross section and polarization fairly well, but only by permitting the spin-orbit diffuseness to take on small values, e.g., a_{so}<0.3 fm. In our preliminary searches on ^{9}Be we found the same result. The fits with a_{so} held fixed at 0.20 fm

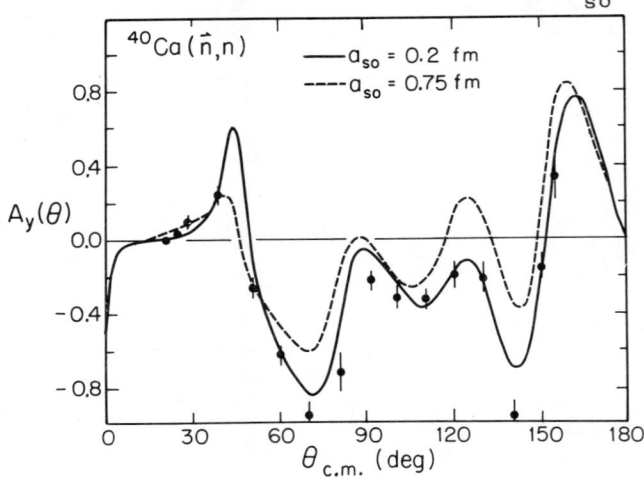

Fig. 15. Data and optical model calculations for ^{40}Ca(n,n).

and the Bechetti-Greenlees value of 0.75 fm are shown in fig. 15 for ^{40}Ca at 12 MeV. Clearly the fit with the smaller a_{so}, the "best-fit" value, is significantly better. The conclusion so far for these lighter nuclei is that if one employs the standard Thomas-type form factor for the real part of the spin-orbit interaction, one must be satisfied with an "unphysically" small diffuseness. A small diffuseness also continually shows up in the optical model for ^{3}He elastic scattering for nuclei between ^{16}O and ^{40}Ca in the work of the Birmingham group[29]. They suggested that since the optical model for ^{3}H has a more "normal" value, the small value observed for ^{3}He might be indicative of a low a_{so} for the neutron well in contrast to the proton well. As we do find this to be true for neutrons, there may be some significance to their proposal.

From our early searches for Fe and Cu we found similar difficulties with a_{so}. However the requirement for a small value disappeared after the final corrections were applied to the cross-section and A_y data. The fit to the ^{54}Fe data shown in fig. 14 uses an a_{so}=0.5 fm, and ^{65}Cu prefers a_{so}=0.6 fm; both are values in the range found for protons in the 15-20 MeV region. The spin-orbit depth is about 6 MeV.

The searches on the ^{54}Fe and ^{65}Cu data at 10 MeV show that the A_y data at forward angles can be fit only by introducing a volume absorption term in addition to the surface term normally employed for neutron scattering in this energy region. Our data are requiring that the two well depths be about equal in magnitude. Another feature we observe is that the data prefer an imaginary spin-orbit depth of about 1 MeV, although this demand must be checked more carefully.

Some closing words of caution are in order at this stage. We have been analyzing here a new generation of data: accurate neutron polarization data for an energy region where optical models should apply. The data are not affected by the long range Rutherford interaction which dilutes proton polarization effects at low energies. The neutron data actually may be showing defects in the form factors for the potentials rather than giving credence to a Thomas-type spin-orbit term that has a structure as thin as 0.3 fm. Furthermore, as pointed out by W.J. Thompson, neutron scattering, as contrasted to charged particle scattering, is sensitive to the tails of the potentials. A breakdown of the Woods-Saxon shape for our energy region may be surfacing. Thompson[30] also is concerned about the sensitivity or accuracy of neutron calculations with the numerical methods currently employed in standard optical model codes. Before the above optical model findings are published in final form, some of these questions will be addressed.

It is a privilege to acknowledge the cooperation of my co-workers at TUNL. This report reviews the progress of an enthusiastic and industrious group. I appreciate the numerous contributions from the students at Duke (C.E. Floyd, P.P. Guss, K. Murphy, S. El-Kadi, and C. Howell), from faculty members associated with TUNL and the project (R.C. Byrd, G. Tungate, T.B. Clegg, S.A. Wender, C. Gould, W.J. Thompson and S.R. Cotanch), and from collaborators from Tübingen (E. Woye, W. Tornow, and G. Mack).

REFERENCES

1. H.H. Barschall in Proceedings of the fourth international symposium on polarization phenomena in nuclear reactions, ed. W. Grüebler and V. König, Birkhäuser Verlag, Basel (1975), p. 427.
2. R.J. Holt, F.W.K. Firk, R. Nath and H.L. Schultz, Nucl. Phys. A213, 147 (1973).
3. M. Ahmed and F.W.K. Firk, contribution to this conference.
4. S.A. Wender, C.E. Floyd, T.B. Clegg and W.R. Wylie, Nucl. Instr. and Methods, in press.
5. G.P. Lawrence, A.R. Roelle, J.L. McKibben, G.Roy and T.B. Clegg, Proc. of the second inter. conf. on ion sources, ed. F. Viehbock, H. Winter and M. Bruck, SGAE, Vienna, Austria (1972), p. 505.
6. G.G. Ohlsen, P.W. Keaton, Jr. and J.E. Simmons in Proc. of the third symp. on polarization phenomena in nuclear reactions, ed. H.H. Barschall and W. Haeberli, Univ. of Wisconsin Press, Madison (1971), p. 415; J.E. Simmons et al. ibid p. 469.
7. R.L. Walter in Proc. of the third symposium, p. 317.
8. W. Tornow et al., contribution to this conference.
9. E. Woye et al., contribution to this conference.
10. R.L. Walter et al., contribution to this conference.
11. B.P. Hichwa et al., in Proceedings of the fourth symposium, p. 655 and references therein.
12. F.D. Brooks et al., contribution to this conference.
13. L.J. Dries et al., Phys. Lett. 80B, 176 (1979); L.J. Dries, Phys. Rev. C21, 475 (1980).
14. W. Drenckhahn et al., contribution to this conference.
15. P.P. Guss et al., this conference. See also references therein.
16. R.C. Byrd, R.L. Walter and S.R. Cotanch, this conference and refs. therein.
17. J. Philpott, this conference and references therein.
18. T.R. Donoghue et al., Phys. Rev. Letters 37, 981 (1976).
19. G. Hale, private communication and to be published.
20. C.E. Floyd et al., this conference.
21. W. Tornow et al., Phys. Rev. Letters 39, 915 (1977); Nucl. Phys. A340, 34 (1980).
22. J. Cote et al., this conference.
23. W. Tornow et al., Nucl. Phys. A296, 23 (1978); P.W. Lisowski et al., Nucl. Phys. A259, 61 (1976); P.W. Lisowski et al., Fourth polarization symposium, p. 534.
24. A. Chisholm, J.C. Duder, and R. Garret, this conference.
25. L. Rosen et al., Ann. of Phys. 34, 96 (1965).
26. E. Zijp and C.C. Jonker, Nucl. Phys. A222, 93 (1974).
27. R.B. Galloway and A. Waheed, Phys. Rev. C19, 268 (1979); see also Phys. Rev. C20, 1711 (1979).
28. W.S. Hogan and R.G. Seyler, Phys. Rev. 177, 1706 (1969).
29. J. Barnwell et al., J. Phys. G: Nucl. Phys. 5, 69 (1979); W. Lui et al., Nucl. Phys. A333, 205 (1980) and references therein.
30. W.J. Thompson, private communication to C.E. Floyd.

DWBA ANALYSIS OF $(\vec{p},d)$: A SPECTACULAR FAILURE

J.R. Shepard, E. Rost and P.D. Kunz
University of Colorado, Boulder, CO 80309

ABSTRACT

We describe a detailed analysis for the $\ell=0$ transition to the 2.36 MeV $1/2^+$ level in ^{23}Mg where the analyzing powers are large and oscillatory at T_p=94 MeV. DWBA calculations completely fail to reproduce this behavior. Qualitative agreement is obtained only by introducing a great deal of absorption artificially, e.g. by using a large radial cutoff.

The distorted wave Born approximation (DWBA) has generally been quite successful in reproducing (p,d) cross-sections over a wide range of bombarding energies extending up to T_p=800 MeV. Analyzing powers depend more sensitively on details of reaction amplitudes and, not surprisingly, the DWBA does not reproduce these quantities as well, although qualitative agreement is frequently achieved. In many cases, however, agreement is quite poor and can only be improved by arbitrary and physically unjustified adjustment of input parameters, such as potential strengths or geometries. Such troubles are frequently encountered for low angular momentum transfer in light nuclei and examples can be found among the contributions to this conference.[1] The origin of the difficulty is not well understood although the possibilities are myriad. Isolating and identifying these problems would provide guidance for developing an improved theoretical understanding of the (p,d) reaction mechanism.

In the present paper, we report on recent measurements for which the DWBA fails in a most spectacular fashion. Specifically, we will discuss cross-section and analyzing power measurements for the ^{24}Mg$(\vec{p},d)^{23}$Mg(2.36 MeV $1/2^+$) transition at T_p=94 MeV.[2] While similar failures have been observed by us for transitions in other light nuclei in this energy regime, this particular one is the most severe and has been the focus of our attention. We have examined the influence on the DWBA of several frequently ignored effects in an attempt to identify the source of the difficulty.

Figure 1 combines the data with "standard" DWBA calculations employing phenomenological proton[3] and deuteron[4] optical model potentials and target wavefunctions tied to electron scattering densities.[5] Both exact-finite range (calculations performed as in Ref. 5) and zero-range results are shown with cross-sections normalized using the theoretical spectroscopic factor of Chung and Wildenthal as reported in Ref. 7. While some differences can be perceived between the two calculations, they are insignificant when compared to the discrepancy with the data. The calculations overestimate the cross-sections by at least an order of magnitude and the pronounced oscillatory structure observed in the asymmetry data is in no way reproduced. These discrepancies constitute the spectacular failure of the DWBA.

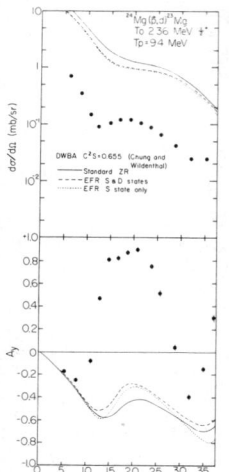

Fig. 1. Zero-range and exact-finite-range DWBA calculations are compared with data.

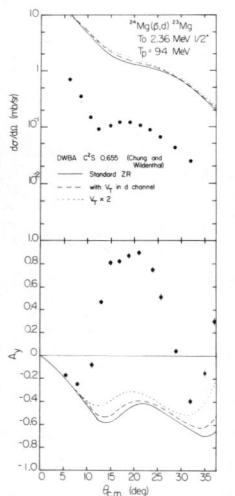

Fig. 2. Zero-range calculations with and without tensor potentials, V_T in the deuteron channel are compared with data.

Several sets of optical potentials were tested and while some sensitivity was observed, especially to deuteron potentials, the variations in cross-section and A_y were miniscule compared to their discrepancy with the data. Since the transition is an $\ell=0$ pickup, the non-zero analyzing powers arise solely from the proton and deuteron spin-orbit potentials which contribute almost equally to the calculated A_y. Reasonable variations of these potentials by themselves, including the addition of reasonable imaginary parts[8] likewise produced small effects.

The influence of a more complicated spin dependence in the deuteron channel was also studied. Figure 2 shows zero-range calculations including a configuration-space tensor interaction, usually designated as T_R, in the deuteron channel. The parameters of T_R were determined by analysis of $d+^{12}C$ elastic scattering data at T=29.5 MeV[9] and, although this extrapolation is likely to be quantitatively unreliable, Fig. 2 shows that the influence of T_R is quite small even when the strength determined in Ref. 9 is arbitrarily doubled.

The effect of the deuteron continuum on (p,d) and (d,p) cross-sections has been frequently studied. Approximate treatments--such as the Johnson-Soper prescription[10]--have resulted in improved agreement with cross-section data in many cases. Full coupled-channels formulation of the three-body problem[11,12] is much more difficult and extended comparisons with data have not yet been made. Farrell, Vincent and Austern[11] suggest that continuum or break-up effects are most pronounced for low partial waves in the nuclear interior. This is just the region emphasized in momentum mismatched reactions such as the one discussed here. We therefore performed coupled channels calculations in the spirit of Ref. 11 using five continuum channels and a separable nucleon-nucleon potential. These calculations appear in Fig. 3. The dashed curves of Fig. 3 correspond to DWBA calculations using unfolded nucleon-nucleus potentials[13] and are therefore equivalent to the Johnson-Soper method.[10] A folded or Watanabe type potential[14] was used in the DWBA to generate the dotted curve; this corresponds to a

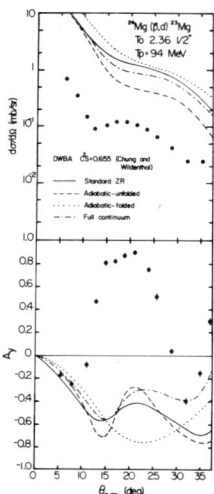
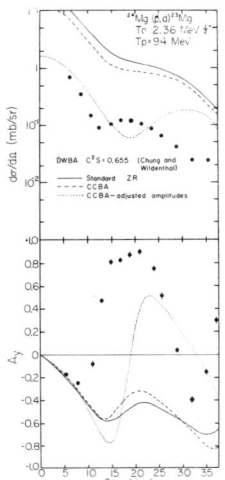

Fig. 3. DWBA calculations with and without continuum contributions are compared with data. See text for explanation.

Fig. 4. Calculations with and without two step contributions are compared with the data. Note that cross-section normalization for the dotted curve is arbitrary.

limiting case of the coupled channels calculations where only the deuteron bound state contribution is retained. Examination of the calculated analyzing powers displayed in Fig. 3 shows that, as asserted in Ref. 11, the Johnson-Soper prescription yields a better approximation to the full calculation than the Watanabe model although no distinct preference can be inferred from the cross-sections. In any case, no matter how the breakup contributions are treated, Fig. 3 shows that they do not appreciably improve agreement with the data.

^{24}Mg and ^{23}Mg are known to be strongly deformed nuclei. Consequently relatively large multistep contributions are possible in single nucleon transfer reactions. These effects have been examined in detail for the ^{24}Mg(d,t) and (d,^{3}He) reactions by Nelson and Roberson.[15] We have used their amplitudes generated from shell model or band-mixed rotational model analyses in a CCBA extension of the DWBA to assess the influence of two-step contributions. The dashed curves of Fig. 4 are typical of the resulting calculations. No significant effect is observed. The dotted curves of Fig. 4 were obtained by arbitrary adjustment of amplitudes to give maximum cancellation of one and two-step contributions. Cross-section normalization is arbitrary. Although the agreement with the A_y data is much improved, it should be noted that the adjusted amplitudes are quite unrealistic - the ratio $C^2S(2^+\to 1/2^+)/C^2S(0^+\to 1/2^+)$ is 160 times greater for the dotted curves than for the dashed--and that the cancellation which gives this improvement is quite delicate.

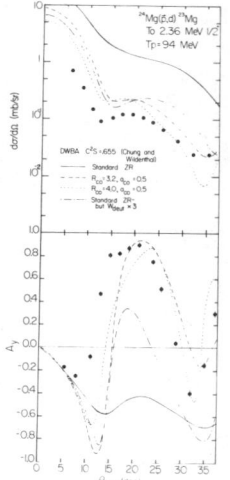

Fig. 5. Zero-range DWBA calculations with and without surface localization are compared with the data. The dashed and dotted curves reflect the use of cutoffs. The dashed-dot curve reflects the use of increased deuteron absorption.

The highly oscillatory A_y angular distribution observed in the ^{24}Mg($\vec{p}$,d) reaction is qualitatively similar to that observed for elastic scattering of strongly absorbed particles which can qualitatively be described using strong absorption models which exploit the localized nature of the reaction.[16] We used several means of artificially introducing such localization into the DWBA calculations. All gave similar results and resulted in vastly improved agreement with the data. Examples appear in Fig. 5 where the dotted and dashed curves were obtained by multiplying the form factor by

$$f(r) = 1 - \{1 + \exp[(r-R_{co})/a_{co}]\}^{-1}$$

with a_{co}=0.5 fm and R_{co}=3.2 and 4.0 fm (or 1.11 x $24^{1/3}$ and 1.39 x $24^{1/3}$) respectively. In effect this amounts to a smoothed lower radial cut-off with cut-off parameters arbitrarily chosen to give the vastly improved agreement with the A_y data shown in the bottom half of Fig. 5. Not surprisingly, a similar effect can be achieved by greatly increasing absorption in the optical potentials. This is illustrated by the dashed-dot curve in Fig. 5 which was generated using the ZR DWBA without a cutoff, but with the imaginary deuteron potential strength increased by a factor of three. The top of Fig. 5 shows that the cutoff chosen to give optimum agreement with the A_y data also greatly improves agreement with the measured cross-sections. Fig. 6 indicates that

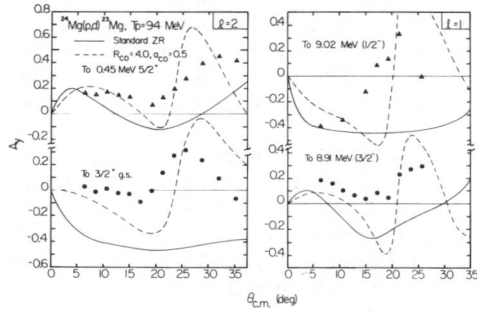

Fig. 6. Zero-range DWBA calculations with and without surface localization are compared with the data. The dashed and dotted curves reflect the use of cutoffs. The dashed-dot curve reflects the use of increased deuteron absorption.

measured A_y values for strong $\ell=2$ and $\ell=1$ transitions are also much better reproduced. Cross section calculations using the Chung and Wildenthal spectroscopic factors are also in better agreement.

Such general improvement using cutoffs suggests that delicate cancellations between one- and two-step amplitudes are not responsible for the poor agreement between the DWBA and experiment for the ^{24}Mg$(\vec{p},d)^{23}$Mg(2.36 MeV 1/2$^+$) reaction. It does suggest that the cutoffs are mocking up important physical effects neglected in the DWBA even when the many refinements discussed above are included. The calculations are obviously very sensitive to contributions from the nuclear interior and apparently the DWBA--even in its more refined formulations--does not treat these contributions correctly. In the present case, this incorrect treatment results in catastrophic disagreement with experiment. The physics which underlies the apparent suppression of interior contributions in the present case should be embodied in an improved reaction theory.

REFERENCES

1. K. Hoshono et al., contribution to this conference.
2. D.W. Miller et al., Bull. Am. Phys. Soc. 25,522 (1980) and to be published.
3. Y.S. Horowitz, Nucl. Phys. A193, 438 (1972).
4. J. Childs and W.W. Daehnick, Bull. Am. Phys. Soc. 20, 626 (1975).
5. J.R. Shepard and P. Kaczkowski, Bull. Am. Phys. Soc. 22, 529 (1977).
6. T.S. Bauer et al., Phys. Rev. C 21, 757 (1980).
7. D.W. Miller et al., Phys. Rev. C 21, 2008 (1979).
8. A. Nadasen et al., to be published.
9. G. Perrin et al., Nucl. Phys. A282, 221 (1977).
10. R.C. Johnson and P.J.R. Soper, Phys. Rev. C1, 976 (1970).
11. J.P. Farrell, Jr., C.M. Vincent and N. Austern, Ann. Phys. 96, 333 (1976) and N. Austern, C.M. Vincent and J.P. Farrell, Jr., Ann. Phys. 114, 93 (1978).
12. G.H. Rawitscher, Phys. Rev. C11, 1152 (1975).
13. H.H. Duhm, Nucl. Phys. A118, 563 (1968).
14. S. Watanabe, Nucl. Phys. 8, 484 (1958).
15. R.O. Nelson and N.R. Roberson, Phys. Rev. C6, 2153 (1972).
16. e.g., R.D. Amado, F. Lenz, J.A. McNeil, and D.A. Sparrow, preprint.

(Editor's Note: The author has provided larger sized figures, and these are appended to the paper.)

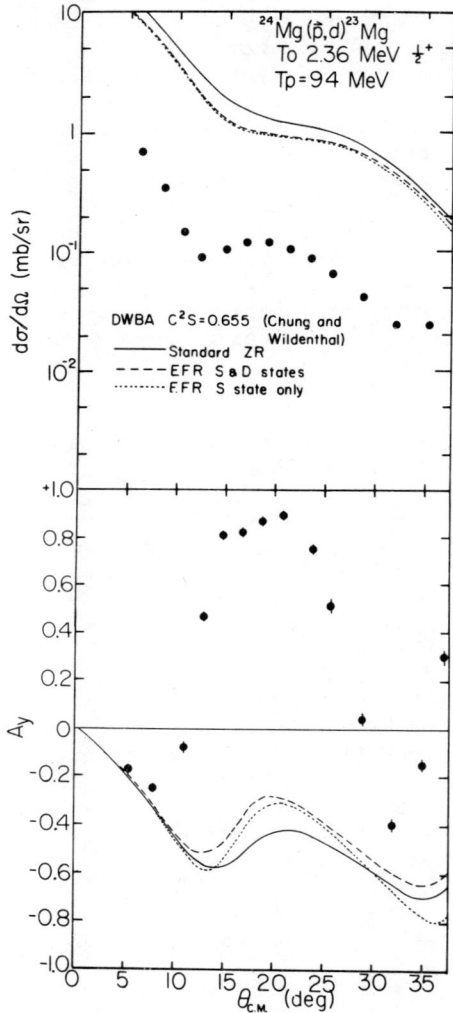

Fig. 1. Zero-range and exact-finite-range DWBA calculations are compared with data.

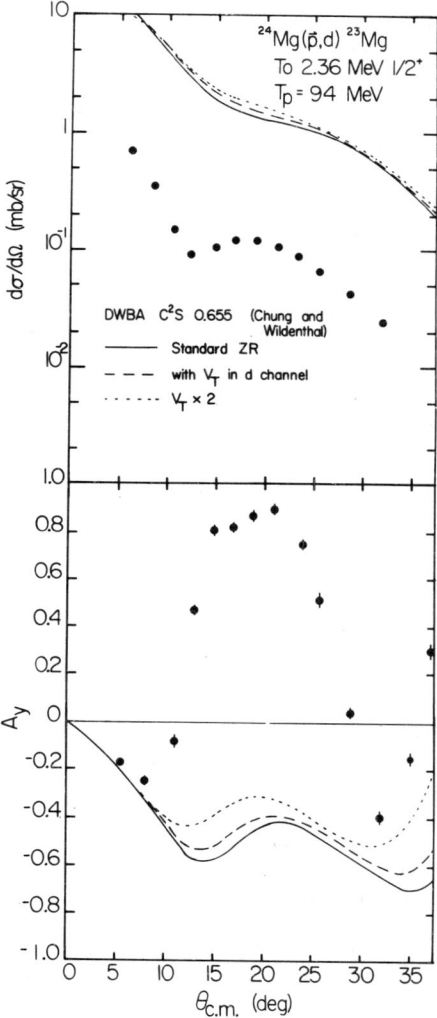

Fig. 2. Zero-range calculations with and without tensor potentials, V_T, in the deuteron channel are compared with data.

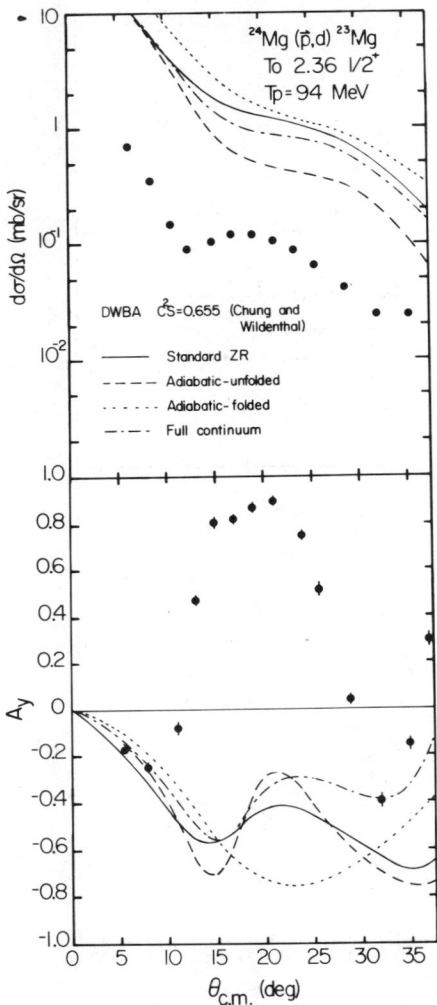

Fig. 3. DWBA calculations with and without continuum contributions are compared with data. See text for explanation.

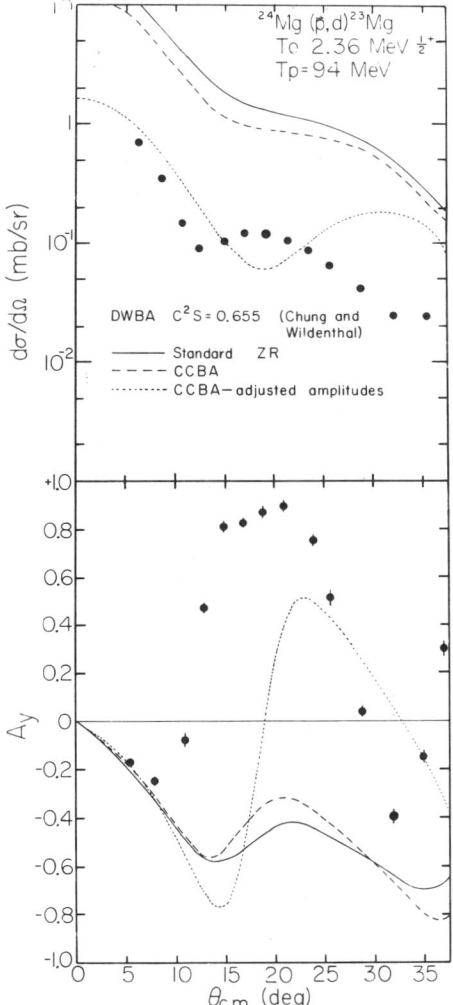

Fig. 4. Calculations with and without two step contributions are compared with the data. Note that cross-section normalization for the dotted curve is arbitrary.

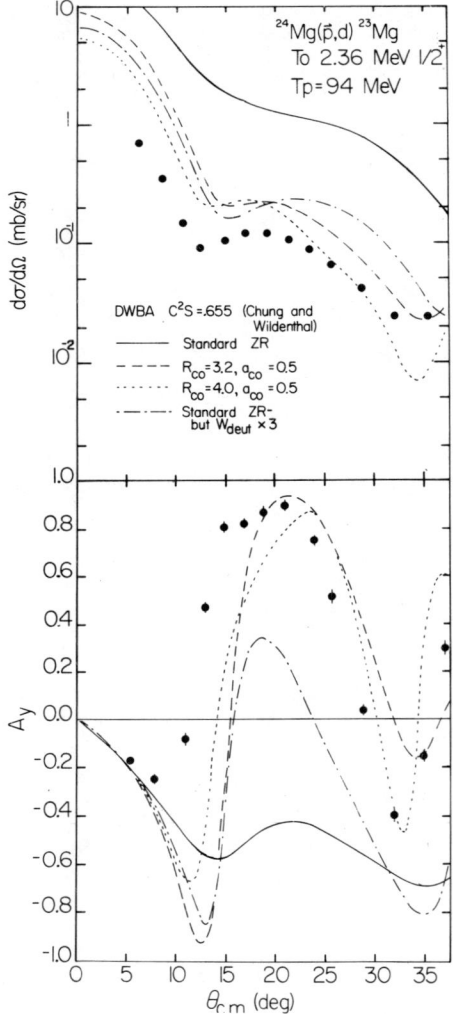

Fig. 5. Zero-range DWBA calculations with and without surface localization are compared with the data. The dashed and dotted curves reflect the use of cutoffs. The dashed-dot curve reflects the use of increased deuteron absorption.

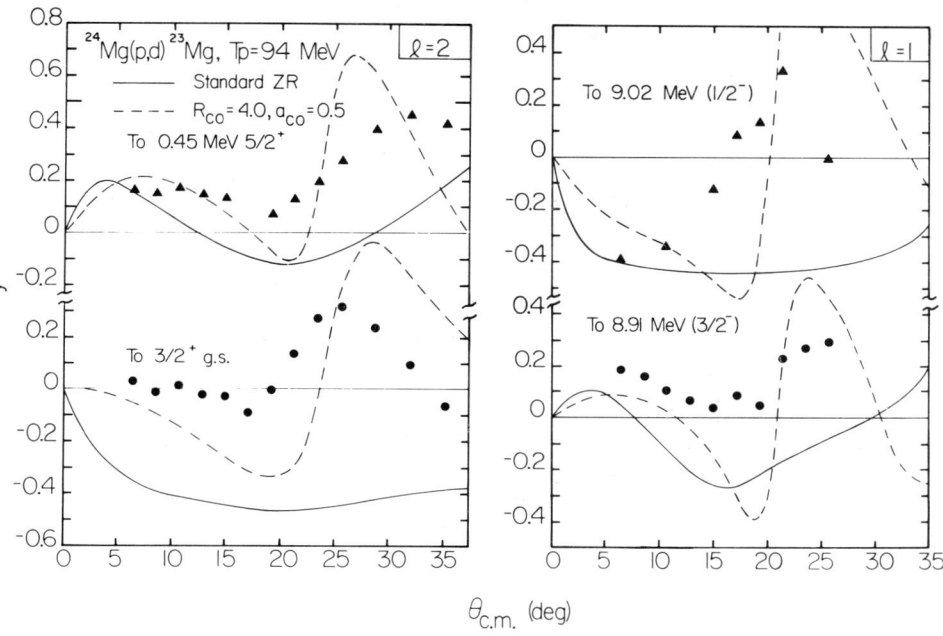

Fig. 6. Zero-range DWBA calculations with and without surface localization are compared with the data for ℓ=1 and 2 transitions.

DISCUSSION

OHNUMA: I'd like to point out that the damping of the nuclear interior region is what we have been asking for at lower energies for a long time, perhaps in a less dramatic way, since the work of Yntema and myself thirteen years ago. Many respectable theorists have come up with various possible explanations, but none of them seems to be good enough. Now we learned that things get disastrous at 200 MeV.

SHEPARD: This is a more extreme effect and we have concentrated on it for this reason.

IGO: Are there other examples besides the Ex. 2.36 KeV., $\ell=0$, $1/2^+$ transition which have large discrepancies with the data? Are those also improved by smooth cut-off calculations?

SHEPARD: The present case is more extreme, but there are similar problems for $^{13}C(\vec{p},d)^{12}C$ at 123, 200 and 400 MeV. Also for other transitions in $^{24}Mg(\vec{p},d)$ involving $\ell=2$ there are problems with the cross-sections and analyzing power agreement is terrible. In the latter case, there is vast improvement when the cut-off is used.

RAWITSCHER: I believe that we are seeing nonlocality effects not only in the deuteron channel (because of breakup, rearrangement and inelastic coupling effects), but also in the proton channel (because of exchange and inherent nucleon-nucleon nonlocality effects). The work reported by the Colorado group is very interesting in that it stimulates us to look for such effects.

SHEPARD: When these effects are included in the local energy approximation, they have little influence. Also, much of the nonlocality in the deuteron channel is expected to come from coupling to the continuium. We have treated this effect explicitly and found it to have little influence. Nonlocality effects should be examined, clearly, though they must very strongly suppress the interior to remedy the disagreement by themselves.

ANTONUK: Are there any particular measurements that should be made to illuminate this problem?

SHEPARD: $d\sigma/d\Omega$ and A_y should be measured at lower energies where the DWBA should work better. Documenting the energy dependence of the failure should be useful in pinpointing its source.

RAPPORTEUR'S REPORT: POLARIZATION IN
NUCLEAR STRUCTURE PHYSICS

R.N. Boyd

The Ohio State University, Columbus, Ohio 43210

INTRODUCTION

In my report I will try to satisfy two major objectives. First of all, I will try to give appreciable coverage to those contributed papers which I feel represent new steps in nuclear structure or reaction physics. There are a number of imaginative new efforts in such traditional areas as low energy particle transfer reactions or medium energy nuclear physics; it is to these that I will devote the most attention, especially when the study indicates promise of new nuclear structure or reaction information. The most meaningful of these studies usually involve both very detailed data and a sophisticated reaction analysis. I am thereby invoking my own philosophical prejudice: Physics is certainly an experimental science, but data are meaningless unless they can be made to impinge in some way on the relevant paradigm.

The other objective I have is to present some notable failures in nuclear reaction studies. The hope here is that focussing on the questions involved will produce sufficient effort, both experimental and theoretical, to solve the prevailing problems.

I will necessarily have to overlook a large amount of excellent physics in my review. This is unfortunate, and I am sorry that it must be so. However, in the Nuclear Structure Physics section of this Symposium, there were 125 papers, nearly half of those submitted to the conference. Thus great selectivity was necessary. I will also not give very much time to work presented in the Nuclear Structure Physics Discussion Session, since that work has already been heard by half the Symposium attendees.

Because time is precious, I don't want to spend much of it comparing the contributions of Polarization Symposium V to those of Symposium IV. However, I do feel that one such comparison is sufficiently noteworthy that it should be made. The contrast in sophistication, both experimental and theoretical, of the Nuclear Structure Physics papers is quite striking. The vector analyzing powers of the deuteron induced reactions which comprised most of the deuteron reaction data of Symposium IV have been supplemented by tensor analyzing powers. The proton elastic scattering studies of Symposium IV have given way to neutron scattering studies. But perhaps most strikingly, the simple DWBA analyses have been replaced by coupled inelastic channel and coupled reaction channel analyses, and the deuteron D-state, a topic treated sparingly in Symposium IV, plays a major role in the theoretical aspects of many of the reaction studies of Symposium V. We are making progress!

What I will do is divide the Nuclear Structure Physics contributions into subsections, and spend a bit of time on each.

ELASTIC AND INELASTIC SCATTERING STUDIES

Both elastic and inelastic scattering have been discussed at some length in the invited talks and the discussion session, so I will make only a few cursory remarks about those topics. Several groups, notably those at TUNL and Stuttgart, have undertaken neutron elastic scattering studies since the last Symposium. While few surprises have arisen from those data, it is nonetheless important to have the detailed information obtained from them. I feel the most significant development in proton elastic scattering since Symposium IV is the addition of the high energy data, i.e., the LAMPF, TRIUMF, and IUCF data. With these data, the analyses have become sophisticated, since they now require proper relativistic treatments and accommodation of pionic effects. A very nice discussion of some of the results of these experiments and the concommitant analyses is given by Schwandt, et al. (2.24).

The set of deuteron elastic scattering data has recently come to include some high energy data from IUCF, and these have been included in the global analysis given by Daehnick (2.34). I should also note the addition of tensor analyzing powers to the considerations of some of the deuteron optical model fits. New triton and helium elastic scattering data also exist: they are presented in the conference contributions. Finally, I should mention the invited talk by Professor Thompson in which he described recent attempts to put the optical model on a more fundamental footing.

Various aspects of proton inelastic scattering to discrete final states have been discussed in invited talks by Professor Bacher, Dr. Moss, and in the Discussion Session by Dr. Comfort (2.54). Thus I will consider that subtopic to have been very adequately covered. I should also note that Dr. Clement, in the Discussion Session (2.64) presented the state of the art in coupled-channels analysis of inelastic

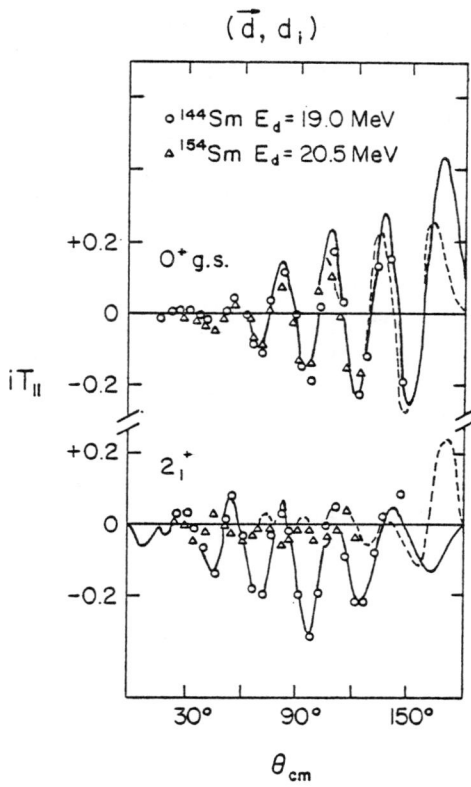

Figure 1. (d,d') on Sm isotopes.

deuteron scattering data. Figure 1 shows a small fraction of those data: one can see the very dramatic effect of including the channel coupling. The reactions for which the data are shown are (d,d') on ^{144,154}Sm. The elastic scattering analyzing powers are very similar. The 2_1^+ analyzing powers for (spherical) ^{144}Sm are well fit by the DWBA. However, for the deformed nucleus ^{154}Sm, the data for which are seen as triangles, the (d,d') analyzing power data to the 2_1^+ state are dramatically different from those for ^{144}Sm, and definitely require the CCBA for representation (dashed curve).

ANALYZING POWERS IN CONTINUUM SPECTRA

Four contributions to this Symposium, three experimental and one theoretical, involve studies of the continuum spectra of light

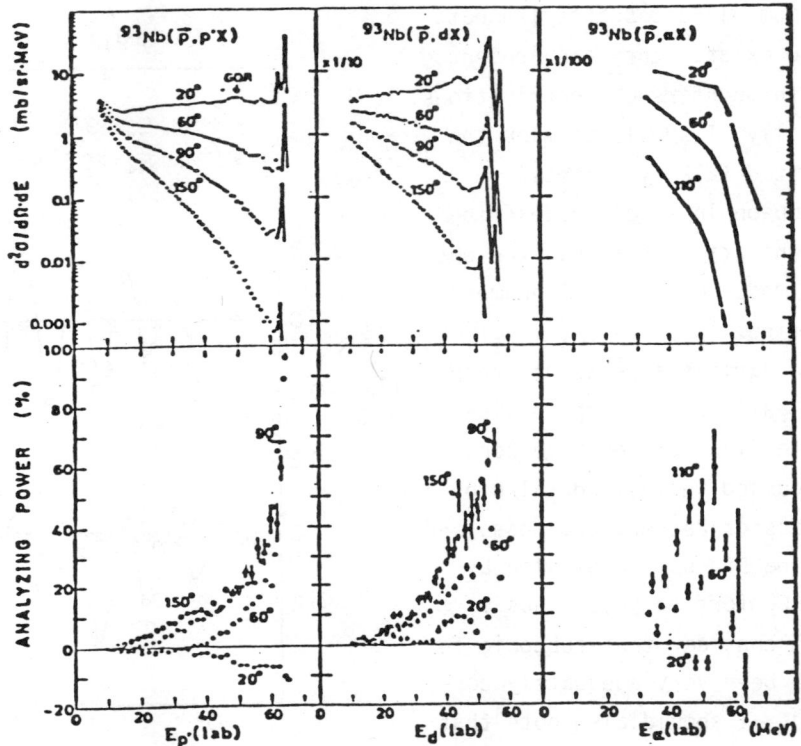

Figure 2. Proton induced reactions on ^{93}Nb.

projectiles emitted when nuclei are bombarded by polarized protons and deuterons. Use of polarized projectiles in such studies affords a beautiful opportunity to study the complexity of the reaction mechanisms involved, i.e., one-step, two-step, and conceivably approaching a traditional compound nuclear reaction, in the different regions of excitation.

Figure 2 shows the sort of behavior exhibited by the outgoing protons, deuterons and alpha particles when ^{93}Nb is bombarded with 65 MeV polarized protons. These data are from the work of Sakai, et al. (2.58). They reveal what would be expected qualitatively, At low excitation (near the right hand side of each of the three sections of the figure) where the reactions would be expected to be dominated by simple one-step processes, the data exhibit the large analyzing powers generally associated with simple direct reactions. At very high excitations the analyzing powers go to zero. This is also reasonable, since the projectiles in this excitation region must undergo a number of energy sharing collisions, thus approaching a compound nuclear process. A similar result is observed for the deuteron induced reactions (2.59).

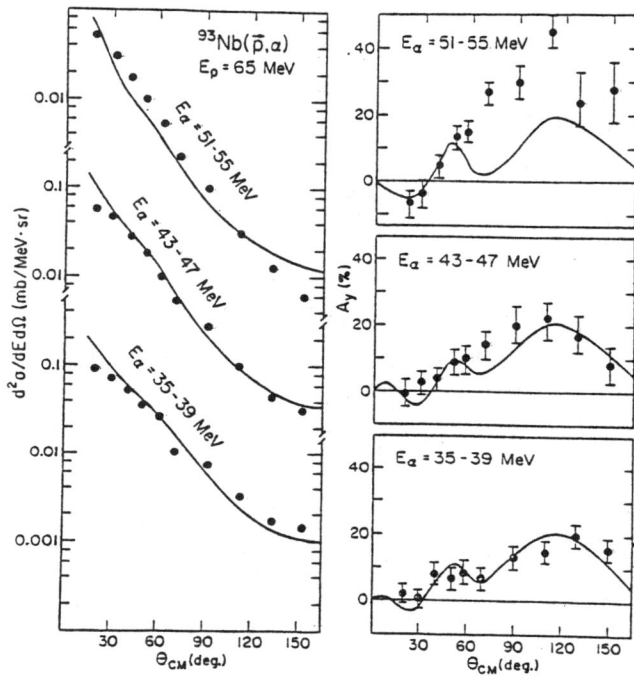

Figure 3. ^{93}Nb($\vec{p},\alpha$) to highly excited regions.

In the middle regions a variety of multistep processes are probably going on. Indeed, Lenske, Tamura and Udagawa (2.60) have attempted fits to some of these data assuming that one- and two-step processes dominate the reaction. The types of processes they include in their calculations for the ^{93}Nb($\vec{p},\alpha$) reaction include simple one-step, and those two-step processes involving an inelastic excitation: (p,p',α) and (p,α,α'). Some simplifying assumptions and a delicate energy averaging are required in order to perform their analyses. Even with those assumptions it can be seen from Figure 3 that their predictions for the different excitation energy regions do reasonably well in reproducing the data.

Another study, that by Weitkamp, et al., (2.42) involves measurement of D, the depolarization parameter, in the ^{63}Cu(p,p') reaction, as a function of excitation energy. Although their study was at a much lower incident beam energy (18 MeV), their results are in qualitative agreement with those of Sakai, et al. The depolarization parameter is found to be about 1.0 at low excitation, and tends toward zero at high excitation.

These continuum region studies are few in number, and the analysis represents a first attempt. It is expected that future developments along these lines will add greatly to our knowledge of the complicated reaction mechanisms in these deeply inelastic collisions.

RADIATIVE CAPTURE STUDIES

Both proton and neutron radiative capture were discussed in invited talks by Prof. Weller and Dr. Snover, so I will not give those subjects much additional attention. I would, however, like to direct your attention to one of the last papers in the Nuclear Structure Physics section, by Wienhard, et al. (2.123). They discuss the $^{16}O(\vec{\gamma},p)$ reaction using linearly polarized photons, and discuss the potential information obtainable from such reaction studies.

SINGLE NUCLEON TRANSFER

Since single nucleon transfer reactions have received much attention in the past decade, I will not deal in detail with most of these studies. However, I do want to note that some excellent nuclear structure physics is represented in the (p,d), (d,p), (d,n), (d,t),

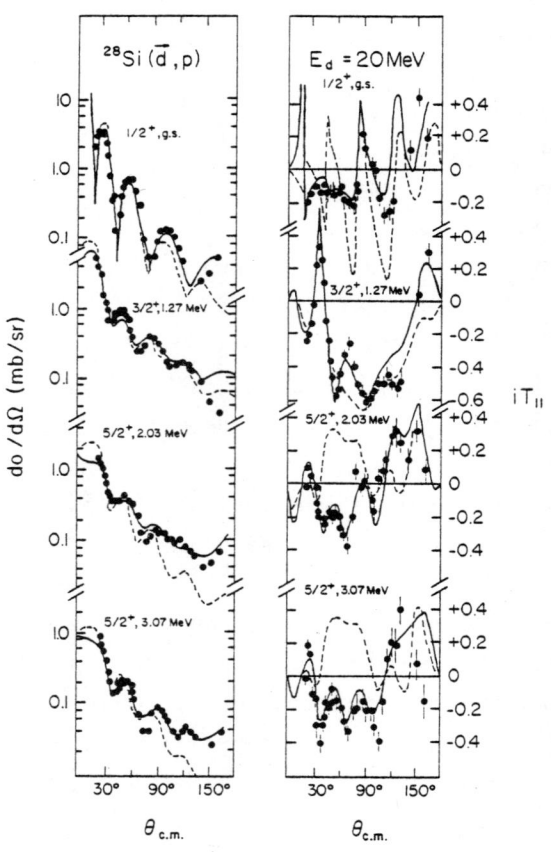

Figure 4. Solid curve is CCBA result, dashed is that of the DWBA.

(d,³He), (t,d), (t,α), (³He,d), and (³He,α) contributions to this Symposium. I will return to some special aspects of some of these studies in subsequent comments. For now, I do want to show you the level of sophistication which has been achieved in analyses of single nucleon transfer reactions. Figure 4 shows results of a (d,p) reaction study on ^{28}Si performed at 20 MeV by Seichert, et al. (2.96). Their analysis involved both inelastic excitations and neutron stripping between various ^{28}Si and ^{29}Si states. The dashed curves are DWBA results. The data representation is indeed remarkable when the coupled channels effects (solid curves) are included.

Several of the single nucleon pickup studies have focussed on deep hole states. In particular, the (p,d) reaction studies done at IUCF at 90-95 MeV by Kasagi, et al. (2.84) and Miller, et al. (2.85) have utilized the analyzing powers to produce some convincing spin-parity assignments of the deeply bound structures. A somewhat more detailed analysis is seen in the (d,³He) study on ^{90}Zr done by Stuirbrink, et al. (2.104). Their results are shown in Fig. 5. Clusters of states at 5.0 and 6.8 MeV are found to exhibit angular distributions characteristic of $f_{7/2}$ proton pickup. In fact, as shown at the bottom, the entire excitation region from 3.3 to 11.3 MeV is characteristic of that same transfer.

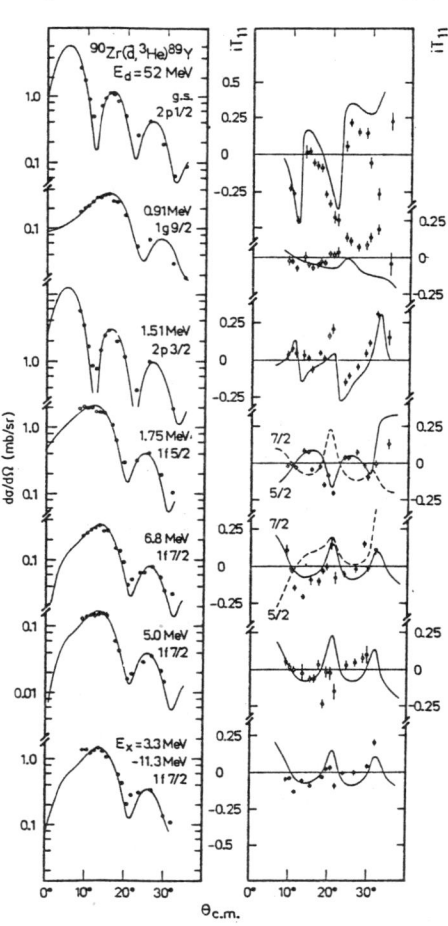

Figure 5. ^{90}Zr(d,³He) to deep hole states.

The same group has similar results for the ^{144}Sm(d,^{3}He) reaction, in which the excitation region from 3.9 to 6.9 MeV is found to have angular distributions characteristic of a $g_{9/2}$ proton pickup.

More work on deep pickup is planned at several laboratories. It is hoped that such studies can be very helpful in mapping out the deeply bound nuclear shells.

The ($^3\vec{\text{He}},\alpha$) and ($\vec{t},\alpha$) reactions are also represented in the conference contributions, and were discussed in the invited talks of Professor Roman and Dr. Flynn. These reactions are of particular

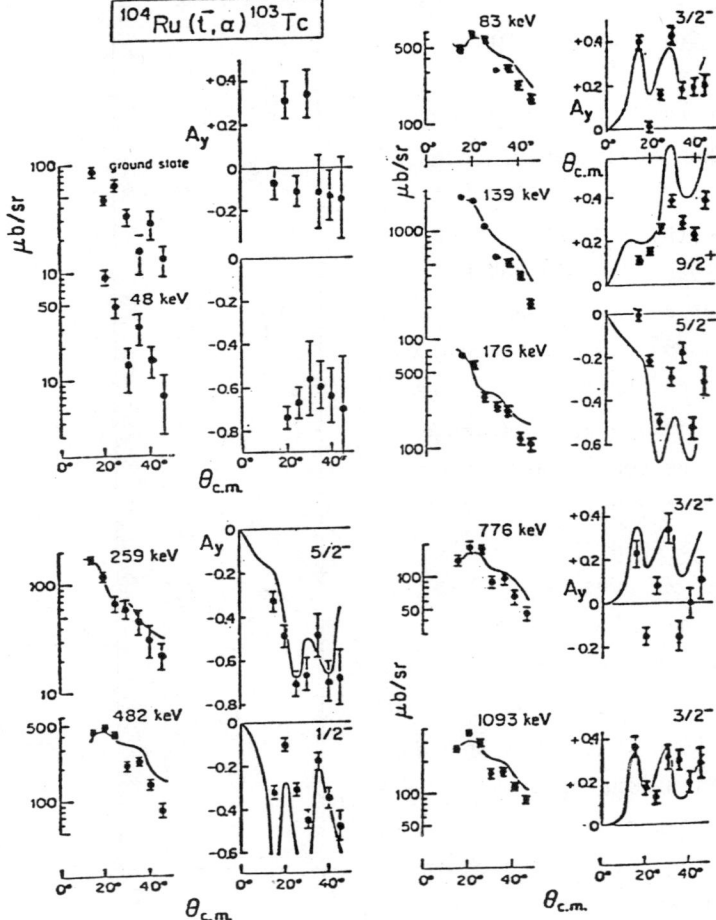

Figure 6. ^{104}Ru($\vec{t},\alpha$) angular distributions.

interest because they favor high spin states, unlike reactions like (d,t). A sample of such work is given by the ^{104}Ru(t,α) data (2.106), shown in Fig. 6. Note the large cross section for the 9/2$^+$ level.

MULTIPARTICLE TRANSFER REACTIONS

The state of the art in two nucleon transfer was well described by Professor Yagi in his invited talk on (p,t) reactions. Some of the data he presented give compelling evidence for the existence of sequential transfers as (p,d,t) in describing two neutron transfer processes. A number of contributed papers deal with this same question, either as (p,t) or (t,p) reaction studies. Both natural and unnatural parity two neutron transfers have been studied.

Several contributions to the conference deal with ($\vec{d}$,α) reaction studies. This particular reaction has had a long history of pathologies. The zero range factor to be used if one assumes a one-step reaction description is known to vary over a wide range. Some investigations have dealt with sequential transfers: such processes appear to be very prominent in this reaction. Extremely detailed data have been obtained for this reaction, and are presented in the contributions to this conference. However, the fits are generally poor, a point to which I will return shortly.

A three-nucleon transfer reaction study is given by Boyd, et al. (2.121), in which they show data from the ^{46,48}Ti($\vec{p}$,α) reactions at 80 MeV. The highest spin

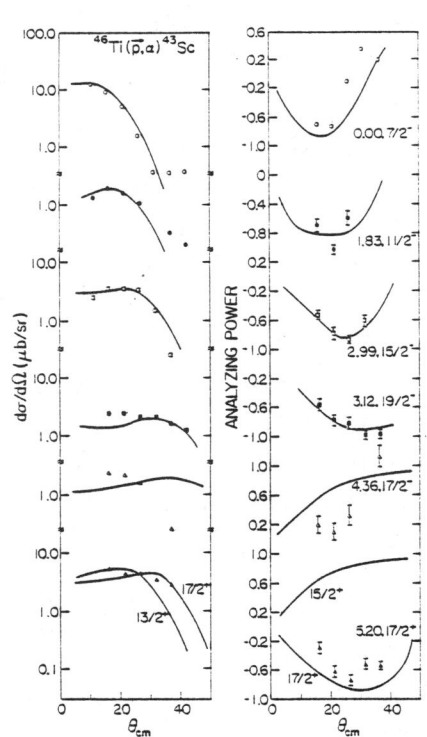

Figure 7. ^{46}Ti($\vec{p}$,α) angular distributions at E = 80 MeV.

state which can be populated in this reaction in a one-step process is 19/2⁻. The highest spin seen, and seen quite strongly, is the 23/2⁻ in ^{45}Sc, a clear indication of a multistep process. The analyzing powers in some cases, notably that for the $17/2^+$ state, have provided clear signatures for spins as shown in Figure 7. The fits assume simple one-step transfer. Note the preference of this reaction for high spin states.

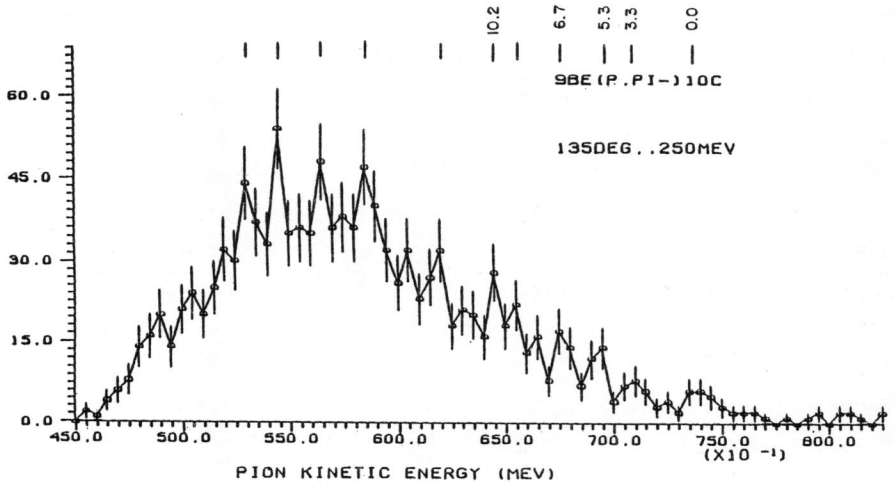

Figure 8. ^{9}Be($\vec{p},\pi^-$) spectrum.

Finally, in this section, I want to mention the recent TRIUMF ($\vec{p},\pi$) reaction results from Lolos, et al. (2.55). The (p,π⁻) spectrum shown in Fig. 8 indicates states, as yet identified only in excitation energy, up to an energy of 20 MeV. The large momentum transfer which occurs in this reaction suggests that the prominent states have high spins. The TRIUMF group, as indicated by Dr. Auld in his invited talk, is hard at work measuring analyzing power angular distributions; their state dependence should be invaluable in spectroscopic identifications and in development of an understanding of the (p,π) reaction. Other groups at other labs, notably IUCF and LAMPF, are also studying this reaction.

SOME OUTSTANDING PROBLEMS

I certainly don't want to leave you with the impression that we

fully understand the
complexities of the
reactions and struc-
tures with which we
have been dealing
for the past decade.
Aside from the danger represented if
our funding agencies
got wind of that, it
just isn't true. So
in my last few minutes I want to present some of the
problems that are
described, either
explicitly or implicitly, in some of
the contributions to
this Symposium.

In Figs. 9 and

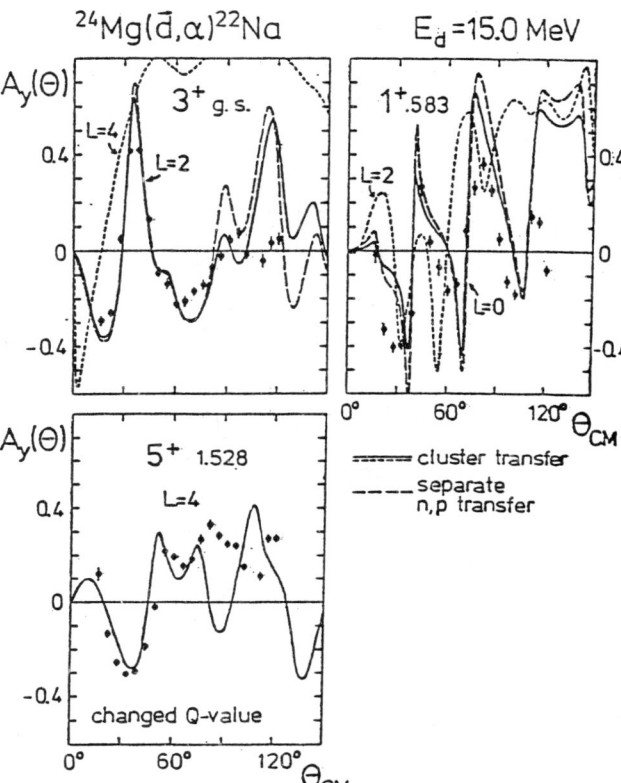

Figure 9. ^{24}Mg($\vec{d},\alpha$) vector analyzing powers.

10 are shown some of the data of Kretschmer, et al. (2.116 and 2.117), from their ($\vec{d},\alpha$) reaction study on ^{24}Mg done at 15 MeV. The data are beautiful, as they include all the vector and tensor analyzing powers. But the DWBA fits generally are poor, even for the 5^+ state, for which no $L_{transfer}$ mixing occurs. These results are fairly typical of the (d,α) work: The DWBA just does not describe that reaction. The failures of the DWBA are most apparent in (d,α), but they are not restricted to it. The other two-nucleon transfer studies, (p,t) and (t,p), also suffer some troublesome unanswered questions. In fitting the (t,p) data it has long been recognized that a sizable increase of the real-central proton distorting potential is essential. When sequential transfer processes through intermediate deuteron channels are involved, it has been necessary to make gross adjust-

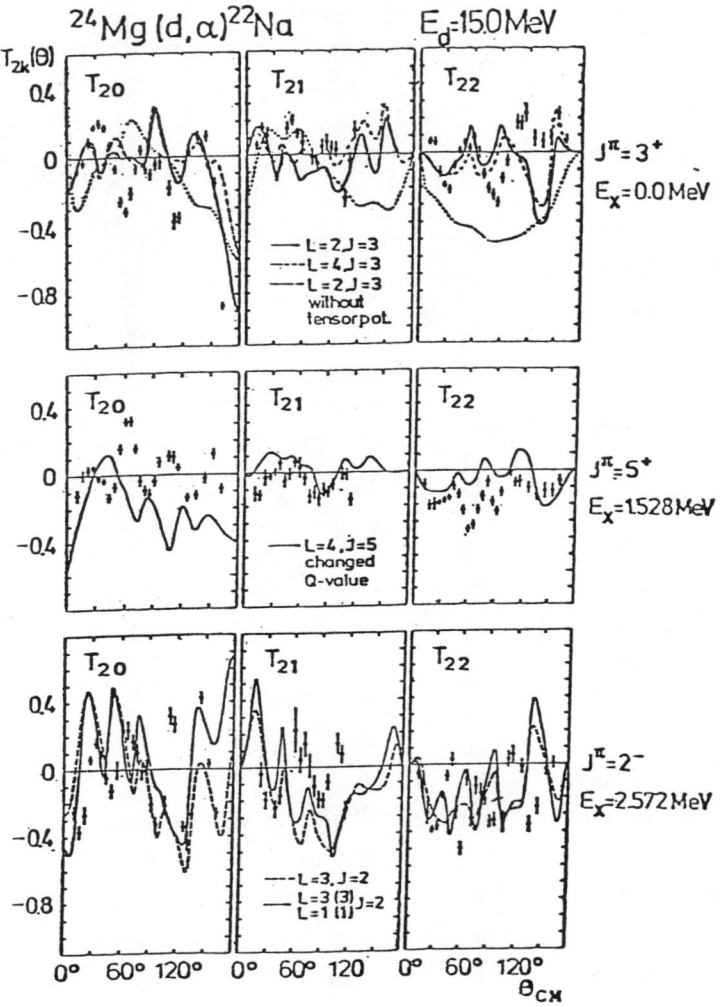

Figure 10. ^{24}Mg($\vec{d},\alpha$) tensor analyzing powers.

ments of the deuteron distorting potentials. This has been done in both (t,p) and (p,t) analyses. While it is nice to fit the data, these parameter modifications are hardly on a fundamental footing, and probably only serve to mask our ignorance of the true reaction process.

These problems are not restricted to two-nucleon transfer either. Several of the contributions to this conference on (p,d) reaction studies express unsettling difficulties with the reaction analysis, and the data representations are often qualitative at best. Dr. Shephard discussed one such case in his invited paper.

So what's wrong? Certainly the deuteron optical potential is a common thread in the difficulties I've mentioned. And we know there are several difficult questions associated with it. Some of these are discussed in conference contributions by Rawitscher and Mukherjee (2.35) who discuss non-locality and breakup effects, and by Tostevin, et al. (2.124), who deal with the effects of antisymmetrization. All of these effects appear to be important, and they don't begin to complete the list.

Several imaginative experiments are reported in this conference which could well provide the information we need to answer the questions we have about reaction mechanisms. Of course the traditional (p,d) and (d,p) reactions provide some information. But important clues may also arise from studies like that of Nakamura, et al. (2.66) who have measured the polarization transfer coefficient of the breakup protons for the $^{90}Zr(d,p)$ reaction at 56 MeV.

Another very pretty experiment is the $^{58}Ni(p,d\gamma)$ reaction study of Kishida, et al. (2.81), who investigate the effects of the deuteron D state on the particle-gamma ray angular correlations. These correlations are related to the magnetic substate populations of the residual nuclei. Some of their results are shown in Fig. 11. Here the

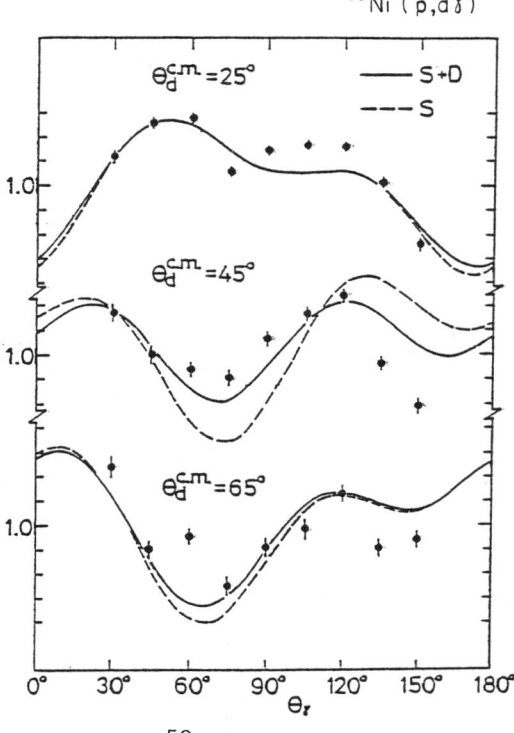

Figure 11. $^{58}Ni(p,d\gamma)$ correlations.

particle detector is at scattering angles of 25°, 45°, and 65°, and the gamma ray detector location is varied in the reaction plane. The solid and dashed curves refer respectively to calculations with and without the deuteron D-state. The sensitivity is seen to be fairly large. Such experiments, with extension to out of plane correlations as well and perhaps using polarized beams, could give the kind of sensitivity to reaction details we need to answer our troublesome reaction questions.

In summary, we've learned a great deal about nuclear structure and reactions with the kinds of detailed studies presented at this Symposium. But we're certainly not done yet!

REFERENCES

2.24 P. Schwandt, A.D. Bacher, W.W. Jacobs, H.-O. Meyer, and S.E. Vigdor, "The Spin Dependence of Intermediate-Energy Proton-Nucleus Elastic Scattering"

2.34 W.W. Daehnick, "Complex $\vec{L}\cdot\vec{S}$ Term in Global Optical Model Potentials for Elastic Deuteron Scattering"

2.54 J.R. Comfort, C.C. Foster, C.D. Goodman, D.W. Miller, G.L. Moake, P. Schwandt, J.R. Rapaport, R.E. Segel, "Analyzing Powers for $^{12}C(p,p')$ ^{12}C at Intermediate Energies"

2.64 H. Clement, R. Frick, G. Graw, I. Oelrich, H.J. Scheerer, P. Schiemenz, N. Seichert, Sun Tsu Hsun, "Inelastic Scattering of Vector Polarized Deuterons from Samarium Isotopes"

2.58 H. Sakai, K. Hosono, N. Matsuoka, S. Nagamachi, K. Okada, K. Maeda and H. Shimizu, "Analysing Powers of the Continuum Spectra (I): 65 MeV Polarized Protons On ^{12}C, ^{28}Si, ^{58}Ni, ^{93}Nb, ^{165}Ho, ^{166}Er and ^{209}Bi"

2.59 H. Sakai, N. Matsuoka, K. Hatanaka, K. Okada and H. Shimizu, "Analyzing Powers of the Continuum Spectra (II): 56 MeV Polarized Deuterons on ^{58}Ni, ^{93}Nb and ^{209}Bi"

2.60 H. Lenske, T. Tamura and T. Udagawa, "Analyzing Power in the Continuum in Light-Ion Induced Reactions"

2.42 W.G. Weitkamp, T.A. Trainor, I. Halpern, H. Bhang, and S.K. Lamoreaux, "Depolarization in the Inelastic Scattering of Protons from Copper"

2.123   K. Wienhard, K. Ackermann, K. Bangert, U.E.P. Berg, C. Bläsing, K. Kobras, W. Naatz, D. Rück, R.K.M. Schneider and R. Stock, "Photonuclear Reactions with Linearly Polarized Photons"

2.96   N. Seichert, H. Clement, R. Frick, G. Graw, P. Schiemenz, Sun Tsu-Hsun, "Inelastic Transfer in $(\vec{d},p)$ Reactions from ^{28}Si and ^{54}Cr"

2.84   J. Kasagi, G.M. Crawley, S. Gales, E. Gerlic, D. Freisel and A. Bacher, "Spin Determination of Deep Hole States From $(\vec{p},d)$ Reactions"

2.85   D.W. Miller, W.W. Jacobs, D.W. Devins, and W.P. Jones, "$(\vec{p},d)$ Analyzing-Power Measurements at 95 MeV"

2.104   A. Stuirbrink, K.T. Knopfle, G. Mairle, H. Riedesel, K. Schindler, G.J. Wagner, V. Bechtold and L. Friedrich, "Spin Determination of Deeply-Bound Hole States from $(\vec{d},^3\text{He})$ Reactions"

2.106   F. Ajzenberg-Selove, E.R. Flynn, Ronald E. Brown, J.A. Cizewski and J.W. Sunier, "Measurements of Masses and Spins of Neutron Rich Nuclei by the $(\vec{t},\alpha)$ Reaction"

2.121   R.N. Boyd, S.L. Blatt, T.R. Donoghue, H.J. Hausman, E. Sugarbaker, and S.E. Vigdor, "The $^{46,48}\text{Ti}(\vec{p},\alpha)^{43,45}\text{Sc}$ Reaction"

2.55   G.J. Lolos, E.L. Mathie, P.L. Walden, E.G. Auld, G. Jones, R.B. Taylor, "New Aspects of the TRIUMF $(\vec{p},\pi)$ Programs"

2.116   W. Kretschmer, E. Heitz, C. Glashausser, A.B. Robbins, J. Duder, D. Melnik, "Investigation of the Reaction $^{24}\text{Mg}(d,\alpha)^{22}\text{Na}$ with Vectorpolarized Deuterons"

2.117   W. Kretschmer, E. Heitz, C. Glashausser, A.B. Robbins, J.C. Duder and D. Melnik, "Tensor Analyzing Powers in the Reaction $^{24}\text{Mg}(d,\alpha)^{22}\text{Na}$"

2.35   G.H. Rawitscher, S.N. Mukherjee, "The Deuteron Optical Potential"

2.124   J.A. Tostevin, M.H. Lopes and R.C. Johnson, "Antisymmetrization Effects in Deuteron-Nucleus Scattering"

2.66   M. Nakamura, H. Sakaguchi, K. Imai, T. Noro, H. Shimizu, H. Sakamoto, S. Kobayashi, S. Kato, N. Matsuoka and K. Hatanaka, "The Polarization of Break-up Protons from Vector Polarized Deuteron Induced Reaction"

2.81   N. Kishida, H. Ohnuma, J. Kasagi, and T. Kubo, M. Yasue, "Effects of the Deuteron D State on the Polarization of the Residual Nuclear State"

(Editors Note: The references are to paper number and are cross-referenced to page number in the Table of Contents.)

POLARIZATION EFFECTS IN THE SMALL-ANGLE SCATTERING OF FAST NEUTRONS BY BISMUTH

M. Ahmed and F. W. K. Firk
Yale University, New Haven, Ct. 06520

ABSTRACT

We have studied $\vec{n}$-^{209}Bi elastic scattering as a continuous function of energy in the MeV region at angles between 3° and 15° using an absolutely calibrated source of polarized neutrons. Calculations using Hogan and Seyler's method with Tanaka's set of optical model parameters have been carried out and compared with our results. The sensitivity of the measurements, and the analyses, to different values of the neutron polarizability has been investigated.

INTRODUCTION

In the past, studies of polarization effects in small-angle scattering of fast neutrons by heavy nuclei have been made using polarized neutrons from charged particle reactions.[1,2] The exact magnitudes of the polarizations in such reactions have often been questionable, and have added to the uncertainties in the interpretation of the results. We have avoided many of the traditional problems in these measurements by using polarized neutrons from the ^{12}C($n, \vec{n}$) reaction, calibrated absolutely in a double-scattering experiment,[3] and by measuring the asymmetry of the polarized neutrons elastically scattered from bismuth as a continuous function of energy between 2 and 5 MeV. The use of the generalized spin precession method further reduced the systematic errors in our measurement. Calculations of multiple scattering corrections (typically less than 15%) were made to the observed data. The errors in our results are dominated by statistical, and not by systematic effects.

EXPERIMENTAL METHOD

Details of the polarized neutron source based on the Yale Electron LINAC have been given in ref. 3: the essential features are the production of an intense pulse of unpolarized photoneutrons with a Maxwellian-like spectrum which scatter from a cylindrical shell of graphite through an angle of 50° to provide a useful flux of polarized neutrons between 2 and 5 MeV. In the present case, the neutrons travelled along a flight path 15m in length, and scattered from a second target into an array of neutron detectors placed at scattering angles ranging from 3° to 15°; the angular resolution was ± 1° for angles <10°. The neutrons passed through a solenoid, set to precess a 2.4-MeV neutron through 180°; the precession angle of all detected neutrons was determined accurately from their measured flight times.

RESULTS AND CONCLUSIONS

The measured analyzing power, $A_y(\theta)$, of polarized neutrons scattered elastically from ^{209}Bi at angles less than 10° revealed no clear energy dependence between 2 and 5 MeV, which is consistent with the expected reduction in the sensitivity of the polarization at very small angles to the details of the nuclear amplitude. Typical results, after corrections for multiple scattering and finite angular resolution effects are shown in Figure 1.

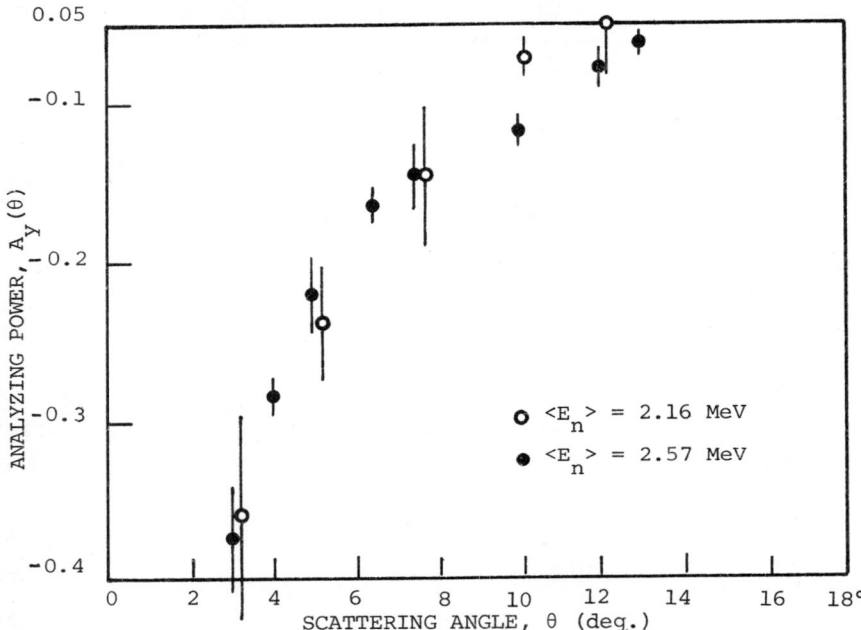

Fig. 1. The point analyzing power $A_y(\theta)$ for n-^{209}Bi elastic scattering in two energy regions. The data have been displaced slightly in angle for display purposes.

The curves shown in Figure 2 represent the results of calculations of Mott-Schwinger scattering using the method of Hogan and Seyler,[4] with Tanaka's optical model and 3 values of the polarizability of the neutron. The present results imply a value of $\alpha \lesssim 0.01$ fm^3;[7,8] an assessment of the error must await further calculations of the influence of slight variations in the optical model parameters.[6] (We note that in refs. 7 and 8, the limits on the value of α were obtained from studies of angular distributions, and not from polarization measurements. In ref. 7, it was assumed that the nuclear scattering was due entirely to potential scattering and that only the $\ell = 1$ partial wave contributed to the interference between nuclear and non-nuclear scattering amplitudes.)

To set an improved limit on α, we are currently studying Mott-Schwinger scattering at low energies in those nuclei in which the

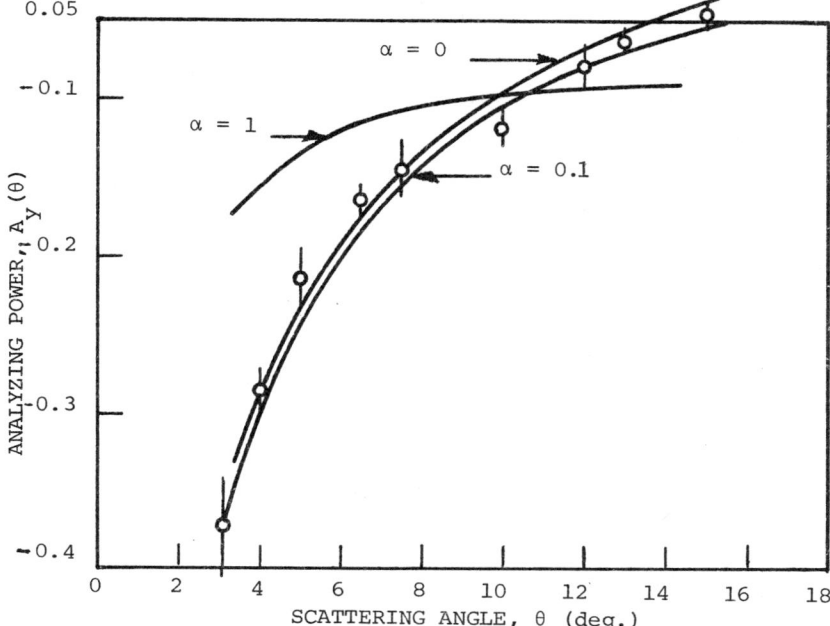

Fig. 2. A comparison between the present results at $\langle E_n \rangle$ = 2.57 MeV and calculations using the optical model of Tanaka et al.[5] for three different values of α (in units of fm^3).

nuclear scattering amplitudes can be calculated with confidence from R-matrix theory.

REFERENCES

1. F. T. Kuchnir, A. J. Elwyn, J. E. Monahan, A. Langsdorf, Jr., and F. P. Mooring, Phys. Rev. **176**, 1405 (1968).
2. A. H. Hussein, J. M. Cameron, S. T. Lam, G. C. Neilson and J. Soukup, Phys. Rev. **15**, C233 (1977).
3. R. J. Holt, F. W. K. Firk, R. Nath and H. L. Schultz, Nucl. Phys. **A213**, 147 (1973).
4. W. S. Hogan and R. G. Seyler, Phys. Rev. **177**, 1706 (1969).
5. S. Tanaka, Y. Tomita, K. Ideno and S. Kikuchi, Nucl. Phys. 179A, 513 (1972).
6. F. D. Becchetti, Jr., and G. W. Greenlees, Phys. Rev. **182**, 1190 (1969).
7. Yu. A. Aleksandrov, G. S. Samosvat, Zh. Sereeter and Tsoi Gen Sor, J. E. T. P. Letters, **4**, 134 (1966).
8. G. V. Anikin and I. I. Kotukhov, Sov. J. Nucl. Phys. **14**, 152 (1972).

ANALYZING POWER OF LANTHANUM USING 7.65MeV NEUTRONS

G. Schleußner, J.W. Hammer, K.W. Hoffmann, D. Kollewe,
W. Kratschmer, E. Speller
Institut für Strahlenphysik der Universität Stuttgart
D-7000 Stuttgart 80

ABSTRACT

Using the Stuttgart neutron scattering facility the analyzing power of natural Lanthanum has been determined for 7.65MeV neutrons.

EXPERIMENT AND EVALUATION

The scattering experiment has been performed by using 4 neutron detectors which were arranged symmetrically to the neutron beam. With the DC-beam of the Dynamitron one has to use pulse height spectroscopy instead of time of flight techniques. The proton recoil spectra have been unfolded by the computer code FERDOR[1] using a response matrix for the detectors. Fig.1a shows a typical proton recoil spectrum measured at a neutron energy E_0 = 7.65MeV and E_1 = 3.2MeV, Fig.1b shows the unfolded neutron spectrum with the two neutron-lines of the $^9Be(\alpha,n)^{12}C$-reaction. The cut off-energy for the proton spectra is about 200 keV. Fig.1c shows the n-γ-discrimination for a dynamic range of 1:50.

The evaluation of the unfolded spectra makes use only of the E_0-peak of the neutrons.

The 50.14%-polarisation of the neutrons has been checked by a scattering experiment using a ^{4}He-high pressure cell in the same collimator-detector-geometry.

Fig.2 shows the analyzing power of natural Lanthanum. Using a coupled channels code we made an optical model analysis using standard parameters [2,3,4] and got preliminary results which are shown in Fig.2 as dashed line.

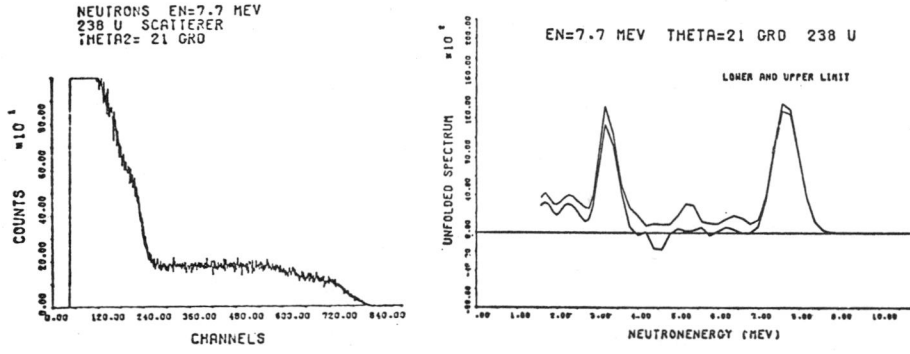

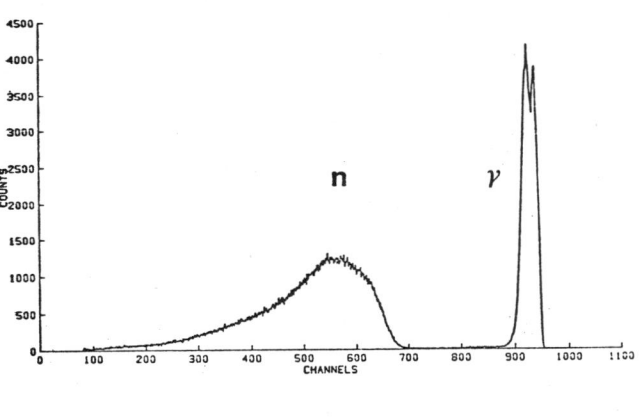

Fig.1a Proton recoil spectrum of $^9Be(\alpha,n)^{12}C$ reaction
 1b With FERDOR-code unfolded spectrum
 1c n-γ-time-spectrum, dynamic range 1:50

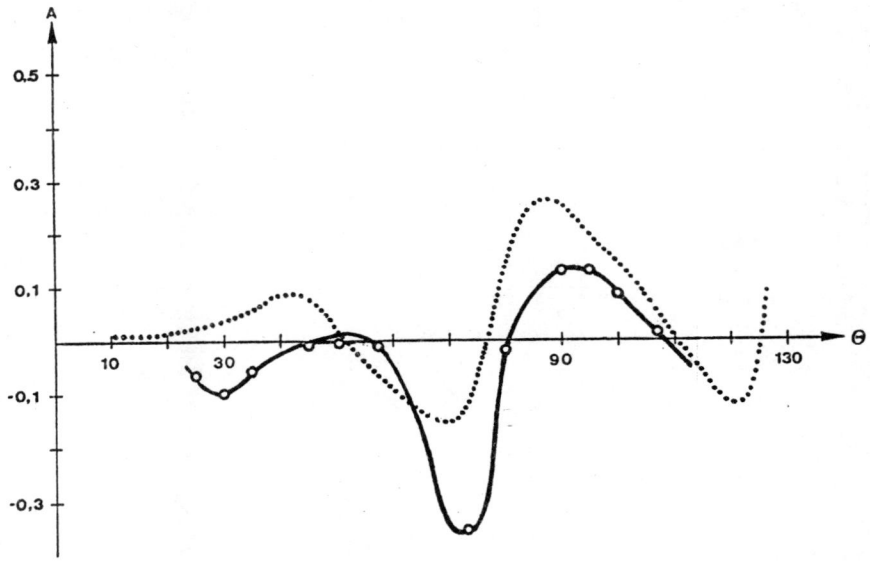

Fig.2 Analyzing Power of Lanthanum (solid line) and optical model calculations (dashed line)

REFERENCES

1. W. R. Burrus and V. V. Verbinski, Nucl. Instr. Meth. <u>67</u>, 181 (1969)
2. L. Rosen et al., Ann. Phys. <u>34</u>, 96 (1965)
3. F. Perey and B. Buck, Nucl. Phys. <u>32</u>, 353 (1962)
4. T. Tamura, Rev. Mod. Phys. <u>37</u>, 679 (1965)

ANALYZING POWER OF URANIUM-238 USING 7.65MeV NEUTRONS

J.W. Hammer, G. Schleußner, K.W. Hoffmann, D. Kollewe,
W. Kratschmer, E. Speller

Institut für Strahlenphysik der Universität Stuttgart

D 7000 Stuttgart 80

ABSTRACT

With a scattering experiment using 50% polarized neutrons of the $^9Be(\alpha,n)^{12}C$-reaction the analyzing power of Uranium-238 has been determined.

THE STUTTGART SCATTERING FACILITY

Fig.1 shows a schematic top view of our scattering facility. The α-beam is delivered by a high current Dynamitron accelerator capable of delivering DC-currents of up to 1mA at an energy of up to 4MeV.

The incoming beam is led via a special beam transport system onto a high power ($10kW/cm^2$) target which is coated with Beryllium of a thickness of 70keV typically.

The neutron scattering facility consists of a shielded source, a spin flip magnet between source and scattering sample, and 4 neutron detectors which are properly shielded too. The reaction angle can be varied between 0 and 70 degrees, the angle of the detectors between 0 and 125 degrees.

Neutron spectroscopy has been performed by unfolding the proton recoil spectra obtaining a resolution of about 8-10%. Gamma events have been nearly totally removed by n-γ-discrimination. Background has been measured and subtracted separately. Measuring left-right asymmetries with symmetrically arranged detectors and with spin up and down we have been able to eliminate apparative asymmetries and to evaluate the analyzing power. Using two monitors for the primary neutron flux one can also evaluate a relative differential cross section.

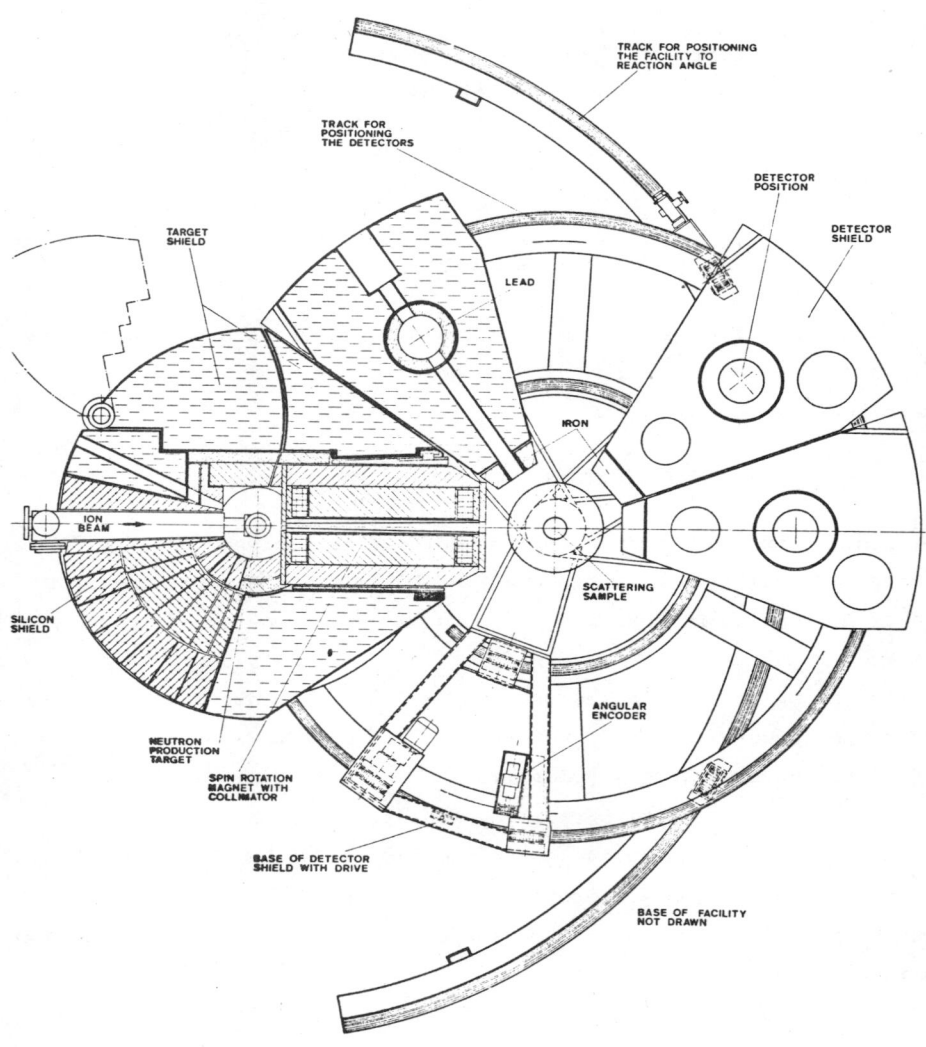

Fig. 1 Schematic top view of the Stuttgart neutron diffraction experiment

The solid line in Fig.2 shows a plot of the analyzing power of Uranium-238.

We have made an optical model analysis for Uranium-238 using a coupled channels code and using standard parameters [1,2] for Uranium-238 and a neutron energy of 7.65MeV:

V_R = 45MeV a_R = 0.62fm r_R = $1.24A^{1/3}$ fm

W_D = 5.74MeV a_D = 0.58fm r_D = $1.26A^{1/3}$ fm

V_{SO} = 7.50MeV a_{SO} = 0.62fm r_{SO} = $1.24A^{1/3}$ fm

Deformation parameters: β_2 = 0.198 β_4 = 0.057

The results are shown as dashed line in Fig.2.

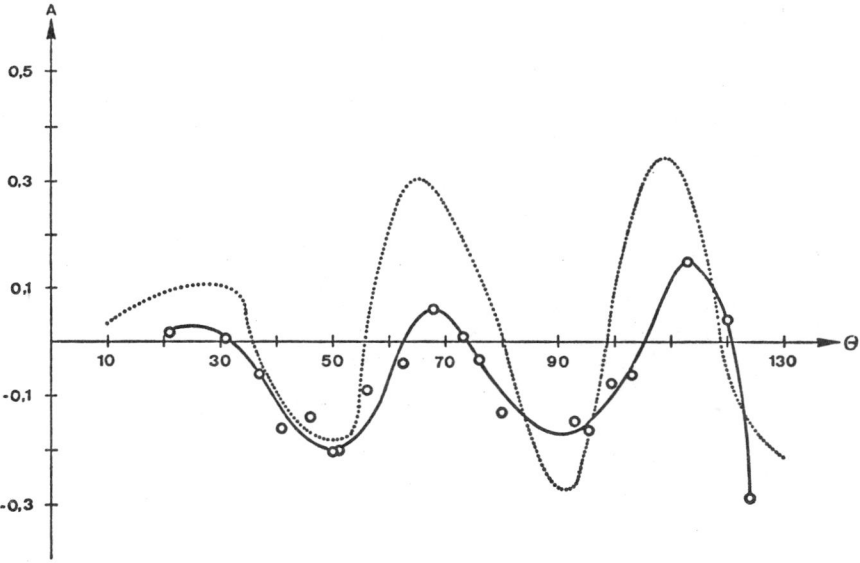

Fig.2   Analyzing Power of Uranium-238 (solid line) and optical model (cc-)calculations (dashed line)

REFERENCES

1. C. H. Lagrange, CEA-Conf. 2852 NEANDC(J) 38 "L" (Feb.1975)
2. G. Haouat et al., NEANDC(E) 196 "L" (Feb.1978)

SCATTERING OF POLARIZED NEUTRONS FROM ^{54}Fe AND ^{65}Cu AT 10 MeV

C.E. Floyd, P.P. Guss, K. Murphy, R.C. Byrd, S.A. Wender, R.L. Walter
Department of Physics, Duke University, Durham, NC 27706 and
Triangle Universities Nuclear Laboratory,* Duke Station, NC 27706

T.B. Clegg
University of North Carolina, Chapel Hill, NC 27514 and TUNL

ABSTRACT

Polarized neutrons have been scattered from ^{54}Fe and ^{65}Cu at 10 MeV. Analyzing powers and cross sections have been fitted with an optical model. The spin-orbit term has a well depth of about 6 MeV, a radius parameter of 1.1 fm, and a diffuseness of 0.5 to 0.6 fm. The need for an imaginary spin-orbit term is indicated. A need is shown for a volume absorptive term in addition to the usual surface term.

The development of a pulsed polarized neutron facility at TUNL has provided a powerful technique for the investigation of spin dependence in neutron scattering processes. Polarized neutrons are produced by polarization transfer in the ^{2}H($\vec{d}$,n)^{3}He reaction at 0°. Pulsing[1] the polarized deuteron beam allows scattered neutrons to be detected with the already existing TUNL time-of-flight facility[2]. Neutron detection with two symmetrically placed scintillators in combination with flipping the incident deuteron spin periodically during each measurement minimizes instrumental asymmetries.

In the present paper we report $A_y(\theta)$ measurements for elastic scattering of 10 MeV neutrons from ^{54}Fe and ^{65}Cu. The data, which have been corrected for finite geometry and double scattering effects, are shown in figs. 1 and 2. Triple scattering effects will be dealt with later. The indicated error bars reflect the propagation of all known uncertainties in the measurement.

These $A_y(\theta)$ values, together with the cross-section data of El-Kadi et al.[3] and the total cross section, have been fitted using the optical-model search code GENOA obtained from F. Perey. For our purposes, the Mott-Schwinger interaction between the neutron magnetic moment and the Coulomb field of the nucleus has been included in the code GENOA and used in all the optical model calculations shown here. The data are compared in figs. 1 and 2 with optical-model calculations based on the global models of Rapaport et al.[4] (RKF) and Wilmore and Hodgson[5] (WH), and a TUNL best fit. While both global sets reproduce the general features of the distributions, neither set predicts the analyzing powers at forward angles where our data has its highest statistical accuracy. Due to the scarcity of accurate polarization data for neutrons, the spin-orbit terms in the neutron model have typically been set to zero as in ref. 5 or set to the proton parameters of Becchetti and Greenlees[6](BG) as in ref. 4. For the present comparison, the BG spin-orbit values were used with the parameter

*Supported in part by U.S. Department of Energy

set of WH.

The results of the searches and the earlier optical model parameter sets are listed in Table 1. Two features of the TUNL parameters should be noted. The forward angle $A_y(\theta)$ requires a volume absorptive term in combination with the surface peaked absorption. This volume term is normally set to zero in low energy neutron scattering. (See

Table 1. OPTICAL-MODEL PARAMETERS FOR ^{54}Fe(UPPER) AND ^{65}Cu(LOWER)

	U	R_r	A_r	W_d	W_s	R_i	A_i	V_{so}	R_{so}	A_{so}	W_{so}
RKF	50.42	1.17	0.67	7.81	0.0	1.30	0.59	6.20	1.01	0.75	0.0
WH	44.22	1.29	0.66	8.99	0.0	1.25	0.48	6.20	1.01	0.75	0.0
Best Fit	45.16	1.28	0.56	4.87	3.31	1.25	0.57	6.02	1.12	0.50	1.00
RKF	48.64	1.20	0.67	6.92	0.0	1.30	0.59	6.20	1.01	0.75	0.0
WH	44.22	1.29	0.66	8.99	0.0	1.25	0.48	6.20	1.01	0.75	0.0
Best Fit	44.56	1.28	0.61	2.95	3.34	1.27	0.61	6.49	1.14	0.62	0.72

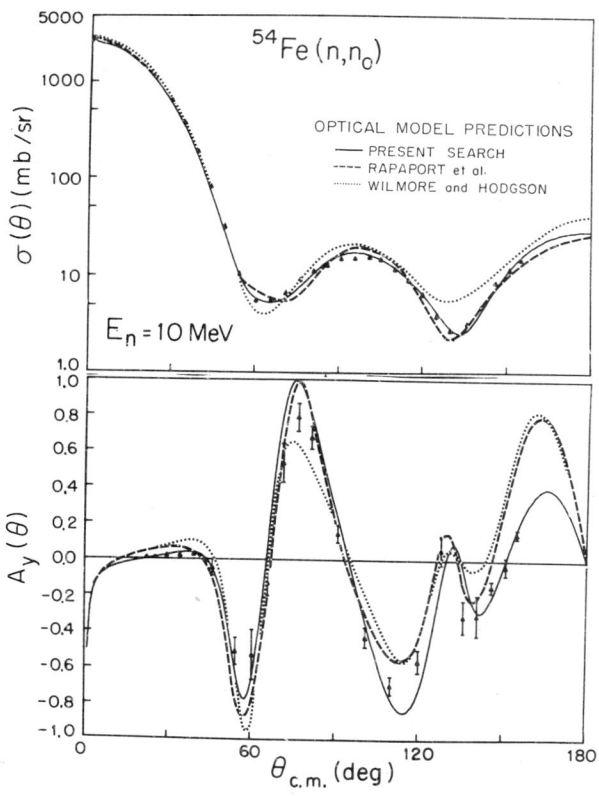

Fig. 1. The data for $\sigma(\theta)$ and $A_y(\theta)$ for ^{54}Fe(n,n). The curves are optical-model calculations.

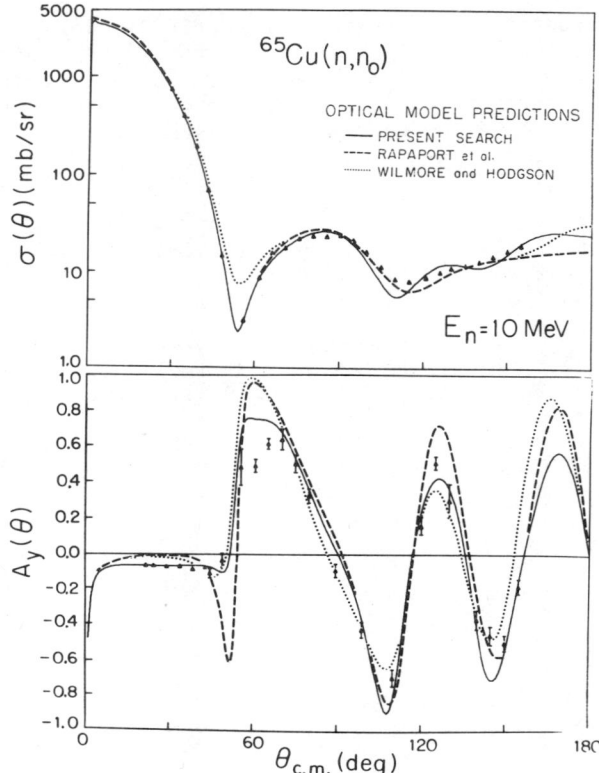

Fig. 2. The data for $\sigma(\theta)$ and $A_y(\theta)$ for ^{65}Cu(n,n). The curves are optical-model calculations.

Table 1.) Secondly, the inclusion of an imaginary spin-orbit term was favored by the data.

Analyzing powers for 14 MeV neutron scattering from Pb, Fe, and Cu have also been obtained. The anomalously high polarizations reported by Galloway and Waheed[7] for angles around 20° are not consistent with either our 14 MeV or our 10 MeV data.

REFERENCES

1. S.A. Wender et al. Nucl. Instr. and Meth. (in press).
2. L.W. Seagondollar et., Bull. Amer. Phys. Soc. 24, 878 (1979).
3. S. El-Kadi et al., Bull. Amer. Phys. Soc. 24, 866 (1979).
4. J. Rapaport et al., Nucl. Phys. A330, 15 (1979).
5. D. Wilmore and P.E. Hodgson, Nucl. Phys. 55, 673 (1964).
6. F.D. Becchetti, Jr. and G.W. Greenlees, Phys. Rev. 182, 1190.
7. R.B. Galloway and A. Waheed, Phys. Rev. C20, 1711 (1979).

Note added in proof: This paper superceeds the contribution presented at the time of the conference where values for A_{so} of 0.4 fm were reported.

NEUTRON SCATTERING FROM ^{58}Ni AND ^{208}Pb AT 10 MeV

P.P. Guss, G. Tungate, C.E. Floyd, E. Woye, K. Murphy, R.C. Byrd,
R.L. Walter and T.B. Clegg
Duke University and University of North Carolina, Chapel Hill and
Triangle Universities Nuclear Laboratory,* Duke Station, NC 27706

ABSTRACT

Cross sections and analyzing powers have been measured for ^{58}N(n,n) and ^{208}Pb(n,n) at 10 MeV. Multiple-scattering corrections are discussed. Comparisons are made to optical-model calculations for ^{58}Ni and to earlier data for ^{208}Pb.

At TUNL we are heavily involved in an ongoing program of neutron elastic and inelastic scattering. The work to be described here involves scattering from ^{58}Ni and ^{208}Pb, and should be viewed as part of the broader neutron scattering program. In this context, an indication is given here of the problem of the unfolding of neutron data to account for multiple scattering corrections.

Both neutron cross sections and analyzing powers are measured at TUNL using time-of-flight methods. The data are obtained in separate experiments, although in both cases the neutrons are produced in the ^{2}H(d,n)^{3}He reaction. For the A_y measurements, the incident deuterons are polarized.

In Fig. 1 the raw data from the cross-section measurement are compared to the cross section obtained after corrections have been applied for attenuation in the sample, for neutrons that are scattered two or more times and then enter the detector, and for geometric effects associated with the finite size of the neutron source, scattering sample, and detector. The cross-section multiple scattering code EFFIGY developed at TUNL was employed to determine the correction factors. The ratio of the multiple scattering events to the neutrons that have only scattered once can be appreciable, as can be seen in the lower half of Fig. 1; note that in the cross-section minimum near 55°, one-half of the detected events are from more than one scattering. The final cross-section data for ^{58}Ni(n,n$_o$) and (n,n$_1$) are shown in Fig. 2 along with polynomial fits.

Also illustrated in Fig. 2 are the ^{58}Ni analyzing powers for both the ground state and the first excited state at 1.454 MeV. (The solid curves for $A_y(\theta)$ are from fitting the product $A_y(\theta) \cdot \sigma(\theta)$ by an associated Legendre polynomial expansion.) Again the data were unfolded by a Monte Carlo multiple scattering code. This code, JANE, developed at TUNL and at the University of Tübingen, provides for finite geometry effects and multiple scattering. Of interest in this instance is the effective analyzing power of the multiple events and the weighting they receive due to the ratio of single-scattered events to multiply-scattered events, as well as the consideration of detected single-scattered events which scatter with

---

*Work supported by U.S. Department of Energy

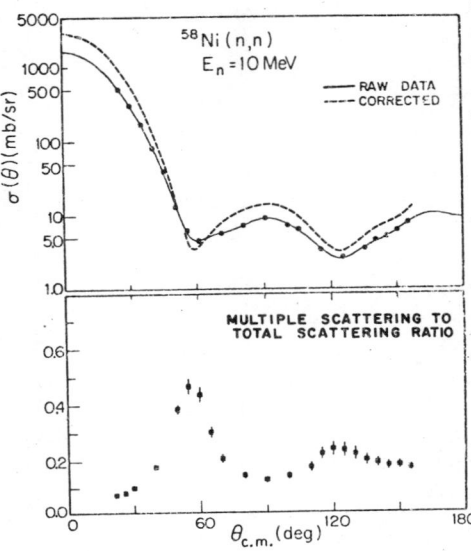

Fig. 1. Multiple scattering corrections for ^{58}Ni(n,n).

various analyzing powers because of the finite size of the scattering sample. From a Monte Carlo simulation of the laboratory process, JANE in one mode of operation estimated for each angle the analyzing power which when smeared for the above effects would yield the experimentally observed analyzing power. In its 2nd mode of calculation, JANE uses for input the point-geometry single-scattering analyzing power (usually that inferred from Mode 1 calculation), and performs a calculation which should have the effect of smearing the input data set to closely resemble the analyzing powers observed in the laboratory. This Mode 2 serves as a consistency check on the corrections obtained in the Mode 1 operation.

Both the cross-section and analyzing power measurements are presently being used to understand the global optical model in a more complete way. We have not progressed to optical-model searching on these data yet, and all we have at the present is a comparison to calculations using the parameters of Rapaport et al.[1] These are shown as dashed curves in Fig. 2. The systematics are reproduced. An optical-model search is expected to give better agreement and thereby better nuclear parameterization for ^{58}Ni(n,n).

We have recently obtained cross-section and analyzing power data for ^{208}Pb at 10 MeV. As yet these data have not been processed through the multiple-scattering codes. Nevertheless, we show the raw $A_y(\theta)$ data to enable a comparison between our data for the isotope ^{208}Pb to the best data available previously,[2,3] that for natural Pb. In Fig. 3 one can see the quality of the new data and the improvement possible with our new time-of-flight techniques.

REFERENCES

1. J. Rapaport, V. Kulkarni and R.W. Finlay, Nucl. Phys. A330, 15 (1979).
2. A.H. Hussein et al., Phys. Rev. C15, 233 (1977).
3. A. Begum and R.B. Galloway, Phys. Rev. C20, 1711 (1979).

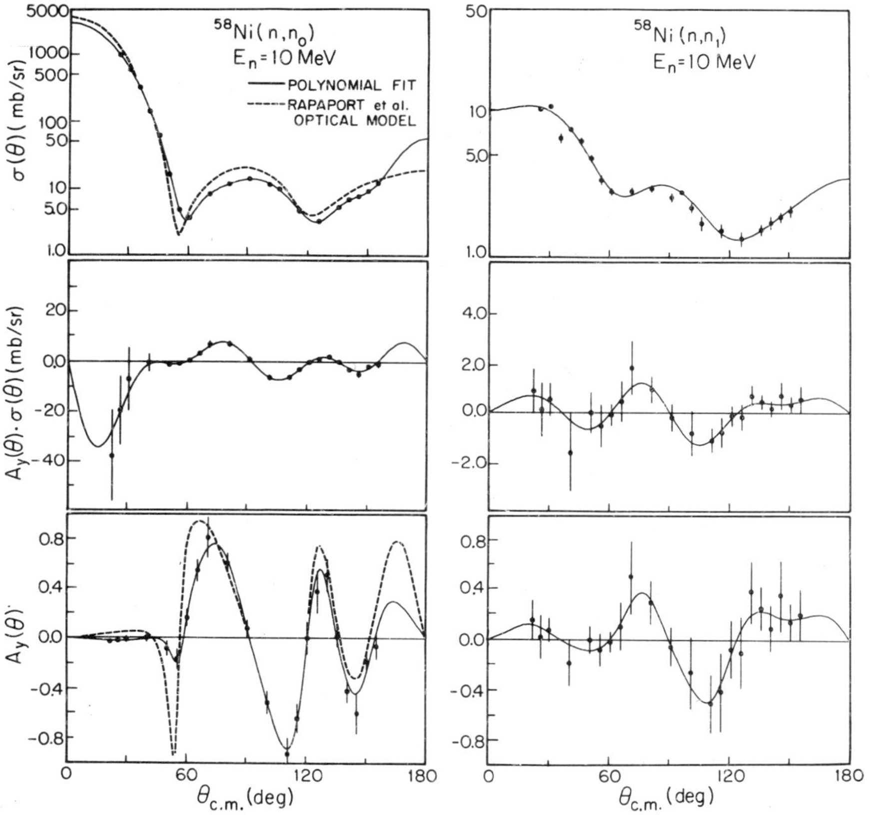

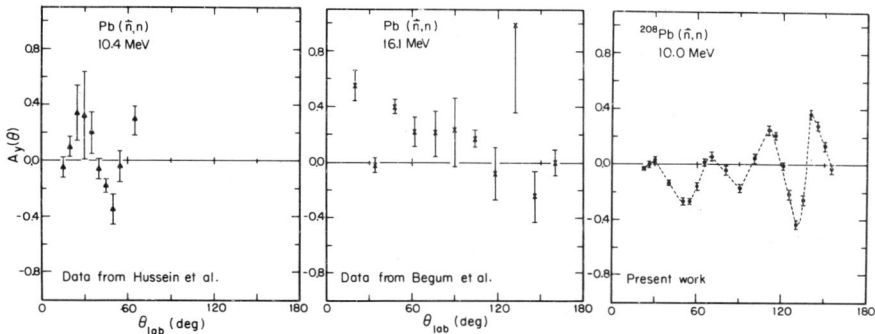

Fig. 2. Data and polynomial fits (solid curves) for ^{58}Ni + n for the functions $\sigma(\theta)$, $A_y(\theta) \cdot \sigma(\theta)$, and $A_y(\theta)$. Dashed curve is optical model prediction with the parameters of Rapaport et al.[1]

Fig. 3. Data of earlier Pb(n,n) measurements and present work.

THE ANALYZING POWER FOR ELASTIC SCATTERING OF 9.9, 11.9 AND 13.9 MeV NEUTRONS FROM Ca

W. Tornow, E. Woye, and G. Mack
Physikalisches Institut, Universität Tübingen,
D7400 Tübingen, West Germany

C.E.Floyd, K.Murphy, P.P.Guss, S.A.Wender, R.C.Byrd, R.L.Walter
Dept. of Physics, Duke University, Durham, NC 27706 and
Triangle Universities Nuclear Laboratory[*], Duke Station, NC 27706

T.B. Clegg
University of North Carolina, Chapel Hill, NC 27514 and TUNL

ABSTRACT

Using the polarization transfer reaction $^2\text{H}(\vec{d},\vec{n})^3\text{He}$ at 0° as a source of polarized neutrons, the analyzing power in elastic neutron scattering from Ca has been measured. Optical-model predictions based on global analyses differ appreciably from our present data. Preliminary optical-model studies indicate a lower strength and a much smaller diffuseness of the spin-orbit potential than usually assumed.

During the last few years neutron differential elastic scattering cross-section data have been obtained at several laboratories for a wide range of nuclei in the energy region from 5 to 25 MeV. In most cases these data have been analyzed in the framework of the optical model. Individual and global optical-model potential parameters have been reported. In the parameter search routines used to obtain best fits to the data, the spin-orbit potential parameters were normally kept constant at some "meaningful" values that were usually obtained from proton scattering and polarization analyses. The searches on the neutron cross-section data therefore gave little information about the true spin-orbit interaction for neutrons because of this constraint and because of the low sensitivity of cross sections to variations in the spin-orbit parameters. In particular, then, one cannot expect good polarization predictions from optical-model parameters derived by the analysis of cross-section data only.
On the other hand, it is well known that polarization observables are more sensitive than cross sections to individual scattering amplitudes. This feature was demonstrated most clearly in the pioneering work of Rosen et al.[1] in the case of proton analyzing power measurements on a wide range of nuclei of 10.5 and 14.5 MeV. Provided that accurate neutron elastic scattering data are available, optical-model analyses performed with neutron data should yield more detailed information than it is possible to obtain with protons as a probe. That is, due to the absence of the Coulomb potential term in the general optical potential, the number of parameters normally kept

[*]Work supported in part by U.S. Department of Energy

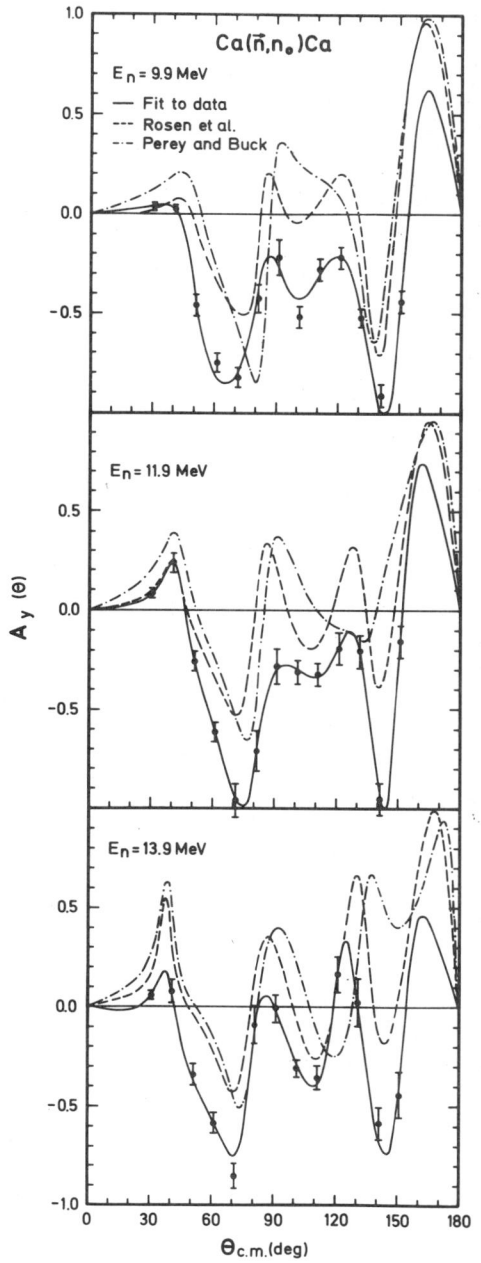

Fig. 1. Analyzing power $A_y(\theta)$ for Ca($\vec{n},n_o$)Ca in comparison to global optical-model predictions.

free in the fitting procedures is reduced, giving rise to a greater constraint on the remaining parameters.

Unfortunately, neutron analyzing power experiments are relatively sparse and imprecise compared to charged particle experiments. The reason for this situation has been that polarized neutron beams are much more difficult to produce and detect. However, since an intense polarized and pulsed deuteron beam is now available at TUNL, accurate neutron analyzing power experiments are possible on nuclei in the energy range 6-18 MeV using the polarization transfer reaction ^{2}H($\vec{d},\vec{n}$)^{3}He at 0° as a source of polarized neutrons.

In this paper we report on one of our first analyzing power experiments. The target was natural Ca (96.97% ^{40}Ca). The choice of Ca was motivated by two facts:

i) ^{40}Ca is considered to be a suitable nucleus for optical-model studies, due to the absence of strong coupling between the ground state and excited states.

ii) Using a T=0 nucleus, the asymmetry parameter ε = (N-Z)/A in the depth of the nucleon optical potential vanishes. By comparison to p-Ca optical-model analyses, the Coulomb correction term, which is usually parameterized as $\Delta U_c = 0.4 \ Z/A^{1/3}$, may be investigated directly.

The experimental setup and data-taking procedure were similar to that described in ref. 3. The TUNL Lamb-shift polarized ion source was used in the pulsed beam operating mode.[2] The Ca sample was a cylinder 2.5 cm in diameter by 2.5 cm in height. The scattered neutrons were detected by a pair of well-shielded detectors using standard time-of-flight techniques.

The raw data have been corrected for finite geometry and multiple scattering using Monte Carlo techniques. Our analyzing power results are given in fig. 1. The error bars shown represent the statistical uncertainty. The solid curves are based on fitting of $A_y(\theta)\sigma(\theta)$ with associated Legendre polynomials. Here $\sigma(\theta)$ is the differential elastic scattering cross section, which was measured at TUNL during the course of the present experiment. In fig. 1 the dashed and dashed-dotted curves represent optical-model calculations performed with the parameters given by Rosen et al.[1] and Perey and Buck[4], respectively. While the Rosen optical-model parameters produce a double-hump structure between 70° and 130° similar to that observed in our data, they fail to reproduce the magnitude. The optical-model parameters of Perey and Buck produce analyzing powers with both the wrong magnitude and a quite different shape between 70° and 130°. The same statement holds for the Becchetti and Greenless[5] optical-model parameters, although here the agreement with our data is even worse than with the predictions using the parameters of Perey and Buck. At 13.9 MeV the agreement between global optical-model predictions and our data is somewhat better than at the lower energies, as may be expected. However, a common feature of all the neutron-nucleus optical-model parameters usually referred to in the literature is that they yield less negative $A_y(\theta)$ values in the 40° to 110° range than the values observed in the present distributions.

Preliminary optical-model studies performed to fit the present data favor a spin-orbit potential which has a diffuseness that is extremely small ($a_s<0.3$fm). In addition the strength of the spin-orbit potential V_{so} is less than 4.5 MeV, lower than usually assumed. The spin-orbit radius parameter ($r_s=0.9$ fm) is in agreement with values expected. Further optical-model studies are in progress at Tübingen using a search code[6] which is capable of fitting simultaneously data sets at different energies.

REFERENCES

1. L. Rosen et al., Ann. Phys. (N.Y.) <u>34</u>, 96 (1965).
2. S.A. Wender et al., Nucl. Instr. and Methods (in press).
3. C.E. Floyd et al., contribution to this conference.
4. F. Perey and B. Buck, Nucl. Phys. <u>32</u>, 353 (1962).
5. F.D. Becchetti and G. Greenless, Phys. Rev. 182, 1190 (1968).
6. H. Leeb, private communication.

THE SCATTERING OF POLARIZED NEUTRONS FROM ^{9}Be BETWEEN 9 AND 15 MeV

C.E.Floyd, P.P.Guss, R.C.Byrd, K. Murphy, S.A.Wender, W.Tornow
and R.L.Walter
Department of Physics, Duke University, Durham, NC 27706 and
Triangle Universities Nuclear Laboratory,* Duke Station, NC 27706

T.B.Clegg and W.J.Thompson
University of North Carolina, Chapel Hill, NC 27514 and TUNL

ABSTRACT

Analyzing powers have been measured between 9 and 15 MeV for neutron scattering to the ground and 2.43-MeV (5/2$^-$) states of ^{9}Be. The $A_y(\theta)$ data vary smoothly with energy. Results at 10 MeV are compared to a Lane optical-model prediction and coupled-channel predictions.

A survey of the literature reveals a meager amount of polarization data for neutron scattering from nuclei at energies above 5 MeV. Our long-standing interest in neutron polarization phenomena has prompted the development of a neutron time-of-flight facility for acquisition of high accuracy analyzing power data. For neutron scattering studies, the polarized neutrons are produced in the ^{2}H(d,n)^{3}He reaction which is initiated with a pulsed polarized deuteron beam.[1] With our tandem Van de Graaff facility, this reaction provides an intense source of highly polarized monoenergetic neutrons over the energy range between 6 and 18 MeV. The first scattering experiment using the pulsed system involved elastic and inelastic scattering from ^{9}Be; reported here are results for $A_y(\theta)$ for incident neutron energies of 9, 10, 11, 13 and 15 MeV for the angular range from 30° to 150°. The distributions will be extended forward to 15° in the future.

In general, neutron data provide unique information concerning the tail of the nuclear potential through scattering at forward angles, whereas the Coulomb force dominates the corresponding proton scattering. In addition, for the present case, the neutron data give new insight into the ^{9}Be + nucleon interaction. In particular, if ^{9}Be is considered to be a pair of α particles with a loosely bound neutron, then the difference between the specifically nuclear force for n and p may involve only the loosely bound neutron. A comparison of proton and neutron scattering data may therefore be expected to provide insight into isospin dependence, as an incident proton may interact through the singlet and triplet states, while a neutron is restricted to the singlet state by Pauli blocking.

The data reported here were obtained using a two-detector spin-flip procedure to eliminate instrumental asymmetries. Neutrons produced in a 400-keV thick gas target were scattered from a cylindrical scattering sample at a distance of 9.5 cm. Flight paths of 3 and 4 m

*Supported in part by the U.S. Department of Energy

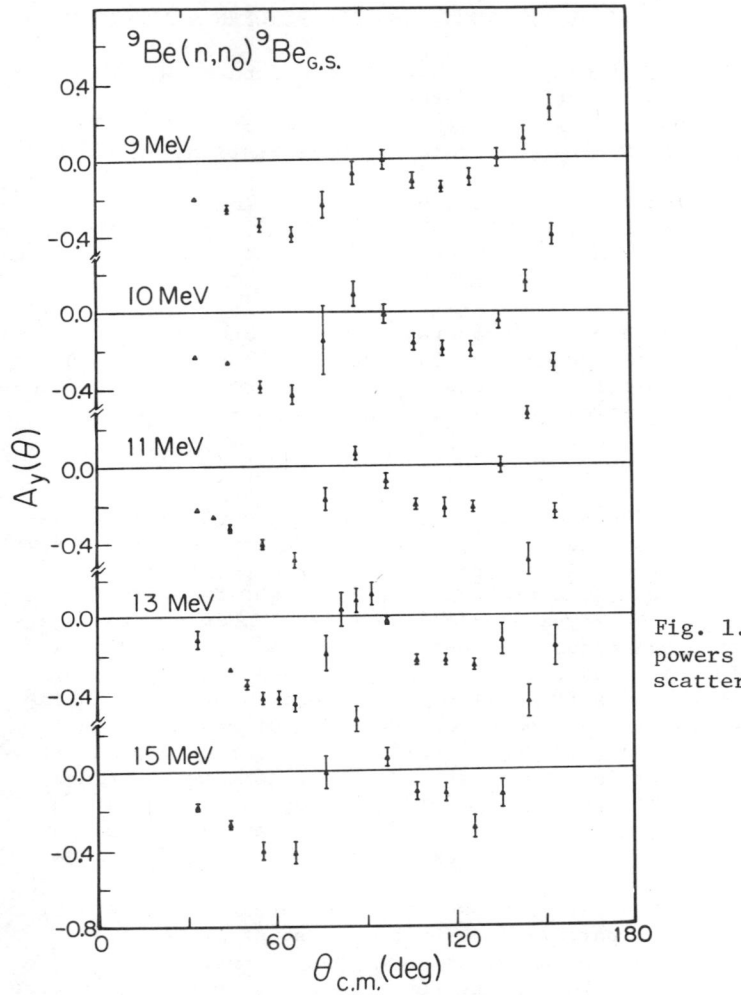

Fig. 1. Analyzing powers for elastic scattering from ^{9}Be.

for the two side detectors provided good separation for neutrons from the ground state and the 2.43 MeV state. The experimental results are shown in fig. 1. All corrections for background and breakup neutrons, finite geometry, and multiple scattering effects are included in these data.

The ^{9}Be + nucleon system has been studied previously by two theoretical techniques at TUNL. A Lane model (which neglects target spin) has been developed to describe the elastic and quasi-elastic processes in the (p,p), (n,n), and (p,n) reaction channels.[2] The model uses one potential to simultaneously fit the (p,p), (n,n), and (p,n) cross sections as well as the (p,p) and (p,n) $A_y(\theta)$ data. Parameters from this analysis were used to generate a prediction for the (n,n)$A_y(\theta)$ distribution. As shown in fig. 2, the prediction

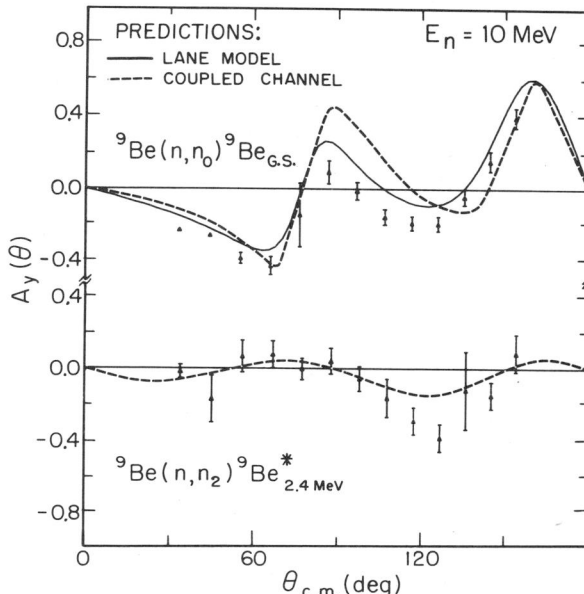

Fig. 2. Analyzing powers for elastic and inelastic scattering from ^{9}Be.

differs from the data at the positive maxima. Inclusion of the data reported here will provide a complete data set for our Lane model analysis.

Proton scattering from the ground and 2.43-MeV (5/2$^-$) states has been analyzed using a coupled-channels analysis. The coupled-channels parameterization is that of Votava et al.[3] except for two nominal modifications which resulted from analysis[4,5] of depolarization data and higher energy data: First, the deformation parameter β_2 is set equal to 1.05 and second, complex form factors are used. Predictions for neutron scattering to the two states are shown in fig. 2. The inelastic $A_y(\theta)$ is reproduced well, although differences exist between calculation and elastic data at the two positive maxima.

Analysis of our data will be extended in order to refine the understanding of the ^{9}Be + nucleon interaction, and to develop further our p-shell studies and the global optical model for the nucleon-nucleus interaction.

REFERENCES

1. S.A. Wender et al., Nucl. Inst. and Meth. (in press).
2. R.C. Byrd et al., Phys. Rev. Lett. <u>43</u>, 260 (1979); see also these proceedings.
3. H.J. Votava et al., Nucl. Phys. <u>A204</u>, 529 (1973).
4. J.S. Blair, M.P. Baker, H.S. Sherif, Phys. Lett. <u>60B</u>, 25 (1975) and private communication.
5. W.J. Thompson (unpublished).

OPTICAL-MODEL AND COUPLED-CHANNEL PREDICTIONS IN COMPARISON TO n-^{12}C ANALYZING POWER DATA

E. Woye, W. Tornow, and G. Mack
Physikalisches Institut, Universität Tübingen,
7400 Tübingen, West Germany

ABSTRACT

Optical model and coupled-channel predictions for the analyzing power $A_y(\theta)$ in n-^{12}C scattering are compared to our data between 9 and 15 MeV neutron energy. It is shown that optical-model and coupled-channel parameters based only on the analysis of cross-section data do not yield a satisfactory description of our present analyzing power data.

As described in another contribution to this conference[1], at Triangle Universities Nuclear Laboratory we have measured the analyzing power $A_y(\theta)$ for the reaction $^{12}C(\vec{n},n_{o,1})^{12}C$ for 7 energies from 9 to 15 MeV. In this paper we present calculations in order to compare with the data. In fig. 1 the dashed dotted curves show optical model calculations obtained by using parameters given by Thumm et al.[2] which were deduced from differential scattering cross-section measurements only. Rather good agreement at forward scattering angles and at the extreme backward angles is obtained. On the other hand the strong negative values around 120° present in the optical model calculation are not reproduced by our data. Around 10 MeV the disagreement observed for the whole backward angle region indicates the existence of resonance effects.

The coupled-channel calculations shown in fig. 1 as solid lines were made using parameters of Hogue et al.[3] which again were based on cross-section data alone. Here the agreement is relatively poor except for 9 MeV. At all the other energies the coupled-channel predictions start with positive values, in disagreement to the trend of our data and the optical model calculations. With increasing energy the strong negative extremum around 60°, which is present in all measured distributions, is slowly decreasing, whereas at around 120° another negative extremum is built up in the calculation much more pronounced than given by the data. As shown in fig. 2 the predictions to inelastic n-^{12}C scattering (dashed curves) also do not give satisfactory agreement with the data. New optical-model and coupled-channel studies are in progress in order to achieve a better fit to the present analyzing power data.

REFERENCES

1. E. Woye et al., contributions to this conference.
2. M. Thumm and H. Lesiecki, Z. Physik A 278, 77 (1976).
3. H.H. Hogue, "Elastic and Inelastic Scattering of Fast Neutrons by ^{6}Li, ^{7}Li, ^{9}Be, and ^{12}C", Ph.D. Thesis 1977, Duke University, unpublished.

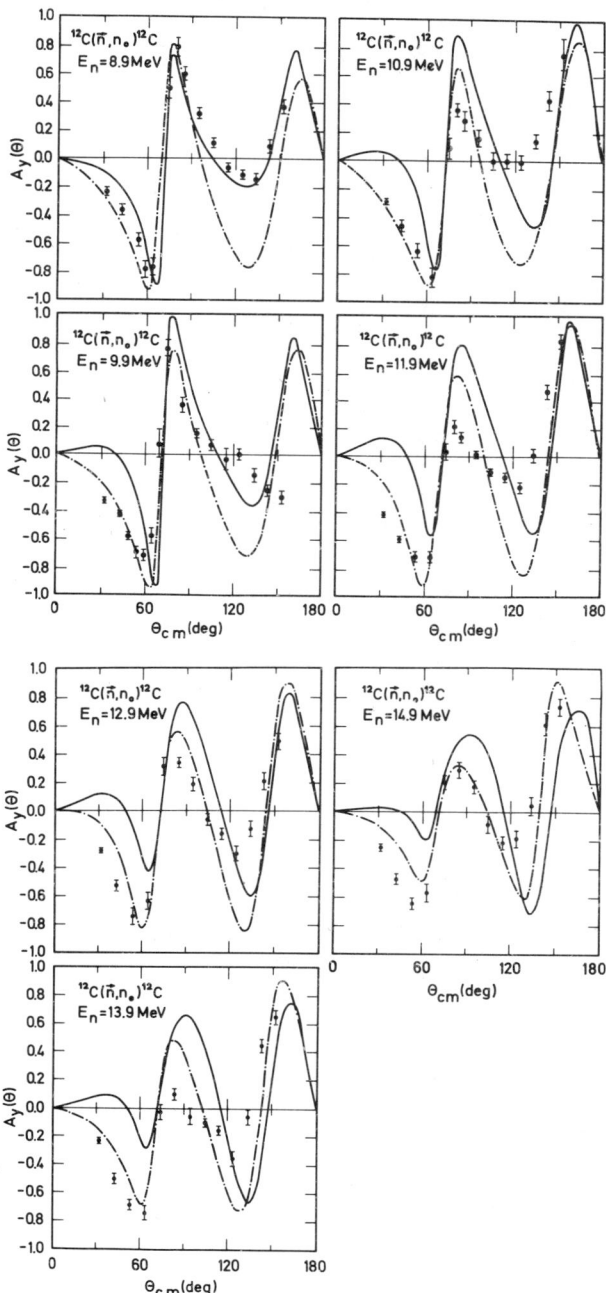

Fig. 1. Coupled-channel (solid lines) and optical model (dashed dotted curves) predictions to $^{12}C(\vec{n},n_o)^{12}C$ elastic scattering.

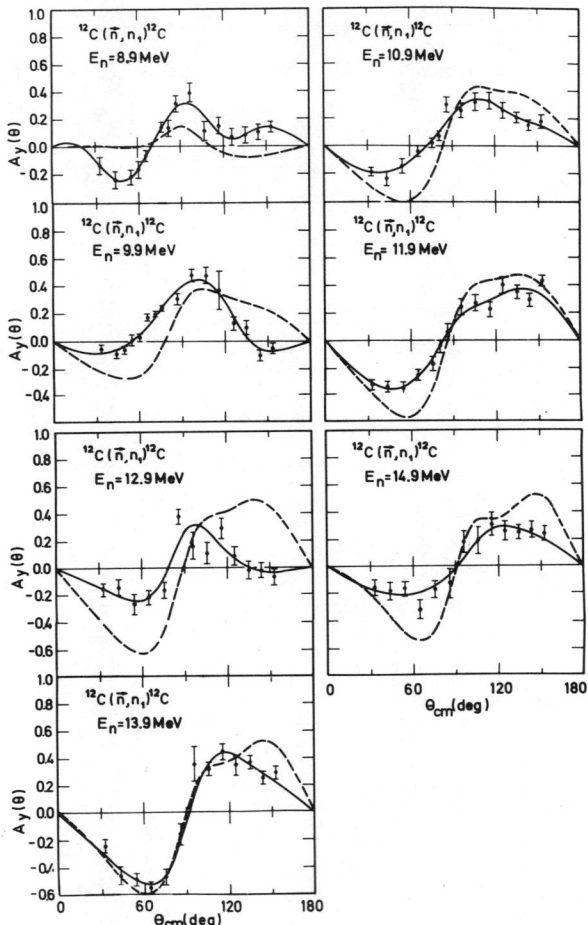

Fig. 2. Coupled-channel predictions (dashed curves) to $^{12}C(\vec{n},n_1)^{12}C$ inelastic scattering. The solid lines are fits with associated Legendre polynomials.

A PROGRAM OF SYSTEMATIC MEASUREMENT AND ANALYSIS FOR THE LOW-ENERGY NUCLEON OPTICAL-MODEL POTENTIAL

R. L. Walter and R. C. Byrd,
Department of Physics, Duke University, Durham, NC 27706 USA,
and Triangle Universities Nuclear Laboratory, Durham, NC

T. B. Clegg, E. J. Ludwig and W. J. Thompson,[*]
Department of Physics and Astronomy, University of North Carolina,
Chapel Hill, NC 27514 USA, and TUNL

ABSTRACT

A program of systematic measurement and analysis of differential cross sections and vector analyzing powers for (n,n), (p,n) and (p,p) reactions at energies below 16 MeV for a wide range of target nuclei is underway at Triangle Universities Nuclear Laboratory.

In more than a decade since the last systematic parameterization of the nucleon optical potential by Becchetti and Greenlees,[1] the techniques for obtaining reliable nucleon scattering data have improved greatly. For example, elastic-scattering analyzing powers in (p,p) are much more extensive, reliable and accurate than are most of the double-scattering polarization data used by Becchetti and Greenlees. Further, the availability of accurate (p,n) cross-section and analyzing-power data provides a strong constraint on the isospin components of the nucleon optical potential. Recently, the theory of the optical potential has advanced considerably,[2] so that extensive data are now needed to enable definitive tests in a bombarding energy region where the Schroedinger equation is adequate.

The experimental goals of the TUNL nucleon-scattering project are to obtain a large data base (~ 5000 data points) of cross section and vector analyzing power data over a wide range of target nuclei for nucleon energies below 16 MeV. Targets of high isotopic enrichment are being used, and as many isotopic sequences as possible are being investigated for both the n and p channels. Consistency of data acquisition procedures, data reduction, statistical accuracy, absolute normalization, solid-angle values, angular steps and ranges, are being maintained. Examples of results obtained so far are given in this proceedings in contributions by Floyd et al. for (n,n) on isotopes of Fe and Cu, by Byrd et al. for (p,n), and by Varner et al. for (p,p) on the Se isotopes.

---

[*]Research supported in part by the U.S. Department of Energy.

Among the goals of the theoretical analysis are to produce a reliable parameterization of the nucleon optical potential in its central, spin-orbit, and isospin components, using recent theories of the optical potential[2] as a guide. Goodness-of-fit criteria, beyond the traditional chi-squared-minimization, are being investigated, which is especially important because of the very large data base expected. In addition to a smooth parameterization as a function of target mass number and Z, shell, pairing, and collective effects are being investigated. The connection of the potential obtained in the low-energy region to the shell-model potential for bound states[3] and to potentials at higher energies is also of interest. The parameterization is also important in the construction of folding-model potentials for composite projectiles in the bombarding energy area of 10 MeV/amu.

The determination of appropriate bombarding energies to allow comparison of neutron and proton potentials is a problem. Perey[4] suggested that one should use a proton bombarding energy exceeding the corresponding neutron energy by the average Coulomb repulsion experienced by the proton. This suggestion can be tested if accurate $(\vec{n},n)$, $(\vec{p},n)$, and $(\vec{p},p)$ data are available. If the energy dependence of the local optical potential is the same for protons and neutrons, then one would not need to know the energy dependence of the potential. For 10-MeV neutrons, as currently being used, the appropriate proton energies exceed 16 MeV (the upper limit of the TUNL tandem accelerator) for all nuclei above calcium. Therefore collaboration with other laboratories with polarized proton beam facilities in the energy range 17 to 30 MeV would be desirable.

REFERENCES

1. F. G. Becchetti and G. W. Greenlees, Phys. Rev. $\underline{182}$, 1190 (1969).
2. "Microscopic Optical Potentials", ed. by H. V. von Geramb, Springer Lecture Notes in Physics, vol. 89, 1979.
3. S. G. Cooper and P. E. Hodgson, J. Phys. G $\underline{6}$, L21 (1980).
4. F. G. Perey, Phys. Rev. $\underline{131}$, 745 (1963).

POLARIZED PROTON BEAM STUDIES OF VERY NARROW RESONANCES

J. F. Wilkerson, W. J. Thompson, E. J. Ludwig and T. B. Clegg*
Department of Physics and Astronomy, University of North Carolina,
Chapel Hill, NC 27514, USA, and
Triangle Universities Nuclear Laboratory, Durham, NC

ABSTRACT

High-resolution polarized proton beam studies of very narrow nuclear resonances are shown to reduce the ambiguities in helicity-amplitude analyses. The examples of $^{12}C(p,p)^{12}C$ near 14.23 MeV and $^{32}S(\vec{p},p)^{32}S$ near 3.4 MeV bombarding energy are shown.

Observation and analysis of nuclear resonances yields valuable information on the properties of nuclear structure. Very narrow resonances, which can be described by a single-level Breit-Wigner amplitude and non-resonant scattering amplitudes, can be analyzed in several different ways; an optical-model analysis of the non-resonant scattering, a phase shift analysis, or a helicity amplitude description. For resonances at high bombarding and excitation energy in the compound nucleus, the optical-model analysis is unsuitable, while the phase-shift analysis becomes unmanageable because of the large number of complex phase shifts required. The helicity-amplitude method is a practical method of analysis for resonances which are sufficiently narrow that the non-resonant scattering amplitudes may be approximated as constant over the energy interval of the resonance excursion.

For scattering of spin-1/2 particles from spin-0 nuclei the differential cross-section, σ, and analyzing power cross section, σA_y, can be expressed in terms of helicity amplitudes as

$$\sigma = |a|^2 + |b|^2 \qquad (1)$$

$$\sigma A_y = 2\mathrm{Im}(a^*b) \qquad (2)$$

where a and b are non-spin-flip (NSF) amplitudes and spin-flip (SF) amplitudes respectively. For narrow resonances

$$a = a_b(\theta) + a_R(\theta,E) \qquad (3)$$

$$b = b_b(\theta) + b_R(\theta,E) \qquad (4)$$

where the off-resonant amplitudes a_b and b_b depend weakly on bombarding energy E. Here the resonance amplitudes

---

* Research supported in part by the U. S. Department of Energy.

$$a_R(\theta,E) = \frac{J+1/2}{k} \left[\frac{\Gamma_p/2}{E-E_R+i\Gamma/2} \right] P_L(\cos\theta) \quad (5)$$

$$b_R(\theta,E) = \frac{(-1)^{J-L-1/2}}{k} \left[\frac{\Gamma_p/2}{E-E_R+i\Gamma/2} \right] P'_L(\cos\theta) \quad (6)$$

where J and L are resonance spin and orbital angular momentum, k is the wave number at bombarding energy E, E_R is the resonance energy, and Γ and Γ_p are total and partial widths for the scattering.

Previous elastic-scattering work [1] using the helicity-amplitude method to analyze narrow nuclear resonances measured the cross-section excitation function σ, including the off-resonance values σ_b but measured A_y only off the resonances, as $(A_y)_b$. These data were sufficient to obtain a reliable analysis, but only because the elastic branching ratio Γ_p/Γ was available from different methods. However, for resonance analyses where Γ_p/Γ is unknown beforehand, the analysis of only σ, σ_b, and $(A_y)_b$ is

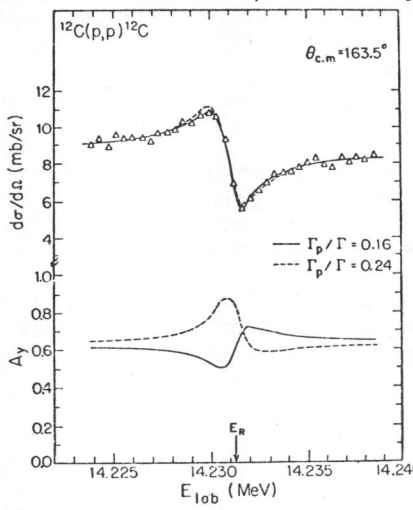

Fig. 1. Best fits to σ for the $^{12}C(p,p)^{12}C$ resonance at 14.23 MeV are compared for two different branching ratios with the off-resonance helicity amplitudes adjusted to optimize the fit for σ. Also shown are the corresponding theoretical predictions for A_y.

ambiguous, as shown in Fig. 1. The calculated fits to σ are equally good, but the predicted A_y are radically different. This increased sensitivity is to be expected, since σ contains cross terms between off-resonance and resonance NSF or SF amplitudes only, but the cross terms in σA_y are between NSF and SF amplitudes only. Thus, if a_b and b_b are adjusted to fit σ as Γ_p/Γ varies, quite different A_y values will be produced. Thus, measurement and analysis of A_y excitation functions adds an additional constraint to the helicity-amplitude analysis.

The acquisition of narrow resonance data with typical widths of 100 eV to 1000 eV requires a high-energy-resolution beam. By using the TUNL tandem accelerator high-resolution system [2] with the

TUNL Lamb-shift polarized ion source, a polarized proton beam of 50-80 nA current with 85% to 90% polarization is available on target with a typical beam resolution (FWHM) of 600 to 800 eV. To obtain polarized high-resolution beam requires the use of thin 2- to 4-μg/cm carbon stripping foils to insure minimum beam straggling with no depolarization. To increase the counting rates, large-area detectors are used. Typical σ and A_y are shown in Fig. 2 along with preliminary fits for the first T=3/2 state in

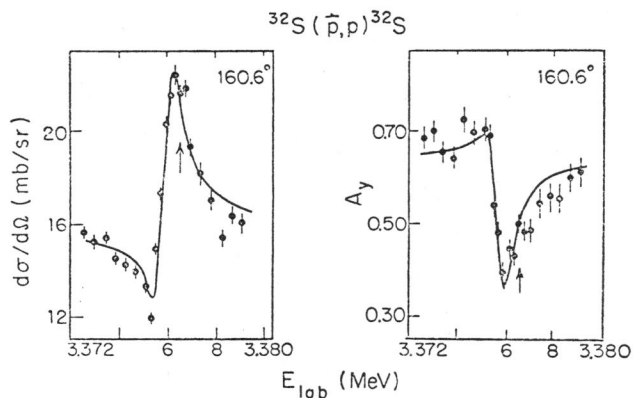

Fig. 2 The first T = 3/2 isospin forbidden resonance from $^{32}S(\vec{p},p)^{32}S$, with E_R = 3.377MeV, Γ = 90eV and Γ_p = 90eV. The arrow indicates E_R.

^{33}Cl populated in $^{32}S(\vec{p},p)^{32}S$. This resonance offers a good test of the helicity-amplitude method since Γ_p/Γ = 1. Currently, a study of higher-excitation, isospin-forbidden T=3/2 resonances in light nuclei is in progress at TUNL. Since Γ_p/Γ is usually unknown for these resonances, polarized high-resolution beams to measure excitation functions of A_y are necessary for an unambiguous analysis. Beam energy resolution, beam energy straggling in the target, Doppler broadening from target-atom vibrations, and atomic-excitation effects[3] also broaden these very narrow resonances, and must be carefully accounted for in the resonance analysis.

REFERENCES

1. P. G. Ikossi et. al., Phys. Rev. Lett. 36, 1357 (1976); Nucl. Phys. A274, 1 (1976).
2. E. G. Bilpuch, in Proc. of the 4th Conf. on the Application of Small Accelerators, (IEEE, New York, 1976), ed. by J. L. Duggan and I. L. Morgan, p. 380.
3. W. J. Thompson et. al., contribution to this conference, and to be published.

THE DEPTH OF THE IMAGINARY POTENTIAL AROUND A = 105 DETERMINED BY PROTON SCATTERING *)

A. Feigel, E. Finckh, B. Rowedder, K. Rüskamp, H. Scheuring,
U. Schneidereit, P. Tröger
Physikalisches Institut der Universität Erlangen-Nürnberg, W.-Germany

A very careful investigation of (p,n) cross sections by Johnson et al.[1] resulted in an anomaly of the imaginary potential. With all other parameters nearly unchanged, the depth of the potential rises from W≈5 MeV at A=89 to values around 30 MeV at A≈105 and falls again to W = 10MeV at A = 120. Other investigations confirmed these results[2] and found another anomaly at lower A-values[3]. All these experiments used sub-Coulomb-energies. An evaluation of neutron data at somewhat higher energies[4] showed systematically smaller values than the average at magic numbers, i.e. A = 90 and A = 120.

A completely independent method to determine the imaginary potential is the evaluation of the analysing power of elastically scattered protons. The absorption of the incoming wave damps the pattern of the analysing power and its amplitude is therefore strongly W-dependent. The investigation of tin isotopes[5] and of ^{109}Ag [6] were in excellent agreement with the (p,n)-data.

Here we report on measurements of ^{103}Rh, ^{107}Ag and ^{106}Cd. They were made with the Erlangen Lamb-shift ion-source, the EN-tandem and the large scattering chamber. It was equipped with 8 movable detectors on each side, 2 monitor-detectors at θ = 20° and a Faraday-cup combined with a polarimeter. The 20 detectors were routed, and the spectra measured in a PDP 11/40 computer using the FORCAM-programme EDDA[7]. The beam intensity on the target was 20 nA, the polarization 75 %. The targets were selfsupporting foils with a thickness of about 1 mg/cm².

The analysing powers were measured at energies around E_p = 6 MeV and the maximum amplitude was about $5 \cdot 10^{-3}$ in the three isotopes (fig. 1).

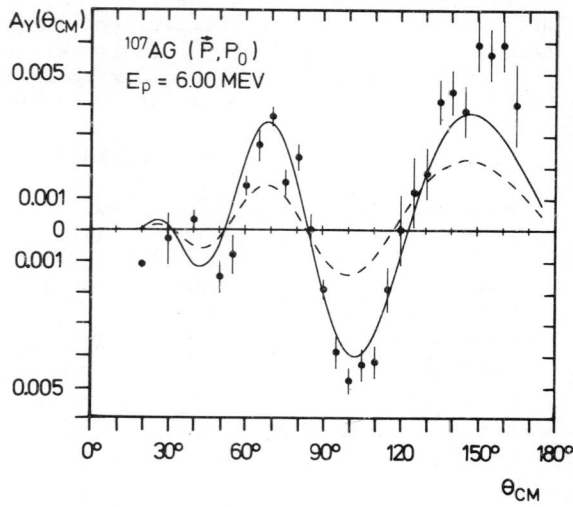

Fig. 1. Analysing power of the reaction ^{107}Ag(p,p$_0$) at E_p^{lab}=6.00 MeV. Dashed curve calculated with the parameters from the (p,n) investigation, full curve W_D=16 MeV, a_D=0.4 fm. (table 1).

The parameters found in the (p,n) investigation give amplitudes of the analysing powers which are definitely to small (dashed curve in fig. 1). The value W = 97 for Rhodium would result in $A_{y,max} \leq 10^{-3}$, experimentally it was found $A_{y,max} = 6 \cdot 10^{-3}$. The result for ^{106}Cd cannot be compared with (p,n)-data since our measurement is below the (p,n) threshold.

The failure in determining W_D and a_D separately in proton scattering seems to occur only for sub-Coulomb energies. Measurements at E_p = 8 and 10 MeV have no longer this ambiguity. At E_p = 10 MeV which is about the height of the Coulomb barrier, the a_D value reaches the usual value $a_{\bar{D}}$ = 0.6 fm.

In fig. 2 we plotted the volume integral of the imaginary potential in the region between A = 89 - 130. Of course, the (p,n) cross section data show the anomaly described above (full curve). The (p,p$_0$) analysing powers also result in smaller values at the magic numbers but have apparently no other systematic departure from an average. The dotted line in fig. 2 gives the volume integral calculated with the parameter set of Becchetti and Greenlees. Outside the magic numbers our data scatter around this line.

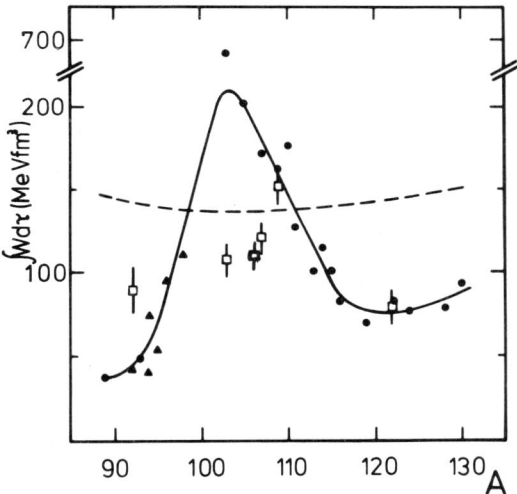

Fig. 2. Volume integral of the imaginary potential versus mass number using the parameters of Johnson et al.[1] (dots), Flynn et al.[7] (triangles), and our parameters (squares) from the (p,p$_0$) scattering.

Table I Optical model parameters as given by Johnson et al. and determined by the analysing power. The real potential is given by the formula $V_R=V_0-0.32\ E+0.4\cdot Z/A^{1/3}+24(N-Z)/A$. (Johnson: $0.45\cdot Z/A^{1/3}$). All potential depths in MeV, all lengths in fm.

	^{103}Rh E_p=5.8 MeV		^{107}Ag E_p=6.0 MeV		E_p=10.5 MeV	^{106}Cd E_p=5.6 MeV
	ref.	(p,p_0)	ref	(p,p_0)	(p,p_0)	(p,p_0)
V_R	60.84	63.56	60.98	62.41	56.0	63.28
r_R	1.2	1.17	1.2	1.20	1.20	1.17
a_R	0.73	0.75	0.73	0.75	0.75	0.75
W_D	97	12.5	22.5	16.07	10.50	13.33
r_I	1.3	1.37	1.3	1.32	1.32	1.36
a_I	0.59	0.37	0.47	0.40	0.60	0.40
V_{SO}	6.4	6.2	6.4	6.2	6.2	6.2
r_{SO}	1.03	1.01	1.03	1.01	1.01	1.01
a_{SO}	0.63	0.71	0.63	0.75	0.75	0.75
r_C	1.22	1.21	1.22	1.21	1.21	1.21

REFERENCES

1. C.H. Johnson, A. Galonsky, and R.L. Kernell, Phys. Rev. C20, 2052 (1979)
2. D.S. Flynn, R.L. Hershberger, and F. Gabbard, Phys. Rev. C20, 1700 (1979)
3. S. Kailas, M.H. Mehta, S.K. Gupta, Y.P. Viyogi, and N.G. Ganguly, Phys. Rev. C20, 1272 (1979)
4. G. Eder, H. Leeb, H. Oberhummer, J. Phys. G3, L 127 (1977)
5. W. Drenckhahn, A. Feigel, E. Finckh, G. Gademann, K. Rüskamp, M. Wangler, and L. Zemło, Nucl. Phys. A339, 13 (1980)
6. A. Feigel, E. Finckh, U. Weise, Phys. Rev. C June 80
7. P. Urbainsky, Jahresbericht 77/78 Phys. Inst. Erlangen-Nürnberg, S. 160

*) Work supported by the Deutsche Forschungsgemeinschaft

INVESTIGATION OF THE REACTION MECHANISM IN THE ELASTIC SCATTERING OF POLARIZED PROTONS ON ^{27}Al *)

W. Kretschmer, J. Jordan, H. Löh, and W. Stach
Physikalisches Institut der Universität Erlangen-Nürnberg, W.-Germany

The enhancement of the compound-elastic cross section σ^{CE} over the Hauser Feshbach[1] cross section σ^{HF}, expressed by $W_{\alpha\alpha} = \sigma^{CE}/\sigma^{HF}$, has been subject of considerable theoretical[2-4] and experimental[5] efforts. In this contribution we report about the determination of $W_{\alpha\alpha}$ for the elastic scattering on ^{27}Al in an energy region where many strongly absorbing channels are open.

Differential cross section $\sigma(\theta)$ and analyzing power $A(\theta)$ were measured[6] for ^{27}Al($\vec{p},p_0$) in the energy range 6.0-10.5 MeV in steps of 250 keV at scattering angles from 50°-160°. The experimental energy averaging by the target (180 keV energy loss for 10 MeV protons) was not sufficient to wash out the Ericson fluctuations and so an artificial energy average was performed. As an example the measured and averaged excitation functions of σ and $\sigma \cdot A$ for $\theta_{Lab}=90°$ and 160° are shown in fig. 1.

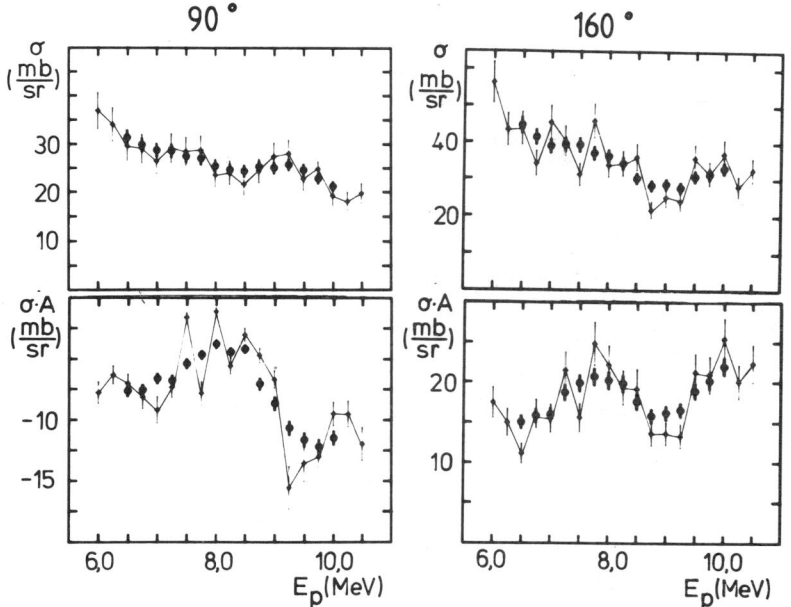

Fig. 1. Excitation functions of σ and $\sigma \cdot A$ for ^{27}Al($\vec{p},p_0$) at θ_{Lab} = 90° and 160°. Small dots: experiment; big dots: average over 1 MeV.

In the energy average the differential cross section $\langle\sigma\rangle$ can be described as a sum of the direct cross section σ^{DI} and the compound-elastic cross section σ^{CE}, whereas the polarization dependent cross section $\langle\sigma \cdot A\rangle$ is independent of compound-elastic contributions[7]. So a separation of σ^{DI} and σ^{CE} is possible, if the direct scattering is known from a description of $\langle \sigma A \rangle$.

The optical potential for the calculation of σ^{DI} has been obtained by a simultaneous fit of all angular distributions $\langle \sigma A \rangle$, where only the real central potential and the imaginary potential are linearly energy dependent. The best fit potential (V_R=58.0, β_R=-0.4, r_R=1.13, a_R=0.75; W_D=4.9; β_I=0.25; r_I=1.21; a_I=0.49; V_{SO}=9; r_{SO}=1.20; a_{SO}=0.62 in the usual units) describes $\langle \sigma \cdot A \rangle$ at least in the lower energy region ($E_p \leq 8$ MeV) quite well (four examples are shown in fig.2).

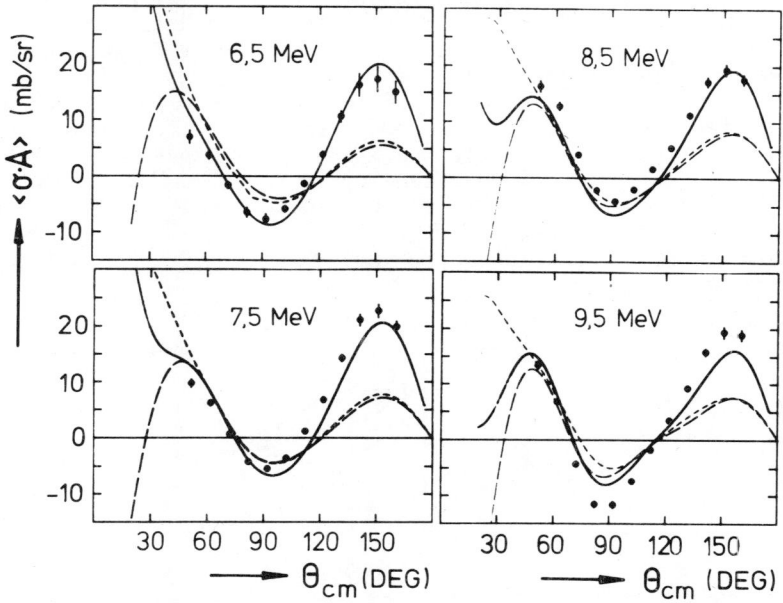

Fig.2. Angular distributions of $\langle \sigma A \rangle$ for ^{27}Al($\vec{p},p_0$) at E_p=6.5-9.5 MeV (best fit——, Perey[8]---, Becchetti and Greenlees[9] — $\cdot$ —).

The compound-elastic cross section σ^{CE}, shown in fig.3 for E_p= 6.5 - 9.5 MeV is obtained by subtracting σ^{DI} from the energy averaged experimental cross section $\langle \sigma \rangle$. The errors for σ^{CE} reflect the experimental uncertainties and optical model ambiguities for the calculation of σ^{DI} as well. The dashed curves in fig.3 are Hauser Feshbach calculations with all exit channels treated explicitly, the transmission coefficients are calculated with the best fit potential (for protons) and with potentials from the literature (for neutrons and alpha particles). The full curves correspond to $W_{\alpha\alpha}$=2. The compound-elastic enhancement factor $W_{\alpha\alpha} = \langle \sigma^{CE}_{exp}(\theta)/\sigma^{HF}(\theta)\rangle$ is obtained as an average over all angles and energies. For the energy region below 8.5 MeV we get $W_{\alpha\alpha}$=2.02±0.07 which is in good agreement with the results of ref.5 for ^{30}Si(p,$\bar{p}_0$) and with theoretical predictions. For the higher energy region $E_p \geq 8.5$ MeV there is a systematic deviation of $\sigma^{CE}_{exp}(\theta)$ from the full curve giving $W_{\alpha\alpha}$=1.51±0.10. This deviation may be due to an intermediate structure[10] which is not explicitely taken into account in the optical model and therefore this latter value of $W_{\alpha\alpha}$ should not be taken too seriously.

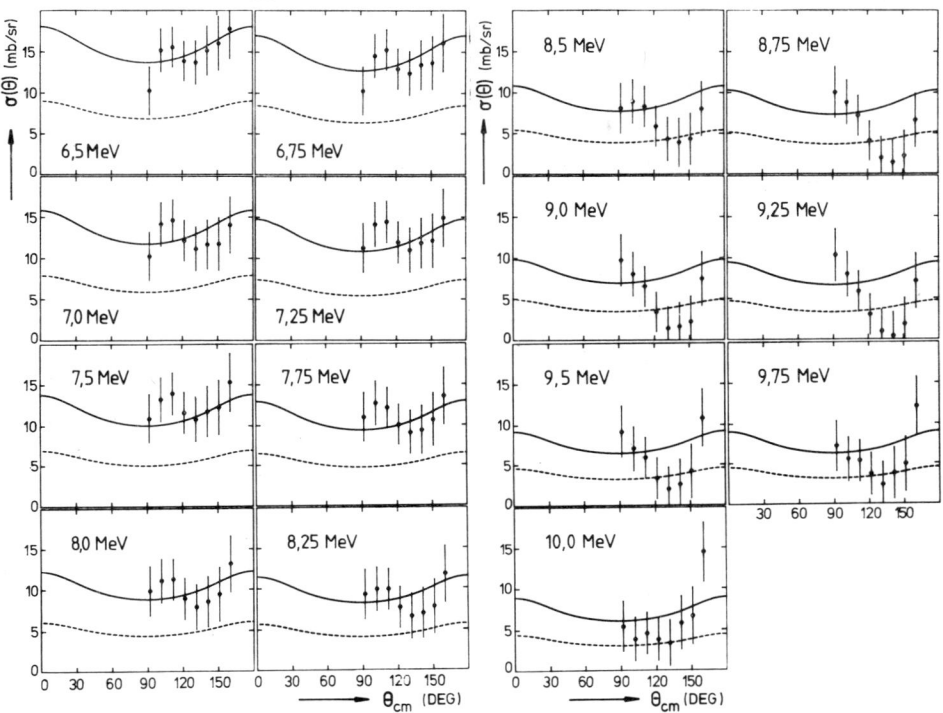

Fig. 3. Compound-elastic cross section for ^{27}Al(p,p$_0$) compared with Hauser-Feshbach calculations.

1. W. Hauser and H. Feshbach, Phys. Rev. <u>87</u>, 366 (1952)
2. H.M. Hofmann, J. Richert, J.W. Tepel, and H.A. Weidenmüller, Ann. Phys. (N.Y.) <u>90</u>, 403 (1975)
3. C. Mahaux and H.A. Weidenmüller, Ann. Rev. Nucl. Part. Sci. <u>29</u>, 1 (1979)
4. P.A. Moldauer, preprint 1980
5. W. Kretschmer and M. Wangler, Phys. Rev. Lett. <u>41</u>, 1224 (1978)
6. J. Jordan, diplom thesis, Erlangen 1979
7. W. Kretschmer and G. Graw, Phys. Rev. Lett. <u>27</u>, 1294 (1971)
8. F.G. Perey, Phys. Rev. <u>131</u>, 745 (1963)
9. F.D. Becchetti and G.W. Greenlees, Phys. Rev. <u>182</u>, 1190 (1969)
10. C. Glashausser, A.B. Robbins, E. Ventura, F.T. Baker, J. Eng, and R. Kaita, Phys. Rev. Lett. <u>35</u>, 494 (1975)

*) Work supported by the Deutsche Forschungsgemeinschaft

PROTON OPTICAL POTENTIAL FOR THE ELASTIC SCATTERING ON MOLYBDENUM ISOTOPES *)

W. Kretschmer, E. Heitz, J. Jordan, H. Löh, W. Schuster, W. Stach,
R. Stingl, P. Urbainsky, and M.B. Wango
Physikalisches Institut der Universität Erlangen-Nürnberg, W.-Germany

For energies below the Coulomb barrier Johnson et al.[1] have found an anomalous behaviour of the optical potential for $89 \leq A \leq 130$ with a resonance-like increase of the absorption W_D near A=103 and an unusually small a_D=0.4 fm. This small imaginary diffuseness is now well established in the sub-Coulomb region[2,3], whereas the strong increase of W_D is confirmed via (p,n)-measurements on Zr and Mo isotopes[2], but not confirmed by low energy analyzing power measurements[4].

It is the aim of the present investigation to extend the energy region above the Coulomb barrier to see whether the absorption still shows an anomalous behaviour as a function of A or whether it agrees with the Becchetti, Greenlees[5] potential. We have measured differential cross section $\sigma(\theta)$ and analyzing power $A(\theta)$ of the elastic scattering of polarized protons on ^{92}Mo, ^{96}Mo, ^{100}Mo at E_p=8.5, 11, and 12 MeV in the angular range $20° \leq \theta_{Lab} \leq 170°$. The measurement was performed in a large 4π scattering chamber with 7 detectors in the forward region on the right side and 8 detectors in the backward region on the left side with one detector overlap at about 90°, the target position was fixed throughout to 45° with respect to the beam, two monitor detectors were placed at θ_{Lab}=±20°. As targets we used rolled foils of isotopically enriched Mo with a thickness of about 1 mg/cm², the exact cross section normalization was obtained from a measurement of the Rutherford scattering at 4 MeV. To avoid false asymmetries due to a movement of the beam on the target, the polarization was switched on and off with a frequency of 10 Hz, the absolute value of it (P~65-70%) was measured continuously with a ^{4}He-polarimeter.

As an example the angular distributions of cross section and analyzing power at E_{Lab}=12 MeV are shown in fig. 1. The effect of increasing A on the elastic scattering is clearly demonstrated both in $\sigma(\theta)$ and $A(\theta)$. For both observables the oscillations are strongly damped and shifted towards forward angles reflecting the increasing strength of the imaginary potential and the increasing radius. The curves in fig. 1 were calculated with the code MAGALI in the following way: starting with the parameters of Becchetti, Greenlees[5] a grid search on a_D with the fit parameters V and W_D was performed, the corresponding X^2 are shown in fig. 2 and the best fit values are listed in table I for all energies. The other parameters didn't improve the fit very much, so the corresponding values of ref.5 were kept fixed.

As a summary we have the following striking results from these experiments above the Coulomb barrier: For this sequence of Mo isotopes the diffuseness a_D and the imaginary volume integral show a definite increase with increasing mass number. A comparison with the corresponding Becchetti, Greenlees values for J_W/A shows an agreement for ^{96}Mo, higher values for ^{100}Mo and lower values for ^{92}Mo which has

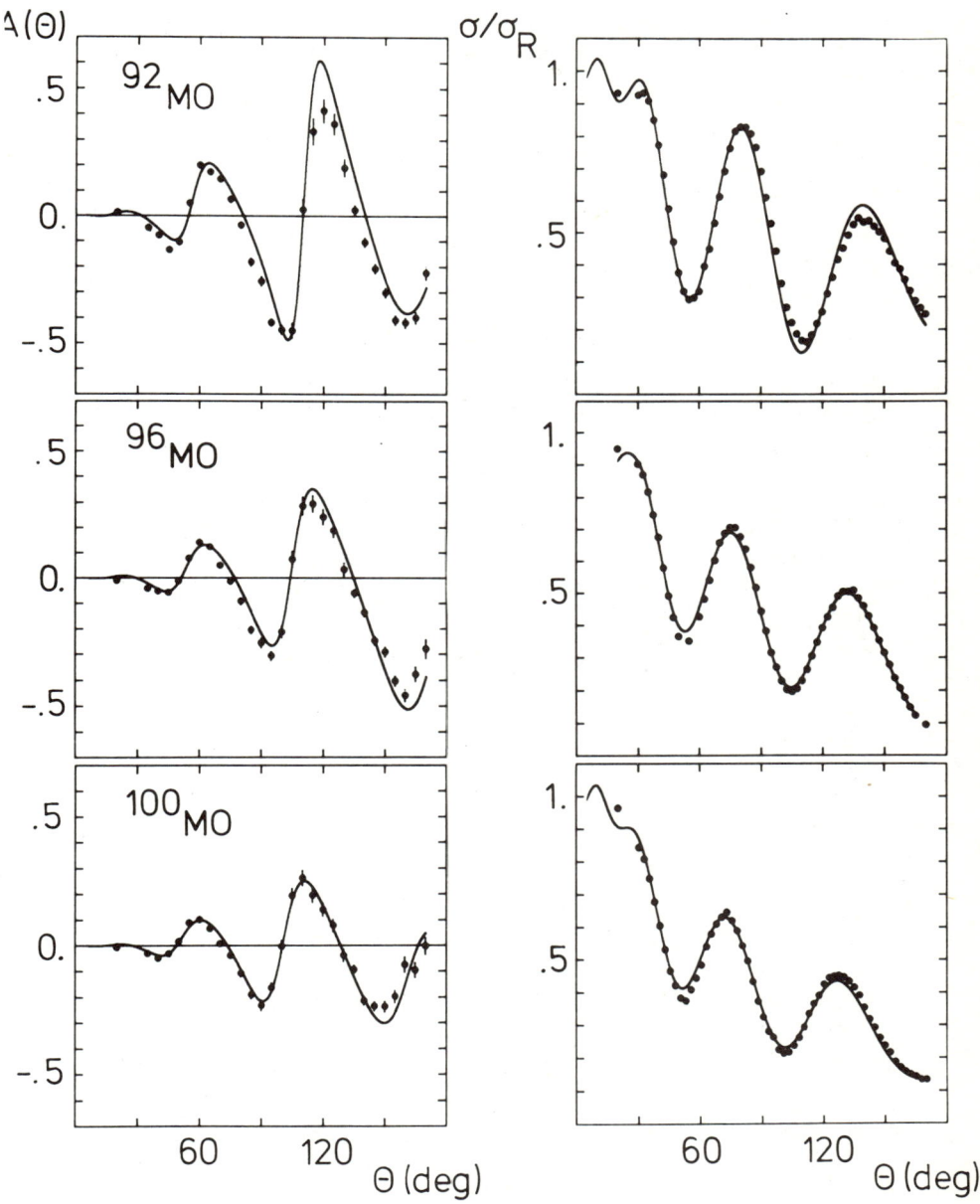

Fig.1: $\sigma(\theta)$ and $A(\theta)$ for the elastic proton scattering on ^{92,96,100}Mo at E_p= 12 MeV with optical model curves.

the closed neutron shell (N=50). This tendency is further supported by a preliminary analysis of ^{108}Pd($\vec{p}$,p$_0$) at 12 MeV giving J_W/A=150 MeVfm^3. This analysis confirms a suggestion by Lane et al.[6] that the imaginary part of the potential may be anomalously small near a closed shell.

Table I Results of a grid search for a_D, V, and W_D and derived magnitudes (real and imaginary volume integrals J_V/A and J_W/A)

		^{92}Mo			^{96}Mo			^{100}Mo		
E_p(MeV)	8.5 a)	11	12	8.5	11	12	8.5	11	12
V (MeV)	59.2	58.4	57.4	58.3	57.8	57.0	58.7	56.5	56.7
W_D(MeV)	10.9	10.2	7.63	7.13	8.27	8.45	8.75	9.4	8.55
a_D(fm)	0.4	0.45	0.55	0.75	0.65	0.70	0.80	0.8	0.8
J_V/A(MeVfm^3)	468	462	454	461	457	451	464	447	448
J_W/A(MeVfm^3)	81	86	79	103	102	113	136	146	132
$J_W/A_{Becc.}$	107.1	100.2	97.7	112.6	105.3	102.8	117.5	109.8	106.8

a) because of the high (p,n)-threshold compound-elastic contributions to $\sigma(\theta)$ cannot be neglected, therefore $\sigma \cdot A$ was fitted.

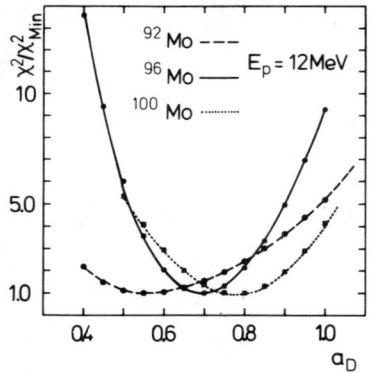

Fig. 2 X^2 for an a_D grid search with V and W_D as fit parameters.

1. C.H. Johnson, A. Galonsky, and R.L. Kernell, Phys. Rev. Lett. 39, 1604 (1977)
2. D.S. Flynn, R.L. Hershberger, and F. Gabbard, C20, 1700 (1979)
3. W. Drenckhahn, A. Feigel, E. Finckh, G. Gademann, K. Rüskamp, M. Wangler, and L. Zemło, Nucl. Phys. A339, 13 (1980)
4. A. Feigel, E. Finckh, B. Rowedder, K. Rüskamp, H. Scheuring, U. Schneidereit, P. Tröger, contribution to this conference
5. F.D. Becchetti and G.W. Greenlees, Phys. Rev. 182, 1190 (1969)
6. A.M. Lane, J.E. Lynn, E. Melkonian, and E.R. Rae, Phys. Rev.Lett. 2 424 (1959)

*) Work supported by the Deutsche Forschungsgemeinschaft

OPTICAL POTENTIAL FOR p + ^{208}Pb SCATTERING BELOW THE COULOMB-BARRIER*)

W. Kretschmer, K.H. Frank, E. Heitz, J. Jordan, H. Löh, W. Schuster,
K. Spitzer, W. Stach, P. Urbainsky, and M.B. Wango
Physikalisches Institut der Universität Erlangen-Nürnberg, W.-Germany

In a previous publication[1] we have measured the analyzing power $A(\theta)$ of proton elastic scattering on ^{208}Pb at 9 and 10 MeV with an accuracy of about 10^{-3}. The combined analysis of these data with cross section ($\sigma(\theta)$) data at 11-14 MeV of Eck and Thompson[2] and of $\sigma(\theta)$ and $A(\theta)$ data at 13 MeV of Rathmell and Haeberli[3] resulted in a stronger energy dependence of the real volume integral per nucleon compared with the higher energy behaviour given by van Oers et al.[4]. It is the purpose of the present investigation to extend the low energy data basis to get a more reliable information about the optical potential in this energy region. We have measured the differential cross section at E_p=9, 10, 11, and 12 MeV from θ_{Lab} = 30° - 170° in steps of 2.5° with a thin rolled target foil fo 1mg/cm² and the analyzing power at E_p=11 and 12 MeV in the same angular range in steps of 5° with a thick rolled target foil fo 10 mg/cm². The experimental method has been described in ref.1 and in another contribution to this conference[5], the essential point for the measurement of these small asymmetries was a fast (10Hz) on-off switching of the beam polarization thus avoiding false asymmetries due to a correlated beam movement on the target.

The data, presented in fig. 1 show the expected behaviour: at low energy the cross section is mainly given by the Rutherford value σ_R and the analyzing power is very small ($< 5\cdot 10^{-3}$ at 9 MeV), whereas with increasing energy higher partial waves contribute, producing more structure in $\sigma(\theta)$ and bigger values in $A(\theta)$ (up to $3\cdot 10^{-2}$ at 12 MeV). This tendency is well reproduced by an optical potential (see table I) whose main features are a small imaginary volume integral J_W/A of about 56 MeVfm³ and an increased energy slope for the real volume integral J_V/A of about -3.6 fm³. This increased slope disagrees with the global set of Becchetti, Greenlees[6] (-2.4 fm³) and with the multienergy analysis at higher energies of van Oers et al.[4] (-1.87 fm³), but it is in accordance with a theoretical prediction of

Table I Volume integrals J_V/A and J_W/A for ^{208}Pb(p,p$_0$) from this analysis and from the global set of Becchetti, Greenlees[6].

E_p (MeV)	this analysis		Becchetti, Greenlees[6]	
	J_V/A (MeVfm³)	J_W/A (MeVfm³)	J_V/A (MeVfm³)	J_W/A (MeVfm³)
9	494.5	59.1	462	120.7
10	491.9	50.6	459.6	118.2
11	488.8	59.6	457.2	115.7
12	483.6	54.6	454.8	113.2

Jeukenne et al.[7]. The anomalously small absorption (factor two smaller than for the global set) may be a hint for shell effects due to the

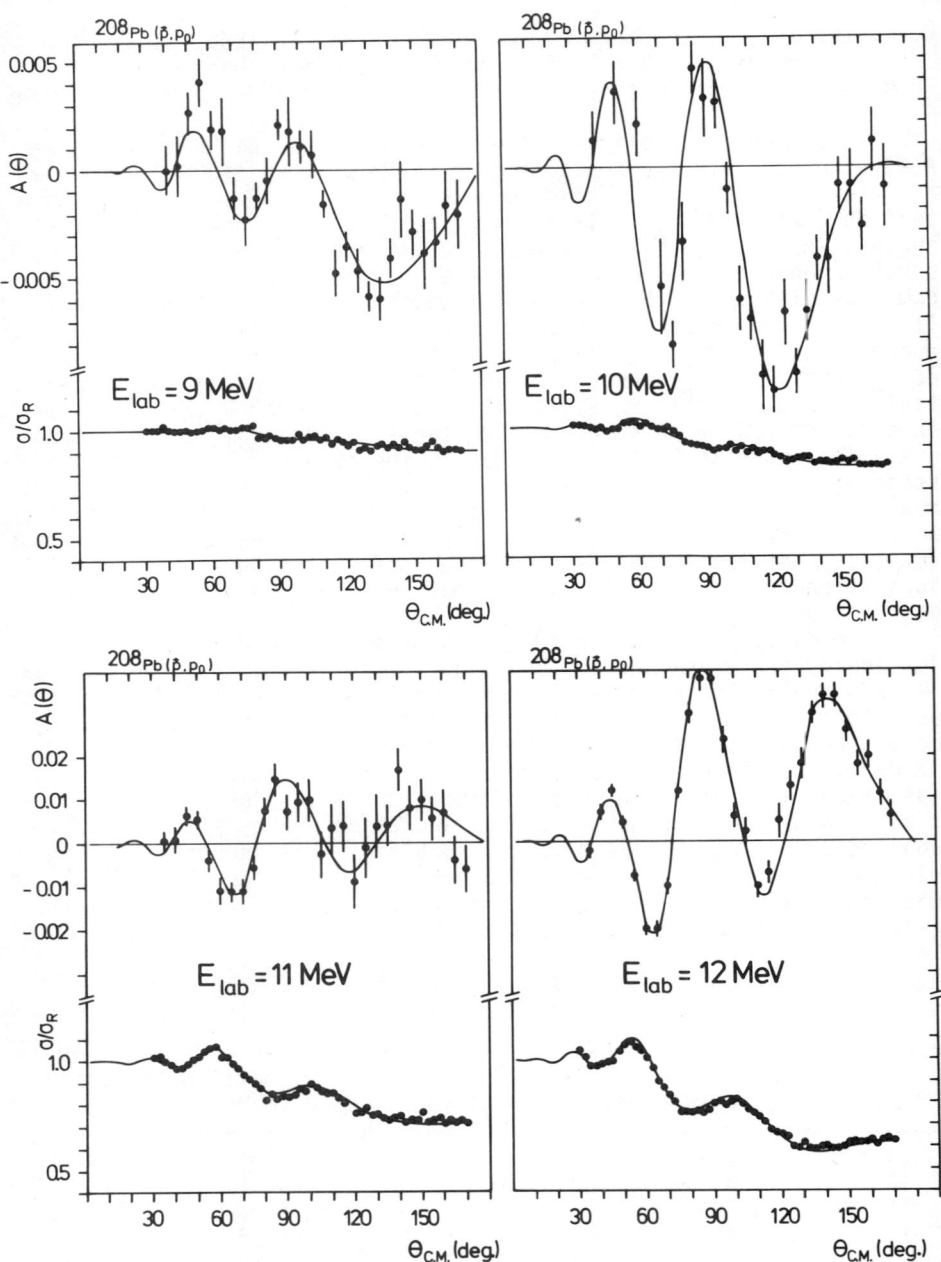

Fig.1. Analyzing power and differential cross section for ^{208}Pb($\vec{p}$,p_0) at E_p = 9, 10, 11, and 12 MeV.

the small level density at magic numbers, which have already been seen by Eder et al.[8] for the neutron-nucleus potential and which have been suggested by Lane et al.[9] for the nucleon-nucleus optical potential.

1. W. Kretschmer, H. Löh, K. Spitzer, and W. Stach, Phys. Lett. 87B 343 (1979)
2. J.S. Eck and W.J. Thompson, Nucl. Phys. A237, 83 (1975)
3. R.D. Rathmell and W. Haeberli, Nucl. Phys. A178, 458 (1972)
4. W.T.H. van Oers, H. Haw, N.E. Davison, A. Ingemarsson, B. Fagerström, and G. Tibell, Phys. Rev. C10, 307 (1974)
5. W. Kretschmer, E. Heitz, J. Jordan, H. Löh, W. Schuster, W. Stach, R. Stingl, P. Urbainsky, and M.B. Wango, contribution to this conference
6. F.D. Becchetti and G.W. Greenlees, Phys. Rev. 182, 1190 (1969)
7. J.P. Jeukenne, A. Lejeune, and C. Mahaux, Phys. Rev. C16, 80 (1977); C15, 10 (1977)
8. G. Eder, H. Leeb, and H. Oberhummer, J. Phys. G: Nucl. Phys. 3, 127 (1977)
9. A.M. Lane, J.E. Lynn, E. Melkonian, and E.R. Rae, Phys. Rev. Lett. 2, 424 (1959)

*) Work supported by the Deutsche Forschungsgemeinschaft

POLARIZED PROTON SCATTERING FROM Se ISOTOPES

R. L. Varner, J. F. Wilkerson, W. J. Thompson, Y. Tagishi,*
E. J. Ludwig, T. B. Clegg and B. L. Burks +
Department of Physics and Astronomy, University of North Carolina,
Chapel Hill, NC 27514 USA, and
Triangle Universities Nuclear Laboratory, Durham, NC

ABSTRACT

Elastic-scattering data for $^{A}Se(p,p)^{A}Se$ with A=76,78,80,82 at 12 MeV are reported for the scattering-angle range 27.5° to 167.5°. The data are compared with predictions from a global optical-model potential and with a deformed-potential calculation which uses parameters consistent with neutron scattering from the same isotopes.

The purpose of these polarized-proton scattering studies is two-fold: (1) To enable comparison of the isotope effects observed for (d,d) scattering from the Se isotopes[1,2] with those for (p,p); (2) To provide high-quality cross-section and analyzing-power data as part of a program of systematic measurement and analysis for the low-energy nucleon optical potential[3], and for comparison with neutron scattering from the same nuclei[4]. Although inelastic-scattering data have also been obtained, we present here only the elastic-scattering data.

The data were obtained by using a polarized beam of protons from a Lamb-shift ion source, accelerated in the TUNL tandem accelerator to impinge on isotopically-enriched (>96%) targets of the even-A Se isotopes from the A=76 to 82. The targets ranged in thickness from 170 to 500 µg/cm. Four detector pairs were placed at angles 17.5° apart, to the left and right of the beam. The data were taken at 2.5° intervals, from 27.5° to 167.5°, with beam spin up and down at each angle. Overlap points were taken at several angles from 45° to 115°. Two monitor detectors at 15° served as an independent check on the beam current integration and target thickness. The quench ratio was used to determine the beam polarization to within $\sim 1\%$ relative accuracy and $\sim 3\%$ overall accuracy. The elastic-scattering cross section, σ, and analyzing power, A_y, data are shown in Fig.1.

The calculations shown in Fig.1 lead us to conclude that: (1) The isotopic dependence of the present (p,p) data is in accord with the global potentials for nucleon scattering[3], and does not show the anomalous behavior reported in Ref.1 for deuteron scattering. (2) The isospin dependence of the nucleon-scattering potential

---

* On leave from University of Tsukuba, Ibaraki 305, Japan.
+ Research supported in part by the U. S. Department of Energy.

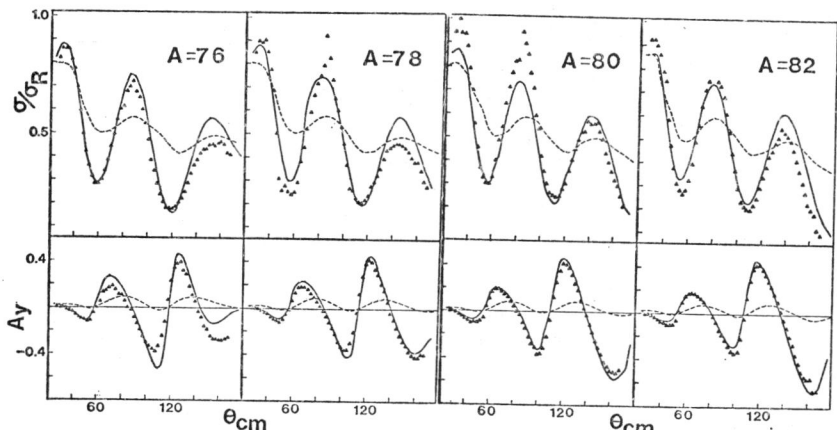

Fig.1. Se(p,p) Se σ and Ay data for the isotopes A=76,78,80 and 82. The σ statistical error is <1% and the normalization error is <5%. The Ay error is <1%. The solid curves are predictions from the global potential of Ref.5. The dashed curves are from coupled-channels calculations using the nucleon-scattering potentials of Ref.4.

deduced in the coupled-channels analysis of (n,n) and (n,n') cross-section data for the same Se isotopes in Ref.4 does not describe our (p,p) data, possibly because the absorption potentials are too large.

We are extending these (p,p) measurements to a wide range of target nuclei, and to bombarding energies between 12 and 16 MeV. Inelastic-scattering data for many of these nuclei are also being obtained and analyzed.

REFERENCES

1. H. R. Bürgi et al., Nucl. Phys. A334, 413 (1980) and references therein.
2. Y. Tagishi et al., "A Test of Collective Effects in Se(d,d)Se", this conference.
3. R. L. Walter et al., "A Program of Systematic Measurement and Analysis for the Low-Energy Nucleon Optical Potential", this conference.
4. J. Lachkar et al., Phys. Rev. C14, 933 (1976).
5. F. G. Becchetti and G. W. Greenlees, Phys. Rev. 82, 1190 (1969).

ANALYSIS OF ELASTIC SCATTERING
WITH AN L-DEPENDENT OPTICAL POTENTIAL[x]

P.J. van Hall, Eindhoven Univ. of Techn., Netherlands
R.S. Mackintosh, Open Univ. Milton Keynes, Great Brittain
A.M. Kobos, Inst. Nucl. Physics, Krakow, Poland
W.M.L. Moonen, Eindhoven Univ. of Techn. Netherlands

It is well known, that second and higher order processes play an important role in nuclear reactions. In the analysis of experimental data they often are accounted for by adjusting parameters in effective or phenomenological interactions. So it has been shown[1] that the intermediate pick-up channels (p,d)(d,p) can give significant conyributions to the elastic scattering. This effect can be simulated by adding a l-dependent part to the conventional optical potential[2]. We have decided to analyze our 20 MeV data obtained for Fe, Ni and Zn nuclei in terms of this potential and in this note we will give some examples.

The l-dependent part has been parametriced as follows:

$$U_1(r) = \{1+\exp((l^2-L^2)/\Delta^2)\}^{-1} \cdot$$
$$\left[V_1 g(r,r_3,a_3) + iW_1 g(r,r_4,a_4) \right]$$
$$g(r,r_x,a_x) = -4a_x \frac{d}{dr} f_{sw}(r,r_x,a_x)$$

in which $f_{sw}(r,r_x,a_x)$ is the Woods-Saxon form factor. The effect of $U_1(r)$ is that it adds to the normal potential an extra surface peaked term for the lower partial waves with $l \lesssim L$. Due to the large number of parameters (19) the searching process converged rather slowly and secondary minima were found. As an example we got for all the Zn isotopes a minimum near $U_1 \sim 0$ (i.e. the "normal" potential).

Here we present our results for ^{56}Fe and ^{70}Zn. Fig. 1 compares the results of calculations with and without a l-dependent part in the optical potential, while table I gives the resulting parameters for U_1. It can be seen from this figure and from the χ^2 values in the table that the quality of the fit, especially for ^{56}Fe, has been improved substantially.

From this work a few conclusions can be drawn. We always found a repulsive V_1 (the minus sign in table I). Though table I suggests a decrease in V_1 when going from ^{56}Fe to ^{70}Zn, we did not find (up till now) evidence for a systematic trend. This potential is felt by most partial waves of interest since $L \sim kR \sim 5$. It therefore is a bit puzzling that the radius of W_1 consequently is very large (~ 2) so it is hardly seen. It would be of interest to make a detailed comparison of the phase shifts from optical model fits with and without an l-dependent part.

[x] Supported in part by F.O.M./Z.W.O.

Table I

	^{56}Fe	^{70}Zn
L	5.2	5.1
Δ	2.6	1.3
V_1	-15.3	-5.3
r_3	1.08	1.02
a_3	0.44	0.56
W_1	0.15	0.16
r_4	1.76	2.38
a_4	0.21	0.67
$\chi^2_\sigma/N\sigma$	0.86(5.1)	5.0(7.5)
χ^2_p/Np	0.85(6.5)	2.3(4.8)

Between parentheses the χ^2 values from an l-independent fit.

REFERENCES

1 R.S. Mackintosh, Nucl. Phys. A230 (1974) 195
2 A.M. Kobos and R.S. Mackintosh, Journ. of Physics G 5 (1979) 97

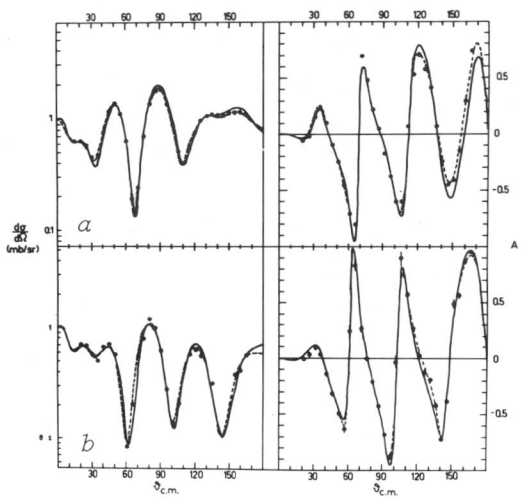

Fig. 1.
Analysis with an L dependent potential
a : ^{56}Fe, b: ^{70}Zn
full curve : L independent
dashed curve: L dependent

NUCLEAR STRUCTURE EFFECTS IN POLARIZED PROTON SCATTERING[x]

P.J. van Hall, W.H.L. Moonen and S.D. Wassenaar,
Eindhoven University of Technology, Eindhoven, Netherlands

In this note we present an effect that emerged when we inspected the optical model parameters obtained from analyses of our $(\vec{p},p)$ experiments around 20 MeV[1]. Moreover we also considered some data from the literature[2] for nuclei with A $\approx$ 90. All nuclei have A $\approx$ 50 and one may expect that no CN effects disturb the general pattern.

It turned out that the geometry of the absorptive potential showed a striking A dependence, which has been visualized in fig. 1. The anticorrelation between r_I and a_I is very clear and also the occurence of closed shells is quite apparent.

First we made sure that this picture was not accidental, due to some ambiguity between r_I and a_I. Such an ambiguity is clearly suggested by the strong anticorrelation, which more or less keeps $r_I^2 a_I$ (the volume integral for surface absorption) constant. At this place it should be said that we did not find a clear indication of an A dependence of the volume integral. We therefore analysed ^{54}Fe with the geometry of ^{70}Zn and vice versa. Fig. 2 shows the results. It is clear the that inclusion of polarization data plays a decisive role to establish that the effect is real. This may be the reason that it was not found in recent work[3] with 35.2 MeV protons, which did not include polarization data.

This A dependence may be connected with the collectivity of a specific nucleus. Though the deformation parameters do not change drastically going e.g. from Cd to Sn, the excitation energy of the 2^+ state, which is related to the softness, changes by a factor of two. A less drastic change of 30% appears when going from Ni to Zn.

Another explanation may be found in the fact that a proton energy of 20 MeV is rather low and thus most of the nonelastic processes take place with the valence nucleons. For the magic numbers 28 and 50 these are in the f 7/2 resp. g 9/2 orbits, which give a density located at the nuclear surface. When e.g. 2p nucleons begin to play a role the absorption takes place more inside the nucleus and over a wider region (r_I smaller, a_I larger).

REFERENCES

[1] S.D. Wassenaar, thesis Univ. Eindhoven, 1980 (to be published)
P.J. van Hall et al. Nuch. Phys. A291 (1977) 63
[2] P.J. van Hall, contribution to this conference
[3] C. Glasshauser et al. Phys. Rev. 184 (1969) 1217
E. Fabrici et al. Phys. Rev. C21 (1480) 844

[x] Supported in part by F.O.M.-Z.W.O.

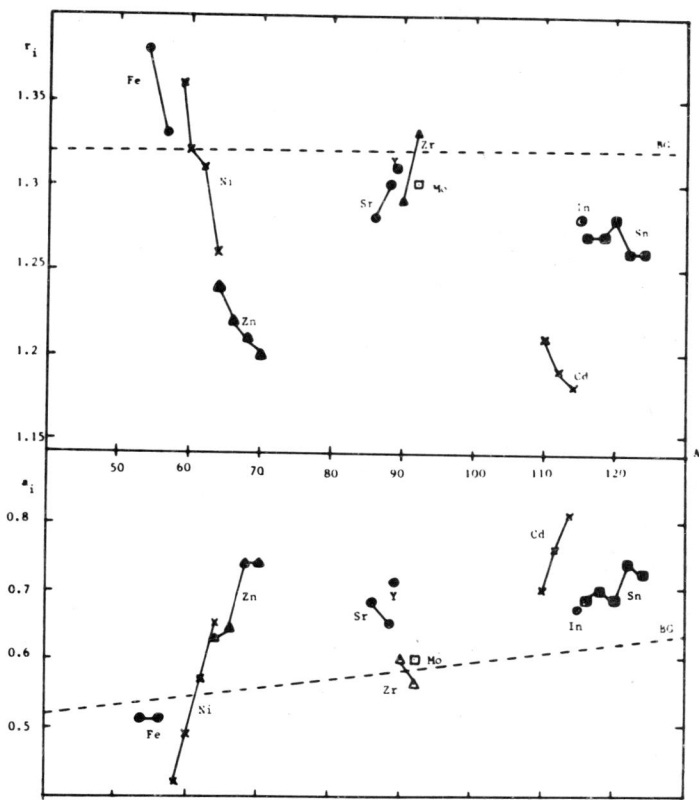

Fig. 1.
Dependence of r_I and a_I on the mass number
dashed curves : dependence according to Becchetti and Greenlees

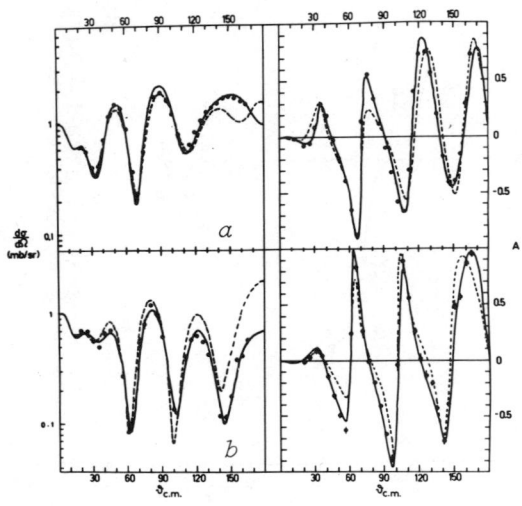

Fig. 2.
a : ^{54}Fe full curve ^{54}Fe geometry
        dashed curve ^{70}Zn geometry
b : ^{70}Zn full curve ^{70}Zn geometry
        dashed curve ^{54}Fe geometry

ANALYZING POWER IN PROTON-NUCLEUS ELASTIC SCATTERING IN THE SMALL ANGLE REGION AT 65 MEV

T. Matsusue, K. Imai, H. Shimizu, J. Shirai, K. Nisimura
Department of Physics, Kyoto University, Kyoto, Japan

K. Hatanaka, T. Saito
Research Center for Nuclear Physics, Osaka, Japan

By the investigation of the proton-nucleus elastic scattering at forward angles including the Coulomb-nuclear interference region, we can expect much information about nuclear forces especially at nuclear surface. However, in order to extract information about nuclear force from experimental data the effect of electromagnetic interaction must be estimated. The effect on analyzing powers, $A(\theta)$ in the neutron-nucleus scattering has been investigated since early time,[1] but that in the proton-nucleus scattering has not been investigated because it has been considered small compared with experimental accuracies. Recently G. Bendiscioli et al. measured $A(\theta)$ and $\sigma(\theta)$ at 36.2 MeV using double scattering method.[2] But their accuracies were not sufficient for detailed analyses and the situation was confusing. We have measured $A(\theta)$ in the proton-nucleus elastic scattering at forward angles with an accuracy of 10^{-3} for several targets and investigated the effect of electromagnetically induced spin-orbit force, electromagnetic LS force.

Analyzing powers have been measured with the polarized proton beam from the AVF cyclotron at the Research Center for Nuclear Physics, Osaka University. Cross sections have been also measured simultaneously. The beam intensity on the target was $\sim$30nA in average and the beam polarization was $\sim$0.7, which was monitored continuously by observing p-C elastic scattering at 47.5°. The sign of spin direction was changed by several Hz. A schematic view of the experimental set-up for solid targets is shown in Fig. 1. Target position 1 was used at larger angles than θ_{lab}=5° and target position 2 was used at smaller angles. NaI(Tl) detectors were located symmetrically to the incident beam. The backward detectors were used as monitors of experimental condition. The angular acceptance of the slits was ±0.13°, but for ^{64}Ni, ^{90}Zr and ^{208}Pb ±0.28° at larger angles than $\theta_{lab}\sim$5°. For gas targets the scattering chamber was isolated by Havar foils of 2.5μm and 5μm at the entrance and the exit, respectively. A double-slit system was used to limit the angular acceptance to ±0.4° maximum. Intrinsic Ge detectors instead of NaI detectors were used for monitors to estimate the level of impurities in the target gas. The pressure was 1.14 and 1.18 atms for ^{16}O and ^{40}Ar, respecively. The beam spread was $\sim$2.5mm FWHM at each target position. The angular divergence due to multiple scatterings in the targets and foils and the incident beam spread was estimated to be 0.16-0.32°. The accuracy of angle setting was better than 0.05°. The spectra of scattered protons with and without ^{90}Zr target at the smallest angles are shown in Fig. 2.

0094-243X/81/690437-3$1.50 Copyright 1981 American Institute of Physics

Angular distributions of $A(\theta)$ are shown in Fig. 3. There are no striking structures of angular distributions and are some smooth dependences on target masses in $A(\theta)$. As the target mass increases, the maximum of $A(\theta)$ at $\theta_{cm} \sim 7°$ decreases and $A(\theta)$ at larger angles than $\theta_{cm} \sim 7°$ decreases and comes to have zero crossing. At smaller angles than $\theta_{cm} \sim 7°$ for ^{208}Pb there is a dip structure. There is no oscillatory behavior as pointed out by G. Bendiscioli et al.

In order to estimate the effect of electromagnetic LS force the nuclear phase shifts were calculated with the optical potential parameters[3] and the phase shifts due to electromagnetic LS force with Coulomb distorted partial wave Born approximation. The electromagnetic LS potential is approximated to $-\frac{Ze^2}{2M_pMc^2}\frac{(g_p-1)}{r^3} \vec{L}\cdot\vec{S}$ on the assumption of point-like particles. Here g_p denotes the magnetic moment of proton in unit of $\frac{e\hbar}{4m_pc}$. The potential is negative just like the nuclear LS potential but has a radial dependence of r^{-3}. For the partial waves between l_{min} and l_{max} the sum of the nuclear phase shift and the electromagnetic LS phase shift was taken as the overall phase shifts. The impact parameter of l_{min} was taken to be the radius of the nucleus and that of l_{max} was taken to be the Bohr radius divided by Z. With these phase shifts, observables were reconstructed and the contribution of electromagnetic LS force were estimated. The results were sensitive to l_{min} but the sensitivity was not so large at forward angles. So the results seem valid at forward angles. The dependence of the results on l_{min} is shown in Fig. 4 and the results for several targets is shown in Fig. 5. The effect of electromagneitc LS force on $A(\theta)$ is sizable and similar for all targets. The effect is additive and maximal around the Coulomb-nuclear interference region. The maximal contribution to $A(\theta)$ reaches ~ 0.03. These results show the effect of electromagnetic LS force should be considered on precise analyses.

We finaly note the possibility of using the proton-nucleus elastic scattering in the Coulomb-nuclear interference region for polarimeters on the view point of figure of merit as shown in Fig. 6.

REFERENCES

1. J. Schwinger, Phys. Rev. 73, 407 (1948).
2. G. Bendiscioli et al., Nucl. Phys. A307, 22 (1978).
3. H. Sakaguchi et al., RCNP Ann. Rep. in 1978 p.12.

FIGURE CAPTIONS

Fig. 1 Schematic view of the experimental set-up.
Fig. 2 Energy spectra of protons at $\theta_{lab}=2.5°$ with and without ^{90}Zr target.
Fig. 3 $A(\theta)$ in the proton-nucleus elastic scattering at forward angles at 65 MeV.
Fig. 4 The effect of electromagnetic LS force on $A(\theta)$ for ^{24}Mg

Fig. 5 target. ΔA denotes the shift of A(θ) when the effect of that force is included in calculation. Three different values of l_{min} were taken as shown in the figure. The effect of electromagnetic LS force on A(θ) for several targets. The solid and the broken lines denote the results of calculation without and with the effect of electromagnetic LS force, respectively.

Fig. 6 Figure of merit for polarimeter, defined to be $\sigma(\theta) A(\theta)^2$.

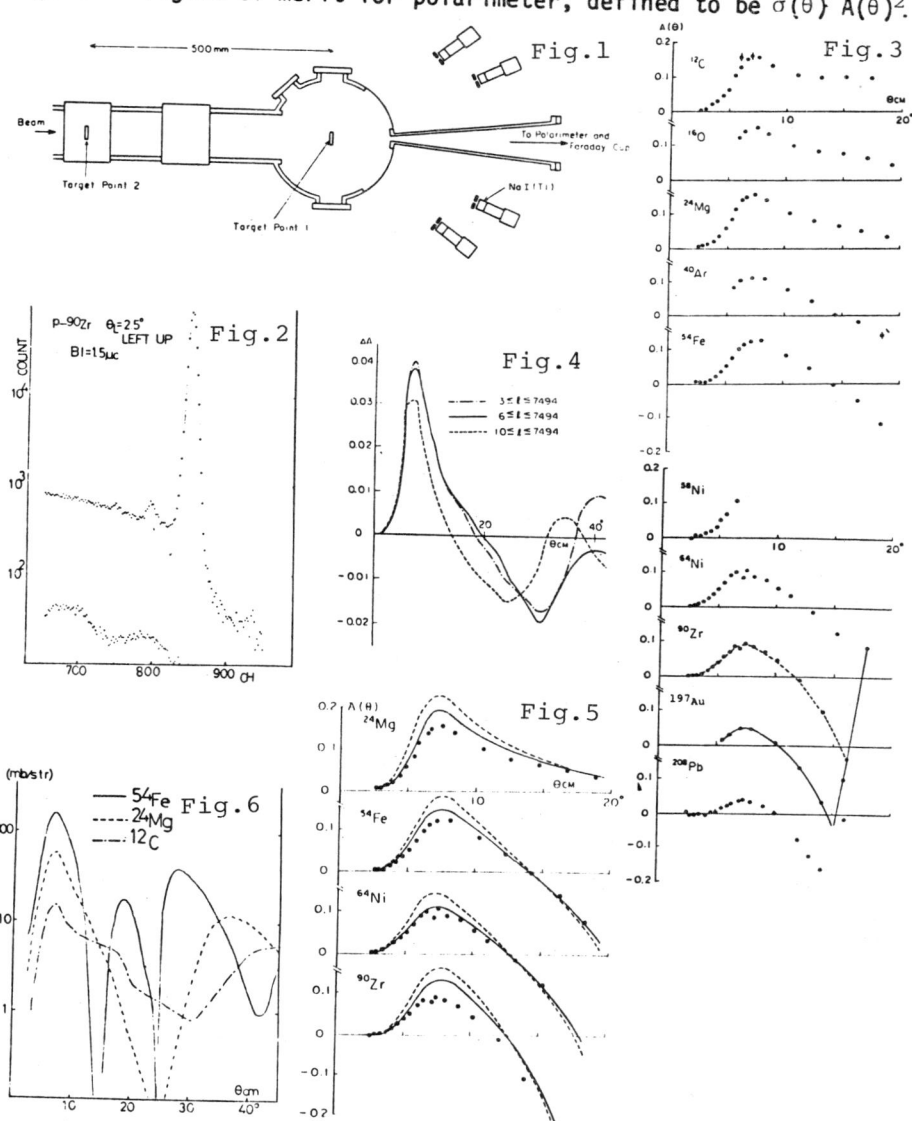

THE SHELL EFFECT OBSERVED IN THE SPIN ORBIT PART
OF THE OPTICAL POTENTIAL

H. Sakaguchi, M. Nakamura, K. Hatanaka, T. Noro,
F. Ohtani, H. Sakamoto and S. Kobayashi

Department of Physics, Kyoto University, Kyoto 606, Japan

ABSTRACT

The elastic scattering of 65 MeV polarized protons on various nuclei from ^{16}O to ^{208}Pb were measured. The real spin-orbit part of the best-fitted optical potential was found to show the nuclear shell effect which cannot be explained by the nuclear size or shape change.

Since Haxel et al.[1] and Mayer[2] pointed out the fundamental importance of the spin-orbit splitting, the search for its origin has been one of the most important problems in nuclear physics. Recently Scheerbaum[3,4,5] has derived the (l·s) potential from a standpoint of the microscopic nuclear matter theory and found good agreement between theory and experiment for splittings in nuclei in which both $j=l+1/2$ and $j=l-1/2$ subshells spin saturated (SS) shells are filled. In SS shell nuclei the effect of the central ($\sigma \cdot \sigma$) and tensor forces on the spin orbit splitting vanishes by spin averaging over the SS shells. The obtained one-body spin-orbit potential explained mainly by the nucleon-nucleon (N-N) spin-orbit interaction in first order (including exchange effect) and effective N-N tensor interaction in second order. The situation is quite different for nuclei having spin-unsaturated (SUS) subshells, that is, for nuclei in which the $j=l+1/2$ subshell is filled, but the $j=l-1/2$ subshell is still empty. Examples of SUS shells are $0f_{7/2}$ neutrons in ^{48}Ca or the $0i_{13/2}$ neutrons and the $0h_{11/2}$ protons in ^{208}Pb. In SUS shell nuclei the central ($\sigma \cdot \sigma$) interaction and the tensor force contribute to the spin-orbit potential or the (l·s) splitting in first order through the exchange. Although a number of workers (Wong,[6] Davies et al.,[7] Stancu et al.,[8] Scheerbaum,[5] ...) have extensively investigated the SUS problem, the (l·s) splitting in SUS shell nuclei remains to be explained by the theory. The calculations invariably underestimate the splitting energy, that is, the existing theory predicts that the (l·s) potential become shallower in SUS shell nuclei than in SS shell nuclei. Very recently applying the SS and SUS shell method to the scattering problem, W.G. Love derived the (l·s) potential for SUS shell targets and predicted in his model calculation of analyzing powers and spin-flip probability that the SUS correction to the (l·s) potential (SUS quenching effect) is large. For open shell nuclei systematic investigation is rare. Only Goodman and Borysowicz[10] have pointed out the SUS problem to explain the mass number dependency of the $0h_{9/2}-0h_{11/2}$ level splitting in Au and Tl isotopes.

Until now all of the experimental values have been calculated or extracted from the level splittings measured by the spectroscopic study – such as one nucleon transfer reactions or $\beta-\gamma$ spectroscopy. From our systematic study[11,12] of the proton optical potential we

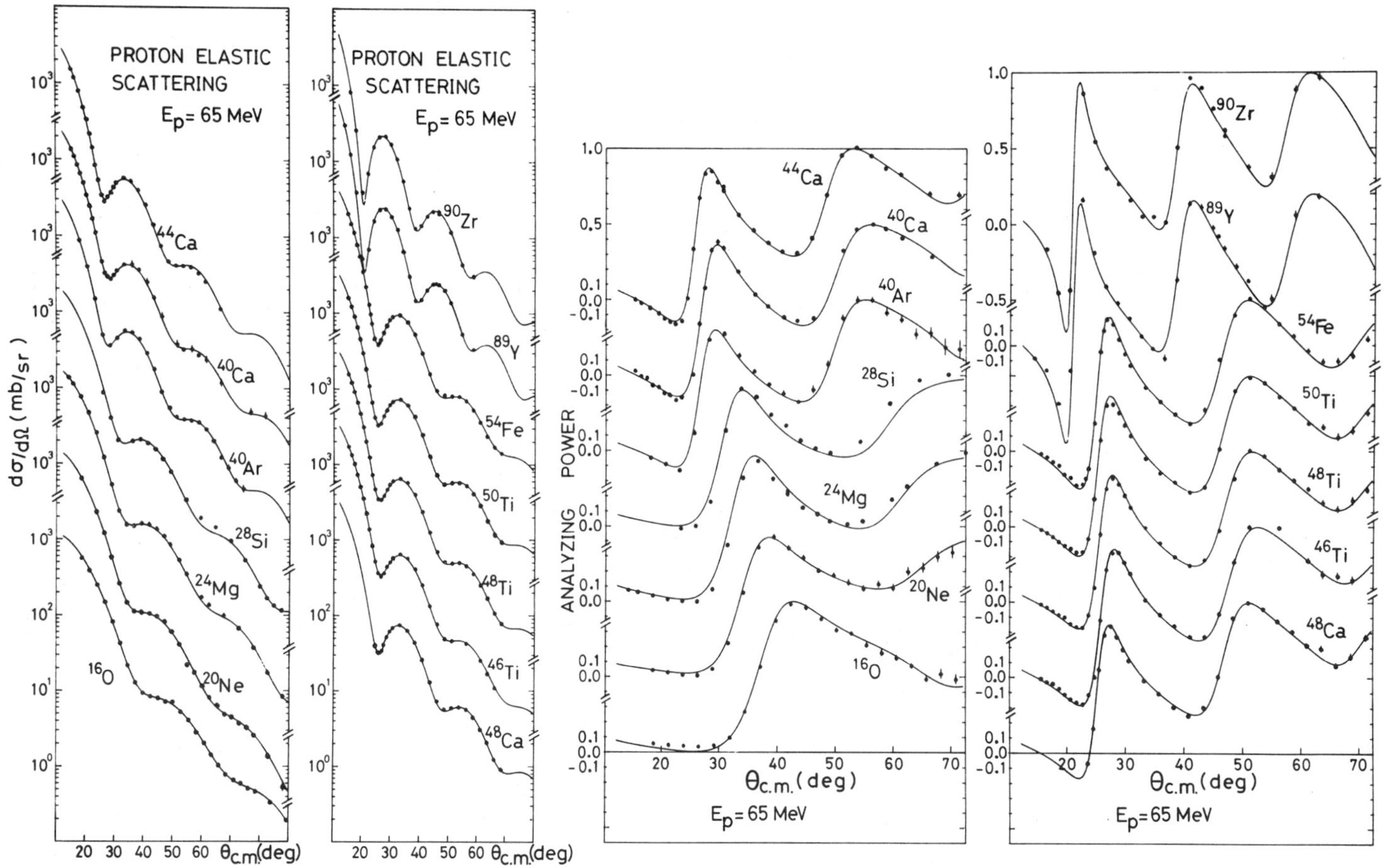

Fig.1 The measured analyzing powers (polarization) and the differential cross sections of the elastic proton scattering. The solid curves show the best fitted optical potential calculations.

propose in a direct way the systematics of the spin orbit-potential at 65 MeV.

In Fig. 1 the analyzing powers (polarization) and the differential cross sections of the elastic scattering are shown. The solid curves in the figure show the best fitted optical potential calculations for each target. The optical potential parameter search was carried out using the Raynal's automatic search code MAGALI. The Becchetti-Greenlees potential was used as the initial parameters of the search. In the phenomenological spin-orbit potential we have three independent parameters V_{1s}, r_{1s} and a_{1s}. In order to reduce the ambiguity, we discuss the volume integral values of the form factor of the (1 s) potential.

$$J_{1s} = -\int V_{1s} \left(\frac{\hbar}{m_\pi c}\right)^2 \left(\frac{1}{r}\frac{d}{dr} f(r; r_{1s}, a_{1s})\right) d\vec{r}$$

$$f(r; r_{1s}, a_{1s}) = 1/(1+\exp(r-r_{1s}A^{1/3}/a_{1s}))$$

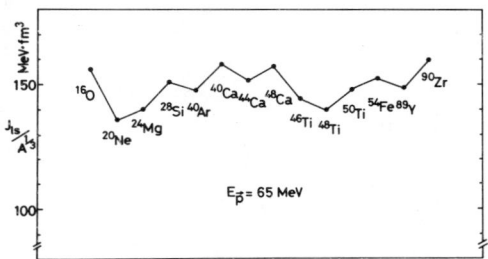

Fig. 2 The volume integral of the (1·s) potential form factor are plotted for each target nuclei. The used potential parameters are listed the in ref. 12.

J_{1s} is calculated as
$$J_{1s} = 4\pi V_{1s} r_{1s} A^{1/3} \left(\frac{\hbar}{m_\pi c}\right)^2$$
$$= 8\pi A \frac{r_{1s}}{} $$
(in fm) V(in MeV) fm^3 MeV

In Fig. 2 $J_{1s}/A^{1/3}$ values are plotted for each target nuclei. A remarkable shell effect is noticed. For the double closed shell nuclei such as ^{16}O, ^{40}Ca, ^{48}Ca and ^{90}Zr the value of $J_{1s}/A^{1/3}$ becomes large. Namely the depth of the (1·s) potential becomes deeper. In the middle of a shell, the depth is the shallowest and it becomes deeper, when the shell becomes more and more closed. For single closed shell nuclei such as ^{54}Fe, ^{50}Ti, ^{89}Y the depth of the (1·s) potential is in the middle between the two utmost. Fig. 3 shows how much a_{1s} can be changed without degrading the χ^2-value too much. For a given a_{1s}-value all other optical potential parameters were free searched to obtain the χ^2-minimum as a function of a_{1s}. Using the potential parameters giving the $\chi^2_{min}(a_{1s})$, $J_{1s}/A^{1/3}$ values are plotted together with $\chi^2_{min}(a_{1s})$ as a function of a_{1s}. This figure shows the followings: 1) With about 10% change of the a_{1s}-value, the fitting does not degrade so much, if we rearrange the other parameters so as to give the minimum χ^2-value. 2) The J_{1s}-value increases, although small, monotonously as the a_{1s} increases. These features can be understood if we think that the form factor of the (1·s) potential is the Thomas-type and its inner part is not so effective due to the imaginary part of the optical potential. So in order to obtain the global systematics of the (1·s) potential it is better to eliminate the ambiguity of the a_{1s}-value. By fixing the a_{1s}-value to one value of 0.6554 fm and by rearranging other parameters, J_{1s}-values are obtained and plotted in Fig. 4. The conclusion from the Fig. 4 are same as the case in the Fig. 2.

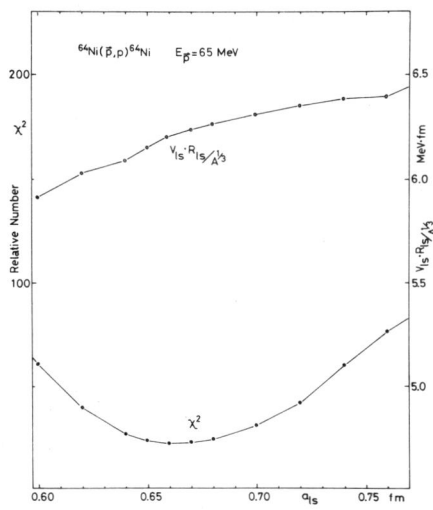

Fig. 3 The $R_{1s} V_{1s}$-values ($\propto J_{1s}$) and the χ^2-values are plotted as a function of a_{1s} in the case of the ^{64}Ni proton elastic scattering. All the potential parameters except a_{1s} and r_c (Coulomb radius) were free searched.

The observed shell effect is about 10% of the total $J_{1s}/A^{1/3}$ in magnitude. But the change of the nuclear radius parameter in one open shell is less than 1 or 2%. In fitting to the experimental data the change of the radius parameter is compensated by the strength of the potential. The net effect to the J_{1s} is minor. The diffuseness parameter of the point nucleon distribution may affect this shell effect. But even for the 10% change of a_{1s}-value in one open shell (which seems to be a too large change) the change of the $J_{1s}/A^{1/3}$ value can be estimated in Fig. 3 to be about 2-3%. So in conclusion the shell effect observed in $J_{1s}/A^{1/3}$-value is small (10%) but it cannot be explained by the nuclear shape or size effect.

From our results (see fig. 4) any remarkable difference between the SS shell nuclei ^{16}O, ^{40}Ca and the SUS shell nuclei ^{48}Ca, ^{90}Zr, ^{208}Pb is not observed, that is, the quenching effect of the SUS shell is not so large as was predicted by W.G. Love. This situation is quite similar to the relation between experimental results and theoretical calculations in the (l·s) splittings of SUS shell nuclei. Nevertheless our data indicates that the quenching effect in the open shell is large. The microscopic calculation including the second order tensor force in the nucleon-nucleon interaction may explain this shell effect. Experimentally we are now in progress to measure elastic scatterings from open shell nuclei and typically deformed nuclei ($150 \leq A \leq 200$ region) using the high resolution spectrograph RAIDEN. The true origin of the shell effect in the (l·s) potential remains to be discovered!

References
1. O. Haxel, J.H.D. Jensen and H.E. Suess, Phys. Rev. 75, 1766 (1949).
2. M.G. Mayer, Phys. Rev. 75, 1969 (1949).
3. R.R. Scheerbaum, Nucl. Phys. A257, 77 (1976).
4. R.R. Scheerbaum, Phys. Lett. 61B, 151 (1976).
5. R.R. Scheerbaum, Phys. Lett. 63B, 381 (1976).
6. C.W. Wong, Nucl. Phys. A108, 481 (1968).

7. K.T.R. Davies and R.J. McCarthy, Phys. Rev. $\underline{C4}$, 81 (1971).
    R.M. Tarbutton and K.T.R. Davies, Nucl. Phys. $\underline{A120}$, 1 (1968).
8. Fl. Stancu, D.M. Brink and H. Flocard, Phys. Lett. $\underline{68B}$, 108 (1977).
9. W.G. Love, Phys. Rev. $\underline{C20}$, 1638 (1979).
10. A.L. Goodman and J. Borysowicz, Nucl. Phys. $\underline{A295}$, 333 (1978).
11. H. Sakaguchi, M. Nakamura, K. Hatanaka, A. Goto, T. Noro, F. Ohtani, H. Sakamoto and S. Kobayashi, Phys. Lett. $\underline{89B}$, 40 (1979).
12. H. Sakaguchi et al., RCNP Annual Report (1978), p.12.

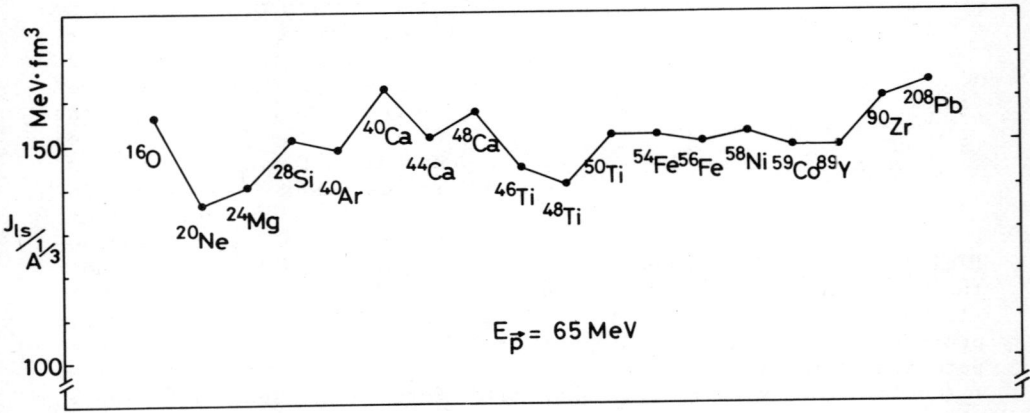

Fig. 4 The modified J_{1s}-values are plotted for various nuclei. In this case a_{1s} is fixed to 0.6554 fm and all other parameters were readjusted to give the χ^2-minimum.

65 MEV POLARIZED PROTON ELASTIC SCATTERING AND THE EFFECTIVE TWO-BODY INTERACTION RANGE

H. Sakaguchi, M. Nakamura, K. Hatanaka, A. Goto,
T. Noro, F. Ohtani, H. Sakamoto, and S. Kobayashi

Department of Physics, Kyoto University, Kyoto, Japan

Abstract

The elastic scattering of 65 MeV polarized protons on various nuclei from ^{16}O to ^{208}Pb were measured. From the systematics of the optical potentials which reproduce the experimental data, the effective 2-body interaction range between the nucleon in the target nucleus and the nucleon in the projectile has been extracted and found to be dependent on the target mass number.

Recently there has been a considerable progress in understanding the nucleon-nucleus optical potential from the standpoint of the nuclear matter theory. The critical check to these global theories seems to be the target mass number dependence and the incident energy dependence of the obtained optical potential. Using the 65 MeV polarized protons of RCNP Osaka cyclotron, we have been measuring systematically the elastic scattering from various target nuclei over the whole periodic table and doing the optical potential fit to the obtained data (Fig. 1). A part of this work[1] was published already.

From the real central part of the obtained optical potential the mean square radius of the potential $<r^2>_{pot}$ was calculated and was plotted as a function of $A^{2/3}$ in Fig. 2. The solid line is a linear fit by the least square method, and is expressed

$$<r^2>_{pot} = (0.928 \pm 0.023)A^{2/3} + 6.57 \pm 0.29 \text{ fm}^2$$

The error bars in Fig. 2 indicate the uncertainty in the optical potential fitting. All the parameters except V_r and r_c were searched so as to obtain the $\chi^2_{min}(V_r)$ as a function of V_r. The error of $<r^2>_{pot}$ was estimated by the equation $\delta<r^2> = |<r^2>_{pot.1} - <r^2>_{pot.2}|/2$, where $<r^2>_{pot.1}$ and $<r^2>_{pot.2}$ were obtained from the parameter sets at $\chi^2_{min}(V_r) = 2\chi^2_{min}$ (best-fit). Thus, the uncertainty dut to the so-called VR^n=const type of ambiguity is included in the error bars.

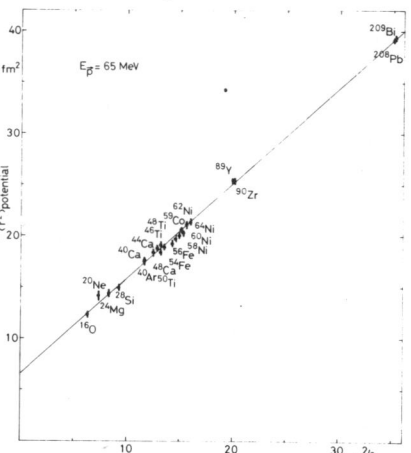

Fig. 2. The mean square radius of the real central part of the obtained optical potential was plotted as a function of $A^{3/2}$. The solid line is a linear fit by the least square method.

Similar relation for the mean square radius of the charge distribution $<r^2>_{charge}$ is obtained by least-square linear fitting to the electron scattering data,

$$<r^2>_{charge} = (0.768\pm0.011)A^{2/3} + (2.60\pm0.23) \text{ fm}^2.$$

Using the folding model, $<r^2>_{pot}$ and $<r^2>_{charge}$ are expressed as

$$<r^2>_{pot} = \frac{3}{5}r_m^2 A^{2/3} + \frac{7}{5}\pi^2 a_m^2 + <r^2>_d .$$

and

$$<r^2>_{charge} = \frac{3}{5}R_p^2 + \frac{7}{5}\pi^2 a_p^2 + <r^2>_{proton} ,$$

where $<r^2>_d$ and $<r^2>_{charge}$ are mean square radii of the effective interaction range and the charge distribution of the proton itself, respectively.

A recent argument[2] based on the experimental data and the Los Alamos experimental results with the small correction from the charge distribution in the neutron itself show that the difference between the root mean square radius of the nuclear matter distribution and that of the point proton distribution is less than 0.1 fm, which is smaller than our $<r^2>_d$-value error. So the average density distribution of the point nucleon is thought to be equal to that of the point proton. By comparing the above two relations the effective two-body interaction range is obtained to be

$$<r^2>_d = 4.49 \pm 0.37 + (0.159 \pm 0.026)A^{2/3} \text{ fm}^2 .$$

This kind of mass number dependence is not strange, if we take into account the Pauli-principle effect on the effective interaction. According to the recent theoretical work on nucleon optical potential from a realistic inter-nucleon interaction the exchange effect on the optical potential mainly reduces the potential depth of its real central part. This does not make any change of $<r^2>_{pot}$ value. The next important Pauli-effect term in $<r^2>_{pot}$ must be the surface type. The possibility of a target dependence of the effective interaction range has been suggested already by B. Sinha.[3] Our $<r^2>_d$ value is larger than the GPT's value[4] of (2.25 ± 0.6) fm^2. GPT's $<r^2>_d$ value was obtained in search for the χ^2-minimum mainly of the cross section data, because of the partial lack of polarization data at that time. By equal weighting the polarization and cross section data we could reduce the VR_n^2 type ambiguity and have found the larger $<r^2>_d$-value. The present $<r^2>_d$-value is as large as the one obtained in Bertsch's calculation[5] (6 fm^2).

1. H. Sakaguchi et al., Phys. Lett. <u>89B</u>, 40 (1979).
2. S. Shlomo and E. Fredman, Phys. Rev. Lett. <u>39</u>, 1180 (1977).
3. B. Sinha, Phys. Lett. <u>C20</u>, 1 (1975).
4. G.W. Greenlees et al., Phys. Rev. <u>C171</u>, 1115 (1968).
5. G. Bertsch et al., Nucl. Phys. <u>A284</u>, 399 (1977).

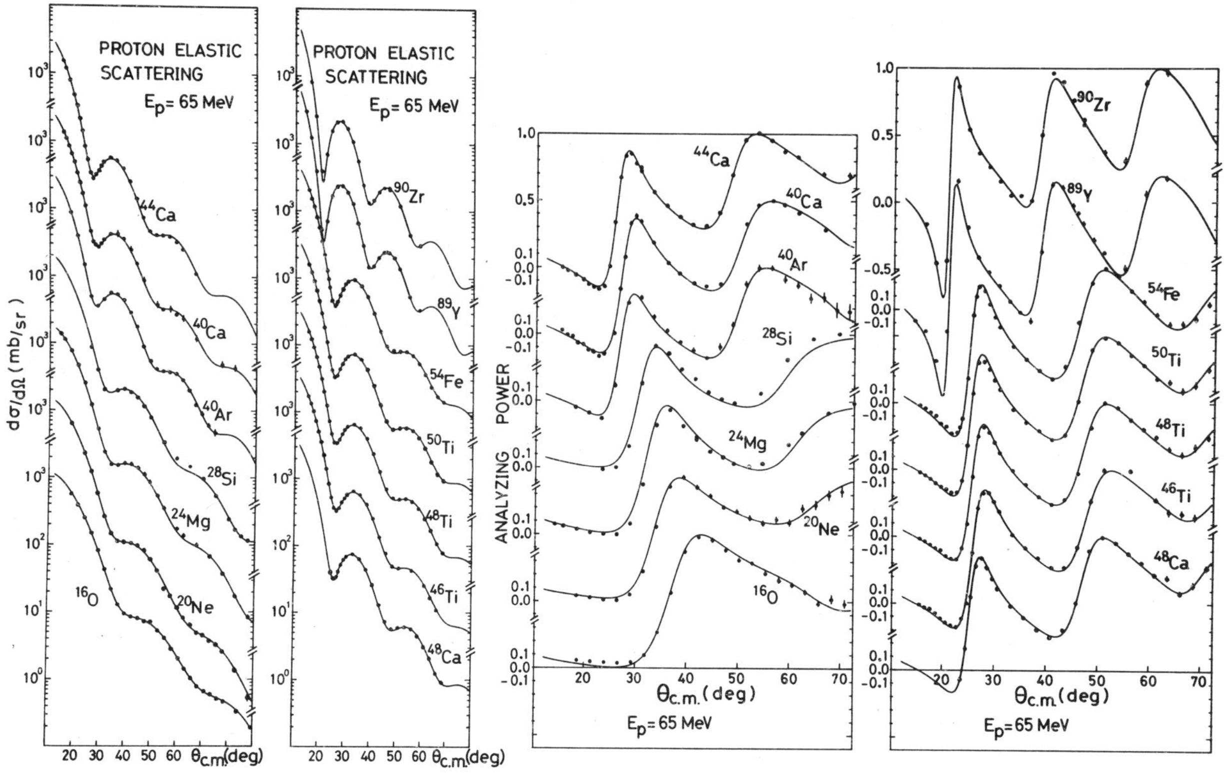

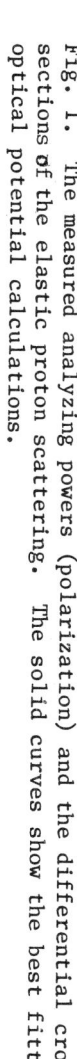

Fig. 1. The measured analyzing powers (polarization) and the differential cross sections of the elastic proton scattering. The solid curves show the best fitted optical potential calculations.

Even-Odd Effect Observed in the Elastic Scattering of Polarized Protons at 65 MeV

M. Nakamura, H. Sakaguchi, K. Hatanaka*, T. Noro,
F. Ohtani, H. Sakamoto and S. Kobayashi

Department of Physics, Kyoto University, Kyoto, Japan
*Research Center for Nuclear Physics, Osaka, Japan

In our systematic measurement using 65 MeV polarized proton beam from the RCNP-OSAKA cyclotron, there can be noticed a differece between the elastic scattering from the even-mass nuclei and the scattering from the neighboring odd-mass nuclei. Fig. 1 shows the angular distributions of the elastic scattering analyzing power for the ^{24}Mg-^{25}Mg-^{26}Mg set and the ^{26}Mg-^{27}Al-^{28}Si set. The points marked by a cross show the odd-mass nucleus. We notice that for the odd-mass nuclei A(θ) decreases at the first peak near θcm=35° and at the second peak near θcm=70° in relative to the neighboring even-mass nuclei. Fig. 2 shows the target dependence of these analyzing power peak values. The points marked by filled and unfilled circles show the first peak value and the second peak value, respectively. For odd-mass nuclei with the target spin I=5/2 such as ^{25}Mg and ^{27}Al, A(θ) decreases appreaciably, but for nuclei with the target spin I=1/2 such as ^{29}Si and ^{31}P the decrease ratio of A(θ) at the peak is small. Such a kind of even-odd effect, that for the target spin I>1/2 the even-odd difference is appreciable, was already suggested by Satchler[1] as the collective effect. Using the Raynal's coupled-channel code ECIS, a rotational model analysis with a deformed optical potential was made.

The spherical part of the used optical potential was same one used in the neighboring even-mass nuclei. A deformed potential was introduced by replacing

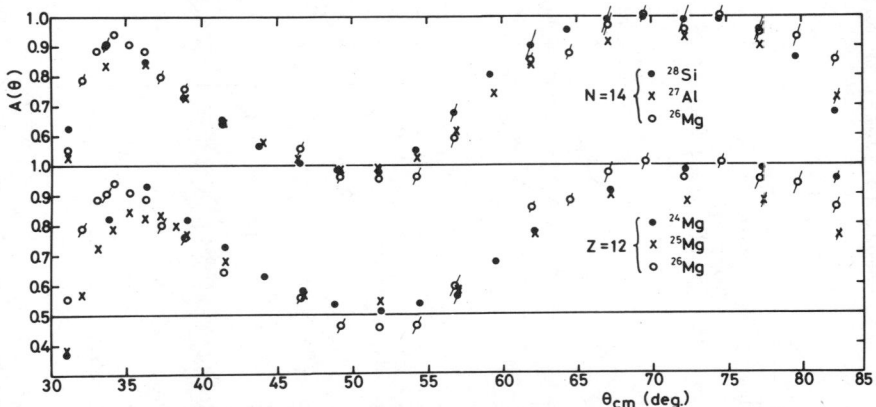

Fig. 1. Angular distributions of the elastic scattering analyzing power for the ^{24}Mg-^{25}Mg-^{26}Mg set and the ^{26}Mg-^{27}Al-^{28}Si set. x points show the data of odd-nuclei.

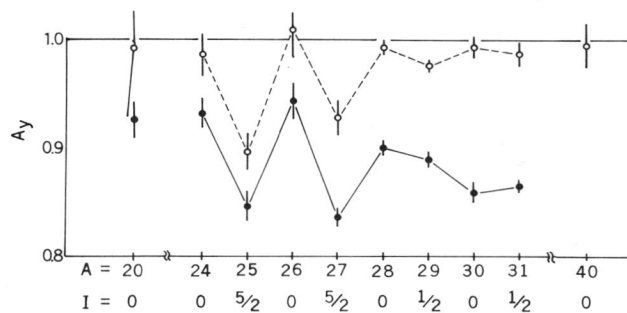

Fig. 2 The target dependence of the analyzing-power peak-values are shown for s-d nuclei. The ● and o mean the first peak value and the second peak value, respectively.

R by R $(1+ \beta Y_2^0(\theta))$, U_{opt} = U (spherical) + ΔU (deformed) thus we took into account only a quadrupole deformation. We neglected the coupling to the excited states and took only the reorientation effect of the ground state, into account. Moreover in order to cut-off the higher order deformation due to β, at first β was reduced to $\beta \times 10^{-3}$ in the calculation of the coupling interaction and later the obtained reduced matrix element was multiplied by 1000. The results are shown in Fig. 3. For ^{25}Mg the best fit was given with $\beta_{1s}/\beta_0 = 1.60$ and $\beta_0 = 0.329$ and for ^{27}Al it was with $\beta_{1s}/\beta_0 = 1.60$ and $\beta_0 = 0.230$. They are shown by solid curves.

The deformed potential is written down as

$$U_0 (def) = R_0 \beta_0 Y_2(\theta) \frac{d}{dR} (V_R/(1+\exp((r-R)/a)))$$

The volume integral and the quadrupole moment of the potential was related to the matter distribution by the folding model.

$$<J>_{pot} = \int U_0(\vec{r}) d\vec{r} \qquad (1)$$

$$<q>_{pot} = \int U_0(\vec{r}_1) r_1^2 Y_2(\theta_1) d\vec{r}_1$$

$$(= \int U_0(def) r_1^2 Y_2(\theta_1) d\vec{r}_1) \qquad (2)$$

Introducing the formula $U_0(\vec{r}) = \int \rho(\vec{r}_0) v_d(|\vec{r}-\vec{r}_0|) dr_0$ into the above definitions, the potential q-moment becomes

$$<q>_{pot} = \int \rho(\vec{r}_2) v(|\vec{r}_1 - \vec{r}_2|) r_1^2 Y_2(\theta_1) d\vec{r}_1 d\vec{r}_2 .$$

Changing the variable $\vec{r}_1 = \vec{s} + \vec{r}_2$ and developing $r_1^2 Y_2(\theta_1)$ by $\vec{s}$ and $\vec{r}_2$, we get

$$<q>_{pot} = \int \rho(\vec{r}_2) Y_2(\theta_2) r_2^2 d\vec{r}_2 \int v(s) d\vec{s} \qquad (3)$$

$$<J>_{pot} = \int \rho(\vec{r}_2) d\vec{r} \int v(s) d\vec{s} \qquad (4)$$

If we assume as our previous paper that the point nucleon distribution is as same as the point proton distribution, we get the similar formula as above by replacing v(s) by the charge distribution function in proton itself. By deviding (3) by (4) we obtain

$$<q>_{pot}/<J>_{pot} = <q>_{matter}/<J>_{matter} = <q>_{charge}/<J>_{charge} \qquad (5)$$

The definition of $<q>_{matter}$, $<J>_{matter}$, $<q>_{charge}$ and $<J>_{charge}$ can be obtained by replacing U(r) in (1) and (2) by the point nucleon distribution function $\rho(\vec{r})$ and the charge distribution function $\rho_{charge}(\vec{r})$ respectively. (5) can be rewritten by

$$\langle q\rangle_{pot}/\langle J\rangle_{pot} = \langle q\rangle_{matter}/A = \langle q\rangle_{charge}/Ze \qquad (6)$$

Then the Q-moment can be calculated by

$$Q = \sqrt{\frac{16}{5} \frac{I(2I-1)}{(I+1)(2I+3)}} \langle q\rangle_{charge} \qquad (7)$$

$$Q = \sqrt{\frac{16}{5} \frac{I(2I-1)}{(I+1)(2I+3)}} Ze\langle q\rangle_{pot}/\langle J\rangle_{pot} \qquad (8)$$

where I is the target spin value. From the best-fit β_0-value $\langle q\rangle_{pot}$ was calculated and using the formula (8) the Q-moment was obtained. For ^{25}Mg the Q-moment was 19 efm^2 and for ^{27}Al 15 efm^2, which are in accord with the value 22 efm^2 for ^{25}Mg and 14.6 efm^2 for ^{27}Al obtained from the electron scattering.

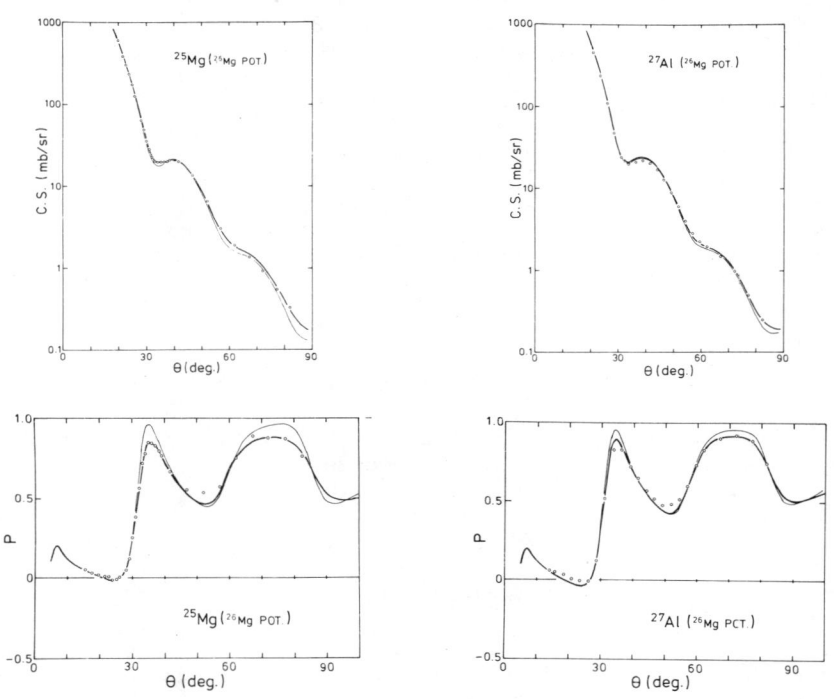

Fig. 3. The best-fitted coupled channel calculation (bold lines) is shown together with the experimental data and the starting optical potential calculation of the neighboring even-nuclei.

References
1. G.R. Satchler, Nucl. Phys. 45, 197 (1963).
2. G.R. Satchler, J. Math. Phys. 13, 1118 (1972).
3. H. Sakaguchi et al., Phys. Lett. 89B, 40 (1979).

THE ISOSPIN DEPENDENCE OF THE 65 MeV PROTON OPTICAL POTENTIAL IN THE f-p SHELL NUCLEI

T. Noro
Department of Physics, Osaka Univ., Osaka 560, Japan

K. Hatanaka
Research Center for Nuclear Physics, Osaka Univ., Osaka 565, Japan

H. Sakaguchi, M. Nakamura, H. Sakamoto, F. Ohtani and S. Kobayashi
Department of Physics, Kyoto Univ., Kyoto 606, Japan

There are several parameter sets of the averaged optical potential which can reproduce well the elastic scattering of nucleons at low energy region (≤ 50 MeV). Almost all of them have the symmetry potential, $V_s(N-Z)/A$, as a part of the real depth parameter and V_s value is set to be about 26 MeV independently of the energy. On the other hand, recent progress in the theoretical and computational means

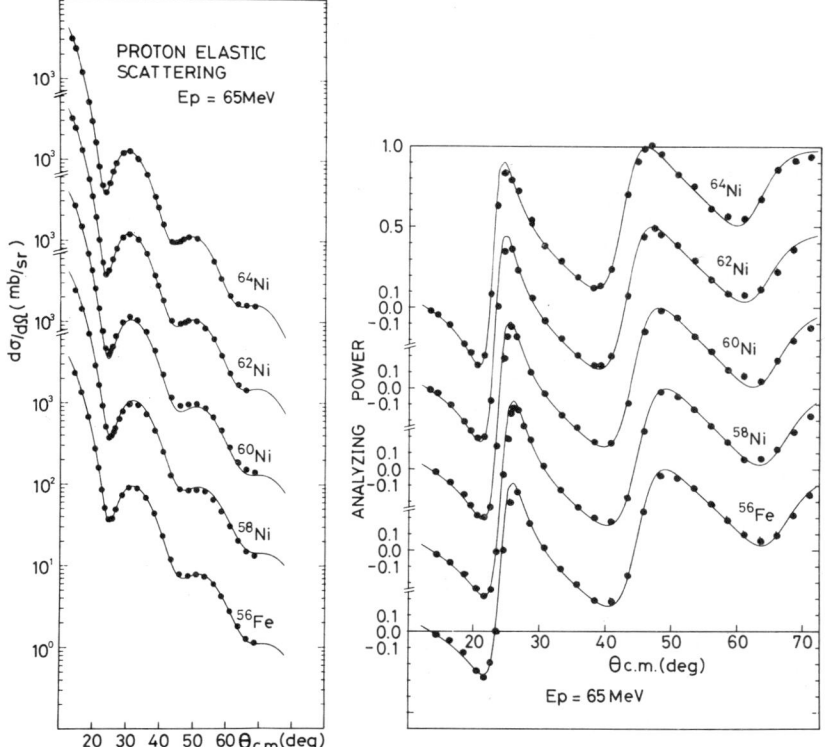

Fig. 1 The elastic scattering data of 65 MeV protons.
The curves are the result of the optical model analysis.

have made it possible to deduce the optical potential which reproduce the experimental data from the realistic nucleon-nucleon interaction quantitatively, and one of these calculation[1] predicts the energy dependence of V_s. The aim of this experiment is to present the precice and systematic data and, especially, to investigate the feature of the symmetry term at the higher energy.

The differential cross section and polarization of the elastic scattering on the various target nuclei were measured using the 65 MeV polarized proton beam accelerated by the RCNP AVF-Cyclotron and a part of these data were already published.[2] The measured angles were from 13.5° to 70° in 1°∼2.5° steps and overall errors (except absolute cross sections) were less than 3% for almost all the angles.

Optical model analysis was performed and the good fit (χ^2/(N-P) = 0.5 ∼ 1.8) was achieved in the usual way except that the geometrical parameters of the surface and volume imaginary potential were searched independently each other. A typical data and the result of the optical model analysis are shown in Fig. 1.

The volume integral per nucleon (J/A) of the real part was used to remove the so-called Vr^n-type ambiguity in the real part of the optical potential. In Fig 2, J/A values of the best fit potentials are plotted as a function of the (N-Z)/A values for ^{40,44,48}Ca, ^{46,48,50}Ti, ^{50,52,54}Cr, ^{54,56}Fe and ^{58,60,62,64}Ni nuclei. It is clear from this figure that the points are scattered and cannot be fitted by a single line, i.e., a simple Lane term cannot reproduce these data. However, it is very interesting to notice that for a single element, the points stands in a straight line and the gradient of the each line itself increases linearly from negative values to positive ones with the charge number of the nuclei in these mass region (Fig. 3).

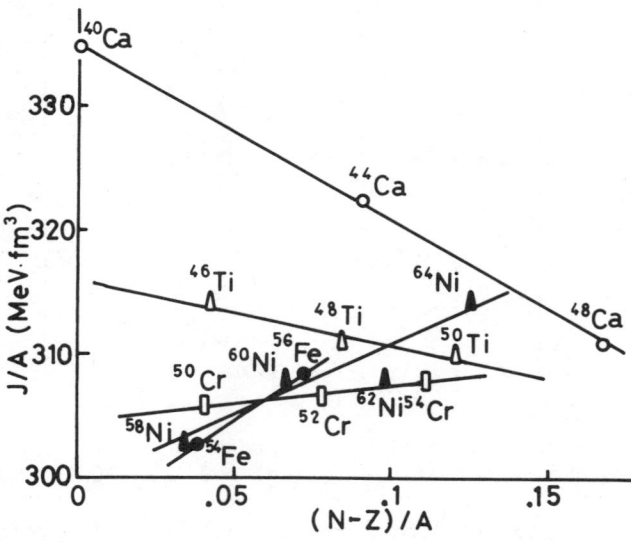

Fig. 2 The volume integral per nucleon of the real part of the 65 MeV proton optical potential.

Though the gradients of these lines for Ca and Ti isotopes are negative, we must not jump to the conclusion that we have here the negative value of V_s, because we also found a strong A dependence of J/A in these mass region. Actually, the comparison between some isobar pairs shows that the J/A values are increased with (N-Z)/A, though the coefficients are very small compared with V_s at low energies.

In the present stage, this is still open problem, but it is probable to attribute those facts to the anomalies in the f-p shell such as Nolen-Schiffer effect or the characteristic behavior of the binding energy in the region of these nuclei.

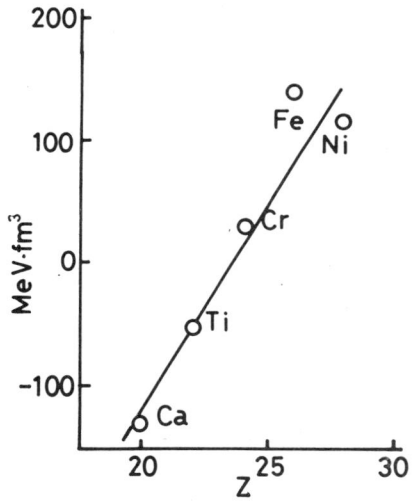

Fig. 3 The gradients of the lines in Fig. 2.

REFERENCES

1. J. -P. Jeukenne, A. Lejeune and C. Mahaux, Phys. Rev. C15, 10 (1977).
2. H. Sakaguchi et al., Phys. Lett. 89B, 40 (1979).

ELASTIC SCATTERING OF POLARIZED PROTONS AT 200 TO 500 MeV

D.A.Hutcheon, J.M.Cameron, R.P.Liljestrand, P.Kitching, C.A.Miller,
W.J.McDonald, D.M.Sheppard, W.C.Olsen, G.C.Neilson, H.S.Sherif,
R.N.MacDonald, and G.M.Stinson
University of Alberta, Edmonton, Alberta T6g 2E1

D.K.McDaniels and J.R.Tinsley
University of Oregon, Eugene, Oregon 97403

L.W.Swensen
Oregon State University, Corvallis, Oregon 97331

P.Schwandt
Indiana University Cyclotron Facility, Bloomington, Indiana 74709

C.E.Stronach
Virginia State University, Petersburg, Virginia 23803

L.Ray
University of Texas at Austin, Austin, Texas 78712

ABSTRACT

We have measured proton elastic scattering cross sections and analyzing power angular distributions on Ca and ^{208}Pb at 200, 300, 400, and 500 MeV. The 400 MeV data from ^{208}Pb have been fitted by an optical model and also compared with results of Impulse Approximation calculations.

EXPERIMENTAL PROCEDURE AND RESULTS

Elastic scattering studies have been done using the TRIUMF polarized proton beam and detecting scattered protons in the 1.5 GeV/c magnetic spectrometer. Cross section and analyzing power angular distributions were measured for targets of natural calcium and 99.1% enriched ^{208}Pb. At 200, 400, and 500 MeV data were obtained from 2.5° to beyond 40°, corresponding to maximum momentum transfers of 3 to 4 fm^{-1}; the 300 MeV distributions, to be completed later this year, extend from 2° to 18°. True zero angle was determined by comparison of yields at positive and negative angles. Beam charge and polarization were monitored by p-p scattering from the CH$_2$ target of an in-beam polarimeter.

The cross section angular distributions show a pronounced diffraction pattern at all four energies. At 200 MeV, however, this pattern is damped between 35° and 50°, a phenomenon which was observed at slightly lower energy in IUCF data[1], and which is a consequence of interference between spin non-flip and spin flip amplitudes.

The analyzing power angular distributions oscillate in step with the diffraction pattern of the differential cross section. The results for ^{208}Pb, shown in figure 1 as a function of momentum transfer, exhibit a decreasing amplitude of oscillation at forward angles as the beam energy increases. At 500 MeV the data for ^{208}Pb are reminiscent of the 800 MeV LAMPF results[2], which never show negative analyzing powers and have an oscillatory pattern superimposed on a trend to large positive analyzing powers as the angle increases.

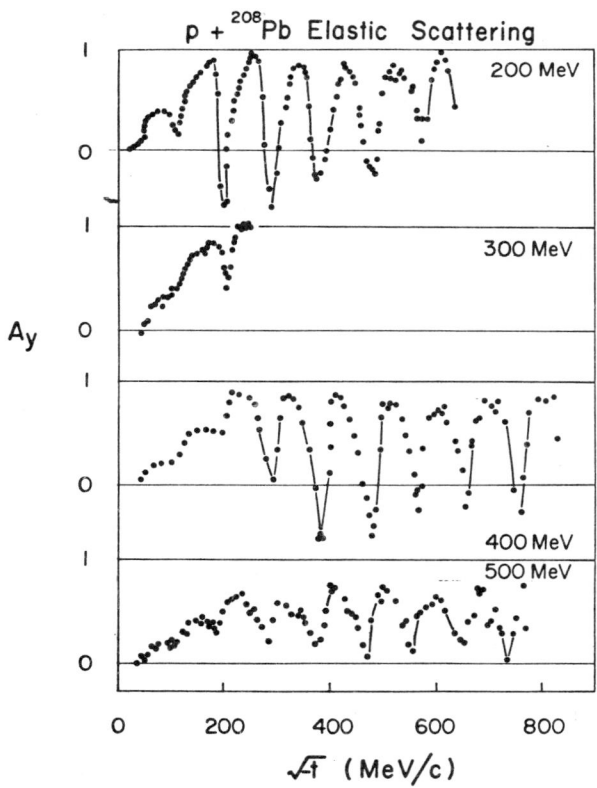

Figure 1. Analyzing power for p + ^{208}Pb elastic scattering as a function of momentum transfer.

DISCUSSION OF RESULTS

The 400 MeV ^{208}Pb data have been fitted using the SNOOPY6 optical model code[1]. The resulting optical model parameters, combined with those from fits to data at other energies, suggest that the real central potential changes sign at about 600 MeV. The volume integrals for the real and imaginary spin-orbit terms are shown in figure 2. The volume integral of the imaginary spin-orbit term changes sign at 400 MeV; there is an abrupt reversal in the slope of this term as a function of beam energy around 200 MeV, behaviour which is not entirely explained by Impulse Approximation estimates of optical model parameters.

Preliminary KMT calculations to second order, using the matter densities for ^{208}Pb deduced from 800 MeV data[3], fail to reproduce many of the deep minima in the analyzing power at 400 MeV. There are indications that corrections for Pauli blocking are important at this energy.

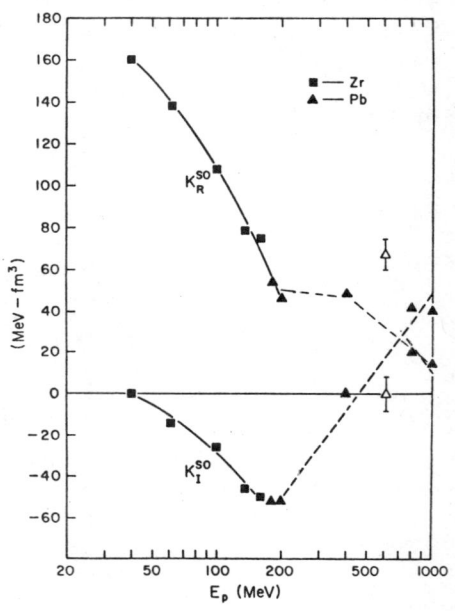

Figure 2. Energy dependence of real and imaginary spin-orbit potential volume integrals.

REFERENCES

1. A.Nadasen et al., Phys. Rev. C, to be published.
2. G.W.Hoffman et al., Phys. Rev. Lett. 40(1979)1256.
3. L.Ray, Phys. Rev. C19(1979)1855.

THE SPIN DEPENDENCE OF INTERMEDIATE-ENERGY
PROTON-NUCLEUS ELASTIC SCATTERING

P. Schwandt, A. D. Bacher, W. W. Jacobs, H.-O. Meyer,
and S. E. Vigdor
Indiana University Cyclotron Facility, Bloomington, IN 47405

Recent measurements [1] at the Indiana University Cyclotron Facility (IUCF) of the differential cross section $\sigma(\theta)$ and the analyzing power $A(\theta)$ for 80-180 MeV polarized proton scattering from ^{40}Ca, ^{90}Zr, and ^{208}Pb were combined with available data at 40 MeV, [2] 65 MeV, [3] 200 and 400 MeV, [4] 800 MeV, [5] and 1 GeV [6] and analyzed in terms of a phenomenological optical model (OM) employing a local, complex, spin-dependent nuclear potential with Woods-Saxon formfactors. The analysis was carried out in a relativistic framework, using a wave equation obtained from the Dirac equation for a massive, energetic fermion in a local potential field.

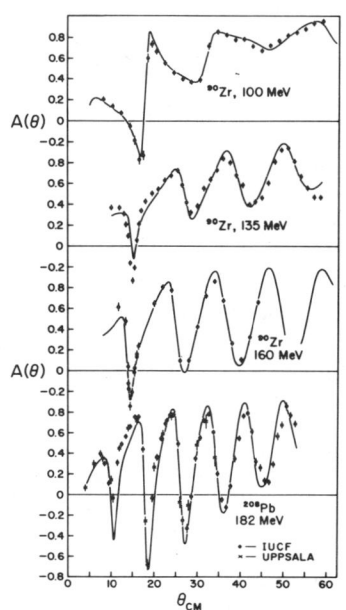

Examples of the IUCF $A(\theta)$ data at energies between 100 and 182 MeV are presented in Fig. 1, along with the results of simultaneous OM fits to the $\sigma(\theta)$ and $A(\theta)$ data. The available $\sigma(\theta)$ and $A(\theta)$ data and OM fits for ^{208}Pb between 120 and 1000 MeV are illustrated in Fig. 2. These figures exhibit the characteristic features of intermediate-energy elastic proton scattering: pronounced damping of diffractive oscillations in $\sigma(40° < \theta < 60°)$ for proton energies around 150-200 MeV, and strong oscillations of $A(\theta)$ (with large negative excursions for the heavier nuclei) at these same energies. In contrast, at both lower and higher energies, $\sigma(\theta)$ is characterized by regular diffractive oscillations and $A(\theta)$ is typically positive and large over most of the angular range. These phenomena observed for $\sigma(\theta)$ and $A(\theta)$ for all nuclei have a common origin: at both low energies (<100 MeV) and high energies (>250 MeV) the partial cross sections $\sigma^+(\theta)$, $\sigma^-(\theta)$ corresponding to channel spins $\vec{\sigma}\cdot\hat{n} = \pm 1$ (where $\hat{n} \equiv \hat{k}_{in} \times \hat{k}_{out}$) are ordered in magnitude as $\sigma^+ \gg \sigma^-$ beyond some moderately forward angle; consequently, the angular structure of $\sigma(\theta) = \sigma^+ + \sigma^-$ is dominated by the oscillatory σ^+, and

Fig. 1. Analyzing powers for $\vec{p}$ + ^{90}Zr, ^{208}Pb measured at IUCF (●). The 182 MeV data labeled (x) are from Ref. 7. The curves are OM fits to $\sigma(\theta)$ and $A(\theta)$ data.

$A(\theta) = (\sigma^+ - \sigma^-)/(\sigma^+ + \sigma^-) \gg 0$. In the "transition" energy range $100 \lesssim E_p \lesssim 250$ MeV, on the other hand, both σ^+ and σ^- oscillate with comparable amplitudes in the forward hemisphere. However, their angular period is slightly different owing to a difference in the radii of the total effective nuclear potentials $U_{centr.} \pm U_{s.o.}$ for $\vec{\sigma}\cdot\hat{n} = \pm 1$. Hence, σ^+ and σ^- move out of phase over a portion of the angular range, leading to a structureless angular distribution for $\sigma(\theta)$ and strong oscillations in $A(\theta)$. This relative phasing of σ^+ and σ^- in the transition energy range is sensitive to the interplay of the complex central and S.O. potentials and accounts for the enhanced sensitivity to the S.O. potential at these energies. It should be noted in this connection that the mechanisms for σ^+ dominance at low and high energies are very different: at low energies the predominantly *real* nuclear potential is more attractive in the $\vec{\sigma}\cdot\hat{n} = +1$ spin state for the important surface region; hence the deflection angles for a given corresponding impact parameter obey $\theta^+ > \theta^-$, causing $\sigma^+(\theta) > \sigma^-(\theta)$ if α decreases with increasing θ. At high energies, $\sigma^+ > \sigma^-$ because of strong absorption [9] by the predominantly *imaginary* potential.

In Fig. 2 the cross section is plotted as a function of momentum transfer $q = 2k \sin(\theta/2)$ in order to demonstrate another significant feature of intermediate-energy proton scattering, namely the reduction in the number of diffraction minima in $\sigma(q)$ over the data range $q < 4$ fm^{-1} from 9 at low energies to 8 at high energies. This change in diffraction structure implies a decrease in the effective radial extent of the nuclear potential with increasing energy which can be qualitatively understood in terms of a progressive reduction in both the Pauli suppression of the potential in the nuclear interior and the range of the fundamental 2-nucleon interaction.

The results of the phenomenological OM analysis are presented here in terms of normalized potential volume integrals $J^C_{R,I}/A$ for the complex central potential and $J^{SO}_{R,I}/A^{1/3} \equiv K^{SO}_{R,I}$ for the complex S.O. potential, where

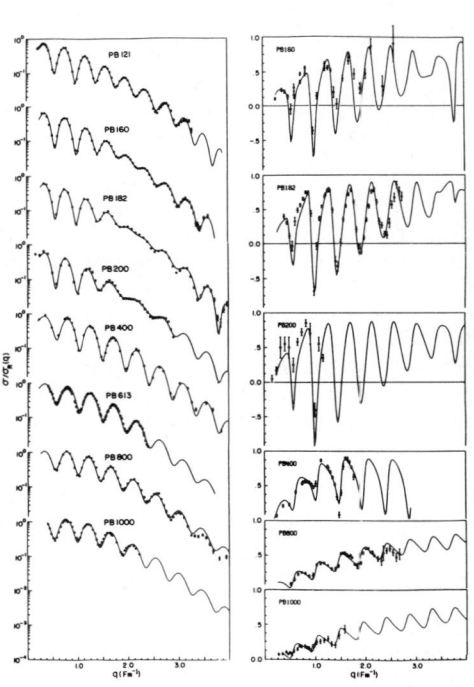

Fig. 2. σ/σ_R and A for $\vec{p} + {}^{208}$Pb as function of momentum transfer q(fm^{-1}) for $E_p = 121$–1000 MeV. The 613 MeV data are from Ref. 8. The curves are OM fits.

$J \equiv 4\pi \int U(r) r^2 dr$. As illustrated in Fig. 3 (left panel) a logarithmic energy dependence is found for the effective strength of the real central potential (which becomes repulsive beyond 600 MeV), while the corresponding imaginary component remains constant up to about 200 MeV but increases rapidly beyond 200 MeV, because of the onset of meson production and nucleon excitation. As displayed in the right panel of Fig. 3, both the real and imaginary S.O. strengths exhibit pronounced energy dependence. Up to 200 MeV the real S.O. strength decreases linearly with $\ln(E_p)$. The imaginary S.O. strength is of opposite sign and *increases* linearly with $\ln(E_p)$ to a maximum near 200 MeV. Beyond 200 MeV the results are less precise because of the paucity of available data, but the present data nevertheless indicate a continuing though somewhat less rapid decrease of K_R^{SO} and a dramatic reversal of the energy variation of K_I^{SO} which in fact changes sign near 400 MeV. Forthcoming additional measurements at TRIUMF and LAMPF between 200 and 800 MeV will allow a more detailed investigation of these trends.

In Fig. 3 the results of the phenomenological OM analysis are also compared to microscopic calculations. At energies below 200, nuclear matter calculations in the BHF approximation by Brieva and Rook [10] represent the most recent and advanced formulation in terms

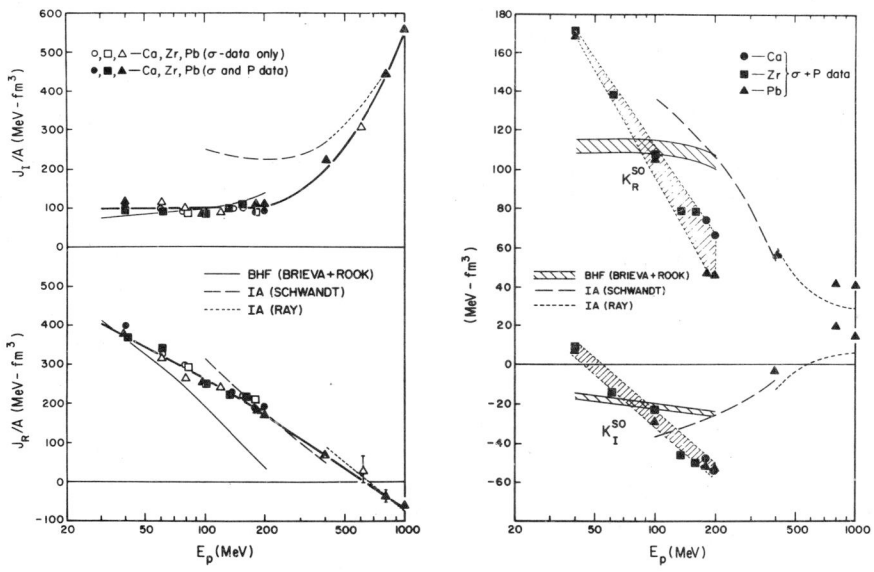

Fig. 3. Normalized volume integrals of the real and imaginary central potentials (left panel) and spin-orbit potentials (right panel). The symbols show the results of phenomenological analyses, the thick solid line is a guide to the eye and the thin solid, dashed, and dotted lines are predictions of microscopic potential models.

of a complex, *effective* 2-nucleon t-matrix (thin solid line). A local energy-dependent optical potential is generated by the antisymmetrized folding model using the local density approximation. Reasonably good average fits to empirical data can be obtained with this model below 100 MeV. Beyond 100 MeV, the real central potential begins to deviate from the empirical values, and the strong energy dependence of the empirical complex S.O. potential is clearly not reproduced by the Brieva-Rook model.

At energies beyond a few hundred MeV multiple-scattering theory in the impulse approximation (IA), in which the reaction matrix reduces to the *free* 2-nucleon t-matrix, is expected to provide a realistic description. Recent application of this model in the KMT approach [11] to medium energy (150 to 400 MeV) proton scattering by one of the present authors (PS) led to moderately satisfactory results when carried out to lowest order, using the Love parametrization [12] of the complex t-matrix. However, by making suitable corrections for Pauli blocking, using physically realistic formfactors with empirically adjusted strengths, the agreement between the IA and experiment is much improved. The potential volume integrals given by the 1st-order (uncorrected) IA (dashes in Fig. 3) agree well with phenomenology above 150 MeV for the real central component and approach the empirical results near 400 MeV for the other components. In the vicinity of 200 MeV, however, the uncorrected IA results for J_I and K_R^{SO} are considerably too large and for K_I^{SO} considerably too small. For energies between 400 and 1000 MeV, Ray [13] has carried out similar IA calculations using 2-nucleon scattering amplitudes with adjusted parameters for the spin-dependent components. Excellent fits are obtained with this model at 800 MeV; the respective central and S.O. volume integrals (dots in Fig. 3) are in good overall agreement with the phenomenological results.

REFERENCES

1. P. Schwandt et *al.*, IUCF Scient. and Techn. Report 1979, p. 47.
2. M. P. Fricke et *al.*, Phys. Rev. 156, 1207 (1967).
3. M. Nakamura et *al.*, RCNP Ann. Rep., Osaka University 1977. p. 129.
4. D. Hutcheon, private communication.
5. L. Ray et *al.*, Phys. Rev. C 18, 1756 (1978).
6. G. D. Alkhazov et *al.*, Leningrad Preprint No. 6 (1977), p. 715.
7. W.T.H. van Oers et *al.*, Phys. Rev. C 10, 307 (1974).
8. G. Bruge, CEN Saclay Intern. Rep. DPh-N/ME/78-1.
9. E. Fermi, Nuovo Cimento II, Suppl. 1, 84 (1955).
10. F. A. Brieva and J. R. Rook, Nucl. Phys. A291, 299 and 317 (1977); *ibid.*, A297, 206 (1978); *ibid.*, A307, 493 (1978).
11. A. K. Kerman, H. McManus, and R. M. Thaler, Ann. Phys. 8, 551 (1959).
12. W. G. Love et *al.*, Phys. Lett. 73B, 277 (1978), and private communication.
13. L. Ray, Phys. Rev. C 20, 1857 (1979), and private communication.

SPIN LENGTH SCALES AND POLARIZATION IN p-NUCLEUS SCATTERING

J. A. McNeil*
Department of Physics, University of Pennsylvania
Philadelphia, Pennsylvania 19104

ABSTRACT

Spin dependent phenomena in p-nucleus scattering at intermediate energies are shown to be dominated by two fundamental lengths. By exploiting their small size compared to the nuclear radius, a simple closed form expression for the spin channel amplitudes is obtained. The closed form permits one to use the unpolarized cross section as input to obtain a data-to-data formula for the polarization in terms of the unpolarized cross section and the spin 'length' parameters. The results generalize to spin phenomena involving scattering to inelastic states.

Intermediate energy proton-nucleus scattering for many nuclei is dominated by the size and surface diffusivity of the target nucleus. This dominance has been exploited in the context of an eikonal amplitude to derive closed form expressions for the elastic scattering amplitudes.[1] Later work extended these results to inelastic processes.[2] By introducing two new length scales associated with the fundamental spin-orbit amplitude, a simple extension to polarization can be made.

From symmetry considerations the forward nucleon-nucleon spin orbit amplitude must vanish like q, the momentum transfer, as q→0. Explicitly extracting this dependence introduces an associated length, w, for p-nucleus applications, which characterizes the forward spin-orbit to central relative strength. A second length, δ, arises from the fact that the fundamental central and spin-orbit amplitudes have different ranges. Upon folding with the nuclear density, the effect of this difference is to shift the spin-orbit optical potential radius with respect to the central radius. The optical potential can thus be written in spin channel form as

$$V^{\pm}(r) = k\gamma \left[\rho(r,a) \pm i w \frac{d}{dr} \rho(r,a') \right]$$

where $\delta = a - a'$ is for simplicity taken to be real and $\gamma = \frac{2\pi A(o)}{k}$. For a Fermi distribution

$$\rho(r,c,\beta) = \rho_0 \left(1 + e^{(r-c)/\beta} \right)^{-1}$$

where the geometry is specified by the radius, c, and diffusivity, β.

*Supported by the National Science Foundation.

By noting that the nuclear radius dominates the character of the scattering, one can us Taylor's theorem to simply shift the radius to first order in $w \ll c$. This just effectively alters c and β:

$$c \rightarrow c \pm v \qquad \beta \rightarrow \beta \mp u/\pi$$

where w = u + iv. At this stage we can use the analytic methods of ADL to evaluate the spin channel amplitudes. To first order in and w, one finds

$$F^{\pm}(q) = \Gamma(q) \, e^{-\pi \beta q \pm qw} \cos(qc \pm q\eta)$$

where small minimum filling terms have been ignored and $\Gamma(q)$ is a slowly varying function of q and

$$\eta \simeq \frac{3 u \delta}{c} \left(\frac{qc}{2\pi \beta \rho_o \gamma}\right)^{2/3} \cos \pi/3$$

arises from the different central and spin-orbit optical potential radii. The spin channel cross sections take the simple form

$$\left(\frac{d\sigma}{dq}\right)_{\pm} = |\Gamma(q)|^2 \, e^{-2\pi \beta^{\pm} q} \cos^2 q c^{\pm} \qquad (1)$$

with $c^{\pm} = c \pm \eta$, $\beta^{\pm} = \beta \mp u/\pi$.

Using (1), the polarization is

$$P = (e^{2qu} - e^{-2qu} \xi^2)/(e^{2qu} + e^{-2qu} \xi^2) \qquad (2)$$

where $\xi = (1 + \tan q\eta \tan qc)/(1 - \tan q\eta \tan qc)$

If we let (1) with w=0 represent the unpolarized cross section (this ignores w^2 contributions), we can eliminate the tan(qc) term as follows

$$\tan qc \simeq \frac{1}{2} \left[F(-\delta q) \frac{d\sigma}{dq}(q - \delta q) - F(\delta q) \frac{d\sigma}{dq}(q + \delta q) \right] / \frac{d\sigma}{dq}(q) \qquad (3)$$

where q =π/4c and F(δq) describes the local variation due to the exponentially falling envelope needed to compensate the δq shift. Ignoring the slow variation due to $\Gamma(q)$ gives F(q)=exp($2\pi \beta q$). The result is a data-to-data formula in the log-derivative tradition but with specific forms for all coefficients. Figure 1 shows the elastic 800 MeV ^{208}Pb(p,p) polarization[3] along with the data-to-data result for w = .15 fm and δ = .25 fm. The data-to-data form is prefered to the closed form because using the elastic scattering data to incorporate the nuclear structure information automatically compensates for approximations made in deriving the closed form.

Extension to inelastic processes is easy. In ALMS closed form amplitudes for collective inelastic excitations have been derived which yield in turn data-to-data relations of the same form as (2) but with the envelope function altered by an additional factor of q^2: $F_{in}(\delta q) = (1-\delta q/q)^2 F(\delta q)$. Figure 2 shows the 800 MeV inel-

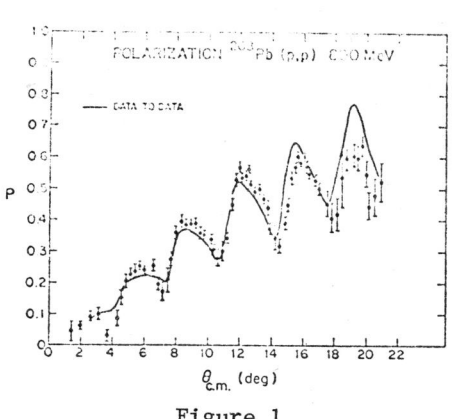

Figure 1

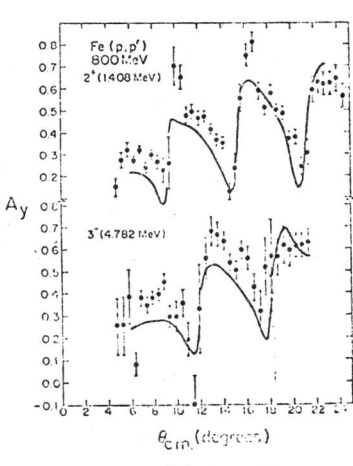

Figure 2

astic ^{54}Fe(p,p') (2_1^+ and 3_1^-) analyzing powers[4] compared with (2) and (3) using the inelastic envelope function and the same values for w and f. The excellent agreement illustrates the power and simplicity of the method and shows that there is little new structure information to be obtained from inelastic polarization experiments.

REFERENCES

1. R. D. Amado, J.-P. Dedonder, and F. Lenz, Phys. Rev. C21, 647 (1980). (Refered to as ADL in the text.)
2. R. D. Amado, F. Lenz, J. A. McNeil, and D. A. Sparrow, to be published in Phys. Rev. C (refered to as ALMS in text)
3. G. W. Hoffman, et al., Phys. Rev. Lett. 40, 1256 (1978).
4. G. S. Adams, et al., to be published in Phys. Rev. C.

ORIGIN OF THE DEUTERON IMAGINARY
SPIN-DEPENDENT POTENTIAL

W.H. Wong and P.A. Quin
University of Wisconsin, Madison, WI. 53706[†]

A complex spin-dependent potential was first suggested by Goddard and Haeberli[1] as a phenomenological way to improve the fit to all four analyzing powers for deuteron elastic scattering. In a systematic study of vector polarized deuteron scattering from the nickel isotopes, the present authors found[2] that as the bombarding energy decreases the vector analyzing power, $iT_{11}(\theta)$, changes from a symmetric oscillation about zero to a predominantly one-signed value in an isotopically systematic way. Quantitative fits to the data could be obtained using a complex spin-orbit potential, but not even qualitative fits could be obtained without an imaginary spin-orbit term. As shown in figure 1, this behavior near the Coulomb barrier is observed for many nuclei in the medium mass region.[2] It is also clear from figure 2 that the predominant sign of $iT_{11}(\theta)$ has a strong dependence on neutron shell closure.

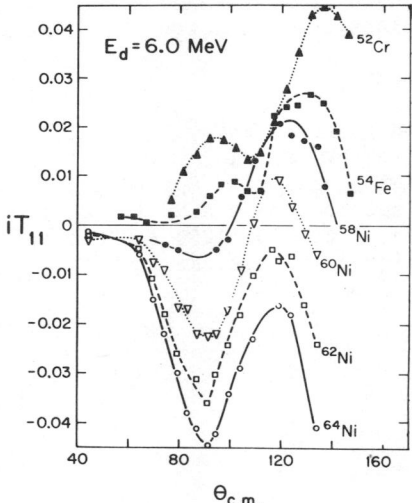

Figure 1. $iT_{11}(\theta)$ for nuclei near mass 60 at a bombarding energy near the Coulomb barrier. The curves are guides to the eye, but for $^{60-64}$Ni the optical model with a complex spin-orbit potential provides a quantitative fit.

[†] Work supported in part by the U.S. Department of Energy.

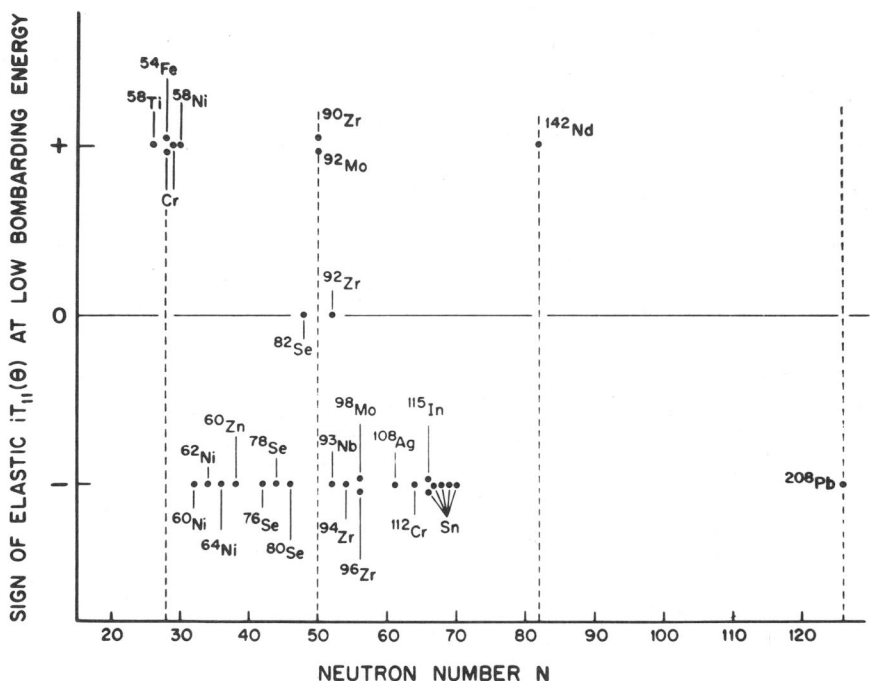

Figure 2. The dominant sign of $iT_{11}(\theta)$ for low bombarding energies as a function of the neutron number.

The imaginary spin-orbit potential we used has a Thomas radial dependence and can be written in the form

$$V_{SO}^{i}(r) = \pm i |V(r)| \vec{L} \cdot \vec{s}. \qquad (1)$$

Here, the positive and negative signs correspond to predominantly positive or negative $iT_{11}(\theta)$. The radius of the potential $r \approx 1.7 A^{1/3}$ is nearly twice as large as that of ref. 1, and V_{SO}^{i} dominates over the real spin orbit potential for bombarding energies near the Coulomb barrier.

It is well known that the dominant absorption mechanism near the Coulomb barrier is stripping to weakly bound or continuum neutron levels. The semiclassical model of Vigdor et al.[3] explains the j-dependence of $iT_{11}(\theta)$ for such transitions in a simple way. Near a closed neutron shell, the weakly bound neutron levels are dominantly $j_n = \ell - 1/2$. Between shell closures $j_n = \ell + 1/2$ dominates.

If such a simple interaction of the absorption mechanism were true, then a coupled-channels calculation which explicitly includes a dominant weakly bound level should simulate the effects of V_{so}^i. The results of such calculations are shown in figure 3. It is clear that the general behavior of $iT_{11}(\theta)$ is described, but the curves would not provide a quantitative fit to measurements. Note that the calculations are very sensitive to the Q-value.

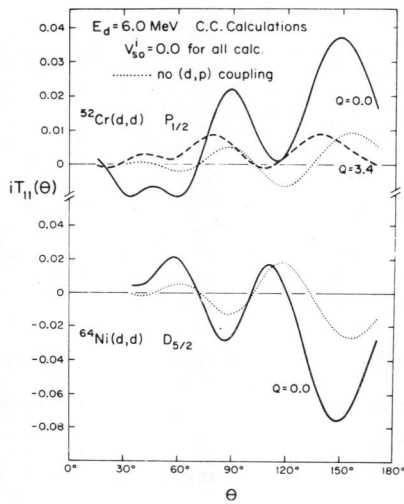

Figure 3. Results of coupled-channels calculations for deuteron elastic scattering which include a (d,p) reaction to a weakly bound state. These calculations should be compared to the data shown in figure 1.

A comment on ^{208}Pb is necessary. ^{208}Pb(d,p) has a small Q-value even for the ground state transition. The largest cross section (d,p) state is a 2d5/2, $j_n = \ell + 1/2$, and therefore the negative $iT_{11}(\theta)$ is expected.

REFERENCES

[1] R.P. Goddard and W. Haeberli, Phys. Rev. Lett. <u>40</u>, 701 (1978).
[2] W.H. Wong, thesis, University of Wisconsin, 1979
[3] S.E. Vigdor, et al., Nucl. Phys. <u>A210</u>, 93 (1973).

A TEST OF COLLECTIVE EFFECTS IN Se(d,d)Se

Y. Tagishi*, B. L. Burks, T. B. Clegg,
E. J. Ludwig, R. L. Varner, J. F. Wilkerson, and W. J. Thompson+
Department of Physics and Astronomy, University of North Carolina,
Chapel Hill, NC 27514 USA, and
Triangle Universities Nuclear Laboratory, Durham, NC

ABSTRACT

Deuteron elastic and inelastic scattering from the Se isotopes with A = 76, 78, 80, 82 has been used to test the suggestion that the change in elastic-scattering vector analyzing powers can be attributed to changing collectivity of these isotopes with A.

An unexpected isotope dependence of the vector analyzing power, $iT_{11}(\theta)$, has been observed experimentally for the ^{A}Se(d,d)^{A}Se, A = 76, 78, 80, 82, elastic scattering at 12 MeV.[1] This isotopic dependence was explained[2] as an effect of smaller imaginary parts of the deuteron optical potentials with decreasing collectivity of the nuclei as A increases. This elastic scattering process is strongly related to coupling to inelastic scattering via the first 2+ state, so an A dependence should also be expected in the inelastic channel. To study such an effect, we measured the differential cross section, σ, and iT_{11} for (d,d) and for (d,d') to the first 2+ excited states of the Se isotopes at 12 MeV.

The experiment was performed with a vector-polarized beam from the TUNL tandem accelerator using a Lamb-shift source. Scattered deuterons were detected by counter telescopes used for mass identification at equal angles to the right and left of the incident beam. Beam polarization was monitored by using the ^{3}He($\vec{d}$,p)^{4}He reaction. Results for σ and iT_{11} are shown in Figs. 1 and 2.

The elastic-scattering data agree well with those in Ref. 1. The magnitudes of the inelastic cross section decrease with increasing A as shown in Fig. 1, as predicted. The cross sections summed between 45° and 160° are approximately proportional to the square of the deformation parameter β_2. As shown in Fig. 2, a systematic A dependence in iT_{11} was also observed for the inelastic scattering. The angular distributions of iT_{11} have quite similar shapes, and the magnitude of iT_{11} increases linearly with increasing A, as observed in iT_{11} for elastic scattering (Fig. 3). Such an A dependence in iT_{11} is more pronounced for the inelastic scattering than for the elastic scattering, as seen in Fig. 2.

----

* On leave from University of Tsukuba, Ibaraki 305, Japan
+ Research supported in part by the U. S. Department of Energy.

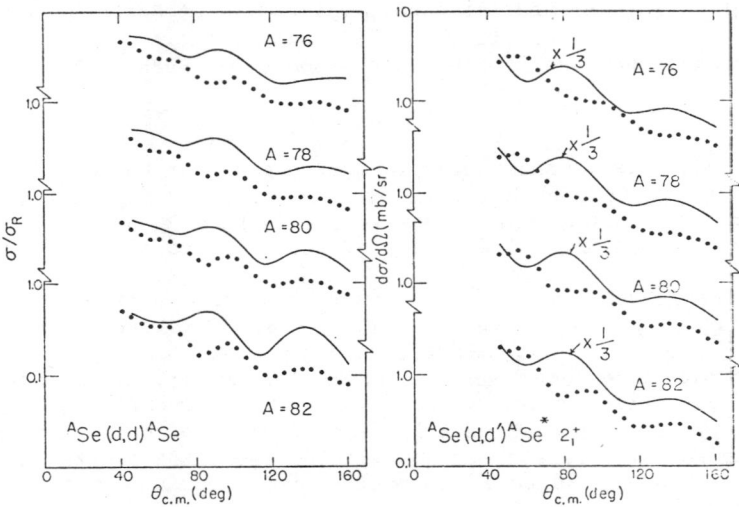

Fig. 1. Angular distributions of the cross section for Se(d,d)Se scattering at 12 MeV. Solid curves are from coupled-channel calculations described in the text. The calculated inelastic cross sections have been reduced by a factor of 1/3.

Bürgi et al.[3], suggested that the A dependence could be described by terms in the strengths V_0 and W_D of the central real and surface-imaginary optical potentials. These terms depend only on nuclear collectivity. Therefore a coupled-channels calculation accounting for this collectivity should require the same V_0 and W_D for all A. The results of such a 0^+-2^+ rotational coupling calculation of the cross section, shown in Fig.1, do not describe the data. We conclude that the observed A dependence is not simply an effect of the collective properties as described in Ref.3. A detailed analysis of both the elastic and inelastic data is in progress.

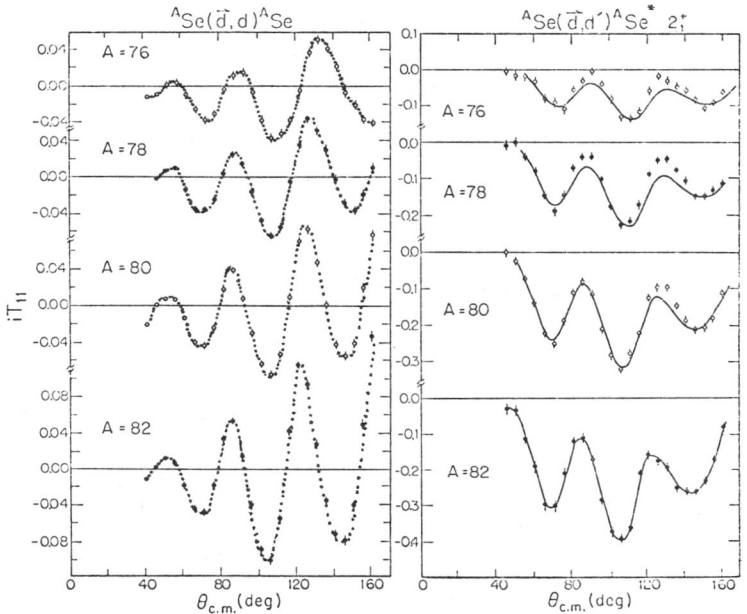

Fig. 2. Elastic and inelastic iT₁₁ for Se(d,d)Se at 12 MeV. The curves through the elastic-scattering data are drawn to guide the eye. The curves for the inelastic-scattering data for A = 76, 78, 80 were obtained by reducing vertically the curve for A = 82 (drawn to connect smoothly the data points) by multiplying by a scale factor to get a good fit to the data. The horizontal scale is also slightly changed from nucleus to nucleus to get a good overlap of the minimum point near 70°. The relative scale factors are shown in Fig. 3.

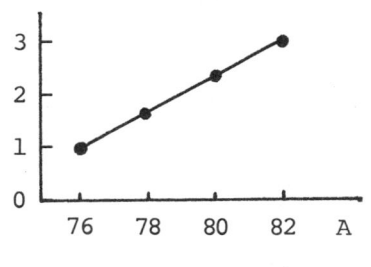

Fig. 3. Relative scale factors for inelastic analyzing power vs A.

REFERENCES

1. J. Nurzynski et al., Phys. Lett. 67B, 23 (1977).
2. W. J. Thompson, Bull. Am. Phys. Soc. 22, 1016 (1977); H. Ohnishi, H. Noya and M. Tanifuji, Phys. Lett. 76B, 256 (1978).
3. H. R. Bürgi et al., Nucl. Phys. A334, 413 (1980).

PHASE SHIFT ANALYSIS OF ^{58}Ni(d,d$_o$) AT 12 MeV

A. Lindner
I. Inst. für Experimentalphysik der Universität Hamburg, W. Germany

ABSTRACT

We have fitted the scattering matrix $<(l_f,1)J|S|(l_i,1)J>$ to the Zürich[1] data σ_o, iT_{11}, T_{20}, T_{21} and T_{22} at 36 angles and gotten standard deviations of real and imaginary parts of S below 0.004. In our approach we rely only on conservation of angular momentum and parity and on time reversal invariance.

GENERAL REMARKS

The interference caused by Coulomb and nuclear interaction enables us to determine the scattering matrix from complete spin correlation experiments[2]. The experiments must be complete in the sense that we need the observables for a variety of angles in order to analyze the partial waves and exploit conservation of angular momentum. Because of this demand there is no contradiction to Simonius[3], who claims that in addition to spin correlation experiments one also needs polarization transfer experiments - when constraining oneself to one scattering angle only.

In the case of scattering by a spinless nucleus, the spin correlation experiment simplifies to a polarization analysis experiment. The relevant formulae are given elsewhere[4]. Instead of choosing the simplest example $\frac{\vec{1}}{2} + 0 \to \frac{1}{2} + 0$ (e.g. nucleon scattering), we have tested a more involved one ($\vec{1} + 0 \to 1 + 0$), since deuteron scattering is not so well known and there are many attempts to derive this from three body approaches[5]. Thus a phase shift analysis of elastic deuteron scattering appears highly desireable.

FITTING PROCEDURE AND RESULTS

The experimental data mentioned consist of N = 177 numbers (iT_{11}, T_{21} and T_{22} were not measured at the most backward angle 175°). Conservation of parity and time reversal invariance are used to reduce the number of parameters to be fitted - conservation of angular momentum is essential for the whole approach. With a maximum l_{max} of orbital angular momenta we have M=8·l_{max} numbers to fit the N data, searching for the minimum of χ^2. (In the search procedure we used the constraint $|S|\leq 1$ because of unitarity.) The search was done for several values of l_{max}. The fit presented has the smallest $\chi^2/(N-M)$ which was achieved with l_{max} = 9. The resulting scattering matrix $S(l_f,l_i)$ is displayed in fig. 1 (note the enlarged scale in the diagram for (l_f,l_i) = (J-1,J+1) and the fit achieved in fig. 2.

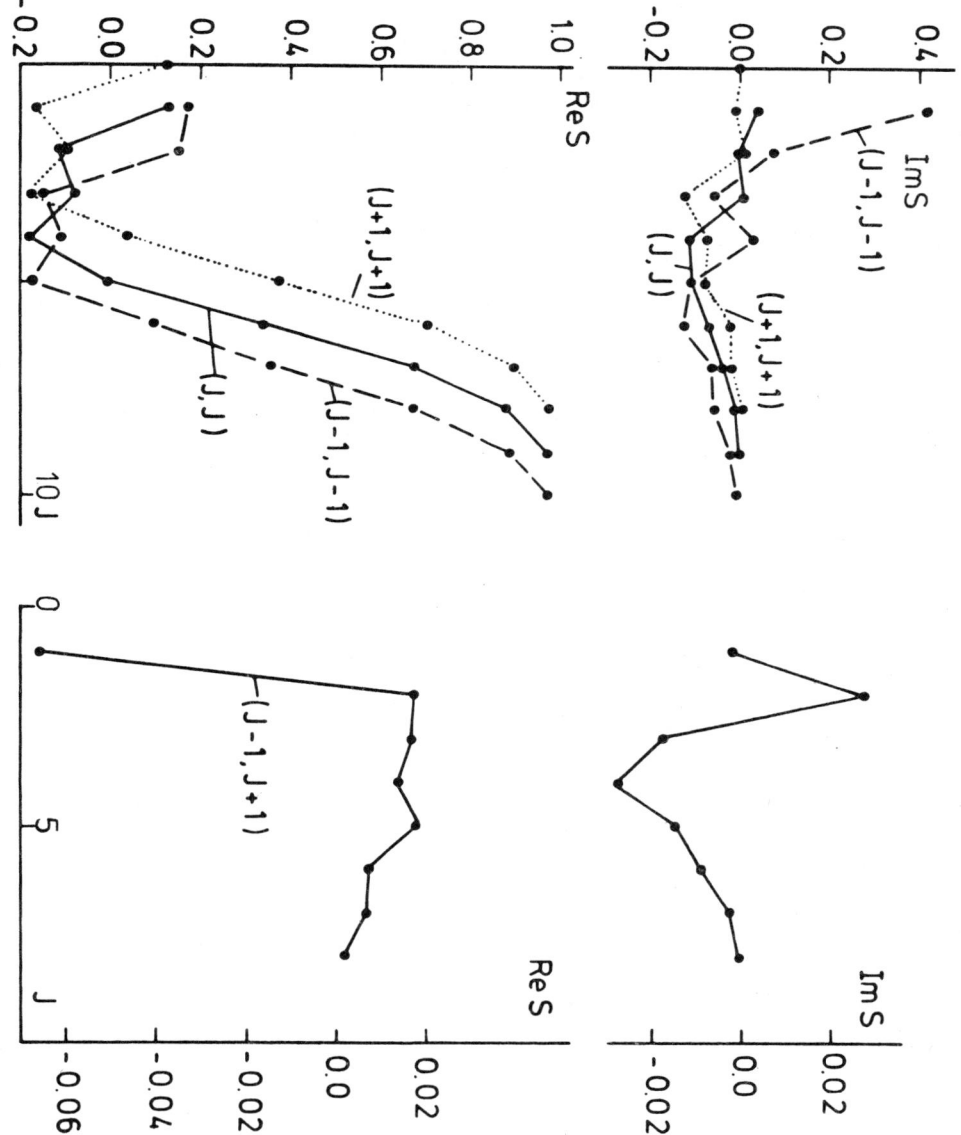

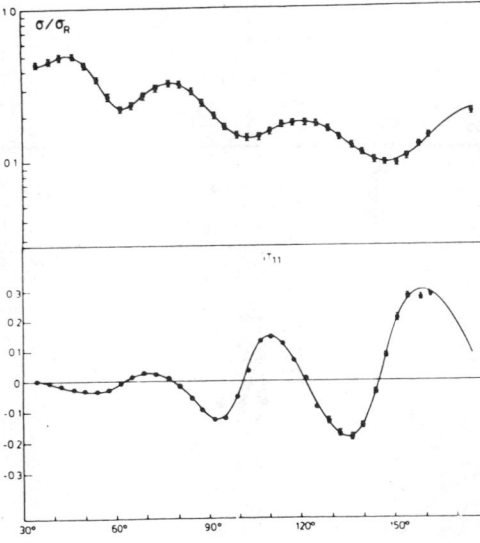
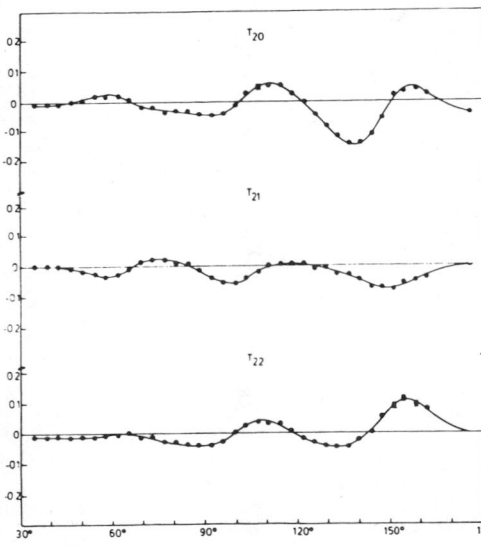

Fig. 2
Fit to the Zürich[1] data achieved

REFERENCES

1. W. Grüebler, priv. communication; cf. H.R. Bürgi, W. Grüebler, J. Nurzynski, V. König, P.A. Schmelzbach, R. Risler, B. Jenny and H.R. Hardekopf, Nucl.Phys. A334 (1980)413

2. A. Lindner in Few Body Systems and Nuclear Forces I, Proc. Graz 1978, eds. H. Zingl, M. Haftel and H. Zankel, Lecture notes in Physics 82 (Springer, Berlin, 1978),p.117

3. M. Simonius, Phys.Rev.Lett. 19(1967)279 and in Polarization Phenomena in Nuclear Reactions, Proc. Third Int.Symp. Madison 1970, eds. H.H. Barschall and W. Haeberli (Univ. Wisconsin, 1971)p. 401

4. A. Lindner, Phase Shift Analysis of Elastic Scattering of Charged Particles with Spin by Spinless Nuclei (submitted to Nucl. Phys.)

5. R.C. Johnson and P.J.R. Soper, Phys.Rev. C1(1970)976; G.H.Rawitscher, Phys.Rev. C9 (1974)2210; J.P. Farrell, C.M. Vincent and N. Austern, Ann. of Phys. 96(1976)333; N.Austern, C.M. Vincent and J.P. Farrell, Ann.of Phys. 114(1978)93; B. Anders and A. Lindner, Nucl. Phys. A296(1978)77; G.H. Rawitscher and S.N. Mukherjee, Ann.of Phys. 123(1979)330; W.M. Wendler, A. Lindner and B. Anders, Influence of Coulomb Polarization on Elastic Scattering of Deuterons by Heavy Nuclei (Submitted to Nucl.Phys.)

MASS DEPENDENCE OF THE T_R TENSOR POTENTIAL FOR ELASTIC SCATTERING OF 20 MeV DEUTERONS[+)]

R. Frick, H. Clement, G. Graw, P. Schiemenz,
N. Seichert and Sun Tsu-Hsun[*)]
Sektion Physik der Universität München, D-8046 Garching

The presence of a tensor term in the optical potential for deuteron scattering is correlated with the existence of a D-state in the deuteron at the place of interaction with the target nucleus. Following the folding model[1)] this tensor potential is expected to be complex in general and only weakly dependent on the target mass. Deviations are expected due to the Pauli principle and the break-up. In consequence of the Pauli principle Ioannides and Johnson[2)] derive a strong reduction and a momentum dependence of the D-state probability, while Rawitscher et al.[3)] discuss an increase of tensor effects by taking into account the break-up. Experimentally a tensor potential shows up most clearly in the tensor analyzing power $A_{xz}(\Theta) = -\sqrt{3}\, T_{21}(\Theta)$. Those experiments[4,5)] had been restricted to deuteron energies below 15 MeV and to heavier nuclei $A \geq 40$.

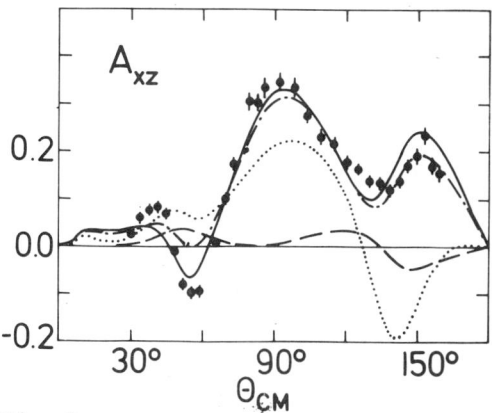

Fig.1

With 20 or 22 MeV deuterons of the Munich tandem accelerator we measured $A_{xz}(\Theta)$ for the elastic scattering on ^{4}He[6)], ^{16}O, ^{24}Mg[7)], ^{28}Si[7)], ^{36}Ar, ^{65}Cu and ^{208}Pb, the results are shown in fig.1 and 2 with calculations including a tensor potential of the T_R type. The strengths of the experimentally observed effects decrease with increasing mass of the target nucleus, they are, however, larger and different in the angular dependence than calculations without a tensor force predict. For the central and spin orbit parts we take potentials[7)] nearly identical to those of ref.[8)], only for ^{16}O we had to increase the radius of the surface absorption from 1.4 fm $A^{1/3}$ to 1.8 fm $A^{1/3}$ as suggested in ref.[9)].

In our folding calculations the radial part $U_T(r)$ of a T_R tensor potential is directly correlated with the phenomenologically determined central part $U_C(r)$ of the deuteron optical potential and the quadrupole moment eQ_d of the deuteron:

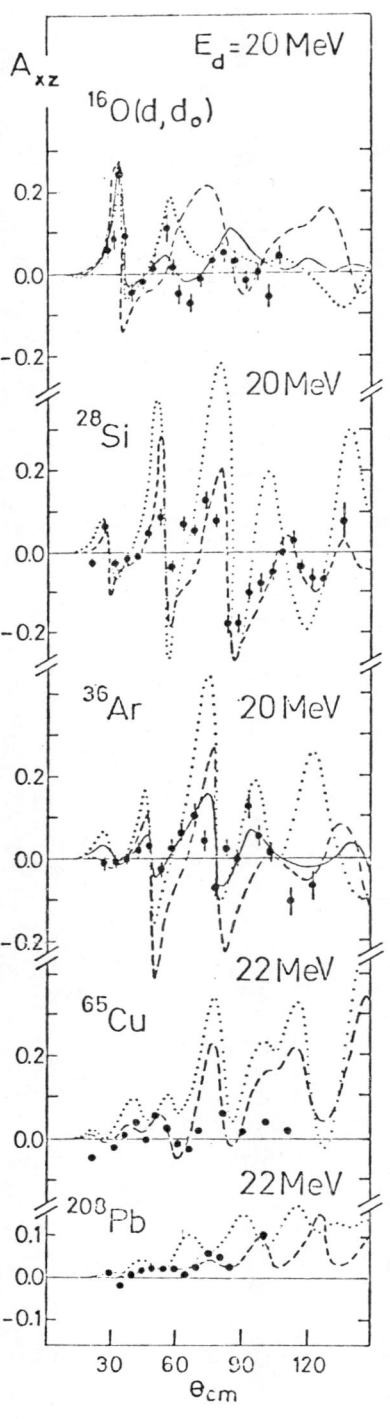

$$U_T(r) = \Omega_d/2 (r \frac{d}{dr} \frac{1}{r} \frac{d}{dr}) U_c(r)$$

This ansatz is an extension of eq. (19) in ref.[10] and accounts in a phenomenological way for break up and higher order terms. The dotted lines in figs. 1 and 2 are calculations with this complex potential, the dashed lines with the imaginary part of $U_T(r)$ only. Solid lines are fits, deviating from this prescription.

We find for ^{4}He the strength of the real part being about twice the folding value[6], for ^{16}O, we have approximately agreement with the folding calculation, if we use the central potential of ref.8 to derive it. For the heavier nuclei the folding potentials overestimate considerably the experimental effects (in agreement with previous investigations for nuclei with A≳40 at E_d<15 MeV [1,4,5]). For ^{24}Mg[7], ^{28}Si and ^{36}Ar it is difficult to separate the relative contributions of the real and the imaginary part. For ^{24}Mg[7] and ^{28}Si[7] the backward angle behaviour favours a pure imaginary T_R-potential, whereas for ^{36}Ar we obtain the best reproduction if both for the real and imaginary part about 50% of the folded T_R potentials are used. For ^{65}Cu and ^{208}Pb a pure imaginary T_R potential reproduces the data. This analysis with a minimal number of adjustable parameters indicates that the imaginary parts of $U_T(r)$ are in approximate agreement with a folding model whereas the real parts show a strong mass dependence: The potentials for the light nuclei are even larger than the folding model predicts and decrease for heavier nuclei to much smaller values.

*) work supported in part by the Bundesministerium für Forschung und Technologie
+) Max-Planck fellow from the Institute of Atomic Energy, Peking, P.R. China
1) P.W. Keaton and D.D. Armstrong, Phys.Rev. C8 (1973)1692
2) A.A. Ioannides and R.C. Johnson, Phys.Rev.C17(1978)1331
3) G.H. Rawitscher and S.N. Muckerjee, Ann.ofPhys. 123 (1979) 330
4) P.P. Goddard and W. Haeberli, Nucl.Phys. A317 (1979)116
5) H.R. Bürgi et al., Nucl.Phys. A321 (1979) 449
6) R. Frick et al., Phys.Rev..Lett. 44 (1980) 14
7) H. Clement et al., Contributions to this conference
8) J.W. Lohr and W. Haeberli, Nucl.Phys. A232 (1974) 381
9) C.E. Busch et al., Nucl.Phys. A223 (1974) 183
10) F.D. Santos, Z. Physik A 295 (1980) 73

ON THE OPTICAL POTENTIAL FOR DEUTERON SCATTERING AT E_d=20 MeV[*]

H. Clement
Sektion Physik der Universität München, D-8046 Garching

For scattering of 20 MeV vector polarized deuterons from various target nuclei we compare the optical potentials obtained in a conventional analysis of the elastic channel (OM) with those of coupled channel analyses, where the coupling to strong collective (2_1^+ and 3_1^-) channels is treated explicitly (CC). The CC calculations show, that strong 2_1^+ excitations influence the elastic scattering channel distinctively. For the nuclei ^{16}O, ^{18}O, ^{24}Mg,

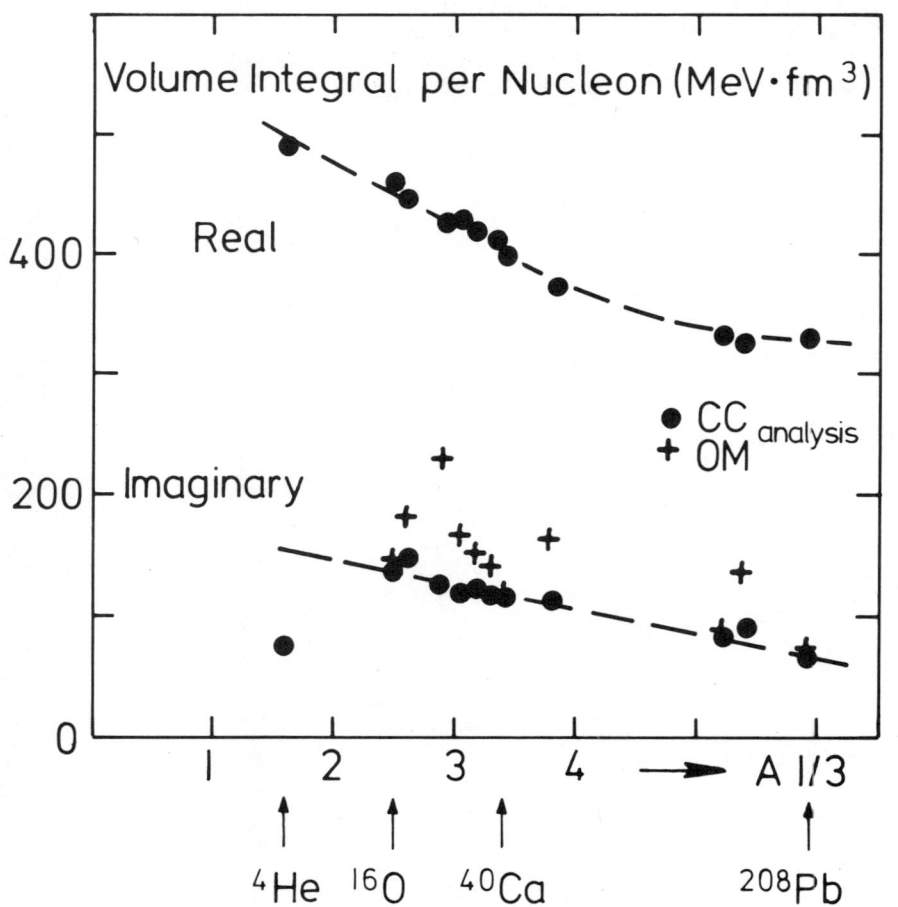

Fig. 1: Volume Integral per nucleon (MeV fm^3) from OM and CC-fits

^{28}Si, ^{32}S, ^{36}Ar, ^{40}Ca, ^{54}Cr, ^{144}Sm, ^{154}Sm and ^{208}Pb the CC analysis gives a satisfying description of the elastic and the inelastic channels[1]) by starting with the global set of Lohr and Haeberli[2]) and adjusting the depth of the real and imaginary central term. In fig.1 the full dots show as result the volume integrals for the real and imaginary central part. The use of other global sets[3,4]) leads to similar results. For the oxygen isotopes the absorption radius has been increased according to the results of ref.5 for very light nuclei. Only for ^{4}He an OM fit[6]) has been performed by optimizing all parameters individually. The OM fits use the same potential, only the depth of the absorption of the central term was adjusted. These results are shown as crosses if they deviate from the CC values. For the deformed nuclei strong deviations are apparent, indicating the strength of collectivity of the 2^+_1 states. For ^{4}He a much lower value is obtained, since here only the break-up mechanism contributes to the imaginary part. The smooth behaviour of all CC results, which is nearly linear for the imaginary part: $J_I = (190-21 \, A^{1/3})$ MeV fm^3, supports very strongly the presence of a global optical potential for 20 MeV deuterons.

*) supported in part by the Bundesministerium für Forschung und Technologie

1) H. Clement et al., contributions to this conference
2) J.H. Lohr, W. Haeberli, Nucl.Phys. A232 (1974) 381
3) G. Perrin et al., Nucl.Phys. A282 (1977) 221
4) WW. Daehnick, J.D. Childs, Z. Vrcelj, Pittsburg 1980, private communication
5) C.E. Busch, T.B. Clegg, S.K. Datta, E.J. Ludwig, Nucl. Phys. A223 (1974) 183
6) R. Frick, H. Clement, G. Graw, P. Schiemenz, N. Seichert, Phys.Rev.Lett. 40 (1980) 14

MEASUREMENTS OF A_{xx} AND A_{yy} FOR ELASTIC SCATTERING OF 56 MeV DEUTERONS FROM ^{28}Si, ^{64}Ni AND ^{144}Sm

K. Hatanaka
Research Center for Nuclear Physics, Osaka University
Suita, Osaka 565, Japan

M. Nakamura, K. Imai, T. Noro, H. Shimizu, H. Sakamoto,
J. Shirai, T. Matsusue and K. Nisimura
Department of Physics, Kyoto University, Kyoto 606, Japan

ABSTRACT

The differential cross sections σ and vector and tensor analyzing powers A_y, A_{xx} and A_{yy} were measured for the elastic scattering of 56 MeV polarized deuterons from ^{28}Si, ^{64}Ni and ^{144}Sm. The optical model analysis including tensor potential was performed. The well depths of the T_R potential were determined from the analysis of the quantity $(2A_{xx}+A_{yy})/\sqrt{3}$.

INTRODUCTION

Recently a comprehensive analysis of σ, A_y and A_{yy} has been performed for the elastic scattering of 56 MeV deuterons over a wide mass region[1]. It has been found that the real part of the T_R potential tends to be zero for all nuclei studied and the inclusion of the imaginary part improves the fit to A_{yy}. A_{yy} was more strongly affected by the tensor potential than at lower energy. It has been suggested, however, that the tensor potential can be better determined in an analysis of the quantity $(2A_{xx}+A_{yy})/\sqrt{3}$ [ref.2].

σ, A_y, A_{xx} and A_{yy} were measured for the elastic scattering of 56 MeV polarized deuterons from ^{28}Si, ^{64}Ni and ^{144}Sm. A comprehensive analysis of these data was made in the framework of the optical model. The quantity $(2A_{xx}+A_{yy})/\sqrt{3}$ was calculated with the best fit parameters and compared with the experimental data.

EXPERIMENTAL PROCEDURES

The experiments were carried out using a 56 MeV polarized deuteron beam from the AVF cyclotron at RCNP. Polarized deuterons were produced by the atomic beam type polarized ion source[3]. The beam intensity on the targets was 5~20 nA. The beam polarizations were monitored continuously during the experiments by a ^{12}C-polarimeter and were 70~80% of the ideal values. The angular distributions of σ, A_y, A_{xx} and A_{yy} were measured for the elastic scattering from ^{28}Si, ^{64}Ni and ^{144}Sm. The experimental methods are described elsewhere[4].

RESULTS AND DISCUSSION

The angular distributions of A_{xx} and A_{yy} are shown in fig. 1. In the first place, the optical model analysis including no tensor potentials were performed. The results of the optical model calculation with the best fit parameters are compared with the experimental data in fig. 1 (solid lines). The tensor analyzing powers were well reproduced without any tensor potentials. This conclusion is consistent with the result of ref. 1. The dashed curves in fig. 1 show the

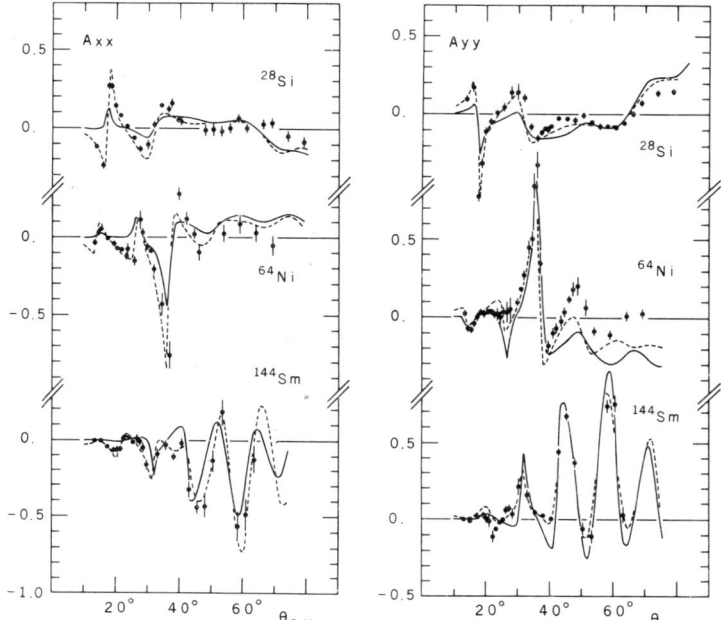

Fig. 1. Angular distributions of the tensor analyzing powers A_{xx} and A_{yy} for the elastic scattering of 56 MeV deuterons from ^{28}Si, ^{64}Ni and ^{144}Sm. The solid curves are results of the optical model analysis with central and spin-orbit terms only. The dashed curves are predictions of the optical model including a complex T_R potential.

results of the optical model analysis with a complex T_R potential. The real parts of the tensor potential were 0.1, 0.03 and 0.8 MeV for ^{28}Si, ^{64}Ni and ^{144}Sm, respectively. The imaginary parts were around 0.8 MeV for all nuclei studied. The radius parameters of the real and imaginary parts were fixed at 1.25 and 1.4 fm respectively. σ and A_y were not strongly affected by the tensor potentials. The inclusion of the complex T_R potential improves the fit to A_{xx} and A_{yy}.

The quantity $(2A_{xx}+A_{yy})/\sqrt{3}$ calculated from the experimental data is shown in fig.2 and is compared with the optical model predictions. The solid and dashed curves correspond to those in fig.1. The predicted values are much smaller than the experimental data, when the tensor potentials are not included. The inclusion of the complex T_R potential much improves the fit to $(2A_{xx}+A_{yy})/\sqrt{3}$.

Hooton et al. derived the relations between the observables and the scattering amplitudes using a perturbation treatment. The equivalent relations can be, however, obtained from the exact treatment of the scattering matrix[5].

$$2A_{xx}+A_{yy} = -\sqrt{2}\mathrm{Re}[(B-D)(A+E-C)^*]\cot\theta/\sigma - 2\mathrm{Re}[(B-D)(B+D)^*]/\sigma,$$

where A∿E are scattering amplitudes and θ is the scattering angle in the center of mass system. It can be easily proved that to the first order the terms A+E-C and B+D depend on the central and spin-orbit

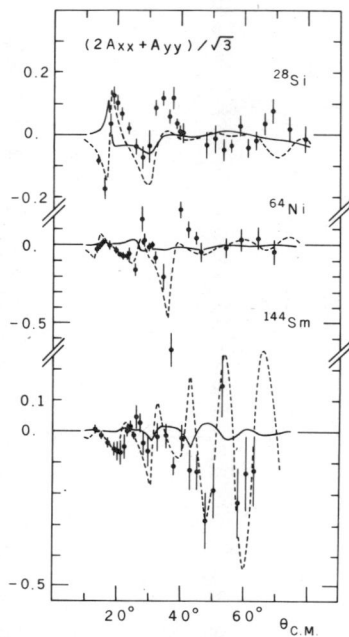

Fig.2. Angular distributions of the quantity $(2A_{xx}+A_{yy})/\sqrt{3}$. The different curves are explained in the caption to fig. 1.

parts of the optical potential, respectively. The combination B-D depends both on the tensor part and the second order of the spin-orbit part. Present analysis shows, however, only the spin-orbit potential can not explain the magnitudes of $(2A_{xx}+A_{yy})/\sqrt{3}$. The optical model predictions including the complex T_R potential can well reproduce both the magnitudes and phases of $(2A_{xx}+A_{yy})/\sqrt{3}$.

The real part of the tensor potential was much smaller than the predictions from the folding model[6]. The deduced dominance of the imaginary T_R potential has not been fully understood, although it was suggested by Rawitscher et al.[7] that the break-up process should be taken into account exactly.

REFERENCES

1) K. Hatanaka et al., Nucl. Phys. A340 (1980) 93.
2) D.J. Hooton and R.C. Johnson, Nucl. Phys. A175 (1971) 583.
3) K. Imai et al., RCNP annual report (1978) p. 154.
4) N. Matsuoka et al., contributed paper to this symposium.
5) H. Ohnishi et al., Private communication.
6) P.W. Keaton and D.D. Armstrong, Phys. Rev. C8 (1973) 1692.
7) G.H. Rawitscher and S.N. Mukherjee, Proc. INS Int. Symp. on nuclear direct reaction mechanism, Fukuoka (1978) p. 236.

FEATURES IN THE ANALYZING POWERS IN DEUTERON ELASTIC SCATTERING NEAR 80 MeV*

E. J. Stephenson, C. C. Foster, and P. Schwandt
Indiana University Cyclotron Facility, Bloomington, IN 47405

D. A. Goldberg
University of Maryland, College Park, MD 20742

Recent measurements at deuteron bombarding energies above 50 MeV show that after a few oscillations the amplitude of the diffractive pattern in the cross section is damped, to be replaced by a nearly smooth exponential decline with angle.[1-2] Measurements of A_y above 45 MeV show a damping of the diffractive oscillations at large angles (> 50°), and a dramatic rise in the average value of the analyzing power to nearly unity.[2-4] This behavior is similar to that observed in the cross sections for ^{4}He projectiles at similar energies/nucleon, and is understood to be an effect of the real nuclear potential (nuclear rainbow scattering).[5] In this contribution we will extend these ideas to the case with spin, and show how they give rise to saturation of the large angle analyzing powers. Calculations will be made for the specific case of 80 MeV deuterons scattering from ^{58}Ni. The optical potentials are from set L of Daehnick et al.[6]

In a semi-classical description of the scattering, one may calculate the deflection function, or angle to which particles are scattered, as a function of impact parameter (or orbital angular momentum). These angles are positive for the Coulomb-dominated large impact parameters, and swing to negative values as the impact parameter decreases and the attraction of the real nuclear potential is felt. For sufficiently high energy (above 35 MeV in this

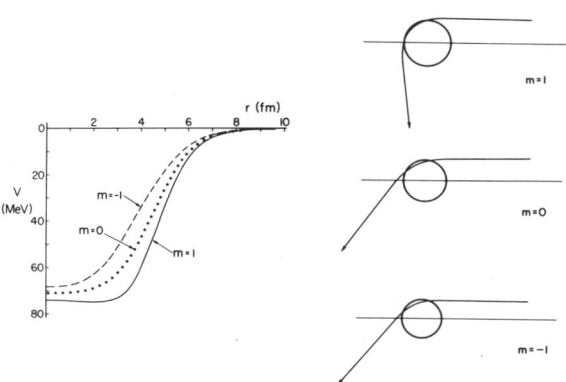

Fig. 1. (left) Sum of central and spin-orbit potentials for the m = 1, 0, and -1 projections of the deuteron spin on the orbital angular momentum axis. (right) Classical trajectories leading to the largest scattering angle for each real potential.

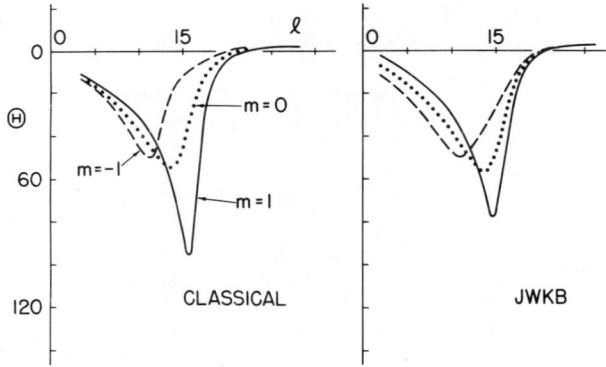

Fig. 2. Deflection functions for deuterons scattering from the three real wells of Fig. 1, calculated using classical and JWKB methods.

example) there is a maximum negative angle. In the optical model calculation, this angle marks the transition in the angular distributions from diffractive structure to an exponential decline.

In the presence of a vector spin-orbit potential, there are three real wells, corresponding to the three projections (m = 1, 0, and -1) of the deuteron spin along the orbital angular momentum axis. These are shown in Fig. 1 at 17 $\hbar$. Deflection functions calculated by integrating the classical equations of motion are shown in Fig. 2. The steepness of the potential for m = 1 deuterons causes these deuterons to experience the largest central force and to scatter to the most negative angles. (The trajectory for scattering to the maximum angle is shown in Fig. 1, superimposed on a circle depicting the half-radius of the potential for each spin state.) Thus, m = 1 deuterons dominate the large angle cross section. Classically, these projections may be associated with pure spin state polarizations of the incoming beam, where they give rise to three polarized cross sections (σ_1 for m = 1, etc.) In terms of these cross sections, the analyzing powers become

$$A_y = \frac{\sigma_1 - \sigma_{-1}}{\sigma_1 + \sigma_0 + \sigma_{-1}} \quad \text{and} \quad A_{yy} = \frac{\sigma_1 + \sigma_{-1} - 2\sigma_0}{\sigma_1 + \sigma_0 + \sigma_{-1}}.$$

At large angles σ_1 dominates, and both A_y and A_{yy} approach unity. The deflection functions on which this conclusion rests are essentially the same when calculated using the JWKB method of Ref. 8, as shown in Fig. 2.

Fig. 3 shows the optical model calculation for deuterons on ^{58}Ni with the cross section decomposed into the contribution from each spin projection in the beam. In accordance with the semiclassical description the three cross sections begin their decline at different angles, with the m=1 at the largest angle. Initially, the differences are emphasized by different slopes as well, with the m=1 cross section falling most slowly. This arises because the range of impact parameters giving large deflections for m=1 is narrow (see Fig. 2), a consequence also of the steepness of the potential slope.[7] At large angles the m=1 cross section dominates by at least an order of magnitude, giving large values of A_y and A_{yy} as shown in Fig. 3. These are general features of deuteron scattering, and should arise for all targets except the lightest at these energies.

Additional calculations were made including the tensor potentials, T_R and T_P. The radial forms were taken from Refs. 8 and 9, respectively. The addition of tensor potentials has almost no effect on the analyzing powers shown in Fig. 3, except for changes in the A_{yy} diffraction pattern at forward angles. In the rainbow region, these potentials create positive or negative shifts in the average values of other analyzing powers, and generally increase the magnitude of any oscillating pattern. The A_{xz} analyzing power is affected only by T_R. Both potentials affect A_{xx}. The sign of the shift changes if the T_P radial form is used with a T_R spin dependence, indicating that at large angles, the ambiguity noted earlier between the effects [10] of these two spin dependences is removed.

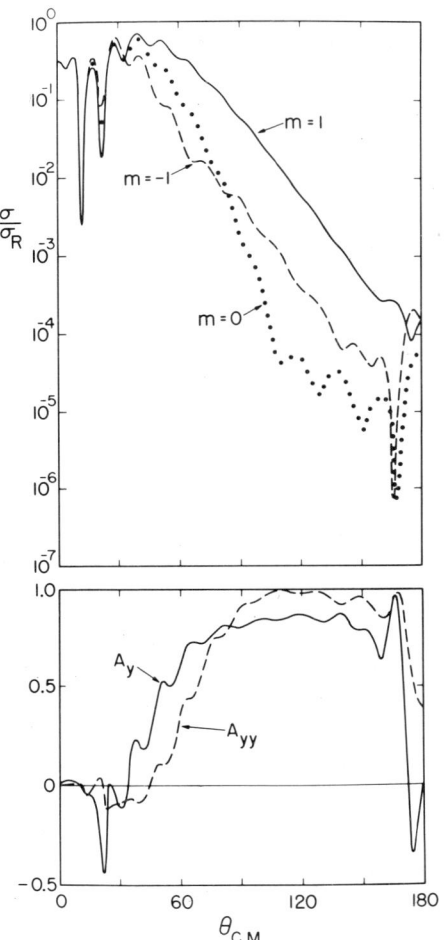

Fig. 3. Optical model calculations for 80 MeV deuterons on ^{58}Ni, with the cross section decomposed by deuteron spin state.

1. F. Hinterberger et al., Nucl. Phys. A111, 265 (1968).
2. C. C. Foster et al., Bull. Am. Phys. Soc. 24, 838 (1979).
3. E. J. Stephenson et al., Bull. Am. Phys. Soc. 22, 587 (1977).
4. G. Mairle et al., Nucl. Phys. A339, 61 (1980).
5. D. A. Goldberg et al., Phys. Rev. C 10, 1362 (1974).
6. W. Daehnick et al., to be published in Phys. Rev.
7. K. W. Ford and J. A. Wheeler, Ann. Phys. 7, 259 and 287 (1959).
8. G. Perrin et al., Nucl. Phys. A282, 221 (1977).
9. A. A. Ioannides and R. C. Johnson, Phys. Rev. C 17, 1331 (1978).
10. R. P. Goddard, Nucl. Phys. A291, 13 (1977).
*Work supported in part by the National Science Foundation.

MEASUREMENTS OF A_y and A_{yy} for 80 MeV DEUTERON ELASTIC SCATTERING ON ^{58}Ni and ^{208}Pb[*]

E. J. Stephenson, J. C. Collins, C. C. Foster, D. L. Friesel
J. R. Hall, W. W. Jacobs, W. P. Jones, S. Kailas
M. Kaitchuck, and P. Schwandt
Indiana University Cyclotron Facility, Bloomington, IN 47405

W. W. Daehnick
University of Pittsburgh, Pittsburgh, PA 15260

D. A. Goldberg
University of Maryland, College Park, MD 20742

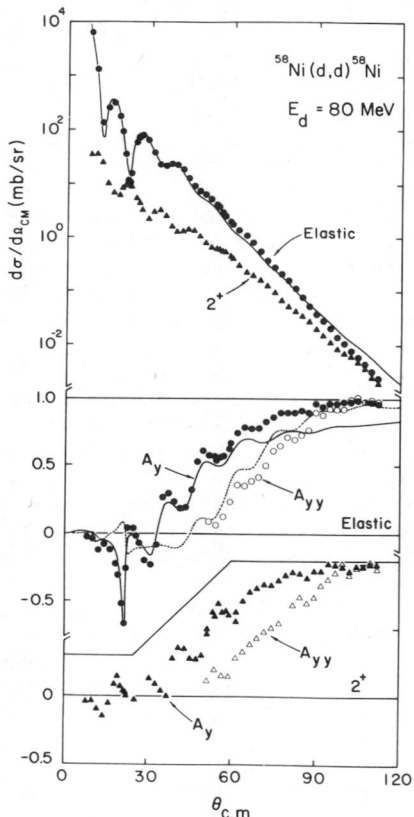

Fig. 1. Differential cross-section, vector (A_y) and tensor (A_{yy}) analyzing power angular distributions for ^{58}Ni $(d,d^*)(0.0, 1.45$ MeV$)$ at 80 MeV.

Measurements have recently been made of the 80 MeV elastic scattering cross section and vector analyzing power (A_y) for ^{58}Ni and ^{208}Pb, and the tensor analyzing power (A_{yy}) for ^{58}Ni. The polarized beam was generated in the Indiana atomic beam source. Both vector and tensor polarization was achieved using combinations of the weak field, and 3-5 and 2-6 intermediate field rf transitions. The beam on target varied between 10 and 50 nA, 15% of which was unpolarized. The polarization was monitored between the two cyclotrons by observing the ^{3}He$(d,p)^4$He reaction at 7.1 MeV using analyzing powers from Ref. 1. Elastically scattered deuterons were observed with the QDDM magnetic spectrograph. A split Faraday cup was used to dynamically center the beam on target, and to measure the integrated charge for computing polarized beam yields. The spin state of the beam was changed automatically about every 30 s during the run.

The results of this experiment are shown in Figs. 1 and 2. The cross sections contain strong oscillations at forward angles and a smooth exponential decline beyond 50°. At forward angles the analyzing powers oscillate

0094-243X/81/690484-3$1.50 Copyright 1981 American Institute of Physics

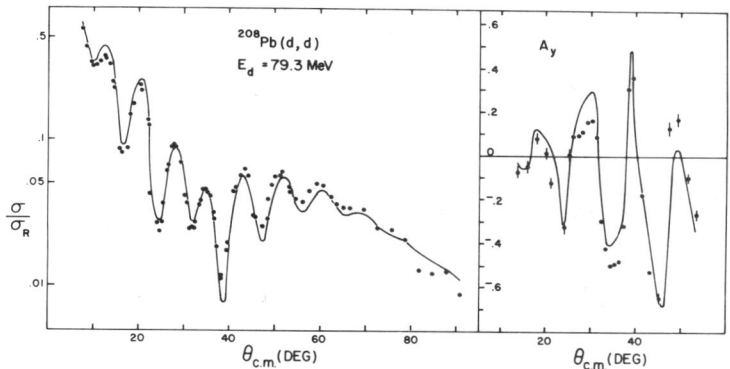

Fig. 2. Angular distributions for the differential cross-section and vector analyzing power for elastic scattering of 79.3 MeV deuterons from ^{208}Pb.

about zero; at back angles the oscillations are damped and the analyzing powers rise to nearly one. These smooth features at large angles are a result of the presence at these energies of a maximum classical angle of scatter from the real potential (nuclear rainbow scattering).[2] At large angles the cross section is produced almost exclusively by the scattering of deuterons with spin along $\vec{k}_{in} \times \vec{k}_{out}$.

The curves in Figs. 1 and 2 are optical model calculations. The potential parameters are taken from a global prescription based on a large range of incident energies and target masses (including analyzing power data).[3] Reasonable reproductions of the forward angle measurements are obtained. At large angles on ^{58}Ni, systematic deviations are noted, including an incorrect slope for the cross section and vector and tensor analyzing powers that are too small and too large respectively. Experience with ^{4}He scattering has shown the angle measurements are sensitive to the detailed radial shape of the optical potential in the nuclear surface.[4] So a complete analysis of these measurements is expected to yield information on both the central and spin orbit radial dependencies for deuterons.

Additional measurements for the 2^+ first excited state of ^{58}Ni are shown as well. While the diffractive oscillations are out of phase with those in elastic scattering, the same smooth trends are seen in both sets of data. Thus, the mechanism of Ref. 2 is also a general feature of inelastic scattering. Preliminary measurements on ^{58}Ni show that the large positive back-angle analyzing powers are common to all reaction channels, including inelastic scattering at excitation energies greater than 10 MeV as seen in Fig. 3. This result is consistent with a similar observation made recently for proton scattering and proton induced reactions to continuum final states.[5]

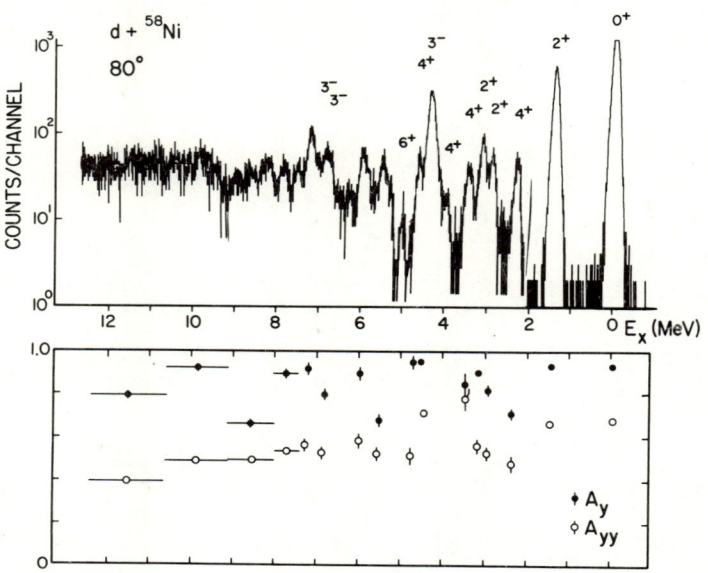

Fig. 3. Spectrum for 80 MeV deuteron inelastic scattering at 80 degrees (upper) and plot (lower) of vector (A_y) and tensor (A_{yy}) analyzing powers for indicated discrete states and continuum regions. All analyzing powers are large and positive.

REFERENCES

1. W. Grüebler et al., Nucl. Phys. A176, 631 (1971); H. R. Bürgi, private communication.
2. E. J. Stephenson et al., to be published and contribution to this proceeding.
3. W. W. Daehnick et al., Phys. Rev. C21, 2253 (1980).
4. D. A. Goldberg, Phys. Lett. 55B, 59 (1975).
5. H. Sakai et al., Phys. Rev. Lett. 44, 1193 (1980).

*Work supported in part by the National Science Foundation.

COMPLEX $\vec{L}\cdot\vec{S}$ TERM IN GLOBAL OPTICAL MODEL POTENTIALS FOR ELASTIC DEUTERON SCATTERING

W. W. Daehnick
University of Pittsburgh, Pittsburgh, PA 15260

ABSTRACT

Simultaneous optical model fits to deuteron scattering and vector polarizations suggest the need for a complex spin orbit term. Individual and global fits to available polarization data in the 15 to 80 MeV range indicate that the imaginary term changes in magnitude with E, and may become positive above 50 MeV.

INTRODUCTION

Utilizing a complex scattering potential of Woods-Saxon form with about 10 free parameters the optical model can give a good account of experimentally observed deuteron scattering cross sections. Furthermore, rather successful "global" prescriptions for a range of targets and energies have been found; only very light targets and energies below $\approx$12 MeV are excluded. Until recently the widest range of validity (12-52 MeV deuteron energy) belonged to the averaged potential obtained by Hinterberger and collaborators[1] from the analysis of their 52 MeV data. However, like other global potentials in the literature it cannot be extrapolated well towards higher energies, and for cross sections and vector polarizations in the 80 to 90 MeV range it is no longer adequate.

NEW GLOBAL POTENTIAL

A potential prescription has been deduced[2] which gives a good account of the available scattering data in the 12 to 90 MeV range. The wider energy range was obtained primarily by introducing a (weak) energy dependence in the diffuseness parameter a_0 and by a gradual change from surface to volume absorption with increasing E. Owing, in part, to these extra degrees of freedom, considerably improved fits (even at 52 MeV) were obtained. Fig. 1 shows a comparison of our global prescription with currently available data[3,4] in the 80-90 MeV range. Agreement at lower energies is of comparable quality.

At 80 MeV the magnitude of the L·S term and its geometric parameters are important factors in the elastic fits. Readjustments of these parameters in order to take better account of vector polarization (where known) lead to poorer fits to the scattering cross sections. In Fig. 2 known vector analyzing powers[5] in the 15 to 80 MeV range are shown. The dotted lines represent predictions of analyzing powers deduced from our global fits[2] to scattering data from 12 to 90 MeV by using real spin-orbit terms of conventional shape and magnitude ($V_{LS} \approx 6$ MeV, $r_{LS}=1.07$ fm, a=0.66 fm). Two deficiencies in these polarization curves are apparent: At

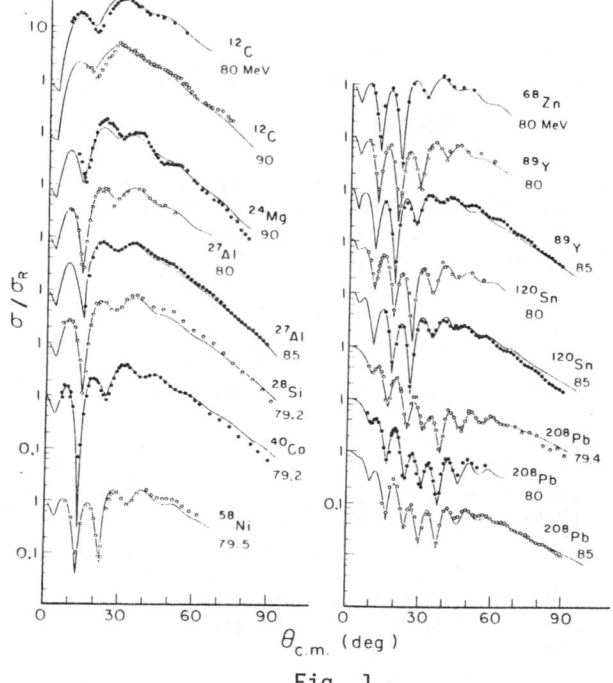

Fig. 1

15 MeV and 30 MeV, the computed $A_y(\theta)$ are shifted to the right with respect to the data. At 79.5 MeV the dotted curves (real L·S term) fail to reach the large positive polarization found in Ni for $\theta > 70°$. We note that such systematic difficulties tend to appear even for simultaneous fits to cross sections and polarization data for a <u>single</u> target and energy.

It has been pointed out[6] that for fits at ≤ 15 MeV the "phase" problem can be reduced considerably by adding an imaginary term W_{LS} (also of Thomas form) to the conventional spin-orbit potential. This approach is also useful in our global analysis at 15 MeV and at higher energies. Our searches indicate that W_{LS} increases with $\sim A^{1/3}$ if the central parameters are kept fixed. At 30 MeV one also needs negative W_{LS} values, but smaller in magnitude than at 15 MeV. For 52 MeV deuterons[7] we find the lowest χ^2 values if $W_{LS}(A)$ straddles zero. At 79.5 MeV W_{LS} must <u>not</u> be negative; for best fits the W_{LS} value for Pb is about 1 MeV less positive than that for Ni.

Generally, the introduction of an imaginary L·S term (characterized by $W_{LS}(E,A)$, $r_{ILS}=0.8$, $a_{ILS}=.25$) produces much improved fits to vector polarizations without hurting the cross section fits. Taking our global presentation, characterized by $r_0=1.17$, $a_0=0.717 + 0.0012$ E, etc., for the central potential (see "Potl. F" in Ref. 2 for parameters) we find that the real L·S term can be kept constant ($V_{LS}=5.0$, $r_{LS}=1.04$, $a_{LS}=0.60$) if W_{LS} is approximated by $W_{LS} = -0.35 A^{1/3} + 0.03$ E(MeV).

For energies at and above 50 MeV, the choice $r_{LS}=1.04$, $a_{LS}=0.60$ may not be optimal. Best fit values as large as $r_{LS}=1.24$ were found in Ref. 7, and $r_{LS}=1.12$ is preferred at 80 MeV, but these values are less useful at low E. The curves from our complex L·S prescription are shown as solid lines in Fig. 2. Visible improvement is found in the relative phase of data and computed analyzing powers at 15 and 30 MeV. (The improvement factor in χ^2 is 0.66.)

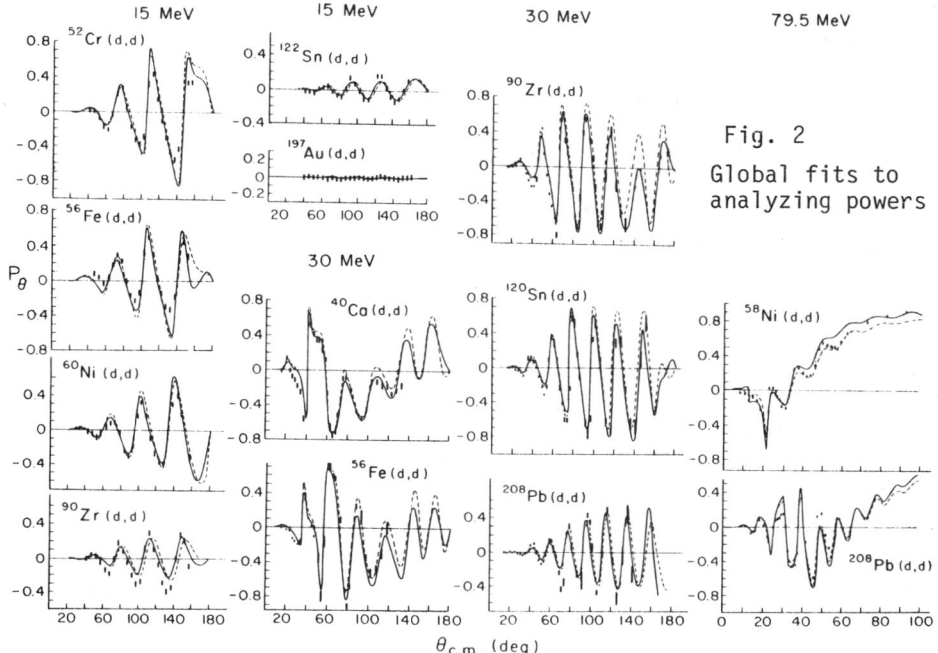

Fig. 2
Global fits to analyzing powers

The improvement at 79.5 MeV depends critically on the weighting of data at angles past 60°. The use of a substantial, positive W_{LS} so far has been the only way to reproduce the $A_y(\theta) > .90$ values which have been found experimentally for ^{58}Ni. The need for a positive imaginary L·S term is reminiscent of intermediate energy proton scattering, and may be understood from a folding construction of the deuteron potential.

REFERENCES

1. F. Hinterberger et al., Nucl. Phys. A111, 265 (1968).
2. W.W. Daehnick, J.D. Childs, and Z. Vrcelj, Phys. Rev. C21 (1980).
3. G. Duhamel et al., Nucl. Phys. A174, 485 (1971); O. Aspelund et al., Nucl. Phys. A253, 263 (1975); A. Kiss et al., Nucl. Phys. A262, 1 (1976).
4. The recent 85 MeV and 79 MeV data have not been reported in final form. Experimental cross sections were taken from the following reports: J. Bojowald et al., Annual Reports KFA Jülich (1974), (1976), (1978). C.C. Foster et al., Bull. Am. Phys. Soc. 24, 594 (1979).
5. R.A. Hardekopf et al., Los Alamos Report LA 5051 (1972); G. Perrin et al., Nucl. Phys. A282, 221 (1979); C.C. Foster, et al., Bull. Am. Phys. Soc. 24, 838 (1979), and private communication.
6. R.P. Goddard and W. Haeberli, Phys. Rev. Lett. 40, 701 (1978).
7. G. Mairle et al., Max Planck Inst., Heidelberg, Report (1980), and private communication.

THE DEUTERON OPTICAL POTENTIAL[*]

G. H. Rawitscher[**]
Center for Theoretical Physics
Laboratory for Nuclear Science and Department of Physics
Massachusetts Institute of Technology
Cambridge, Massachusetts 02139

S. N. Mukherjee
Banaras Hindu University
Varanasi, India

All is not well with the deuteron optical potential. Good global parameters for $d\sigma/d\Omega$ and iT_{11} do exist[1] (see however [2]) but problems arise in the standard DWBA treatment of rearrangement reactions. Not well fitted are the j dependence of stripping cross sections[3], the asymmetries and proton polarizations for ^{116}Sn(d,p)^{117}Sn near the Coulomb barrier[4] and the analyzing power in the ^{24}Mg($\vec{p}$,d) reactions at 95 MeV[5]. Both the Wisconsin[5] and the ETH groups[2] suspect that the central part of the deuteron optical potential is mainly at fault. Since both the (d,p) cross sections as well as the analyzing powers are better fitted when the adiabatic potential of Johnson and Soper is used[6] rather than a elastic optical potential, part of the problem may be due to the presence of breakup effects, which are inadequately taken into account in the DWBA.

Breakup has two effects: 1) the elastic optical potential becomes non local, 2) the rearrangement amplitude can occur both via transitions from the bound (the elastic) channel, as well as from breakup channels. The present note concerns itself with 1).

In order to investigate the breakup corrections to the deuteron optical potential, we illustrate in this note the "Trivially Equivalent Local Potentials" (TELP) for 45 and 80 MeV deuterons incident on the nucleus of Nickel. The TELP's are obtained by solving the elastic-breakup coupled equations, and then taking the coupling terms which are present in the elastic channel and dividing them by the elastic wave function. The TELP thus replaces the effect of the coupling terms by means of a correction to the deuteron optical potential. The TELP is local but it depends both on L, the deuteron-nucleus orbital angular momentum, as well as J the total angular momentum ($\vec{J}=\vec{L}+\vec{S}$). We express the J dependence by decomposing the TELP into a central, a spin orbit and a tensor component[7]. The spin orbit and tensor parts have been given previously[8] for incident energies of 13 and 21.6 MeV. However, since we do not solve the coupled equations exactly, but only to second order in the breakup potential, our results

---

[*] Supported in part through funds provided by the U.S. Department of Energy under contracts EG-77-5-02-4444.A002, DE-AC02-76ERO3069 and by the Council for the International Exchange of Scholars, Washington, D.C.

[**] On leave of absence 1979-80 from the University of Connecticut

are only approximate.

In Figs. 1 and 2 the central and spin orbit parts of the TELP's are shown for deuteron incident on ^{58}Ni for 45 and 80 MeV incident energies. With the exception of the L=12 curves for 80 MeV, only results for the "surface" partial waves are shown. The "interior" partial waves represent the transmission of a pair of nucleons through nuclear matter, which necessitates the Pauli exclusion principle, not included here. The assumptions for the calculation are as described in Refs.7,8. Only breakup momentum bin 1 is included here. The Watanabe or folding potential is used for the deuteron potential before breakup corrections.

The conclusions one can draw from the figures are
1) The TELP's markedly decrease with increasing deuteron energy.
2) The real parts of the TELP's are more L dependent than the imaginary parts.
3) The bulk of the imaginary part of the central TELP is located at larger radial distances than the real part. Indeed, beyond 7 fm the former is nearly equal to the Watanabe value, thus causing a near doubling of the overall imaginary central potential. In view of 2) it is reasonable that a local, L independent complex optical potential should be able to represent the major portion of the breakup effect.
4) The real part of the central TELP is quite "ondulatory" as a function of R.

One is tempted to dismiss this ondulation as an arifact of the definition of the TELP, which contains in the denominator the elastic wave and its respective zeros. However, a Green's function which represents the propagation of breakup components contains wavelengths which are different from the elastic channel (in view of the breakup Q value) and hence ondulations in the TELP are to be expected on physical grounds. Interestingly, as shown by Kobos and MacKintosh[9] an L dependent potential gives rise to an equivalent L independent potential which is also very ondulatory.

Is this ondulation the signature of nonlocality? If so, then the possibility exists that a ondulatory term introduced ad hoc into the deuteron optical potential may simulate the breakup effects, at least for the surface partial waves.

REFERENCES

1. W.W. Daehnick, J.D. Childs and Z. Vrcelj, Phys. Rev. C, to be published.
2. H.R. Burgi, W. Gruebler, J. Nuzyuski, V. Konig, P.A. Schmelzbach, R. Risler, B. Jenny and RA. Hardekopf., Nucl. Phys. A334, 413(1980).
3. N. Kishida and H. Ohnuma, J. Phys. Japan 46, 1375(1979) and in the Proc. INS Int. Symp. on Nuclear direct reaction mechanisms, Fukuoka, Japan (1978) Ed. M. Tanipuji and K. Yazaki, p. 221.
4. R.R. Cadmus, Jr., and W. Haeberli, Nucl. Phys. A327, 419(1979).
5. D.W. Miller, D.W. Devins, W.W. Jacobs, W.P. Jones, J.R. Shepard, Bull. Am. Phys. Soc. 25, 522 (1980).
6. R.C. Johnson and P.J.R. Soper, Phys. Rev. C1, 976(1970).

7. G.H. Rawitscher and S.N. Mukherjee, Phys. Rev. Lett. <u>40</u>, 1486 (1978).
8. G.H. Rawitscher and S.N. Mukherjee, Ann. Phys. (N.Y.) <u>123</u>, 330 (1979), also, Nucl. Phys. to be published.
9. A.M. Kobos and R.S. MacKintosh, Ann. Phys. (N.Y.) <u>123</u>, 296 (1979).

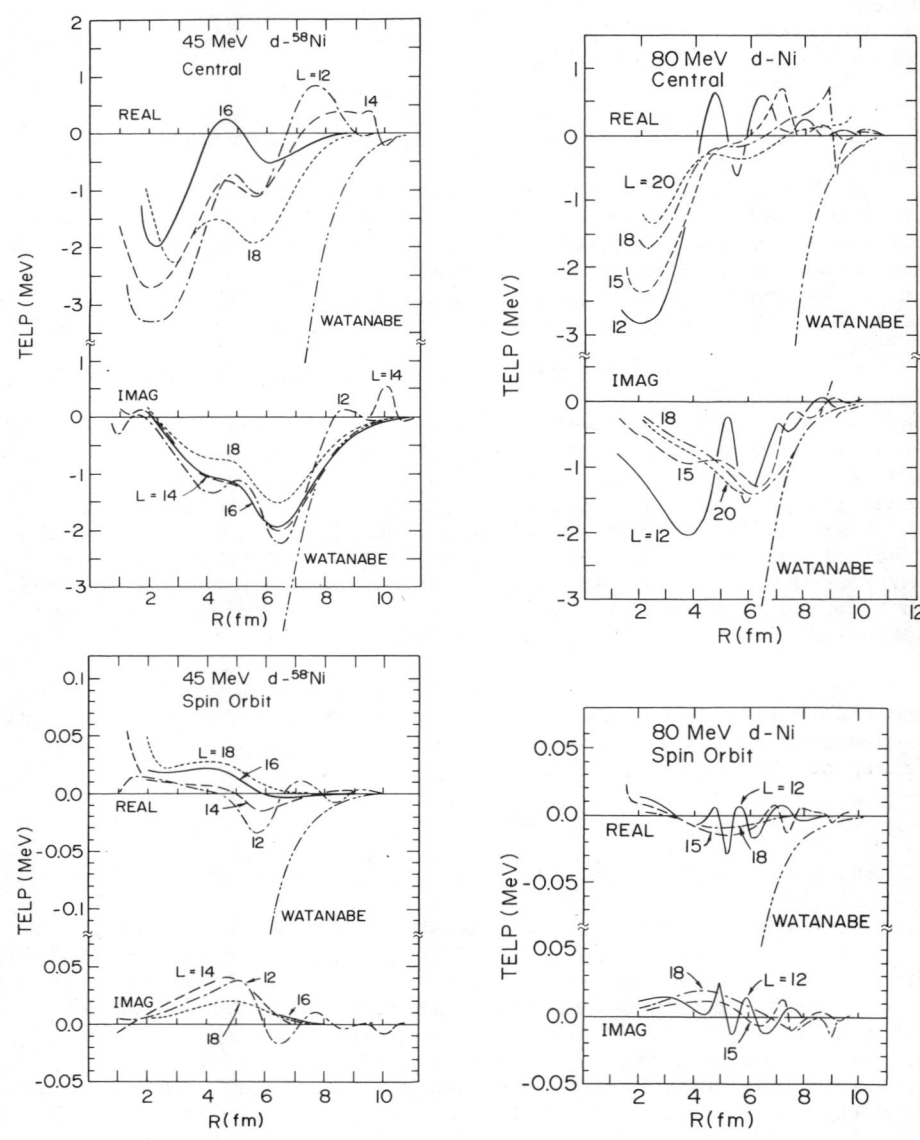

NEW REPRESENTATION OF ANALYZING POWERS IN ELASTIC SCATTERING OF POLARIZED DEUTERONS

H. Ohnishi and M. Tanifuji
Hosei University, Fujimi, Chiyoda, Tokyo

H. Noya
Western Michigan University, Kalamazoo, Mich. 49008

ABSTRACT

New linear combinations of tensor analyzing powers of polarized deuterons are proposed to identify separately individual effects of spin-dependent interactions in elastic scattering from 0^+ target nuclei. One of them, T_{ST} is shown to be particularly sensitive to tensor interactions, real contributions of which is identified in analyses of experimental data.

INTRODUCTION

Tensor terms in deuteron-nucleus interactions have been investigated in many ways, where tensor analyzing powers of polarized deuterons in elastic scattering have been found to be influenced strongly by the tensor terms. However, the tensor analyzing powers contain contributions both of the tensor term and the spin-orbit term in a complicated way and thus pure effects of the tensor term can hardly be seen in these quantities. In reference 1, it has been pointed out that a linear combination of the tensor analyzing powers is useful to identify the contribution of the tensor interaction but the theoretical development depend inevitably on the perturbational treatment of the spin-dependent interactions and its applicability is limited to particular angular ranges.

This note proposes new linear combinations of the analyzing powers in the case of spin-zero target nuclei, which can discriminate the effects of the tensor interaction from those of other interactions. The theoretical formulation is free from the above restrictions in ref. 1.

THEORETICAL FORMULATION

Scattering amplitudes $M(\theta)$ for the deuteron elastic scattering from spin-zero targets can be described by[2]

$$M(\theta) = \begin{pmatrix} A & B & C \\ D & E & -D \\ C & -B & A \end{pmatrix} \qquad (1)$$

where

$$C = A - E - \sqrt{2}(D+B)\cot\theta.$$

The cross section $\sigma(\theta)$, the vector analyzing power $iT_{11}(\theta)$ and the tensor analyzing powers $T_{2K}(\theta)$ are expressed as

$$\sigma = [2(|A|^2 + |B|^2 + |C|^2 + |D|^2) + |E|^2]/3, \qquad (2)$$
$$iT_{11} = -i[(A^*D - AD^*) + (B^*E - BE^*) - (C^*D - CD^*)]/\sqrt{6}\sigma,$$
$$T_{20} = \sqrt{2}[|A|^2 + |B|^2 + |C|^2 - 2|D|^2 - |E|^2]/3\sigma,$$
$$T_{21} = -[(A^*D + AD^*) + (B^*E + BE^*) - (C^*D + CD^*)]/\sqrt{6}\sigma$$

and $T_{22} = [(A^*C + AC^*) - |B|^2]/\sqrt{3}\sigma.$

By close examinations of A~E, it is found that the following S, T and U have characteristic features concerning the contribution of each type of the interaction, the central, spin-orbit and tensor ones.

$$S \equiv B + D, \qquad (3)$$
$$T \equiv B - D$$
$$\text{and} \qquad U \equiv A + E - C,$$

where the dominant contributions arise from the first order of the spin-orbit interaction in S, the first order of the tensor interaction and the second order of the spin-orbit interaction in T and the central interaction in U. The new representations of the analyzing power which we postulate are

$$T_{SU} = \text{Im}(SU^*)/\sigma, \qquad (4)$$
$$T_{ST} = \text{Re}(ST^*)/\sigma$$
$$\text{and} \qquad T_{TU} = \text{Re}(TU^*)/\sigma.$$

These quantities are related to the conventional analyzing powers as

$$T_{SU} = -\sqrt{6}\,iT_{11}, \qquad (5)$$
$$T_{TU} = \sqrt{3/2}\,[(\sqrt{3/2}\,T_{20} - T_{22})\cos\theta - 2T_{21}\sin\theta]\sin\theta$$
$$\text{and} \qquad T_{ST} = \sqrt{3/2}\,[(\sqrt{3/2}\,T_{20} - T_{22})\sin\theta + 2T_{21}\cos\theta]\sin\theta.$$

Eqs. (3) and (4) suggest that T_{SU} depend strongly on the spin-orbit interaction but very weakly on the tensor interaction and T_{ST} and T_{TU} depend on both of the spin-orbit interaction and the tensor one.

NUMERICAL RESULTS

Numerical calculations justify the above speculation, which are shown in fig.1, where both of T_{ST} and T_{SU} vanish in the limit of V_{so} (spin-orbit strength)=0, while T_{ST} vanishes in the limit of V_T(tensor strength)=0. The last quantity T_{ST} is proportional to the strength of the tensor interaction. From these results, it should be emphasized that T_{ST} is a good measure of the strength of the tensor interaction. Fig.2 shows T_{ST} derived from the experimental data[3,4] on T_{20}, T_{21} and T_{22}, where the non-vanishing values of T_{ST} prove the real contribution of the tensor interaction.

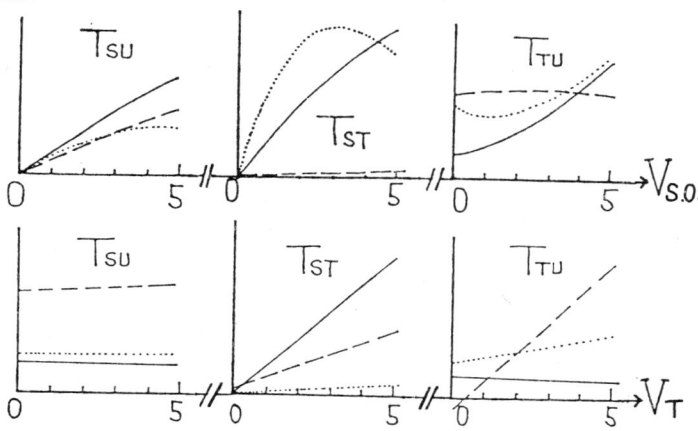

Fig.1. Behaviours of T_{SU}, T_{TU} and T_{ST} for ^{90}Zr at E_d=15 MeV. The dashed, solid and dotted lines are for θ=45°, 90° and 135°, respectively. The unit of abscissa is MeV.

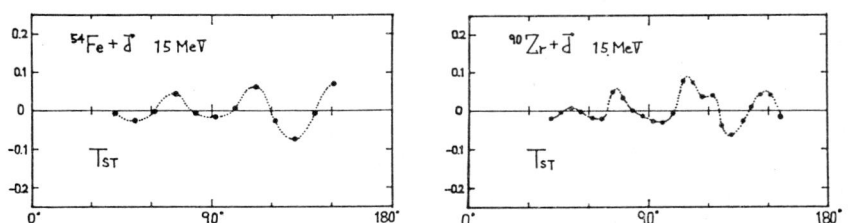

Fig.2. Angular distributions of T_{ST}. The lines are for guiding eyes.

REFERENCES

1. D.J.Hooton and R.C,Johnson, Nucl.Phys.A175,583 (1971)
2. B.A.Robson,The Theory of Polarization Phenomena (Clarendon Press Oxford,1974)
3. F.T.Baker et al., Nucl.Phys.A233,409(1974)
4. H.R.Burgi et al., Nucl.Phys.A321,445(1979)

Polarized Triton Scattering from ^{26}Mg, ^{27}Al and ^{28}Si at 17 MeV*

R. A. Hardekopf, Ronald E. Brown, F. D. Correll, and G. G. Ohlsen
Los Alamos Scientific Laboratory, Los Alamos, NM 87545 USA

P. Schwandt
Indiana University, Bloomington, IN 47401 USA

We have measured differential-cross-section and analyzing-power angular distributions for 17 MeV tritons elastically scattered from targets of ^{26}Mg, ^{27}Al, and ^{28}Si in the angular range 20° to 160°. The experiment was performed at the Los Alamos Scientific Laboratory Van de Graaff facility using the Lamb-shift polarized triton source[1] and the supercube[2] scattering chamber. A pair of detector telescopes with angular resolutions of ± 0.4° detected the reaction products, with mass identification and storage performed by an on-line computer. The triton beam intensity available at the target was about 70 nA with a polarization of 0.77. The target thicknesses were about 3 mg/cm^2, although thinner targets were used for the ^{27}Al forward-angle data.

We recently completed a study[3] of the triton optical model (OM) for targets with masses in the range 40 ≤ A ≤ 208. The present purpose was to extend that study to somewhat lighter nuclei. In addition, we wanted to obtain OM parameters in this lighter mass range for reaction calculations and for comparison with the results of polarized ^{3}He (helion) scattering.[4] The polarized triton data are shown in Figs. 1-3 along with curves from the OM calculations.

Of the 3 nuclei studied, the ^{26}Mg data showed the most regular pattern of oscillations in both the differential cross-section and analyzing power distributions, and most of our calculation efforts attempted to fit these data. Although the fit is acceptable for the cross sections, it is only marginal for the polarization data, a feature that we have also noted in other polarized

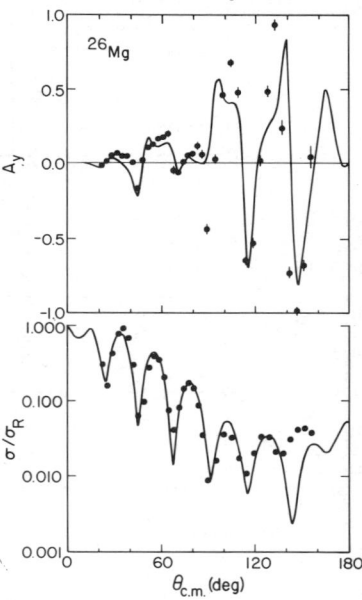

Fig. 1. Analyzing powers A_y and differential cross-sections σ/σ_R (ratio-to-Rutherford) for 17 MeV triton elastic scattering from ^{26}Mg. The curves are from OM calculations using the parameters in Table I.

---

*Work supported by the U.S. Department of Energy.

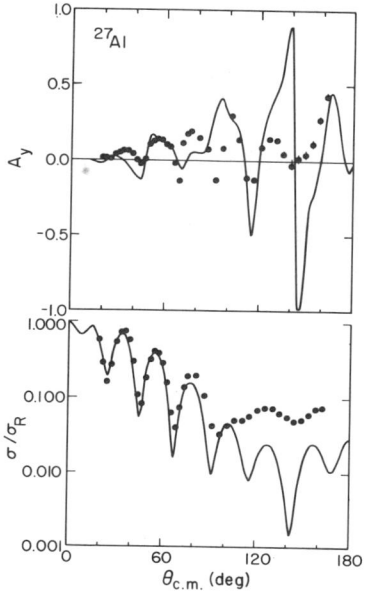

Fig. 2. Triton elastic scattering from ^{27}Al at 17 MeV. See caption to Fig. 1. The curves are from OM calculations using the ^{26}Mg parameters.

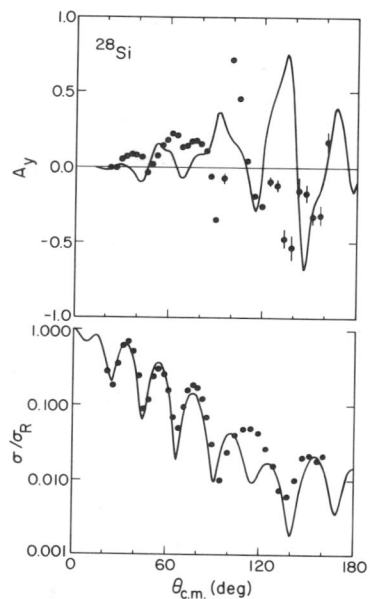

Fig. 3. Triton elastic scattering from ^{28}Si at 17 MeV. See caption to Fig. 1. The curves are from OM calculations using the ^{26}Mg parameters.

triton studies. The analyzing power data seem to provide a severe test of the applicability of the OM in this mass and energy range and thus constrain the acceptable parameters more than do the cross section data alone. The curves shown with the ^{27}Al (Fig. 2) and ^{28}Si (Fig. 3) data use the same parameters found for the ^{26}Mg data. Although some additional calculations were made for these nuclei, no significant improvement could be achieved in the fits.

It can be noted for all of the cross-section distributions that a departure from the OM calculations occurs at backward angles, past 135° for ^{26}Mg and past 100° for ^{27}Al and ^{28}Si. In the latter cases, the analyzing power magnitudes are also reduced, pointing to the importance of compound nuclear contributions to the scattering in this angular range. A Hauser-Feshbach compound-elastic contribution was added to the ^{26}Mg calculation, improving the cross-section fit near 150°, but the overall improvement was slight. We could find no combination of parameters that give a satisfactory fit to the middle-angle analyzing power data for ^{26}Mg while maintaining a good representation of the cross sections.

Table I gives the OM parameters for the curves shown in the figures along with parameters reported by the Birmingham group[4]

TABLE I. Optical model parameters found in the present study for ^{26}Mg(t,t) and from Ref. 4 for ^{26}Mg(^{3}He,^{3}He). Well depths are in MeV and other parameters are in fm.

	V_o	r_o	a_o	W_v	r_w	a_w	V_{so}	r_{so}	a_{so}
Triton	160	1.16	0.71	28.0	1.38	0.92	14.5	1.04	0.25
Helion	160	1.12	0.68	31.7	1.37	0.95	7.2	0.96	0.25

for helion scattering at 33.4 MeV. In spite of the energy difference in the two data sets and the fact that the helion polarization data extend only to 77°, the parameter agreement is quite good. This is the first case we have studied with polarized tritons in which there is clear evidence for a small spin-orbit radius, a feature often noted in the analyses of helion data. The real volume integral per nucleon for the triton potential is about 495 MeV fm^3, which is in the same family as the heavier nuclei we have studied.[3] Although a real potential radius of 1.16 fm resulted from the automatic searches carried out, a radius of 1.20 fm, as used in Ref. 3, gives an equally good fit if V_o is decreased to 150 MeV, maintaining roughly the same volume integral.

The spin-orbit well depth of 14.5 MeV is higher than the average of 6 MeV found for heavier nuclei, and much higher than the folding-model prediction of about 2.5 MeV. In the present study, the analyzing-power fit is not good enough to warrant a firm conclusion, but the result does support our earlier indications[3,5] that there is something wrong with the assumptions of the simple folding-model.

---

1. R. A. Hardekopf, in Proc. 4th Int. Symp. on Polarization Phenomena in Nuclear Reactions, Zürich, 1975, ed. by W. Grüebler and V. König (Birkhäuser, Basel, 1976), p. 865.
2. G. G. Ohlsen and P. A. Lovoi, in Proc. 4th Int. Symp. on Polarization Phenomena in Nuclear Reactions, Zürich, 1975 (see Ref. 1) p. 907.
3. R. A. Hardekopf, R. F. Haglund, Jr., G. G. Ohlsen, W. J. Thompson, and L. R. Veeser, Phys. Rev. C 21, 906 (1980).
4. M. D. Cohler, N. M. Clarke, C. J. Webb, R. J. Griffiths, S. Roman and O. Karban, Journal of Physics G 2, L151 (1976).
5. R. A. Hardekopf, L. R. Veeser, and P. W. Keaton, Jr., Phys. Rev. Lett. 35, 1623 (1975).

IMAGINARY SPIN-ORBIT OPTICAL POTENTIAL IN TRITON ELASTIC SCATTERING ON ^{9}Be

J. Meyer and E. Elbaz
Institut de Physique Nucléaire (and IN2P3)
Université Claude Bernard Lyon-I
69622 Villeurbanne Cedex, France

Recently T.F. Hill and W.E. Frahn [1] have interpreted $^3\vec{\text{He}}$ elastic scattering data on various nuclei using a spin-orbit potential with an imaginary part. Such a part appears necessary to obtain good results for experimental cross sections and analyzing powers. We have investigated this effect for the $\vec{t} + ^9$Be system at $E(t) = 15$ and 17 MeV. The experimental data are taken from P.A. Schmelzbach et al. [2]. These authors have built up two sets BA and BB which give a quite well agreement with the data but only at forward angles. The top of Fig. 1 shows these results up to 180 degrees and exhibits then some large deviations at backward angles for the cross sections and for the analyzing powers. Using the Optical Model MAGALI Code of J. Raynal, we have built up two other sets of optical parameters. These new sets are listed in Table 1 and their results shown at the bottom of Fig. 1.

The central imaginary part has a surface form (Saxon Wood derivation) and all the potentials are defined by J. Raynal [3]. The first set (BC 15 and BC 17), for the two energies respectively, uses only a real spin-orbit term ; it gives very satisfying results especially for the analyzing powers in the whole angular range. The second set (BD 15 and BD 17) uses an imaginary spin-orbit term and reduces the χ^2 - square without major modifications to the fits. These sets of optical parameters use an absorptive surface term and are characterized by a rather large V_{so}. The differences between BC and BD sets are not very important for the two energies and it seems very difficult to conclude about an evidence for a strong imaginary term in the spin-orbit optical potential for these elastic scatterings.

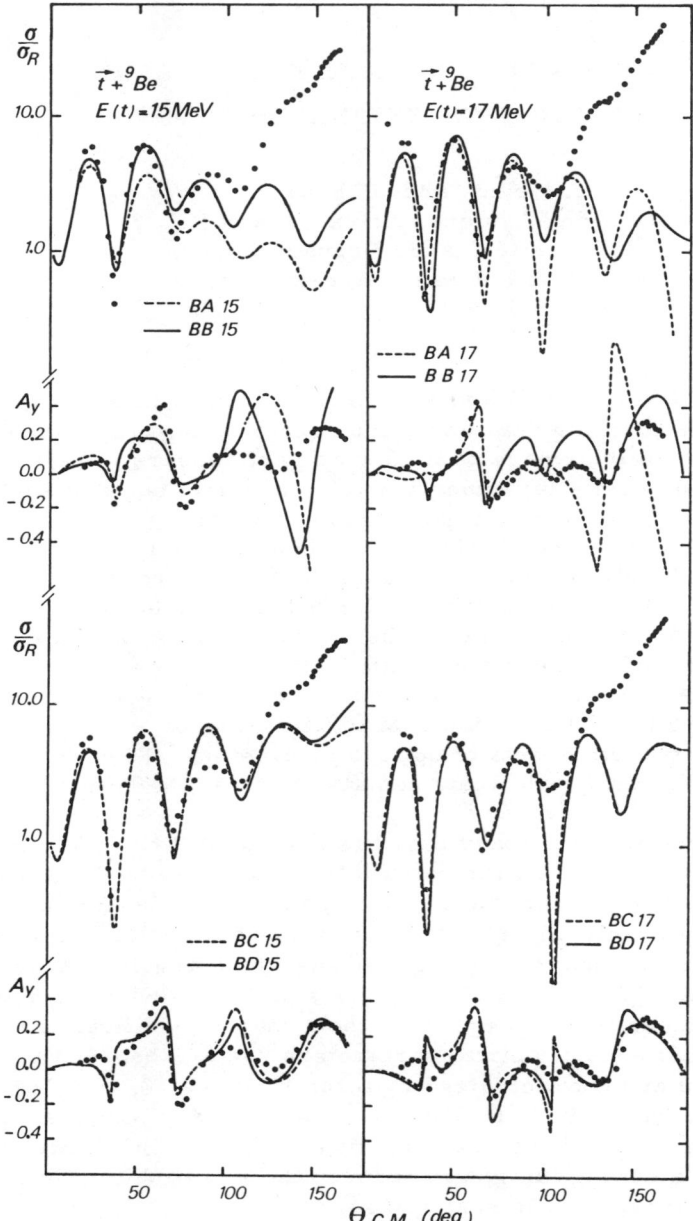

Fig. 1. $\vec{t}$ + ^{9}Be elastic scattering at E(t) = 15 and 17 MeV.

Table I. Optical parameters (All depths are in MeV ; all radii and diffusenesses are in fm - r_c = 1.3 fm).

Energy	Set		V	r_o	a_o	W	r_w	a_w
15 MeV	BC 15	Central	76.71	1.906	0.511	70.28	1.921	0.265
		Spin-orbit	13.30	1.624	0.390	0.	-	-
	BD 15	Central	77.43	1.882	0.525	70.36	1.921	0.265
		Spin-orbit	20.38	1.705	0.334	1.28	1.210	0.267
17 MeV	BC 17	Central	76.64	2.000	0.551	81.25	1.935	0.246
		Spin-orbit	15.35	1.597	0.356	0.	-	-
	BD 17	Central	76.64	2.000	0.551	81.25	1.935	0.246
		Spin-orbit	15.35	1.597	0.356	0.14	2.439	0.256

REFERENCES

1. T. F. Hill, W. E. Frahn, Ann. Phys. (N. Y.) **124**, 1 (1980).

2. P. A. Schmelzbach, R. A. Hardekopf, R. F. Haglund Jr., G. G. Ohlsen, Phys. Rev. C 17, 16 (1978).

3. M. A. Melkanoff, T. Sawada, J. Raynal, in Methods in Computational Physics. Vol. 6 : Nuclear Physics, edited by B. Adler, S. Fernbach, M. Rotenberg, Academic Press, New York, 1966.

ANALYSES OF THE SCATTERING OF POLARISED HELIONS FROM ^{32}S

J.M. Barnwell, N.M. Clarke and R.J. Griffiths
Wheatstone Physics Laboratory, King's College, London, England.

ABSTRACT

^{32}S($\vec{^{3}\text{He}}$,^{3}He)^{32}S and ^{32}S($\vec{^{3}\text{He}}$,α)^{31}S cross section and analysing power measurements have been made using the University of Birmingham polarised ^{3}He beam. The ^{3}He spin orbit potential has been investigated through an optical model analysis of the eleastic scattering, and also through both DWBA and CCBA analyses of the reaction data.

INTRODUCTION

The ^{3}He spin orbit potential is of interest because of the small value of its diffuseness parameter ($a_s \sim 0.2$ fm) required in all optical model analyses of ^{3}He polarisation data to date. This is in contrast with the results for tritons and the predictions of folding models where a more conventional geometry is indicated. It has been shown that asymmetries can be produced in the elastic scattering channel even in the absence of a spin orbit potential because of feedback from reaction channels[1]. If coupling effects are important for (^{3}He,α) reactions it is possible that the small values of a_s are unphysical in that they are simulating these feedback effects. If so, additional information about the potential might be obtained from reaction data. (^{3}He,α) reactions involving l=0 transfer are of special value in determining spin orbit potentials since, in the DWBA formalism, the reaction analysing powers are determined solely by the spin orbit distortion of the entrance channel wavefunction. Such data therefore provide a stringent test of the optical potential determined from the elastic polarisation data.

The ^{32}S nucleus was chosen for study because the ground state ^{32}S(^{3}He,α)^{31}S reaction proceeds via an l=0 transfer. Elastic scattering cross sections and polarisations were measured along with the cross sections and analysing powers for the ground and first two excited states in ^{31}S. The data were taken using the 33 MeV polarised ^{3}He beam from the University of Birmingham Radial Ridge Cyclotron.

OPTICAL MODEL ANALYSIS

The optical model analysis of the elastic scattering data[2] gave rise to ambiguities in both the central and spin orbit potential parameters. The analysis yielded six real central potential "families" corresponding to the discrete ambiguity and within each family it was usually possible to find several different spin orbit potentials which each gave comparable fits to the polarisation data. However, although the radius parameters of these ambiguous spin orbit potentials varied quite considerably, they all had diffuseness parameters which were small (≈0.2 fm). Figure 1 illustrates the fit to the polarisation data obtained with one of the potential parameter sets.

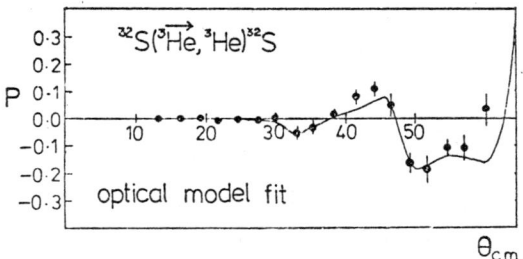

Figure 1.

DWBA ANALYSIS OF ^{32}S($^3\vec{\text{He}},\alpha$)^{31}S DATA

The DWBA calculations were performed with the zero range code DWUCK4[3] and the fits to the analysing power data are shown as dashed lines in Figure 2. The $3/2^+$ (j=l-1/2) and $5/2^+$ (j=l+1/2) states are quite well predicted and show the characteristic j-dependence of (^{3}He,α) reactions. This j-dependence is independent of the ^{3}He spin orbit potential and the predictions remain almost unchanged if the spin orbit potential is set to zero. This contrasts with the $1/2^+$ (l=0) data where the prediction becomes zero if the spin orbit potential is turned off.

The $1/2^+$ (l=0) analysing power data are not predicted by the DWBA using any of the sharply peaked spin orbit potentials required to fit the polarisation data. However, a reasonable fit to the l=0 data may be obtained if a more diffuse spin orbit potential is used (a_S>0.7) although this diffuse potential is unable to fit the elastic polarisations. CCBA analyses have been performed in an attempt to explain this discrepancy.

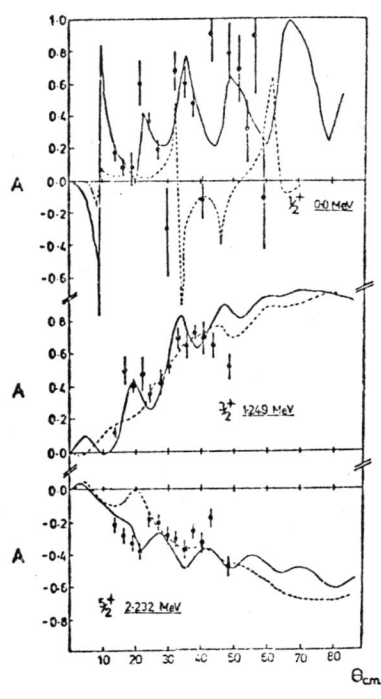

Figure 2   - - - - - - DWBA
                    ———— CCBA

CCBA ANALYSIS OF ^{32}S($^3\vec{\text{He}},\alpha$)^{31}S DATA

The CCBA calculations were performed with the King's College London version of CHUCK[3] using the coupling diagram shown in Figure 3. Initially coupled channels parameters were obtained for the entrance and exit channels and then the magnitudes and phases of the

various transition amplitudes were adjusted
to obtain the best fits to all the data sets.
The fits to the analysing powers are given
by the continuous lines in Figure 2, and show
a marked improvement over the DWBA fits. It
was found that the oscillatory structure of
the 1/2+ analysing power is determined largely
by interference between the inelastic coupling
to the 3/2+ and 5/2+ states in the exit
channel. The effect of two-step processes
via the 2+ state in ^{32}S is simply to shift the

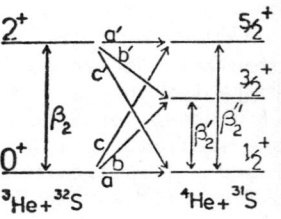

Figure 3.

trend of the prediction in a positive or negative direction. When
these couplings are included the prediction is fairly insensitive to
the ^{3}He spin orbit geometry.

REFERENCES

1. N.M. Clarke, M. D. Cohler, and R.J. Griffiths; J.Phys. G5 1233 (1979).
2. J.M. Barnwell, N.M. Clarke, M.D. Cohler, R.J. Griffiths, J.S. Hanspal, O. Karban and S. Roman; J.Phys. G5 L69 (1979).
3. P.D. Kunz; unpublished.

ELASTIC AND INELASTIC SCATTERING OF POLARIZED PROTONS THROUGH
ISOBARIC ANALOG RESONANCES IN ^{207}Bi AND ^{209}Bi.

N. L. Back, H. C. Bhang, J. G. Cramer,
T. A. Trainor, and R. Von Lintig
Nuclear Physics Laboratory
University of Washington, Seattle, WA 98195

ABSTRACT

Cross-section and analyzing-power excitation functions have been measured for elastic and inelastic scattering of polarized protons from ^{206}Pb and ^{208}Pb. These measurements are made at θ_{lab}=120°, 135°, 150°, and 165°, in the energy range E_p=14.25-18.00 MeV. Isobaric analog resonances in ^{207}Bi and ^{209}Bi were observed in the elastic scattering and in the inelastic scattering to the 2$^+$(0.803 MeV) and 3$^-$(2.647 MeV) states of ^{206}Pb and to the 3$^-$(2.615 MeV) state of ^{208}Pb. The elastic data has been analysed and total and partial widths extracted for each resonance. Analysis of the inelastic data is in progress.

We are continuing our program of polarized beam studies of the isobaric analog resonances in the lead region for the purposes of high-precision nuclear spectroscopy. Here we report the measurement of excitation functions of the scattering of polarized protons from ^{206}Pb and ^{208}Pb[1]. Cross-section and analyzing power excitation functions have been measured for these targets between 14.25 and 18.00 MeV, i.e., in the region of the single-particle isobaric analog resonances (IAR) in ^{209}Bi*. These resonances are observed in the elastic scattering and the scattering to the first 3$^-$ state in each target, and in the scattering to the first 2$^+$ state in ^{206}Pb.

The University of Washington Lamb-shift polarized ion source, operated in the fast-flip mode with the spin flipped every 100 msec, produced a beam of polarized protons. This beam was accelerated in the UW FN tandem accelerator and used to bombard self-supporting targets of isotopically enriched ^{206}Pb (1.1 mg/cm^2) and ^{208}Pb (0.6 mg/cm^2). The scattered protons were detected by an array of silicon detectors and their signals were routed into spin-up and spin-down arrays by the computer. Thus, systematic effects resulting from the passage of time between spin-up and spin-down runs were eliminated.

In addition to the excitation function data, runs were taken at 4.00, 6.00, 8.00, 10.00, 12.00, and 13.75 MeV with unpolarized beam to aid in the normalization of the cross-section data. Absolute normalization of the polarized-beam data was performed by employing the unpolarized data, after correcting for multiple scattering effects in the target at the lowest energies. Cross section and analyzing power angular distributions were also measured for each target at 13.75, 15.50, and 18.00 MeV, for use in determining optical-model parameters needed for the analysis of the excitation-function data.

An example of the data obtained for $\vec{p} + {}^{206}Pb$ are shown in the figure below. The smooth curves shown here represent a fit to the data using a parameterized-background approach, in which each background amplitude is expressed as a Taylor series with terms up to second order in $(E - E_o)$, with the non-spin-flip amplitude also containing an E^{-1} term. Here E_o is the average of the highest and lowest energies being considered The background coefficients and resonance parameters are allowed to vary simultaneously to fit the data.

The results of this analysis of the elastic $\vec{p} + {}^{208}Pb$ are completely consistent with previous work of our group[2,3] on this system. The resonance structure in ${}^{207}Bi^*$ is considerably more complicated. Here there is considerable mixing between single-particle states built on a ${}^{206}Pb$ ground-state core and those built on a 2^+ core. As a result, the single-particle strength for each (ℓ,j) is split among several IARs. This mixing is also the reason for the appearance of IARs in the scattering to the 2^+ state of ${}^{206}Pb$.

Preliminary results for ${}^{207}Bi^*$ are listed in Table 1. The resonance energies (relative to the $g_{9/2}$ IAR) were determined by identifying the IARs with some of the states observed in the ${}^{206}Pb(d,p)$ reaction[4]. IARs corresponding to many of the weaker states have not yet been identified, but some of them will be required to improve the fit to the data. The column labeled "Previous Spin Assignment" contains the spin-parity assignments given in Ref. 4. Where the numbers are in parentheses, the assignments are uncertain.

TABLE 1

Resonance Parameters for IARs in ${}^{207}Bi^*$ (Preliminary)

ℓ_j	Previous Spin Assignment[4]	Γ_p (keV)	Γ (keV)	E_R (MeV-c.m.)
$g_{9/2}$	$g_{9/2}$	19.7	191	14.851
$i_{11/2}$	$(i_{11/2})$	2.2	338	15.632
$d_{5/2}$	$d_{5/2}$	5.0	229	15.757
$d_{5/2}$	$(d_{5/2})$	8.7	269	16.442
$d_{5/2}$	$d_{5/2}$	32.5	269	16.510
$s_{1/2}$	$s_{1/2}$	44.7	275	16.750
$g_{7/2}$	$(g_{7/2})$	15.6	288	17.304
$d_{3/2}$	$(d_{3/2})$	43.2	279	17.342
$g_{7/2}$	$(d_{5/2}, g_{7/2})$	8.3	288	17.551

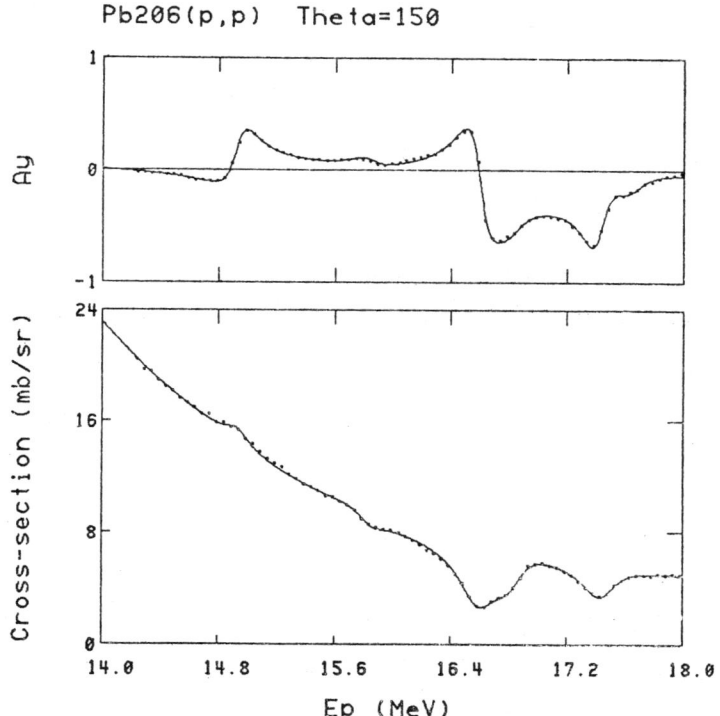

Cross-section and analyzing-power excitation functions for ^{206}Pb$(\vec{p},p_0)$. The curves are a fit using nine IARs and a parameterized background.

REFERENCES

1. Nuclear Physics Laboratory Annual Report, University of Washington (1979), p. 53.
2. P. von Brentano and J. G. Cramer, in Nuclear Spectroscopy and Reactions, Part B, J. Cerny, editor, pp.101-104, Academic Press, New York (1974).
3. M. P. Baker, J. S. Blair, J. G. Cramer, E. Preikschat, and W. Weitkamp, in Proceedings of the 4th International Symopsium on Polarization Phenomena in Nuclear Reactions, W. Grüebler and V. König, eds., p.781, Birkhaüser Verlag, Basel, (1975).
4. Table of Isotopes, 7th Edition, edited by C. M. Lederer and V. S. Shirley, John Wiley and Sons, New York, (1978).

GIANT RESONANCE ANALYSIS OF THE ^{54}Fe($\vec{p},p'$) REACTION[x]

P.J. van Hall, S.D. Wassenaar and J.P.M.G. Melssen[xx]
Eindhoven University of Technology, Eindhoven, Netherlands

In this note we present an analysis of data for the ^{54}Fe($\vec{p},p'$) reaction between 10 and 20 MeV. It is well known[1,2] that the analysing powers show anomalies, which can be accounted for only by an unphysically large spin orbit deformation. As this enhancement is clearly energy dependent[1], we have tried to analyse these data in terms of the giant resonance model of Von Geramb[3]. In order to get a clear picture we have performed additional experiments at 12.6 and 15.3 MeV.

The inelastic data were analysed first with the collective DWBA, from which we extracted the enhancement factor λ. As can be seen from fig. 1 the data at 17.2 MeV and 15.3 MeV clearly need a $\lambda \gtrsim 3$, while at 12.6 MeV $\lambda \gtrsim 1.5$ suffices. Though the descriptions of the 15.3 and 12.6 MeV data along these lines are far from perfect, they nevertheless give strong evidence for an energy dependence of 1 as given in fig. 2. Here we also present the values at other energies extracted by a similar analysis[1,2,4]. The value of $\lambda = 2$ even at 30 MeV supports the suggestion of Raynal[5] that proton excitations play a dominant role.

Next we performed an analysis with the code MEPHISTO[6] in which the excitation of a giant resonance can be included as an intermediate step. As our attempts to describe the direct part of the T-matrix microscopically failed badly[7] we used for it the collective DWBA with $\lambda = 2$. We then assumed simple $(f_{7/2})^{-2}$ proton configurations for both ground and excited states and used this for the resonance calculations. We tried several multipolarities for the intermediate GR but only an L = 3 assumption proved to be acceptable. By varying the complex coupling constant y_3 we tried to fit our data. The results of these atttempts can also be found in fig. 1. Again the fits are far from perfect, but some qualitative features of the data are certainly reproduced i.e. the large positive analyzing powers around 25 deg. and 90 deg. as well as the raising of the cross section at backward angles. The fitted value of y_3 shows the behaviour in amplitude and phase as can be expected from a resonance (fig.3), indicating an L=3 GR around 20 MeV with a width of 3 MeV, which exceeds the EWSR by a factor of 2. This factor of 2 can be explained by realizing that the assumed configuration is much too simple.

Recently evidence has been found for an L = 3 GR by direct excitation[8] and therefore we feel confident to conclude that in our experiment we also see the 3 $\hbar\omega$ part of the octupole strength.

---

[x] Supported in part by F.O.M./Z.W.O.
[xx] Present address: Computer Centrum Limburg, Heerlen.

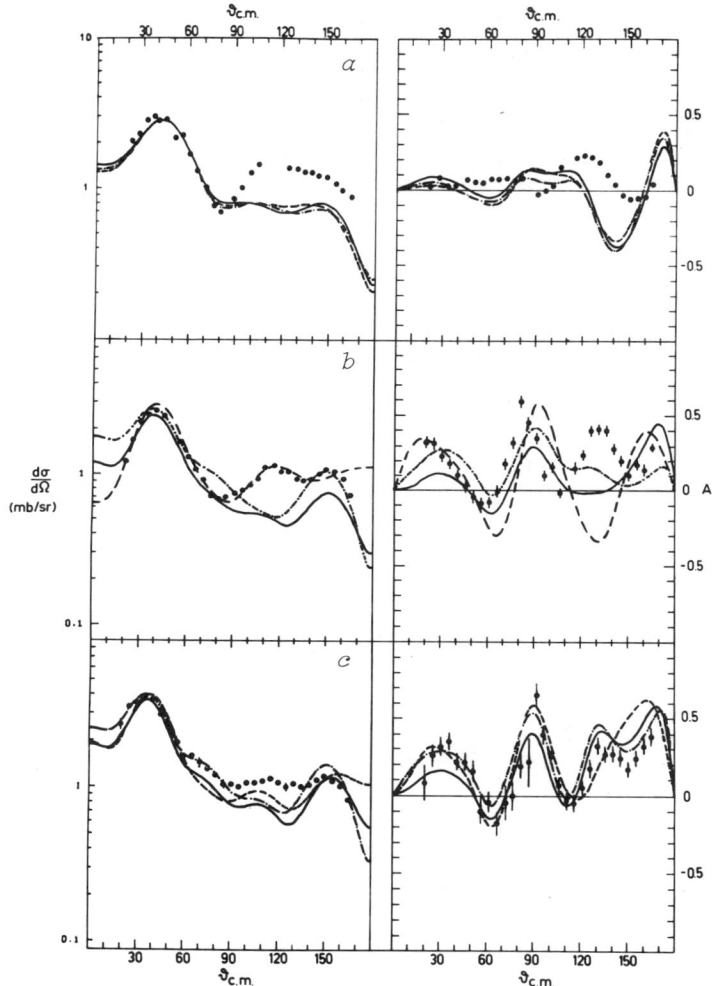

Fig. 1.
Analysis of the $^{54}Fe\ (\vec{p},p')\ ^{54}Fe\ (2_1^+)$ reaction
a : 12.6 MeV, b : 15.3 MeV, c : 17.2 MeV
Full curve : collective DWBA with $\lambda = 2$
Dot dashed curve : collective DWBA with enhanced λ
Dashed curve : collective DWBA with GR (L = 3)

REFERENCES

1. P.J. van Hall et al. Nucl. Phys. A291 (1977) 63
2. C. Glasshauser et al. Phys. Rev. 164 (1967) 1437
3. D.L. Hendrie et al. Phys. Rev. 186 (1969) 1188
4. H.V. von Geramb et al. Nucl. Phys. A199 (1973) 545
5. O. Karban et al. Nucl. Phys. A147 (1970) 462
6. J. Raynal, Trieste Lectures 1971 p. 75
7. H.V. von Geramb, unpublished
8. P.J. van Hall, contribution to this conference
9. R. Pitthan et al. Phys. Rev. C21 (1980) 147.

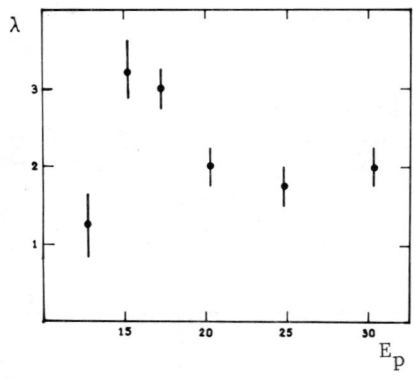

Fig. 2.
Energy dependence of enhancement factor λ

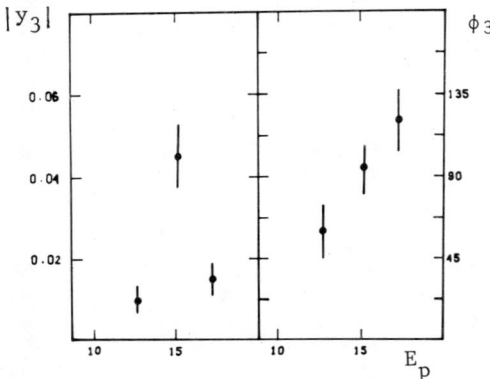

Fig. 3.
Energy dependence of GR coupling strength y_3

DEPOLARIZATION IN THE INELASTIC SCATTERING OF PROTONS FROM COPPER[*]

W.G. Weitkamp, T.A. Trainor, I. Halpern, H. Bhang, and S.K. Lamoreaux
University of Washington, Seattle WA 98195

ABSTRACT

The proton depolarization resulting from elastic and inelastic scattering from copper has been measured with an 18 MeV polarized proton beam at scattering angles of 45°, 90° and 135° and excitation energies up to 12 MeV. Outgoing proton polarization was measured by scattering from ^{4}He in a high pressure gas cell with a resolution of 1 to 1.7 MeV. The depolarization at all angles drops from unity at the lowest excitation energies to near zero above 8 MeV excitation energy.

Measurement of polarization observables such as analyzing power and depolarization in inelastic scattering to high excitation energies in medium weight nuclei may provide new insight into the process by which a projectile transfers energy to a nucleus. Recent measurements of the analyzing power[1,2] show small, systematic effects in the excitation energy region just above the excitation energy where discrete states begin to overlap significantly. We report here on the first measurement of the depolarization parameter D (or $K_y^{y'}$) in inelastic scattering to relatively high excitation energies.

The nucleus ^{63}Cu was chosen for this study because it gives a high yield of inelastic protons and because inelastic scattering to the the evaporation region has been extensively studied[3]. In this nucleus the depolarization parameter D is essentially the ratio of outgoing to incident proton polarization.

To measure the polarization of the protons leaving the nucleus, we use a broad range helium polarimeter designed to maximize the counting rate and minimize the background. To maximize the counting rate we make the target thick, 14 atm, and the angular spread large. This leads to rather poor resolution, from 1 to 1.7 MeV FWHM. To minimize background, we use a counter telescope consisting of two proportional counters and a silicon detector. The entire polarimeter is made of iron to reduce the flux of neutrons from the target striking the detectors. The pulses from the detectors are calibrated in energy by observing pulse heights corresponding to protons entering the polarimeter after scattering from carbon, leaving the carbon in the ground, 4.439 or 9.632 MeV states.

Fig. 1 shows typical depolarization data. Note that the relatively poor resolution of our polarimeter completely wipes out any struture. In interpreting these data, it is convenient to divide the excitation energy into three regions: the direct reaction region from 0 to about 5 MeV excitation, the evaporation region above 10

MeV, and the pre-equilibrium region from about 5 to 10 MeV. The depolarization is expected to show different behavior in these three regions.

Direct Reaction Region. In the region of lowest excitation energy, the yield is dominated by scattering to discrete states. It is clear from the data that spin flip is not an important process in exciting these states, especially at forward angles. This is expected since in those cases where spin flip in inelastic scattering to discrete states has been measured, it is generally quite small, i.e., D is approximately unity.[4]

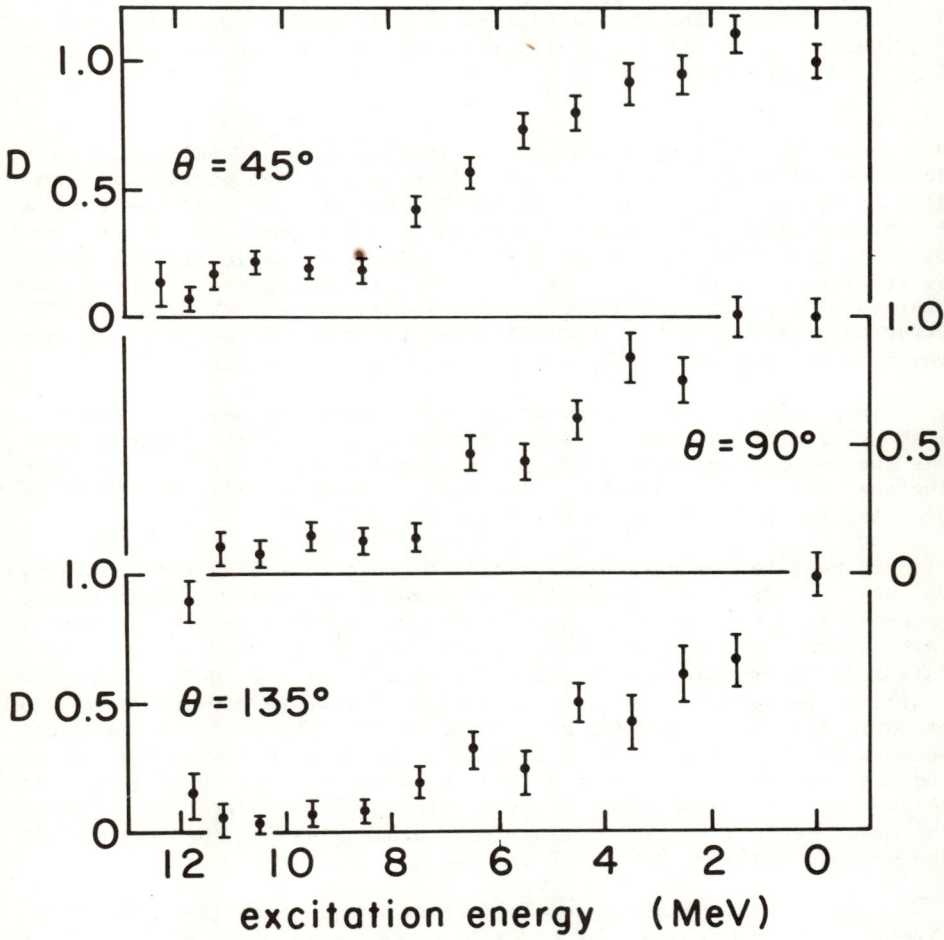

Fig. 1. Depolarization vs excitation energy for 18 MeV proton scattering from ^{63}Cu.

The Evaporation Region. At the highest energies at which a proton can still escape from the nucleus, the cross section is well described by statistical theory.³ One can make a simple minded prediction for the depolarization parameter in the evaporation region. If Z_e is the number of protons in the target nucleus which can evaporate, in the absence of spin flip the depolarization should be approximately $1/(Z_e + 1)$, since half of the target protons which can evaporate have their spins opposite to the incident proton. Our data give a hint that the depolarization parameter doesn't quite go to zero at high excitation energies.

The Pre-equilibrium Region. In the pre-equilibrium region the depolarization decreases with increasing excitation energy. The reaction mechanism here is not well understood. There are theoretical calculations predicting the cross section, but no predictions of the depolarization.

The decrease in polarization of the outgoing protons with increasing excitation energy, i.e. with increasing energy transfer, would arise from two expected effects. First, the probability for spin flip increases with the intimacy of the collision, and second, a collision between protons where considerable energy is tranferred results in the emission of an (unpolarized) target proton in place of the incident proton even when there is no spin exchange. We are currently working on quantitative models of these effects to explain the observed depolarization as a function of energy transfer and angle.

REFERENCES

* Work supported in part by the U.S. Department of Energy
1. H.C. Bhang et al., Bull. Am. Phys. Soc. 24, 829 (1979).
2. H. Sakai et al., to be published in Nucl. Phys.
3. A. Sprinzak et al., Nucl. Phys. A203, 280 (1973).
4. J.M. Moss, W.D. Cornelius and D.R. Brown, Phys. Lett. 71B, 87 (1977).

SCATTERING OF POLARIZED PROTONS FROM ^{64,66,68,70}Zn[x]

P.J. van Hall, J.F.A.G. Ruyl[xx], J. Krabbenborg,
W.H.L. Moonen and H. Offermans
Eindhoven University of Technology, Eindhoven, Netherlands.

This investigation is a part of our program to investigate the scattering of 20 MeV polarized protons from medium weight and heavy nuclei. Previous experiments on Zn isotopes with polarized protons[1,2] yielded contradictory results concerning the deformation of the spin-orbit potential, when the data were analysed with collective DWBA.

We have measured the elastic and inelastic scattering to the 2_1^+ and 3_1^- states for the even Zn isotopes at 20.4 MeV. The data have been analysed with the conventional optical model and the collective DWBA. Results are shown in fig. 1 and fig. 2. The agreement between experiment and theory is quite satisfactory and no anomalous spin-orbit deformation has been found. Only one comment can be made. Like for the Ni isotopes[3] the experimental cross sections for the 2_1^+ states fall below the theoretical predictions at backward angles for ^{64,66,68}Zn. It turned out however, that this was not the case for ^{70}Zn.

In addition to these strongly excited collective states we obtained also data for the two-phonon states in ^{64}Zn and ^{68}Zn. Preliminary calculations with the code ECIS[4] were done with as ingredients the parameters obtained in the analyses mentioned above including the β_2 value. A pure 2-phonon character was assumed and the deformation of the optical potential was taken into account up to the second order. It can be seen from fig. 3 that the agreement with the data is encouraging and that especially the analysing power for the 0^+ state is described surprisingly well.

REFERENCES

[1] W.H. Tait et al, Nucl. Phys. A203 (1973) 193
[2] M.J. Throop et al., Nucl. Phys. A283 (1977) 475
[3] P.J. van Hall et al. Nucl. Phys. A291 (1977) 63
[4] J. Raynal, unpublished.

[x] Supported in part by F.O.M./Z.W.O.
[xx] Present address: ECN, Petten.

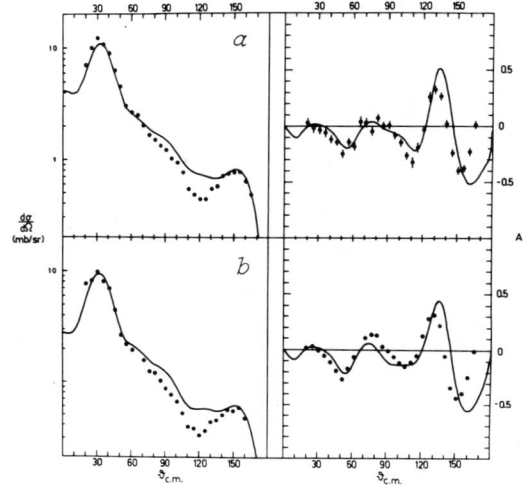

Fig. 1.
Inelastic scattering from 2^+ states   a : ^{66}Zn, b : ^{68}Zn.

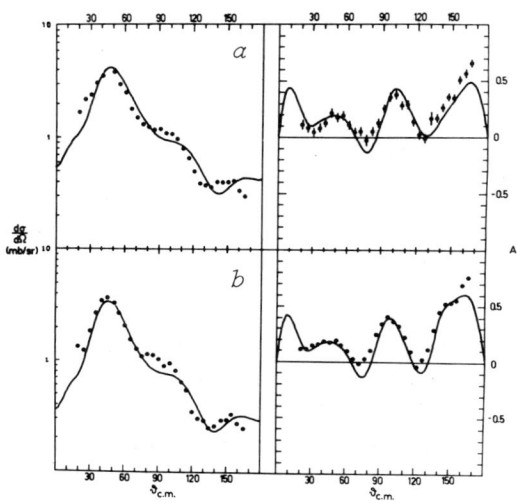

Fig. 2.
Inelastic scattering from 3^- states   a : ^{66}Zn, b : ^{68}Zn

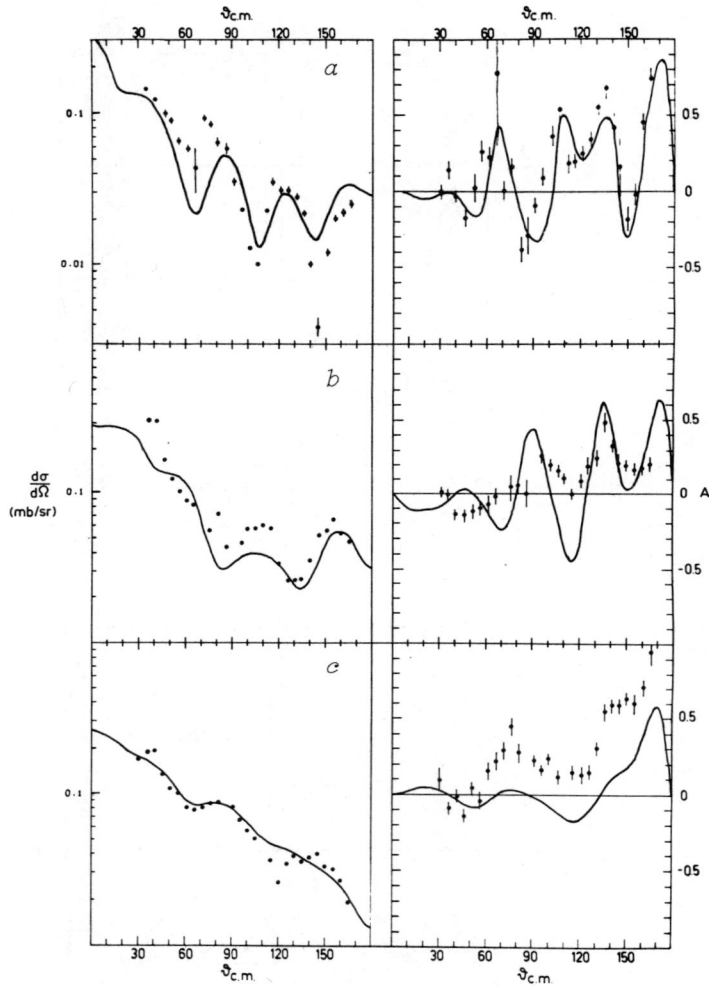

Fig. 3.
Scattering from the two-phonon multiplet in ^{68}Zn
a : 0^+ (1.656), b : 2^+ (1.833), c : 4^+ (2.417)

STUDY OF ^{68,64}Ni AND ^{86,88}Sr BY SCATTERING OF POLARIZED PROTONS[x]

S.D. Wassenaar, P.J. van Hall, S.S. Klein, G.J. Nijgh
O.J. Poppema, W.F. Feix, J.H. Polane and J.F.J. Dautzenberg.

Cyclotron laboratory, Eindhoven University of Technology,
The Netherlands.

The Ni isotopes and ^{88}Sr are nuclides of interest because of their closed shells (Z=28, N=50 respectively). For the experiment on ^{60,64}Ni the proton energy was 20.4 MeV, while for ^{86,88}Sr we used 24.6 MeV. The statistics of the measurements on ^{60}Ni were better than the previous results[5], so that we could extract data for some of the two-phonen states. With conventional optical model and DWBA analysis we could fit the scattering from the ground states and the collective 2^+ and 3^- states quite well, except for the differential cross section of ^{60}Ni at backward angles. This same problem also exists for the other Ni isitopes and for ^{64,66,68}Zn, but not for ^{70}Zn[4]. We have performed calculations with the CO code CHUCK[1] in an attempt to find a better description for the cross section of ^{60}Ni at backward angles. The coupling to the two phonon multiplet around 2.5 MeV, indeed lowers the cross section, but not enough to give a satisfactory result. With the code ECIS[2] we performed some calculations for the 4^+ state at 2.51 MeV. In a second order vibrational model we mixed the two-phonon L = 2 and the direct L = 4 contributions. Nearly equal amounts of both contributions gave a good result, better than the DWBA did.

The inelastic scattering from the Sr isotopes has been analysed with collective DWBA. The quantity of interest is the ratio $\lambda = \beta_{so}/\beta_C$. Table 2 lists the λ-values deduced from polarized proton scattering. We see that λ is energydependent and that it increases with Z. So the observed anomaly of large λ-values is not the same for the N = 50 isotones. Experiments with lower bombarding energies would be of interest to complete table 2. Perhaps for lower energies the ^{88}Sr 2^+ excitation needs a larger spin orbit deformation, as it is a semi-closed shell nucleus. Experiments on ^{92}Mo have been performed in our laboratory at lower energies and are currently being analysed.

Table 1 Deformation parameters.

	2^+	3^-		J	Ex(MeV)	
^{60}Ni	0.255	0.209	^{60}Ni	2	2.15	0.030
^{64}Ni	0.203	0.203	^{60}Ni	2	3.12	0.051
^{86}Sr	0.158	0.153	^{60}Ni	4	2.51	0.127
^{88}Sr	0.114	0.166	^{60}Ni	4	3.67	0.066

[x] Supported in part by Z.W.O./F.O.M.

Table 2 $\lambda = \beta_{so}/\beta_C$ values.

Energy (MeV)	^{86}Sr	^{88}Sr	^{89}Y	^{90}Zr	^{92}Zr	^{92}Mo
20			2.0^4	3.0^3	2.5^3	
25	1.0	1.0				
30				1.5^6	1.5^6	1.5^6
40				1.0^6	1.0^6	

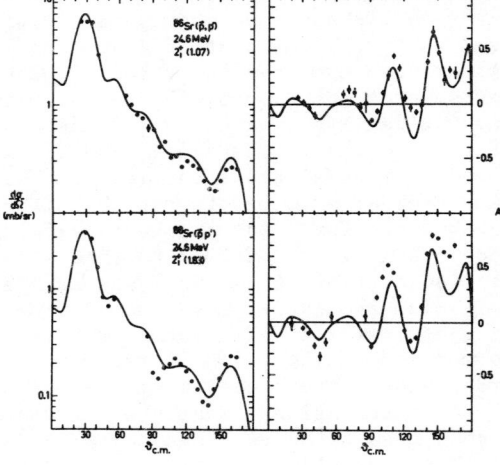

Fig. 1. DWBA curves for ^{86,88}Sr, 2^+.

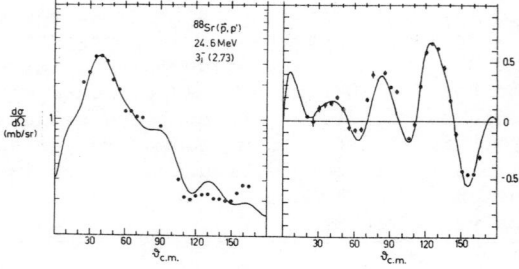

Fig. 2. DWBA curve for ^{88}Sr, 3^-.

REFERENCES

1. P.D. Kunz, unpublished
2. J. Raynal, unpublished
3. C. Glashauser, et al., Phys. Rev. 164 (1967) 1437
4. P.J. van Hall et al., contributions to this conference
5. P.J. van Hall et al., Nucl. Phys. A291 (1977) 63.
6. R. de Swiniarski et al., Phys. Lett. 79B (1978) 47; Can. Journ. Phys. 57 (1979) 540.

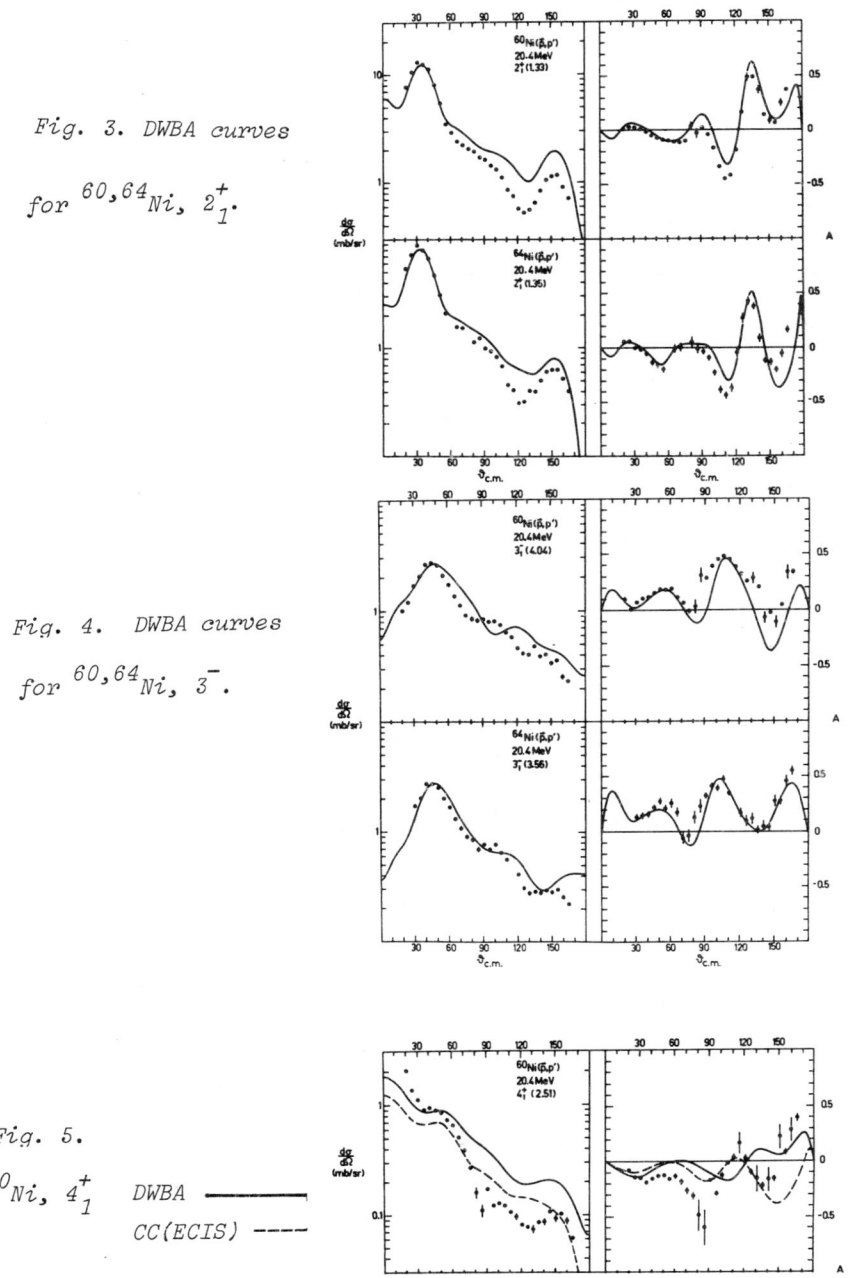

Fig. 3. DWBA curves for $^{60,64}Ni$, 2_1^+.

Fig. 4. DWBA curves for $^{60,64}Ni$, 3^-.

Fig. 5. ^{60}Ni, 4_1^+ DWBA ———
   CC(ECIS) - - - -

SCATTERING OF POLARIZED PROTONS FROM ^{110,112,114}Cd AND ^{115}In.[x]

S.D. Wassenaar, J.F.J. Dautzenberg, J.H. Polane, P.J. van Hall,
S.S. Klein, G.J. Nijgh and O.J. Poppema.

Cyclotron laboratory, Eindhoven University of Technology,
The Netherlands.

The elastic and inelastic scattering of 20.4 MeV polarized protons from the even Cd isotopes and from ^{115}In has been investigated. A reason for chosing these nuclides is that they are lying in the neighbourhood of the tin isotopes which also have been studied[1]. In Cd and In, however, the proton shell is not closed and this results in appreciably lower excitation energies for the 2^+ and 3^- collective states. Since ^{115}In is an odd nuclide with one proton hole in the Z = 50 core, the weak coupling model can be applied. We fitted the elastic scattering with the conventional optical model. Table I lists the best fit parameters. The global fit parameter set of tin, as given in a companion paper[1], also gives good results. The scattering from the 2^+ and 3^- states of Cd and from the L = 2 and L = 3 multiplets in ^{115}In is described very well by collective DWBA calculations with the full Thomas spin-orbit term included. The deduced deformation parameters are listed in table II. According to the weak coupling model we can expect to find five L = 2 and seven L = 3 transitions in ^{115}In, with weighted mean energies equal to the energies of the parent states in ^{116}Sn. The deformation parameters should show a 2J+1 dependence:

$$B_f = (2J_i+1)(2L+1)(\beta_f/\beta_{parent})^2 = 2J_f + 1$$

for each final state. Their sum should equal the deformation parameter of the parent state. In our spectra we could see only three L = 2 and about five L = 3 transitions. The main cause for this is our energy resolution of about 80 keV. Only two L = 3 transitions were strong enough to be analysed. The total strenght of the analysed transitions amounted to be 79 and 76% of the parent 2^+ and 3^- strenghts of ^{116}Sn respectively. From a weak coupling model calculation by Smits[2] the main overlap of the J = 0 and J = 2 states results in 73%, which is in good agreement with our value of 79%. The weighted mean energies are only 3% too high which is also a satisfactory result. Our B_f values for the L = 2 transitions agree very well with the J^π values of Smits[2]. In the L = 3 case a ratio of the two strongest excitations of 3/4 is found, but J^π values are not known. A high resolution experiment with polarized protons should be of interest for a further study of ^{115}In.

---

[x] Supported in part by F.O.M.-Z.W.O.

Table I Best fit optical model parameters, at 20.4 MeV

Cd	V_o	r_o	a_o	W_V	W_D	r_i	a_i	V_{so}	r_{so}	a_{so}
110	52.12	1.213	.691	.63	9.78	1.210	.700	6.26	1.128	.571
112	53.21	1.200	.688	.63	9.36	1.194	.760	6.11	1.110	.582
114	54.38	1.181	.732	.63	9.70	1.183	.811	6.09	1.097	.569

Table II Deformation parameters[x]

	E_x(MeV)	L=2	B_f	$2J_f+1$	E_x(MeV)	L=3	B_f
^{110}Cd	0.66	0.168			2.08	0.146	
^{112}Cd	0.62	0.165			1.97	0.154	
^{114}Cd	0.59	0.128			1.96	0.145	
^{115}In	1.13	0.089	17.5	12+6	2.13	0.092	22.7
^{115}In	1.29	0.075	12.4	14	2.46	0.106	30.2
^{115}In	1.48	0.066	9.5	10			

[x] Without the statistical factor $\sqrt{(2J_i+1)(2L+1)/(2J_f+1)}$

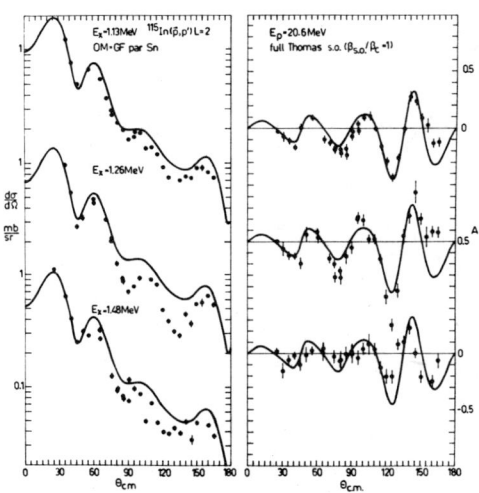

Fig. 1. L=2 excitations in ^{115}In, DWBA curves.

REFERENCES

1. S.D. Wassenaar et al, these conference proceedings
2. J.W. Smits, thesis, KVI, University of Groningen.
   J.W. Smits and R.H. Siemssen, Nucl. Phys. A261 (1976) 385.

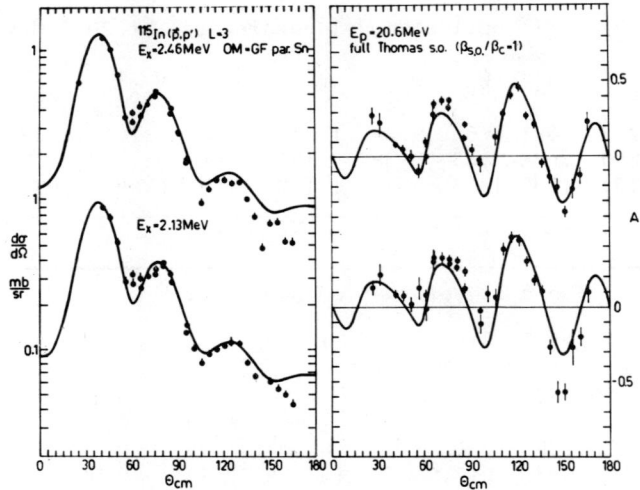

Fig. 2. L=3 excitations in ^{115}In, DWBA curves.

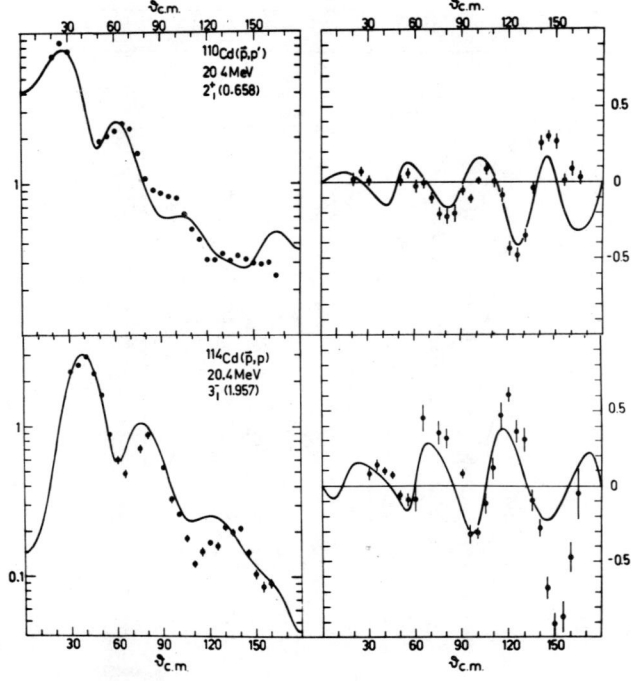

Fig. 3. DWBA curves for ^{110}Cd, 2^+ and ^{114}Cd, 3^-.

SCATTERING OF POLARIZED PROTONS FROM EVEN TIN ISOTOPES [x]

S.D. Wassenaar, P.J. van Hall, S.S. Klein, G.J. Nijgh,
O.J. Poppema, J.H. Polane and J.F.J. Dautzenberg.

Cyclotron laboratory, Eindhoven University of Technology,
The Netherlands.

Since tin has a more or less closed proton shell a lot of stable isotopes is available. So the influence of the increasing neutron number on the optical model parameters, the angular distribution of the cross sections and analysing powers and the deformation parameters can be investigated. In accordance with our program of scattering of polarized protons we have chosen the energy of 20.4 MeV. The cross section and analysing power of the elastic scattering have been fitted simultaneously with the optical model, for every isotope separately, "best fit", and also for the five isotopes together, "global fit". The best fit parameters did not differ much from the global fit parameters that are listed in table I. An unexpected feature of the global fit set is that only the imaginary surface absorption depends on the neutron number. Other isospin dependences e.g. for the real central term turned out to be negligible. Also an extra real surface term, as has been used by Sinha[2], was not significant in our case. Due to the large amount of data points that have been fitted simultaneously the global fit optical model parameters are very well fixed, and can be used for DW and other calculations. A comparison with microscopic optical model calculations should be interesting.

The global optical model parameters of Becchetti and Greenlees did give curves that were nearly equal though the chi squared values were higher by a factor of four.

The geometry given by Satchler[4], that also has been used by Beer[5] was less satisfactory in describing the elastic and inelastic data.

The curves shown in figures 1-3 have been calculated with the collective DWBA code of Verhaar and Tolsma[3]. No large spin-orbit deformation is needed for describing the collective inelastic scattering, it could be equal to the deformation of the central well. From the cross sections of the 2^+ states one can see that the minimum around 120 degrees is described better for the heavier isotopes, whereas for the 3^- states the opposite is true.

For ^{122,124}Sn data for the 5^- states could be extracted from the spectra. Also here the collective DWBA gives a good description. The deduced deformation parameters are listed in table II, no large deviations with those of Beer[5] are found.

[x] Supported in part by F.O.M.- Z.W.O.

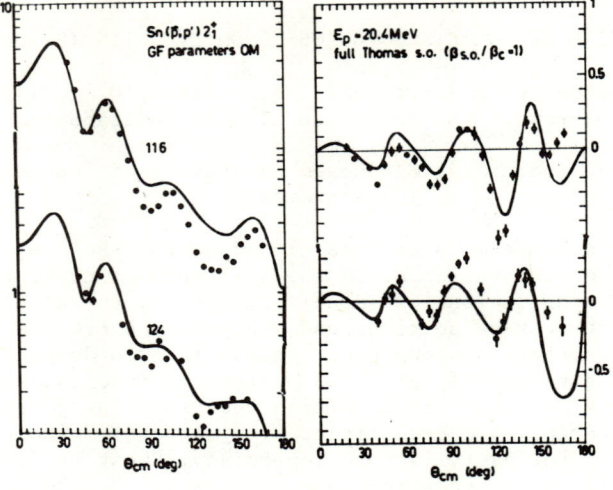

Fig. 1. DWBA curves for $^{116,124}Sn$, 2^+.

Table I Global fit optical model parameters

V_0	r_0	a_0	W_v	W_d	r_I	a_I
54.97	1.178	0.730	1.21	5.93+21(N-Z)/A	1.266	0.695

		V_{so}	r_{so}	a_{so}		
		5.65	0.970	0.699		

Table II Deformation parameters

A	116	118	120	122	124
2^+	0.151	0.138	0.138	0.127	0.109
3^-	0.164	0.158 x		0.141	0.123
5^-				0.084	0.088

x after correction for the 5^- overlap

REFERENCES

1. P.J. van Hall et al., Nucl. Phys. A291 (1977) 63.
2. B.C. Sinha et al., Nucl. Phys. A183 (1972) 401.
3. B.J. Verhaar et al., Nucl. Phys. A195 (1972) 379
4. G.R. Satchler, Nucl. Phys. A82 (1967) 273.
5. O. Beer et al., Nucl. Phys. A147 (1970) 401.

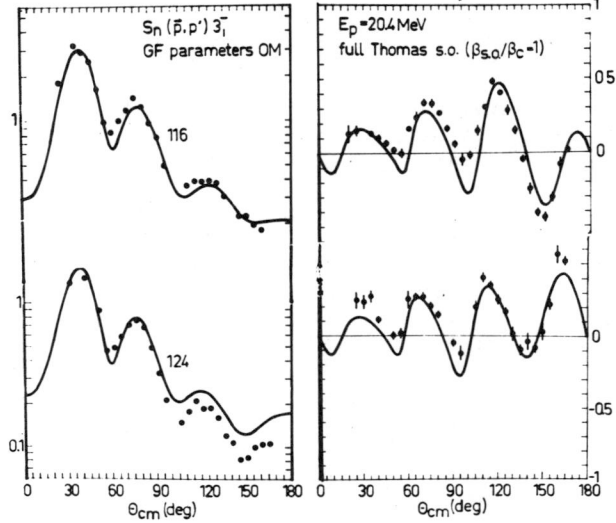

Fig. 2. DWBA curves for $^{116,124}Sn$, 3^-.

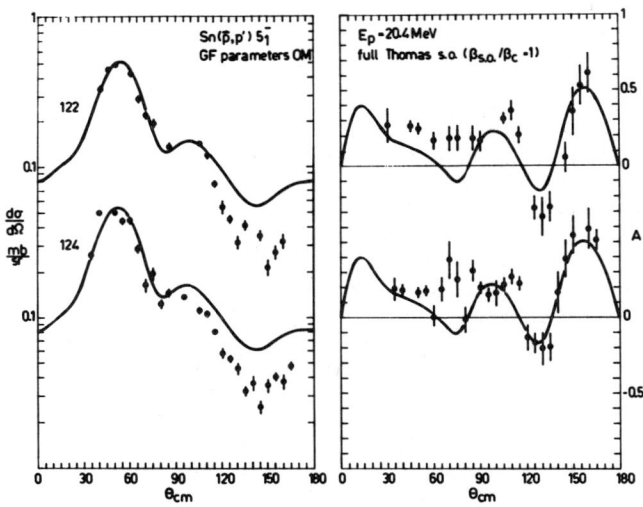

Fig. 3. DWBA curves for $^{122,124}Sn$, 5^-.

MICROSCOPIC ANALYSIS OF POLARIZED PROTON SCATTERING FROM TIN[x]

S.D. Wassenaar and P.J. van Hall,
Cyclotron laboratory, Eindhoven University of Technology,
The Netherlands.

The inelastic scattering from the collective 2^+ and 3^- states of ^{116}Sn can be described by macroscopic or microscopic models. The macroscopic method has been applied in a companion paper[1], from which the global fit optical model and deformation parameters have been used. We have calculated the central, tensor and spin-orbit contributions for all single-particle transitions, using the program MEPHISTO[2]. For the effective N-N interaction we have taken the long-range part of the Hamada-Johnston potential as central force together with the tensor and spin-orbit interactions of Eikenmeier and Hackenbroich. In some cases we have used the phenomenological central interaction of Austin[3] and the tensor interaction of Sprung[4]. The differences, however, in cross section and analysing power with the previous mentioned interactions turned out to be small. The bound state wave functions have been generated in the usual Woods-Saxon well. The theoretical reduced transition probabilities B(EL) have been calculated with the same wave functions. Every single-particle transition is scaled by a reduced spectroscopic amplitude. We have used those calculated by Allaart[5] who has done number projected BCS calculations including proton particle-hole excitations. In this calculations the inert proton and neutron shells were Z=28 and N=50 respectively.

The sum of the neutron and proton excitations is too low to match the absolute value of the experimental cross section. Following Terrien[7] the effective charges of protons and neutrons can be adjusted to fit the experimental cross section and the experimental value of B(EL). The effective charge of the protons is the square root of the ratio of the experimental[9] and theoretical B(EL). From fitting the experimental cross section at the first maximum the effective charge of the neutrons is deduced. When we add to the microscopic part the full imaginary collective form factor[8], with a deformation parameter as found from the collective model analysis[1], the description is much better than the pure microscopic one. The collective Coulomb excitation has been added too though the effect is small. It can be seen from fig. 2 that the 2^+ fit is rather poor but the 3^- one is good. The spectroscopic amplitudes of Gillet[6] that do not differ much from those of Allaart, have been used too. For the cross section of the 3^- we found a factor of 1.5 difference, while the analysing powers are nearly the equal, as can be seen in fig. 3. Some differences with the work of Terrien are: (i) the inclusion of non-central components in the effective interaction, which of course are important especially for the analysing power, (ii) somewhat more difference between neutron and proton contribu-

---

[x] Supported in part by F.O.M./Z.W.O.

tions and between direct and exchange, as can be seen in fig. 1.

We arrive at the conclusion that the effective charges (table I) needed to describe the data are unexpected high, especially for the 3^- states. This points to deficiencies in the wave functions. The inclusion of a phenomenological imaginary interaction is certainly needed.

Table I. Effective charges of protons and neutrons, obtained with spectroscopic amplitudes of Allaart.

	2^+		3^-	
	λ_p	λ_n	λ_p	λ_n
proton + neutron = valence (V)	9.43	2.01	3.52	5.58
V + imaginary + Coulomb (V + I + C)	9.43	1.48	3.52	4.02

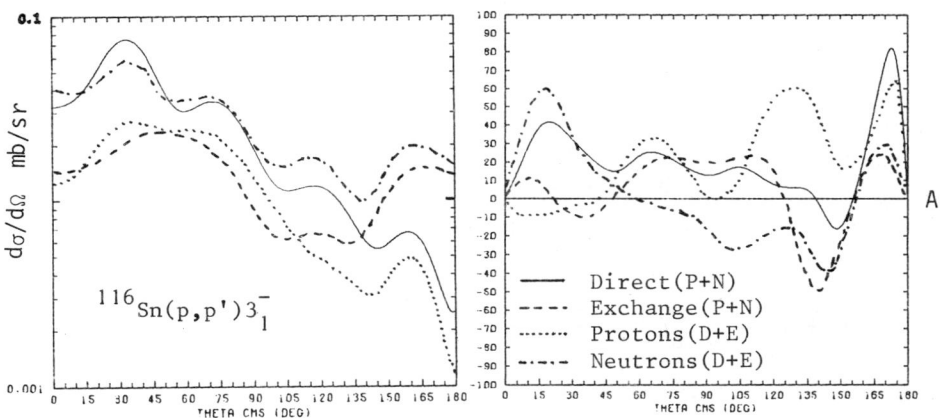

Fig.1. Diverse parts compared with $\lambda_p = \lambda_n = 1$.

REFERENCES

[1] S.D. Wassenaar, contribution to this conference
[2] H.V. von Geramb, unpublished
[3] S.M. Austin, MUSCL-302, (1979), Michigan State
[4] D.W.L. Sprung, Nucl. Phys. A182 (1972) 97.
[5] K. Allaart, private comm., Vrije Universiteit Amsterdam.
[6] V. Gillet et al., Le Journal de Phys. 37 (1976) 189.
[7] Y. Terrien, Nucl. Phys. A199 (1973) 65
[8] Y. Terrien, Nucl. Phys. A215 (1973) 29.
[9] P.H. Stelson et al., Phys. Rev. C2 (1970) 2015.
D.G. Alkhazov et al., Bull. Acad. Sci. USSR Ser.Phys. 28(1965) 149.

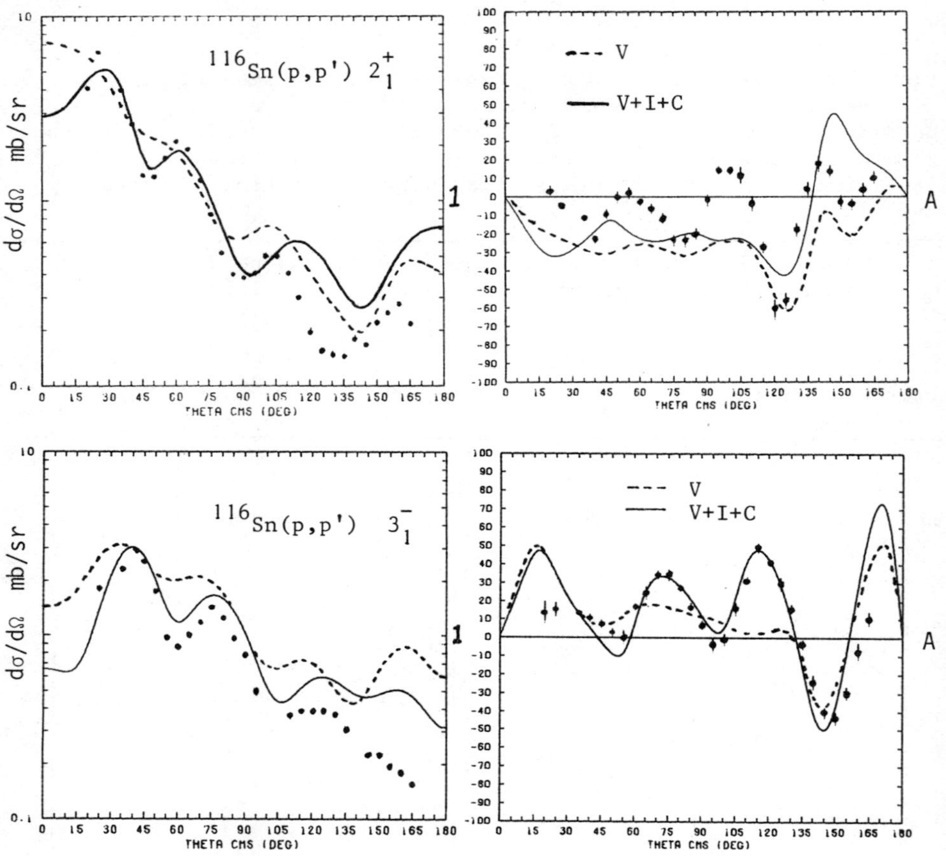

528

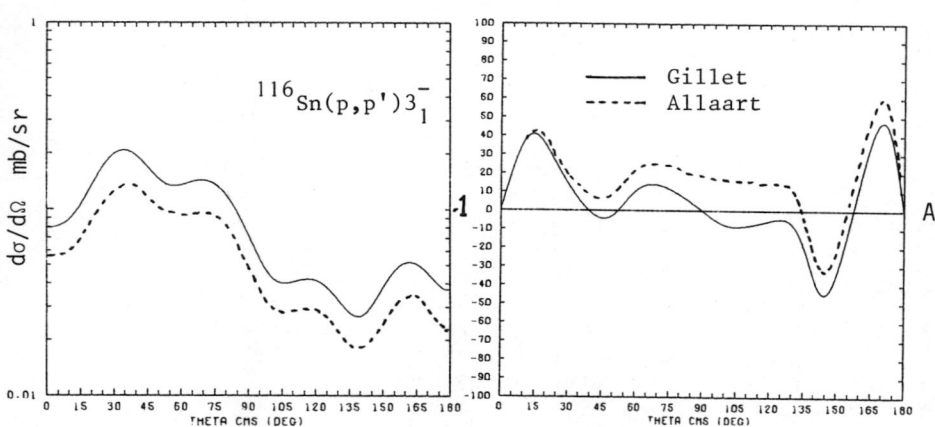

Fig.2. Valence calculations with effective charges.

Fig.3. Gillet and Allaart amplitudes compared with $\lambda_p = \lambda_n = 1$.

ONE- AND TWO-STEP ANALYSIS OF ^{14}N($\vec{p}$,p')^{14}N(2.31 MeV)
REACTION AT 21.0 MeV AND THE EFFECTIVE INTERACTION

Y. Aoki, K. Nagano, Y. Toba, S. Kunori and K. Yagi
Institute of Physics and Tandem Accelerator Center,
The University of Tsukuba, Ibaraki, 305, Japan

ABSTRACT

One-step and two-step DWBA analyses via one-nucleon transfer channel have been made for spin- and isospin-flip reaction of ^{14}N($\vec{p}$,p')^{14}N(2.31 MeV) at 21.0 MeV. The one-step DWBA predicts the pattern of differential cross section reasonably well, but not the vector analyzing power. Some one-fifth of the differential cross section is predicted by finite-range two-step calculation via the ground state of ^{13}N. Inclusion of the (p,d)(d,p') process reproduces the experimental data and reduces the strength of the effective interaction of the inelastic scattering by a factor of 2.

INTRODUCTION

It is of interest to study the spin- and isospin-dependent part of the effective interaction responsible for inelastic scattering of nucleons by nuclei. Inelastic excitation of the first excited state in ^{14}N by protons offers a good base for this purpose and many authors[1,2,3] agree the importance of tensor force for this reaction. The differential cross section for the reaction, however, is an order of magnitude smaller compared to the ^{14}N(p,p')^{14}N(3.95 MeV) or ^{14}N(p,d)^{13}N(gs) reactions. Two-step contribution should be assessed under these circumstances to reduce reliably the information on the effective interaction. Oregon group[3] predicted that (p,n)(n,p') effect was not a main contributor to the reaction. We know that Wigner-type interaction is the strongest among all types of nuclear forces. This force causes nucleon transfer reactions. Spin and isospin are transferred by nucleon in the transfer reactions so that second-order DWBA term of Wigner-type interaction can compete with the first-order DWBA term of the weak spin- and isospin-dependent interaction. So the largest second-order DWBA term should be due to the (p,d)(d,p') channel.

Experimental data of ^{14}N($\vec{p}$,p')^{14}N(2.31 MeV) and ^{14}N($\vec{p}$,d)^{13}N(gs and 3.51 MeV) reaction at proton energy of 21.0 MeV ara analyzed by using a zero- and finite-range one- and two-step program TWOFNR[4].

ANALYSIS

Optical potentials of protons are taken from the work by Baron

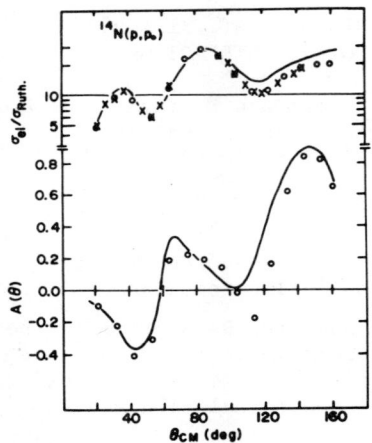

Fig. 1. $^{14}N(\vec{p},p)^{14}N$(gs) reaction. Circles are obtained by using a magnetic spectrograph, while crosses are obtained by using a counter telescope. Solid lines are due to optical model.

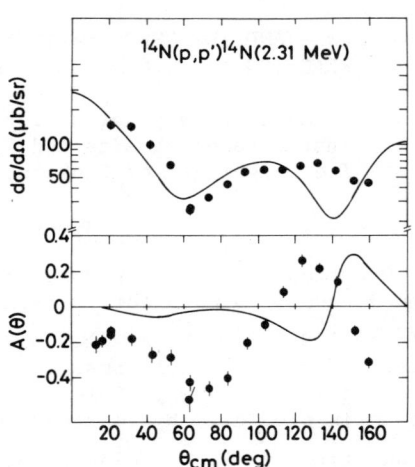

Fig. 2. $^{14}N(\vec{p},p')^{14}N$(2.31 MeV) reaction. Solid lines are due to DWBA with the interaction strength of $V_c=-3.55$ and $V_t=-3.45$ MeV.

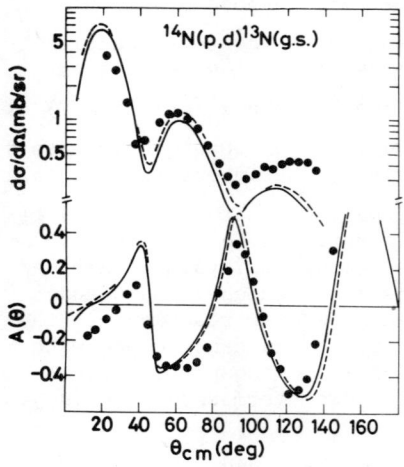

Fig. 3. $^{14}N(\vec{p},d)^{13}N$(gs) reaction. Solid lines are due to finite-range DWBA, while dashed lines are due to zero-range DWBA.

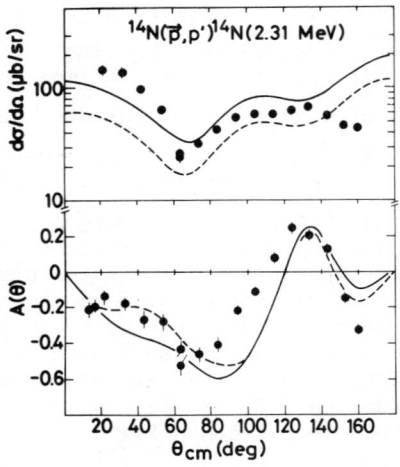

Fig. 4. $^{14}N(\vec{p},p')^{14}N$(2.31 MeV) reaction. Dashed lines are due to finite-range two-step calculation via ^{13}N(gs). Solid lines are the final result. See the context.

et al.[5], which reproduces our elastic data well (Fig. 1). Spectroscopic amplitudes are obtained from the p-shell matrix elements of 8-16 POT by Cohen and Kurath[6]. Types and geometric parameters of the effective interaction are takne from ref. 1. Interaction strengths of around 3. 5 MeV for both central and tensor terms are adjusted to reproduce the experimental data. Figure 2 compares the experimental data with DWBA predictions. Except at around 130°, DWBA reproduces the differential cross section well. Analyzing power data, however, are not reproduced at all within the one-step calculation.

Deuteron optical potentials should be settled before going into (p,d)(d,p') two-step analysis. Phenemenological optical potential parameters are introduced by modifying the adiabatic prescription[7]. Figure 3 shows the zero- and finite-range DWBA curves for $^{14}N(p,d)^{13}N$ (gs) reaction with this potential. The zero-range two-step calculation via the ground and the second excited state in ^{13}N shows that the two processes give the same angular distribution both in differential cross section and in analyzing power. The two processes interfere constructively to double the contribution via the ground state in ^{13}N. Interference nature of one- and two-step processes is examined by using finite-range option of the program, because zero-range two-step calculation is reported to over-estimate the magnitude[8]. Figure 4 shows the final calculation, where central and tensor interaction strengths of -1.74 MeV and -1.31 MeV are used. Two-step amplitude via the $^{13}N(gs)$ was multiplied by 1.27 and the one via the second excited state was neglected. The fit to the experimental data is reasonable except for the forward angle region.

We believe the importance of (p,d)(d,p') two-step process still remains if knock-on exchange effect is taken into account.

REFERENCES

1. G. M. Crawley, S. M. Austin, W. Benenson, V. A. Madsen, F. M. Schmidttroth and M. J. Stomp, Phys. Letts. 32B, 92 (1970).
2. T. H. Curtis, H. F. Lutz, D. W. Heikkinen and W. Bartolini, Nucl. Phys. A165, 19 (1971), J. L. Escudie, A. Tarrats and J. Raynal, Proc. 3rd Int. Symp. on Polarization Phenomena in Nuclear Reactions, (the University of Wisconsin Press 1971), p. 705, H. F. Lutz, D. W. Heikkinen and W. Bartolini, Nucl. Phys. A198, 257 (1972) and S. H. Fox and S. M. Austin, Phys. Rev. C21, 1133 (1980).
3. L. F. Hansen, S. M. Grimes, J. L. Kammerdiener and V. A. Madsen, Phys. Rev. C8, 2072 (1973).
4. M. Igarashi, program TWOFNR unpublished.
5. N. Baron, R. F. Leonard and D. A. Lind, Phys. Rev. 180, 978 (1969).
6. S. Kohen and D. Kurath, Nucl. Phys. 73, 1 (1965).
7. R. C. Johnson and J. R. Soper, Phys. Rev. C1, 976 (1970).
8. P. D. Kunz and L. A. Charlton, Phys. Letts. 61B, 1 (1976).

THE ^{89}Y $(\vec{p},p')$ REACTION AT 21.1 MEV[x]

J.P.M.G. Melssen[xx], P.J. v. Hall, S.D. Wassenaar,
O.J. Poppema, S.S. Klein, G.J. Nijgh,

Eindhoven Univ. of Techn., Eindhoven, Netherlands.

Microscopic DWBA calculations of the analyzing powers in $(\vec{p},p')$ reactions at energies below 50 MeV have been performed with various degrees of success[1]. Most of these calculations for heavier nuclei bore on rather collective states because for obvious reasons for these shades experimental data are available. They generally yield fits inferior as compared with fits obtained by simple collective model DWBA. To perform these calculations rather complicated wave functions have to be used. Therefore in case of failure it is difficult if not impossible to ascertain whether the reaction theory or the spectroscopy (i.e. the wave functions) is to be blamed (or even both).

In an attempt to bring this dilemma nearer to its solution we have performed an inelastic scattering experiment on ^{89}Y at 21.1 MeV. ^{89}Y was chosen because its three lowest excited states have a rather pure single particle character[2]. This nucleus has been a test case for microscopic calculations for quite a long time[3] and so it seemed worthwhile to add analysing powers to the existing cross section data.

We have obtained data for the 0.91 MeV ($9/2^+$), 1.51 MeV ($3/2^-$) and 1.74 MeV ($5/2^-$) states. These data have been analysed with microscopic DWBA calculations using the code MEPHISTO[4]. The long-range part of the Hamada-Johnston potential[5] with the Eichenmeier-Hackenbroich[6] tensor and spin-orbit forces was used for the N-N interaction. Generally it was necessary to include core polarization, we used for it the collective prescription of Love and Satchler[7].

THE 0.91 MEV ($9/2^+$) STATE

This state can be considered as an excitation of a proton from the $2p_{1/2}$ orbit into the $1g\ 9/2$ orbit. In the calculations a core polarization term with $y_5 = 4.2 \times 10^{-4}$ MeV^{-1} had to be added to account for the absolute value of the cross section. From the analysing power it is suggested that the effective tensor force is a bit too strong. It is, however, premature to draw definite conclusions in view of the large core contribution.

---

[x] Supported in part by F.O.M./Z.W.O.
[xx] Present address: Computer Centrum Limburg, Heerlen

THE 1.51 MEV ($3/2^-$) STATE

We assume that this state is excited by promoting a proton from the $2p^{3/2}$ orbit into the $2p^{1/2}$ orbit. We therefore expect a large contribution from the LSJ = 011 triad. To account for the electromagnetic transition rate a core polarization of $y_2 = 3.2 \times 10^{-4}$ MeV^{-1} has to be added. We now overestimate the cross section. We think that this difficulty may be overcome by a microscopic calculation of the core polarization, including an S = 1 pout which in some cases has been shown to reduce the core polarization effects. When inspecting the analysing power we see that the agreement is only qualitative though the main features are described quite well.

THE 1.74 MEV ($5/2^-$) STATE

This state is considered to be excited by lifting a proton from the $1f^{5/2}$ orbit into the $2p^{1/2}$ orbit. As shown in fig. 3 this single particle excitation alone hardly contributes to the differential cross section. We therefore need a core coupling parameter of 1.0×10^{-3} MeV^{-1}. Because of this large core polarization the information about the effective interaction from this transition is rather poor. Nevertheless it can be observed that the analysing power is described better at the maxima, when the microscopic term is added. This is mainly due to the inclusion of the spin-dependent parts.

Our conclusions are that due to the large core polarization contributions even in this case only rather general statements can be made. The microscopic calculations certainly do not fail completely; some features, especially of the analysing powers, are predicted correctly.

REFERENCES

[1] See e.g.
P.J. van Hall et al. contribution to this conference
S.D. Wassenaar et al. contribution to this conference
J.L. Escudié et al. Zurich Conf. 1975, pg. 721
[2] B.M. Preedom et al. Phys.Rev. 166 (1968) 1156
[3] M.M. Stautberg et al. Phys. Rev. 157 (1967) 977
See e.g.
[4] H.V. von Geramb and K.A. Amos, Nucl. Phys. A163 (1971) 337
[5] H.V. von Geramb, unpublished
[6] T. Hamada and I.D. Johnston, Nucl. Phys 34 (1962) 382
[7] H. Eikenmeier and H.H. Hackenbroich, Nucl. Phys. A169 (1971) 407
W.G. Love and G.R. Satchler, Nucl. Phys. A101 (1967) 977

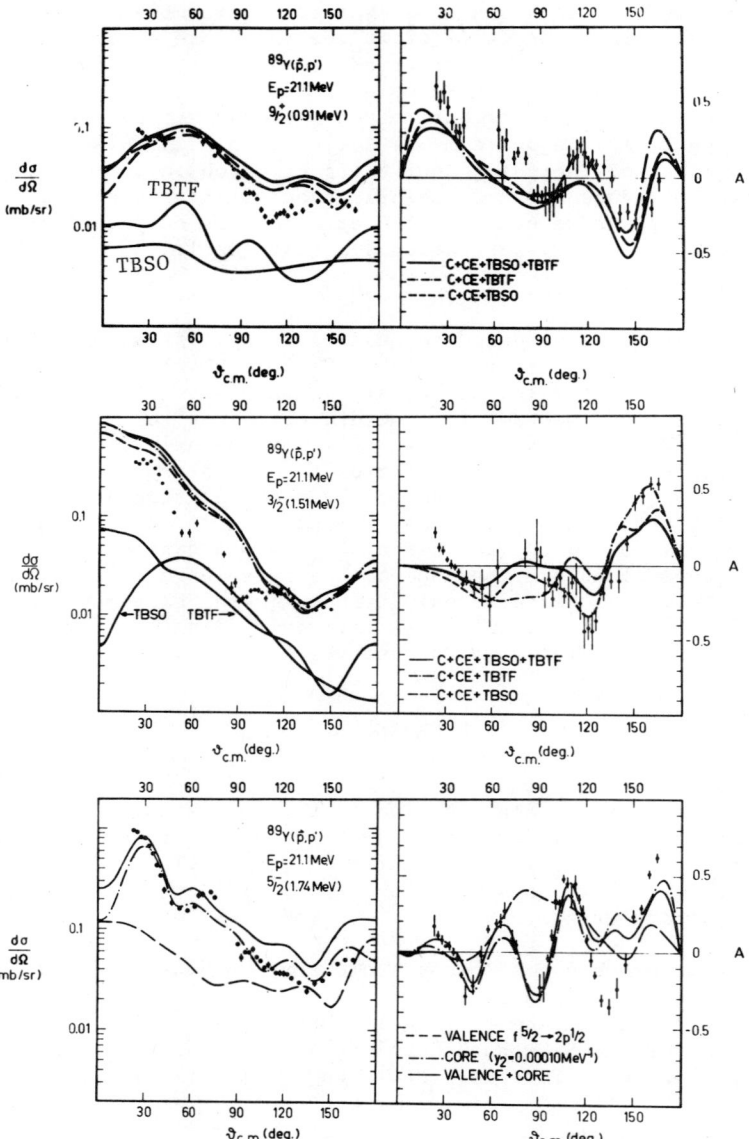

Fig. 1, 2, 3.
Inelastic scattering from the single-particle states in ^{89}Y

SPIN FLIP ASYMMETRY IN THE INELASTIC SCATTERING OF PROTONS ON ^{12}C AT ENERGIES FROM 22.0 TO 29.0 MEV

T. Fujisawa, N. Kishida, and T. Kubo
The Institute of Physical and Chemical Research,
Wako-shi, Saitama, 351 Japan

T. Hasegawa, M. Sekiguchi, N. Ueda, and M. Yasue
Institute for Nuclear Study, University of Tokyo,
Tokyo 188, Japan

Y. Wakuta and A. Nagao
Kyushu University, Fukuoka 812, Japan

ABSTRACT

The spin flip asymmetry (SFA) in the inelastic scattering of protons on ^{12}C (2^+ 4.43 MeV state) was determined by measuring the spin flip probability of polarized protons at incident energies from 22.0 to 29.0 MeV. The SFA shows a pronounced interference-like energy dependence.

INTRODUCTION

A large number of studies for elastic and inelastic scatterings of protons on ^{12}C has been performed over the incident energy range of about 20 to 30 MeV but the reaction mechanism has not been well understood[1~5]. Spin-dependent observables are more sensitive to a reaction mechanism than cross-sections. Particularly it is important to measure a difference between analyzing power(A) and polarization(P)[6,7]. Therefore, we measured the σ_0, A, SF and SFA simultaneously and deduced the A-P at incident energies from 22 to 29 MeV.*

---

* If the z-axis is chosen along the normal to the scattering plane the observables for nucleon scatterings are expressed as follows;

$\sigma_0 = (\sigma_{++} + \sigma_{+-} + \sigma_{-+} + \sigma_{--})/2$, $A = (\sigma_{++} + \sigma_{+-} - \sigma_{-+} - \sigma_{--})2\sigma_0$,

$P = (\sigma_{++} - \sigma_{+-} + \sigma_{-+} - \sigma_{--})/2\sigma_0$, $SF = (\sigma_{+-} + \sigma_{-+})/2\sigma_0$,

$SFA = (\sigma_{+-} - \sigma_{-+})/(\sigma_{+-} + \sigma_{-+}) = (A-P)/(2 \cdot SF)$,

where σ_0 and SF are the differential cross section and the spin flip probability of unpolarized protons, respectively. σ_{+-} is the partial differential cross section for scattering from an incoming spin-up state (+) to a final spin-down state(-). In the present report $(\sigma_{+-} - \sigma_{-+})/(\sigma_{+-} + \sigma_{-+})$ was called a spin flip asymmetry but sometimes $1/2(A-P)$ was called[8].

EXPERIMENTAL PROCEDURE AND RESULTS

Direct measurement of the polarization is very difficult

because it requires a double scattering. So the A-P and SFA were deduced by measuring the spin-flip probability of polarized protons. The spin flip probability was measured by the (pp'γ) method, i.e., by measuring the angular correlation between protons scattered inelastically from the first 2^+ excited state and the E2($2^+\to 0^+$ ground) de-excitation γ-rays emitted perpendicularly to the scattering plane [9]. The polarized proton beam was extracted from an atomic beam type polarized ion source constructed at the Institute of Physical and Chemical Research and was accelerated by the SF Cyclotron at the Institute for Nuclear Study, University of Tokyo. The beam current

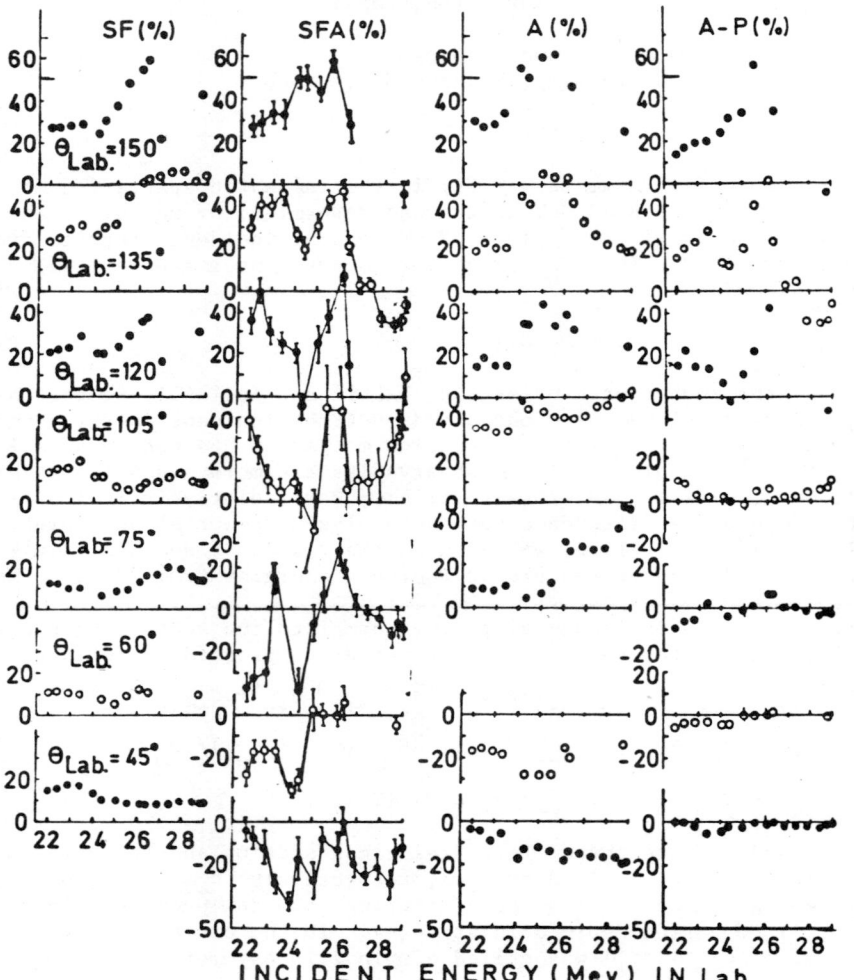

Fig. 1. Excitation curves of the Spin flip probability(SF), spin flip asymmetry(SFA), analyzing power(A) and A-P in the inelastic scattering of protons on ^{12}C at the lab. scattering angles 45°, 60°, 75°, 105°, 120°, 135° and 150°. The line is to guide the eye.

on the target was about 10 nA and the beam polarization was about 50 %. The beam polarization was reversed about every one minute by the RF transition units of the ion source which were controlled automatically with signal from a beam current integrator to eliminate systematic errors. The ^{12}C target was a self-supporting foil of natural carbon with thickness of 5.65 mg/cm^2. The γ-rays were detected with a shielded 7.6 x 7.6 cm NaI(Tℓ) crystal mounted on a 56 AVP photomultiplier. This detector had a lead slit of 10 cm thick with an aperture of 57 mmϕ at a position 30 cm apart from the target. The scattered protons were detected by four pairs of solid state detectors which were set symmetrically on both sides of the beam direction. A defining slid of 10 mmϕ in front of each detector was placed at a position 10 cm from the target.

In Fig. 1 the experimental data are shown. The values of SF have been corrected for non-spin-flip contributions due to the off-z axis γ-rays. The solid bars in Fig. 1 indicate the statistical errors. It was surprising that the SFA and the (A-P) have such a large values and change drastically with the incident energy in contrast with a rather monotonous change of the inelastic scattering cross section.[1] A coupled-channel calculation with Breit-Wigner terms is in progress.

The authors thank M. Nakamura, F. Soga, K. Hatanaka, T. Tanaka, Y. Toba, S. Motonaga, and T. Wada for their helps with the experiment. They also thank H. Kamitsubo and Y. Chiba for their encouragements during the course of this work.

References

[1] J.K. Dickens, D.A. Haner, and C.N. Waddell, Phys. Rev. <u>132</u> 2159 (1963).
[2] T. Tamura and T. Terasawa, Phys. Lett. <u>8</u> No 1, 41 (1964).
[3] R.M. Craig, J.C. Dore, G.W. Greenlees, J. Lowe and D.L. Watson, Nucl. Phys. <u>83</u> 493 (1966); ibid <u>79</u> 177 (1966).
[4] H.V. Geramb, K. Amos, R. Sprickman, K.T. Knöfle, M. Rogge, D. Ingham and C. Mayer-Boricke, Phys. Rev. C <u>12</u> No 6 1679(1975).
[5] R. DE. Leo, G. D'Erasmo, F. Ferrero, A. Pantaleo and M. Pingnanelli, Nucl. Phys. A <u>254</u> 156 (1975).
[6] G.R. Satchler, Phys. Lett. <u>19</u> 312 (1965); H. Sherif, Canadian Jur. of Phys. <u>49</u> 983 (1971).
[7] C. Glashausser, Proc. of the 4th Inter. Symp. on Pol. Phen. in Nucl. Reac., 333 (Zurich 1975).
[8] R.N. Boyd, D. Slater, R. Avida, H.F. Glavish, C. Glashauser, G. Bissinger, D. Davis, C.F. Haynes and A.B. Robbins, Phys. Rev. Lett. <u>29</u> 955 (1972).
[9] F.H. Schmidt, R.E. Brown, J.B. Gerhart and W.A. Kolasinski, Nucl. Phys. <u>52</u> 353 (1964).

MICROSCOPIC ANALYSIS OF THE ^{54}Fe($\vec{p}$,p') REACTION BETWEEN 20 AND 30 MEV [x]

P.J. van Hall, J.P.M.G. Melssen[xx] and S.D. Wassenaar
Eindhoven University of Technology, Eindhoven, Netherlands.

We have analyzed our previously obtained data[1] for the inelastic scattering of polarized protons from ^{54}Fe together with the data taken at a higher energy by the Birmingham group[2]. Our aim in this analysis was to investigate the effects of differences in the microscopic structure[3] between the 2_1^+ (1.41) and 2_2^+ (2.96) states. Experimentally significant effects in the analysing powers are observed.

For our calculations we used the code MEPHISTO written by Von Geramb[4] with the spectroscopic amplitudes as given by Amos[3]. To keep computer time within reasonable limits we omitted the very small components of the wave functions. The N-N force was the long range part of the Hamada-Johnston potential with the Eikenmeier-Hackenbroich tensor and spin-orbit forces. To obtain the correct absolute transition strength we added the part due to the imaginary potential of the collective T-matrix to the microscopic one. Moreover, the latter was multiplied with an enhancement factor ε. A second method is to add a core polarization term, which in fact is a collective transition amplitude.

The results of these calculations are shown in fig. 1 and fig. 2 where for comparison also the collective model DWBA calculation is given. It can be seen that especially the description of the analysing power is very poor. The core polarization calculation of course does the better job as it resembles more the collective DWBA. It is indeed true that the results for the 1.41 MeV state are somewhat less negative than those for the 2.96 MeV state, but no real conclusion should be drawn at all.

Concluding we may say that these microscopic DWBA calculations with rather complete forces seriously fail to reproduce especially the analysing power data. This of course can be attrituted to the use of incomplete wave functions as reflected by the enhancement factor $\varepsilon \approx 2$. We feel, however, that the tendency of the calculated analysing powers being too negative is rather representative for this type of calculations and that it would not change drastically by the inclusion of a large number of small components. It would be of interest to perform similar calculations with a "realistic" N-N force as calculated by Brieva and Rook[5].

REFERENCES

[1] P.J. van Hall et al., Nucl. Phys. A291 (1977) 63
[2] O. Karban et al., Nucl. Phys. A147 (1979) 461
[3] K. Amos et al., Nucl. Phys. A304 (1978) 191
[4] H.V. von Geramb, unpublished
[5] F.A. Brieva and J.R. Rook, Nucl. Phys. A291 (1977) 299.

[x] Supported in part by F.O.M.-Z.W.O.
[xx] Present address: Computer Centrum Limburg, Heerlen.

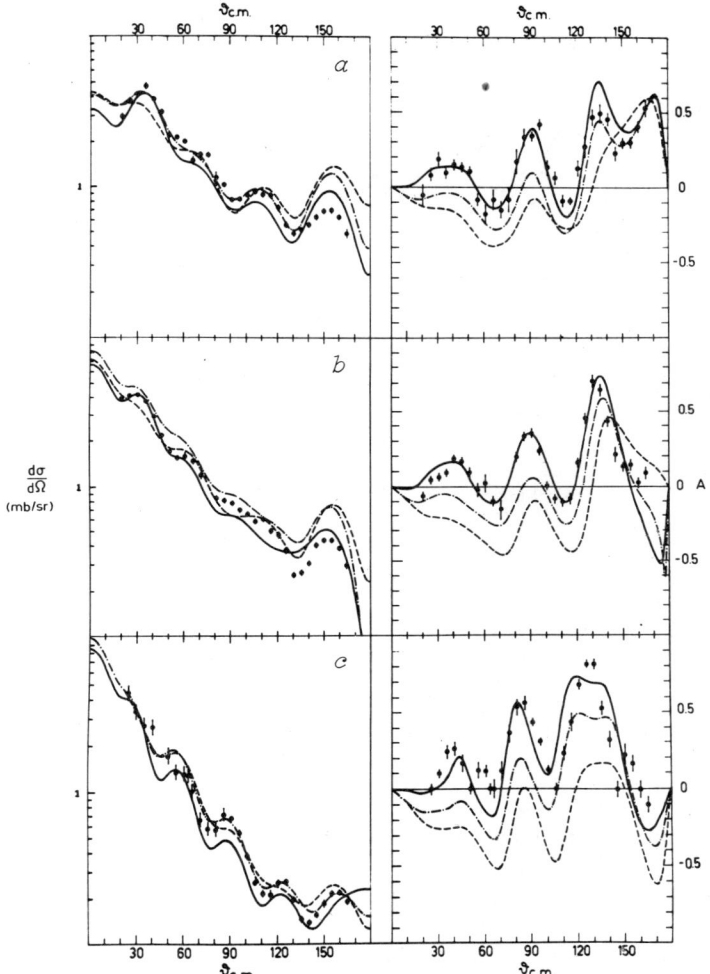

Fig. 1.
Analysis of the $^{54}Fe\ (\vec{p},p')\ ^{54}Fe\ (2^+_1)$ reaction
a : 20.4 MeV,   b : 24.6 MeV,   c : 30.3 MeV
full curve : collective model DWBA
dashed curve : microscopic + imaginary form factor
doth dashed curve : microscopic + core polarization

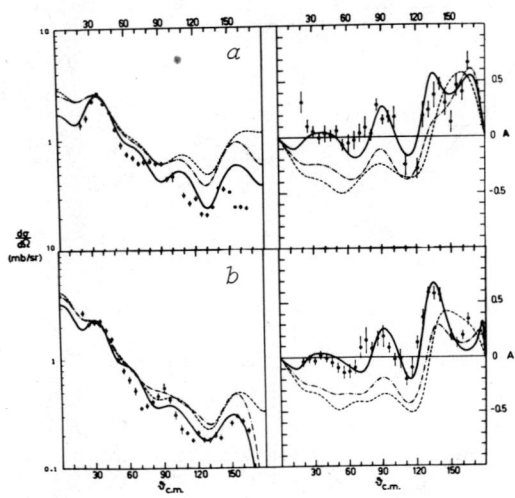

Fig. 2.
Analysis of the $^{54}Fe\,(\vec{p},p')\,^{54}Fe\,(2_1^+)$ reaction
a : 20.4 MeV, b : 24.6 MeV
curves as in fig. 1.

Enhancement factors
Microscopic+Imaginary $\varepsilon V+I$

2_1^+ $\varepsilon = 2.25, 2.25, 2.25$ (20.4, 24.6, 30.3 MeV)
2_2^+ $\varepsilon = 1.64, 1.47$ (20.4, 24.6 MeV)

Microscopic +Core polarisation $V+y_2 C$

2_1^+ $y_2 = 1.7, 1.8, 1.5 \times 10^{-3}$ (20.4, 24.6, 30.3 MeV)

2_2^+ $y_2 = 1.4, 1.2 \times 10^{-3}$ (20.4, 24.6 MeV)

NEW EVIDENCES FOR AN IMAGINARY SPIN-ORBIT POTENTIAL IN THE INELASTIC SCATTERING OF POLARIZED PROTONS FROM ^{12}C AND ^{16}O

R. de SWINIARSKI and Dinh-Lien PHAM
Institut des Sciences Nucléaires (IN2P3-U.S.M.G.)
53, avenue des Martyrs, 38026 Grenoble Cédex, France

ABSTRACT

Analyzing powers and cross sections for elastic and inelastic scattering of 20, 30 and 40 MeV polarized protons from ^{12}C and ^{16}O have been analyzed in the coupled-channels collective model using the full Thomas deformed spin-orbit term. Evidence is given that there is an imaginary spin-orbit potential in the proton optical potential for these nuclei.

Some time ago, inelastic scattering of polarized protons on ^{40}Ca between 15 and 25 MeV [1] have been interpreted as requiring an imaginary spin-orbit term ($W_{LS} = 1.5 - 2.0$ MeV) in the optical potential in order to explain the polarization mainly around 20 MeV. However, another experiment using double scattering around 20 MeV [2] has not confirmed this requirement. In addition, measurements of the inelastic scattering of 18.6 MeV polarized protons on closed shell nuclei in the $f_{7/2}$ region [3] have shown that the large analyzing power obtained by exciting the first 2^+ state in these nuclei could be reproduced by introducing a large imaginary and negative spin-orbit term in the optical potential ($W_{LS} = -2$ MeV). Recent theoretical work by Brieva and Rook [4] suggested that the real and imaginary parts of the spin-orbit potential have different shapes and different energy dependences. In the framework of the impulse approximation, Jackson and Abdul-Jalil[5] were indeed able to reproduce elastic proton cross sections and polarizations between 49 and 185 MeV using this new form for the imaginary spin-orbit interaction. These calculations confirmed the need for a W_{LS} term in the phenomenological optical potential at these energies. Recently, Mackintosh and Kobos [6] analyzing elastic proton polarization data on ^{40}Ca and ^{16}O with an ℓ-dependent optical model potential have also clearly demonstrated the presence of a definite contribution of an imaginary spin-orbit potential which was found to improve the fits to the data around 30 MeV.

Analyzing powers and cross sections for elastic and inelastic scattering of 20, 30 and 40 MeV polarized protons from ^{12}C and ^{16}O have been therefore analyzed in the coupled-channels (CC) collective model using the code ECIS 78. For these calculations using the full Thomas deformed spin-orbit term, the 0^+ (g.s), 2^+ (E_x = 4.43 MeV) and 4^+ (E_x = 14.08 MeV) states in ^{12}C were coupled together using the rotational model with the following deformation parameters $\beta_2 = -0.70$ and $\beta_4 = 0.05$ [7]. Also in these calculations but with the vibrational model, the 0^+(g.s), 3^- (E_x = 6.13 MeV) and 2^+(E_x=6.92 MeV) states in

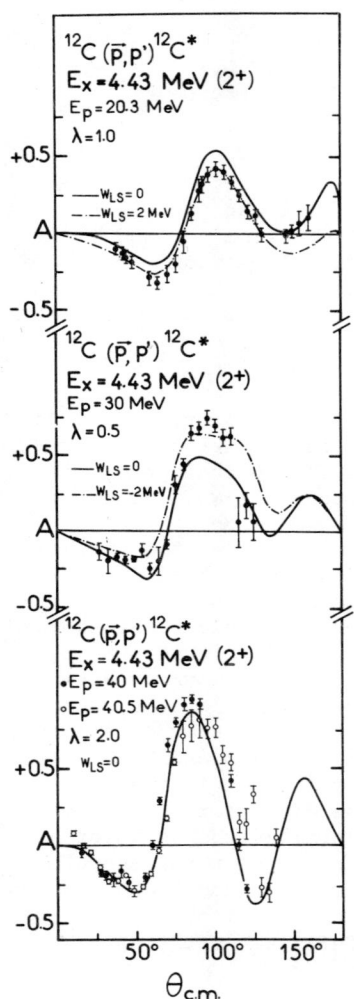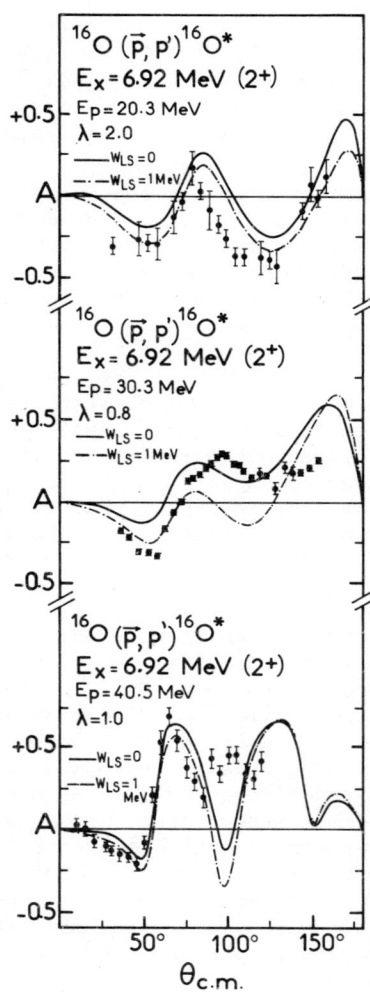

Fig. 1
Coupled-channels calculations for the 2^+ state in ^{12}C. The data are taken from ref. 7.

Fig. 2
Coupled-channels calculations for the 2^+ state in ^{16}O. The data are taken from ref. 7.

^{16}O were coupled together with the following deformation parameters $\beta_3=0.5$ and $\beta_2=0.2$ [7]. The interaction potential arises from the deformation of the Coulomb potential, the complex central potential and the complex spin-orbit potential. Optical potential parameters from ref.7 were used, they have standard definition.

Good agreement between the CC calculations using optical model parameters ($W_{LS}\equiv 0$) in ref. 7 and the cross section and analyzing power data had been obtained by varying the ratio $\lambda = \beta_{LS}/\beta_{central}$ (β_{LS} is the spin-orbit deformation and $\beta_{central}$, the central deformation) [7].

Somewhat improved agreement between the CC calculations using the same optical model parameters but with an imaginary spin-orbit potential W_{LS} different from 0 and the data has been obtained.

Figure 1 shows the better agreement obtained with the ^{12}C data with W_{LS} equal to 2, -2 MeV respectively for the energies E_p equal to 20, 30 MeV. For E_p = 40 MeV, no value of W_{LS} was needed in the case of ^{12}C. Figure 2 shows the better agreement obtained with the ^{16}O data at E_p=20 MeV with W_{LS}=1 MeV. This last figure shows also a better agreement namely at the forward angles obtained with the ^{16}O data at 30 and 40 MeV with W_{LS}=1 MeV.

In conclusion, there is evidence for a contribution of the imaginary deformed spin-orbit potential in the inelastic scattering of polarized protons from ^{12}C and ^{16}O at around 30 MeV, this is in agreement therefore with recent works [5,6].

REFERENCES

1. D.G. Baugh et al., Phys. Lett. 13, 63 (1964) ; R.M. Craig et al., Nucl. Phys. 58, 515 (1964)
2. R. de Swiniarski et al., Bull. Am. Phys. Soc. 13, 1663 (1968)
3. C. Glashausser, R. de Swiniarski, J. Thirion and A.D. Hill, Phys. Rev. 164, 1437 (1967)
4. F.A. Brieva and J.R. Rook, Nucl. Phys. A297, 206 (1978)
5. D.F. Jackson and I. Abdul-Jalil, J. Phys. G 6, 481 (1980)
6. R.S. Mackintosh and A.M. Kobos, J. Phys. G 4, L135 (1978)
7. Dinh-Lien Pham and R. de Swiniarski, Z. Physik A291, 327 (1979), and references therein.

MEASUREMENTS OF ANALYZING POWERS FOR 6⁻ STATES IN ^{28}Si and ^{24}Mg BY INELASTIC SCATTERING OF 65 MeV POLARIZED PROTONS

K. Hosono, N. Matsuoka, T. Saito, K. Hatanaka and M. Kondo
Research Center for Nuclear Physics, Osaka University,
Suita, Osaka 565, Japan

T. Noro and H. Shimizu
Department of Physics, Kyoto University, Kyoto 606, Japan

S. Kato and K. Okada
Laboratory of Nuclear Studies, Faculty of Science, Osaka University
Toyonaka, Osaka 560, Japan

K. Ogino and Y. Kadota
Department of Nuclear Engineering, Kyoto University, Kyoto 606, Japan

ABSTRACT

The angular distributions of cross sections and analyzing powers for proton inelastic scattering leading to the 11.58 MeV and 14.36 MeV states in ^{28}Si and the 15.15 MeV state in ^{24}Mg have been measured with 65 MeV polarized proton beam using the spectrograph RAIDEN. Preliminary microscopic DWBA calculations including exchange process have been performed. The results are discussed.

INTRODUCTION

The angular distributions of cross sections and analyzing powers for the ^{28}Si(p,p')^{28}Si and ^{24}Mg(p,p')^{24}Mg inelastic scatterings have been measured using 65 MeV polarized protons from the RCNP AVF cyclotron. We are interested in the scattering to the high spin and unnatural parity states of (E_x, j^π, T)=(11.58 MeV, 6⁻, 0) and (14.36 MeV, 6⁻, 1) in ^{28}Si and (15.15 MeV, 6⁻, 1) in ^{24}Mg. These states have been studied recently by (e,e')[1], (π,π')[2] and (p,p')[3,4] scatterings. The (p,p') scattering data were used to provide informations on the nucleon-nucleon effective interactions. We also intend to investigate the effective interactions in the (p,p') scattering. If these states are represented in the one-particle, one-hole model, we may extract informations of spin-flip and spin-isospin-flip interactions separately from the measurements of a pair of T=0 and T=1 states. Present work is a part of systematic studies of inelastic scatterings to unnatural parity states.

EXPERIMENTAL PROCEDURES

The scattered protons were measured using the spectrograph RAIDEN[5]. The overall energy resolution in the measurements was about 35 keV. The main part of the resolution came from the target thickness. A ^{12}C-polarimeter which monitored the polarization of the beam continuously was placed 100 cm upstream the scattering chamber. The beam polarization was about 70% during the measurements.

RESULTS AND DISCUSSIONS

0094-243X/81/690544-3$1.50 Copyright 1981 American Institute of Physics

Figures 1 and 2 show the angular distributions of the cross sections and analyzing powers leading to the 6⁻ states. The shapes of the angular distributions of the analyzing powers leading to a pair of T=0 and T=1 states are similar to each other. This is different from the case of the 1+ pair of ^{12}C in which the angular distributions of the analyzing powers show an out-of-phase pattern[6].

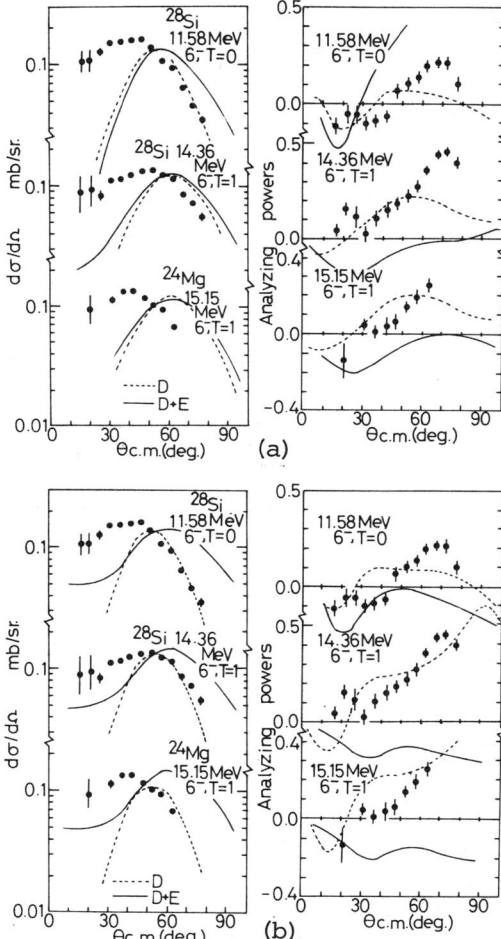

Fig. 1. Angular distributions of the cross sections and analyzing powers leading to the 11.58 MeV and 14.36 MeV states in ^{28}Si and the 15.15 MeV state in ^{24}Mg. The solid and dotted curves show DWBA predictions. Curves in fig. 1a are calculated with an interaction derived by Bertsch et al. and curves in fig. 1b are calculated with that by Hinricks et al,.

Preliminary microscopic DWBA calculations including exchange process have been performed. We used the code DWBA74[7] in the calculations. The predominant particle-hole configuration for these 6⁻ states is considered to be $(1f_{7/2})(1d_{5/2})^{-1}$. In a ^{27}Al($^{3}He,d$)^{28}Si reaction, the $f_{7/2}$ spectroscopic factor for the T=0 state was found to be 0.9 of the maximum possible value which was referred in ref. 2. On the other hand, the implied spectroscopic amplitude ranges from 0.5 to 0.7 times pure particle-hole value, which was obtained from the ratio of the experimental cross sections to theoretical ones in the electron scattering by Lindgren et al.[4] The $f_{7/2}$ spectroscopic factors for proton and neutron derived from ^{27}Al($^{3}He,d$)^{28}Si reactions and from $^{27}Al(d,p)$$^{28}Al$ reaction leading to the parent analog state are 0.2 (T=0 and T=1)[8] and 0.32(T=1)[9] respectively. These facts may attribute to more complicated configurations of these states. However in our calculations, the configuration of these states was assumed to be pure $(1f_{7/2})(1d_{5/2})^{-1}$. The results are shown in figs. 1a and 1b. In the figures, D+E means an inclusion of both direct and

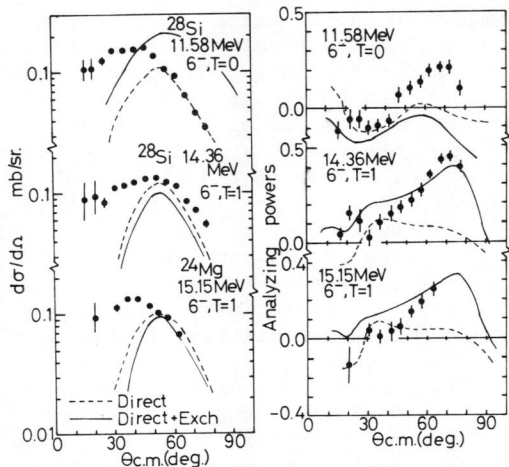

Fig. 2. Angular distributions of the cross sections and analyzing powers leading to the 6⁻ states. Curves are examples of the results of the calculations with an interaction obtained by the wide range of parameter searches.

exchange contributions. Figures 1a and 1b represent the results of the calculations using effective interactions derived by Bertsch et al.[10] and by Hinricks et al.[11] respectively. The calculated cross sections are normalized to the experimental data. The cross sections deduced from the DWBA calculations are small by factor of 0.2-1. compared with the experimental data. Theoretical cross sections and analyzing powers for these states obtained with both interactions do not so well reproduce the shapes of the experimental angular distributions.

We tried to adjust the strength of the effective interactions in order to obtain good fits to the experimental data. However we failed to obtain a good fit to the experimental ones for a pair of the T=0 and T=1 states simultaneously in spite of the wide range of parameter searches. An example of the results are displayed in fig. 2. The failure of the calculations to reproduce the experimental data may be due to the inadequacy of the description of these states in terms of one-particle one-hole configuration or to a deficiency in the reaction calculations.

REFERENCES

1. H. Zarek et al., Phys. Rev. Lett. $\underline{42}$ (1979) 1524.
2. C. Olmer et al., Phys. Rev. Lett. $\underline{43}$ (1979) 612.
3. G. S. Adams et al., Phys. Rev. Lett. $\underline{38}$ (1977) 1387.
4. R. A. Lindgren et al., Phys. Rev. Lett. $\underline{42}$ (1979) 1524.
5. H. Ikegami et al., RCNP Annual Report (1978) p.113.
6. K. Hosono et al., Phys. Rev. Lett. $\underline{41}$ (1978) 621.
7. J. Raynal, unpublished.
8. S. Kato et al., unpublished data
9. R. M. Freeman et al., Phys. Rev. $\underline{C11}$ (1975) 1948.
10. G. Bertsch et al., Nucl. Phys. $\underline{A284}$ (1977) 399.
11. R. A. Hinricks et al., Phys. Rev. $\underline{C7}$ (1973) 1981.

ANALYZING POWERS FOR $^{12}C(\vec{p},p')^{12}C$ AT INTERMEDIATE ENERGIES

J. R. Comfort
Physics Department, University of Pittsburgh, Pittsburgh, PA 15260

C.C. Foster, C.D. Goodman, D.W. Miller, G.L. Moake and P. Schwandt
Indiana University Cyclotron Facility, Bloomington, Indiana 47401

J. R. Rapaport
Physics Department, Ohio University, Athens, Ohio 45701

R. E. Segel
Physics Department, Northwestern University, Evanston, IL 60204

ABSTRACT

Data are presented for the analyzing powers for the $^{12}C(\vec{p},p')^{12}C$ reaction at 120 and 200 MeV to 1^+ and 2^+ states with T=0 and 1. The isovector transitions do not change much with energy, while the isoscalar transitions are strongly sensitive to it. In particular, $A(\theta)$ for the 12.71-MeV (1^+,T=0) transition changes sign. Microscopic DWIA calculations for the 120-MeV data, which describe the cross sections rather well, exhibit numerous disagreements with the analyzing power data.

INTRODUCTION

The interesting nucleus ^{12}C is very suitable for studying and testing effective nucleon-nucleon (N-N) interactions in (p,p') reactions. Not only are good wave functions available, but individual transitions are selectively sensitive to different components of the effective interaction. To the extent that the impulse approximation is valid, data at intermediate energies should provide new insights into the free N-N interaction. Polarization data are especially sensitive to spin and tensor contributions, but such data are generally scarce.

This work is an extension of an earlier study of the cross sections for the $^{12}C(p,p')^{12}C$ reaction at 122 MeV.[1] Analyzing powers have now been obtained for the same transitions. In addition, a full set of data has been obtained at 200 MeV in order to examine the energy dependence of reaction and its theoretical description.

THE EXPERIMENTS

Data for the $^{12}C(\vec{p},p')^{12}C$ reaction were obtained at I.U.C.F. at energies of 120 and 200 MeV. The beam polarization was typically about 70% and the emitted protons were detected in a helical-cathode proportional chamber in the focal plane of a magnetic spectrograph. The data extend to above 21 MeV excitation at 200 MeV and, except for a small gap, to about 17 MeV excitation at 120 MeV. They span the angular range 6-60°, typically in steps of 2-3°.

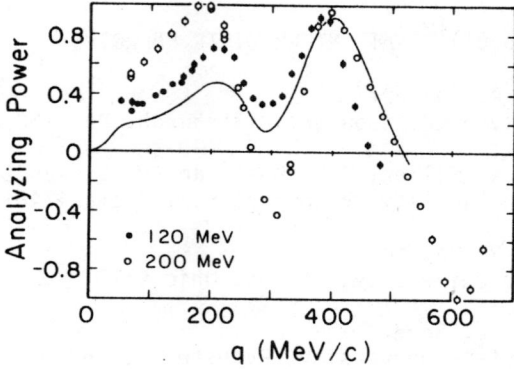

Fig. 1. Elastic analyzing powers for proton scattering from ^{12}C. The optical model curve is for 120 MeV.

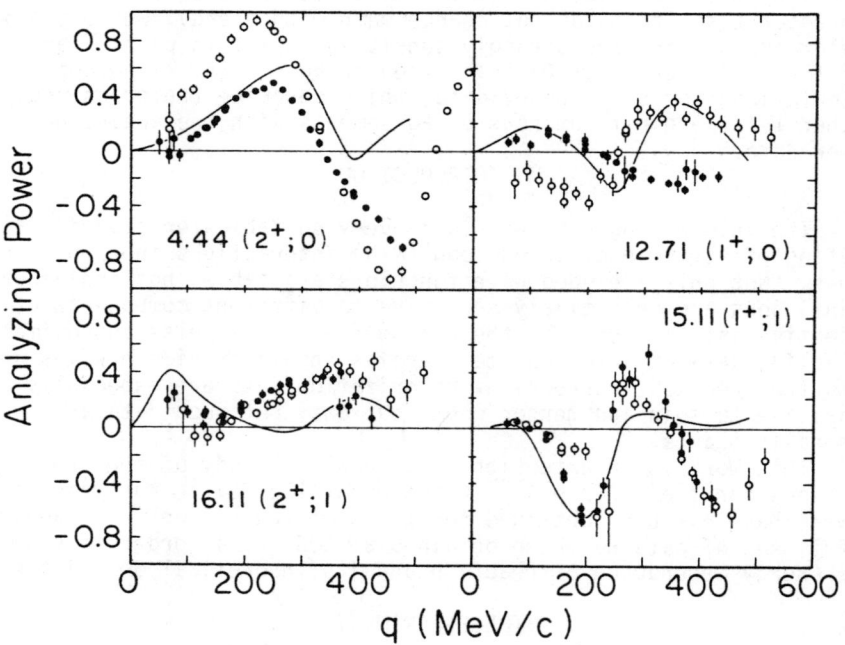

Fig. 2. Analyzing powers for ^{12}C($\vec{p},p'$)^{12}C at 120 MeV (solid points) and 200 MeV (open points). States are labelled by $E_x(J^\pi;T)$. The curves are DWIA calculations at 120 MeV.

RESULTS AND DISCUSSION

Analyzing powers for the elastic scattering and several inelastic transitions are plotted against momentum transfer q in Figs.1 and 2. The most striking feature of the energy comparison is the stability of the features for the isovector transitions while the isoscalar transitions, including elastic scattering, undergo substantial changes. Many of the observed isoscalar transitions oscillate sharply between about ±1. A second sharp change is the reversal in sign of $A(\theta)$ for the 12.71-MeV transition. It seems that the signs of $A(\theta)$ for the two 1^+ states are not signatures of their isospins[2] but instead reflect the energy dependence of the reaction dynamics.

Data at 50 and 65 MeV for the 1^+ states[2] accentuate the trends indicated here, while data at 800 MeV[3] for both $d\sigma/d\Omega$ and $A(\theta)$ are similar to the present 200-MeV data. This suggests that the impulse approximation is taking on greater validity as the energy increases and that it should be most useful at 200 MeV and beyond.

An optical-model calculation is shown for the 120-MeV elastic data in Fig. 1. Although this represents a best fit to both $d\sigma/d\Omega$ and $A(\theta)$ with conventional potentials,[1] it is not very satisfactory. Detailed microscopic calculations for the inelastic transitions, based on the free N-N t matrix and with an exact treatment of knockon exchange, are shown in Fig. 2 for the 120-MeV data. Although these calculations generally reproduce the cross section data below 300 MeV/c rather well,[1] numerous discrepancies are seen for $A(\theta)$.

The best agreement is for the 15.11-MeV transition where the one-pion exchange process is known to dominate.[1] Nevertheless, the calculations have insufficient structure beyond 250 MeV/c. No satisfactory explanation of this discrepancy has yet been found. It should be noted that this region could be affected by precritical phenomena related to the threshold for pion condensation.[4] However, our preliminary calculations show that while the cross sections can be strongly affected by this, $A(\theta)$ is very insensitive to it.

CONCLUSIONS

The analyzing power data for proton scattering from ^{12}C are providing very valuable and unique guides to the nature of the reaction mechanism at intermediate energies. It appears that the impulse approximation should be reasonably valid above 200 MeV. Considerable effort is being directed towards understanding the discrepancies with the $A(\theta)$ data. It is expected that the 200 MeV data will be especially beneficial, particularly for revealing more features of the short-range two-body correlations.

REFERENCES

1. J. R. Comfort, et al., Phys. Rev. C21, 2147 (1980).
2. K. Hosono et al., Phys. Rev. Letters 41, 621 (1978).
3. J. M. Moss et al., Phys. Rev. Letters 44, 1189 (1980).
4. H. Toki and W. Weise, Phys. Rev. Letters 42, 1034 (1974).

NEW ASPECTS OF THE TRIUMF ($\vec{p},\pi$) PROGRAM[†]

G.J. Lolos, E.L. Mathie, P.L. Walden, E.G. Auld, and G. Jones
University of British Columbia, Vancouver, B.C., V6T 2A6

R.B. Taylor
James Cook University of North Queensland, Townsville,
Queensland, Australia

INTRODUCTION

Recent experimental efforts to study the $(p,\pi^{\pm})$ reaction on light nuclei leading to low excited, discrete states of the residual nuclei[1,2] have been implemented to delineate the nature of the reaction mechanism. An understanding of the reaction mechanism is necessary before attempting to exploit the $(\vec{p},\pi)$ reaction for investigating nuclear structure. Studies of the (p,π^-) reaction, do however, provide immediately useful spectroscopic information due to the double charge exchange nature of the reaction. The first $(\vec{p},\pi^-)$ studies at TRIUMF, using the newly commissioned 65 cm pion spectrograph, the Resolution, are discussed in this contribution. A second focus of attention is the observation of $(\vec{p},\pi^+)$ reactions leaving the residual nucleus in highly excited, discrete states.

EXPERIMENTAL TECHNIQUES

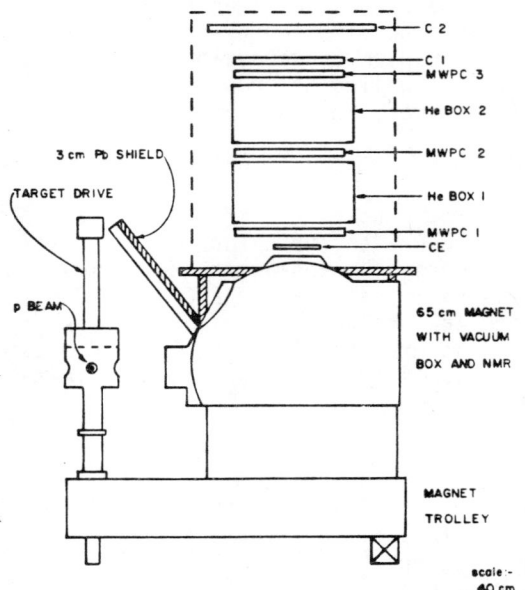

The measurements were done using a 65 cm Browne-Buechner spectrograph incorporating a 3 counter (CE, C1, C2) telescope and three helically wound multi-wire proportional chambers (MWPC) (see Fig. 1).[3] The pions are identified by appropriate pulse height cuts, particle propagation times to the counters, as well as track analysis. The spectrograph was calibrated using the $^1H(p,\pi^+)^2H$ reaction. A proton polarimeter based on the polarization dependence of proton-proton elastic scattering (at 26° laboratory angle) is used both as the beam polarization monitor and the

Fig. 1. "Resolution Spectrometer: with the three counter telescope (CE, C1, C2) and the three helically wound MWPC separated by two helium bags.

and the beam monitor. Typical beam intensity was 5 namp with a polarization of 75%.

RESULTS

A) $^9\text{Be}(\vec{p},\pi^-)^{10}\text{C}$ reaction: A sample of our $(\vec{p},\pi^-)$ data is shown in Figure 2(a,b). The states which have been clearly identified above

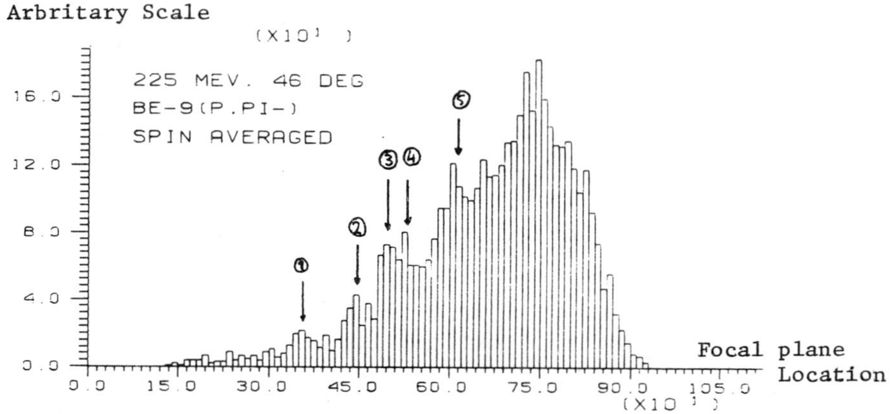

Fig. 2a   The $^9\text{Be}(p,\pi^-)^{10}\text{C}$ spectrum for $\theta_{lab} = 46°$ at $T_p = 225$ MeV. The numbers correspond to: 1, g.s.; 2, 3.3 MeV; 3, 5.3 MeV; 4, 6.6 MeV; 5, 10.6 MeV.

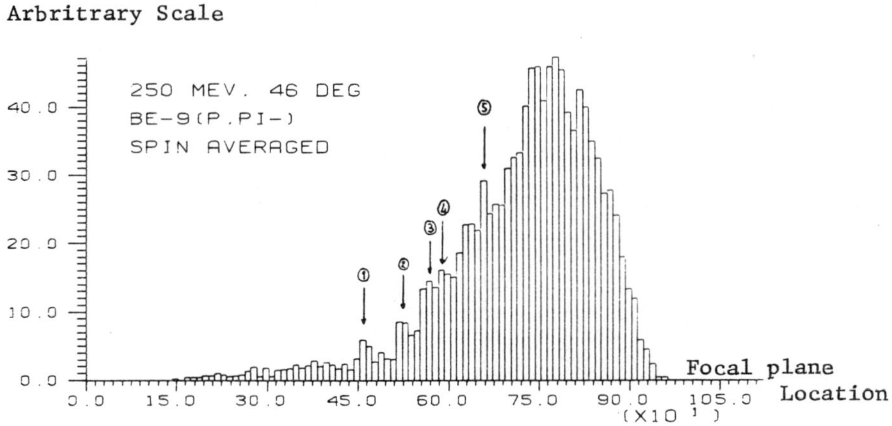

Fig. 2b   The $^9\text{Be}(p,\pi^-)^{10}\text{C}$ spectrum for $\theta_{lab} = 46°$ at $T_p = 250$ MeV. The numbers correspond to the same states as Fig. 2a.

the pion continuum are at excitation energies of 0.0, 3.4, 5.3, 6.6 and at 10.6 ± .3 MeV. We also see evidence for a 9.1 ± .2 MeV ^{10}C state. This level structure can be compared with Figure 3, which is a pion spectrum for the $^9\text{Be}(p,\pi^+)^{10}\text{Be}$, ^{10}Be being a mirror nucleus to ^{10}C. The energy levels inferred for ^{10}C may be compared to previous $^9\text{Be}(p,\pi^-)$

data at 185 MeV[4], and to ^{10}B(p,n), ^{12}C(p,t) and ^{10}B(^{3}He,t) data[5]. The analyzing powers, never previously reported are given in Table 1.

B) (p,π$^+$) High Excitation States in ^{10}Be and ^{13}C: States at very high excitation have been observed in the reactions ^{9}Be($\vec{p}$,π$^+$)^{10}Be (see Figure 3), and ^{12}C($\vec{p}$,π$^+$)^{13}C (see Figure 4). In many cases it is diffi-

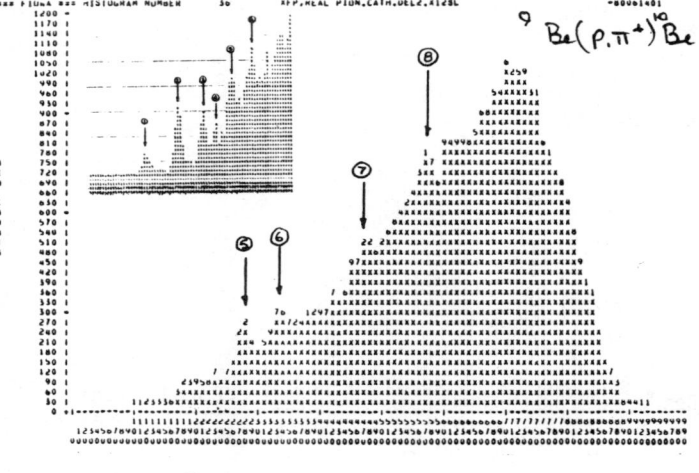

Figure 3. The ^{9}Be(p,π$^+$)^{10}Be high excitation spectrum for θ_{lab} = 46° at T_p = 250 MeV. The low excitation spectrum is shown in the inset. The numbers correspond to the following states in ^{10}Be: 1, g.s; 2, 3.4 MeV; 3, 6.0 MeV; 4, 7.3; 5, 9.4; 6, 12.0; 7, 17.8; 8, 22.8.

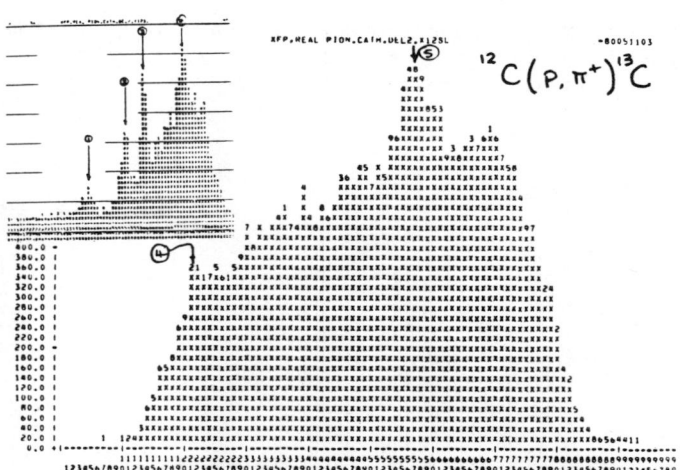

Figure 4. The ^{12}C(p,π$^+$)^{13}C high excitation spectrum for θ_{lab} = 46° at T_p = 200 MeV. The low excitation spectrum is shown in the inset. The numbers correspond to the following states in ^{13}C: 1, g.s.; 2, 3.9 MeV; 3, 6.1; 4, 10.5; 5, 18.0.

cult to distinguish the states above the pion continuum, which has an average analyzing power opposite in sign to that of the reactions to the discrete states. States at an excitation of 16 to 23 MeV have been identified in ^{13}C with a variety of reactions[6] but have never been observed with the (p,π) reaction. The analyzing powers for the excited states are given in Table I.

The uncertainty in the positions of the energy levels arises primarily from statistical uncertainties in the peak position along the focal plane. The uncertainty in the analyzing power for any particular state is primarily statistical but also includes the uncertainty of the incident beam polarization ($\leq 5\%$).

Table I   Analyzing powers for the $(\vec{p},\pi)$ reaction

$^9Be(p,\pi^-)^{10}C$		$\theta = 46°$		
^{10}C states =	g.s.	3.4	5.3	6.6
$T_p = 225$	-0.7 ± 0.4	-0.8 ± 0.2	-0.2 ± 0.2	-0.4 ± 0.3
$T_p = 250$	-0.5 ± 0.2	-0.4 ± 0.3	-0.5 ± 0.2	-0.6 ± 0.4

$^9Be(p,\pi^+)^{10}Be$ (High Excitation) $\theta = 46°$		
^{10}Be states =	17.8 MeV	22.8 MeV
$T_p = 250$	?**	?**

$^{12}C(p,\pi^+)^{13}C$ (High Excitation) $\theta = 46°$	
^{13}C states =	18.0 MeV
$T_p = 200$	$-.06 \pm .10$

** Analyzing power not calculable as pion continuum tends to mask the peak in the spin down condition.

REFERENCES

1. E.G. Auld, A. Haynes, R.R. Johnson, G. Jones, T. Masterson, E.L. Mathie, D. Ottewell, P. Walden, and B. Tatischeff, Phys. Rev. Lett. 41, 462-465 (1978).
2. P.H. Pile, R.D. Bent, R.E. Pollock, P.T. Debeuec, R.E. Marrs, M.C. Green, T.P. Sjoreen, and F. Soga, Phys. Rev. Lett. 42, 1461 (1979).
3. D.M. Lee, S.E. Sobottka, and H.A. Thiessen, NIM 120, 421-428 (1976).
4. S. Dahlgren, P. Grafström, B. Höistad, and A. Åsberg, Nucl. Phys. A204, 53 (1973).
5. W. Benenson, G.M. Crawley, J.D. Dreisbach and W.P. Johnson, Nucl. Phys. A97, 510 (1967).
6. F. Ajzenberg-Selove, Nucl. Phys. A152, 1 (1970).

^{90,92}Zr$(\vec{p},p')$ REACTIONS AT 800 MeV[*]

F. T. Baker[a], C. Glashausser[b], A. Scott[a],
G. Adams[c], M. Grimm[a], G. Hoffmann[d], G. Igo[c],
W. G. Love[a], J. Moss[d], V. Penumetcha[a],
W. Swenson[e], and B. E. Wood[f]

Angular distributions of differential cross sections and analyzing powers have been measured for elastic and inelastic scattering of 800 MeV protons from ^{90}Zr and ^{92}Zr. The data, acquired in 0.3° steps for the angular range 2°-20°, were obtained using the High Resolution Spectrometer at LAMPF. Excited states for which data have been obtained are the 2^+ (2.18), 5^-, 3^-, 4^+, 2^+ (3.31) 6^+, and 8^+ states for ^{90}Zr and the 2^+ (0.93), 4^+, 2^+ (1.83), 3^-, and 5^- states for ^{92}Zr. The data for elastic scattering were well fitted using an optical potential previously determined[1] for 800 MeV proton elastic scattering from ^{90}Zr; the fits for ^{92}Zr are shown in Figure 1. Data for inelastic scattering to collective 3^- and 5^- states are virtually identical for the two nuclei. Collective-model DWBA analyses provide excellent fits to these data and, particularly from the fits to the analyzing power

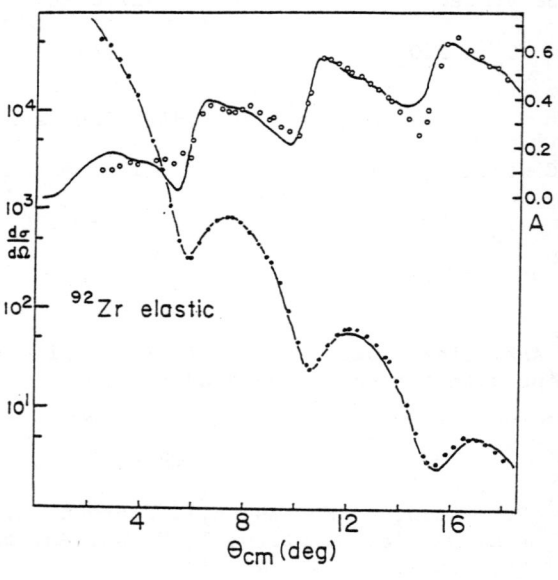

Figure 1

[*]This work was supported by the National Science Foundation and the U. S. Department of Energy.

[a]The University of Georgia, Athens, GA 30602
[b]Rutgers University, New Brunswick, NJ
[c]University of California at Los Angeles, Los Angeles, CA
[d]Los Alamos Scientific Laboratory, Los Alamos, NM
[e]Oregon State University, Corvallis, OR
[f]The University of Oregon, Eugene, OR

data, strongly indicate the need for deforming the spin-orbit potential; an example of this is shown in Figure 2 where the dashed curves are with no spin-orbit deformation and the full-drawn curves include a full-Thomas deformation with the same deformation parameter as for the central potential. Significant differences have been observed for excitation of the less collective first 2^+ and 4^+ states. The data for the 4^+ states are compared to collective-model DWBA predictions in Figures 3 and 4. Although microscopic-model analyses of these data have not yet been completed, a possible origin of these differences would be the differences in the microscopic structure of these states: it is generally assumed that the dominant configurations of these $J=2,4$ states are $(\pi g_{9/2})^2$ for ^{90}Zr and $(\nu d_{5/2})^2$ for ^{92}Zr. We therefore expect that microscopic-model impulse-approximation analyses of these data will yield interesting results concerning the sensitivity of intermediate-energy proton scattering to the microscopic nature of the states excited.

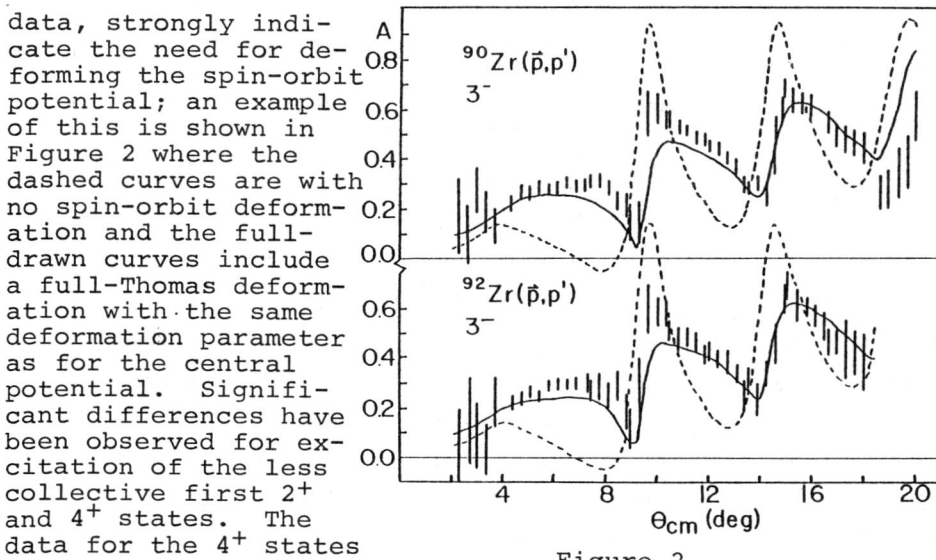

Figure 2

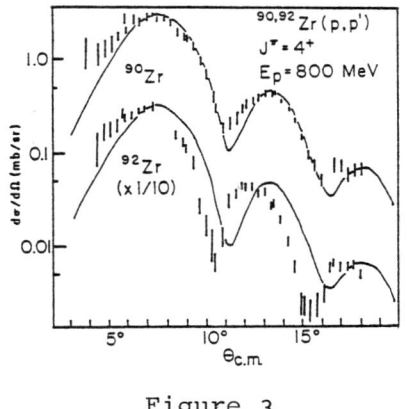

Figure 3

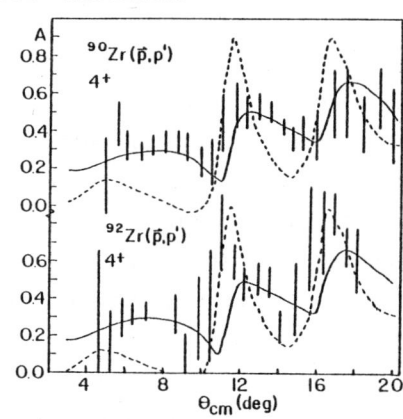

Figure 4

[1] L. Ray, private communication

NUCLEAR INFORMATION FROM THE WOLFENSTEIN PARAMETERS
IN INELASTIC PROTON NUCLEUS SCATTERING [*]

E. Bleszynski, M. Bleszynski and Ch. A. Whitten Jr.
Physics Department, UCLA, Los Angeles, Ca 90024

We study the structure of the collision matrices describing inelastic proton-nucleus scattering, and verify which suitable sets of experiments are necessary to achieve their complete determination. As a specific example we discuss the $^{12}C(p,p')$ reactions to the $1^+(12.72$ MeV,T=0) and $1^+(15.11$ MeV,T=1) levels. We show that by taking an appropriate combination of the Wolfenstein parameters, proton polarization and unpolarized differential cross section, it is possible to extract a selective information on the nuclear longitudinal and transverse formfactors and the isovector and isoscalar NN spin-flip amplitudes.

Very accurate experimental data for inelastic proton-nucleus scattering at intermediate energies are becoming available. Since at these energies the Glauber's diffraction approximation provides us with a very accurate and transparent description of nuclear scattering, a question arises how the data can be used to extract information on the structure of the excited states.

To date only data on unpolarized differential proton-nucleus cross sections and proton polarizations are available. Therefore, in order to obtain more precise and unambiguous information on these processes (which are, in general, described by a number of various amplitudes) some additional polarization experiments are necessary.

The aim of the present work is to analyze, in the framework of the Glauber model, the structure of the collision matrices for inelastic proton-nucleus reactions and to find the suitable sets of polarization experiments allowing their complete determination.

Here, we present the results of our preliminary analysis of the $^{12}C(p,p')$ reaction to the 1^+(12.72 MeV, T=0) and 1^+(15.11 MeV, T=1) states.

From the general symmetries with respect to rotations, parity and time reversal transformations, it follows that each of these processes is described by four independent complex scalar amplitudes: $F_o^T(q), F_x^T(q), F_y^T(q), F_z^T(q)$, being the functions of the momentum transfer q. The complete (up to an arbitrary phase) determination of these amplitudes can be achieved by doing 7 independent polarization measurements. Some of these measurements would require the determination of the spin alignement of the excited target state, which would cause substantial experimental problems. Therefore we focused our discussion here on the processes which are more feasible from the experimental point of view, namely those in which one can analyze the incident and scattered proton spin projections, averaging over the magnetic substates of the target.

We found, that in these circumstances it is possible to extract only 5 independent observables. These are: the absolute values of the amplitudes $F_j^T(q), j=0,x,y,z$ and the relative phase between $F_o^T(q)$ and $F_y^T(q)$. We have related these observables to the unpolarized differential cross section I_o, proton polarization P, and 3 (out of 4) arbitrarily chosen Wolfenstein parameters, e.g. A, D and R. The fourth parameter R' is related to A,D,R and P through the following relation:

$$R' = -R - \sin(2\theta_L)A - \sqrt{(1+D)^2 + 4P^2 + 4A^2\cos^2\theta_L(1+\sin^2\theta_L)} ,$$

where θ_L is the proton scattering angle in the target laboratory frame.

In order to check what physical information is contained in the above mentioned observables, we have calculated them in the framework of the Glauber model, assuming the one particle - one hole excitation mechanism.

It turns out that the single scattering contributions to the amplitudes $F_j^T(q)$, j=0,x,y,z are proportional to the different and only one spin-flip isoscalar (for T=0 state) or isovector (for T=1 state) part of the NN amplitude. Furthermore the amplitude $F_x^T(q)$ is proportional to the longitudinal formfactor (i.e. reduced matrix element of the axial longitudinal multipole operator L_{11}^5) , while the other three amplitudes are proportional to the transverse electric formfactor (i.e. reduced matrix element of the axial transverse electric multipole operator $T_{11}^{el}5$).[1]

We also checked that this interesting, selective character of the amplitudes is not affected after the higher order terms of the multiple scattering expansion (distortion effects) are taken into account.

We conclude that the measurement of the three Wolfenstein parameters, polarization and differential cross section will allow one to isolate specific parts of the $^{12}C(p,p')$ reactions to the 1^+(12.72 MeV,T=0) and 1^+(15.11 MeV,T=1) states collision matrix, and study separately contributions from the two different formfactors, as well as the isoscalar, or isovector spin flip-parts of the NN amplitude.

Such an information might be useful also for studies of the pion precondensation effects in inelastic proton-nucleus scattering, because their possible importance can be checked by extracting the contributions proportional to the ("effective") longitudinal formfactor.

Finally we point out that in order to complete the information on 4 amplitudes describing this reaction two additional measurements are necessary. These might be, e.g. measurements of the polarization of the excited target state corresponding to the transverse (in the scattering plane) and the longitudinal (along the incident beam direction) polarizations of the incident proton. These measurements would allow one to extract the relative phases between the $F_o^T(q)$ and $F_x^T(q)$, and $F_y^T(q)$ and $F_z^T(q)$ respectively.

*Work supported in part by US DOE

References:

1. J. D. Walecka, "Semi-Leptonic Interactions in Nuclei", in Muon Physics, V.H. Hughes and C. S. Wu, editors, Academic Press 1973.

ANALYSING POWERS OF THE CONTINUUM SPECTRA (I) :
65 MeV POLARIZED PROTONS ON
^{12}C, ^{28}Si, ^{45}Sc, ^{58}Ni, ^{93}Nb, ^{165}Ho, ^{166}Er and ^{209}Bi

H.Sakai, K.Hosono, N.Matsuoka and S.Nagamachi[*]
Research Center for Nuclear Physics, Osaka University,
Suita, Osaka 565, Japan

K. Okada
Dept. of Physics, Osaka University, Toyonaka, Osaka Japan

K. Maeda
Dept. of General Education, Tohoku University, Sendai, Japan

H. Shimizu
Dept. of Physics, Kyoto University, Kyoto, Japan

Analyzing powers A_y of the continuum spectra were measured in order to gain an insight into the reaction mechanism of the pre-equilibrium process. 65 MeV polarized protons were used to bombard 8 target nuclei (^{12}C, ^{28}Si, ^{45}Sc, ^{58}Ni, ^{93}Nb, ^{165}Ho, ^{166}Er and ^{209}Bi) spanning the periodic table. The A_y of the continuum spectra of ($\vec{p}$, pX) and ($\vec{p}$, dX) reactions were measured for all target nuclei and those of ($\vec{p}$, αX) reaction were also measured for ^{93}Nb and ^{209}Bi target nuclei. Attention was directed to measurements made over a wide range of excitation energy extending from 10 MeV up to the maximum kinematically allowed and to a wide angular range (20°-150°). Some of the results have been previously reported[1].

The 65 MeV polarized proton beam was provided by the AVF cyclotron at the Research Center for Nuclear Physics, Osaka University. The beam currents used varied from 10 to 60 nA depending upon the angle of observation and the target thickness. The beam spot size was approximately 2 mm × 2 mm with negligible beam halo. The emitted p, d and α particles were detected by a pair of counter telescopes positioned at symmetric angles to the beam. Each counter telescope consisted of a Si detector of 400 μm thickness and a high purity Ge detector of 15 mm thickness. The emitted particles were identified by the ΔE and E signals. The solid angle of two detector systems were 1.01 msr, respectively. The beam polarization was reversed every 300 msec and its polarization was monitored continuously with a ^{12}C-polarimeter during the measurements. The beam polarization was ∼70%.

Typical double differential energy spectra $d^2\sigma/d\Omega dE$ and analyzing powers $A_y(\theta)$ at $\theta_L = 20°, 60°, 90°$ and $150°$ for (p, p'X), (p, dX) and (p, αX) reactions are shown in Fig. 1 for ^{93}Nb. Since the energy spectra are summed over 1 MeV, 2 MeV and 4 MeV bins for (p, p'X), (p, dX) and (p, αX) reactions respectively, the structure due to discrete levels in the low excitation region is smeared out. The results for other target nuclei are in general similar for the ^{93}Nb target.

The following results were obtained from the A_y of the continuum spectra. (1) The A_y of the (p, p'X), (p, dX) and ($\vec{p}$, αX) reactions have a characteristic behavior with excitation energies and angles which is similar for all target nuclei except the ^{12}C(p, p'X) reaction. (2) The A_y are small at forward angle where the pre-equilibrium process is important. However there is no systematic tendency on the A_y; ^{58}Ni and ^{93}Nb(p, p'X) reactions have negative A_y values at 20° while ^{165}Ho, ^{166}Er and ^{209}Bi(p, p'X) reactions have positive A_y values at 20°. This fact clearly indicates that the spin dependent interaction as well as the nuclear structure effect have to be taken into account in the understanding of the continuum spectra. (3) One of the most striking features is that the A_y are large and positive at backward angle where the shape of the continuum spectra resembles that of a conventional evaporation spectrum. This fact may indicate that the direct interaction process is very important and contributes a large fraction of the total continuum spectra even at the backward angle. (4) The maximum values of the A_y in the backward hemisphere for ^{93}Nb(p, p'X), (p, dX) and (p, αX) reactions at E_x = 20 MeV are 15%, 20% and 35% for p, d and α emitted particles, respectively. (5) These large values are due mainly to the entrance channel effect. This conclusion is based on the fact that the (p, αX) reaction has a large A_y. (6) There is no appreciable even-odd mass effect, since the A_y values of ^{165}Ho and ^{166}Er agree with each other at almost all angles and excitation energies. (7) The magnitude of the A_y in the light mass region, depends on the target mass number, while that for the nuclei heavier than ^{58}Ni does not depend on the target mass number strongly.

The present features of the A_y of the continuum spectra are new and suggest the importance of the inclusion of the spin-dependent interaction in the pre-equilibrium reaction models. A more sophisticated model is clearly required to account for both angular distributions and complex particle spectrum.

REFERENCE

1. H. Sakai et al., Proc. Int. Symposium of Continuum Spectra of Heavy Ion Reactions, San Antonio, 1979, to be published.
   H. Sakai et al., Phys. Rev. Lett. <u>44</u> (1980) 1193.

* Hitachi Medical Corporation, Chiba, Japan

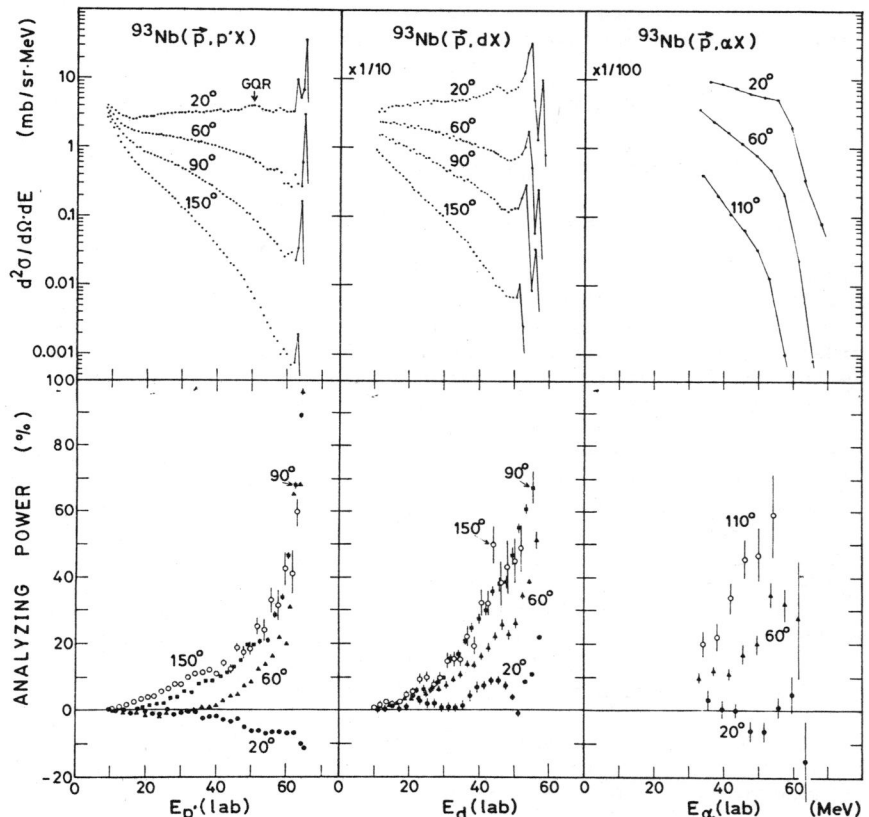

The double differential energy spectra (upper) and the analyzing powers (lower) of p, d and α particles resulting from the bombardment of 65 MeV polarized protons on ^{93}Nb.

ANALYZING POWERS OF THE CONTINUUM SPECTRA (II):
56 MeV POLARIZED DEUTERONS ON ^{58}Ni, ^{93}Nb AND ^{209}Bi

H. Sakai, N. Matsuoka and K. Hatanaka
Research Center for Nuclear Physics, Osaka University, Suita, Osaka, Japan

K. Okada
Department of Physics, Osaka University, Osaka, Japan

H. Shimizu
Department of Physics, Kyoto University, Kyoto, Japan

Vector and tensor analyzing powers of the continuum energy spectra for ^{58}Ni, ^{93}Nb, ^{209}Bi(d, d'X) and (d, pX) reactions at E_d = 56 MeV have been measured, for the first time, over the angular range from 20° to 137° and the energy range from ~12 MeV to maximum energy which is kinematically allowed. The polarized deuteron beam was provided by the RCNP AVF cyclotron at Osaka University. Fig. 1 shows the energy spectra, the vector analyzing power A_y and the tensor analyzing powers A_{xx} and A_{yy} at various angles for the ^{93}Nb(d, d'X) and (d, pX) reactions. Fig. 2 shows the angular dependences of the cross-sections and vector and tensor analyzing powers for ^{93}Nb target against 10 MeV excitation energy bins. The A_y, A_{xx} and A_{yy} for the ^{58}Ni and ^{209}Bi targets are in general similar for the ^{93}Nb target.

The most prominent features are; (1) at forward angle where the pre-equilibrium process dominates, the A_y, A_{xx} and A_{yy} are small but not zero for almost all energy range. (2) At backward angle where the spectrum shape resembles to that of an evaporation spectrum, they are very large in magnitude. (3) At the lower energy end where the equilibrium process dominates, they are generally consistent with zero.

These features on A_y are also observed in the reaction induced by 65 MeV polarized protons[1]. Thus these A_y data may indicate the importance of the spin dependent interaction in the understanding of the continuum spectra and also the importance of the direct reaction process at backward angle as well as forward angle.

The tensor analyzing power A_{xx} (A_{yy}) can be expressed by the change between the cross section due to a beam aligned along x (y) axis and that due to an unpolarized beam as follows[2]

$$A_{xx} = 2(\sigma_x - \sigma_0)/\sigma_0, \quad A_{yy} = 2(\sigma_y - \sigma_0)/\sigma_0.$$

If we apply the same mode of thinking as was made for the tensor analyzing power in the heavy-ion (^{7}Li) scattering[3], we could connect the deuteron shape with a tensor analyzing power. Then we expect σ_x (σ_y) is larger (smaller) than σ_0. Thus we get $A_{xx} > 0$ and $A_{yy} < 0$. The signs of A_{xx} and A_{yy} are in accord with our experimental results of the (d, pX) reaction. However in the absence of a quantitative theoretical understanding of the reaction mechanism, we can only state that the data are very suggestive of a deuteron D-state effect. Therefore it is very interesting to apply a more sophisticated method such as the Statistical Multi-Step Nuclear Reaction theory[4] or Multi-Step Direct Reaction theory[5] to our data.

REFERENCES

1. H. Sakai et al., contribution to this symposium.
   H. Sakai et al., Phys. Rev. Lett. <u>44</u> (1980) 1193
2. M. Simonius, Lecture Notes in Physics, Vol. 30, ed. D. Fick (Springer, Berlin)
   W. Haeberli, Lecture Notes in Physics, Vol. 30, ed. D. Fick (Springer, Berlin)
3. W. Dreves et al., Phys. Lett. <u>78B</u> (1978) 36,
   P. Zupranski et al., Phys. Lett. <u>91B</u> (1980) 358
4. H. Feshbach, A. Kerman and S. Koonin, Ann. of Phys. <u>125</u> (1980)
5. T. Tamura, T. Udagawa, D.H. Feng, and K.-K. Kan, Phys. Lett. <u>66B</u> (1977) 109, T. Tamura and T. Udagawa, Phys. Lett. <u>71B</u> (1977) 273

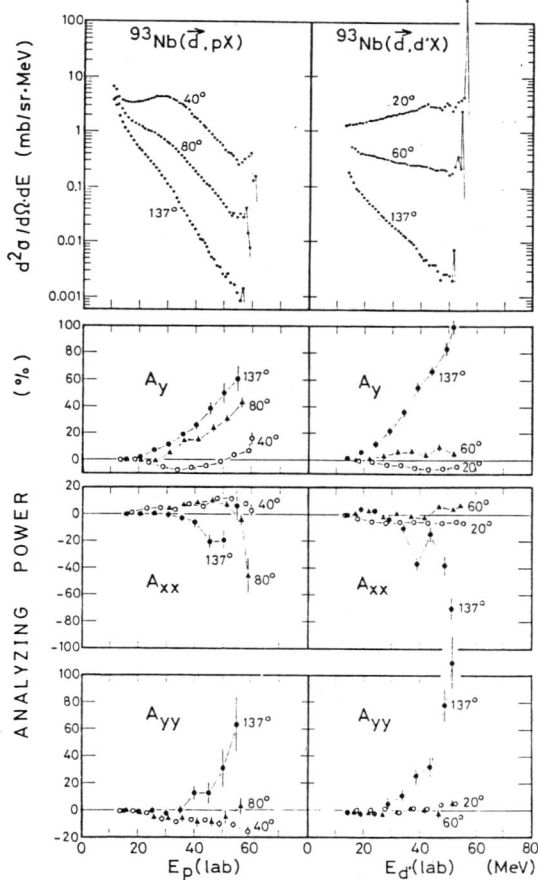

Fig. 1 The double differential energy spectra (upper), vector (A_y) and tensor analyzing powers (A_{xx} and A_{yy}) (lower) for ^{93}Nb($\vec{d}$,dX) and ($\vec{d}$,pX) reactions at 65 MeV.

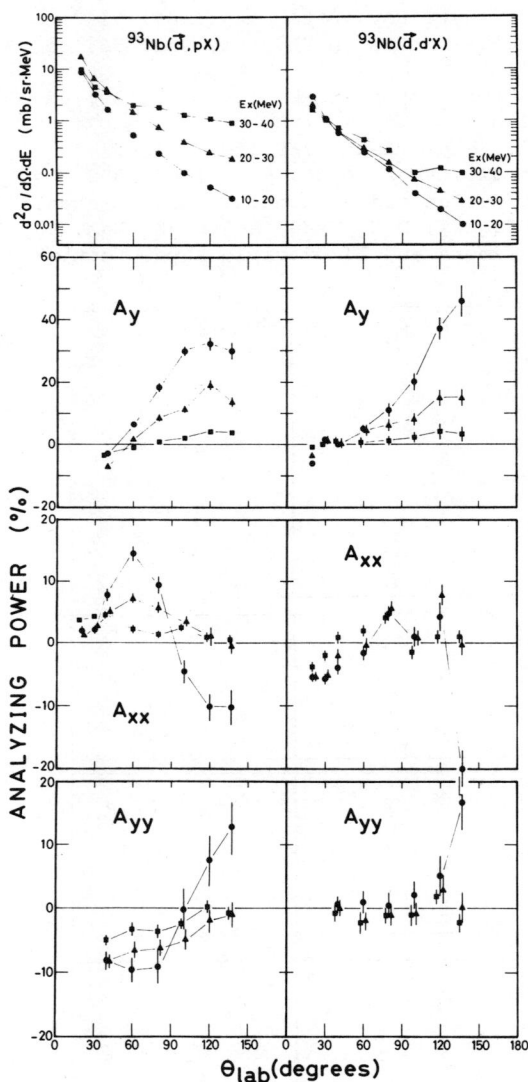

Fig. 2 Angular distributions of various energy bins and vector and tensor analyzing powers. Solid lines are drawn to guide the eye. The errors quoted are statistical.

ANALYZING POWER IN THE CONTINUUM IN LIGHT-ION
INDUCED REACTIONS

H. Lenske, T. Tamura and T. Udagawa
Department of Physics, University of Texas, Austin, Texas 78712

We showed earlier that continuum spectra observed in both light-ion[1,2] and heavy-ion[3] induced reactions can successfully be explained in terms of a multi-step direct reaction (MSDR) theory. The work was based on the belief that the MSDR theory, known to work rather well for discrete state transitions, must also be applicable to continuum state transitions. Successful fits to a variety of experimental data[1-3] appear to have confirmed the validity of this approach.

Recently, Sakai et al.[4] measured the analyzing power (A_y) in reactions leading to the continuum, induced by 65 MeV polarized protons. Here we report on MSDR analyses of their data. As it is the case for reactions leading to discrete states, the analysis of A_y data is expected to add new informations on the reaction mechanisms, beyond what one gets by just analyzing the cross section data.

We first took the ^{58}Ni($\vec{p},p'$) data of ref.4 and performed a calculation very similar to that of ref.1. We found that the theoretical A_y, if multiplied by an overall factor of 1/3, fit very well the angular and E_p' dependence of the observed A_y. This result[5] was considered rather encouraging, because it had been thought previously that the contributions from various components to the continuum cross section might wash out the overall A_y.

In spite of the success achieved in fitting the experimental angular and energy distributions, the analysis of ref.1 encountered a difficulty in that the theoretical cross sections were too small by an overall factor of 3 or so. In order to solve this problem we have been engaged in a work to redo the analysis of ref.1, by using the microscopic form factors (MFF)[6], rather than the collective form factor (CFF)[1]. We found that the magnitude problem was all but removed this way. We thus decided to compare further the MFF and CFF predictions of A_y in the continuum (p,p') reactions.

The comparison made for the case of the ^{208}Pb($\vec{p},p'$) reaction is presented in Fig.1, where the data of ref.4 are also given. The MFF result (solid line) is in good accord with experiment. On the other hand the (envelope of the) CFF result (dotted line) overpredicts the experimental A_y, repeating the trouble as encountered in ref.5.

The reason why A_y(CFF)>A_y(MFF) is easy to understand. In the CFF method the radial form factor is assumed to be the same for all particle-hole (ph) excitations; in practice the first derivative of the optical potential (hence the name CFF). Thus different ph transitions result in very similar A_y. Therefore the washing-out effect is too weak. With the MFF the amount of washing-out is just right, the resultant overall A_y agreeing rather well with experiment.

The result in Fig.1 is incomplete in that only one-step contributions have been considered so far. To use MFF for higher-step calculations is too time consuming, however. We are thus working on

0094-243X/81/690565-3$1.50 Copyright 1981 American Institute of Physics

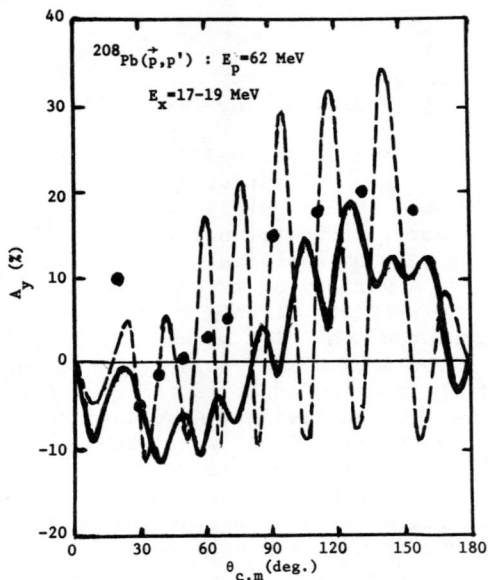

Fig.1

Analyzing power in the ^{208}Pb($\vec{p},p'$) reaction for E'_p= 17-19 MeV. The data were taken from ref.4. The theoretical lines are explained in the text.

modifying the CFF method in such a way as to take into account the crucial features embodied by the MFF method, but to avoid increasing the computation time drastically.

In Fig.2, we show results of analyzing the ^{93}Nb($\vec{p},\alpha$) data[4]. Here, both one-step, ($\vec{p},\alpha$), and two-step, ($\vec{p},\alpha',\alpha$) and ($\vec{p},p',\alpha$), processes were taken into account. (See ref.2 for details.) It is seen that very good agreement with experiment has been obtained, except possibly that the theoretical A_y is somewhat too small for E_α=51-55 MeV. This last difficulty may, however, be removed by dropping a simplifying assumption that states with $j=\ell\pm1/2$ are degenerate, where ℓ is the orbital angular momentum transferred in the (p,α)-step. Thus the fit shown in Fig.2 may be considered sufficiently good as to convince one that the MSDR method does explain the A_y data in the continuum.

We discussed above, regarding Fig.1, the subtle feature pertaining to the shape of the form factors, not detected while fitting only the cross section data[1]. Similarly, the A_y's given in the lower-right box in Fig.2 show that the (p,p',α) process is crucial in explaining the bump of A_y in the θ=70°-160° region, a feature which was not detected earlier[2], when only cross sections were fitted.

We are very much indebted to Dr.H. Sakai, for informing us of his data prior to publication, and for useful discussions.

This work was supported in part by the U.S. Department of Energy and the Deutsche Forschungsgemeinschaft.

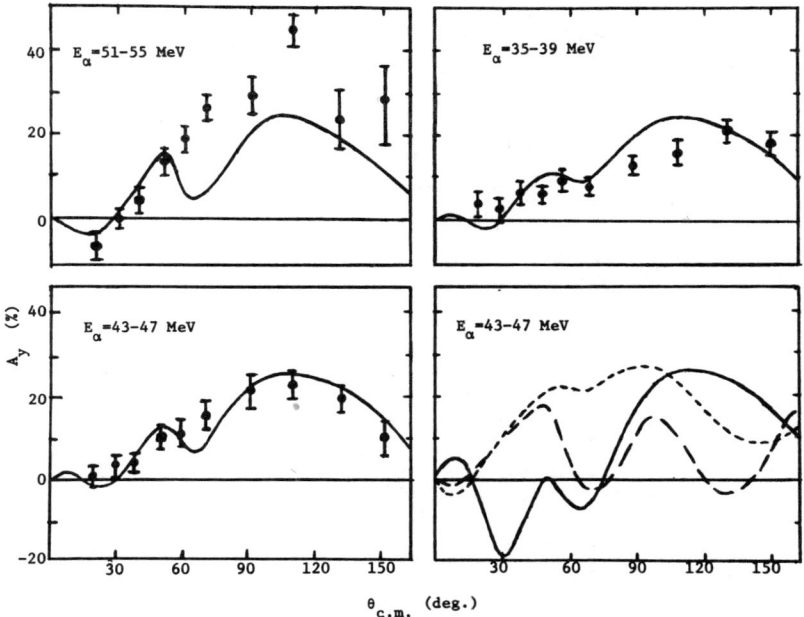

Fig.2 Analyzing power in the ^{93}Nb(p,α) reaction. The data were taken from ref.4. The lower-right box shows A_y obtained when (p,α), (p,α',α) and (p,p',α) processes are considered seperately. They are denoted, respectively, by dashed, dotted and solid lines.

REFERENCES

1. T. Tamura, T. Udagawa, D.H. Feng and K.K. Kan, Phys. Lett. <u>66B</u>, 109 (1977).
2. T. Tamura and T. Udagawa, Phys. Lett. <u>71B</u>, 273 (1977).
3. T. Udagawa and T. Tamura, in "Continuum Spectra of Heavy Ion Reactions", ed. by T. Tamura, J.B. Natowitz and D.H. Youngblood (Harwood Acad. Pub., N.Y.) in press.
4. H. Sakai, K. Hosono, N. Matsuoka, S. Nagamachi, K. Okada, M. Maeda and A. Shimizu, Phys. Rev. Lett. <u>44</u>, 1193 (1980); H. Sakai, private communications.
5. Annual report, Theoretical Nuclear Physics Group, University of Texas (1979) p.8
6. See e.g.: V.A. Madsen, "Nuclear Spectroscopy and Reactions", Part D (Academic Press, N.Y., 1975) p.249

QUASI-ELASTIC ^{40}Ca($\vec{p}$,2p) SCATTERING AT 200 MEV AT TRIUMF

P. Kitching, L. Antonuk, C. A. Miller, D. A. Hutcheon,
W. J. McDonald, W. C. Olsen, G. C. Neilson, G. M. Stinson
Physics Department, University of Alberta,
Edmonton, Canada, T6G 2J1

A. W. Stetz
Oregon State University
Corvallis, Oregon, 97331, U.S.A.

Previous ($\vec{p}$,2p) experiments on ^{16}O have demonstrated j dependence in the analysing powers for $\ell = 1$ states.[1,2,3] In a recently completed ^{40}Ca($\vec{p}$,2p) experiment at 200 MeV, unpolarized cross sections and analysing powers for a wide range of final state proton angle pairs, separation energies, and recoil momenta were measured using ~70% polarised beam at TRIUMF. The experimental arrangement consisted of four detector telescopes, each detector set consisting of x and y multiwire planes, a passing detector providing timing, and dE/dx information, and a NaI(Tℓ) stopping counter for measuring the outgoing energies. With two telescopes on the left and right of the beam, left-right coincidence measurements provided data for 4 angle pairs simultaneously with an energy resolution of 3 to 4 MeV (FWHM).

While each of the $1d_{3/2}$ and $2s_{1/2}$ strengths is believed to be concentrated in a single peak, other reactions such as (p,d) and (d,^{3}He) show the $1d_{5/2}$ strength to be spread out over at least 5 MeV in separation energy.[4,5] Using spectroscopic factors from the analysis of a ^{40}Ca(d,^{3}He) experiment,[4] we fit the separation energy spectra. In this way, cross sections and analysing powers as a function of energy-sharing between final state protons were extracted for 11 angle pairs with recoil momenta ranging from 0 to 250 MeV/c for the $1d_{3/2}$, $2s_{1/2}$, and $1d_{5/2}$ single particle states.

The $\ell = 2$ states demonstrate the marked j dependence in their analysing powers predicted by Jacob and Maris.[6] In Fig. 1

Fig. 1
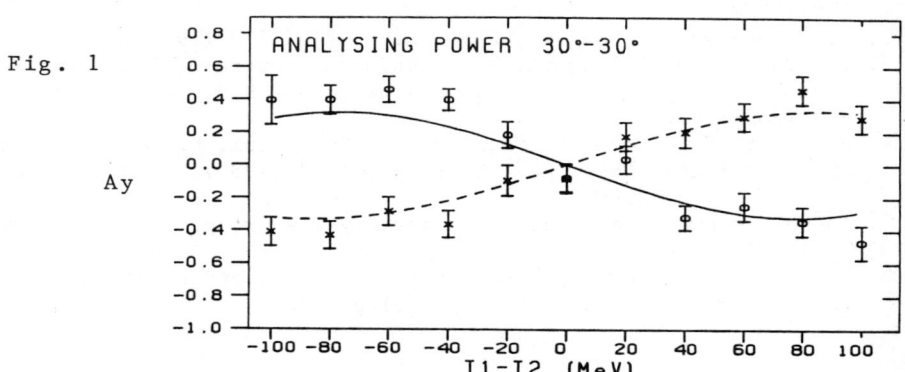

analysing power versus energy sharing is plotted for the $1d_{3/2}$ (circles) and $1d_{5/2}$ (crosses) states. Within the framework of DWIA and in the absence of spin-orbit distortion, the cross section for $(\vec{p},2p)$ reduces to a product of a phase space factor, the free proton-proton scattering cross section and the distorted momentum distribution of the struck nucleon. The j dependence of analysing powers for $\ell \neq \emptyset$ states arises from:
- the target nucleon being effectively polarized by nuclear spin-orbit coupling in the initial bound state and distortion in the scattering channels.
- the free polarized proton-proton scattering cross section being 3 to 5 times larger for parallel spins than for anti-parallel spins as a result of the large positive value of C_{NN} at these energies where:

$$\frac{\partial \sigma}{\partial \Omega P_o} = \frac{\partial \sigma}{\partial \Omega_o} \left[1 + (\vec{P}_o + \vec{P}_{eff}) \cdot \vec{A}_y + \vec{P}_o \cdot \vec{P}_{eff} \cdot C_{NN} \right] \quad (1)$$

$\vec{P}_o$, $\vec{P}_{eff}$ are the polarizations of the incident and struck protons. The analysing power for an $\ell = \emptyset$ state is just that for free p-p scattering with no contribution from the C_{NN} term. If spin-orbit distortion is included, the cross section factorization is destroyed and there is a further contribution to the analysing power for $\ell = \emptyset$ as well as $\ell \neq \emptyset$ states.

Comparisons with DWIA calculations were made using the University of Maryland code THREEDEE which allows the inclusion of spin-orbit dependence in the optical potentials for incoming and outgoing protons. In our calculations we used the bound state wave functions of Elton and Swift,[7] a spin-orbit dependent optical model potential incorporating recent 181 MeV (Indiana[8]) and 200 MeV (TRIUMF[9]) p-^{40}Ca elastic scattering data, and a half-off-shell prescription for p-p scattering. The predictions for analysing powers are in good agreement with the data for equal and near-equal forward angles (Fig. 1) but overestimate the cross sections (Fig. 2). In Fig. 2, where the units are mb/sr^{2}MeV, DWIA cross section predictions for $1d_{5/2}$ (dashed line) and $1d_{3/2}$ (solid line) have been reduced by a factor of 2.

Our results for the $2s_{1/2}$ analysing power at or near zero recoil (Fig. 3) were insufficiently precise to distinguish between the inclusion (solid line) and exclusion (dashed line) of spin-orbit distortion in the DWIA calculations. Peaks in the cross section at ~20 and ~30 MeV separation energies suggest the possible presence of $1p_{1/2}$ and $1p_{3/2}$ single particle states. Further experiments with improved resolution are underway.

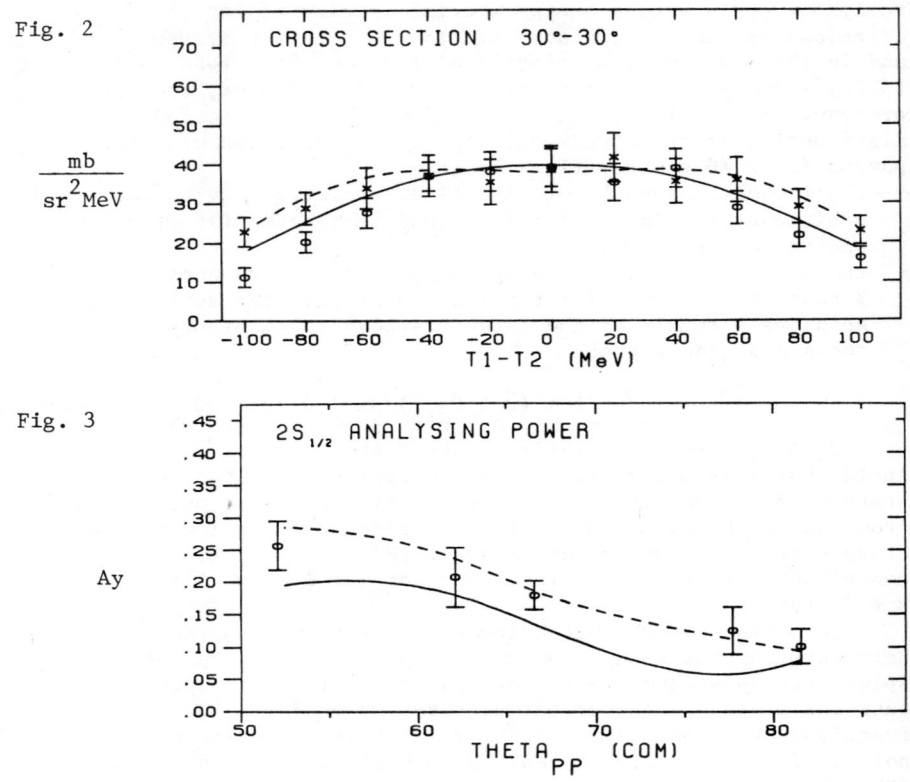

Fig. 2

Fig. 3

REFERENCES

1. P. Kitching, C. A. Miller, D. A. Hutcheon, A. N. James, W. J. McDonald, J. M. Cameron, W. C. Olsen, G. Roy. Phys. Rev. Lett. 37, 1600 (1976).
2. P. Kitching, C. A. Miller, W. C. Olsen, D. A. Hutcheon, W. J. McDonald, A. W. Stetz. Nucl. Phys. A340, 423 (1980).
3. G. Jacob, Th.A.J. Maris, C. Schneider, M. R. Teodoro. Phys. Lett. 45B, 181 (1973).
4. P. Doll, G. J. Wagner, K. T. Knopfle, G. Mairle. Nucl. Phys. A263, 210 (1976).
5. P. Martin, M. Buenerd, T. Dupon, M. Chabre. Nucl. Phys. A185, 465 (1972).
6. G. Jacob, Th.A.J. Maris, C. Schneider, M. R. Teodoro. Nucl. Phys. A257, 517 (1976).
7. L.R.B. Elton and A. Swift. Nucl. Phys. A94, 52 (1967).
8. P. Schwandt (private communication).
9. D. A. Hutcheon (private communication).

THE NUCLEAR QUADRUPOLE-QUADRUPOLE INTERACTION IN THE INELASTIC SCATTERING OF TENSOR POLARIZED DEUTERONS[+]

H. Clement, R. Frick, G. Graw, F.D. Santos[x], P. Schiemenz,
N. Seichert, Sun Tsu-Hsun[*]
Sektion Physik der Universität München, D-8046 Garching

With aligned (tensor polarized) deuterons we have measured at 20 MeV the elastic and inelastic scattering from two extremely deformed nuclei ^{28}Si (oblate, $\beta_2 = -.38$) and ^{24}Mg (prolate, $\beta_2 = +.45$). The figure shows the results for the observable $A_{xz}(\Theta)$, which exhibits most clearly the effect from the alignment of the deuteron quadrupole moment. While the elastic scattering data at forward angles behave

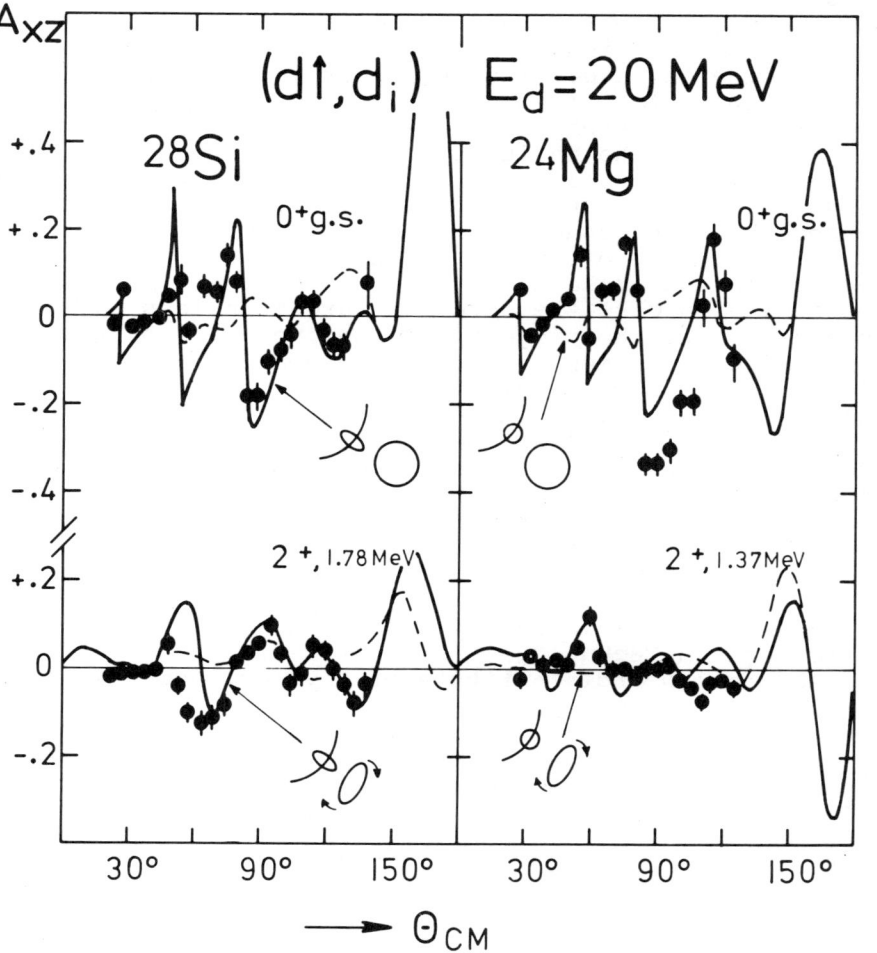

similar for ^{28}Si and ^{24}Mg, differ the inelastic data considerably in their angular distribution.

The drawn lines show CC-calculations with (full) and without (dashed) tensor interaction. The tensor potential at present is not determined in a unique way from the experimental data. For heavier nuclei, there are indications that the real part is reduced strongly[1,2], however, the interplay of real and imaginary part of the tensor potentials and the determination of the radial shapes are still an open problem. In the calculations shown in fig.1 we used a folding potential derived from the deuteron quadrupole moment and the central part of the deuteron nucleus interaction[2] and neglect the real part. The other parameters of the calculation are fixed by the analyses of measurements with vector polarized deuterons[2] resulting thus in a quasi parameter free calculation of $A_{xz}(\Theta)$.

The dashed lines simulate the scattering of a spherical deuteron, i.e. without tensor interaction. The analyzing powers for the 0_1^+ and the 2_1^+ states in both nuclei are calculated hereby to be small and mostly out of phase with the data. The inclusion of the tensor potential in the elastic scattering leads to a phase correct, nearly quantitative reproduction of the data. The pronounced difference between dashed and full lines demonstrates the strong influence of the deuteron quadrupole moment on $A_{xz}(\Theta)$. In the inelastic scattering to the 2_1^+-state due to reorientation also the deformed target nucleus is aligned in addition to the projectile. The inclusion of the deuteron quadrupole moment and the quadrupole-quadrupole interaction by means of a deformed tensor potential in the formalism of ref.3, used in the code CHUCK, leads to a description for the 2_1^+-scattering (full lines), which reproduces the characteristic features of the data. The difference between full and dashed lines in the inelastic scattering accounts in part for a significant contribution of the quadrupole-quadrupole interaction in the scattering of two deformed and aligned nuclei.

+) work supported in part by the Bundesministerium für Forschung und Technologie
x) DAAD-fellow from Institute of Physics, University of Lisbon, Portugal
*) Max-Planck fellow from the Institute of Atomic Energy, Peking, P.R. China
1) R.P. Goddard, W. Haeberli, Nucl. Phys. A316 (1979) 116
2) see further contributions to this conference by H. Clement et al. and R. Frick et al.
3) F.D. Santos, Z. Physik, Z 295 (1980) 73

MEASUREMENT OF QUADRUPOLE MOMENTS AND P_3-TERMS BY INELASTIC SCATTERING OF VECTOR POLARIZED DEUTERONS[+)]

H. Clement, R. Frick, G. Graw, F. Merz, P. Schiemenz
N. Seichert, Sun Tsu-Hsun[*)]
Sektion Physik der Universität München, D-8046 Garching

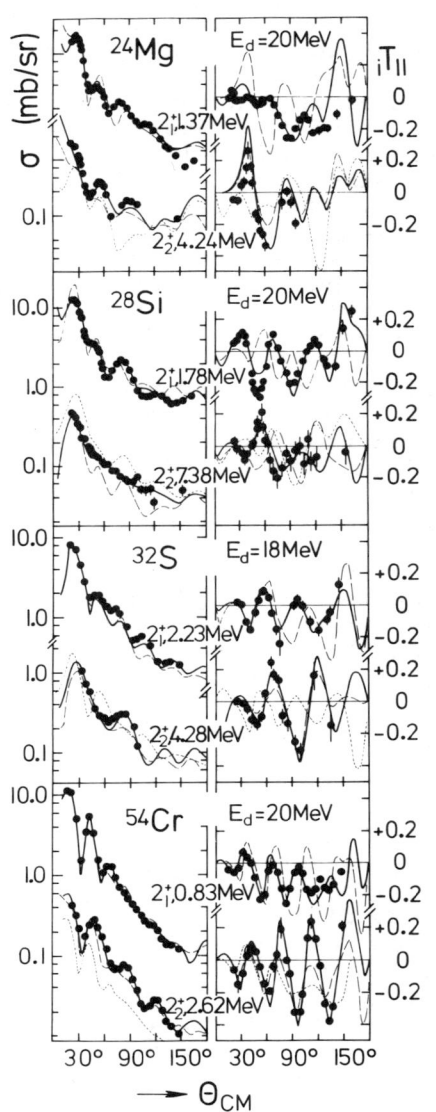

The high sensitivity of the scattering of 20 MeV vector polarized deuterons to interference terms in the scattering process has been used to study in detail the structure of excited states. Special attention has been paid to 2^+_1 and 2^+_2-states, which play a dominant role in the determination of the nuclear shape. Fig.1 shows the cross sections and the analyzing powers for the first two 2^+-states in ^{24}Mg, ^{28}Si, ^{32}S and ^{54}Cr. In all four cases the angular distributions for the 2^+_2-states behave very differently from those of the 2^+_1-states exhibiting thus the sensitivity of the reaction mechanism to differences in the structure of the excited states. The drawn full curves are the result of CC-analyses with the code ECIS[1)], where cross sections and analyzing powers for the level system $0^+_1 - 2^+_1 - 2^+_2 - 4^+_1$ and others specific for the individual nucleus have been analyzed simultaneously. In all cases a global set[2)] of optical model parameters has been used for the description of the spherical part of the deuteron-nucleus interaction. For the parametrization of the transition amplitudes we use the collective model. In the analyses the data show

0094-243X/81/690573-03$1.50 Copyright 1981 American Institute of Physics

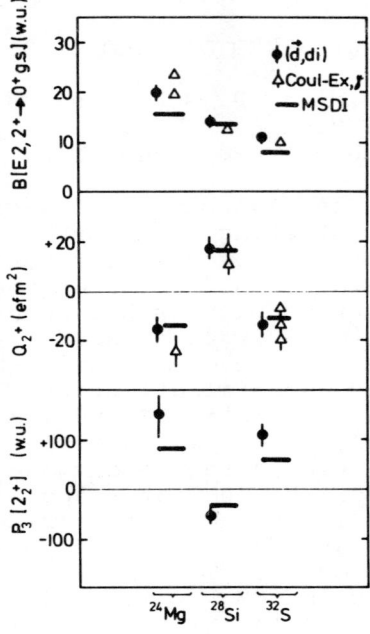

Fig. 2

strong sensitivity to the phases of nuclear quadrupole interference terms in the scattering process, which correspond to the Q_2 quadrupole moments and the P_3 term in the case of electromagnetic transitions; the P_3 terms describe for the 2_2^+-states the interference of the direct transition from the groundstate with the indirect one via the 2_1^+-state.

Reversing in the calculations the sign of the quadrupole moments (solid and dashed lines in fig.1) gives strong effects of $Q_{2_1^+}$ on the 2_1^+ data, especially in $iT_{11}(\theta)$ and somewhat smaller effects of $Q_{2_2^+}$ on the 2_2^+ data. (The 2_2^+-scattering is influenced only very weakly by $Q_{2_1^+}$ and the effect of $Q_{2_2^+}$ on the 2_1^+-scattering is negligible.) For the 2_2^+ excitation much stronger effects result from the P_3-term. This can be seen from the comparison of the solid and the dotted lines in fig.1, which show for the 2_2^+-scattering the results of reversing the sign of P_3 in the calculations. For all four nuclei we have pronounced effects, which lead to a determination of phase and magnitude of the nuclear matrix elements equivalent to $Q_{2_1^+}$ and P_3 and, with lower accuracy, of at least the phase of $Q_{2_2^+}$. The latter quantity to our knowledge has not been measured before in a direct way, it is important for the test of models for these nuclei.

For $Q_{2_1^+}$ our values agree with the ones obtained from Coulomb excitation measurements. For the phase of P_3 of these nuclei there are no measurements else known to compare with.

Our results agree favourably with microscopic calculations for the sd-shell using the MSDI[3]. This is shown in fig. 2 for the $2_1^+ \to 0_1^+$ transition strength, the quadrupole moments $Q_{2_1^+}$ and the P_3-values of ^{24}Mg, ^{28}Si and ^{32}S, where also the results from Coulomb excitation and from γ-ray-studies are included.

For all four nuclei these quantities support remarkably well also geometric collective models. This points to a dominance of simple symmetries in the low energy spectrum of these nuclei[4]. For ^{54}Cr, this is supported

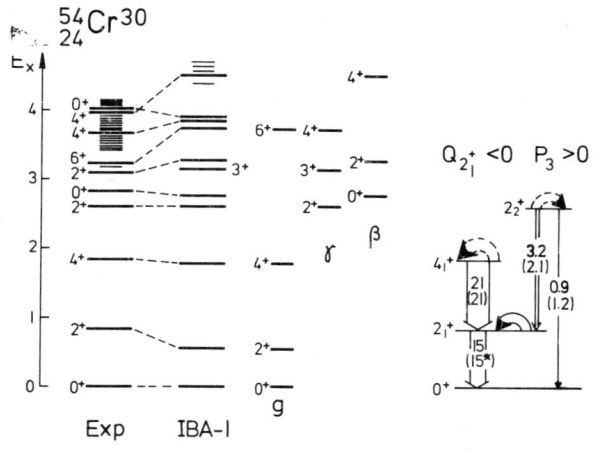

Fig. 3

also from a different point of view, fig.3 shows our calculations of ^{54}Cr in terms of the Interacting Boson Model[4] with the code PHINT:[5] the energy spectrum (left side) and our experimental values for the transition amplitudes and moments (shown schematically on the right side) are described satisfactorily in this semi-microscopic calculation. An investigation of the ß-band members, as a further test of this model, is currently under way.

+) supported in part by the Bundesministerium für Forschung und Technologie
*) Max-Planck-fellow from the Institute of Atomic Energy, Peking, P.R. China
1) J. Raynal, C.E.N. Saclay, code ECIS
2) J.M. Lohr, W. Haeberli, Nucl. Phys. A232 (1974) 381
3) Knüpfer, Stumm, private communication, Erlangen 1980
4) see e.g.: O. Scholten, F. Iachello, A. Arima, Ann. Phys. 115 (1978) 325
5) O. Scholten, Groningen, code PHINT

INELASTIC SCATTERING OF VECTOR POLARIZED DEUTERONS FROM SAMARIUM ISOTOPES[+)]

H. Clement, R. Frick, G. Graw, I. Oelrich[x)],
H.J. Scheerer[x)], P. Schiemenz, N. Seichert, Sun Tsu-Hsun[*)]
Sektion Physik der Universität München, D-8046 Garching

From inelastic scattering of vector polarized deuterons at E_d = 19 and 22 MeV we determine static and dynamic nuclear quadrupole transition matrix elements in ^{144}Sm and ^{154}Sm. Comparing with electromagnetic transition quantities and assuming geometrical collective models we observe differences between mass and charge deformation.

In fig.1 the data for ^{144}Sm exhibit a strict phase correlation between the elastic and the inelastic channels, which shows up most clearly in the analyzing power. While the 2_1^+-scattering is exactly in phase with the elastic channel, the 3_1^- angular distribution is just opposite in sign. This characteristic feature is reproduced by the CC-calculations with the code ECIS[1)] assuming collective, pure vibrational excitations (solid lines). Fig.2 compares the analyzing power of the 0_1^+ and the 2_1^+-scatte-

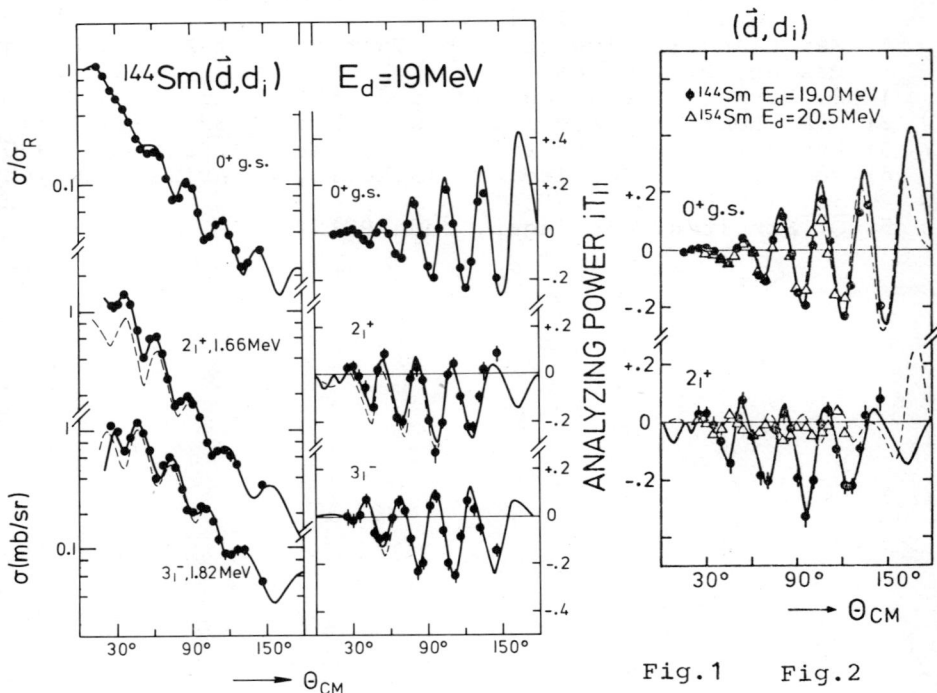

Fig.1   Fig.2

0094-243X/81/690576-03$1.50 Copyright 1981 American Institute of Physics

ring of ^{144}Sm with ^{154}Sm. In the elastic scattering the amplitudes of the oscillations are nearly the same, the deuteron incident energies have been chosen to give approximately an identical pattern for both nuclei.

The inelastic data exhibit pronounced differences: The diffraction pattern of ^{154}Sm is strongly damped and more narrow than that of ^{144}Sm. Both these effects result from (destructive) interference with the nuclear scattering amplitudes correlated with the large negative quadrupole moment of the 2^+_1 state of ^{154}Sm. They are reproduced quantitatively by the CC-analysis assuming for ^{154}Sm static deformation of prolate shape (dashed lines). In

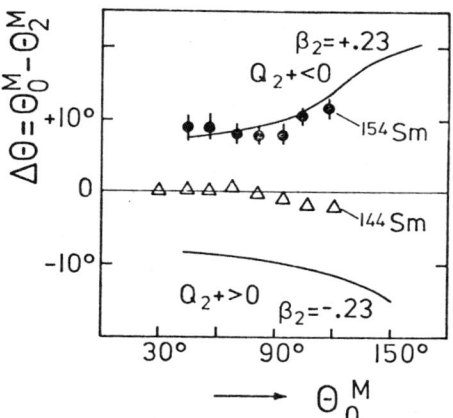

Fig. 3

fig.3 the angular shifts are displayed as differences of the scattering angle between the maxima (and minima) of the elastic (Θ^M_0) and the inelastic scattering (Θ^M_2). For ^{144}Sm these values are close to zero, while for ^{154}Sm they show a large shift of $\Delta\Theta \cong +10°$ on the average. The drawn lines represent results of CC-calculations assuming prolate or oblate shape, the shifts are approximately linear with the static nuclear deformation, fig.3 illustrates the sensitivity to determine this quantity.

Fig.4 shows for ^{154}Sm the data for the groundstate band up to the 6^+ state, the Q3D-magnetic spectrograph was used to resolve the transitions. All data are reproduced by the CC-calculations (including the 8^+-channel and assuming static deformation of the nuclear potential with $\beta_2 = +.225$ and $\beta_4 = +.05$ (solid line). (Using constant deformation lengths $\delta_2 = +1.51$ fm, $\delta_4 = +0.34$ fm gives a fit of the same quality.) The dashed lines show the effect of reversing the sign of β_2. In all calculations (fig.1-4) for the optical model the global set of Daehnick et al.[2] has been used. With other potentials we see that the results are not very dependent on the choice of the potential. Both for ^{144}Sm and ^{154}Sm the deduced nuclear quadrupole transition elements are about 30% smaller than the corresponding B(E2)-values. This is in contrast to lighter nuclei[3], where exactly the same procedure gives full agreement. We suggest to take this result for two heavy nuclei as evidence for a deviation from the geometrical collec-

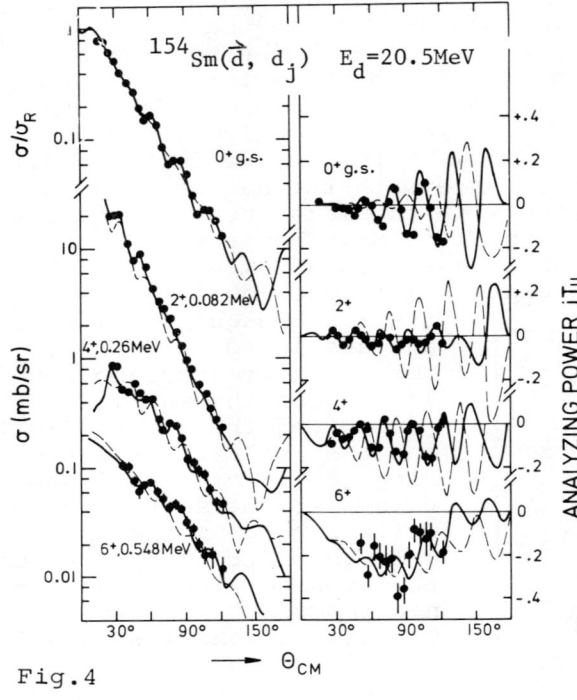

Fig.4

tive model. For states of high collectivity as those in ^{154}Sm this deviation is quite interesting. Of course, further investigation on other nuclei is needed to establish this behaviour.

The 2_1^+-cross sections in the forward angle region are very sensitive to Coulomb excitation (dashed lines in fig. 1: calculation with nuclear excitation only), we need indeed larger Coulomb deformations in the analyses in order to fit the forward angle region (full lines in fig.1 and 4). This mass charge discrepancy is observed also for the hexadecapole component β_4 in ^{154}Sm, where larger charge deformations have been found in Coulomb excitation measurements[4].

+) supported in part by the Bundesministerium für Forschung und Technologie
x) Technische Universität München, 8046 Garching
*) Max-Planck fellow from the Institute of Atomic Energy, Peking, P.R. China
1) J. Raynal, C.E.N. Saclay, code ECIS
2) W.W. Daehnick, J.D. Childs, Z. Vrcelj, Pittsburg, private communication
3) H. Clement, R. Frick, G. Graw, F. Merz, P. Schiemenz, N. Seichert, Sun Tsu Hsun, further contribution to this conference
4) F.S. Stephens, R.M. Diamond, J. de Boer, Phys.Rev.Lett. 27 (1971) 1151

OCTUPOLE-QUADRUPOLE COUPLING OBSERVED IN THE EXCITATION OF 3^- STATES WITH POLARIZED DEUTERONS *)

H. Clement, R. Frick, G. Graw, P. Schiemenz, N. Seichert
Sektion Physik der Universität München, D-8046 Garching

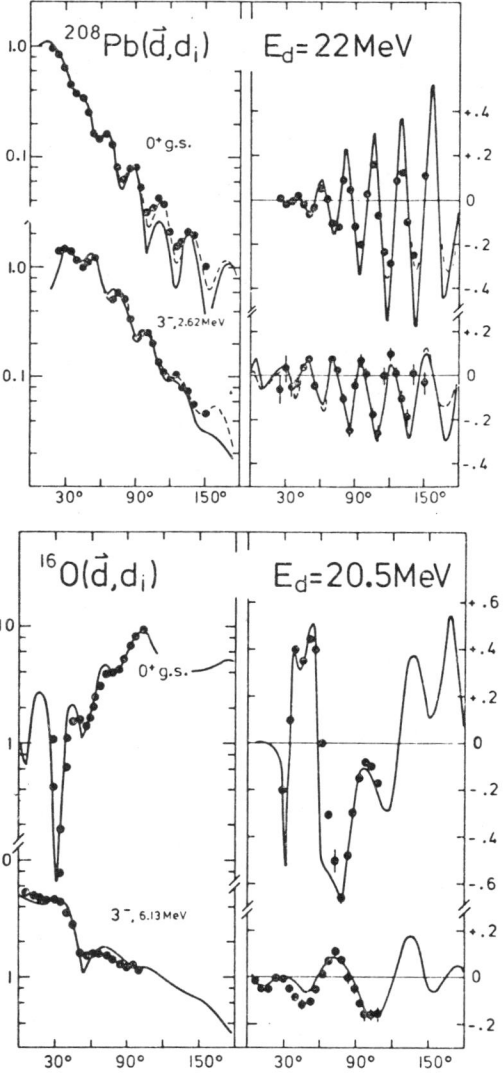

Figs. 1 and 2: Elastic cross section as $\sigma(\theta)/\sigma_R(\theta)$, inelastic in mb/sr, analyzing power as $iT_{11}(\theta)$

In contrast to the collective quadrupole excitation, the octupole excitation is most pronounced for magic nuclei. In figs. 1 and 2 we show our data for $\sigma(\theta)$ and $iT_{11}(\theta)$ for the 0_1^+ and 3_1^- scattering channels of ^{208}Pb+) and ^{16}O with the results of a coupled channels calculation, assuming pure octupole vibrational excitation. Data for ^{40}Ca [1] and ^{144}Sm are reproduced with similar quality. In fig. 3 we show in addition our 3^- data for the sd-shell nuclei ^{18}O, ^{24}Mg, ^{28}Si, ^{32}S and ^{36}Ar. The solid lines are coupled channel calculations, assuming again pure octupole vibration. The optical potentials are taken from the CC analyses of even parity states[2] and resemble closely those of ref.3. The data are reproduced satisfactorily only at the beginning (^{16}O, ^{18}O) and at the end (^{36}Ar, ^{40}Ca) of the s,d shell. For ^{32}S the angular distribution of $iT_{11}(\theta)$ deviates totally from the calculation, though $\sigma(\theta)$ is described well. For ^{28}Si the situation is similar, here, however, $\sigma(\theta)$ fails to be reproduced too. (For ^{28}Si and for ^{24}Mg the 4_1^+ excitation could not be separated completely from the 3_1^- excitation in the

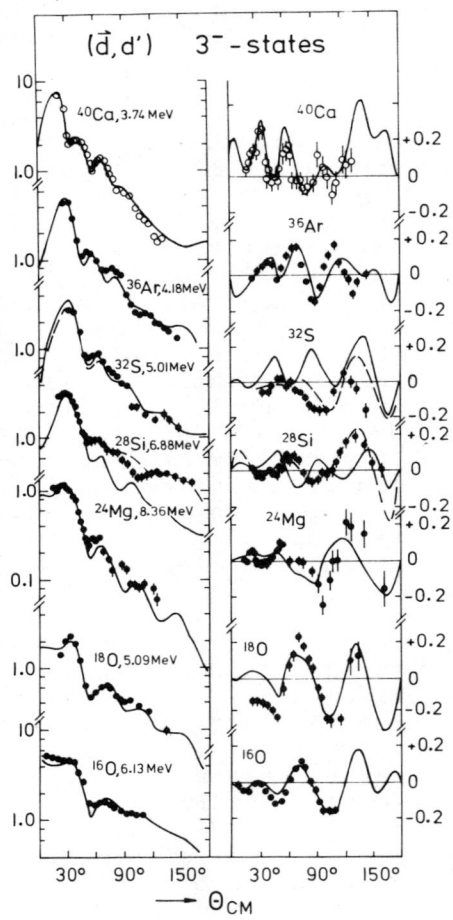

Fig. 3: Differential cross section in mb/sr and vector analyzing power as $iT_{11}(\theta)$

spectrum. All estimates, however, show, that this contribution is negligible for our discussion). Since both nuclei show strong quadrupole excitations, we interprete the anomalous behaviour of the analyzing power data for the 3^- states in ^{28}Si and ^{32}S as evidence for an octupole-quadrupole type coupling. This is supported by the fact that higher lying negative parity states with spin 1 to 5 which would be a consequence of such an assumption, indeed are known to exist in these nuclei. The dashed lines in fig.3 show CC-calculations which strongly improve the reproduction of the data. In the calculation we assume about half of the 3^- states to be of pure octupole-phonon nature, populated directly by an octupole excitation of the target ground state. The other part of the 3^- wave-function is assumed to be an octupole-quadrupole configuration, excited in the scattering process in two steps via the the 2_1^+-state. A consequence of such a configuration mixing is, that the 3^--states then can have a static quadrupole moment.

*) supported in part by the Bundesministerium für Forschung und Technologie
1) The data for ^{40}Ca are taken from G. Perrin, thesis Grenoble, 1974
2) see further contribution
3) J.M. Lohr and W. Haeberli, Nucl.Phys. A232 (1974) 381
4) W.W. Daehnick, J.D. Childs, Z. Vrcelj, Pittsburg 1980, private communication
+) the dashed curve in fig.1 is calculated with the potential of Ref.4.

THE POLARIZATION OF BREAK-UP PROTONS
FROM VECTOR POLARIZED DEUTERON INDUCED REACTION

M. Nakamura, H. Sakaguchi, K. Imai, T. Noro, H. Shimizu,
H. Sakamoto and S. Kobayashi
Department of Physics, Kyoto University, Kyoto, Japan

S. Kato
Laboratory of Nuclear Studies, Osaka University, Osaka, Japan

N. Matsuoka and K. Hatanaka
Research Center for Nuclear Physics, Osaka University, Osaka, Japan

Recently it has been recognized that the polarization measurement of emitted particles in nuclear reaction may play a significant role in the exploration of nuclear reaction mechanism, where the reaction may be induced by either polarized or unpolarized beam and even heavy ion beam. The polarization measurement should extend to not only the emitted particles leading to the low-lying states but also to those leading to the highly excited states of residual nuclei. The latter due to, for example pre-compound process, the break-up reaction process, giant resonances, quasi elastic process, deeply inelastic process, deep hole states, isobaric analog states and so.

Matsuoka et al.[1] studied the break-up process of medium energy deuterons through the continuum spectra of the (d,p) reaction over a wide range of target masses. A typical energy spectrum of protons from the (d,p) reaction is shown in Fig. 1 for the ^{90}Zr(d,p) reaction at θ_{lab}=9.5°. The spectrum shows a prominent bump, the peak energy of which is about a half of the deuteron incident energy. Similar bump spectra are observed for all target nuclei. The bumps should be closely concerned with the deuteron break-up process.

In the present experiment, we have tried to measure the polarization of emitted proton corresponding to the bump occured in the energy spectra of the ^{40}Ca(d,p) reaction induced vector polarized deuteron. The experimental layout is shown in Fig. 2. The experiment were carried out using a 56 MeV vector polarized deuteron beam from the AVF cyclotron with atomic type polarized ion source at RCNP. A Silicon proton polarimeter combined with high purity Ge detectors was used. The details of the polarimeter was described elsewhere.[2] Since the Ge detectors were susceptible to radiation damage, it was desirable to set it far apart from the first target and beam duct. A pair of quadrupole magnets was installed between the Calcium target and the polarimeter. Thus we succeeded in the collection of scattered proton without reduction of solid angle. Of course in the present case, the transmission efficiency through quadrupole magnet due to the energy dependence of scattered proton was variable.

The experimental result is shown in Table I, in which the data are presented at the emitted proton energy of 24.6 MeV (FWHM~4 MeV). The further experiment is planning to measure at several separated energy bins, taking account of the co-existence of various reaction mechanisms each of which may contribute to a part of the energy spectra in a different way. The results should be still understood as preliminary ones because the estimation of the analyzing power of the polarimeter due to inelastic process in Si itself has an uncertainty, although the deduced maximum values are included in error bars. The separate experiment should be able to reduce the uncertainty by a great deal. Nevertheless it is very interesting to notice the considerable reduction of polarization comparing with the expected values (K∿2/3) based on the simple spectator model.

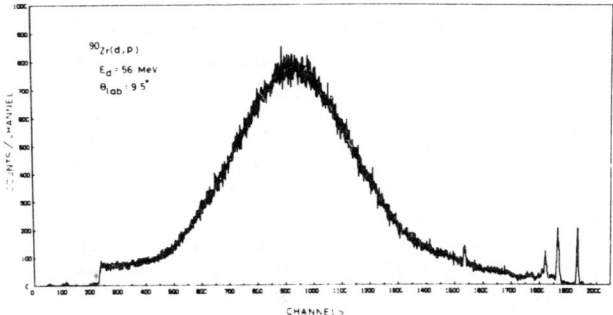

Fig. 1 A typical proton energy spectrum of the ^{90}Zr(d,p) reaction at θ_{lab}=9.5° with 56 MeV deuteron.

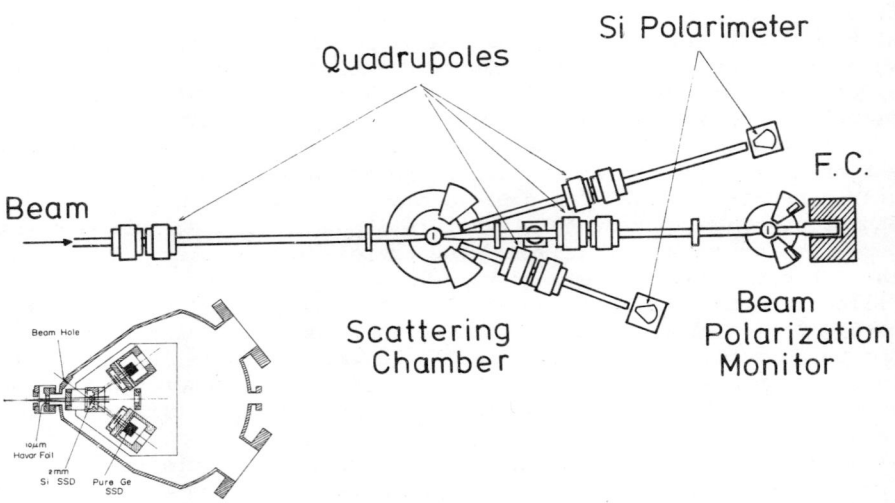

Fig. 2 Schematic view of the experimental arrangement.

Table I Measured value of the polarization transfer coefficient $K(\theta)$.

E_d (MeV)	θ_{lab} (deg.)	E_p (MeV)	K
56	15.5	24.6±2.0	0.58±0.06

REFERENCES

1. N. Matsuoka et al., RCNP Annual Report 1978 and Nucl. Phys. to be published.
2. M. Nakamura et al., RCNP Annual Report 1978, K. Kato et al., RCNP Annual Report 1979 and contribution to this conference.

ON DEFORMED TENSOR POTENTIAL FOR INELASTIC DEUTERON SCATTERING

by

Jacques RAYNAL
DPh-T - CEN - Saclay, B.P. N°2, 91190 Gif-sur-Yvette, France

Tensor analysing powers for inelastic deuteron scattering have been measured since a long time around 12 to 15 MeV [1,2]. The aim of such measurements for elastic scattering was to obtain informations on a tensor potential which comes from the interference [3] between the S and D parts $u(r)$ and $\omega(r)$, of the deuteron in a crude folding model:

$$V_T(r_d) = f_T(r_d) T_r \qquad T_r = \frac{\sqrt{8\pi}}{3} [Y_2(\hat{r}_d) T_2]^0_0 \qquad (1)$$

$$f_T(r_d) = \int \frac{18\omega(r)}{r^2\sqrt{2}} \left(u(r) - \frac{\omega(r)}{\sqrt{8}}\right)[V(r_p) + V(r_n)] P_2(\cos \widehat{r_d r}) \, d\vec{r}$$

where the deuteron wave function is :

$$\varphi_d(r) = \frac{1}{r\sqrt{4\pi}} \left[u(r) + \frac{\omega(r)}{\sqrt{8}} S_{np}\right] \qquad S_{np} = 3(\vec{\sigma}_p \hat{r})(\vec{\sigma}_n \hat{r}) - 1 \qquad (2)$$

There is a similar term generated by the spin-orbit nucleon-nucleus potential.

There is no problem to use such a tensor potential for the excited states in coupled channels calculations. However, for transition potentials, form factors are very different if $f_T(\vec{r}_d)$ [4] or $V(\vec{r}_p)$ is assumed to be deformed : in the last case the form factors of $[Y_L T_2]_\lambda$ for $L = \lambda \pm 2, \lambda$ are different.

Using a nucleon nucleus potential $V_C(\vec{r}_p) + \vec{\nabla}_p V_{LS}(\vec{r}_p) \wedge \frac{\vec{\nabla}_p}{i} \cdot \vec{\sigma}_p$, the contribution of the central potential to the folded one is [1] :

$$V_C(\vec{R},\vec{S}) = \int \varphi_d(r) V_C(\vec{r}_p) \varphi_d(r) d\vec{r} = \frac{1}{4\pi} \int V^C(\vec{r}_p) \left\{ \frac{u^2+\omega^2}{r^2} + \frac{\omega}{r^2\sqrt{2}}\left(u - \frac{\omega}{\sqrt{8}}\right) S_{np} \right\} d\vec{r} \qquad (3)$$

Using $\vec{\nabla}_p = \frac{1}{2} \vec{\nabla}_R + \vec{\nabla}_r$, the contribution of the spin-orbit potential is splitted into a spin-orbit term :

$$V_{LS}(\vec{R},\vec{S}) = \frac{1}{2} \vec{\nabla}_R \int \varphi_d(r) V_{LS}(\vec{r}_p) \wedge i\vec{\sigma}_p \varphi_d(r) d\vec{r} \cdot \vec{\nabla}_R$$

$$= \frac{1}{8\pi} \vec{\nabla}_R \wedge i \int V_{LS}(\vec{r}_p) \frac{1}{r^2} \left\{ \left(u^2 - \frac{u\omega}{\sqrt{2}} + \omega^2\right) \vec{\sigma}_p + \frac{3\omega}{\sqrt{2}}\left(u - \frac{\omega}{\sqrt{2}}\right)(\sigma_n \hat{r})\hat{r} + \frac{9\omega^2}{4}(\sigma_p \hat{r})\hat{r} \right\} d\vec{r} \cdot \vec{\nabla}_R$$

$$= \frac{1}{8\pi} \vec{\nabla}_R \int V_{LS}(\vec{r}_p) \left\{ \frac{1}{r^2}\left(u^2 - \frac{u\omega}{\sqrt{2}} + \omega^2\right) - \int_a^r \frac{3\omega(r')}{r'^3\sqrt{2}}\left[u(r') + \frac{\omega(r')}{\sqrt{2}}\right]dr'\right\} d\vec{r} \wedge \frac{\vec{\nabla}_R}{i} \cdot \vec{S} \qquad (4)$$

and a central plus tensor potential:

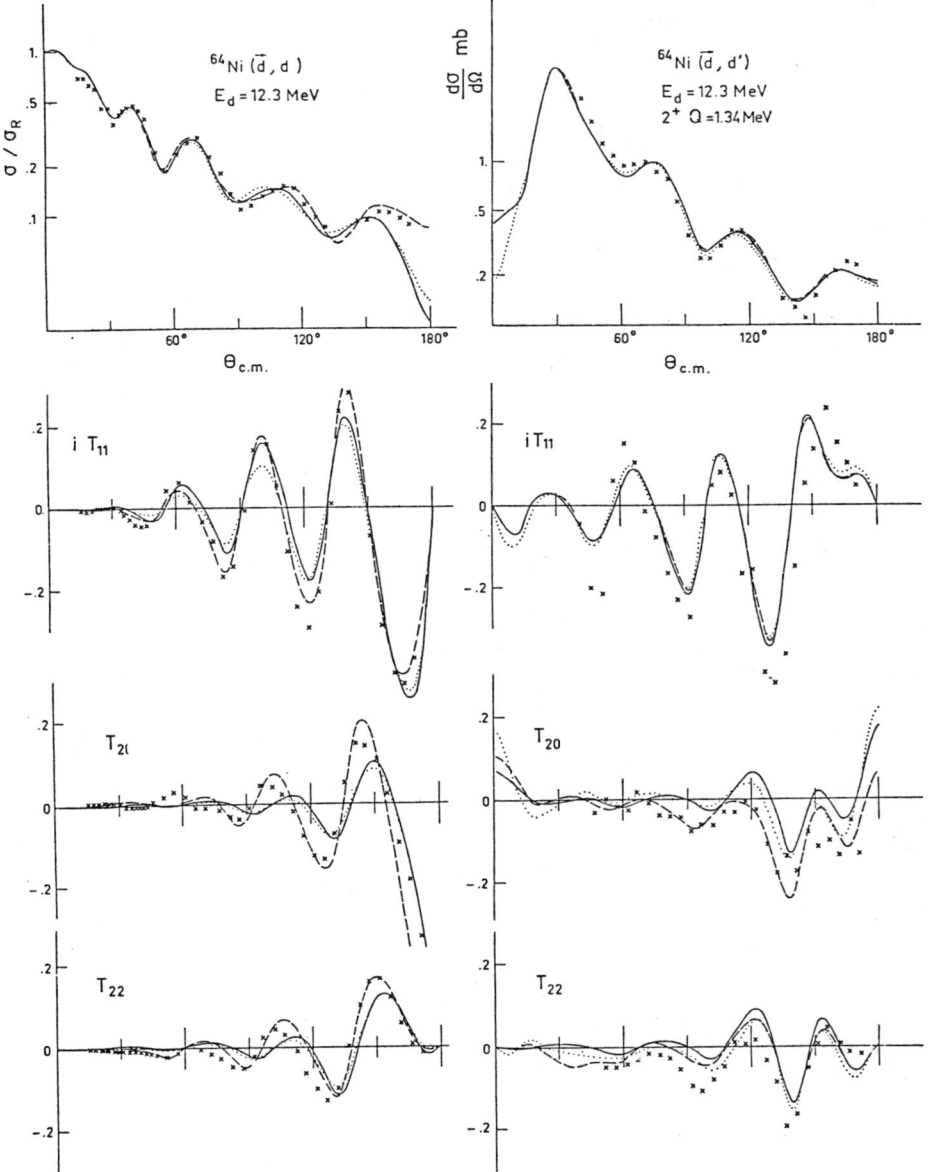

$$V'_{LS}(\vec{R},\vec{S}) = -i \int [\vec{\nabla}_r \varphi_d(r) \wedge \vec{\sigma}_p \cdot \vec{\nabla}_r \varphi_d(r)] V_{LS}(\vec{r}_p) d\vec{r} \qquad (5)$$

$$= \frac{1}{4\pi} \int V_{LS}(\vec{r}_p) \left\{ \frac{3\omega}{r^4} [2r\omega' - \omega] + \frac{3\omega}{2r^4} [\sqrt{2}ru' - \sqrt{2}u - r\omega'] S_{np} \right\} d\vec{r}$$

Due to the simple expressions in r, these integrals are easily evaluated, changing $d\vec{r}$ into $d\vec{r}_p$ and $\vec{r} = 2(\vec{r}_p - \vec{R})$. The central form-factor of a transfert λ is obtained from the multipole λ of $\frac{1}{r^2}(u^2+\omega^2)$ in (3) and of $\frac{3\omega}{r^4}[2r\omega'-\omega]$ in (5). Similarly, the form-factor of $[Y_L T_2]_\lambda$ involves the multipoles of $f_1(r) = \frac{\omega}{r^4 \sqrt{2}} \left(u - \frac{\omega}{\sqrt{8}} \right)$ in (3) and $f_2(r) = \frac{3\omega}{2r^4} [\sqrt{2}ru' - \sqrt{2}u - r\omega']$ in (5). The results are

$$\frac{64}{\sqrt{10\pi}} \frac{2L+1}{2\lambda+1} <L200|\lambda 0> \int [V^\lambda(r_p) f_1(r) + V^\lambda_{LS}(r_p) f_2(r)] \mathbb{R}^2 P_\lambda(\cos\theta)$$
$$+ r_p^2 P_L(\cos\theta) - 2Rr_p P_{\frac{L+\lambda}{2}}(\cos\theta)] r_p^2 dr_p d\cos\theta$$
$$L = \lambda \pm 2$$

$$\frac{64}{\sqrt{10\pi}} <\lambda 200|\lambda 0> \int [V^\lambda(r_p) f_1(r) + V^\lambda_{LS}(r_p) f_2(r)] \left[(R^2+r_p^2) P_\lambda(\cos\theta) \right. \qquad (6)$$
$$\left. - \frac{Rr_p}{2\lambda+1} \{(2\lambda-1) P_{\lambda+1} + (2\lambda+3) P_{\lambda-1}\} \right] r_p^2 dr_p d\cos\theta$$
$$L=\lambda$$

To study the importance of such terms, a fit has been done with the first order vibrational model for ^{64}Ni(dd')^{64}Ni*, 2$^+$ at 1.344 MeV. The plain curve shows the best fit obtained without tensor potentials. The dashed elastic curves are an optical model fit with complex spin-orbit interaction and some tensor potential. These curves were obtained, using primarily cross-sections and elastic vector polarization in the χ^2. In the optical model, the tensor polarizations are quite well fitted as soon as the fit of the vector one is good. In coupled channel calculations the spin-orbit interaction became real but the fit of the elastic vector analyzing power cannot be maintained. The dotted curve, obtained with a tensor potential with different strengths for the elastic and inelastic channels, shows that a tensor potential cannot fill the gap between calculation and experiment. The inelastic dashed curves are obtained with deformed tensor potentiels (6) of which the strength was varied. These results show that a tensor potential for the 2$^+$ or a tensor transition potential, chiefly L = 4 can give a better fit for inelastic tensor analyzing powers. However, they show primarily the importance of central and spin-orbit terms : without a very good fit for cross-sections and vector polarizations, such a search is not conclusive.

References
1) O. Karban et al. Nucl. St. Annual Report, Birmingham (1976) 30
2) F.T. Baker et al, Nucl. Phys. A 233 (1974) 409
3) J. Raynal, Thesis (Orsay 1964) ANL-TRANS-258
4) O. Karban, Nucl. St. Annual, Report Birmingham (1975) 60

INTERACTIONS OF POLARIZED ^{3}He PARTICLES WITH ^{24}Mg

F. Entezami, A. K. Basak, O. Karban, P. M. Lewis and S. Roman

Department of Physics, The University of Birmingham, England

ABSTRACT

Differential cross sections and analysing powers have been measured for the elastic and inelastic scattering of polarized ^{3}He particles by ^{24}Mg. Simultaneously, data have been extracted for (^{3}He,d) and (^{3}He,α) reactions leading to the g.s. (5/2$^+$), 0.451 MeV ($\frac{1}{2}^+$) and 0.945 MeV (3/2$^+$) states of ^{25}Al and to the g.s. (3/2$^+$) and 0.451 MeV (5/2$^+$) states of ^{23}Mg. The obtained results are compared with the optical model, DWBA and coupled reaction channel (CRC) calculations.

The elastic scattering data, shown in Fig. 1, could be very well reproduced by an optical-model potential, together with the cross-section data which are not presented here. The "best fit" potential has the following parameter values 171.6, 1.096, 0.729; 20.51, 1.173, 0.913; 3.38, 1.255, 0.211 respectively for V_R, r_R, a_R; W_D, r_W, a_W ; V_{so}, r_{so}, a_{so}. It is interesting to note that the ^{3}He spin-orbit potential is characterized by a very small diffuseness parameter $a_{so} \leqslant 0.3$ fm, consistent with all previous measurements for a wide range of nuclei.

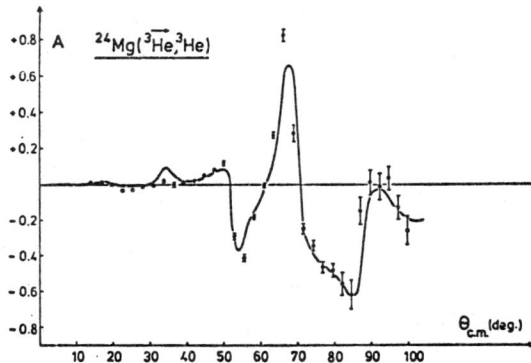

Fig. 1 Polarization in the elastic scattering of 33.3 MeV ^{3}He particles by ^{24}Mg. The solid curve was obtained with the "best fit" optical-model potential.

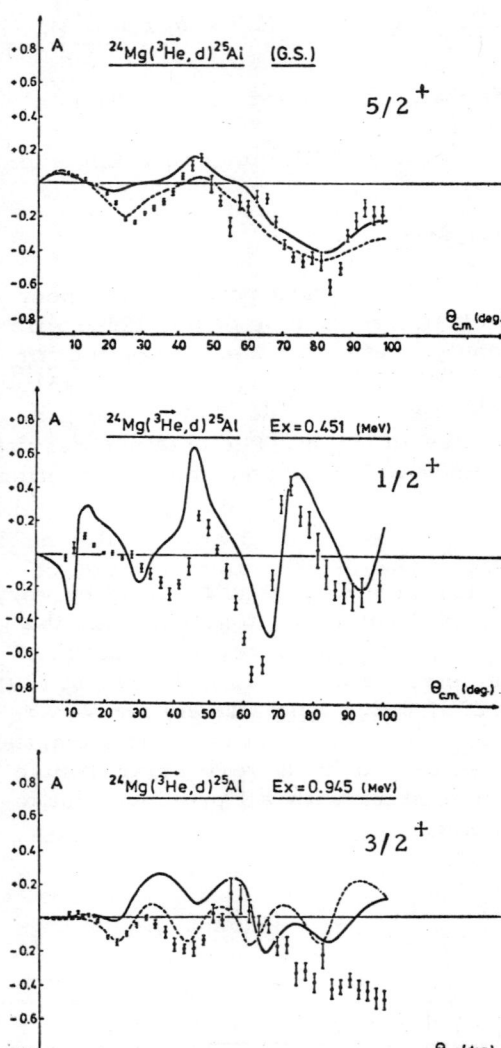

Fig. 2 Analysing power of the ^{24}Mg(^{3}He,d) reactions induced by 33.3 MeV ^{3}He particles compared with DWBA predictions using different imaginary potential depths.

The analysing power of the (^{3}He,d) reactions leading to three excited states of ^{25}Al are shown in Fig. 2. The solid curves are DWBA predictions using the "best fit" ^{3}He potential and an average set of deuteron parameters of ref. 1. Clearly, these calculations are unable to describe the polarization data, except for the 0.451 MeV state. At the same time the cross-sections were well reproduced. Investigation of the sensitivity of the predictions to different optical model parameters showed that changes of imaginary potential depth in both channels produced significant changes of the shape of the predicted analysing power, except for the 1st excited state in ^{25}Al ($\frac{1}{2}^+$). The dashed curve for the g.s. transition was obtained with reduced imaginary potential depth by 30 % and 60 % for the incident and outgoing channel respectively. For the 0.945 MeV state the dashed curve was obtained increasing both imaginary potential depths by 80 %. Predictions for the 0.451 MeV state did not improve by changing the imaginary potential.

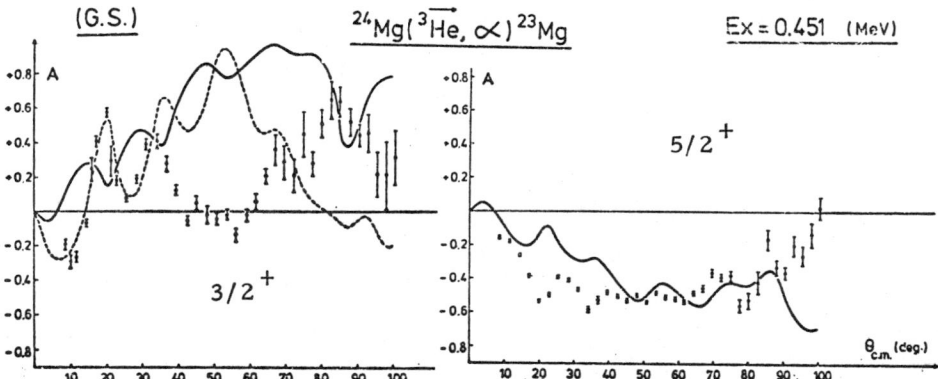

Fig. 3 Analysing power of the ^{24}Mg(^{3}He,α) reactions induced by 33.3 MeV ^{3}He particles compared with DWBA calculations. The dashed curve for the g.s. is a CRC prediction.

For the (^{3}He,α) reactions, shown in Fig. 3, the analysing power predictions were found to be very sensitive to the choice of the optical-model potentials in both channels. Clearly, calculations fail to describe the analysing power for the g.s. transition, but the results for the 1st excited state are qualitatively reproduced by DWBA. The calculations were carried out in zero-range, since the addition of finite range corrections had only a slight effect.

An attempt has also been made to reproduce the ^{24}Mg(^{3}He,α) ^{23}Mg g.s. results by coupled-reaction-channel(CRC) analysis using the code CHUCK, with shell-model spectroscopic amplitudes given in ref. 2, assuming a rotational deformation of ^{24}Mg with $\beta_2 = 0.46$. It is encouraging to note that in the first attempt the experimental data have been well reproduced at small angles (Fig.3, dashed curve). In conclusion, the strong j-dependent analysing powers of the reactions on ^{24}Mg were not reliably reproduced by DWBA. The improved description of the (^{3}He,α) reaction by CRC suggest that the difficulties may be due to multistep processes which should be taken into consideration.

1. R.J. Peterson and R.A. Ristinen, Nucl. Phys. A246 402 (1975)
2. R.O. Nelson and N.R. Robertson, Phys. Rev. C6 2153 (1972)

THE SCATTERING OF POLARIZED ^{3}He BY OXYGEN ISOTOPES

P.M. Lewis, O. Karban, A.K. Basak, E.C. Pollacco and S. Roman
Department of Physics, University of Birmingham, England

ABSTRACT

Differential cross sections and analysing powers were measured for elastic and inelastic scattering of ^{3}He by ^{17,18}O. Large differences in the elastic data observed for all three oxygen isotopes can be accounted for by variations in the absorption term of the optical model potential. Deformation parameters of $\beta_2 = 0.30$ and 0.28 for ^{17}O and ^{18}O respectively were deduced from a rotational-model coupled-channels analysis of both the elastic and inelastic scattering data.

The elastic and inelastic scattering cross sections and analysing powers were measured using enriched (98%) gas targets of ^{17}O and ^{18}O. A comparison with previous measurements[1] on ^{16}O shows significant differences in the magnitude and shape of both the differential cross sections and analysing powers for the three isotopes at $\theta_{cm} > 55°$. An optical-model analysis of the elastic data was carried out using the code RAROMP[2] and the best-fit parameters together with the χ^2-values are given in the table. The results of this analysis show the following three main features of the scattering of ^{3}He by the oxygen isotopes:

(i) The optical model describes best the ^{18}O data while ^{16}O has the highest χ^2-values. The experimental differences are accounted for mainly by changes in parameters of the imaginary term. It is possible to fix all parameters except W and r_W for the three isotopes on some average values without significant deterioration in the fit. The spin-orbit strength is 2 MeV, i.e. one third of the nucleon potential as expected from the folding model, while the diffuseness $a_{so} = 0.23$ fm follows the trend observed for other nuclei.

(ii) Two potential families[1] were used in the analysis, characterised by the strength of the real central potential, the shallow having $V \sim 122$ MeV and the deep $V \sim 180$ MeV. For the ^{16}O data there is an ambiguity between these potential, i.e. they both give equally good fits (χ^2-values). However, this is not the case for ^{17}O and ^{18}O: there is about 50% increase in χ^2 for ^{17}O and 80% for ^{18}O when going from the shallow to the deep potential. Thus, the heavier oxygen isotopes resolve the discrete ambiguities and favour the shallow potential characterised by the volume integral $J_R \sim 400$ MeV fm^3.

(iii) According to Becchetti and Greenlees[3] the real central potential has a positive (N-Z)/A dependence (assuming that the isospin for ^{3}He is equivalent to that of a proton), i.e. $V(^{16}O) < V(^{17}O) < V(^{18}O)$. However, from the present analysis we see that $V(^{16}O) \sim V(^{17}O) > V(^{18}O)$. On the other hand, for the imaginary potential, the expected trend[3] is seen: $W(^{16}O) < W(^{17}O) < W(^{18}O)$. The large

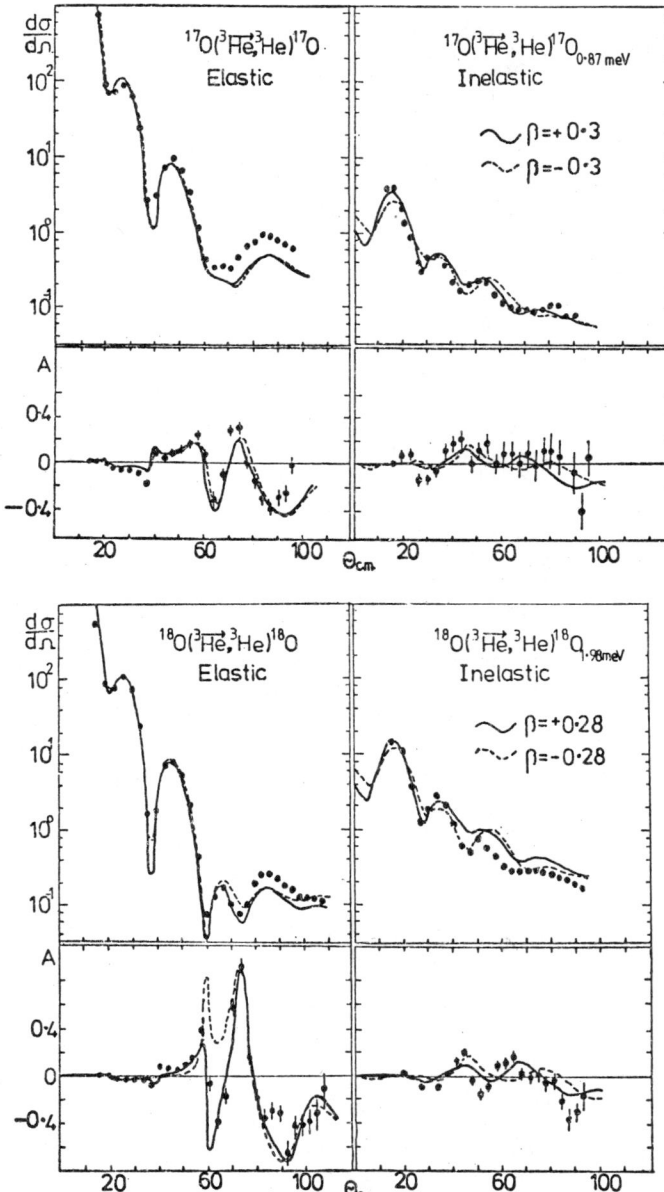

Fig.1. Differential cross sections and analysing powers for elastic and inelastic scattering of ^{3}He by ^{17}O and ^{18}O compared with rotational-model coupled-channels calculations.

difference between W(^{18}O) and W(^{16,17}O) apparently reflects the presence of strong inelastic channels in the ^{18}O case.

Data were also taken for the inelastic scattering to the 0.87 MeV state in ^{17}O and the 1.98 MeV state in ^{18}O. At $\theta_{cm}>60°$ the differential cross section for the 1.98 MeV state is greater than for the elastic scattering. An attempt was made to account explicitly for the inelastic scattering using a coupled-channels code ECIS[4]. The rotational model was used to describe the ground and first excited state data assuming the K=0 band for ^{18}O and K=1/2 band for ^{17}O. The average optical-model potential obtained from the analysis of the ^{16}O elastic scattering was first used in the hope that the coupling could account entirely for the differences in the absorption term found for ^{17,18}O. However, this procedure was not successful. In the next step, β, W and r_W were all adjusted in order to predict the magnitude and shape of the inelastic scattering without affecting the fit to the elastic scattering data. A subjectively best fit was obtained with the following parameters (fig.1):

^{17}O: W = 11.0 MeV, r_W = 1.45 fm, $β_2$ = 0.30
^{18}O: W = 11.0 MeV, r_W = 1.45 fm, $β_2$ = 0.28

The phase of the inelastic analysing powers selected the sign of β: it is positive for both ^{17}O and ^{18}O. The calculations also show that deformation of the spin-orbit term reduces the predicted magnitude of the inelastic scattering analysing power, resulting in an inferior fit in the case of ^{18}O. The above deformation parameters are in good agreement with values published in literature[5,6]. It is worth noting that from the present analysis β(^{17}O)>β(^{18}O), contrary to expectation, but in agreement with preliminary results for the elastic and inelastic scattering of deuterons on ^{17,18}O.

Table I. Optical model parameters for ^{16,17,18}O

	V_0	r_0	a_0	W	r_W	a_W	V_{so}	r_{so}	a_{so}	χ^2
^{16}O	122.8	1.065	0.80	8.35	1.55	0.85	1.49	1.07	0.30	25.0
^{17}O	122.9	1.07	0.80	9.65	1.51	0.84	2.20	1.098	0.21	12.5
^{18}O	121.0	1.07	0.79	13.8	1.35	0.83	2.16	1.076	0.23	3.5

REFERENCES

1. Y.-W. Lui et al., Nucl. Phys. A333 205 (1980)
2. G.J. Pyle, Code RAROMP, Univ.of Minnesota Rep. COO-1246-64(1964).
3. F.O. Becchetti and G.W. Greenlees, 3rd Symp. on Pol. Phen., p.682, Madison, 1970.
4. J. Raynal, Code ECIS, private communication.
5. H.F. Lutz et al., Nucl.Phys. 81, 423 (1966).
6. J.L. Escudie et al., Phys. Rev. C10 1645 (1974).
7. O. Karban et al., Birmingham Prog.Rep.1979, p.44, unpublished.

ANALYSING POWERS OF ($^3\vec{\text{He}}$,t) REACTIONS ON LIGHT NUCLEI

A.K. Basak, O. Karban, P.M. Lewis, G.C. Morrison and S. Roman
Department of Physics, University of Birmingham, England

ABSTRACT

Analysing powers of the ($^3\vec{\text{He}}$,t) reactions on ^{6,7}Li, ^{11}B, ^{13}C ^{17}O and ^{18}O targets were measured. Considerable variations in observed angular distributions indicate sensitivity to the (LSJ) transferred combinations allowed in the reaction.

The (^{3}He,t) reaction has been studied[1-10] extensively in recent years as a spectroscopic tool for investigating analogue and non-analogue states. Both macroscopic and microscopic DWBA calculations have been used to describe the cross section data of the reaction on various nuclei. The strengths of the isospin dependent parts of the effective interaction have been deduced. It has also been found that the angular distribution of the reaction cross section determines the orbital angular momentum transfer L, provided the reaction mechanism is predominantly a one-step process. The analysing powers (AP) of the reaction, on the other hand, are also expected to be sensitive to the transferred J-value, where J=L±S with S=0 or 1.

Measurements have been made of the analysing powers of the ($^3\vec{\text{He}}$,t) reaction on ^{6,7}Li, ^{11}B, ^{13}C, ^{17}O and ^{18}O nuclei. The reactions were initiated by the 33 MeV polarized ^{3}He beam of the University of Birmingham Radial Ridge Cyclotron. The average beam intensity on target was 1.0 nA with a polarization of about 55%. The method of extracting the data is described in ref.[11]. The AP data for the various final states are shown in Fig.1. The measured AP values are significantly different from zero and the distributions have distinct structures.

DWBA analyses in the framework of both macroscopic and microscopic models are in progress using the code DWUCK4[12]. Initial calculations were performed separately for each of the allowed (LSJ) triads. Preliminary calculations suggest that transitions to the analogue states are not necessarily represented by the (LSJ)=(000) transfer, which would be the simplest mechanism. The reactions to the ^{7}Be$_{g.s.}$ and ^{11}C$_{g.s.}$ are $3/2^- \rightarrow 3/2^-$ analogue transitions, yet the observed analysing powers for the two cases are completely different (Fig.1c). The cross section distributions for the reactions populating the ground ($3/2^-$) and 2.0 MeV ($1/2^-$) states are identical (Fig.1b). Nevertheless, the AP data for the two states are again different. This is a clear indication that different J-transfers are involved in the two reactions or, in other words, that the reaction analysing powers are strongly J-dependent.

For most of the final states populated by the reaction, several combinations of the (LSJ) triads are allowed by the selection rules. This makes an analysis of the J-dependence of analysing powers

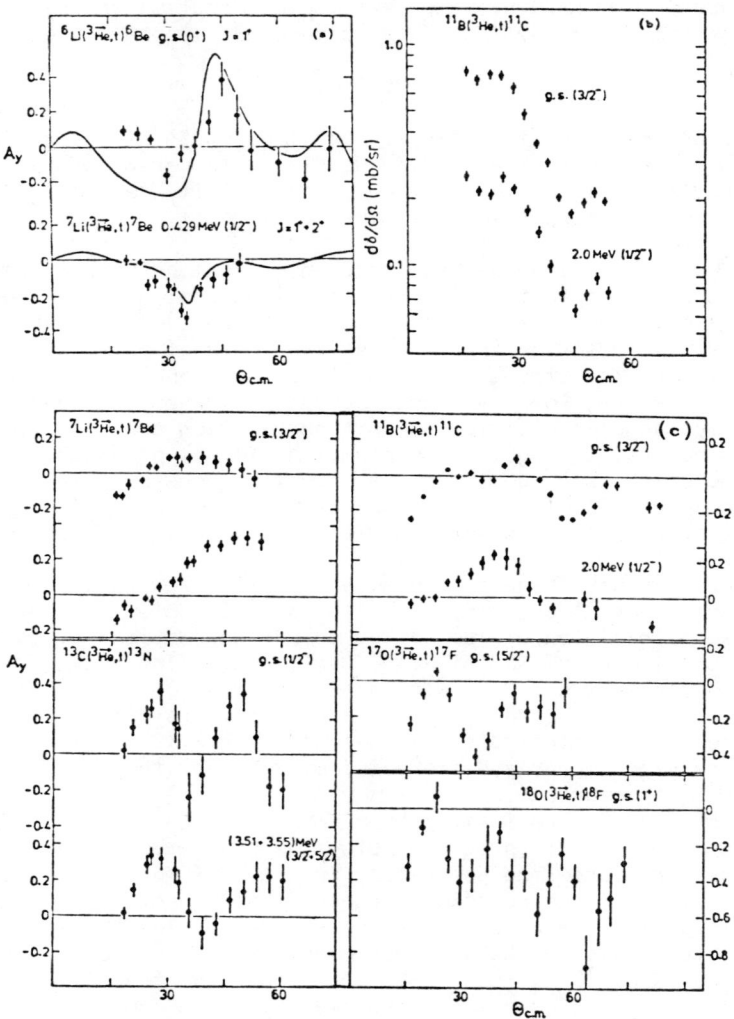

Fig.1. (a) Analysing powers and the microscopic DWBA predictions (solid curves) for the ^{6}Li($\vec{^3\text{He}}$,t_0)^{6}Be and ^{7}Li($\vec{^3\text{He}}$,t_1)^{7}Be reactions;

(b) Cross sections for the ^{11}B(^{3}He,t)^{11}C reactions leading to the ground (3/2$^-$) and 2.0 MeV (1/2$^-$) states;

(c) Analysing powers for the ($\vec{^3\text{He}}$,t) reactions on ^{7}Li, ^{11}B, ^{13}C, ^{17}O and ^{18}O.

difficult. However, the microscopic calculations (solid curves in Fig.1a) reproduce the features of the AP data for the ^{6}Be$_{g.s.}$ where $J=1^+$ is the only value allowed. Also, the AP data for the 0.429 MeV ($1/2^-$) state of ^{7}Be are fitted well by the coherent sum of the $J=1^+$ and 2^+ contributions (solid curve in Fig.1a). To obtain a clear picture of the J-dependence of analysing powers more experiments, preferably with zero-spin targets are needed.

REFERENCES

1. J.J. Wesolowski, E.H. Schwarcz, P.G. Roos and C.A. Ludeman, Phys.Rev. 169 (1969) 878.
2. P.D. Kunz, E. Rost, R.R. Johnson, G.D. Jones and S.I. Hayakawa, Phys. Rev. 185 (1969) 1528.
3. S.I. Hayakawa, W.L. Fadner, J.J. Kraushaar and E. Rost, Nucl. Phys. A139 (1969) 465.
4. L.F. Hansen, M.L. Stelts, J.G. Vidal, J.J. Wesolowski and V.A. Madsen, Phys. Rev. 174 (1968) 1155.
5. E. Rost and P.D. Kunz, Phys. Lett. 30B (1969) 231.
6. S.I. Hayakawa et al., Phys. Lett. 29B (1969) 327.
7. R.C. Bearse, J.R. Comfort, J.P. Schiffer, M.M. Stautberg and J.C. Stiltzfus, Phys. Rev. Lett. 23 (1969) 864.
8. W.L. Fadner, L.C. Farwell, R.E.L. Green, S.I. Hayakawa and J.J. Kraushaar, Nucl. Phys. A162 (1971) 239.
9. W.L. Fadner, J.J. Kraushaar and L.C. Farwell, Nucl. Phys. A178 (1972) 385.
10. R.J. Peterson and R.A. Ristinen, Nucl. Phys. A276 (1979) 61.
11. W.E. Burcham, J.B.A. England, R.G. Harris, O. Karban and S. Roman, Nucl. Phys. A246 (1975) 269.
12. J.R. Comfort, Univ. of Pittsburg, private communication.

INTERMEDIATE STRUCTURE IN THE GIANT E1 RESONANCE OF ^{20}Ne STUDIED BY POLARIZED PROTON CAPTURE

P. M. Kurjan, G. A. Fisher,* J. R. Calarco, and S. S. Hanna
Department of Physics, Stanford University, Stanford, CA 94305

It has long been recognized that the giant dipole resonance (GDR) of ^{20}Ne should provide a severe test of giant resonance theory. As can be seen in Figs. 1 and 2 the GDR of ^{20}Ne, as observed in the ^{19}F(p,γ_0)^{20}Ne reaction, breaks up into a series of nearly isolated structures. The ^{19}F(p,γ_0)^{20}Ne reaction was studied with both polarized and unpolarized proton beams. Gamma-ray spectra were recorded with the Stanford 24 cm × 24 cm NaI spectrometer. Data were accumulated over an energy range encompassing the entire GDR, E_p = 3.5 to 10 MeV. With unpolarized protons data were accumulated in 100 keV steps at seven angles in the range 40 to 140° to determine the coefficients a_k, k = 1 - 4, which describe the angular distribution $4\pi d\sigma/d\Omega = \sigma[1 + \Sigma a_k P_k(\theta)]$, where σ is the total (p,γ) cross section. Polarized protons at the same incident energies were used to measure $4\pi A d\sigma/d\Omega = \sigma[\Sigma b_k P_k^1(\theta)]$, where A is the analyzing power, at three angles, 45°, 90° and 135°. From these polarization measurements and the a_k coefficients from the unpolarized measurements, the coefficients b_k were determined for k = 1 and 3 and for the particular linear combination b_{24} = b_2 + 0.417 b_4. This special combination is significant because it is independent of E2 radiation. Figure 1 shows the 90° cross section for ^{19}F(p,γ_0)^{20}Ne along with the extracted coefficients.

The a_k and b_k coefficients can be related to the L-S coupling amplitudes for proton capture ^{1}P and ^{3}P, leading to E1 radiation, the amplitudes, ^{1}D and ^{3}D, leading to E2 radiation, and their relative phases.

The derived E1 amplitudes and relative phase are shown in Fig. 2. The dominant am-

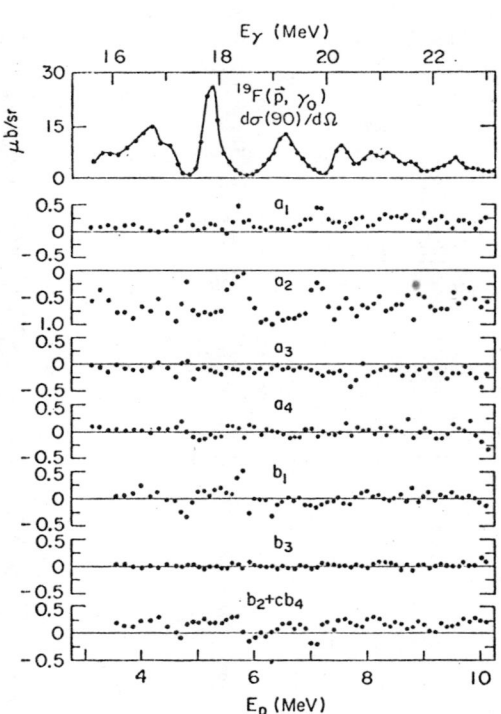

Fig. 1: The differential cross section and coefficients for the reaction ^{19}F(p,γ_0)^{20}Ne.

* San Francisco State University, San Francisco, CA 94132.

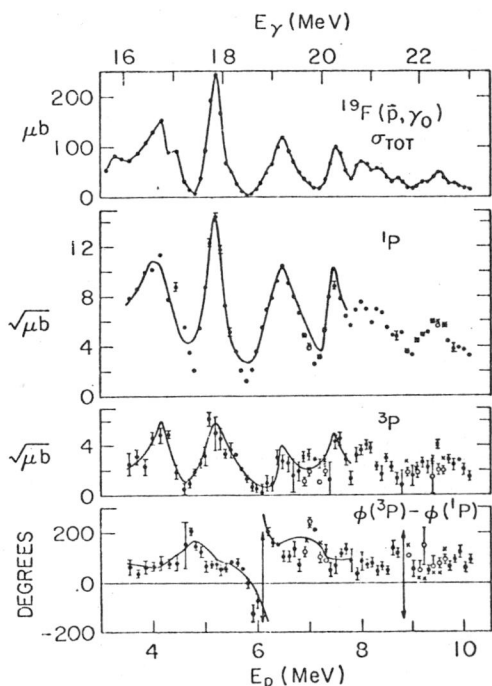

Fig. 2: The total (p,γ_0) cross section, the ^1P and ^3P amplitudes, and the relative phase.

plitude is that for ^1P capture. However, the capture of ^3P waves is also significant and shows similar resonance behavior, particularly in the region from E_x = 16 to 19 MeV. The subsequent $^3P \to {}^1S$ radiative transition represents a so-called "spin-flip" transition and is often said to be negligible for E1 transitions. However, its relative strength here and in other well-established cases shows that neglecting it in analysis of E1 transitions is not justified.

We have made a fit to the data with the doorway-state model. Four primary doorways having simple Breit-Wigner shapes were used, one for each of the observed peaks at E_x = 16.7, 17.8, 19.1 and 20.2 MeV. The parameters of the resonances were adjusted to these observed resonances. Figure 2 shows the results of the fit with the four Breit-Wigner resonances. The agreement is quite good for both the amplitudes and the relative phase, lending support for the picture of the resonances as different, slightly overlapping doorway states.

In order to test whether or not an alternative picture might equally well account for the data we have tried to fit the data with a single broad GDR interfering with a set of underlying secondary doorways. We find that such a picture, which describes very well the intermediate structure in ^{16}O, is unable to fit the ^{20}Ne data, except in the limit that the strength of the primary GDR becomes very small with all the strength going into the "secondary doorways." But this is just the picture used successfully above.

In summary, our analysis supports the doorway-state model which ascribes the prominent structure in the GDR of ^{20}Ne to reasonably well separated primary particle-hole configurations based on the ground state of ^{20}Ne. It would be interesting to see if the doorway-state calculations can also accommodate the actual microscopic configurations of this model.

This paper was supported in part by the National Science Foundation.

POLARIZED PROTON CAPTURE TO THE FIRST EXCITED STATE OF ^{60}Ni*

K. Sparks, J. D. Turner, N. R. Roberson, D. R. Tilley and H. R. Weller
Triangle Universities Nuclear Laboratory
Duke Station, Durham, North Carolina 27706

ABSTRACT

Polarized protons have been used to measure the analyzing powers for the reaction ^{59}Co(p,γ)^{60}Ni leading to the first excited 2+ state of ^{60}Ni in the region of the giant dipole resonance. The extracted a_2 and b_2 coefficients are compared to a direct-semidirect calculation.

We have used the reaction ^{59}Co($\vec{p}$,γ_1)^{60}Ni to study the giant dipole resonance built on the first excited (2$^+$) state of ^{60}Ni.[1,2] The γ-rays were detected with a 25.4 × 25.4 cm NaI detector assembly incorporating a plastic anticoincidence shield. The target was a 4.2 ± 0.4 mg/cm^2 self-supporting natural Co foil. A standard line-shape fitting procedure was used to strip the γ_0 and γ_1 peaks from the spectra.

Yield curve measurements were taken at 90° in 100 keV steps for E_p of 5.7 to 16.5 MeV. When these data are averaged over a 500 keV interval, the curve shown in Fig. 1 is obtained. The two peaks seen in this curve near E_γ of 15.8 and 19.0 MeV are shifted up by about 0.4 MeV in ^{60}Ni with respect to two peaks observed in the ^{59}Co(p,γ_0)^{60}Ni reaction. In the latter case these two peaks have been previously associated with the T< and T> components of the GDR of the ground state of ^{60}Ni.[1]

Angular distributions of cross section and analyzing power were measured at several energies across the region of the experiment. Data were taken at five angles for each proton beam energy. Typical results are shown in Fig. 2. The differential cross section data were fitted to an expansion of Legendre polynomials:

$$\sigma(\theta) = A_0 \left[1 + \sum_{k=1}^{2} a_k Q_k P_k (\cos \theta) \right] ,$$

while the product of the cross section and the analyzing power were fitted by an expansion in associated Legendre polynomials:

$$\frac{A(\theta)\sigma(\theta)}{A_0} = \sum_{k=1}^{3} b_k Q_k P_k^1 (\cos \theta) .$$

The coefficients Q_k correct for finite geometry. The solid lines in Fig. 2 are the result of this procedure.

The resulting a_2 and b_2 coefficients are shown in Fig. 3. Since the angular distributions of cross sections and analyzing powers for the ^{59}Co(p,γ_0)^{60}Ni reaction could be reasonably well accounted for by the direct-semidirect model with a form factor $\propto$ r, a similar calculation was performed

* Partially supported by the U. S. D. O. E.

for this case.[2]

The results of this calculation which takes the 2^+ state of ^{60}Ni to be an $f_{7/2}$ single particle state are shown as the solid curves in Fig. 3. The overall trend of the coefficients is in fair agreement with the predictions of the calculation. Stripping reactions have indicated a significant $p_{3/2}$ single particle spectroscopic strength in the first excited state of ^{60}Ni.[3] However, attempts to improve the agreement of the DSD calculation by including this strength were unsuccessful.

REFERENCES

1. E. M. Diener, J. F. Amann, P. Paul, and S. L. Blatt, Phys. Rev. C3, 2303 (1971).
2. J. D. Turner, C. P. Cameron, N. R. Roberson, H. R. Weller and D. R. Tilley, Phys. Rev. C17, 1853 (1978).
3. R. L. Auble, Nuclear Data Sheets 28 No. 2, 154 (1979).

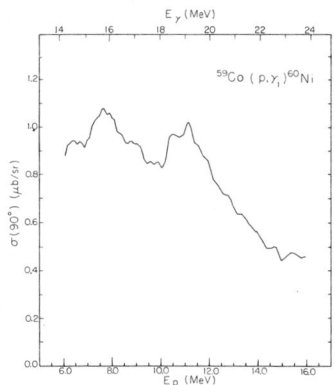

Fig. 1. Energy Averaged Yield Curve

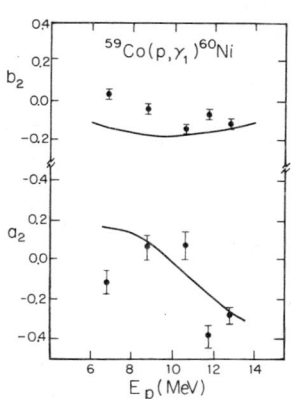

Fig. 3. Angular Distribution Coefficients are shown along with the results of the DSD calculation (solid lines). Errors are statistical.

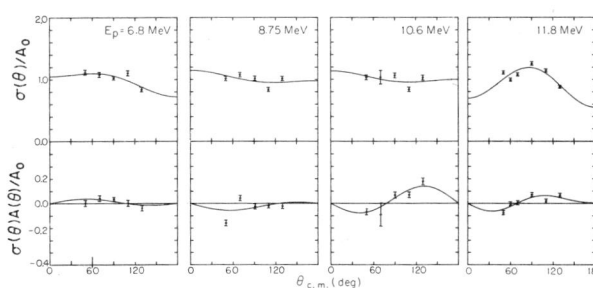

Fig. 2. Angular Distribution Data

DECAY MECHANISMS OF THE GIANT E1 RESONANCE IN ^{90}Zr STUDIED BY POLARIZED PROTON CAPTURE

J. R. Calarco, P. M. Kurjan, G. A. Fisher,* and S. S. Hanna
Department of Physics, Stanford University, Stanford, CA 94305

The study of two strong E1 $T_>$ analogue resonances (IAR) superposed on the main ($T_<$) giant E1 resonance (GDR) in ^{90}Zr, by ^{89}Y($\vec{p},\gamma_0$)^{90}Zr, provides a means of determining the nature of the decay of the GDR. The region of interest in ^{90}Zr is E_x = 14-17 MeV which includes the peak of the GDR at 16.5 MeV as well as the E1 IAR's at E_x = 14.4 and 16.3 MeV. Complete angular distributions of yields and analyzing powers have been obtained throughout this region. The quantities important to this study A_0, a_2, and b_2 are shown in Fig. 1.

Since ^{89}Y has a ground state with $J^\pi = 1/2^-$, only two T-matrix elements, which involve $s_{1/2}$ and $d_{3/2}$ proton waves, are responsible for radiative capture through 1^- dipole states. Analysis of the angular distributions give two solutions, one predominantly s-wave, the other mainly d-wave, as shown in Figs. 2 and 3. If the s-wave dominated solution is preferred at E_x = 14.4 MeV (since this IAR has been assigned an s-wave character from elastic scattering) and if the d-wave dominated solution is preferred in the main part of the GDR, then a crossing over of the two solutions is expected in the region in between. However, the measurements are in energy steps about equal to the target thickness, but no such crossing is observed.

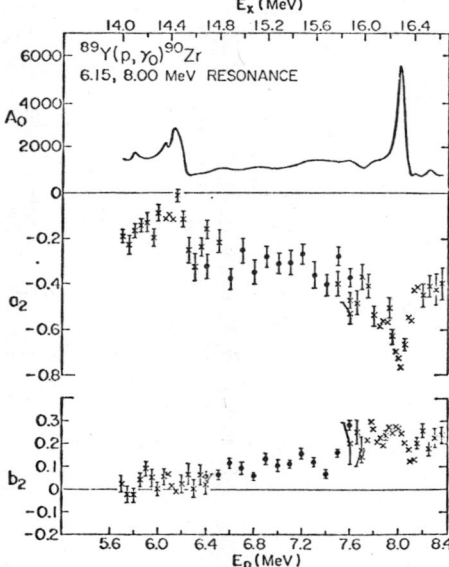

Fig. 1: The coefficients A_0, a_2, and b_2 for the reaction ^{89}Y(p,γ_0)^{90}Zr.

A resolution of this dilemma is to suppose that a direct-semidirect (DSD) mechanism dominates the capture in the main part of the GDR ($E_x = \geq 16.5$ MeV) but that a statistical mechanism dominates around $E_x \simeq 14.4$ MeV. In the former case the angular correlation coefficients are

$$A_0 = |s|^2 + |d|^2$$
$$A_0 a_0 = -0.5|d|^2 + \sqrt{2}|s||d|\cos(\phi_d-\phi_s)$$
$$A_0 b_2 = (1/\sqrt{2})|s||d|\sin(\phi_d-\phi_s)$$

which lead to the desired d-wave solution discussed above. In the statistical case these equations become

$$\langle A_0 \rangle \cong -0.5|d|^2$$
$$\langle A_0 a_2 \rangle \cong -0.5|d|^2$$
$$\langle A_0 b_2 \rangle \cong 0$$

given that the amplitudes vary

* San Francisco State University, San Francisco, CA 94132.

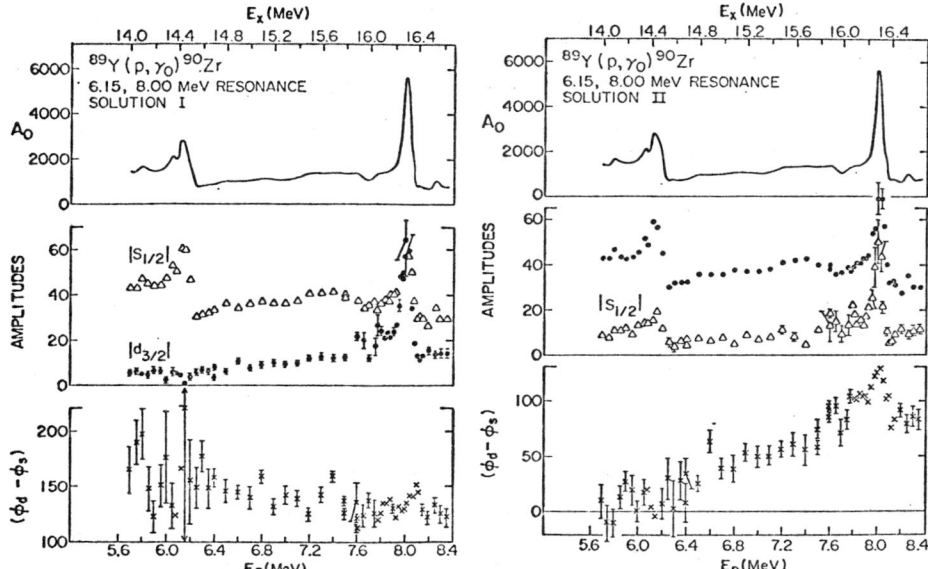

Fig. 2: The $s_{1/2}$ dominated solution in the DSD model.

Fig. 3: The $d_{3/2}$ dominated solution in the DSD model.

smoothly and the major statistical fluctuations are restricted to the phases. The last equation accounts for the result $\langle A_0 b_2\rangle \simeq 0$ observed around $E_x \simeq 14.4$ MeV. Optical-model estimates obtained from ABACUS give $|d|^2/(|s|^2 + |d|^2) \simeq 0.3$ for a statistical mechanism. This would account for the small, but non-zero, value of $A_0 a_2$ seen in the same region.

Thus, we can explain the trends seen in the coefficients a_2 and b_2 if we assume a statistical mechanism near $E_x \simeq 14.4$ MeV that decreases to zero near $E_x \simeq 16.5$ MeV as the DSD mechanism takes over. This model supports the observation (in other mass regions) that thte decay of the amin part of the GDR to low-lying hole states is definitely not statistical.

We wish to thank Frank Dietrich for helpful discussion and Mamiko Sasao for aid in carrying out the calculations. This work was supported in part by the National Science Foundation.

POLARIZED PROTON CAPTURE TO THE FIRST EXCITED STATE IN ^{31}P

C. Fitzpatrick, C. P. Cameron, Hideo Kitazawa, N. R. Roberson,
D. R. Tilley, and H. R. Weller

Triangle Universities Nuclear Laboratory,* Duke Station, Durham,
North Carolina 27706

ABSTRACT

Angular distributions of cross section and analyzing power for the reaction ^{30}Si$(\vec{p},\gamma_1)^{31}$P* (1.27 MeV) were measured for excitation energies in the region of the giant dipole resonance. The extracted a_k and b_k coefficients are compared to direct-semidirect calculations.

The reaction ^{30}Si$(\vec{p},\gamma_1)^{31}$P has been used to study the excitation energy region in ^{31}P from 13 to 23 MeV. The γ-rays were detected with a 25.4 x 25.4 cm NaI crystal surrounded by a plastic anticoincidence shield. The target was a 1.15 mg/cm^2 self-supporting silicon dioxide foil. The beam currents used during the experiment were typically 40 nA with polarizations of 0.80 ± 0.02. The extraction of the γ_1-peak from the spectra has been described previously.[1]

When the 90° yield curve (see Fig. 2 below) is averaged with a 1.1 MeV energy interval, two broad peaks are observed near E_γ of 16 and 20 MeV respectively. These peaks may be identified with the $T_<$ and $T_>$ components of the GDR built on the first excited state of ^{31}P. Potokar[2] has carried out a detailed analysis of this yield curve.

Angular distributions of cross section and analyzing power were measured at seven angles with θ_γ = 42° to 142°; at some energies the cross sec-

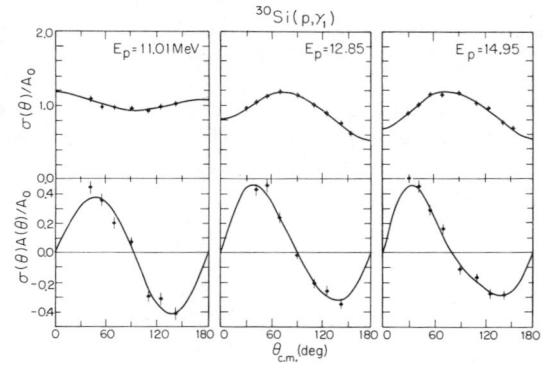

Fig. 1. Typical angular distributions

*Supported in part by the U. S. Department of Energy

0094-243X/81/690602-03$1.50 Copyright 1981 American Institute of Physics

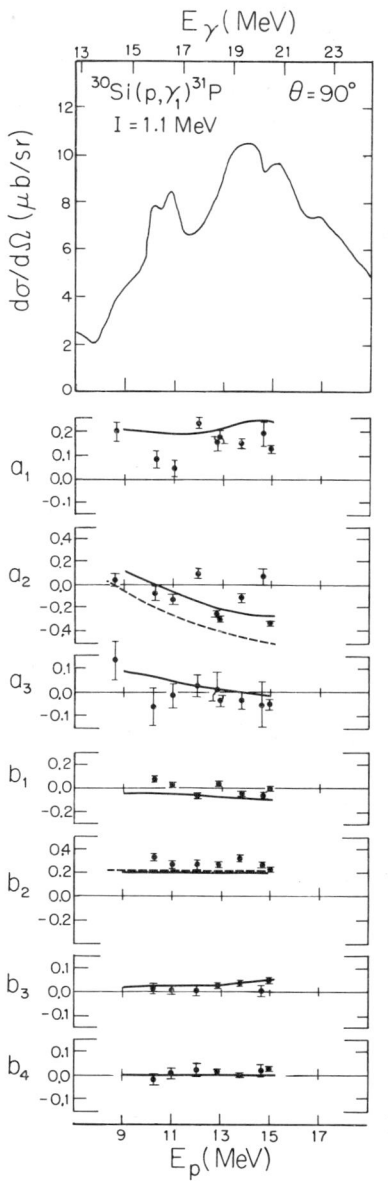

Fig. 2. The experimental a_k and b_k coefficients and results of DSD calculation.

tions were measured at nine angles (30° to 154°). Typical results are shown in Fig. 1. The solid lines represent a fit of the data to an expansion of Legendre polynomials.

$$\sigma(\theta) = A_0 [1 + \sum_{k=1}^{3} a_k Q_k P_k (\cos\theta)] ,$$

and associated Legendre polynomials,

$$\frac{A(\theta)\sigma(\theta)}{A_0} = \sum_{k=1}^{4} b_k Q_k P_k^1 (\cos\theta) .$$

The coefficients Q_k correct for finite geometry. The a_k and b_k coefficients obtained from the fits are shown in Fig. 2. Also shown at the top of Fig. 2 is the energy averaged 90° yield curve.

We first assumed that only E1 is important. Since the first excited state spin of ^{31}P is $3/2^+$, the complex T-matrix elements which contribute to the E1 strength are $p_{1/2}\exp(i\phi_{1/2})$, $p_{3/2}\exp(i\phi_{3/2})$ and $f_{5/2}\exp(i\phi_{5/2})$. Having to determine five unknowns with only three measured quantities A_0, a_2 and b_2 means that some simplifying assumptions must be made. Calculations using the direct-semidirect theory[3] with a form factor $\propto r$ showed that the $p_{3/2}$ T-matrix elements (spin-flip) should be small and that the $(\phi_{1/2}-\phi_{3/2})$ relative phases are determined primarily by the spin-orbit term in the optical-model potential. The analysis was carried out by setting the ratio $\sigma(p_{3/2})/\sigma(p_{1/2})$ and the relative p-wave phases equal to the DSD calculated values. Two sets of solutions were found: one corresponds to having $\sigma(f_{5/2})/\sigma(E1) \approx 0.7$ and the other to ≈ 0.25. The DSD model calculations suggest that the solutions with $f_{5/2}$ dominant are the physical ones.

A calculation[4] based on an extended DSD model[2,5] which includes amplitudes corresponding to direct electric dipole and quadrupole capture, semidirect $T_<$ and $T_>$ GDR components, and semidirect isoscalar and isovector GQR components has been performed. The effect of the isovector GQR is quite small in the excitation energy region studied here. The parameters used for the calculation were taken from our earlier study of the ^{30}Si$(\vec{p},\gamma_0)$ reaction.[1] The results are shown as the solid lines in Fig. 2. Also shown as dashed lines are the pure E1 calculations discussed above. The overall agreement with the experimental data is good, especially at the higher energies, but there are some substantial deviations.

The calculated $\sigma(p,\gamma_1)$ E2 cross sections were converted by the method of detailed balance to $\sigma(\gamma,p)$ cross sections and used to determine the fraction of the E2 EWSR for $E_p = 8.5$ to 15.5 MeV. When compared to the prediction of Nathan and Nilsson,[6] we find 1.5% and 1.6% for the full calculation and for the case of pure direct E2 capture, respectively.

The authors acknowledge helpful conversations with M. Potokar during early stages of this work.

REFERENCES

1. C. P. Cameron, M. Potokar, D. G. Rickel, N. R. Roberson, H. R. Weller, and D. R. Tilley, Phys. Rev. (1980) in press.
2. M. Potokar, Triangle Universities Nuclear Laboratory Report, 1978 (unpublished), p. 68.
3. G. E. Brown, Nucl. Phys. $\underline{57}$, 339 (1969).
4. Computer Code HIKARI, Hideo Kitazawa, Tokyo Institute of Technology, Tokyo, Japan.
5. M. Potokar, Stanford University Progress Report, 1977 (unpublished), p. 99.
6. O. Nathan and S. G. Nilsson, in Alpha-, Beta- and Gamma-ray Spectroscopy, ed. by K. Siegbahn (North Holland, 1965) p. 601.

POLARIZED PROTON CAPTURE ON ^{88}SR

R. D. Ledford, C. Cameron, M. Potokar, N. R. Roberson,
D. R. Tilley, and H. R. Weller
Triangle Universities Nuclear Laboratory,* Duke Station,
Durham, North Carolina 27706

ABSTRACT

Cross sections for ^{88}Sr$(p,\gamma_0)^{89}$Y and ^{88}Sr$(p,\gamma_1)^{89}$Y have been measured at 90° for E_p = 5.5 to 27.0 MeV. Angular distributions of cross section and analyzing power for (p,γ_0) were measured at 14 proton energies in the GDR region for E_p = 6.05 to 15.0 MeV. The resulting a_k and b_k coefficients are compared with direct-semidirect calculations.

EXPERIMENT

The excitation region from 12 to 22 MeV in ^{89}Y has been studied[1] with the ^{88}Sr$(p,\gamma)^{89}$Y reaction. A 2.1 mg/cm 99.84% enriched Strontium-88 target was used for the measurements, and gamma rays were detected with a 25.4 × 25.4 cm NaI crystal surrounded by a plastic anticoincidence shield. Average beam currents on target were 40 na and typical beam polarization was 0.80 ± 0.02.

The (p,γ_0) and (p,γ_1) 90° yield curves are shown in Fig. 1. We note that photonucleon work of Lepretre et al.[2] determined that the GDR in ^{89}Y peaks at 17 MeV and extends from 12 to 23 MeV. Both (p,γ) yield curves exhibit considerable structure. The (p,γ_0) yield does not drop off below the GDR region, and has a good deal of intermediate structure below 18 MeV. The (p,γ_1) yield diminishes to very small values near 12 MeV, but has additional clusters of strength near 20 MeV and 25 MeV. This structure was also apparent in the data of Cue et al. which was published along with the shell model work of Vergados and Kuo.[3]

Angular distributions of cross section and analyzing power were measured at 14 proton energies corresponding to the excitation region from 13 to 22 MeV. These data were fit to the Legendre polynomial expansion,

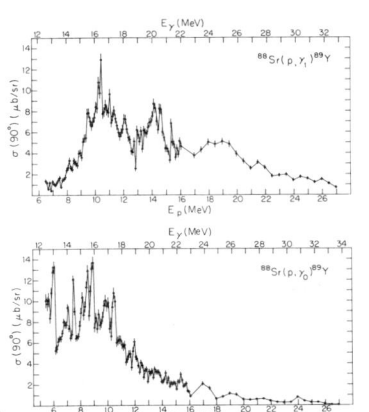

Fig. 1. Yield Curves

$$\sigma(\theta) = A_0 [1 + \sum_{k=1}^{3} a_k Q_k P_k (\cos \theta)] \quad (1)$$

*Supported by U. S. Department of Energy

and the associated Legendre polynomial expansion,

$$\frac{A(\theta)\sigma(\theta)}{A_0} = \sum_{k=1}^{3} a_k Q_k P_k^1(\cos\theta) \quad (2)$$

The coefficients Q_k correct for finite geometry. Fits were made only through k=3. Data at additional angles would have been necessary to justify the inclusion of higher order terms. Some of these data and the fits are shown in Fig. 2. The a_k and b_k coefficients are plotted in Fig. 3.

ANALYSIS AND CONCLUSIONS

Although the $J = 1/2$ spin of the ^{88}Sr ground state enables the extraction of dipole and quadrupole amplitudes in a model-independent way, the approach here is to use the direct-semidirect model to calculate the a_k and b_k coefficients.

Transition amplitudes used to calculate the (p, γ_0) cross sections were of the form given in eqn. (3) in which $|\phi_{n\ell jm}(x)\rangle$ is the single particle bound state, $|X_i^{(f)}(x)\rangle$ the scattering states, and $d^\nu(x)$ and $q^\nu(x)$ are the single-particle dipole and quadrupole operators respectively. The summation index T refers to the two possible isospin components of the GDR which are labeled as T< and T>. The index τ labels the isoscalar ($\tau = 0$) and isovector ($\tau = 1$) transition operators, the E1 strength being pure isovector. The dipole and quadrupole form factors, $V_1^\nu(x)$ and $V_2^\nu(x)$, respectively, appear in the terms which represent the

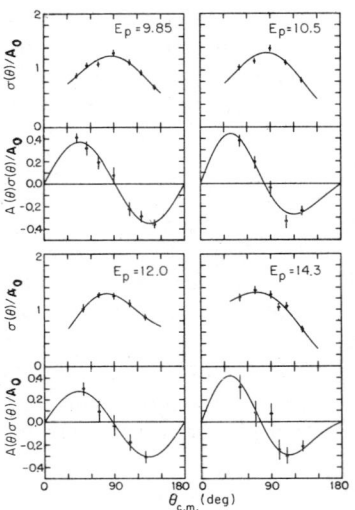

Fig. 2. Typical Angular Distributions

$$t^\nu(p\to\gamma) = \langle \phi_{n\ell jm}(x) | d^\nu(x) + \sum_T \frac{V_{1,T=1}^\nu(x)}{E - E_{11}(T) + \frac{1}{2} i \Gamma_{11}(T)} |X_i^{(f)}(x)\rangle$$

$$+ \langle \phi_{n\ell jm}(x) | q^\nu(x) + \sum_{\tau=0,1} \frac{V_{2,\tau}^\nu(x)}{E - E_{2\tau} + i/2 \Gamma_{2\tau}} |X_i^{(+)}(x)\rangle \quad (3)$$

possible presence of the GDR and GQR. Except for the dipole coupling interaction, the parameters were obtained from other experiments. The spectroscopic factor is from Ref. 4. The T< and T> GDR parameters are from Ref. 5, and the GQR parameters are from Ref. 6. The Bechetti Greenless optical potential was used.

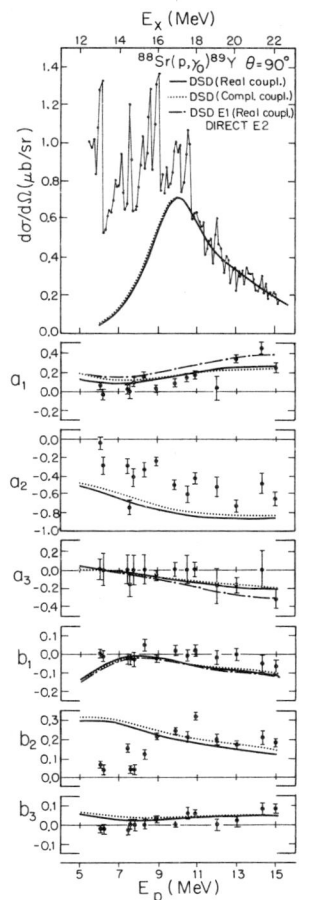

The results are shown in Fig. 3. The a_k and b_k coefficients are well reproduced on the high-energy side of the calculated resonance, where the DSD mechanism would be expected to dominate. These results indicate that direct E2 radiation is sufficient to account for the essential non-E1 effects observed in the data.

Fig. 3. The ^{88}Sr$(p,\gamma_0)^{97}$Y 90° yield and a_k and b_k coefficients and comparisons with DSD calculations.

REFERENCES

1. R. D. Ledford, Ph.D. Thesis, Duke University, 1976 (unpublished)
2. A. Lepretre, H. Beil, R. Bergere, P. Carlos, A. Veyssiere, and M. Sugawara, Nuc. Phys. A175, 609 (1971)
3. J. D. Vergados and T. T. S. Quo, Nuc. Phys. A168, 225 (1971)
4. J. Picard and G. Bissard, Nucl. Phys. A131, 636 (1969)
5. R. Pitthan, F. R. Buskirk, E. B. Dally, J. D. Shannon, and W. H. Smith, Phys. Rev. C16, 970 (1977)
6. Compilation by F. E. Bertrand, Ann. Rev. Nucl. Sci. 26, 457 (1976)

THE UNIQUE EXTRACTION OF E2 STRENGTH
IN POLARIZED PROTON CAPTURE REACTIONS

J. Sowinski and D.G. Mavis
University of Wisconsin, Madison, WI. 53706[†]

Extensive measurements have recently been made[1-3] which attempt to measure the collective E2 strength in nuclei by observing the radiative capture of polarized protons. For these reactions which proceed via a simple spin sequence (such as for $^{15}N(\vec{p},\gamma_0)^{16}O$) there are only a total of four complex T-matrix elements that need to be considered if only E1 and E2 radiations are present. In principle, polarized angular distribution measurements for such reactions provide a sufficient number of experimentally determined quantities to completely determine the T-matricies and, in turn, the E2 cross section. In practice, however, the T-matrix values must be extracted through a non-linear fitting procedure which often results in multiple solutions.

One class of multiple solution is known to exist from earlier polarized capture measurements[4]. There are only two solutions in this class and they are characterized by an interchange of the spin-flip and non-spin-flip capture amplitudes along with a slight adjustment of the relative phases. Since both solutions provide the same fit to the data, this class of multiplicity is truly a mathematical artifact and reflects the quadratic nature of the equations used in the fitting procedure. Fortunately, the resulting E2 strength is nearly identical for two such solutions[3] and this type of multiplicity will not be considered further.

A second type of multiple solution has been recently observed for the reactions $^{12}C(\vec{p},\gamma_0)^{13}N$ (ref. 1) and $^{13}C(\vec{p},\gamma_1)^{14}N$ (ref. 2). Unlike the multiple solutions of the first type, two solutions of this type result in E2 cross sections which often differ by an order of magnitude. Furthermore, since all solutions within this class are characterized by a dominant E1 non-spin-flip amplitude, there is no straightforward theoretical justification for choosing one solution over the other.

We have investigated this second class of multiplicity in order to determine if there exists some procedure by which one can isolate the true physical solution or, at least, extract a unique value for the E2 cross section. To do this, we have performed computer simulations of capture experiments and have been able to study in detail the standard analysis procedures currently used. By knowing the "physical" solution beforehand, we are then able to isolate the various mathematical and statistical properties of the extraneous solutions.

For the computer simulations, sets of E1 and E2 T-matrix values were chosen to provide a variety of E2 cross sections and were selected to be representative of the various solutions found in refs. 1 and 2. For each T-matrix set, parent distributions were computed

† Work supported in part by the U.S. Department of Energy.

0094-243X/81/690608-03$1.50 Copyright 1981 American Institute of Physics

for $\sigma_+(\theta)$ and $\sigma_-(\theta)$ (the capture cross sections for proton polarization +1 and -1 respectively). A Poisson random distribution generator was then used to select several hundred random samples $\tilde{\sigma}_+(\theta)$ and $\tilde{\sigma}_-(\theta)$ of each parent distribution over a variety of angle ranges and for a variety of counting statistics.

The sums and differences of $\tilde{\sigma}_+(\theta)$ and $\tilde{\sigma}_-(\theta)$ were first fitted with Legendre and associated Legendre polynomial expansions in order to extract the usual A_k and B_k coefficients[1,2]. E2 cross section values were then extracted using a non-linear iterative search routine which incorporated the quadratic expressions of ref. 2 relating the T-matrix values to the A_k and B_k coefficients. Another fit in which the T-matricies were varied to fit the $\tilde{\sigma}$ values directly was also performed and yielded equivalent results.

Extraneous solutions with varying E2 strengths were found for each parent distribution. The existence of these solutions, however, depended ultimately upon the statistical accuracy of the $\tilde{\sigma}$ samples and upon the angular range over which their values were taken. It was found that for $\tilde{\sigma}_+(\theta)$ and $\tilde{\sigma}_-(\theta)$ values within a given angle range, the χ^2 of the fit at the extraneous solution scaled in proportion to the total number of counts taken in the angular distribution. Also, for a given number of total counts, the χ^2 of this solution depended sensitively upon the angular range.

The latter of these properties is illustrated in Fig. 1 for the case of 2200 total counts in the $\tilde{\sigma}$ distributions. Figure 1(a) shows the behavior of χ^2 for the extraneous solution as a function of angular range centered about 90°. Complete χ^2 vs. E2 strength contours are shown in Fig. 1(b) for $\tilde{\sigma}$ values taken over two separate angle ranges. The "physical" solution in this example has $\sigma(E2) = 0.02\sigma(TOT)$ and the extraneous solution disappears (at the level of 1% confidence) for "data" taken to sufficiently wide angles.

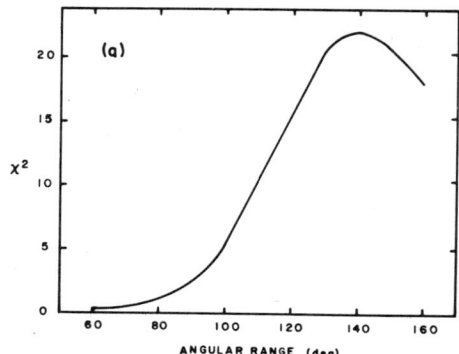 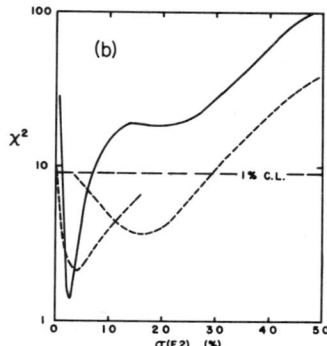

Fig. 1(a) χ^2 for the extraneous solution as a function of angular range for 2200 total counts. (b) χ^2 vs. E2 strength for the two angular ranges 40°-140° (dashed curve) and 25°-155° (solid curve).

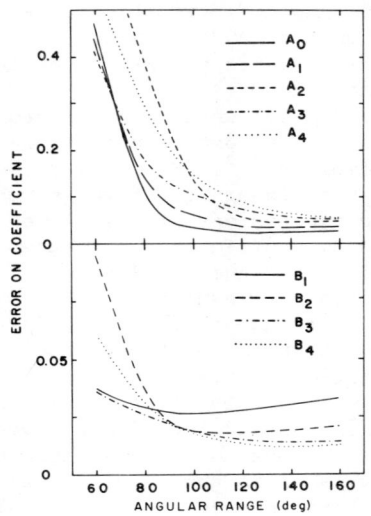

Fig. 2. Errors on A_k and B_k coefficients for 2200 total counts in the $\tilde{\sigma}$ distributions. The curves scale as the inverse of the square root of the total number of counts. The values shown are normalized to $A_0 = 1.0$.

The information contained in the "data" as a function of angular range is reflected in the errors on the resulting A_k and B_k coefficients. The χ^2 behavior of Fig. 1(a) is clearly a consequence of the A_k and B_k uncertainties shown in Fig. 2. Figure 2 also illustrates the importance of obtaining data beyond the outermost zero crossings of the highest order polynomial required to fit the data.

In all cases examined, the fit generated at the extraneous solution differed from the fit at the "physical" solution by a redistribution of the χ^2 among the "data". As the "data" improved (either due to improved statistics or due to wider angle range) this redistribution became impossible and the extraneous solution disappeared, indicating that the mathematical form of the extraneous fit provided an inadequate description of the parent distribution.

In no case did we find any extraneous solution capable of describing "data" of an arbitrary accuracy. This indicates that the multiple E2 solutions found in refs. 1 and 2 are not a necessary consequence of the non-linear fitting procedure, but can be eliminated by taking sufficiently precise data.

REFERENCES

[1] R.L. Helmer, et al., Nucl. Phys. A336, 219 (1980).
[2] J.D. Turner, et al., Phys. Rev. C21, 525 (1980).
[3] R. LaCanna, Ph.D. thesis, Stanford University (1977), (unpublished)
[4] S.S. Hanna, et al., Phys. Rev. Lett. 32, 114 (1974).

ANALYZING POWERS AND CROSS SECTIONS FOR (p,d) REACTIONS ON NUCLEI OF N=50-82

K. Nagano, Y. Aoki, H. Iida, S. Kunori, Y. Toba and K. Yagi
The University of Tsukuba, Ibaraki 305, Japan

ABSTRACT

Analyzing powers and differential cross sections have been measured for (p,d) reactions on nuclei of N=50-82 at a proton energy of 22.0 MeV. DWBA calculations employing deuteron optical potential parameters obtained from adiabatic model qualitatively reproduce the measured angular distributions. The effect of the deuteron D-state on DWBA calculations has been studied for $\ell=0$ and $\ell=2$ transitions.

EXPERIMENT AND L=0,2 TRANSITIONS

Angular distributions of analyzing powers $A(\theta)$ and differential cross sections $\sigma(\theta)$ were measured for (p,d) reactions on nuclei of N=50-82 at E_p=22.0 MeV, with simultaneous measurements of $A(\theta)$ and $\sigma(\theta)$ for the corresponding (p,t) reactions. Deuterons and tritons are momentum analyzed with a magnetic spectrograph with about 50 keV energy resolution and detected with several position-sensitive solid state detectors arranged at the focal plane.

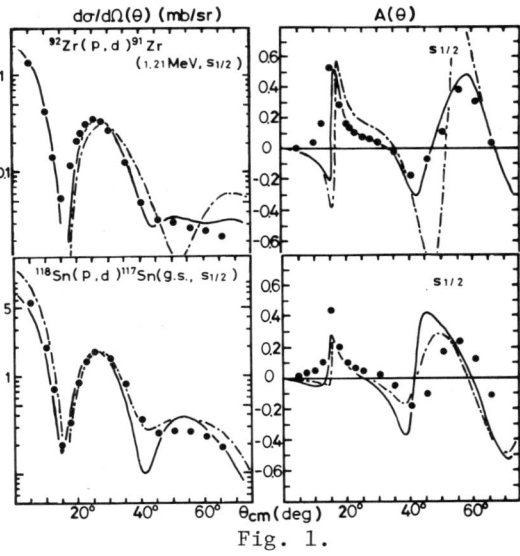

Fig. 1.

Data were obtained for transferred angular momenta from $\ell=0$ to $\ell=5$. These data were analyzed in terms of DWBA by using the code FNR-TWOSTP[1]. Proton optical potentials of Becchetti and Greenlees type[2] and deuteron optical potentials obtained from adiabatic model (Johnson-Soper model)[3,4] are used. Solid curves in all figures except Fig. 4 show zero range DWBA calculations obtained by using above-mentioned optical potentials.

$A(\theta)$ and $\sigma(\theta)$ for $\ell=0$ transitions are shown in Fig. 1. $A(\theta)$ show almost same oscillatory behaviors,

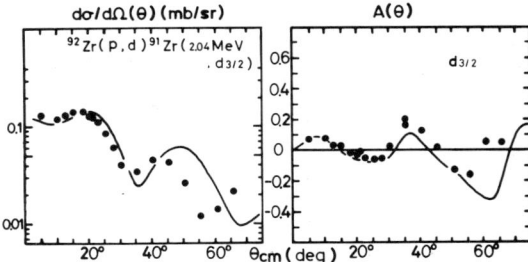

Fig. 2.

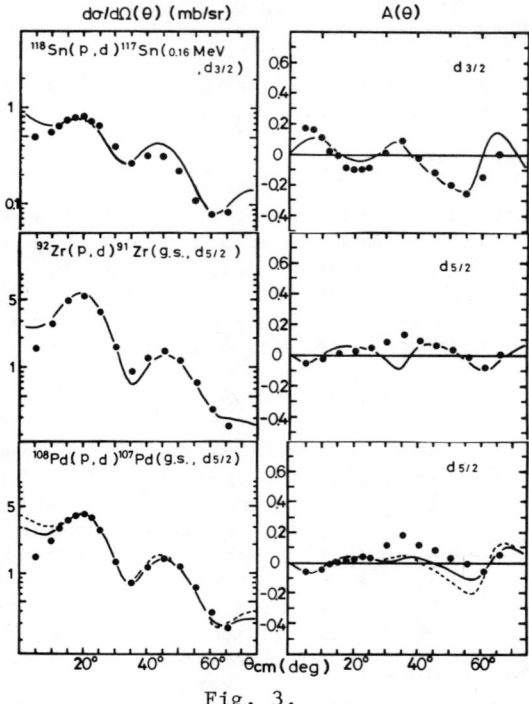

Fig. 3.

independent of the targets. However, there exist a little differences among $\sigma(\theta)$ at about θ_{cm} =45°. The sign of the data of $A(\theta)$ disagrees with the logarithmic derivative of $\sigma(\theta)$ at θ_{cm} =5°-15°. In addition, DWBA predictions misfit to the data at the same angles. DWBA predictions with deuteron optical potential obtained from an analysis of the elastic scattering data[5] (dot-dash curves) are shown in Fig. 1, which are compared with the adiabatic model calculations (solid curves).

Figs. 2 and 3 show $A(\theta)$ and $\sigma(\theta)$ for ℓ=2 transitions. $A(\theta)$ for d5/2 display quite similar angular distributions with each other. DWBA predictions of $A(\theta)$ for d5/2 disagree with the data at about θ_{cm} =35°. Theoretical curves for d5/2, including the second order process, p-p'(2^+)-d, are shown in Fig. 3 (dashed curves).

DEUTERON D-STATE EFFECT FOR s1/2 AND d5/2

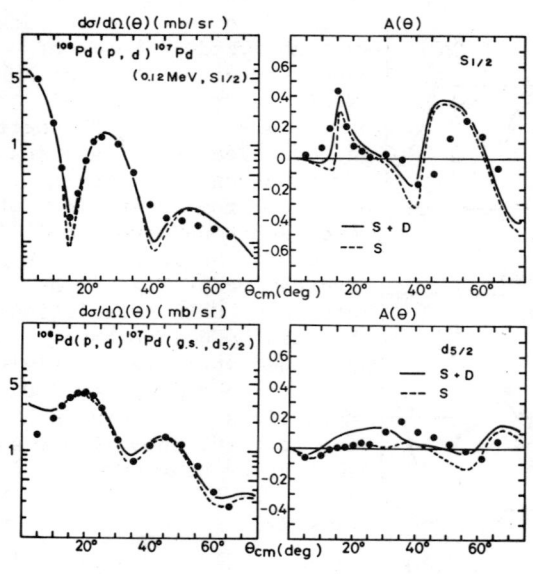

Fig. 4.

Fig. 4 shows the effect of the deuteron D-state on DWBA calculations for s1/2 and d5/2. Solid curves are finite range (FR) DWBA[6] calculations including deuteron D-state. Dashed curves are FR DWBA calculations for S-state only. These show that FR DWBA calculations including the deuteron D-state improve fits to the data, particularly at θ_{cm} =5°-15° for s1/2 and at about θ_{cm} =30° for d5/2 in $A(\theta)$. The effect of the deuteron D-state for d5/2, in $A(\theta)$, is more than that for s1/2.

The deuteron D-state effect for $\ell=3,4$ and 5 is not studied but expected to be much.

L=4 AND L=5 TRANSITIONS

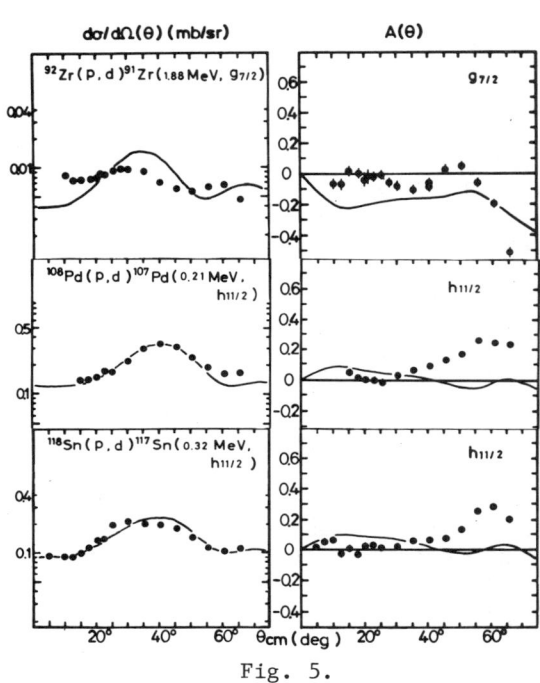

Fig. 5.

Fig. 5 shows $A(\theta)$ and $\sigma(\theta)$ for g7/2 and h11/2. DWBA calculations reproduce the data of $\sigma(\theta)$ but do not the data of $A(\theta)$. While, in ^{142}Nd(p,d)^{141}Nd(0.76MeV, h11/2) reaction (see Fig. 6), there is a large disagreement between the data and DWBA calculation, particularly at forward angles. The oscillation of the experimental $\sigma(\theta)$ for Nd is flat and different from that for Pd or Sn (see Figs. 5 and 6), at the same angles (forward angles). However, by changing the radius parameter of the real part of the proton optical potential from 1.25 fm to 1.4 fm, DWBA calculation well reproduce the data of $\sigma(\theta)$. Dashed curves in Fig. 6 show DWBA calculation with r_R=1.4 fm. This result is intimately connected to momentum mismatch.

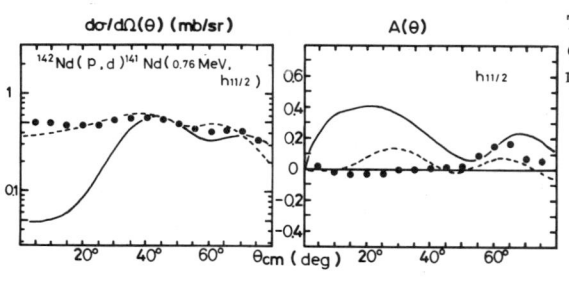

Fig. 6.

REFERENCES

1. M. Toyama and M. Igarashi, computer code FNRTWOSTP.
2. F. D. Becchetti,Jr. and G. W. Greenlees, Phys. Rev. 182 (1969) 1190.
3. R. C. Johnson and P. J. R. Soper, Phys. Rev. C1 (1970) 976.
4. J. D. Harvey and R. C. Johnson, Phys. Rev. C3 (1971) 636.
5. M. Irshad and B. A. Robson, Nucl. Phys. A218 (1974) 504.
6. M. Igarashi, private communication.

ANALYZING POWERS FOR (p,d) REACTIONS ON ^{208}Pb AND ^{142,144}Nd EXCITING NEUTRON-HOLE STATES

Y. Toba, K. Nagano, Y. Aoki, S. Kunori and K. Yagi
Institute of Physics and Tandem Accelerator Center
The University of Tsukuba, Ibaraki 305, Japan

ABSTRACT

Vector analyzing powers $A(\theta)$ and cross sections $\sigma(\theta)$ have been measured for (p,d) reactions on ^{208}Pb and ^{142,144}Nd by the use of a polarized proton beam of 22.0 MeV and a magnetic spectrograph. Excitations of neutron-hole states of $p_{1/2}$, $p_{3/2}$, $f_{5/2}$, $f_{7/2}$, $h_{9/2}$ and $i_{13/2}$ have been observed in the ^{208}Pb($\vec{p}$,d)^{207}Pb (N=125) reaction, those of $f_{7/2}$ and $p_{3/2}$ in the ^{144}Nd($\vec{p}$,d)^{143}Nd (N=83) reaction and those of $d_{3/2}$, $s_{1/2}$ and $h_{11/2}$ in the ^{142}Nd($\vec{p}$,d)^{141}Nd (N=81) reaction.

EXPERIMENAL RESULTS AND DWBA ANALYSES

Angular distributions of $A(\theta)$ and $\sigma(\theta)$ have been measured from $\theta=5°$ to $65°$ in $5°$ or $2.5°$ steps. Good agreement between the measured and calculated analyzing powers and cross sections is obtained for the $\ell=0$, 1, 2, 3 (p,d) transitions employing zero-range DWBA theory, as shown in Figs. 1 to 6. Strong j-dependence of analyzing powers has been observed for the pairs of transitions of $p_{1/2}$-$p_{3/2}$ (Fig. 1) and $f_{5/2}$-$f_{7/2}$ (Fig. 2). The (p,d) transitions with large angular momentum transfer $\ell=5$ and 6, however, are not interpreted in terms of the DWBA calculations because of large angular momentum mismatch (Figs. 7, 8 and 9).

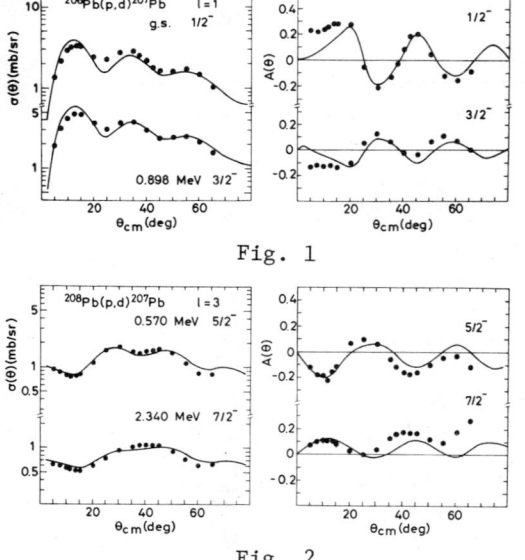

Fig. 1

Fig. 2

The experimental analyzing powers for the $p_{3/2}$ and $f_{7/2}$ neutron pick-up in the ^{208}Pb(p,d)^{207}Pb reaction (Figs. 1 and 2) have very different angular distribution from the corresponding analyzing powers for the ^{144}Nd(p,d)^{143}Nd

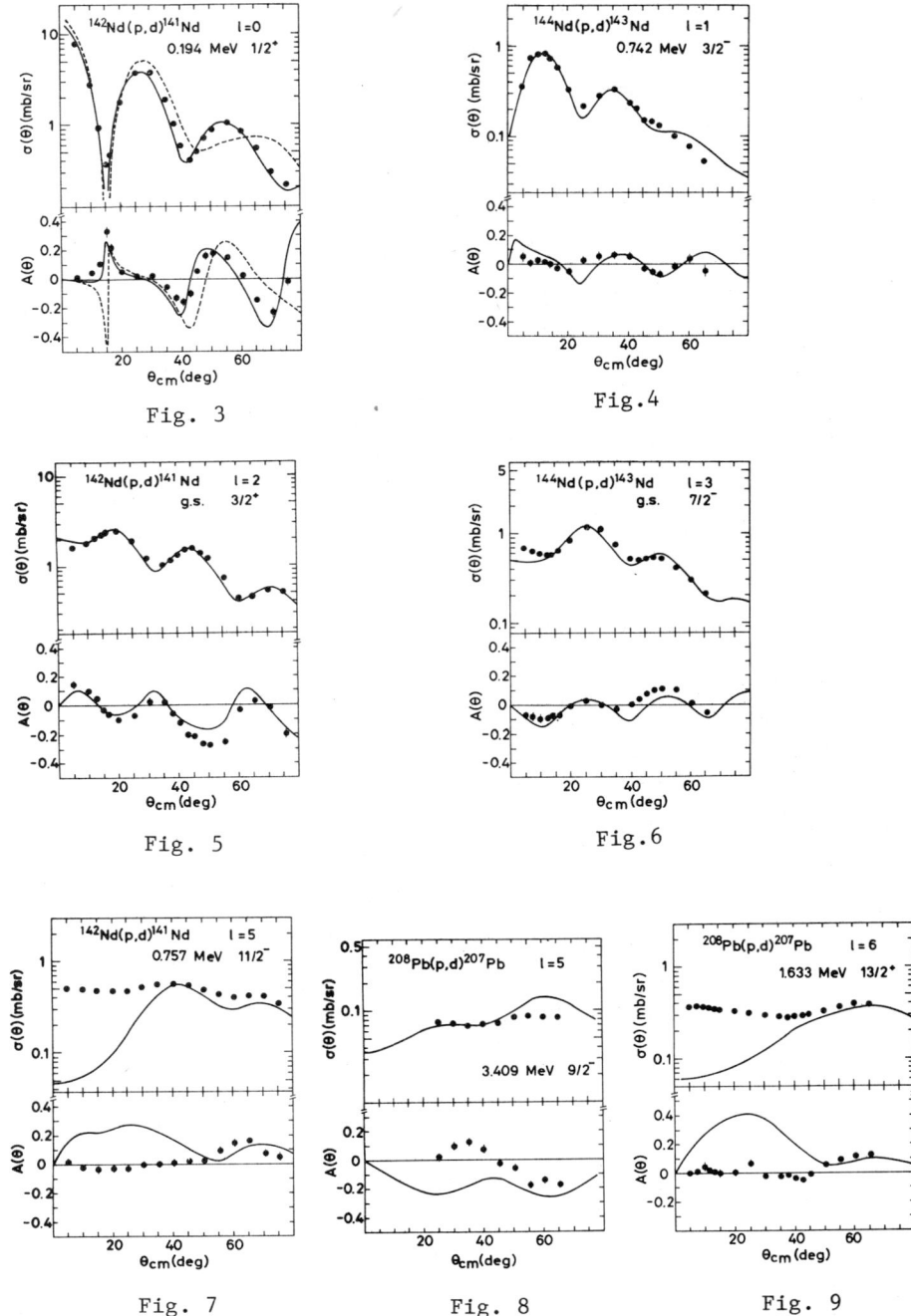

Fig. 3 Fig. 4

Fig. 5 Fig. 6

Fig. 7 Fig. 8 Fig. 9

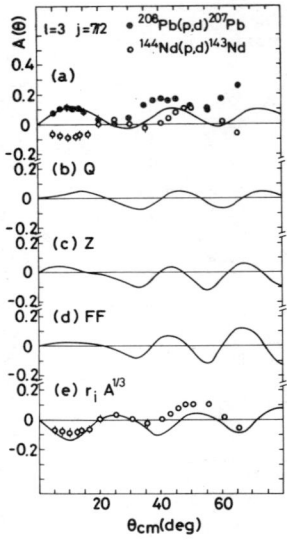

Fig. 10

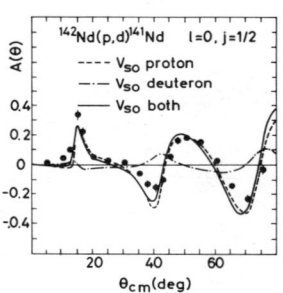

Fig. 11

reaction (Figs. 4 and 6). As shown in Fig. 10, the oscillation of the analyzing powers for the $f_{7/2}$ transitions is completely out of phase. Solid curves in Fig. 10 correspond to the DWBA calculations in which several parameters involved are changed by turns from (a) to (e). The curve (b) is obtained from (a) by changing the reaction Q-value from -7.485 MeV (^{208}Pb) to -5.593 MeV (^{144}Nd). In the similar way, the curves (c), (d) and (e) are obtained from the curves (b), (c) and (d) by changing, respectively, Coulomb potential of Z from 82 to 60, the form factor for the ^{208}Pb(p,d)^{207}Pb($f_{7/2}$) to for the ^{144}Nd(p,d)^{143}Nd($f_{7/2}$) reaction, and the nuclear radii of the optical potentials from $r(208)^{1/3}$ to $r(144)^{1/3}$. The curves of (b) to (d) show only small change in shape and there occurs no phase change. On the contrary, a drastic sign change occurs in going from (d) to (e). Therefore, the opposite phase oscillations of the $f_{7/2}$ analyzing powers observed in the two reactions are ascribed to the change in the nuclear radii in the distorting potentials.

The analyzing power for an $\ell=0$ transfer reaction is sensitive to the spin-orbit interaction in the distorting potentials. Indeed, Fig. 11 shows that the effect of spin-orbit distortion in the proton channel on the analyzing power of the $\ell=0$ transfer is greater than that in the deuteron channel over the whole angular range. Another interesting feature in the $\ell=0$ transfer is the fact that DWBA fitting to the experimental analyzing power is improved by using the adiabatic model for deuteron optical potential (solid curve in Fig. 3) instead of using the optical potential obtained from the study of elastic scattering of deuteron (dashed curve).

THE REACTION ^{56}FE $(\vec{p},d)$ AT 24.6 MEV[x]

J.H. Polane, P.J. van Hall
Cyclotron laboratory, Eindhoven University of Technology
Eindhoven, the Netherlands

We measured the differential cross sections and analysing powers of the following nine levels of ^{55}Fe: g.s., 0.41, 0.93, 1.32, 1.41, 2.94, 4.51, 4.88 and 7.79 MeV excitation energy. The polarized proton beam was produced by the Eindhoven A.V.F. Cyclotron. The beam current was 10 nA, with a degree of polarisation of 75%. The deuterons were detected by four telescopes consisting of 0.2 mm ΔE and 2 mm E detectors. The reaction ^{56}Fe $(\vec{p},t)$ was measured simultaneously and the deuterons and tritons were identified by analogue techniques[1].

Like in the ^{58}Ni $(\vec{p},d)$ reaction[2] we performed two-step DWBA calculations (code CHUCK) with inelastic scattering in the proton channel. We used the following scheme:

2.66 $\beta = 0.10$ 2_2^+ ─────────

0.85 $\beta = 0.24$ 2_1^+ ───────── ^{55}Fe

                       0_1^+ ─────────

The spectroscopic amplitudes were computed[3] from wave functions of Vennink and Glaudemans[4]. These wave functions have two $7/2^-$ proton holes (closed proton core in the Ni case) and at most one $7/2^-$ neutron hole. The particle states are $p_{3/2}$, $p_{1/2}$ and $f_{5/2}$. A modified Kuo-Brown interaction has been used. The inelastic scattering is calculated with a collective model, so one must find the sign of β by comparison with the experiment.

The influence of two-step transitions was found to be chiefly in the analysing power and the magnitude of the cross section. The effects are of the same order of magnitude as in the reaction ^{58}Ni $(\vec{p},d)$. The two-step processes raise the cross section at the forward maximum with about 15% for the $3/2^-$(g.s.) and $1/2^-$(0.41). There is no change in magnitude for the $5/2^-$(0.93). The spectroscopic factors are then, after correction for the two-step contribution:

$S(3/2) = 0.70$ $S(1/2) = 0.28$ $S(5/2) = 0.33$

To appreciate these numbers one should know that we used corrections for the nonlocality of the proton, the neutron and the deuteron and for finite range (LEA). Especially the nonlocality of the neutron is important as this lowers the spectroscopic factors by 30%.

[x] Supported in part by F.O.M.-Z.W.O.

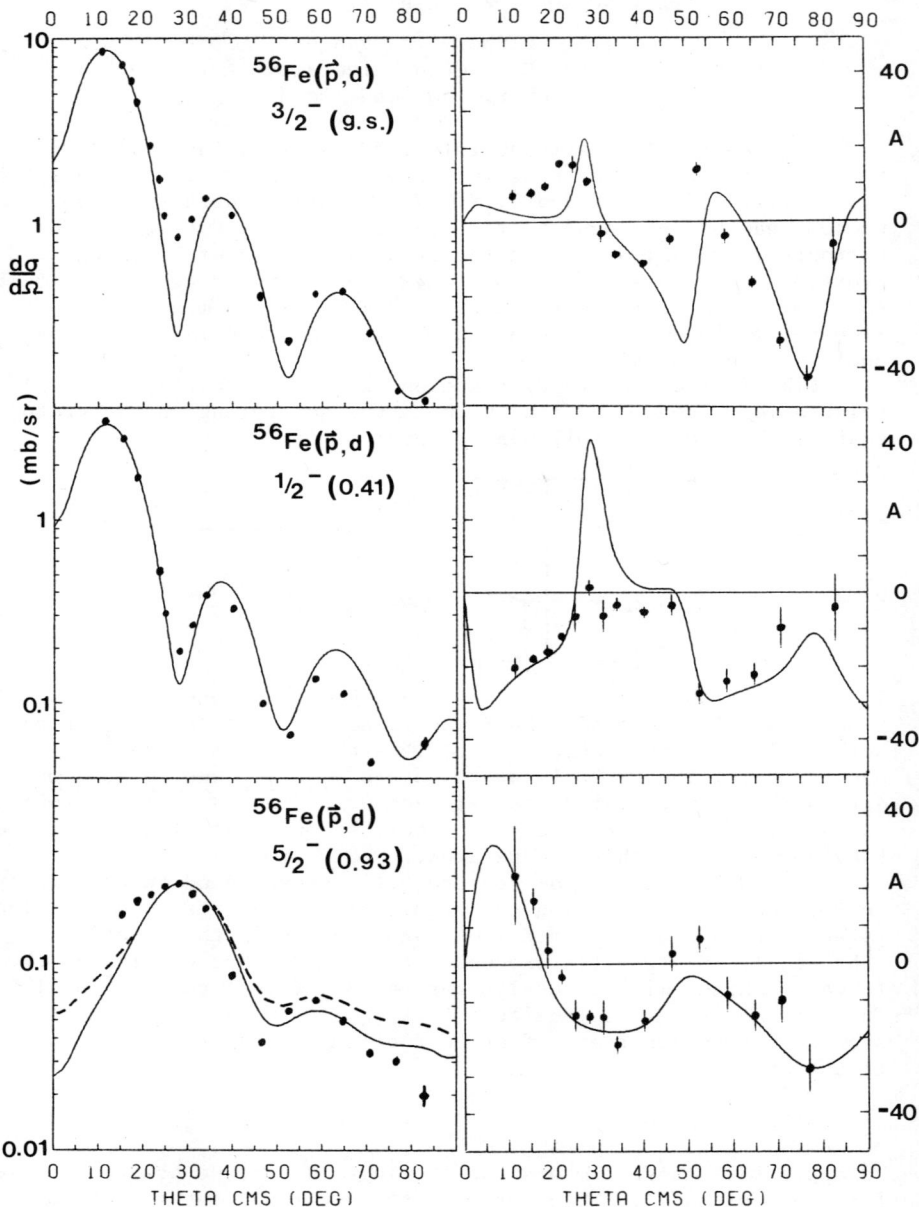

fig. 1. Two-step DWBA calculations of the first three levels of ^{55}Fe
full curves with deuteron potential D1
dashed curve with deuteron potential D2

Like in the ^{58}Ni $(\vec{p},d)$ reaction[2] we used a modified adiabatic deuteron potential (D1). The reduction of the real well depth V_R is 7.5% instead of 10% if one employs nonlocality, maybe because part of the exchange is already incorporated. The imaginary well had to be deepened with respect to the adiabatic potential and it took the form $W = 11.8 + 1.25$ E.

In fig. 1 ($5/2^-$) we show the difference between D1 and a potential D2 with an adiabatic imaginary potential and 5% reduction of V_R. From the better fit of D1 we infer the need for more absorption than in the ordinary adiabatic potential.

The calculations of the analysing power of $7/2^-$ states are very sensitive to the deuteron energy and the real well depth V_R as shown in fig. 2, where we display one-step calculations to the $7/2^-$ (2.94) state with 0%, 8% and 16% reduction of V_R. So one must be on guard against such instabilities when comparing experiment with DWBA calculations.

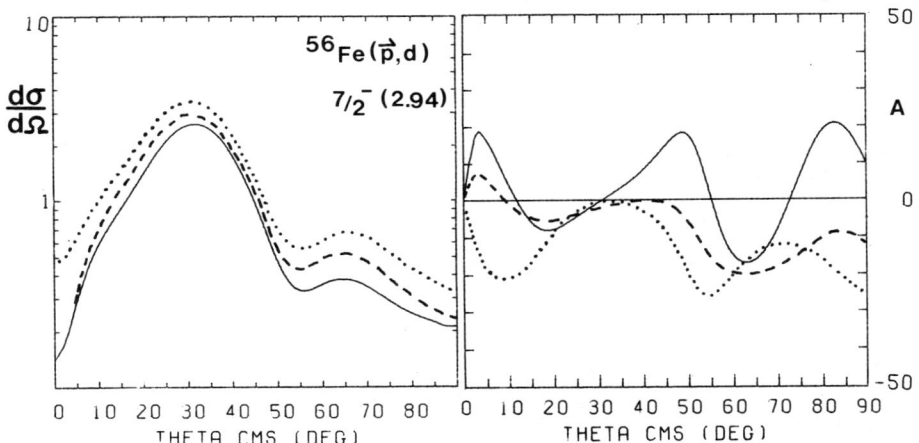

fig. 2. Full curve   0% reduction of V_R
     Dashed curve 8%   "   "   "
     Dotted curve 16%   "   "   "
     Cross-section in arbitrary units

REFERENCES

1. J.E. Sluiters et al., Nucl. Instr. and Meth. <u>120</u>, 305 (1974)
2. J.H. Polane et al., these proceedings.
3. F. v. Hees, University of Utrecht, the Netherlands (priv.comm.)
4. R. Vennink, University of Utrecht, doctoral thesis (1979).

TWO-STEP PROCESSES IN THE REACTION ^{58}NI $(\vec{p},d)$ AT 24.6 MEV[x]

J.H. Polane, P.J. van Hall, O.J. Poppema, S.S. Klein,
G.J. Nijgh, S.D. Wassenaar, J.F.J. Dautzenberg, W. Feix,
Cyclotron laboratory, Eindhoven University of Technology,
Eindhoven, the Netherlands

The analysing power of a $(\vec{p},d)$ reaction is much more j-dependent than the differential cross section. As a general rule one can say that the sign of the analysing power at the pick-up peak angle is + or − depending on whether $j = 1 + \frac{1}{2}$ or $j = 1 - \frac{1}{2}$. By comparing the measured analysing power with DWBA calculations one thus obtains a reliable spin assignment of the final state in the case of spin-zero target nuclei[1]. This is a good procedure for levels which are reached mainly by direct pick-up. If two-step processes are important then l and j assignments become less clear because of more parameters in the DWBA calculations.

The nucleus ^{58}Ni has a 2^+ state at 1.45 MeV which can be reached easily by inelastic proton scattering ($\beta = 0.22$). Therefore we expected some influence from the process of inelastic scattering followed by pick-up. In fact we tried the following coupling scheme (inelastic scattering in the deuteron channel was found to give negligible contributions):

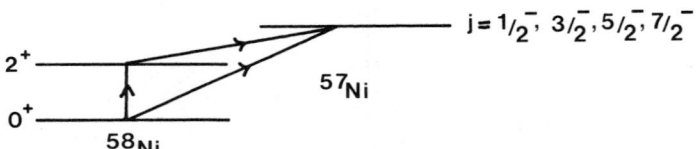

In a simple model with only 2 neutrons in the $p3/2$, $p1/2$ and $f5/2$ subshells outside an inert ^{56}Ni core this scheme already implies 4 different spectroscopic amplitudes. So one needs structural calculations of these amplitudes (at least as starting values). We compared several sets of amplitudes for the $3/2^-$, $5/2^-$ and $1/2^-$ states and did not find significant differences in the shape of the differential cross sections and analysing powers. On the other hand there is a distinct difference between the one-step and the full calculation (fig. 1, $3/2^-$).

We had some difficulty in finding a good deuteron potential. The Lohr-Haeberli (LH) potential which is derived from elastic deuteron scattering is certainly not sufficient (fig. 1, $5/2^-$). But also the adiabatic potential fails (fig. 1, $7/2^-$). The adiabatic potential was constructed from the BG potentials following the prescriptions of Satchler[2]. In getting better fits the most sensitive parameter appeared to be the depth of the real potential V_R (or its radius). We had to reduce the magnitude of V_R by about 10%. The need for this reduction is supported by several authors [2,3] and is probably connected with neglecting exchange effects in the calcula-

[x] Supported in part by F.O.M.-Z.W.O.

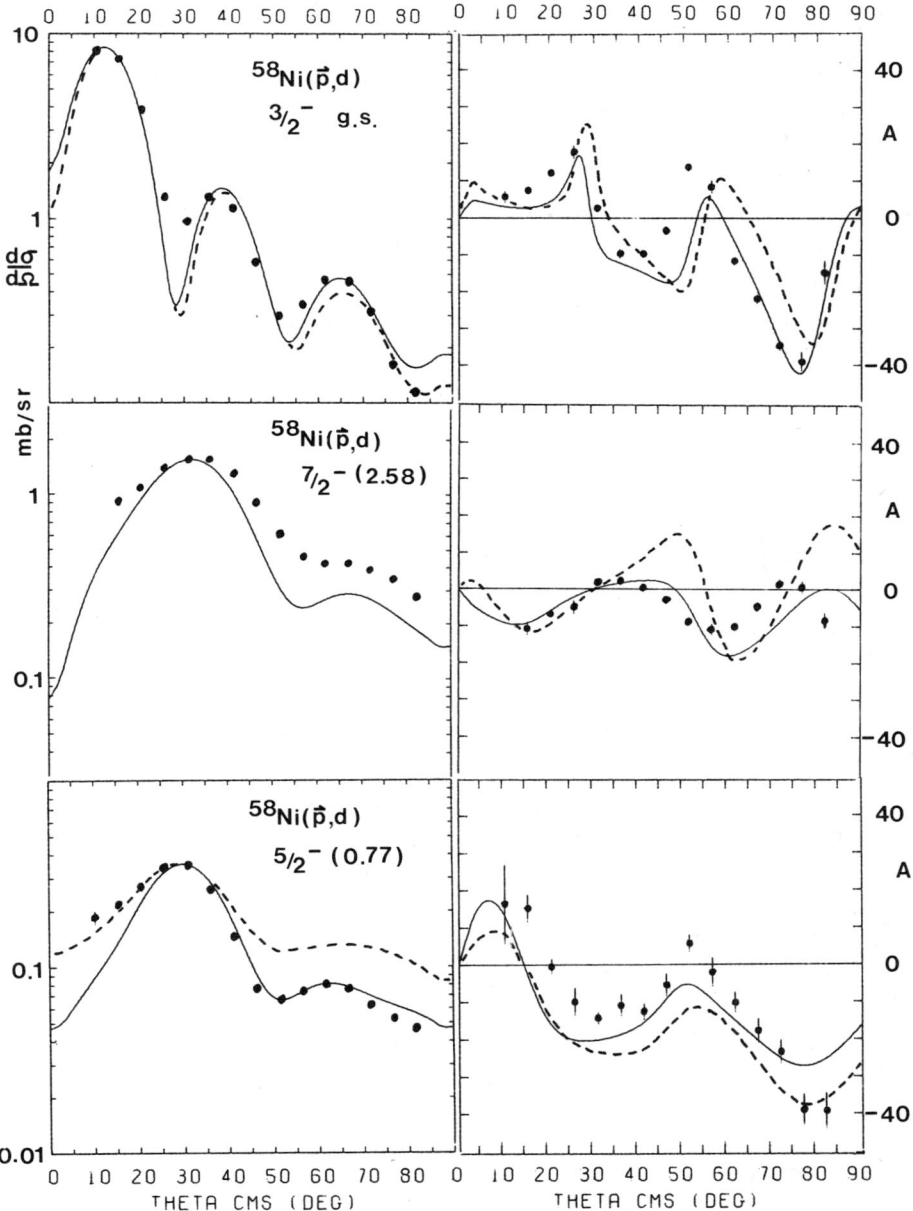

Fig. 1. full curve: full calculation with modified adiab. potential
dashed curve ($3/2^-$): pure one-step
dashed curve ($7/2^-$): adiabatic deuteron potential
dashed curve ($5/2^-$): LH deuteron potential

tion of the adiabatic potential.

This reduction of V_R was not sufficient and in addition we were forced to deepen the well of the imaginary surface potential. That means more absorption.

In fig. 2 we have displayed a level which has a big two-step component. In the two-step process inelastic scattering is followed by pick-up of a $7/2^-$ neutron from the core. In this case the change is chiefly in the differential cross section.

Final remarks: we measured 19 levels of ^{57}Ni. Most of these were described by direct pick-up. These results will be published in a forthcoming paper. The full calculations in fig. 1 were performed with amplitudes derived from wave functions of Koops and Glaudemans[4]. The inelastic scattering was described with a collective model. For more details of the calculations see the companion paper on ^{56}Fe $(\vec{p},d)$ in these proceedings.

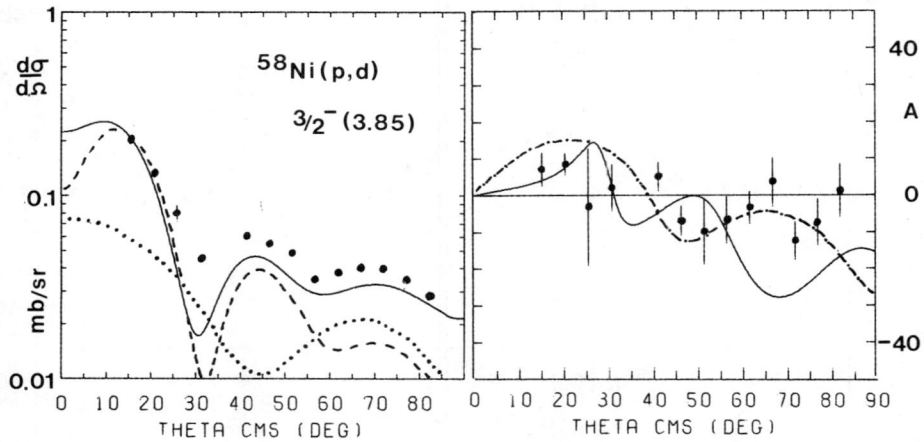

Fig. 2. two-step calculation of $3/2^-$ (3.85)
   ful curve: full calculation
   dashed curve: pure one-step
   dotted curve: pure two-step (cross section)
   dash-dot curve: pure two-step (analysing power)

REFERENCES

1. B. Mayer et al., Nucl. Phys. A 177, 205 (1971).
2. G.R. Satchler, Phys. Rev. C4, 1485 (1971).
3. P.J. Blankert, Investigation of the (p,p'), (p,d) and (p,t) reactions on some light Sn isotopes (doctoral thesis, V.U., Amsterdam, 1979), p. 56.
4. P.J. Brussaard, P.W.M. Glaudemans, Shell-model applications in nuclear spectroscopy (North-Holland, 1977), p. 440.

EFFECTS OF THE DEUTERON D STATE ON THE POLARIZATION OF THE RESIDUAL NUCLEAR STATE

N. Kishida, H. Ohnuma, J. Kasagi and T. Kubo
Department of Physics, Tokyo Institute of Technology, Meguro-ku, Tokyo, Japan

M. Yasue
Institute for Nuclear Study, University of Tokyo, Tanashi, Tokyo, Japan

ABSTRACT

A clear indication of the effects of the deuteron D state on the polarization of the residual nuclear state has been observed in the ^{58}Ni(p,dγ) angular correlation measurement at an incident energy of 30 MeV.

In general, residual nuclei are left polarized in nuclear reactions, and this polarization gives information independent of the other quantities such as cross sections and analyzing powers. Experimental information on the polarization, or more strictly the spin statistical tensors, of residual nuclei may be obtained from the correlation functions between the emitted particles and de-exciting gamma rays.

We have measured the correlation functions for the ^{58}Ni(p,dγ)^{57}Ni reaction using a 30-MeV proton beam from the INS cyclotron, and obtained for the first time a clear indication of the deuteron D-state effects on the polarization of the 2.58-MeV (7/2$^-$) state in ^{57}Ni. The target used was 2.4 mg/cm^2 thick metallic foil of enriched ^{58}Ni. Four sets of Si counter telescopes, each consisting of a 200 μm thick ΔE detector and a 2 mm thick E detector, and four 2"×2" NaI(Tl) detectors were used to detect deuterons and gamma rays. Energy signals and timing signals from these detectors were fed into an on-line computer TOSBAC-40C to sort out 4×4=16 combinations of coincidence events.

The angular correlation functions measured at θ_d^{cm} = 25° and 65° are shown in Fig. 1, and that obtained at 45° in Fig. 2. These angles correspond to the first and the second maxima and the first minimum, respectively, of the differential cross sections shown in Fig. 3. The solid and dashed curves represent exact-finite-range (EFR) DWBA calculations with and without the D state. The n-p interaction, which is used to describe the transfer reaction as well as to generate the deuteron internal wave function, is taken to be the soft-core potential of Reid[1]. Integration range is enough to cover the interaction range, the D_2 value[2] obtained in the present calculation being 0.483 fm^2, which is to be compared with the exact value of 0.484 fm^2. In Table I are given the potential parameters used in the calculations. Finite-solid angle corrections, although small, have been made to the calculated results to facilitate di-

Table I Potential parameters used in the calculations. The deuteron parameter set D3 is that of an adiabatic potential.

set	V_o	r_o	a_o	W	W_D	r_i	a_i	V_{so}	r_{so}	a_{so}	r_c	ref.
P1	48.1	1.20	0.712	1.8	5.7	1.235	0.702	5.8	1.20	0.712	1.20	3
D1	108.5	1.05	0.85		15.7	1.34	0.69	4.88	0.70	0.41	1.30	4
D2	96.3	1.12	0.735		12.63	1.261	0.842	6.37	1.12	0.735	1.30	5
D3	103.5	1.17	0.779	0.8	17.96	1.29	0.583	6.2	1.01	0.79	1.25	6,7

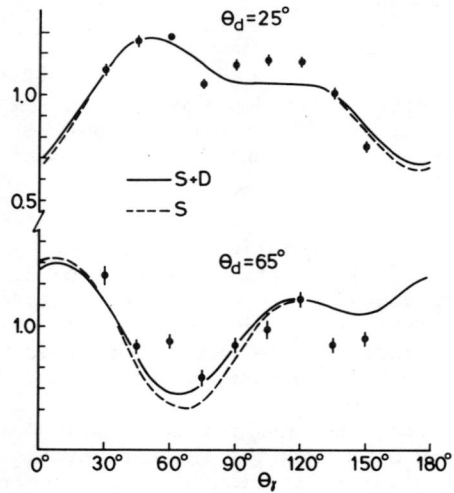

Fig. 1. Angular correlations measured for $\theta_d=25°$ and $65°$ compared with EFR DWBA calculations with the sets P1-D1.

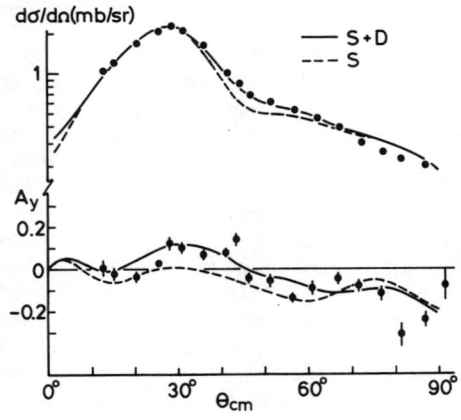

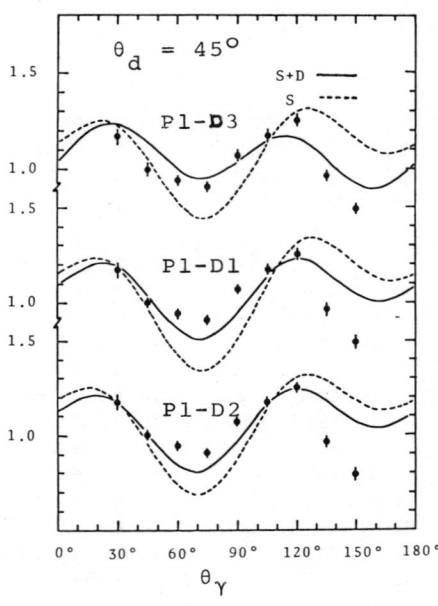

Fig. 2. Angular correlation measured for $\theta_d=45°$ compared with EFR calculations with various parameter sets.

Fig. 3. Differential cross sections and analyzing powers for the ^{58}Ni(p,d)^{57}Ni (2.58 MeV, $7/2^-$) reaction compared with EFR DWBA calculations with the sets P1-D1.

rect comparison with the data.

The D-state effects increase the calculated cross sections at minima as was first pointed out by Johnson and Santos[8], but the effects are small at this incident energy. The D-state effects on the vector analyzing power are also rather small, but somewhat improve the fit to the data. This can be seen in Fig. 3. On the other hand the D-state effects are appreciable on the correlation function at 45°, and significantly improve the fit to the data. Near the maxima of cross sections, i.e. at 25° and 65° the D-state effects are small even on the correlation functions, although slight improvement of the fit to the 65° data is seen.

The D-state effects observed in the fit to the correlation data at 45° are not related to a specific choice of DWBA parameters. This can be seen in Fig. 2, where examples of the correlation functions calculated with various parameter sets are compared with the data. The best fit to the data is obtained with the adiabatic potential, although the fits to the differential cross sections and analyzing powers are slightly worse. Irrespective of the potential parameters used, the 45° correlation data could not be explained without the deuteron D state.

REFERENCES

1. R. V. Reid, Ann. Phys. 50, 411 (1968).
2. R. C. Johnson and F. D. Santos, Particles and Nuclei 2, 285 (1971).
3. G. W. Greenlees and G. J. Pyle, Phys. Rev. 149, 836 (1966).
4. F. T. Barker, S. Davis, C. Glashausser, and A. B. Robins, Nucl. Phys. A233, 409 (1974).
5. C. M. Perey and F. G. Perey, Phys. Rev. 152, 923 (1966).
6. F. D. Becchetti and G. W. Greenlees, Phys. Rev. 182, 1190 (1969).
7. G. L. Wales and R. C. Johnson, Nucl. Phys. A274, 168 (1976).
8. R. C. Johnson and F. D. Santos, Phys. Rev. Lett. 19, 364 (1967).

MEASUREMENT OF THE ANALYZING POWERS FOR THE FRAGMENTED $f_{7/2}$
STATES BY THE ^{58}Ni$(\vec{p},d)^{57}$Ni REACTION AT 65 MeV

M. Fujiwara, Y. Fujita, S. Morinobu, I. Katayama, T. Yamazaki
and H. Ikegami
Research Center for Nuclear Physics, Osaka University
Suita, Osaka 565, Japan

K. Imai
Department of Physics, Kyoto University, Kyoto 606, Japan

ABSTRACT

Levels in ^{57}Ni were populated by the (p,d) reaction using 65 MeV polarized protons. A striking similarity was observed in the angular distribution of the analyzing powers for all the 7/2$^-$ states in ^{57}Ni observed in this experiment.

INTRODUCTION

Recently Ikegami et al.[1] have reported an observation of grouping of weak 7/2$^-$ states around the $(f7/2)^{-1}$ analog states in the (p,d) reactions on Ni isotopes. They claimed that the excitation of these weak states can be interpreted in terms of small $T_>$ admixtures in the $T_<$ states of presumably complicated configuration.
In order to check their interpretation, it would be of importance to investigate whether these weak levels might be excited only through the component of the f7/2 single neutron hole configuration or partly through the other reaction mechanisms.
In ^{57}Ni, several $J^\pi = 7/2^-$ levels exist below the isobaric analog (5.23 MeV) of the ^{57}Co ground state[2~6] and have been well studied from the theoretical view point[7]. In the present study, the analyzing powers of these 7/2$^-$ states, including both the $T_<$ and $T_>$ states, were measured using polarized proton beam.

EXPERIMENTAL PROCEDURE AND RESULTS

Polarized protons from an atomic beam type ion source were accelerated to 65 MeV by the RCNP AVF cyclotron. The spin direction of the proton beam was inverted every second by switching the low and high field RF spin units alternatively. The beam current of about 5 nA was obtained at the target position.
During the experiment, the beam polarization was monitored by a carbon polarimeter system[8], which consisted of two NaI(Tl) scintillators to detect the asymmetries of elastic protons from a carbon target at 47.5 degrees. The beam polarization was about 69 % throughout the present experiment.
The target used was an enriched ^{58}Ni foil of 2 mg/cm^2. Scattered deuterons were analyzed by the spectrograph "RAIDEN"[9] and detected with a 1.5 m position sensitive gas proportional counter.

Particle identification of deuterons was carried out using the energy loss signal in the two ΔE counters and the energy signal from a plastic scintillator. Angular distributions of the deuterons were measured in the range of 5°-80°(lab.) with a solid angle of 3.2 msr. The overall resolution obtained in the present experiment was dominated by the target thickness, giving the value of about 50 keV.

An example of the momentum spectra of the ^{58}Ni($\vec{p}$,d)^{57}Ni reaction is shown in Fig. 1. Energy values are only given for the 7/2$^-$ states obtained in the previous works[2~6]. The analyzing powers were taken for these 7/2$^-$ states.

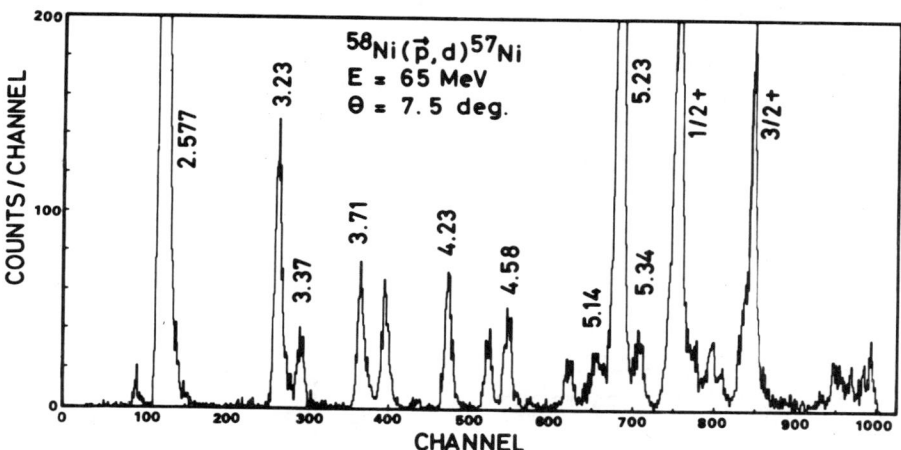

Fig. 1 Momentum spectrum of deuterons at θ_L= 7.5°. Energy values are only given for the 7/2$^-$ states.

Angular distributions of analyzing powers are shown in Fig. 2. They showed quite a similar pattern to each other. This observation is compatible with the previous results obtained for relatively intense levels in the ^{58}Ni($\vec{p}$,d)^{57}Ni reaction at lower beam energy (30 MeV)[10]. It is of interent that the present experiment shows virtually complete agreement of the distribution even among the weak states at 5.14 and 5.34 MeV.

It should be noted that each 7/2$^-$ levels in ^{57}Ni is expected to have different components in the wave function. However, the difference in configuration of the state seems to little affect on the angular distribution of the analyzing power in the present ^{58}Ni(p,d)^{57}Ni reaction at E_p=65 MeV. These facts should be understood to show that all of the observed 7/2$^-$ states in ^{57}Ni are excited purely by picking up a particle in the f7/2 neutron orbit and the strength may correctly represent the f7/2 component included in the wave function of complicated configuration. The difference in the analyzing power between the $T_>$ and $T_<$ states also seems to be absent.

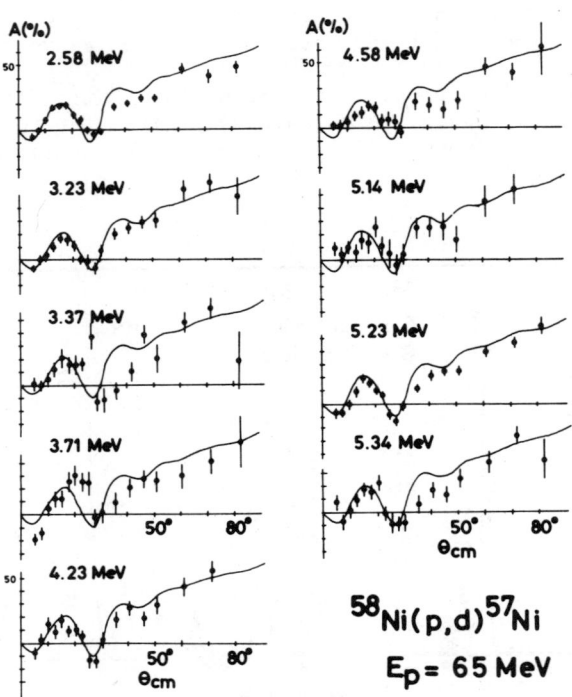

Fig.2 Analyzing powers for the $7/2^-$ states in ^{57}Ni with DWBA prediction.

REFERENCE

1) H. Ikegami et al., Nucl. Phys. A329 (1979) 84.
2) F.M. Edwards, J.J. Kraushaar and B.W. Ridley, Nucl. Phys. A199 (1973) 463.
3) H. Ohnuma et al., J. Phys. Soc. Japan, 36 (1974) 1236.
4) C. Glashausser and M.E. Rickey, Phys. Rev. Lett. 12 (1964) 420.
5) R. Sherr et al., Phys. Rev. 139B (1965) 1272.
6) B. Mayer et al., Nucl. Phys. A177 (1971) 205.
7) T. Motoba and K. Ogawa, Prog. Theor. Phys. 51 (1974) 173.
8) K. Hosono, N. Matsuoka, T. Saito, M. Fujiwara, S. Nagamachi and M. Kondo, RCNP annual report (1978) 191.
9) H. Ikegami, S. Morinobu, I. Katayama, M. Fujiwara and S. Yamabe, Nucl. Instr. and Meth. in print.
10) T. Hasegawa, N. Ueda, N. Kishida, H. Ohnuma, T. Fujisawa, T. Wada K. Iwatani, T. Tanaka and Y. Wakuta; INS Int. Symposium on Nucl. Direct Reaction Mechanism, (1978) 105.

THE (p, d) REACTIONS ON A=12-94 NUCLEI
BY 65 MeV POLARIZED PROTONS

K. Hosono, M. Kondo, T. Saito, N. Matsuoka and S. Nagamachi
Research Center for Nuclear Physics, Osaka University,
Suita, Osaka 565, Japan

T. Noro and H. Shimizu
Department of Physics, Kyoto University, Kyoto 606, Japan

S. Kato and K. Okada
Laboratory of Nuclear Studies, Faculty of Science, Osaka University,
Toyonaka, Osaka 560, Japan

K. Ogino and Y. Kadota
Department of Nuclear Engineering, Kyoto University, Kyoto 606, Japan

ABSTRACT

Differential cross sections and analyzing powers have been measured for (p,d) reactions on 1p,2s,1d,1f,2p,1g and 2d shell nuclei with 65 MeV polarized protons. The analyzing powers clearly show a j-dependence which depends on the numbers of nodes in the radial wave function of the neutron. The shapes of the angular distributions of the analyzing powers can be classified into five types for simple neutron pickup reactions from these shells. The pickup from the $1p_{1/2}$ shell is singular. The analyzing powers for the $1p_{1/2}$ shell neutron pickup can be reproduced by DWBA calculations only when large values of the radius and diffuseness parameters are used in the bound state potential.

INTRODUCTION

Differential cross sections and analyzing powers have been measured for (p,d) reactions on 1p,2s,1d,1f,2p,1g and 2d shell nuclei with 65 MeV polarized protons from the RCNP AVF cyclotron. The major emphasis in the present work is to point out the systematics of the j-dependence of the analyzing powers for the simple pickup reaction in the wide mass region and to provide a systematic basis for DWBA calculations.

EXPERIMENTAL PROCEDURES

The experiments have been carried out on the targets of ^{12}C, ^{13}C, ^{16}O, ^{24}Mg, ^{28}Si, ^{29}Si, ^{40}Ca, ^{56}Fe, ^{58}Ni, ^{90}Zr and ^{94}Zr with 65 MeV polarized proton beam. The deuterons were detected by a pair of counter telescopes positioned at symmetric angles to the beam axis. Each telescope consists of a 500 μm thick transmission type Si detector and a 15 mm thick high-purity Ge detector cooled by liquid nitrogen.

RESULTS AND DISCUSSIONS

The transitions which can be considered to proceed via a simple pickup reaction were selected. Cross sections and analyzing powers were measured for these transitions, which are listed in Table 1. There is not so distinct feature in the angular distributions of the

	Excitation Energy (MeV)	Q-value (MeV)	J^π	nlj assumed (neutron shell)
$^{12}C(p,d)^{11}C$	0.00	-16.496	$3/2^-$	$1p_{3/2}$
	2.00	-18.496	$1/2^-$	$1p_{1/2}$
$^{13}C(p,d)^{12}C$	0.00	-2.722	0^+	$1p_{1/2}$
	12.71	-15.432	1^+	$1p_{3/2}$
	15.11	-17.832	$1^+(T=1)$	$1p_{3/2}$
	16.11	-18.832	$2^+(T=1)$	$1p_{3/2}$
$^{16}O(p,d)^{15}O$	0.00	-13.444	$1/2^-$	$1p_{1/2}$
$^{24}Mg(p,d)^{23}Mg$	0.451	-14.310	$5/2^+$	$1d_{5/2}$
$^{28}Si(p,d)^{27}Si$	0.00	-14.954	$5/2^+$	$1d_{5/2}$
$^{29}Si(p,d)^{28}Si$	0.00	-6.253	$1/2^+$	$2s_{1/2}$
$^{40}Ca(p,d)^{39}Ca$	0.00	-13.504	$3/2^+$	$1d_{3/2}$
$^{56}Fe(p,d)^{55}Fe$	0.00	-8.986	$3/2^-$	$2p_{3/2}$
	0.411	-9.397	$1/2^-$	$2p_{1/2}$
	0.931	-9.917	$5/2^-$	$1f_{5/2}$
$^{58}Ni(p,d)^{57}Ni$	0.00	-9.700	$3/2^-$	$2p_{3/2}$
	0.769	-10.496	$5/2^-$	$1f_{5/2}$
	1.113	-10.813	$1/2^-$	$2p_{1/2}$
	2.577	-12.277	$7/2^-$	$1f_{7/2}$
	5.23	-14.930	$7/2^-$ (IAS)	$1f_{7/2}$
$^{90}Zr(p,d)^{89}Zr$	0.00	-9.724	$9/2^+$	$1g_{9/2}$
	0.588	-10.312	$1/2^-$	$2p_{1/2}$
	1.096	-10.820	$3/2^-$	$2p_{3/2}$
$^{94}Zr(p,d)^{93}Zr$	0.00	-5.800	$5/2^+$	$2d_{5/2}$

Table 1

Transitions for which the cross sections and analyzing powers were measured.

cross sections. On the other hand, the angular distributions of the analyzing powers of the pickup reaction from same orbit are very similar and j-dependence and n-dependence are quite pronounced. The shapes of the angular distributions of the analyzing powers could be classified into five types (the n=1, $j_>$; the n=1, $j_<$; the n=2, $j_>$; the n=2, $j_<$ and $1p_{1/2}$ transition group) as seen in fig. 1. The analyzing powers for the neutron pickup from $1p_{1/2}$ shell are expected naturally to belong to the n=1,$j_<$ transition group. However they show oscillatory pattern and are quite different from those of the n=1,$j_<$ transition group. The pickup from $1p_{1/2}$ shell appears to be singular.

In order to explain these characteristics, DWBA calculations have been carried out for the (p,d) reactions by TWOSTP code[1]. The proton optical potentials were determined using the cross sections and polarizations of the elastic scatterings measured in this experiment. The deuteron potentials were obtained by the adiabatic approximation[2] using nucleon optical potentials[3]. The geometrical parameters of neutron bound state potential are 1.25 fm for the radius and 0.65 fm for the diffuseness (case A). The well depth was adjusted to match the neutron separation energy corresponding to the state. The DWBA calculations for the n=1,$j_>$, the n=1,$j_<$, the n=2,$j_>$ and the n=2, $j_<$ transition groups can reproduce the shapes of the experimental angular distributions and j-dependence which depends on the n-value. An agreement between calculations and experimental data for the $1p_{1/2}$ transition group is very poor. An agreement between calculations and

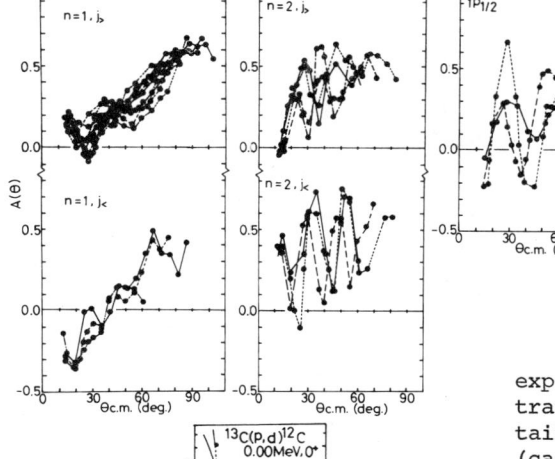

Fig. 1. angular distributions of the analyzing powers summarized for each transition groups. The lines are drawn only to guide eyes.

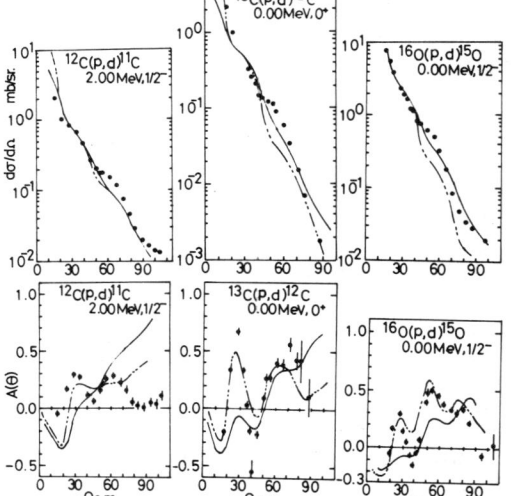

Fig. 2. Angular distributions of the cross sections and analyzing powers for the $1p_{1/2}$ transition group in the (p,d) reactions. The solid lines are the results of DWBA calculations in the case A. The two dots-dash curves show the DWBA calculations in the case B.

experimental data for the $1p_{1/2}$ transition group could be obtained only when large values (case B) of radius and diffuseness parameters of the neutron bound state potentials were used. They were 1.98 fm and 0.98 fm for ^{11}C, 1.78 fm and 0.90 fm for ^{12}C and 1.75 fm and 0.95 fm for ^{15}O respectively. These results are shown by two dots-dash curves in fig. 2. A better treatment of the form factor might facilitate more reliable predictions for $1p_{1/2}$ shell.

REFERENCES

1. M. Toyama and M. Igarashi, unpublished.
2. R. C. Johnson and P. J. R. Soper, Phys. Rev. C1 (1970) 976
3. C. M. Perey and F. G. Perey, Atomic Data and Nucl. Data Table 17 (1976) 1.

SPIN DETERMINATION OF DEEP HOLE STATES FROM $(\vec{p},d)$ REACTIONS[*]

J. Kasagi and G.M. Crawley
Cyclotron Lab., Michigan State Univ., East Lansing, MI 48824

S. Gales and E. Gerlic
IPN, BP No. 1 - 91406, Orsay, France

D. Friesel and A. Bacher
IUCF, Indiana University, Bloomington, IN 47405

ABSTRACT

Analyzing power measurements in $(\vec{p},d)$ reactions on ^{90}Zr, ^{120}Sn and ^{208}Pb have been performed using a 90 MeV polarized proton beam. The spin of the broad structure near 5 MeV excitation energy in ^{119}Sn has been unambiguously determined to be 9/2 by comparing the analyzing power with that observed for the ground state of ^{89}Zr as well as with a DWBA calculation.

INTRODUCTION

In previous studies of deep hole states in medium mass nuclei, the ℓ value of the transferred particle has been determined by angular distribution measurements. The j of the deep hole states, however, has been assigned only on the basis of theoretical expectations of the positions of particular orbits. It is therefore highly desirable to determine the j unambiguously.

A strong j dependence is expected from DWBA calculation for the analyzing power (Ay) of $(\vec{p},d)$ reactions even with rather high incident energies. The Ay measurement in $(\vec{p},d)$ reactions on some light nuclei at E_p = 65 MeV has shown the experimental j dependence of the analyzing power.[1] Spin determination of deeply-bound proton hole states has been done by Ay measurements in $(\vec{d},^3\text{He})$ reactions.[2] Measurements of the (p,d) reaction at 90 MeV on a series of Sn isotopes showed that the deep hole states near 5 MeV and 8 MeV excitation energy, were strongly excited.[3]

EXPERIMENT

Differential cross sections and analyzing powers have therefore been measured for $(\vec{p},d)$ reactions on ^{90}Zr, ^{120}Sn and ^{208}Pb using a 90 MeV polarized proton beam from the Indiana University Cyclotron. The beam polarization was measured frequently and was found to be about 70%. The spin direction of the incident beam was flipped automatically every minute during the data taking runs to reduce systematic errors. The outgoing particles were detected by two Si/Ge detector telescopes and were identified using hard wired particle identification boxes to minimize computer dead time. The energy resolution was about 100 keV FWHM.

[*]Supported by N.S.F. under Grant No. Phy 78-22696.

RESULT AND DISCUSSION

The measured Ay for several low-lying states in ^{89}Zr and ^{119}Sn are shown in Fig. 1. The Ay for the low-lying $p_{1/2}$ (0.595 MeV) and $p_{3/2}$ (1.1 MeV) states in ^{89}Zr are completely out of phase. The Ay for the $g_{9/2}$(g.s., ^{89}Zr) and $h_{11/2}$ (0.09 MeV, ^{119}Sn) states with $j=\ell+1/2$ are very similar with a dip at about 20° and increase rather steadily with angle beyond 20°. On the other hand, the Ay for the $f_{5/2}$ (1.46 MeV, ^{89}Zr) and $g_{7/2}$ (0.79 MeV, ^{119}Sn) states with transferred $j=\ell-1/2$ have negative values at forward angles and smaller values than the $j=\ell+1/2$ cases.

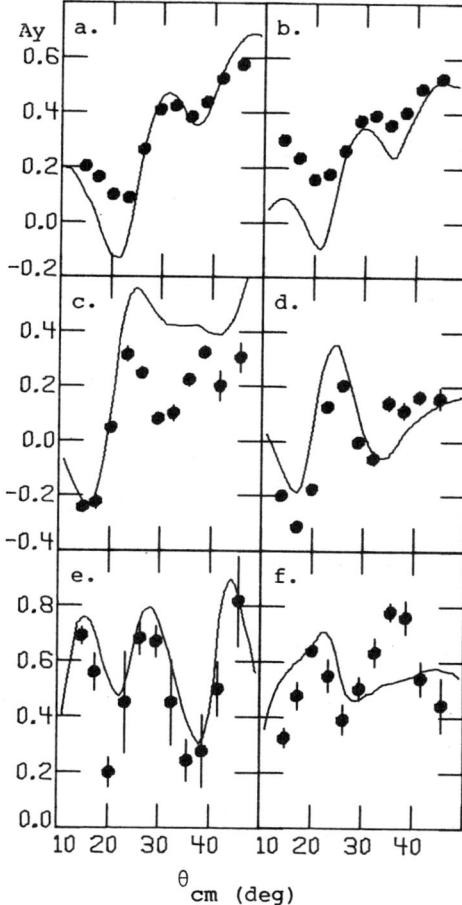

FIG. 1. Experimental and calculated Ay for the following states; (a) ^{89}Zr (g.s., $9/2^+$), (b) ^{119}Sn (0.09 MeV, $11/2^-$), (c) ^{89}Zr (1.46 MeV, $5/2^-$), (d) ^{119}Sn (0.79 MeV, $7/2^+$), (e) ^{89}Zr (0.595 MeV, $1/2^-$) and (f) ^{89}Zr (1.1 MeV, $3/2^-$).

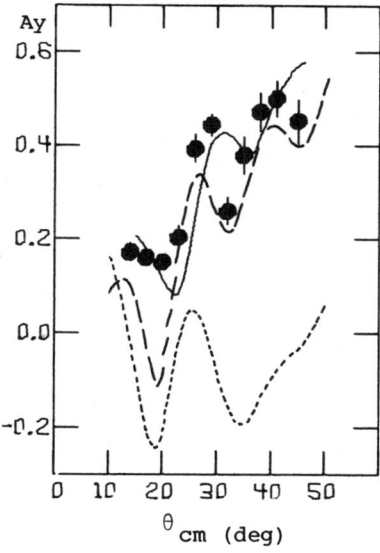

FIG. 2. Analyzing power of the broad structure near 5 MeV excitation energy in ^{119}Sn compared with the g.s. transition of ^{89}Zr (solid line). DWBA calculations with the $g_{9/2}$ transfer (dashed line) and with the $g_{7/2}$ transfer (dotted line) are also shown.

Distorted wave Born approximation (DWBA) calculations have been carried out using the code DWUCK with finite range corrections. Optical potentials used in the calculation are almost the same as in ref. 4) and ref. 5) except the Vs.o. of the proton channel which is slightly modified to obtain better fits to the data.

The calculated angular distributions of the cross sections reproduce the experimental angular distributions rather well for all the transitions. The spectroscopic factors of the low-lying states deduced in the present analysis are also in good agreement with the previously reported values.

The calculated Ay's are shown as solid lines in Fig. 1. Reasonable fits are obtained for the $p_{1/2}$ (0.595 MeV, ^{89}Zr) and $g_{7/2}$ (0.79 MeV, ^{119}Sn) transitions. The calculations also reproduce the experimental data fairly well for the $g_{9/2}$ (g.s., ^{89}Zr) and $h_{11/2}$ (0.09 MeV, ^{119}Sn) transitions except for the forward angles. However, for the $p_{3/2}$ (1.1 MeV, ^{89}Zr) and $f_{5/2}$ (1.46 MeV, ^{89}Zr) transitions, somewhat worse fits are obtained. The calculated Ay for the analog states around 9 MeV excitation in ^{89}Zr are also in good agreement with experiment. Thus, in general conventional DWBA calculations reproduce the Ay, as well as the differential cross sections, even at incident energies as high as 90 MeV.

The Ay of the broad structure near 5 MeV excitation energy in ^{119}Sn is shown in Fig. 2. It is similar to other $j=\ell+1/2$ transitions and is compared to the ground state of ^{89}Zr with a known j of 9/2 in Fig. 2 (solid line). Considering the small angle shifts which come from the momentum transfer differences, the two measurements agree very well. This provides a unique spin determination of 9/2 for the broad structure of ^{119}Sn. In Fig. 2 the results of a DWBA calculation are also shown. The long dashed curve is the calculation for the $g_{9/2}$ transition and the dotted curve for the $g_{7/2}$ transition. The $g_{9/2}$ calculation gives a good fit to the data except at forward angles, where the calculation cannot reproduce the data for the low-lying state. The $g_{7/2}$ calculation is quite dissimilar to the data. This comparison gives a further confirmation of the spin assignment of this neutron hole state.

REFERENCES

1. K. Hosono et al., Proc. 1978 INS International Symposium on Nuclear Direct Reaction Mechanism, ed. M. Tanifuji and K. Yazaki (INS, Univ. Tokyo, 1979), p 115.
2. A Stuirbrink et al., Contribution for International Simposium on Highly Excited States in Nucleus, Ohsaka, 1980.
3. G.M. Crawley, in Structure of Medium-Heavy Nuclei 1979, ed. Demokritos Group, Athens (The Institute of Physics, 1980) p 127.
4. R.E. Anderson, J.J. Kraushaar, J.R. Shepard and J.R. Comfort, Nucl. Phys. A311 (1978) 93.
5. G. Duhamel, L. Marcus, H. Langevin-Joliot, J.P. Didelez, P. Narboni and C. Stephan, Nucl. Phys. A174 (1971) 485.

MSU is an affirmative action/equal opportunity institution.

$(\vec{p},d)$ ANALYZING-POWER MEASUREMENTS AT 95 MeV[†]

D.W. Miller, W.W. Jacobs, D.W. Devins, and W.P. Jones
Indiana University Cyclotron Facility, Bloomington, IN 47405 USA

ABSTRACT

Analyzing-power angular distributions have been measured for prominent states excited by the ^{24}Mg$(\vec{p},d)$ reaction at 95-MeV bombarding energy. The distributions for the previously-known states below 6 MeV exhibit a clear j-dependence which is most pronounced near 6° for $\ell=1$ transitions, and near 35° for $\ell=2$ transitions. The analyzing power observed for $\ell=0$ pickup reaches a value $\simeq$ 90% near 20°. These spin signatures allow the identification of "deep-hole" p states at 8.91, 9.67, and 10.57-MeV excitation in ^{23}Mg as $p^{-1}_{3/2}$ and of a state at 9.02 MeV as $p^{-1}_{1/2}$.

INTRODUCTION

The (p,d) reaction has proved to be a useful spectroscopic tool below about 60-MeV bombarding energy when a simple pickup mechanism is involved. Differential cross section measurements are sensitive to the ℓ-dependence of the transfer, and fairly reliable spectroscopic information can be extracted at the lower bombarding energies with the use of DWBA calculations. Furthermore, analyzing-power measurements are sensitive to the j-dependence of the transfer. There is considerable current interest in finding out whether or not these features of the (p,d) reaction will persist at the higher bombarding energies where, in particular, the reaction could be useful for identifying "deep-hole" states inaccessible at the lower energies. The present paper reports on analyzing-power angular distribution measurements for the ^{24}Mg$(\vec{p},d)^{23}$Mg* reaction at E_p=95 MeV, carried out at the Indiana University Cyclotron Facility(IUCF); a companion paper submitted to this symposium[1] presents DWBA analyses of some of the results. Cross-section measurements for the states in ^{23}Mg up to 13.28-MeV excitation, carried out at the same bombarding energy, have been reported previously.[2]

EXPERIMENTAL METHOD AND RESULTS

Polarized protons from an atomic-beam source located in the large 800-kV electrostatic terminal at IUCF were accelerated through the injector and main cyclotrons to 94.8 MeV. Reaction products were momentum analyzed by a QDDM magnetic spectrograph. Overall resolution was about 70 keV. Beam polarizations were monitored by a ^{4}He polarimeter, periodically inserted directly after the injector cyclotron (E_p=8.3 MeV). No depolarization could be detected after acceleration in the main cyclotron. Typical beam polarizations were about +71% and -68% in the two spin orientations, with beam intensities on target of 40-100 nA.

[†]Work supported in part by the National Science Foundation.

Figure 1 shows the analyzing-power angular distributions obtained for known $\ell=0$ and $\ell=2$ pickup reactions to several low-lying states in ^{23}Mg. The two known $\ell=0$ transfers in the left panel show essentially identical angular distributions, the large oscillation reaching $\simeq$ 90% at about 20°. Transfers with $\ell=0$ are especially important for DWBA studies, since the analyzing power would be zero in the absence of spin-dependent distortions. The $\ell=2$ transitions shown in the right panel of Fig. 1 exhibit a substantial j-dependence at angles around 35°; little difference is seen at forward angles where the cross sections are much larger.

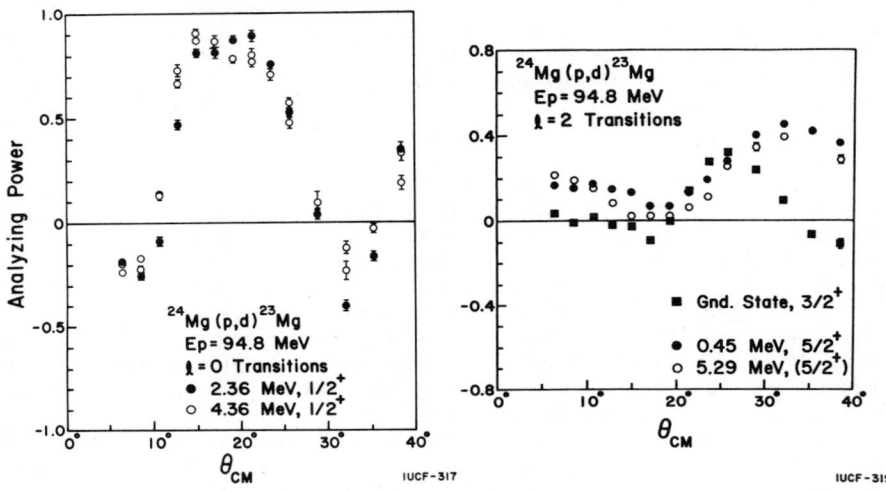

Fig. 1. Analyzing power distributions for $\ell=0$ and $\ell=2$ transitions.

A very characteristic spin signature for $\ell=1$ transitions at forward angles is shown for known states in the left panel of Fig. 2. The pronounced negative analyzing power observed for $p_{1/2}$ pickup at very forward angles has also been observed in this energy range with ^{13}C targets at 123 MeV at IUCF, and with ^{13}C and ^{16}O targets at 200 MeV.[3] The distributions presented in the right panel of Fig. 2 show the analyzing powers obtained in the present experiment for four deep-hole states previously assigned as $\ell=1$ pickup in 95-MeV (p,d) cross section studies.[2] A comparison of these angular distributions with the $p_{1/2}$- $p_{3/2}$ spin signatures observed for the known states results in a $p_{1/2}^{-1}$ assignment for the 9.02-MeV and $p_{3/2}^{-1}$ assignments for the 8.91, 9.67, and 10.57-MeV deep-hole states. The latter spin assignments suggest a concentration of $p_{3/2}^{-1}$ strength at an excitation energy which is consistent with the predictions of a shell-model calculation[4] for the nearby nucleus ^{27}Si; little $p_{1/2}^{-1}$ strength is predicted by this calculation to lie this high in excitation.

These same four deep-hole states have been studied at IUCF using the ^{24}Mg(d,t) reaction at E_p= 76 MeV.[5] It is interesting to note that in the observed spectrum for the mirror reaction ^{24}Mg(d,^{3}He),

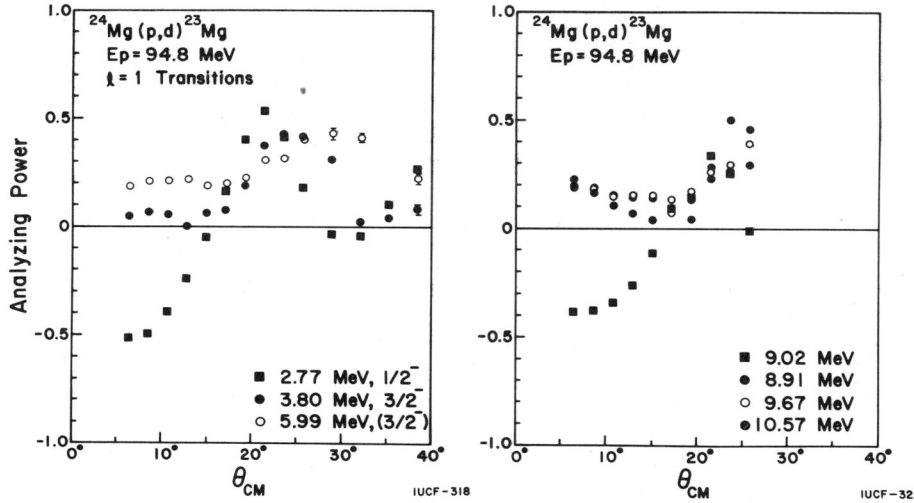

Fig. 2. Analyzing power distributions for $\ell=1$ transitions.

also studied at 76 MeV, the only ^{23}Mg state for which there is no obvious mirror counterpart is the 9.02-MeV $p_{1/2}^{-1}$ state. Systematic $(\vec{d},^{3}\text{He})$ analyzing-power measurements at lower energies for other sd shell targets [6] also fail to reveal any $p_{1/2}^{-1}$ strength except in the lowest-lying p state.

Analyzing powers for low-lying states believed to be excited in two-step processes [2] in the ^{24}Mg(p,d) reaction were observed to vary slowly with angle and did not exceed 30%.

CONCLUSIONS

$(\vec{p},d)$ analyzing-power measurements at bombarding energies around 100 MeV can serve as a very useful spin filter in light nuclei due to the characteristic j-dependence observed for $\ell=1$ and $\ell=2$ transitions. These measurements are particularly useful in establishing the systematics of the $p_{1/2}^{-1}$ and $p_{3/2}^{-1}$ deep-hole states in sd shell nuclei. Furthermore, analyzing-power measurements can provide stringent additional experimental tests for DWBA analyses [1] of the (p,d) reaction mechanism at intermediate energies.

REFERENCES

1. J. R. Shepard, E. Rost, P. D. Kunz; submitted to this symposium.
2. D. W. Miller et al., Phys. Rev. C20, 2008 (1979).
3. J. M. Cameron (private communication).
4. S. Maripuu (private communication).
5. W. W. Jacobs et al., Bull. Am. Phys. Soc. 23, 539 (1978).
6. H. Breuer et al., Annual Report, Max Planck Institut für Kernphysik, Heidelberg, 1977, p. 98.

POLARIZATION TRANSFER IN THE REACTION ^{56}Fe($\vec{d}$,p) $^{57}\vec{Fe}^*$ BY IN-BEAM MÖSSBAUER MEASUREMENT[†]

B.J. vom Feld, Th. Müller, C. Günther, H. Hübel
Institut für Strahlen- und Kernphysik, Universität Bonn, D-5300 Bonn, West-Germany

H. Paetz gen. Schieck
Institut für Kernphysik, Universität Köln, D-5000 Köln, West-Germany

ABSTRACT

All three tensor moments $t_{k0}(k=1,2,3)$ of the polarization of the 14.4 keV ($3/2^-$) state of ^{57}Fe produced by polarization transfer from a vector-polarized deuteron beam have been measured at $E_d=6.85$ MeV by the in-beam Mössbauer technique.

INTRODUCTION

Griffith et al.[1] showed in 1972 that Mössbauer techniques can be used to obtain information on polarization transfer in a ($\vec{d}$,p) reaction leading to a suitable Mössbauer transition. With the presently available polarized-beam currents these experiments can be done with much higher accuracy, so that all tensor moments of the Mössbauer state may be determined.

EXPERIMENT AND RESULTS

Fig. 1. shows the experimental arrangement. An average vector-polarized deuteron beam of 200nA on target from the Cologne polarized ion source LASCO was accelerated to 6.85 MeV by the Cologne Super FN tandem accelerator. The conventional Mössbauer apparatus consisted of a target foil of ^{56}Fe with a thickness of 5 mg/cm^2 (99.9% enriched) and tilted at 74° against the beam direction. It was magnetized to saturation by an electromagnet. The absorber was made of stainless steel of 1mg/cm^2 (^{57}Fe) and was moved sinusoidally. The γ's were detected by a Si(Li) detector (300 mm^2, 3 mm thick) at 90° to the beam.

The beam polarization was measured between up and down runs every 4 hours in a seperate ^{4}He polarimeter and found to be on the average $t_{10}^B=0.418\pm0.018$ and constant within this error.

Fig. 2. shows the Mössbauer spectra obtained and compared to a spectrum from a ^{57}Co source. The effects of the beam polarization on the occupation numbers are clearly visible. After background subtraction the areas under the four lines considered were determined and gave the occupation numbers of the magnetic substates. From these the tensor moments $t^{k0}(k=1,2,3)$ of the $3/2^-$ state were calculated according

[†]Supported in part by Bundesministerium für Forschung und Technologie D-5300 Bonn, West-Germany

Fig. 1. Schematic of the
experimental arrangement.

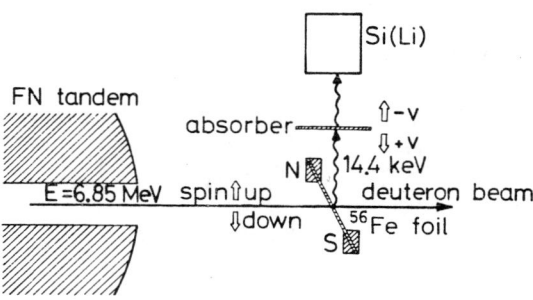

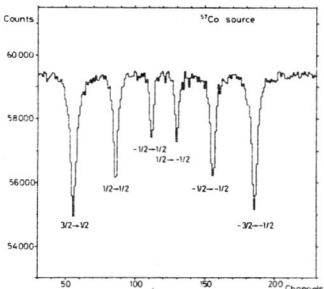
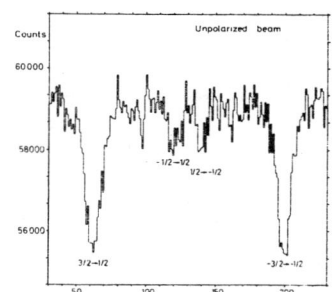

Fig. 2. Mössbauer spectra with ^{57}Co source
and in-beam with the beam unpolarized
and vector-polarized up and down. In the
in-beam measurements the two $\pm 1/2 \to \pm 1/2$
transitions are very weak due to the angle
of 16° between the target foil and the
γ-emission direction. Differences in the
line width are due to the different experimental conditions in the two cases.

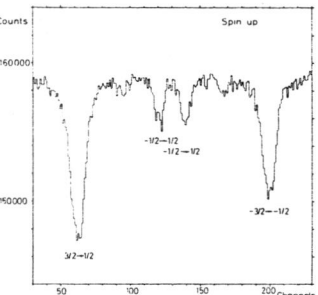

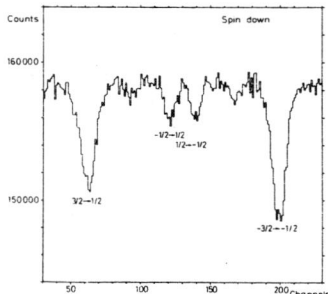

to ref.[2] (they are equal to T_{kq} there) for spin up and spin down together with their purely statistical errors. These values are given in the first two rows of Table I. The absolute values of t^{10} and t^{30} agree within the errors for the two cases whereas t^{20} shows a systematic deviation.

Defining the polarization transfer coefficients as ratios of the average of these tensor moments over spin up and spin down runs and the beam vector polarization t^B_{10}

$$t^{k0}_{10} = t^{k0}/t^B_{10}$$

we obtain the results in the third row of Table I (the error of t^{20}_{10} was increased to include the systematic deviation between the spin up/spin down runs):

		k=1	k=2	k=3
t^{k0}	Spin up	0.164±0.003	0.380±0.010	0.034±0.009
	Spin down	-0.155±0.010	0.293±0.010	-0.028±0.010
t^{k0}_{10}	average	0.382±0.024	0.806±0.104	0.073±0.023

Table I. Tensor moments and transfer coefficients of the reaction $^{56}Fe(\vec{d},p)^{57}Fe^*$.

CONCLUSION

The results show significant polarization transfer in the reaction $^{56}Fe(\vec{d},p)^{57}Fe^*$ and demonstrate the usefulness of the in-beam Mössbauer method in selected cases. Since the protons have not been detected the results are quantities averaged over all proton angles.

REFERENCES

1. J.A.R. Griffith, G.R. Isaak, S. Roman, Phys. Rev. Lett. **28**, 375 (1972).
2. G.G. Ohlsen, Rep. Prog. Phys. **35**, 717 (1972).

DETERMINATION OF j-MIXING IN ^{53}Cr(d,p)^{54}Cr FROM TENSOR ANALYZING POWER MEASUREMENTS

J.E. Kammeraad, J.A. Bieszk, L.D. Knutson, and W. Haeberli
University of Wisconsin, Madison, WI. 53706†

For (d,p) reactions on odd-A targets it is generally the case that more than one value of j, the total angular momentum of the transferred neutron, is allowed. Within the context of conventional DWBA theory, it can be shown that for a mixed-j transition the contributions to the cross section and analyzing powers arising from the different j-values add incoherently. In the case of the two allowed j-values we obtain at each angle

$$\sigma(\text{mixed}) = S(j_1)\sigma(j_1) + S(j_2)\sigma(j_2) \tag{1}$$

$$T_{kq}(\text{mixed}) = \frac{S(j_1)\sigma(j_1)T_{kq}(j_1) + S(j_2)\sigma(j_2)T_{kq}(j_2)}{S(j_1)\sigma(j_1) + S(j_2)\sigma(j_2)} \tag{2}$$

Here $\sigma(j)$ and $T_{kq}(j)$ are the cross sections and analyzing powers for the corresponding pure-j transitions, and the $S(j)$ are spectroscopic factors for the mixed-j transition. Since the vector analyzing power is strongly j-dependent, it is found[1] that measurements of $\sigma(\theta)$ and $iT_{11}(\theta)$ for a mixed-j transition often make it possible to determine the spectroscopic factors, provided that $\sigma(j)$ and $iT_{11}(j)$ are known, a priori, for each allowed j-value.

Measurements of the tensor analyzing power T_{22} show that this quantity is also strongly j-dependent. The purpose of the present experiment is to investigate the usefulness of T_{22} measurements in the analysis of mixed-j transitions. In particular, we wish to compare the spectroscopic information obtained from the T_{22} measurements with that obtained from the vector measurements. Agreement of the results would further increase our confidence in this kind of analysis. Additional interest in the use of T_{22} measurements follows from the observation[2] that DWBA calculations seem to be more reliable for T_{22} than for iT_{11}. In the case of the vector analyzing power, it has been necessary to determine $iT_{11}(j)$ empirically, by measuring the vector analyzing power for the appropriate pure-j transitions. However, for T_{22} it is possible that DWBA calculations rather than empirical calibration curves might be used. One purpose of the present experiment is to investigate this possibility.

The reaction we have chosen to study is ^{53}Cr(d,p)^{54}Cr(0.83 MeV). The initial and final states have $J^\pi = 3/2^-$ and 2^+ respectively, so that for this transition $\ell = 1$ and $\ell = 3$ are allowed. Previous measurements[1] of $\sigma(\theta)$ show that only $\ell = 1$ occurs with significant probability and thus the transition may be treated as a mixture of $j^\pi = 1/2^-$ and $j^\pi = 3/2^-$. Empirical calibration curves for the two j-values were obtained by measuring the analyzing powers for the pure-j transitions listed in Table I.

† Work supported in part by the U.S. Department of Energy.

Table I. Reactions studied in this experiment

Reaction	j^π	Q-value
^{53}Cr(d,p)^{54}Cr(0.83 MeV)	Mixed	6.662 MeV
^{52}Cr(d,p)^{53}Cr(G.S.)	$3/2^-$	5.717
^{54}Fe(d,p)^{55}Fe(G.S.)	$3/2^-$	7.074
^{54}Fe(d,p)^{55}Fe(0.41 MeV)	$1/2^-$	6.663

The measurements were taken at a bombarding energy of 10 MeV using deuterons from the Wisconsin crossed-beam polarized ion source. For the measurement of the tensor analyzing powers, the orientation of the spin-alignment axis was chosen to make one of the tensor beam moments (t_{20}, t_{21}, t_{22}) non-zero. During the run, the sign of the tensor polarization was changed every 2 seconds by switching RF transitions at the ion source, and the corresponding spectra were routed to separate areas of the computer memory. The beam polarizations were determined by a polarimeter located downstream of the scattering chamber. The targets were enriched foils of about 2 mg/cm^2 thickness. Reaction protons were detected by solid state detectors located to one side of the beam. Thin sheets of graphite, thick enough to just stop elastically scattered deuterons, were placed over each detector slit making it possible to extend the

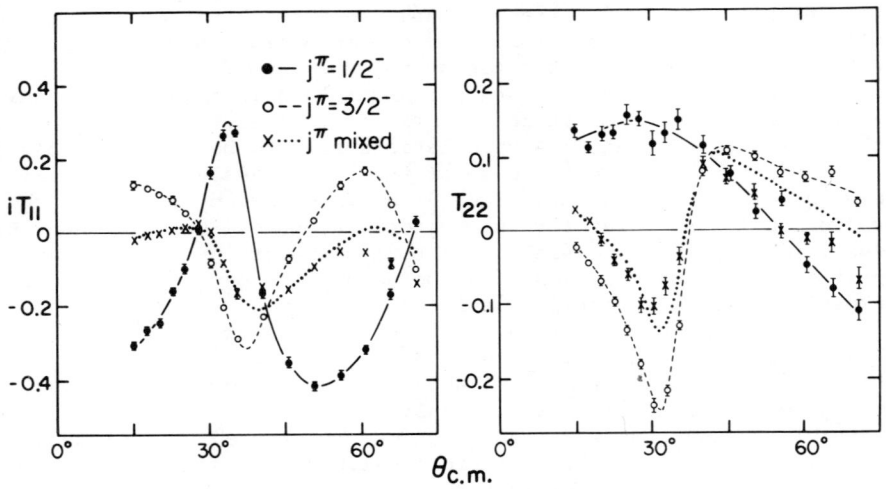

Figure 1. Measured analyzing powers. The solid and dashed lines are smooth curves through the data. The dotted line is a fitted curve using Eq. (2).

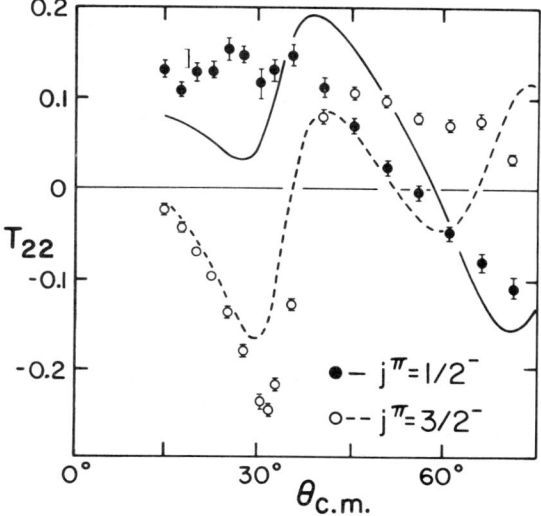

Figure 2. Measured analyzing powers for the pure-j transitions. The solid and dashed lines are DWBA calculations.

measurements to forward angles. The FWHM of the proton peaks was approximately 260 keV.

Measurements of the analyzing powers iT_{11} and T_{22} for the mixed-j transition are shown in Fig. 1 along with calibration data for the pure-j transitions. The calibration data shown for $j^{\pi}=3/2^-$ correspond to a Q-value of 6.662 MeV and were obtained by taking a weighted average of the measurements for the two $3/2^-$ transitions listed in Table I. The solid and dashed lines are smooth curves drawn through $1/2^-$ and $3/2^-$ calibration data.

By using Eq. (2) along with the empirically determined values of $\sigma(j)$ and $T_{kq}(j)$ it is straightforward to extract the relative contribution of the two j-values. This was done separately for iT_{11} and T_{22} using only the nine most forward angles. In both cases the best fit to the mixed-j analyzing power data was obtained for $S(3/2^-)\sigma(3/2^-)/S(1/2^-)\sigma(1/2^-) = 2.0\pm0.2$. The dotted curves in Fig. 1 show the corresponding fits. The fact that the iT_{11} and T_{22} measurements give consistent results indicates that the j-mixing determinations are probably quite reliable.

In Fig. 2 the T_{22} calibration data are shown along with DWBA calculations (which include the deuteron D-state) for the pure-j transitions. The DWBA calculations do not represent the data accurately and it is clear that if one were to use these curves rather than the empirical curves of Fig. 1 in the analysis it is likely that incorrect results would be obtained. We conclude that the DWBA calculations are not reliable enough for determining the j-mixing parameters from either iT_{11} or T_{22}.

REFERENCES

[1] D.C. Kocher et al., Nucl. Phys. A196, 225 (1972).
[2] E.J. Stephenson et al., Nucl. Phys. A277, 374 (1977).

TENSOR ANALYSING POWER OF THE REACTION ^{64}Ni(d,p)^{65}Ni *)

K. Rüskamp, W. Drenckhahn, A. Feigel, E. Finckh, G. Gademann, M. Wangler
Physikalisches Institut der Universität Erlangen-Nürnberg, W.-Germany

All three tensor analysing powers are known only for a few nuclei with medium A-values. Therefore, we have measured the ^{64}Ni(d,p)^{65}Ni reaction which allows to evaluate several transitions with different ℓ - and j-values. The comparison with calculation shall show the reliability of the reaction models and the parameters. The j-dependent effects can be used to decide a tentative spin asignment of a final state.

The tensor polarized deuteron beam of the Erlangen Lamb shift source was accelerated in the EN-tandem to E_d = 10 MeV, energy analysed and focused through an aperture with 2 mm diameter into the 4π-scattering chamber. Four slits in front of the collimator measured the beam position and corrected it with magnetic steerers. Eight proton detectors were placed left, right, up and down from the target. The tensor polarization of the beam was switched on and off with a frequency of 0.1 Hz, it was measured with a ^{3}He(d,p)^{4}He polarimeter. The beam intensity on the target was 60 nA, the polarization $\hat{t}_{20}$ = -0.44. The eight proton spectra were measured via a CAMAC-interface in a PDP 11/40 computer. A biased amplifier suppressed the scattered deuterons and enlarged the high energy part of the proton spectrum. The FORCAM programme EDDA written by Urbainsky[1] was used. It allowed a direct evaluation of the analysing powers for the different transitions.

The data were compared with DWBA calculations using the DW code which includes the deuteron D-state[2]. Many different deuteron parameters were tested, the best overall agreement for all transitions was obtained with the set of Lohr and Haeberli[3]. The parameters of Becchetti and Greenlees were used for the protons[4].

For the T_{20} analysing power, the transition to 1/2$^-$-states have larger negative values at small angles and a more pronounced structure than the transition to 3/2$^-$-states (fig. 1). For larger ℓ-values the effect becomes smaller. That is in agreement with the results of other nuclei[5].

The D-state admixture in the deuteron strongly influences[6] T_{20} at small angles and the amplitude of T_{21}. These effects should decrease for smaller Q-values and should be seen better for j = ℓ+1/2. Fig. 2 compares a j = 3/2$^-$-transition in the ^{52}Cr(d,p)^{53}Cr reaction[7], Q = 5.7 MeV with a j = 3/2$^-$-transition in ^{64}Ni(d,p)^{65}Ni, Q = 3.2 MeV. The smaller effects in ^{64}Ni are clearly seen.

The T_{22} component is j-dependent[8]. The angular distribution has a definite minimum for a j = ℓ+1/2-transition at an angle slightly above the stripping maximum. Transitions with j = ℓ-1/2 have a positive value of T_{22} at this angle. Fig. 3 shows T_{22} angular distributions for j = 3/2$^-$, 5/2$^+$ and 9/2$^+$. The minimum clearly moves to larger angles and becomes deeper as described by the model of Santos[9].

The j-dependent effects in the tensor components confirm the

0094-243X/81/690644-03$1.50 Copyright 1981 American Institute of Physics

tentative spin assignment of the state at E* = 1.013 MeV, I = 9/2+, as shown in fig. 4.

*) Work supported by the Deutsche Forschungsgemeinschaft

1. P. Urbainsky, Jahresbericht 77/78, Phys.Inst.Univ.Erlangen-Nürnberg, S. 160
2. J.P. Harvey and F.D. Santos (1970) Internal Report, University of Surrey, Guildford, England
3. J.M. Lohr and W. Haeberli, Nucl.Phys. A232, 381 (1974)
4. F.D. Becchetti and G.W. Greenlees, Phys.Rev.182, 1190 (1969)
5. A.K. Basak, J.A.R. Griffith, M. Irshad, O. Karban, E.J. Ludwig, J.M. Nelson, S. Roman, and G. Tungate, Nucl.Phys.A278,217 (1977)
6. R.C. Johnson and F.D. Santos, Nuclei and Part. 2, 285 (1971)
7. N. Rohrig and W. Haeberli, Nucl. Phys. A206, 225 (1973)
8. E.J. Stephenson and W. Haeberli in Proc. of the Fourth Intern. Symp. on Polarization Phenomena in Nuclear Reactions, 1975
9. F.D. Santos, Phys. Rev. C13, 1145 (1976)

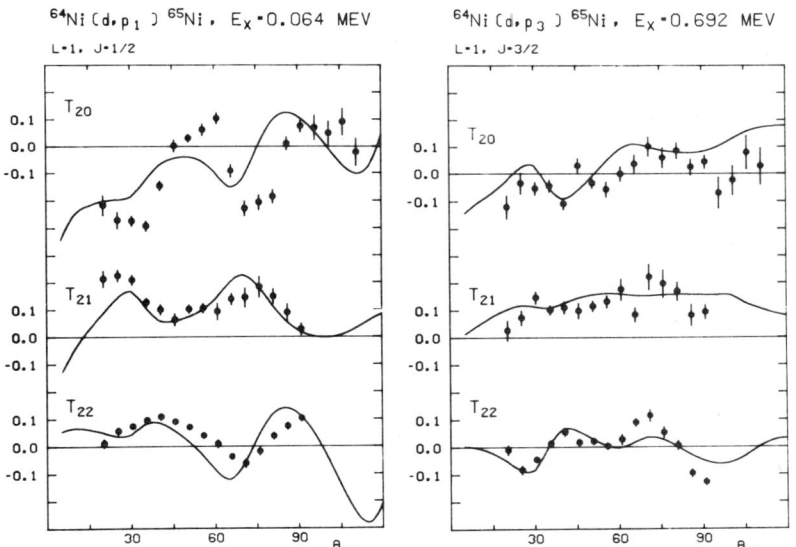

Fig.1. Tensor analysing powers of the ^{64}Ni(d,p$_1$) and ^{64}Ni(d,p$_3$) reaction compared with calculations using the deuteron parameters of Lohr and Haeberli^3.

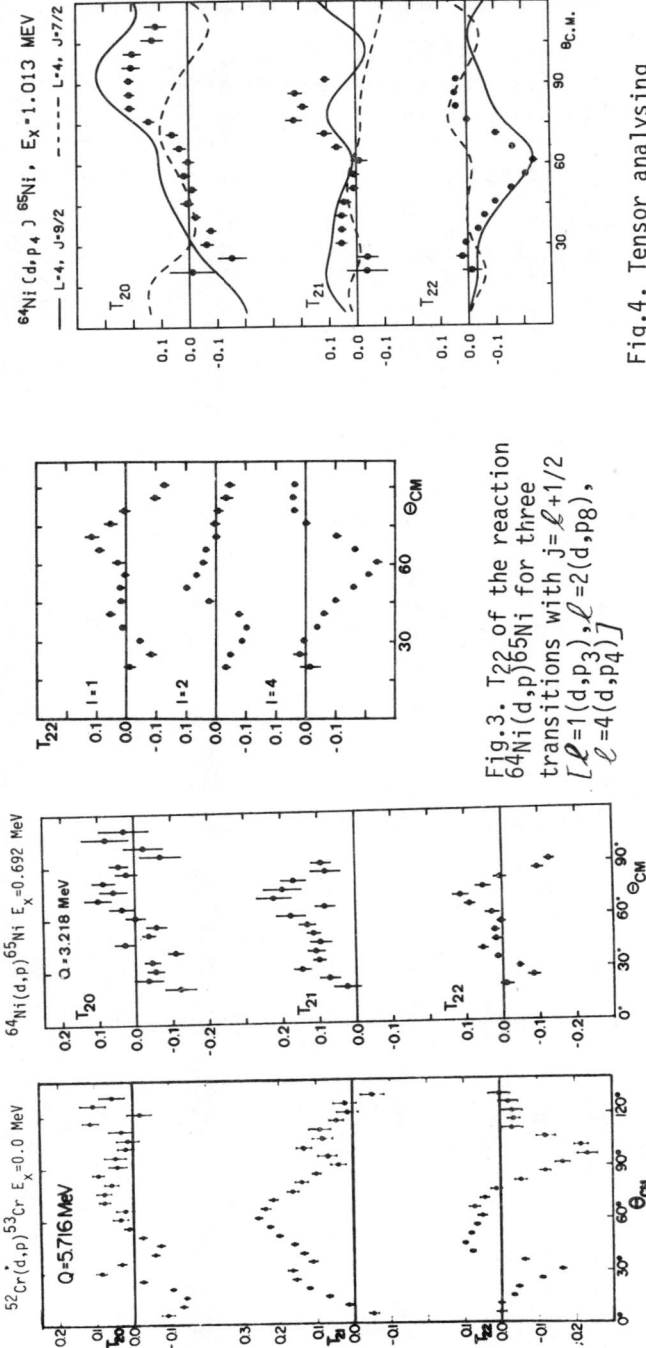

Fig.4. Tensor analysing powers of the reaction ^{64}Ni(d,p$_4$)^{65}Ni, confirming the spin asignment j=9/2$^+$.

Fig.3. T_{22} of the reaction ^{64}Ni(d,p)^{65}Ni for three transitions with j=ℓ+1/2 [ℓ=1(d,p$_3$), ℓ=2(d,p$_8$), ℓ=4(d,p$_4$)].

Fig.2. Comparison of tensor analysing powers for two ℓ=1, j=ℓ+1/2 transitions in ^{52}Cr(d,p) and ^{64}Ni(d,p) with different Q-values.

MEASUREMENT AND DWBA ANALYSIS OF THE ^{78}Kr$(\vec{d},p)^{79}$Kr REACTION[†]

B. L. Burks, R. R. Cadmus, Jr.[††], T. B. Clegg, E. J. Ludwig
Department of Physics and Astronomy, University of North Carolina
Chapel Hill, NC 27514 and
Triangle Universities Nuclear Laboratory, Durham, NC

ABSTRACT

The structure of ^{79}Kr has been investigated via the ^{78}Kr$(d,p)^{79}$Kr reaction using an isotopically enriched gas target and an 11-MeV vector-polarized deuteron beam. Differential cross sections, σ, and vector analyzing powers, A_y, have been measured from 25° to 95° for 10 proton groups below 2.5 MeV. Comparisons of these distributions to DWBA calculations were made to extract spectroscopic factors and values of spin (J) and parity (π) for these states.

The experimental arrangement was the same as that described in a recent study of ^{78}Kr$(\vec{d},t)^{77}$Kr.[1] The target was 99% enriched in ^{78}Kr and experimental resolution was about 55 keV FWHM. Rutherford scattering at 7.0 MeV provided a reference for absolute measurement of the cross sections to within a 5% uncertainty. An optical model analysis of deuteron elastic scattering at 11 MeV was fitted to σ and A_y data from 25° to 125° as shown in Fig. 1. The resulting optical-model potentials are presented in Table I along with proton potentials deduced from proton scattering on natural krypton at 12 MeV.[2]

Table 1. Optical-model parameters used in DWBA calculations. Potential depths given in MeV; geometrical parameters given in fm.

Particle	V	r_o	a_o	W_D	r_W	a_W	V_{SO}	r_{SO}	a_{SO}	r_c
p	50.48	1.26	0.65	13.77	1.20	0.53	6.8	1.26	0.72	1.27
d	89.10	1.21	0.82	23.30	1.35	0.66	6.9	1.00	0.55	1.30

Using the optical-model potentials of Table I, DWBA calculations were performed and are shown in Fig. 2 along with the experimental data. The value of ℓ for each transfer was determined by fits to σ. For each possible j transfer (= $\ell \pm 1/2$), A_y angular distributions were calculated and compared to the A_y data to

---

[†]Support in part by the U.S.D.O.E.
[††]Present address: Physics Department, Grinnell College, Grinnell, Iowa 50112.

0094-243X/81/690647-03$1.50 Copyright 1981 American Institute of Physics

determine J as in Fig. 2.

A previous ^{78}Kr(d,p)^{79}Kr study utilized cross section angular distributions only and relied on shell-model arguments to make J determinations. In Table 11 values for the spin and parity of these states from the present work are compared to the previous (d,p) assignments[3] as well as to ^{79}Rb decay measurements[4]. Of the ten states studied, eight have been assigned firm values of J^π. There may be 2 or more states not resolvable by this experiment near the 0.147 MeV and 0.384 MeV levels. The strongest state in each of these 2 groups has been assigned the values of $J^\pi = 7/2^+$, and $1/2^-$, respectively.

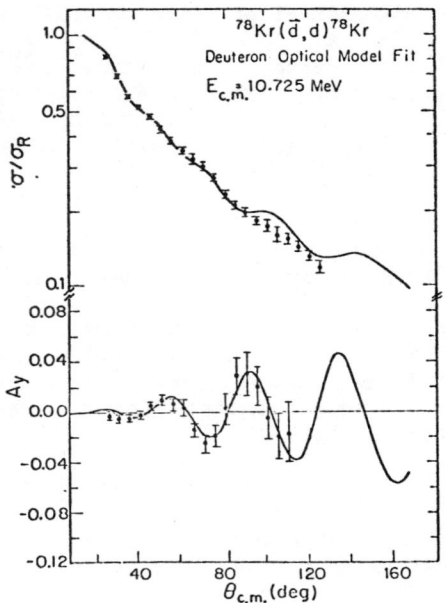

Fig. 1 Cross section and vector analyzing power distributions for elastic deuteron scattering from ^{78}Kr and the resulting optical model fits.

Table 11. Summary of spectroscopic information for states in ^{79}Kr.

| E_x(MeV) | J^π | | | | S_j |
	this work	Ref. 3	Ref. 4	Ref. 5	this work
0.0	$1/2^-$	$1/2^-$	$1/2^-$	$(1/2^-)$	0.26
0.147*	$7/2^+$	$9/2^+$	$(5/2^-,9/2^+)$	$(5/2^+)(9/2^+)$	0.98
0.384*	$1/2^-$	$(1/2^-)$	$3/2^-$	$(1/2^-,3/2^-)$	0.33
0.533	$1/2^+$	$1/2^+$	$1/2^+$	$(1/2^+)$	0.15
0.639	$5/2^+$	$5/2^+$	$5/2^+$	$(3/2^+,5/2^+)$	0.09
0.688	$3/2^+$	$(5/2^+)$	$3/2^+$	$(1/2^- - 5/2^-)$	0.11
0.752	$3/2^+$		$5/2^+$		0.04
0.810	$1/2^-$	$3/2^-$	$9/2^-$	$(1/2^-,3/2^-)$	0.12
1.912	$1/2^+$	$1/2^+$	$13/2^-$	$(1/2^+)$	0.07
2.060	$5/2^+$	$(5/2^+)$	$15/2^-$	$(3/2^+,5/2^+)$	0.09

*Two or more states may be unresolved at this energy.

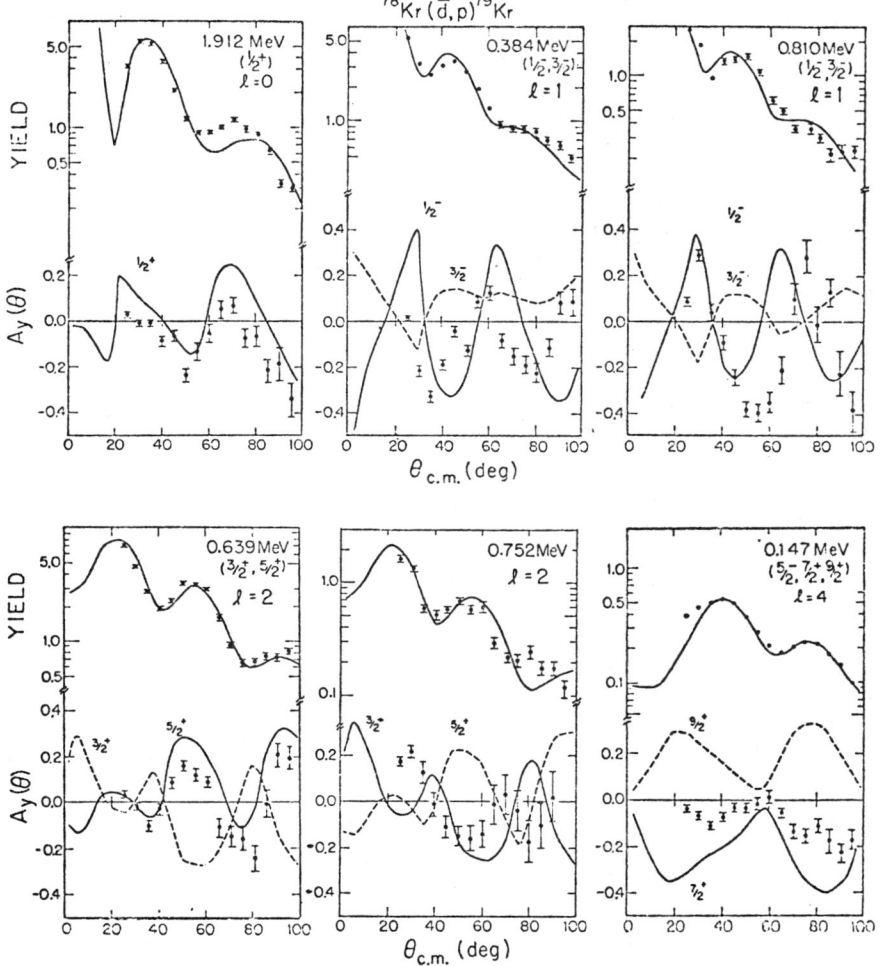

Fig. 2. DWBA calculations compared to cross section and analyzing power data.

REFERENCES

1. R. R. Cadmus, Jr., T. B. Clegg and E. J. Ludwig, Nucl. Phys. A319 (1979) 165.
2. C. P. Browne, D. K. Ohlsen, J. Chao, P. J. Riley, Phys. Rev. C, 9 (1974) 1831.
3. J. Chao, D. K. Ohlsen, C. Newsom and P. J. Riley, Phys. Rev. C, 11 (1975) 1237.
4. J. Liptak and J. Kristiak, Nucl. Phys. A311 (1978) 421.
5. Nuclear Data Sheets 15, 3 (1975) 257.

AN INVESTIGATION OF CONFIGURATION MIXING IN ^{210}Bi USING POLARIZED DEUTERONS

C.A. Gossett, L.D. Knutson and P.A. Quin
University of Wisconsin, Madison, WI. 53706[†]

The reaction ^{209}Bi$(\vec{d},p)^{210}$Bi has been used to study the structure of the ground state multiplet in ^{210}Bi. This multiplet consists of ten levels with spins from 0^- to 9^- which are formed by the coupling of an $h_{9/2}$ proton with a $g_{9/2}$ neutron outside the doubly closed ^{208}Pb core. Shell model calculations[1,2] predict extremely pure two particle configurations for these levels.

Although previous measurements are, for the most part, consistent with the shell model predictions, some evidence for configuration mixing, not predicted by the calculations, has been observed[3,4]. In particular, significant fragmentation of the 8^-, 580 keV level has been reported by Cline et al.[3] and by Kolata and Daehnick[4].

In the present experiment, cross section and vector analyzing powers have been measured for the ^{209}Bi(d,p) transitions leading to members of the ground state multiplet at a deuteron bombarding energy of 12 MeV. In the absence of configuration mixing (and neglecting possible Q-value dependence) one would expect the vector analyzing powers and the shapes of the cross sections to be identical for all ten final states. Deviations from this rule can result from small admixtures of configurations in which the neutron occupies an orbital other than $g_{9/2}$ (e.g. $i_{11/2}$). It is expected that the iT_{11} measurements should be more sensitive than the cross section to the presence of these admixtures, since for low bombarding energies the cross section angular distributions for different ℓ-values are all rather featureless. In previous ^{209}Bi(d,p) experiments, only the cross sections have been measured.

Vector polarized deuterons were produced with the Wisconsin crossed-beam polarized ion source. Beam currents on target were typically 200-300 nA and the typical beam polarization was $|it_{11}|$ = 0.55. To measure the vector analyzing power, the sign of it_{11} was reversed every 0.25 seconds by switching RF transitions at the source, and the corresponding spectra were stored in separate areas of the computer memory. Since the ten states of interest lie below 580 keV of excitation, good resolution is required. The scattered particles were detected with surface barrier detector telescopes which were cooled to -15°C in order to reduce noise. Improved resolution was obtained by using magnets and collimating slits to prevent recoil electrons from reaching the detectors. To reduce pileup, we used ΔE detectors thick enough to stop elastically scattered deuterons. The E detectors were selected to be just thick enough to stop the protons. Resolutions of 22-26 keV FWHM were obtained. This allowed us to resolve all of the levels except for the 5^- and 7^- which are separated by only 6 keV.

The analyzing power measurements for the transitions to the

[†] Work supported in part by the U.S. Department of Energy.

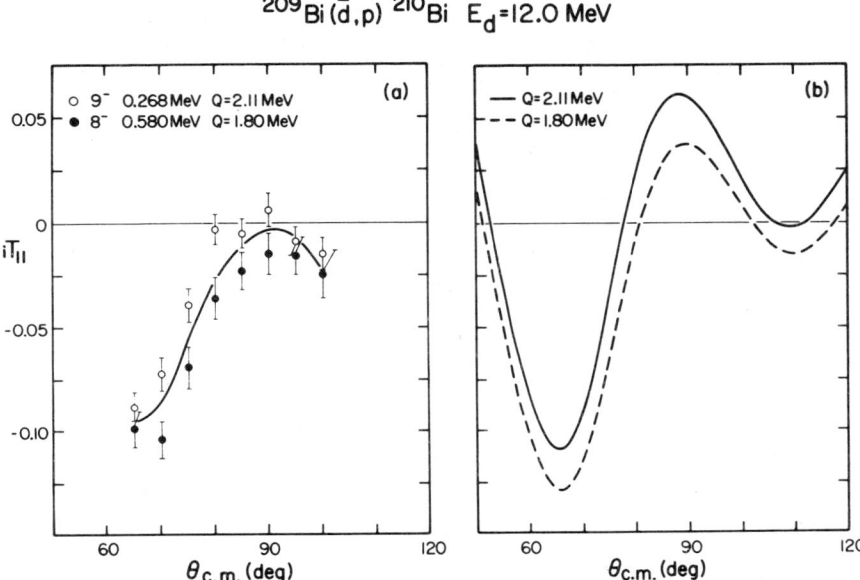

Figure 1 (a) Vector analyzing power measurements for the ^{209}Bi(d,p) transitions to the 8⁻ and 9⁻ states in ^{210}Bi induced by 12 MeV deuterons. The solid curve is a smooth fit to the "average" analyzing power of the ground state multiplet. (b) DWBA calculations which illustrate the Q-value dependence of the vector analyzing power for ^{209}Bi(d,p)^{210}Bi.

9⁻, 268 keV level and the 8⁻, 580 keV level are shown on the left-hand side of Fig. 1. The solid line is a smooth curve drawn through the measured analyzing power points for the entire ground state multiplet and thus represents the average analyzing power. Note that the measurements for the 8⁻ transition are consistently more negative than the average while those for the 9⁻ state are consistently more positive. The shell model calculations of Kim and Rasmussen[2] predict that the 9⁻ level has a virtually pure (99.9%) $\pi h_{9/2}\, \nu g_{9/2}$ configuration while the 8⁻ level contains small (~ 1%) admixtures of $\pi h_{9/2}\, \nu i_{11/2}$ and $\pi f_{7/2}\, \nu g_{9/2}$. In the angular range studied the vector analyzing power is more positive[5] for an $i_{11/2}$ transition than for a $g_{9/2}$ transition, and thus one might have expected the iT_{11} measurements to be slightly more positive for the 8⁻ state, in disagreement with the measurements. In addition, the magnitude of the change in iT_{11} resulting from a 1% admixture is much smaller than the observed discrepancy between the 8⁻ and 9⁻ measurements.

Distorted-wave calculations suggest that the differences between the two data sets may simply be a result of the difference in Q-value. This is illustrated on the right-hand side of Figure 1. The solid and dashed curves show the DWBA predictions for pure $g_{9/2}$ transitions with Q-values of 2.11 and 1.80 MeV, corresponding to the 9^- and 8^- levels, respectively. Although the calculations do not reproduce the measurements accurately in the angular region studied, it appears that the changes in iT_{11} which result from changing the Q-value in the DWBA calculation are comparable to the differences between the measurements for the 8^- and 9^- levels. The measurements for the other members of the ground state multiplet also appear to be consistent with a gradual trend toward more negative iT_{11} values with decreasing Q-value. Thus, no evidence for configuration mixing is seen in the vector analyzing power measurements for the ground state multiplet.

The cross section measurements confirm the observation[3,4] that the spectroscopic factor for the 8^- state is approximately 15% smaller than expected for a pure $\pi h_{9/2} \nu g_{9/2}$ configuration, thus indicating the presence of some configuration mixing for this level. It has been suggested[3] that the admixture is primarily $\pi f_{7/2} \nu g_{9/2}$. The analyzing power measurements appear to be consistent with this conclusion, since the presence of such an admixture would have no effect on iT_{11}.

REFERENCES

[1] G.H. Herling and T.T.S. Kuo, Nucl. Phys. A181, 113 (1972).
[2] Y.E. Kim and J.O. Rasmussen, Nucl. Phys. 47, 184 (1963).
[3] C.K. Cline, W.P. Alford, H.E. Gove and R. Tickle, Nucl. Phys. A186, 273 (1972).
[4] J.J. Kolata and W.W. Daehnick, Phys. Rev. C5, 568 (1972).
[5] S.E. Vigdor, R.D. Rathmell, H.S. Liers and W. Haeberli, Nucl. Phys. A210, 70 (1973).

ANALYZING POWER OF THE $^{12}C(d,p)$ and $^{12}C(d,n)$ REACTION *)

W. Drenckhahn, A. Feigel, E. Finckh, G. Gademann, K. Rüskamp, M. Wangler, L. Zemło **)
Physikalisches Institut der Universität Erlangen-Nürnberg, W.-Germany

The tensor analyzing power T_{20} and T_{22} of the $^{12}C(d,p)$ reaction was measured[1] at E_d = 12.3 MeV. The data could only be reproduced by a calculation which includes the deuteron-D-state and uses a very deep spin orbit potential, V_{SO} = 19 MeV. To see whether this is due to a resonance in the $^{12}C+d$ channel and whether the T_{21} analyzing power is fitted with the same parameters we investigated the $^{12}C(d,p)^{13}C$ and the $^{12}C(d,n)^{13}N$ reaction at several energies between 9 and 11 MeV.

Both experiments were carried out with the Erlanger Lamb shift source and the EN-tandem. The experimental arrangement for the $^{12}C(d,p)$ reaction was the same as for $^{64}Ni(d,p)$ [2]. The $^{12}C(d,n)$ reaction was measured in a 2π-geomtry. Eight liquid scintillators (NE 213) placed symmetrically left and right of the beam axis were used. To get all analyzing powers the orientation of the spin alignment axis had to be varied with Wien-filter and solenoid, according to the coordinate system of the target. For the determination of the tensor analyzing powers, three independent runs were necessary.

The data of both reactions at E_d = 10 MeV are shown in fig. 1

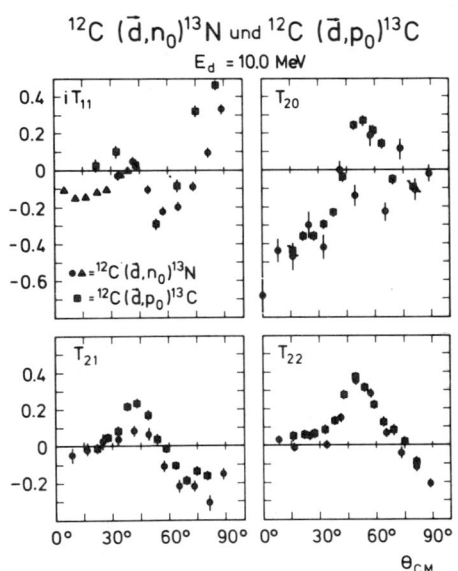

Fig. 1. Analyzing power data of the $^{12}C(d,p_0)^{13}C$ and $^{12}C(d,n_0)^{13}N$ reaction at E_d = 10 MeV.

All analyzing powers are nearly equal. For iT_{11} only the width of the maximum differs, the steep rise of T_{20} at forward angles is identical, the agreement in T_{22} is very good and in T_{21} only the amplitude differs.

Both reactions were analyzed with the DWBA-Code which uses the D-state of the deuteron in a local energy approach[3]. For the (d,p) reaction, the analyzing powers are reproduced quite well (fig. 2). The steep rise of T_{20} at forward angles is only given when the D-state of the deuteron is included in the calculation. The parameter set for the deuteron channel with the deep spin orbit potential is necessary for all energies, especially to describe the amplitude T_{22}.

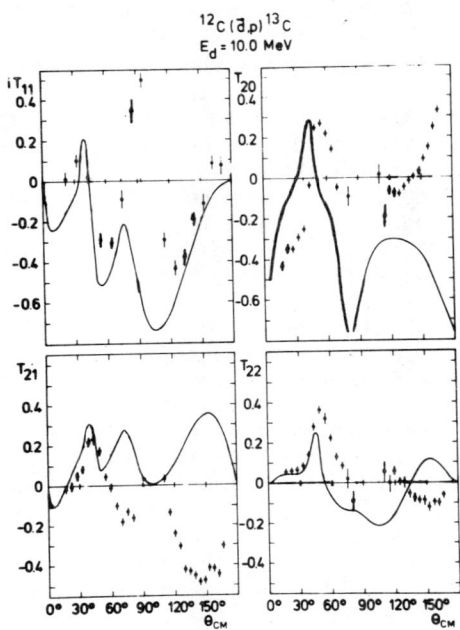

Fig. 2. Analyzing powers of the $^{12}C(d,p)^{13}C$ reaction. Calculation with a DWBA code using the deuteron-D-state.

Fig. 3. Analyzing powers of the $^{12}C(d,n_0)^{13}N$ reaction. Calculation with the same programme and parameter set as in fig. 2.

A decription of the analyzing powers for the (d,n) reaction fails (fig. 3). With no parameter sets the steep rise in T_{20} can be reproduced, but the maximum of T_{22} needs the deep spin orbit potential, too.

To improve the description of the analyzing powers for the $^{12}C(d,p)$ and (d,n) reaction and to avoid such an unusual spin orbit potential, we calculated with a coupled channel programme. We didn't get any improvement for the description of the (d,n) data. The quality for the (d,p) analyzing powers is nearly the same as with the D-state programme of Harvey and Santos, however, the larger spin orbit potential can be avoided. Only the steep rise of T_{20} cannot be reproduced, for this we probably need a coupled channel programme which includes the deuteron D-state or a tensor potential.

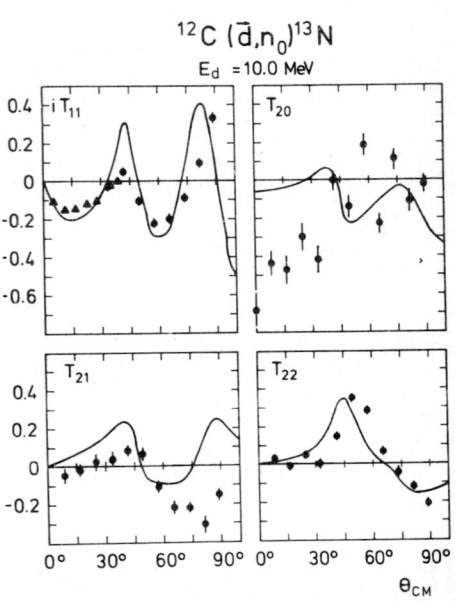

REFERENCES

1. A.K. Basak, J.A.R. Griffith, O. Karban, S. Romana, G. Tungate, and F.D. Santos, Nucl. Phys. A 286, 420 (1977)
2. Fifth International Symp. on Polar. Phen. in Nucl. Phys., Santa Fe, USA, Aug. 80
3. J.D. Harvey and F.D. Santos (1970) Internal Report, University of Surrey, Guildford, England

*) Work supported by the Deutsche Forschungsgemeinschaft
**) Visitor from Institute of Nuclear Research, Warsaw, Poland

VECTOR ANALYSING POWERS FOR THE ^{12}C($\vec{d}$,n)^{13}N, ^{9}Be($\vec{d}$,n)^{10}B AND ^{28}Si($\vec{d}$,n)^{29}P REACTIONS

F.D. Brooks
Dept. of Physics, Univ. of Cape Town, Rondebosch 7700, South Africa

P.M. Lister, J.M. Nelson and K.S. Dhuga
Dept. of Physics, Univ. of Birmingham, Birmingham, B15 2TT, England

ABSTRACT

Vector analysing powers have been measured for the ^{12}C($\vec{d}$,n)^{13}N, ^{9}Be($\vec{d}$,n)^{10}B and ^{28}Si($\vec{d}$,n)^{29}P reactions at E_d = 12.3 MeV. The results obtained for transitions to the ground state of ^{13}N and the 1.38 and 1.95 MeV states of ^{29}P agree well with DWBA predictions at forward angles and demonstrate the j-dependence of the analysing power.

INTRODUCTION

As a spectroscopic tool in the study of proton transfer reactions the ($\vec{d}$,n) reaction is a potentially strong competitor to the (^{3}He,d) reaction because of the more intense polarized beams which are available for deuterons than for ^{3}He ions. However, this advantage may be offset by limitations imposed by the efficiency and energy resolution of available neutron spectrometers, especially when neutron energies exceeding 10 MeV are involved. We report here on some ($\vec{d}$,n) studies made using a new type of deuteron recoil spectrometer in conjunction with the 12.3 MeV vector polarized deuteron beam of the Birmingham University Radial Ridge Cyclotron.

EXPERIMENTAL

The neutron spectrometer consisted of a deuterated anthracene scintillation crystal (10 mm diam. × 21 mm length) placed 10-20 cm away from the target under study. Recoil deuterons from D(n,n)D elastic scattering in the crystal were internally detected and were separated from breakup protons (from D(n,2n)H reactions in the crystal) and gamma background by pulse shape discrimination. The spectrum of recoil deuterons produced by monoenergetic neutrons (E_n > 5 MeV) contains a prominent forward recoil peak with a sharp cutoff at its high energy limit, thus giving the deuterated scintillator a lineshape which is particularly suitable[1] for neutron spectrometry. By folding in the correct lineshape the neutron spectum may be extracted from the observed recoil deuteron spectrum. Fig. 1 shows an example of a neutron spectrum obtained from measurements of the ^{9}Be(d,n)^{10}B reaction. Vector analysing powers were determined from neutron spectra obtained simultaneously for the "up" and "down" spin states of the incident polarized deuteron beam.

0094-243X/81/690656-03$1.50 Copyright 1981 American Institute of Physics

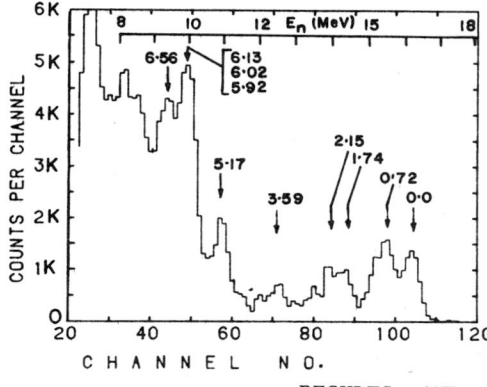

Fig.1. Neutron spectrum obtained from the ^{9}Be(d,n)^{10}B reaction at $\theta_{lab}=20^0$. Arrows show the expected positions of peaks corresponding to transitions to different levels in ^{10}B.

RESULTS AND DISCUSSION

Fig.2 shows analysing powers obtained for transitions to different final states in the ^{12}C($\vec{d}$,n)^{13}N reaction, together with data from refs.[2-4]. The data for the ground states are in good overall agreement at forward angles but diverge at $\theta_{cm} > 45^0$. The DWBA predictions shown by the solid line were obtained using the computer code NELMAC[5] which includes corrections for non-locality and finite range. Optical-model parameters were obtained from refs.[6,7]. The agreement between theory and the present data is very good for the ground state transition in this reaction.

The central and lower sections of fig.2 show results for the reactions leading to the unbound 2.37 MeV state and the 3.51-3.56MeV doublet in ^{13}N. For the 2.37 MeV state the DWBA calculation was performed in the weakly-bound-nucleon approximation where the unbound proton is treated as being bound by a small (10 keV) amount. The fit to the data is reasonable at forward angles, both for our data and for those of Tenhaken and Quin[4].

Figs.3(a) and (b) show analysing powers for transitions to different final states in the reactions ^{9}Be($\vec{d}$,n)^{10}B and ^{28}Si($\vec{d}$,n)^{29}P respectively. Optical model para-

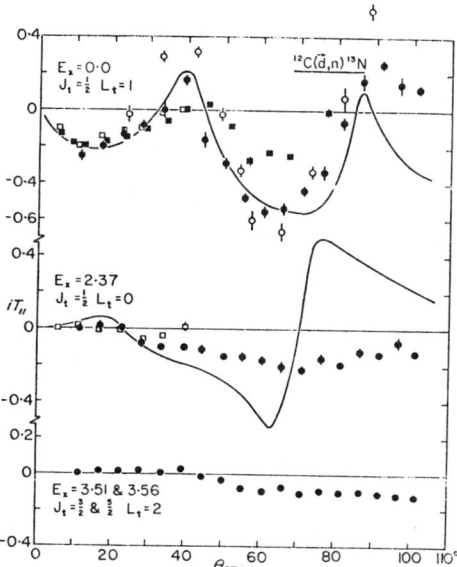

Fig.2. Analysing powers for the ^{12}C($\vec{d}$,n)^{13}N reaction leading to the ground, 2.37 MeV and (3.51 & 3.56) MeV states of ^{13}N.
Data shown are:
● present work (at 12.3 MeV);
○ Krämmer et al[2] (at 10.0 MeV);
■ Hilscher et al[3] (at 8.5 MeV);
□ Tenhaken and Quin[4] (at 9.7 MeV).
Curves show DWBA predictions.

eters for the DWBA calculations were taken from refs.[8,9] and from refs.[6,10]. Since no experimental spin-orbit parameters were available for ^{10}B+n, the spin-orbit parameters of Bechetti and Greenlees[6] were taken in conjunction with parameters from ref.[9]. The DWBA describes the overall features of the data quite well and achieves satisfactory fits to the data for the 1.38 MeV and 1.95 MeV states of ^{29}P (fig.3(b)) at forward angles. The analysing powers for these transitions show a j-dependence at forward angles similar to that found in many neutron transfer reactions. Vector analysing powers have also been determined for other final states in the ^{9}Be($\vec{d}$,n)^{10}B and ^{28}Si($\vec{d}$,n)^{29}P reactions and DWBA analyses of these data and the data shown in fig. 3 are in progress.

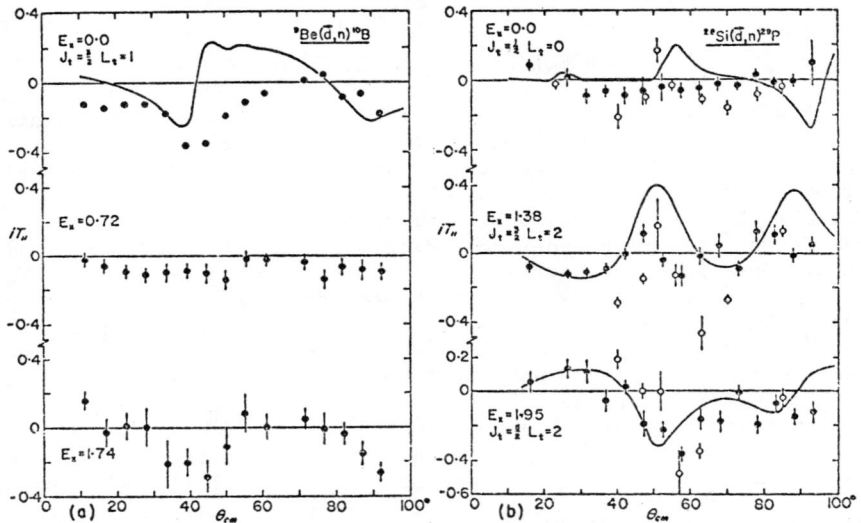

Fig. 3. Analysing powers for transitions to different final states in : (a) the ^{9}Be($\vec{d}$,n)^{10}B reaction; and (b) the ^{28}Si($\vec{d}$,n)^{29}P reaction. Data shown are: ● present work; ○ Krämmer et al[2] (taken at 10.0 MeV). Curves show DWBA predictions.

REFERENCES

1. F.D. Brooks, Nucl. Instr. and Meth., 162, 477 (1979).
2. P. Krämmer et al, Proc. 4th Int. Symp. on Polarization Phenomena (Birkhauser Verlag, Basel, 1976) p.651.
3. D. Hilscher, J.C. Davis and P.A. Quin, Nucl. Phys., A174, 417(1971).
4. R.K. Tenhaken and P.A. Quin, Nucl.Phys., A271, 173 (1976).
5. J.M. Nelson and B.E.F. Macefield, Birmingham Univ. Rept.No.74-9 (unpublished).
6. F.D. Bechetti and G.W. Greenlees, Phys. Rev., 182, 1190 (1969).
7. C.E. Busch et al., Nucl. Phys., A223, 183 (1974).
8. K.I. Zaika et al., Sov. J. Phys., 13, 553 (1971).
9. J.A. Cookson and J.G. Lock, Nucl. Phys., A146, 417 (1970).
10. A.K. Basak et al., Nucl. Phys., A295, 111 (1978).

SPECTROSCOPY OF ^{62}Ni FROM THE ^{61}Ni(d,p)^{62}Ni REACTION

O. Karban, A.K. Basak, F. Entezami and S. Roman
Department of Physics, University of Birmingham, England

ABSTRACT

States in ^{62}Ni were populated using the ^{61}Ni(d,p) reaction initiated by a polarized deuteron beam. Measurements of vector and tensor analysing powers allowed to deduce dominant j-transfer values for 32 levels in ^{62}Ni. Below 3.5 MeV excitation the reaction is dominated by 3/2$^-$, 1/2$^-$ and 5/2$^-$ transfers whereas higher levels are populated mainly by 5/2$^+$ and 9/2$^+$ transfers. No strong 7/2$^-$ transition was observed.

The level scheme of the ^{62}Ni nucleus has been studied in a number of experiments which are summarized in ref.[1]. In the neutron stripping reaction many levels are populated but the values of the transferred orbital angular momentum are well established only for states below 3.5 MeV. The spin of most of the higher levels are also not known. The present study with the 12.3 MeV polarized deuteron beam was undertaken to determine the dominant total momentum transfer in the ^{61}Ni(d,p)^{62}Ni reaction which can provide information on the wave functions of the ^{61}Ni ground state and the final states in ^{62}Ni.

Vector and tensor analysing powers iT_{11}, T_{20} and T_{22} together with differential cross sections were extracted for 32 proton groups which were identified with ^{62}Ni states using high-resolution spectra of ref.[2,3] and also results of the (p,t) reaction of ref.[4]. Below 3.5 MeV excitation the (d,p) reaction is dominated by the j^π = 3/2$^-$, 1/2$^-$ and 5/2$^-$ transfers. These can be easily identified from the shape of the iT_{11}: the ground state, 2.05, 2.89, 3.16 and 3.85 MeV states are populated by the 3/2$^-$ neutron, while the 1.17 and 3.36 MeV states are characterised by the 1/2$^-$ transfer. The j^π = 5/2$^-$ pattern can be seen in the VAP for the 2.34, 3.26 MeV data and, less clearly for the 3.06 and 4.40 MeV states. The proton group at 3.50 MeV corresponds to an unresolved doublet of 3.464 (ℓ=3) and 3.518 MeV (ℓ=1) states[1] which are populated by 5/2$^-$ and 1/2$^-$ transfers respectively. All these j-values are confirmed by the tensor analysing power data.

The j^π = 5/2$^+$ pattern of the VAP observed in ref.[5,6] is clearly visible in the data for the 5.33, 5.63, 5.83, 6.13, 6.32 and 7.4 MeV proton groups. This transfer value also follows from the position of the minimum[7] in the T_{22}. The proton groups at 3.76, 4.50 and 5.54 MeV have the VAP similar to the 9/2$^+$ transitions in Ge isotopes[6]. Analogous behaviour of the VAP has been seen for the 4.19, 4.87 and 5.03 MeV groups. Again, the dominance of the 9/2$^+$ transfer is confirmed by the position of the minimum in the T_{22} angular distribution.

The two strongest states populated in the (d,p) reaction are at

8.13 and 8.46 MeV excitation, both observed in ref.[3]. The latter has the VAP similar to the 6.74 MeV one, while the former resembles that of the group of states around 7.8 MeV. It is suggested that these states and the 8.13 MeV state are predominantly populated by a 5/2+ transfer. The 6.74 MeV state was observed in several reactions and in ref.[3] it was associated with an ℓ=0 transfer. The present data support this transfer value[8] for both 6.74 and 8.46 MeV states, implying possible 1⁻ or 2⁻ assignments.

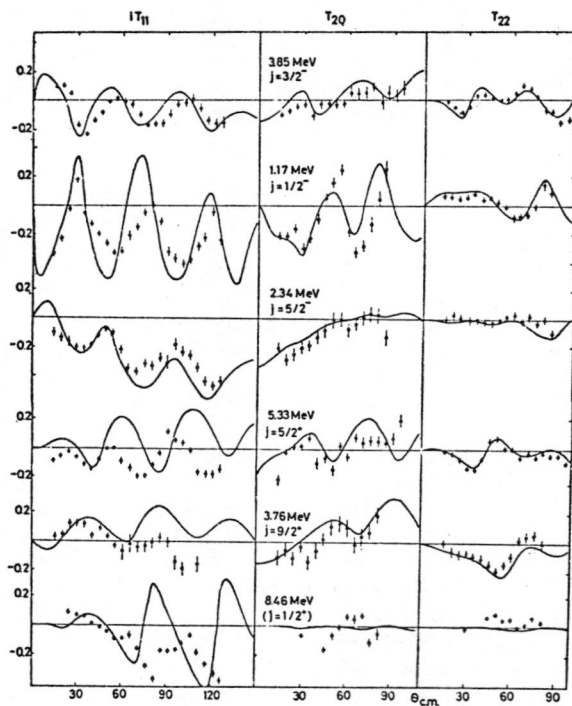

Fig.1. Typical analysing powers of the ^{61}Ni(d,p) reaction populating states in ^{62}Ni with dominant j-values given in the figure. Predictions of the DWBA theory including the deuteron D-state are shown as solid lines.

Examples of the measured analysing powers for each of the above-mentioned j-transfer values are shown in Fig.1 and compared with predictions of the DWBA theory (solid lines). The calculations were performed using the programme DWCODE[9] with the deuteron D-state parameter D_2=0.484. An inclusion of the latter was essential to reproduce the shape and magnitude of the T_{20} and T_{22}. Moreover, the best agreement between the experiment and theory was achieved for the T_{22}, therefore the position of the first minimum can be reliably used to identify the j=ℓ+1/2 transfers[7].

Koops and Glaudemans[10] performed extensive calculations for the Ni and Cu isotopes in an unrestricted ($2p_{3/2}$, $1f_{5/2}$, $2p_{1/2}$) shell-model space. As the present polarization experiment can distinguish between the 3/2 and 1/2 transfer, a more detailed comparison with the theoretical wave functions is possible. The expected[11] dominance of the 1/2⁻ transfer to the first 2⁺ state is confirmed by our data. Apart from the pure 5/2⁻ transfer to the first 4⁺ state at 2.34 MeV two dominant 5/2⁻ transitions are predicted at 2.75 (2⁺) and 3.10 MeV (1⁺). They could be associated with states observed at 3.06 and 3.26 MeV respectively.

The first negative parity state in ^{62}Ni was established from several experiments[1] at 3.76 MeV, in contradiction with earlier (d,p) results[3]. From our data a 9/2⁺ transfer to this state (3⁻) was deduced, suggesting an anti-parallel configuration of the 3/2⁻ and 9/2⁺ neutrons. Generally, the ^{61}Ni(d,p) reaction proceeds to a large extent by a neutron transfer to the empty 9/2⁺ and 5/2⁺ shells, which apparently should be included in shell-model calculations. There is no evidence from the present data for a 7/2⁻ transfer, therefore possible $f_{7/2}$ hole configurations in the ^{61}Ni ground state could be omitted.

REFERENCES

1. Nucl. Data Sheets 26 (1979) 5.
2. R.G. Tee and A. Aspinall, Nucl. Phys. 98 (1967) 417.
3. R.H. Fulmer and A.L. McCarthy, Phys.Rev. 131 (1963) 2137.
4. D.H. Kong-A-Sion and H. Nann, Phys. Rev. C11 (1975) 1681.
5. A.A. Debenham et al., Nucl. Phys. A167 (1971) 289.
6. W.A. Yoh et al., Nucl. Phys. A263 (1976) 419.
7. F.D. Santos, Proc. 4th Pol. Symp., Zurich, p.675
8. J.A. Ayman et al., Nucl. Phys. A207 (1973) 596.
9. J.D. Harvey and F.D. Santos, Univ. of Surrey Report, 1970.
10. J.E. Koops and P.W.M. Glaudemans, Z. Phys. A280 (1977) 181.
11. P.W.M. Glaudemans, private communication.

SPIN-TENSOR INTERACTION IN POLARIZATION TRANSFER REACTIONS

J. W. Hugg* and S. S. Hanna
Department of Physics, Stanford University, Stanford, CA 94305

Beta-emitting nuclei have been polarized in $(\vec{p},n)$, $(\vec{d},n)$, $(\vec{d},p)$, and $(\vec{p},\alpha)$ polarization transfer reactions in thick crystalline targets. All recoil nuclei are stopped in the thick target which is placed in a strong magnetic field to preserve polarization. The asymmetric beta decay of the implanted recoil nuclei is detected to measure the polarization transferred. We have observed polarization transfer in 17 reactions: ^{7}Li$(\vec{d},p)^8$Li, ^{11}B$(\vec{d},p)^{12}$B, ^{16}O$(\vec{d},n)^{17}$F, ^{19}F$(\vec{p},n)^{19}$Ne, ^{19}F$(\vec{d},p)^{20}$F, ^{23}Na$(\vec{p},n)^{23}$Mg, ^{25}Mg$(\vec{p},n)^{25}$Al, ^{28}Si$(\vec{p},\alpha)^{25}$Al, ^{27}Al$(\vec{p},n)^{27}$Si, ^{27}Al$(\vec{d},p)^{28}$Al, ^{28}Si$(\vec{d},n)^{29}$P, ^{29}Si$(\vec{p},n)^{29}$P, ^{31}P$(\vec{p},n)^{31}$S, ^{32}S$(\vec{d},n)^{33}$Cl, ^{35}Cl$(\vec{p},n)^{35}$Ar, ^{39}K$(\vec{p},n)^{39}$Ca, and ^{40}Ca$(\vec{d},n)^{41}$Sc.

A simple spectator model (Fig. 1) describes the deuteron stripping polarization transfer reactions. The addition of a single spin-polarized nucleon to the valence orbit of the unpolarized target nucleus determines that the polarization transfer to the product nucleus will be positive if the valence orbit is a stretched configuration and negative if it is jackknifed.

Similarly, a simple quasi-elastic model (Fig. 2) describes the charge-exchange polarization transfer reactions. The proton and neu-

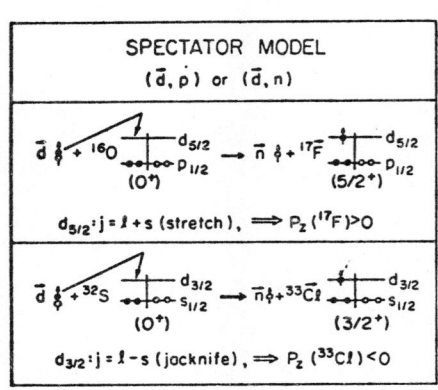

Fig 1: Schematic illustration of the spectator model of polarized deuteron stripping reactions.

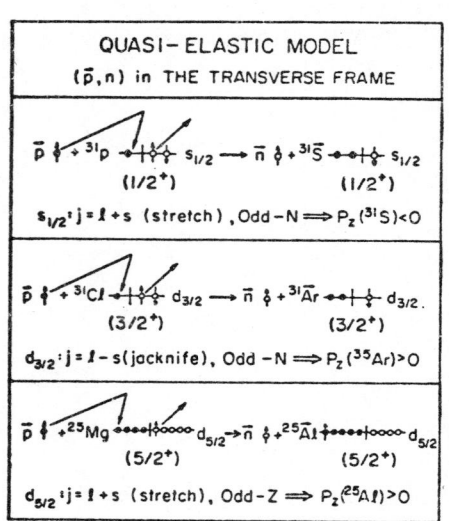

Fig 2: Schematic illustration of the quasi-elastic scattering model of polarized proton charge-exchange reactions.

*Present address: Shell Development Co., Houston, Texas 77001.

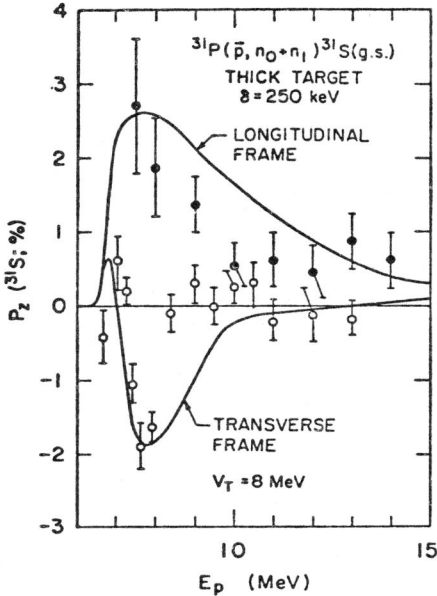

Fig 3: Ground state polarizations generated by DWUCK compared to experimental measurements for the $^{31}P(\vec{p},n)^{31}S$ reaction.

tron exchange charge but rarely spin, so the recoil neutron spin is preferentially parallel to the projectile proton spin. If the product nucleus is an odd proton mirror of the target nucleus, the polarization transfer will be positive if the valence orbit is stretched and negative if jackknifed. If the product nucleus is an odd neutron mirror of the target nucleus, the unpaired neutron will have spin preferentially antiparallel to the projectile proton and the polarization transfer will be negative if the valence orbit is stretched and positive if jackknifed.

A DWBA calculation of the $^{31}P(\vec{p},n)^{31}S$ reaction can describe the sign and general shape of the polarization transfer (Fig. 3). The nucleon-nucleon interaction includes a charge-exchange potential $V_T = 9$ MeV, a spin-flip potential $V_{\sigma T} = 6$ MeV, and a spin-tensor potential $V_{T} = 8 \pm 4$ MeV. This result is consistent with recent determinations of the tensor interaction in polarized nuclear reactions.

This work was supported in part by the National Science Foundation.

SPECTROSCOPY OF ^{145}Sm AT $E_x \leqslant 3.2$ MeV VIA THE $(\vec{d},p)$ REACTION AT $E_d = 19$ MeV [+)]

Sun, Tsu-Hsun[*)], H. Clement, R. Frick, G. Graw, F. Merz, F. Riess, P. Schiemenz, N. Seichert
Sektion Physik der Universität München, D-8046 Garching

The structure of N=83 nuclei is dominated by the coupling of the neutron single particle motion with collective modes in the excitation of the core. Theoretically this aspect is investigated in the work of K. Heyde and others[1)]. Experimentally for ^{145}Sm ℓ values had been assigned to all states up to E_x=3.2 MeV observed by Booth et al.[2)] with a magnetic spectrograph at E_d = 19 MeV. In the present work, we measured this reaction at the same energy with polarized deuterons using solid state detectors. The figs. 1 to 3 show for well separated transitions $\sigma(\theta)$ and $A_y(\theta)$ data for the transfer of $p_{1/2}$, $p_{3/2}$, $f_{5/2}$, $f_{7/2}$, $h_{9/2}$ and $i_{13/2}$ neutrons, respectively. In all cases DWBA calculations with standard potentials reproduce at forward angles those features that are essential to determine ℓ and $j = \ell \pm 1/2$. A better reproduction of the backward angle data (not shown in the figs.) was obtained only by using different potentials for different transitions.

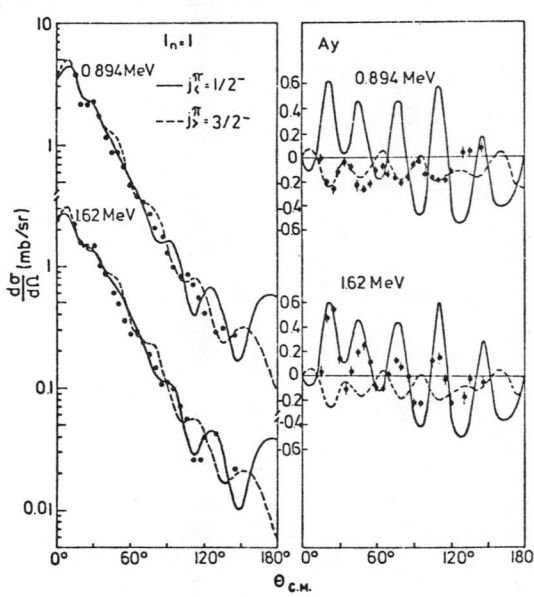

Fig. 1

Our energy resolution was about 60 keV. For overlapping levels we used the information of Booth et al.[2)] on the energy, ℓ-value and cross section at forward angles to obtain j assignment for overlapping pairs of transitions. To construct the reference angular distributions we used our experimental angular distributions of $\sigma(\theta)$ and $A_y(\theta)$ of isolated lines with the corresponging quantum numbers. In most cases this

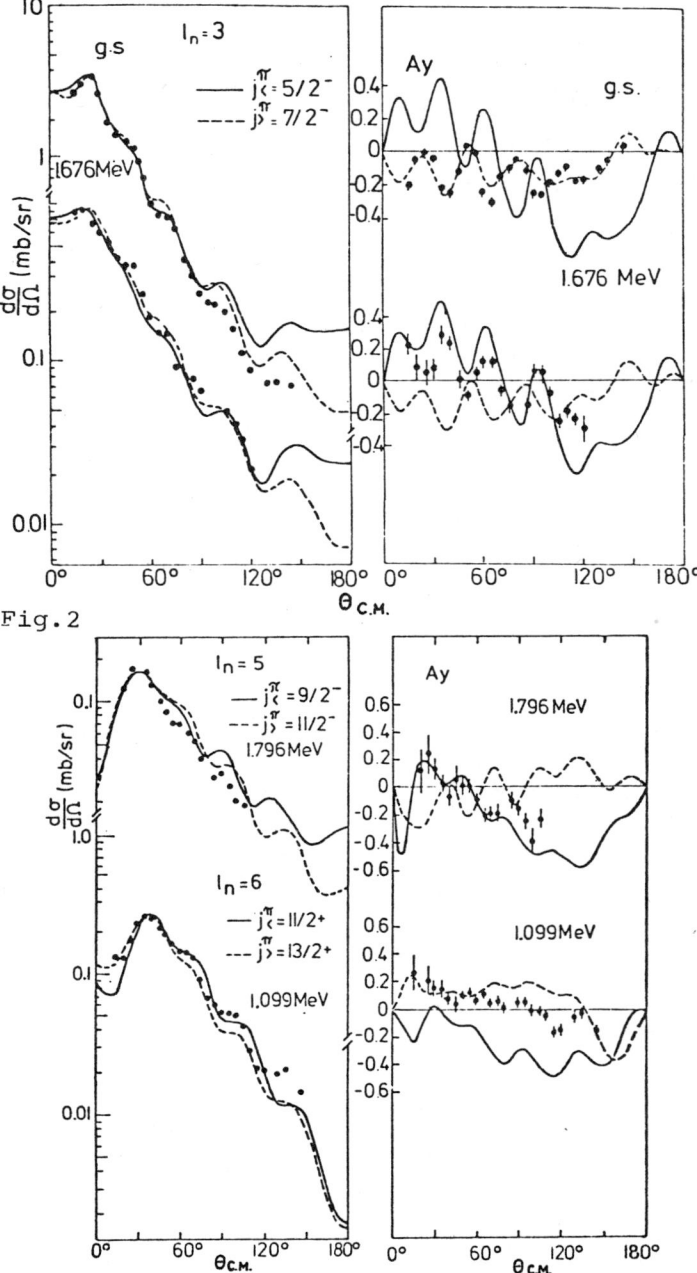

Fig. 2

Fig. 3

Table 1: Spectroscopic data of ^{145}Sm

Nr.	E_x/MeV	J^π	S
0	0	$7/2^-$	0.78
1	0.894	$3/2^-$	0.50
2	1.099	$13/2^+$	0.38
3	1.430	$9/2^-$	0.65
4	1.620	$1/2^-$	0.58
5	1.676	$5/2^-$	0.19
6	1.796	$9/2^-$	0.28
7	1.850	$(1/2^-)$	(0.06)
8	1.876	$7/2^-$	0.09
9	1.977	$3/2^-$	0.16
10	2.000	$5/2^-$	0.38
11	2.138	$(3/2^-)$	(0.06)
12	2.162	$(1/2^-)$	(0.06)
13	2.287	$(9/2^+)$	–
14	2.350	$5/2^-$	0.05
15	2.429	$5/2^-$	0.09
16	2.674^-	$3/2^-$	0.05
17	2.674^+	$13/2^+$	0.21
18	3.096	$7/2^-$	0.07
19	3.131	$1/2^-$	0.11

provided a unique j-assignment. The results are listed in table 1, the labeling and the energy of the levels are taken from ref.2. Up to $E_x = 2.5$ MeV the results agree with a study of isobaric analog resonances with polarized protons[3]). The spectroscopic factors indicate, that the observed levels do not exhaust the 1i and the $2f_{5/2}$ strength, comparing with structure calculations[1]), the over all agreement is good, for the weak transitions discrepancies are apparent.

+) work supported in part by the Bundesministerium für Forschung und Technologie
*) Max-Planck fellow from the Institute of Atomic Energy, Peking P.R. China
1) G. vanden Berghe and M. Waroquier, Nucl.Phys. A196 (1972) 303 and P. von Isaker et al., Phys.Rev. C19 (1979) 498 and K. Heyde, Gent, private communication (1979)
2) W. Booth and S. Wilson, Nucl.Phys. A247 (1978)126
3) H. Clement et al., Nucl. Phys. A285 (1977) 109

INELASTIC TRANSFER IN ($\vec{d}$,p) REACTIONS FROM ^{28}Si AND ^{54}Cr[+)]

N. Seichert, H. Clement, R. Frick, G. Graw, P. Schiemenz
Sun Tsu-Hsun[*)]
Sektion Physik der Universität München, D-8046 Garching

With vector polarized deuterons of 20 MeV we measured the (d,p) reaction for a larger number of transitions from the target nuclei ^{16}O, ^{18}O, ^{24}Mg, ^{28}Si, ^{36}Ar, ^{54}Cr, ^{144}Sm and ^{208}Pb. For nuclei near closed shells the main features of cross section and vector analyzing power have been reproduced by DWBA. This is true at least for the strong transitions. For reactions from deformed nuclei as ^{24}Mg, ^{28}Si and ^{54}Cr we observe even for strong transitions significant deviations of the data from the DWBA pattern. The figs. 1 and 2 show the excitation of $1/2^+$, $3/2^+$ and two $5/2^+$ states in ^{29}Si and of $1/2^+$, $5/2^-$ and $7/2^-$ states in ^{55}Cr. DWBA calculations are presented as dotted lines, CCBA-fits as solid lines.

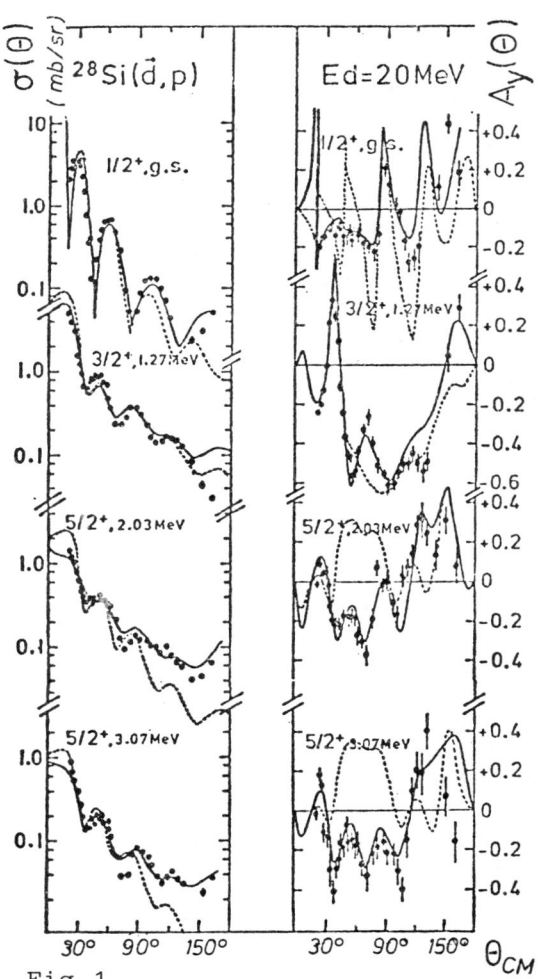

Fig. 1

It is dependent on the individual transition, whether the deviation from the DWBA pattern is most apparent from $\sigma(\Theta)$ or $A_y(\Theta)$ or from both. It should be noted, that for all transitions with an orbital angular momentum transfer larger than zero DWBA reproduces approximately the pattern at the most forward angles and thus allows for ℓ and $j=\ell\pm 1/2$ assignment, if those data are available.

For the CCBA calculations we expand the wave function of the residual states $|\lambda\rangle$ in a series of neutron single particle states coupled to excited states

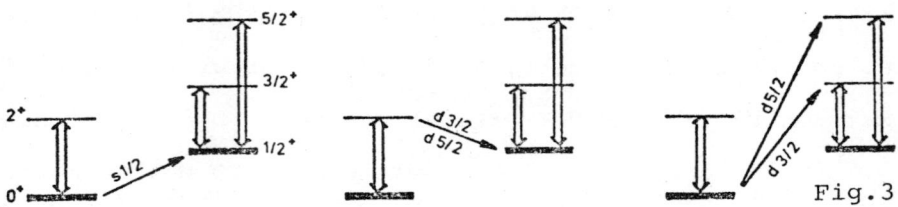

Fig. 3

of the even even target nucleus using a spherical basis and restrict this to strong excited states as the 2_1^+ in ^{28}Si and ^{54}Cr.

$$|\lambda, J^\pi\rangle = \alpha^\lambda_{J_o} |J \otimes 0^+\rangle + \sum_j \alpha^\lambda_{j_2} |j \otimes 2^+\rangle_{J^\pi}$$

Fig. 3 shows schematically the transitions, included in the calculation for the excitations of a $1/2^+$ state. Most important are the channel coupling in the entrance channel and the interference of direct and inelastic transfer. Channel coupling in the exit channel due to protons is comparatively weaker. For (normalized) configurations these transition amplitudes are caculated with the code CHUCK[1] only once. A code BARBARA calculates $\sigma(\theta)$ and $A_y(\theta)$ from these transition amplitudes weighted with the spectroscopic coefficients

$\alpha^\lambda_{J_o}$ and $\alpha^\lambda_{j_2}$.

In a least square search routine variation of these coefficients and comparison with the data gives fitted spectroscopic coefficients and an error estimate (Tab.1). For ^{29}Si this analysis indicates considerable deviations from the Nilsson coupling scheme. For some of the coefficients the errors are large. The reason is

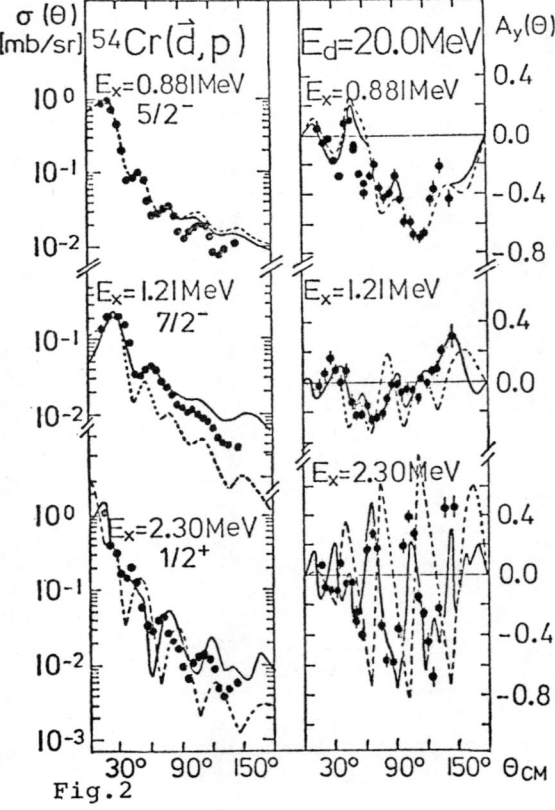

Fig.2

not the insensitivity to individual variations of these coefficients but that the same χ^2 may be obtained by rather different combinations of spectroscopic coefficients. We think, the present investigation is promising, if coefficients from a theoretical model can be used and taken as a prediction for the data. Especially the analyzing power of 1/2⁺ states is very sensitive to weak admixtures.

J^π	$1/2^+$	$3/2^+$	$5/2^+$
E_x/MeV	0.0	1.27	2.03
α_o	.61±.05	−.48±.05	.16±.06
$\alpha_{5/2}$	.78±.13	−.30±.22	−.35±.27
$\alpha_{1/2}$	---	.80±.35	.68±.19
$\alpha_{3/2}$	−.62±.12	−.79±.32	−.17±.25

Table 1: Spectroscopic coefficients for ^{29}Si

+) work supported in part by the Bundesministerium für Forschung und Technologie
*) Max-Planck fellow from the Institute of Atomic Energy, Peking, P.R. China
1) P.D. Kunz, code CHUCK, University of Colorado

D-STATE EFFECTS IN (d,p), (d,t) AND (d,^{3}He) REACTIONS AT E_d = 22 MeV *)

N. Seichert, H. Clement, R. Frick, G. Graw, S. Roman+),
F.D. Santos++), P. Schiemenz, Sun Tsu-Hsun+++)
Sektion Physik der Universität München, D-8046 Garching

DWBA calculations[1,2] for the nucleon transfer reactions (d,p), (d,t) and (d,^{3}He) show, that tensor analyzing powers, especially $T_{21}(\theta)$, provide an experimental information on the $\ell=2$ (D-state) amplitude $u_2(r)$ of the transfered nucleon in the light fragment. In the local energy approximation (LEA)[1] the D-state effects depend only on the quantity D_2, given by

$$D_2 = \frac{1}{15}\int u_2(r) r^4 dr / \int u_0(r) r^2 dr$$

where $u_0(r)$ and $u_2(r)$ are the radial parts of the S and D-state components, respectively. Comparison of experimental data with those calculations determine only the asymptotic D-state to S-state ratio in the wave function of n relative to p in ^{2}H and n(p) relative to d in ^{3}H(^{3}He), respectively.

Previous experimental information originates from low (E_d<13 MeV) or subcoulomb energies of the deuteron where in the analysis the restrictions due to the LEA are considered to be of minor importance. For the (d,^{3}He) reaction, values for D_2 near -0.24 fm² [3,4] and -0.37 fm² [5] have been obtained, thus further investigation of this reaction, even on a relatively low level of accuracy is demanded. Also it is important to have data at higher energies to study the validity of the LEA.

With 22 MeV tensor polarized deuterons we measured $T_{21}(\theta)$ for the reactions ^{208}Pb(d,p), ^{208}Pb(d,t) and ^{65}Cu(d,^{3}He), some of the transitions are presented in the figs.1,2 and 3. Pronounced effects are observed, which qualitatively agree with the data at lower energy. The drawn curves show the results of a preliminary analysis.

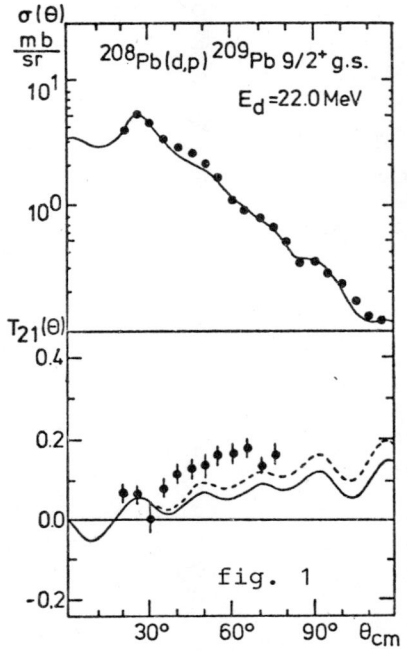

fig. 1

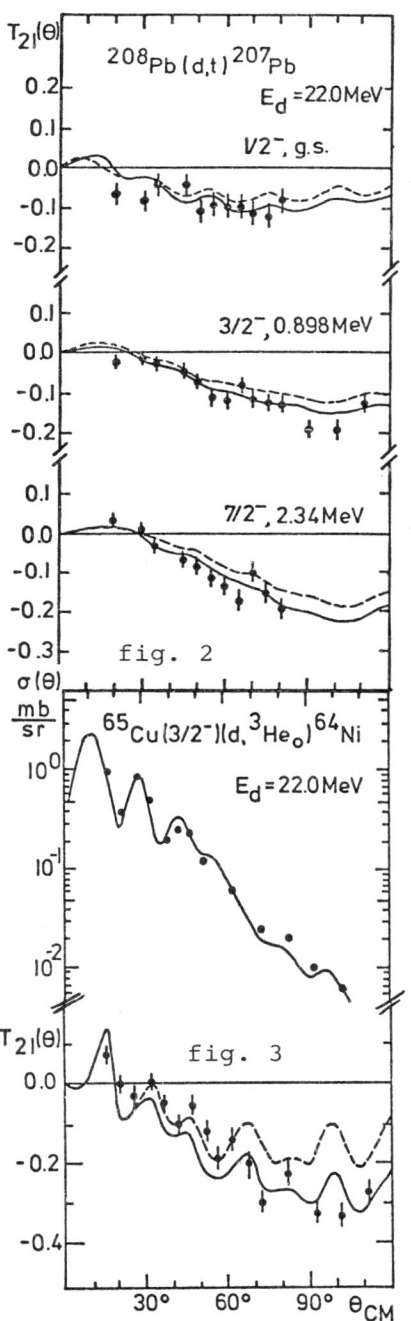

fig. 2

fig. 3

The dashed lines show DWBA calculations[6] where D_2 is taken into account using the LEA. For the optical potentials, global sets have been used. The solid curves contain corrections originating from the inclusion of a tensor potential of the T_R type, which have been calculated separately with the zero range DWBA code CHUCK. They yield a 10% to 20% correction to $T_{21}(\Theta)$, while for deuteron energies near $E_d=12$ MeV the corrections obtained are much smaller. Neglecting both T_R and D_2 results in very small analyzing powers $T_{21}(\Theta)$. The T_R tensor potential was taken from a preliminary analysis of elastic scattering[7], where for these nuclei a pure imaginary T_R potential has been obtained. Fig.1 shows the results for the (d,p) reaction, taking the established value[8] of $D_2=0.484$ fm^2. The calculated values for $T_{21}(\Theta)$ are about 50% smaller than the measured ones. The difference even increases in case of a (d,p) reaction, if a T_R correction is included.

We compared the results of these calculations with those of the code PTOLEMY[9]. This full finite range code treats explicitely the radial dependence of the s- and d-state parts of the deuteron wave function (parametrized as a Reid soft core wave function) and complex tensor coupling in the entrance channel. The calculations done with PTOLEMY reproduce nearly exactly the curves shown in figure 1. This supports the validity of the LEA and also the approximation to add in the $T_{21}(\Theta)$ analyzing power the effects resulting from the T_R potential and the D_2 contribution, obtained in separate calculations. The discrepancy of these calculations with the experimental results seems to indicate the presence of a dynamical change of the deuteron wave function in a (d,p) reaction at this energy.

The calculations for the pick up reactions (dashed curves in figs. 2 and 3) have been done in Birmingham with the DTCODE[6], an expansion of DWCODE, using for D_2 the most recent values from a Fadeev calculation for ^{3}He and ^{3}H[10]: $D_2 = -0.24$ fm^2 (see also ref.11), which are also in agreement with the experimental observations of ref. 3 and 4. Including the T_R corrections, we obtain a reasonable reproduction of the experiment. For the (d,t) reaction, the analysis and the result agree with the experimental work at $E_d = 12$ MeV[3]. For the (d,^{3}He) reaction we find no evidence for a value as large as $D_2 = -0.37$ fm^2 reported recently[5]. In the pick up reactions the effects from a dynamic change of the wave function, as observed in (d,p) stripping at this energy, should be weaker because of two reasons:
(a) the binding energy of the transferred nucleon is much stronger for the A=3 nuclei than for the deuteron, (b) the absorption is much stronger for the A=3 nuclei than for protons, thus the transfer takes place at larger distance from the target nucleus.

We thank Dr. Steven Pieper for his advice to use the recent version of the program PTOLEMY, which he installed at the Leibniz Rechenzentrum in München and Dr. A. Feigel for his help in doing test calculations with the DWCODE in Erlangen.

*) supported in part by the Bundesministerium für Forschung und Technologie
+++) Max-Planck fellow from the Institute of Atomic Energy, Peking, P.R. China
++) DAAD Guest from the University of Lisbon, Portugal
+) University of Birmingham, Great Britain
1) R.C. Johnson and F.D. Santos, Part.and Nucl. 2, 285 (71)
2) G. Delic and B.A. Robson, Nucl.Phys. A156 (1970) 97
3) L.D. Knutson, B.P. Hickwa, A. Barroso, A.M. Eiro, F.D. Santos and R.C. Johnson, Phys.Rev.Lett.35 (75) 1570
4) S. Roman, A.K. Basak, J.B.A. England, J.M. Nelson, N.E. Anderson, F.D. Santos and A.M. Eiro, Journal Phys.Soc.Japan 44 (1978) Suppl. 635
5) M.E. Brandon and W. Haeberli, Nucl.Phys. A287 (77) 213
6) J.D. Harvey, F.D. Santos: DWCODE, Univ. of Surrey Report, 1970, unpublished
7) R. Frick et al., contribution to this conference
8) K. Stephenson and W. Haeberli in Europ.Symp. on Few Body Probl.in Nucl. and Part.Phys., Sesimbra, Portugal, 1980
9) code PTOLEMY by M.H. Macfarlane and Steven C. Pieper, Argonne, Illinois, version Febr. 1980
10) Y.E. Kim and Muslim, Phys.Rev.Lett. 42 (1979) 1328
11) F.D. Santos, A.M. Eiro, A. Barroso, P.R. C19 (1979) 238

CONTRIBUTIONS OF SPIN-DEPENDENT EFFECTS TO THE VECTOR ANALYZING POWER AND PROTON POLARIZATION FOR $\ell_n=0$ (d,p) REACTION

T. Hasegawa and N. Ueda
Institute for Nuclear Study, Univ. of Tokyo, Tanashi-shi, Tokyo

T. Kubo, N. Kishida and H. Ohnuma
Tokyo Institute of Technology, Meguro-ku, Tokyo

T. Fujisawa and T. Wada
Institute of Physical and Chemical Research, Wako-shi, Saitama

and K. Iwatani
Faculty of Engineering, Hiroshima University, Hiroshima-shi

ABSTRACT

Both the vector analyzing powers and the proton polarizations have been measured for the ^{116}Sn(d,p)^{117}Sn(grd., 1/2$^+$) and ^{29}Si(d,p)^{30}Si(grd.,0$^+$) reactions at 22.2 MeV. The separation of the contributions from the proton and deuteron spin-orbit distortions has been performed for these $\ell_n=0$ transfers using the method of Johnson. The obtained results have been compared with DWBA and the adiabatic calculations.

INTRODUCTION

The vector analyzing powers and proton polarizations in an $\ell_n=0$ (d,p) reaction are particularly sensitive to the effects of spin-dependent interactions, since these polarization quantities for $\ell_n=0$ transfers are not affected by the absorbing potentials within the frame work of DWBA theory. It has been pointed out by Johnson[1] that the contributions of the proton and deuteron spin-dependent distortions can be separately estimated by measuring both the analyzing powers and polarizations for an $\ell_n=0$ transfer. Namely, the contributions from the proton and deuteron spin-orbit potentials are given to first order by the quantities $S_p=2(P_y-A_y)$ and $S_d=3A_y-2P_y$, respectively, where A_y and P_y are a vector analyzing power and a polarization. Recently the first experimental determination of S_p and S_d was made for the ^{116}Sn(d,p) reaction at 8.22 MeV[2].

We have measured the vector analyzing powers for the ^{116}Sn($\vec{d}$,p) and ^{29}Si($\vec{d}$,p) reactions at E_d=22.2 MeV. The polarizations of protons have also been obtained from the analyzing power measurements of the inverse ^{117}Sn($\vec{p}$,d) and ^{30}Si($\vec{p}$,d) reactions at 27.0 and 30.0 MeV, respectively. The $S_p/2$ and $S_d/2$ values have been determined from these results.

EXPERIMENTAL METHODS

The vector polarized deuteron and proton beams were provided by an atomic beam type polarized ion source and INS SF-cyclotron.

The average beam polarizations were about 40% for deuterons and 55% for protons, where a carbon polarimeter was used for the determination of the beam polarization. The analyzing powers of the polarimeter were determined by comparing with previous works[3,4]. The systematic errors due to this procedure were estimated to be within ±10%. The beam intensity was 30-50 nA on taget.

The targets were self-supporting foil of enriched isotopes of $^{116}Sn(2.5mg/cm^2)$, $^{117}Sn(5.7mg/cm^2)$ and $^{29}Si(0.7mg/cm^2)$, and $^{30}SiO_2$ (2.2mg/cm^2) supported by Mylar backing. The reaction particles were detected with four silicon counter telescopes. The polarizations for the elastic scatterings were also measured.

RESULTS AND DISCUSSION

1). $^{116}Sn(d,p)^{117}Sn(grd., 1/2^+)$ reaction

The vector analyzing powers(A_y) and the proton polarizations (P_y) obtained from the inverse ($\vec{p},d$) reaction are shown along with the differential cross sections in Fig. 1. The error bars shown are statistical errors only.

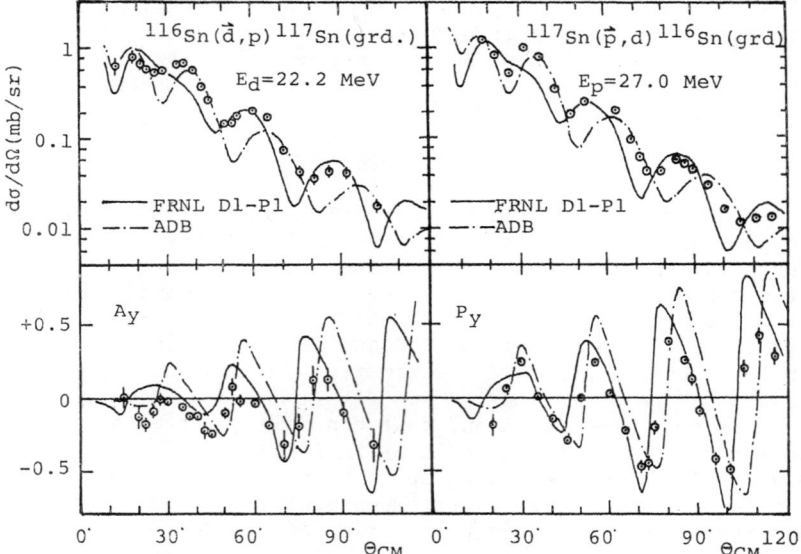

Fig. 1. Angular distributions of differential cross sections, vector analyzing powers and proton polarizations for $^{116}Sn(d,p)$ $^{117}Sn(grd.,1/2^+)$ reaction at 22.2 MeV. The solid and dashed curves are calculated using DWBA and ADB, respectively.

The obtained angular distributions of both (d,p) and (p,d) cross sections are almost identical as expected from the detailed balance. The **derivative** rule is clearly seen by comparing the zero crossing points of the polarization data with the peaks and valleys in the cross sections.

The quantities $S_p/2$ and $S_d/2$ calculated from the A_y and P_y are shown in Fig. 2. As was the case at 8.22 MeV, S_p and S_d are opposite in sign at 22.2 MeV. However, the magnitudes at 22.2 MeV are considerably larger than those at 8.2 MeV.

These data were compared with FRNL DWBA calculations. The adiabatic calculations of Johnson and Soper(ADB)[5] were also made. These curves are compared with the data in Figs. 1 and 2.

The experimental A_y and P_y are mostly determined from the proton spin-orbit potential, but affected only little by the deuteron spin-orbit potential. The fits to the data is good for $S_p/2$,

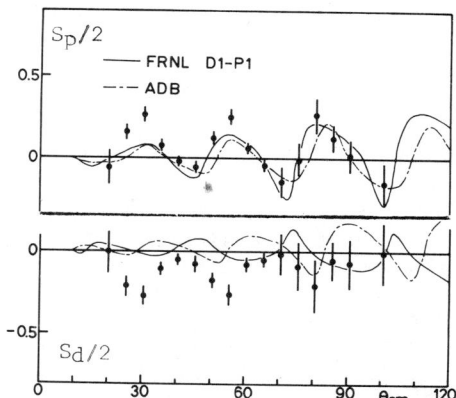

Fig. 2. Angular distributions of $S_p/2$ and $S_d/2$ for $^{116}Sn(d,p)^{117}Sn$ reaction at 22.2 MeV. The curves are same with Fig. 1.

but is very poor for $S_d/2$. This is not improved by changing the strengths of the spin-orbit forces. These circumstances can be improved by including the deuteron D-state effects as is described in another contribution to this symposium.

2). $^{29}Si(d,p)^{30}Si$ reaction

Obtained quantities $S_p/2$ and $S_d/2$ are shown in Fig. 3. Again, $S_p/2$ and $S_d/2$ are opposite in sign, but the magnitudes are considerably smaller than the latter. Preliminary DWBA and ADB calculations are also shown in Fig. 3. More detailed analyses are in progress.

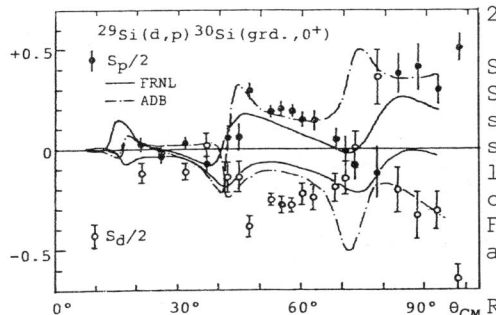

Fig. 3. Angular distributions of $S_p/2$ and $S_d/2$ for $^{29}Si(d,p)$ reaction at 22.2 MeV. Preliminary calculations are made using DWBA and ADB methods.

REFERENCES

1. R. C. Johnson, Nucl. Phys. <u>35</u>, 654 (1962).
2. R. R. Cadmus and W. Haeberli, Nucl. Phys. <u>A327</u>, 419 (1979).
3. R. M. Craig et al., Nucl.Phys. <u>83</u>, 493 (1966).
4. G. Perrin et al., Nucl. Phys. <u>A193</u>, 215 (1972)
5. R. C. Johnson and P. J. R. Soper, Phys. Rev. <u>C1</u>, 976 (1970).
6. H. Ohnuma, contribution to this symposium.

ON SOME NEW EFFECTS OF THE DEUTERON D STATE OBSERVED IN LOW ENERGY (p,d) AND (d,p) REACTIONS

H. Ohnuma
Department of Physics, Tokyo Institute of Technology, Meguro-ku, Tokyo, Japan

ABSTRACT

Studies of the energy dependence of the j dependence seem to rule out the possibility of explaining the j dependence in terms of the deuteron D state. On the other hand two previously unexplored experiments have revealed the importance of the deuteron D state effects on (d,p) and (p,d) reactions at low energies. They are the vector polarization/analyzing power measurements for the $^{116}Sn(d,p)^{117}Sn$(g.s., $1/2^+$) reaction at E_d = 22 MeV and the d-γ angular correlation measurement for the $^{58}Ni(p,d)^{57}Ni$(2.58 MeV, $7/2^-$) reaction at E_p = 30 MeV. Physical implications of these results are discussed.

The deuteron D-state effects (DSE) on (d,p) and (p,d) reactions were first studied by Johnson and Santos[1]. They showed that DSE could give a partial account of j dependence of differential cross sections raising j=ℓ+1 cross sections at minima. Most of the later calculations[2,3], however, indicate smaller DSE on cross sections. Studies of the energy dependence of the (d,p) j dependence between 14 MeV and 30 MeV have shown[4] that the "oscillation type" j dependence observed at lower incident energies becomes less pronounced with increasing energy, and the transition to the "phase type" j dependence takes place around 20 MeV. Typical examples are found in Fig. 1. Since DSE are more important at higher energies, because of the larger momentum transfer involved in the reaction[5], it would be expected that DSE, if they ever explain j dependence, could explain the "phase type" j dependence. Actually DSE may to some extent reproduce "oscillation type" j dependence, but not the "phase type" j dependence. No theory at the moment can reproduce such variation of the j dependence with energy in a consistent manner. It is no wonder that the cross sections, or scalar quantities, are insensitive to DSE at low energies, since primary contributions to the cross sections come from the spin-independent part of the distorting potentials and the S component of the n-p interaction.

Vector polarizations and analyzing powers, or rank-one tensor quantities, are largely affected by the spin-orbit parts and the absorptive parts of the distorting potentials through the modification of the distorted waves. In order to see DSE on these quantities, therefore, one has to eliminate the contributions from the spin-orbit potentials and the absorption. This can be achieved in some special cases. A recent study of the $^{116}Sn(d,p)^{117}Sn$(g.s.,$1/2^+$) reaction at E_d = 22 MeV[6] clearly indicates the importance of DSE on these quantities. In this experiment both the vector polarization P_y and vector analyzing power A_y were measured. The absorption of the distorted waves has no effect on P_y and A_y because of an $\ell=0$ transfer in this case. Then the quantities $S_p=2(P_y-A_y)$ and $S_d=3A_y-2P_y$ were calculated from the measured P_y and A_y. It has been pointed out by Johnson[7] that the separation of proton and deuteron spin-orbit distortion can be approximately achieved by taking such linear combinations. The S_d values thus obtained from the experiment are large and negative, while DWBA without DSE gives only small S_d although a fair fit is obtained for S_p. One might ask whether larger S_d could be obtained in DWBA with a stronger spin-orbit potential for deuterons. Increasing the spin-orbit potential for deuterons however does not improve the fit to the experimental S_d. The reason is that both P_y and A_y are mostly determined by the spin orbit potential in the proton channel in this case. Thus a better agreement with the data could be expected for S_p by slightly adjusting the spin-orbit potential for protons, but S_d are little affected by changing either proton spin-orbit or deuteron spin-orbit potential. Thus the inclusion of DSE is the only natural way to obtain large S_d in DWBA. Indeed exact-finite-range calculations using the Reid soft-core potential give nice fits to the experimental S_d as seen in Fig. 2, while the quality of fit to S_p is not much different from that without DSE. Whether the present conclusion can be extended to all $\ell=0$ transfers and why the spin-orbit potential for deuterons does not contribute to P_y and A_y are yet to be investigated.

Significant SDE have been predicted for tensor analyzing powers and tensor polarizations in (d,p) and (p,d) reactions and experimentally observed by many authors. Spin transfer has also been found[8] to be sensitive to DSE. Usually the term tensor polarization refers to that of the outgoing deuterons in (p,d) reactions. Recently the tensor polarizations of the other outgoing particles, or rather the residual nuclei, have been studied in a (p.d) reaction[9]. In this experiment the d-γ angular correlation functions were measured in the ^{58}Ni(p,d)^{57}Ni(2.58 MeV,7/2$^-$) reaction at E_p = 30 MeV. Since the decaying gamma ray is pure E2 in this particular case, we are looking at the ρ_{2q} and ρ_{4q} parts of the spin statistical tensors of the residual nuclear state. The angular correlation function obtained for $\theta_d = 45°$, at the minimum of the differential cross sections, could not be explained without DSE. It is not difficult to understand why DSE are important to correlation functions. In the DWBA formalism the angular momenta of initial and residual nuclei enter the theory through the form factor, the product of the residual interaction and the initial and final state wave functions. The inclusion of the D component in the residual interaction and the deuteron wave function directly modifies the form factor and the angular momentum coupling therein, in contrast to the spin-orbit potentials which modify distorted waves. Thus the tensor polarization of the final nuclear state can be sensitive to DSE, giving entirely different information from other quantities.

The author gratefully acknowledges many discussions with Drs. N. Kishida, J. Kasagi, T. Kubo and T. Hasegawa.

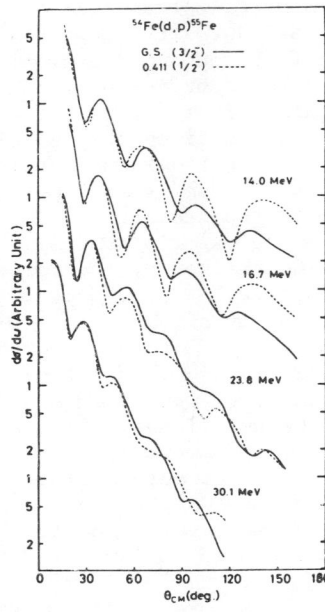

Fig. 1. Variation of ℓ=1 j dependence with energy.

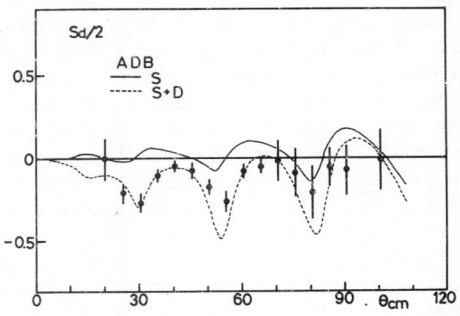

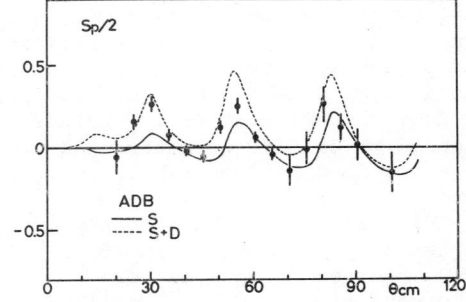

Fig. 2. Experimentally obtained S_d and S_p compared with EFR calculations with (dashed curves) and without (solid curves) DSE.

REFERENCES

1. R. C. Johnson and F. D. Santos, Phys. Rev. Lett. 19, 364 (1967).
2. G. Delic and B. A. Robson, Nucl. Phys. A156, 97 (1970).
3. N. Kishida and H. Ohnuma, J. Phys. Soc. Jpn. 46, 1375 (1979).
4. N. Kishida and H. Ohnuma, Proc. 1978 INS Int. Symp. Nuclear Direct Reaction Mechanism, ed. M. Tanifuji and K. Yazaki (INS, Univ. Tokyo, 1979), p.211.
5. E. Rost and J. R. Shepard, Phys. Lett. 59B, 413 (1975).
6. T. Hasegawa et al., contribution to this conference.
7. R. C. Johnson, Nucl. Phys. 35, 654 (1962); A90, 289 (1967).
8. A. K. Basak et al., Nucl. Phys. A275, 381 (1977).
9. N. Kishida, H. Ohnuma, J. Kasagi, T. Kubo and M. Yasue, contribution to this conference and to be published.

MIXED-j TRANSFER IN ^{55}Mn$(\vec{d},t)^{54}$Mn

J.A. Cameron, E. Habib and A.A. Pilt
Tandem Accelerator Laboratory, McMaster University,
Hamilton, Ontario, Canada

ABSTRACT

The $(\vec{d},t)$ reaction on ^{55}Mn has been studied at 17 MeV. Mixed $p_{1/2}-p_{3/2}$ neutron pick-up was observed to 1p-3h states near the ground state. Above 1 MeV $f_{7/2}$ transfer to a group of 2p-4h states was found.

INTRODUCTION

Neutron pick-up on even-even N=30 targets leads to states with $J^{\pi}=1/2^-, 3/2^-, 5/2^-$ and $7/2^-$ in the N=29 final nucleus. The former three are thought to be one particle states, the last a hole state relative to a core with N=28. In odd Z nuclei, these states couple with the odd protons to form many levels. Calculations have suggested [1,2] that little mixing should occur between particle and hole states in such cases.

Strong $\ell=3$ pick-up has been reported [3] to three states above 1 MeV in the ^{54}Mn$(d,t)^{54}$Mn reaction. These states have been tentatively identified as the expected $f_{7/2}$ hole states.

EXPERIMENTAL

Triton groups were observed up to 3.5 MeV excitation by using a particle identifying focal plane counter in a magnetic spectrograph. Cross sections and analyzing powers were observed at 5° intervals from 25° to 50°. Few transitions in the reaction can be clearly assigned unique j_n so an analyzing power calibration was made using the reactions ^{54}Cr$(\vec{d},t)^{53}$Cr and ^{56}Fe$(\vec{d},t)^{55}$Fe. The Q dependence was not well reproduced by any DWBA calculations so the empirical analyzing powers were used to estimate the degree of mixing shown in Table 1.

DISCUSSION

The observed spectroscopic factors are in good agreement with shell model calculations, except for the 839 keV 4^+ level, which theory associates with the $f_{5/2}$ particle state. The summed spectroscopic factors are similar to those observed in odd-A nuclei. The data and calibrations are not sufficiently precise to allow a quantitative measure of the 1p-3h - 2p-4h mixing, though it would seem to be smaller than the 20% forseen by Ref 2.

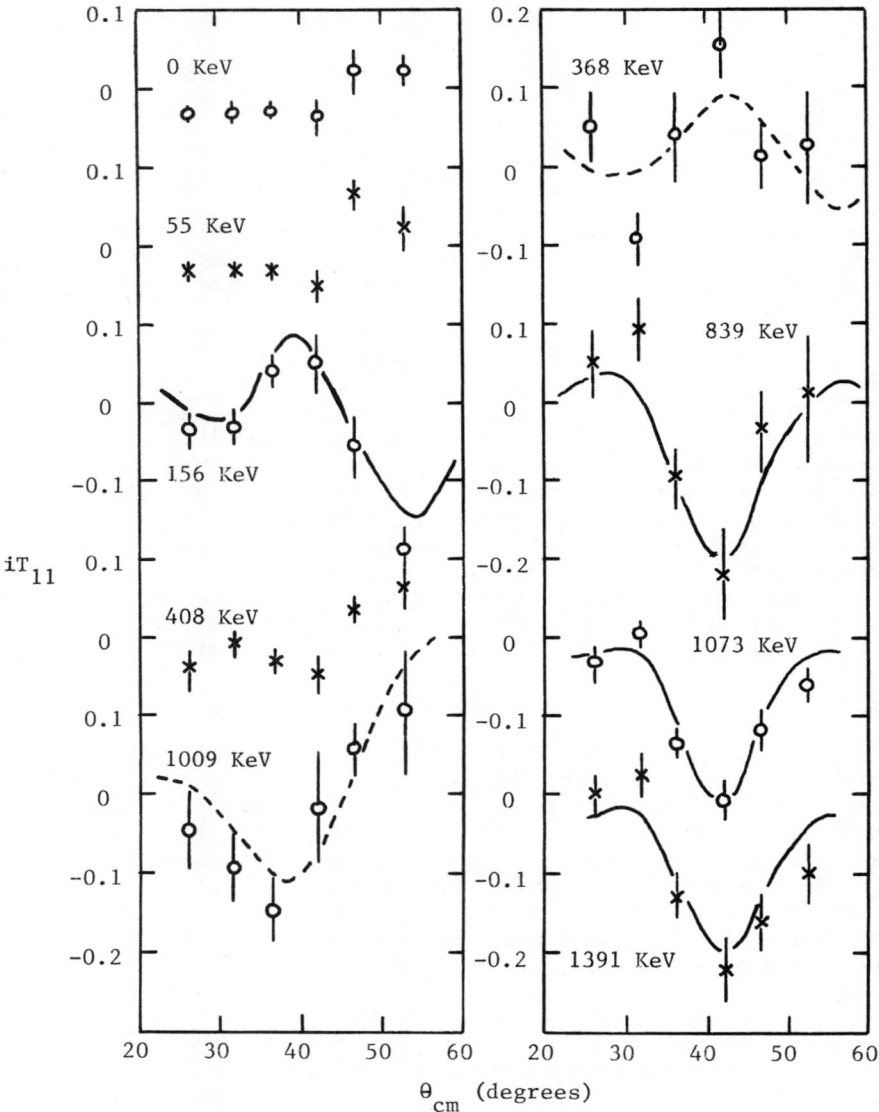

Fig. 1. Vector analyzing powers for ^{55}Mn($\vec{d}$,t)^{54}Mn at 17 MeV.
Left: $\ell = 1$; Right: $\ell = 3$
Solid lines: $j = \ell + \frac{1}{2}$; Dashed lines: $j = \ell - \frac{1}{2}$.

TABLE 1. Spectroscopic factors for ^{55}Mn$(\vec{d},t)^{54}$Mn

E_x (keV)	J^π	C^2S $p_{1/2}$	$p_{3/2}$	$f_{5/2}$	$f_{7/2}$
0	3+	0.1	0.3		
55	2+	0.1	0.2		
156	4+		0.13		
368	5+			0.09	
408	3+	0.07	0.07		
839	4+				0.14
1009	3+	0.02	0.00		
1073	6+				1.2
1391	1+				0.35
1508	2+				0.32
1850	3+				0.30
3032					0.20
Summed strength		0.3	0.7	0.1	2.5
^{54}Cr(d,t)^{53}Cr		0.23	0.64	0.40	2.8
^{56}Fe(d,t)^{55}Fe		0.48	1.0	0.54	4.7

REFERENCES

1. I.P. Johnstone and H.G. Benson, J. Phys. G: Nucl. Phys. 3 (1977) L69 and private communication.
2. P.W.M. Glaudemans, private communication.
3. T. Taylor and J.A. Cameron, Nucl. Phys. A257 (1976) 427.

CORE-COUPLED STATES EXCITED IN THE ^{208}Pb($\vec{d}$,t)^{207}Pb REACTION

E. Sugarbaker
University of Colorado, Boulder, CO 80309

W. P. Alford
University of Western Ontario, London, Ontario

R. N. Boyd
Ohio State University, Columbus, OH 43212

J. Cameron
McMaster University, Hamilton, Ontario

E. Flynn and J. Sunier
Los Alamos Scientific Laboratory, Los Alamos, NM 87545

ABSTRACT

A study of known core excited states in ^{207}Pb shows that the cross section and analyzing power data are not well reproduced by two-step mechanisms.

DISCUSSION

States up to an excitation energy of 3.8 MeV in ^{207}Pb have been studied with the ^{208}Pb($\vec{d}$,t) reaction at a bombarding energy of 17 MeV. Among the known core excited states in this region are the (^{208}Pb(3$^-$) $\otimes$ p$_{1/2}^{-1}$) 5/2$^+$, 7/2$^+$ doublet near 2.6 MeV and the (^{206}Pb(g.s.) $\otimes$ g$_{9/2}^2$) state at 2.74 MeV. The former would be excited by a two-step process involving inelastic excitation while the latter should be excited by a (d,p)(p,t) sequential transfer two-step reaction. Given the relatively simple structure of these states, it was thought that measurements of analyzing power as well as cross section should provide a sensitive test of reaction theories for two-step processes.

The polarized deuteron beam from the LASL Van de Graaff accelerator was used in this investigation. The tritons were momentum analyzed in the Q3D spectrometer and detected in a helical wire focal plane detector. Differential cross sections, vector analyzing power (Ay), and tensor analyzing powers (Ayy) were measured for θ_{lab}=10-60°. The Ayy data are of reasonable quality for only the single-hole states.

Data for the single-neutron hole states in ^{207}Pb provided a test of the optical model parameter sets to be employed in the subsequent two-step calculations. While the fits to these cross sections were fairly insensitive to the choice of parameter set, the fits to the analyzing powers provided a more stringent requirement. The one-step DWBA analysis was performed with the code DWUCK[1] and included first-order corrections for finite range and nonlocality. The best fit was obtained using the "best fit" deuteron set of

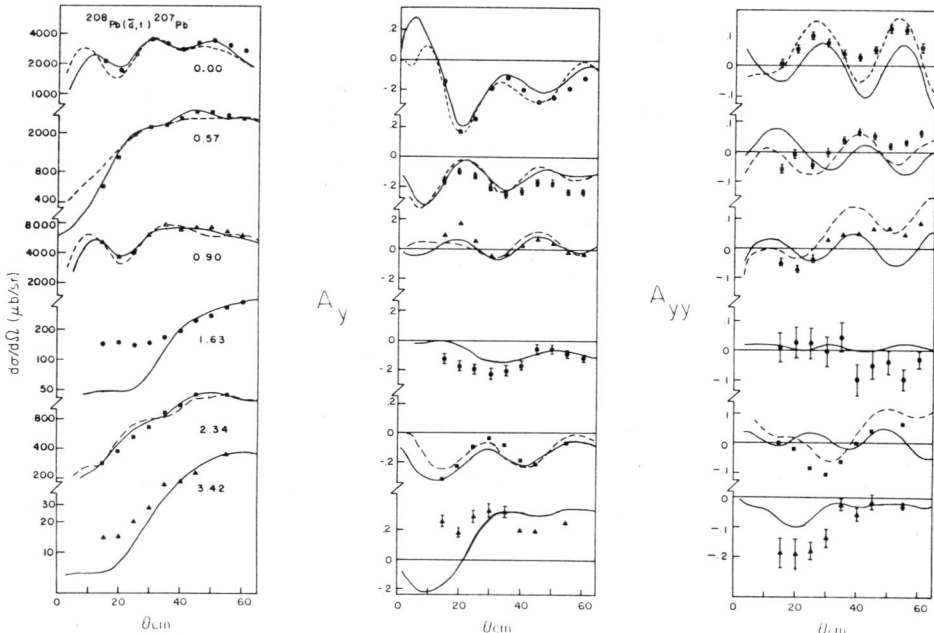

Fig. 1. Angular distributions for the single-hole states.

Childs et al.[2] and the global triton set of Becchetti and Greenlees.[3] These fits are shown in Fig. 1. The dashed lines indicate the predictions of the finite-range code DWUCK5[1] using the same parameters. The agreement with the Ayy data is improved by including the full finite range effects, while the quality of the fits to the cross section and Ay data is not significantly modified.

The assignment of the doublet near 2.6 MeV as (^{208}Pb(3^-)$\otimes p_{1/2}^{-1}$) is based on its strong L=3 excitation in inelastic scattering. Accordingly, a two-step calculation involving inelastic excitation plus $p_{1/2}$ pickup is expected to reproduce the data. The solid curves in Fig. 2 show the two-step predictions calculated with the code CHUCK[1] assuming inelastic excitation of the 3^- state in ^{208}Pb followed by a $p_{1/2}$ transfer. The agreement with the data is very poor. If the Johnson-Soper prescription is used for the deuteron potential, the results are only slightly improved. The dot-dash curves in Fig. 2 present the result of such a calculation in which inelastic excitation of the 3^- state is permitted in both the target and residual nuclei. Further calculations including plausible one-step contributions were unable to provide better agreement.

The 2h-1p character of the $9/2^+$ state at 2.74 keV is established by its strong excitation with ℓ=4 in the ^{206}Pb(d,p) reaction. In this case it was expected that the reaction should be described by sequential (d,p)(p,t) transfer, plus a one-step component due to the 2p-2h excitations in the ^{208}Pb ground state. The pure (d,p)(p,t) sequential calculations using reaction strengths from the literature,

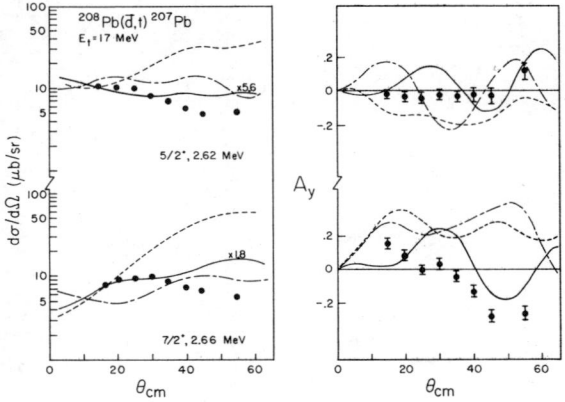

Fig. 2. Angular distributions for the ^{208}Pb($\vec{d}$,t)^{207}Pb reaction to the multiplet at 2.6 MeV of excitation. Dashed curves are 1-step predictions normalized to the data.

yield the dot-dash curves in Fig. 3. The fits (dashed line) to back angle data assuming only a 1-step $g_{9/2}$ pickup gives $C^2S \simeq 0.012$. This value is lower than that obtained in previous studies.[4] The 1-step prediction appears to provide better agreement with the Ay than does the pure sequential calculation. The combined 1 and 2-step calculations (solid line) (using the 1-step strength of Ref. 4) over-predicts the cross sections and provides only a marginal fit to the data. The 2-step contribution appears again to be too large.

In conclusion, we have been unable to obtain good agreement with measured cross sections using obvious reaction models based on the known properties of the states. Further calculations involving additional reaction channels are in progress and will be reported.

REFERENCES

1. P. D. Kunz, unpublished.
2. Childs et al., Phys. Rev. C 10, 217 (1974).
3. F. Becchetti and G. Greenlees, in Polarization Phenomena in Nuclear Reactions (University of Wisconsin Press, Madison, WI, 1971) p. 682.
4. Moyer et al., Phys. Rev. C 2, 1898 (1971).

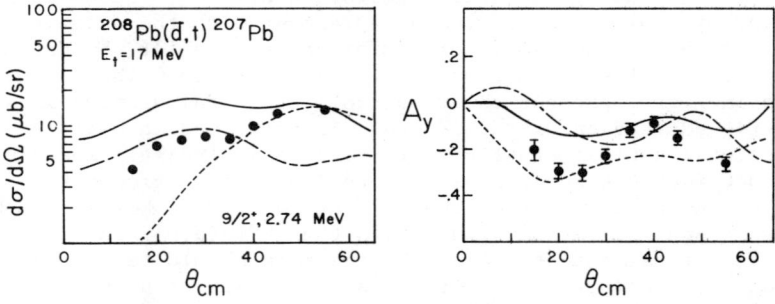

Fig. 3. Angular distributions for the ^{208}Pb($\vec{d}$,t)^{207}Pb (2.74 MeV) reaction. Curves as described in the text.

CONFIGURATION MIXING OF PARTICLE-HOLE STATES IN A = 16 NUCLEI STUDIED BY THE $^{17}O(\vec{d},t)^{16}O$ AND $^{17}O(\vec{d},\tau)^{16}N$ REACTION

G. Mairle, K.T. Knöpfle, H. Riedesel, K. Schindler and G.J. Wagner
Max-Planck-Institut für Kernphysik, 6900 Heidelberg, W.-Germany

V. Bechtold and L. Friedrich
Kernforschungszentrum Karlsruhe, IAK, 7500 Karlsruhe, W.-Germany

ABSTRACT

The reactions $^{17}O(\vec{d},t)^{16}O$ and $^{17}O(\vec{d},\tau)^{16}N$ have been measured simultaneously at E_d = 52 MeV. The analyzing powers for ℓ=1 pick-up, which differ characteristically for $1p_{1/2}$ and $1p_{3/2}$ transfer, allow to determine the degree of j-mixing in 2^- and 3^- states in ^{16}O(T=0 and T=1) and in ^{16}N. All observed strongly excited states appear to have rather pure particle-hole configurations.

INTRODUCTION

The simultaneous measurement [1] of the $^{17}O(d,t)^{16}O$ and $^{17}O(d,\tau)^{16}N$ reactions at 52 MeV has yielded a wealth of spectroscopic information on states which have a large overlap with $(1d_{5/2} \cdot 1p_{1/2}^{-1})_{2^-,3^-}$ and $(1d_{5/2} \cdot 1p_{3/2}^{-1})_{1^-,2^-,3^-,4^-}$ multiplets both for T=0 and T=1. A remaining problem were the relative $1p_{1/2}$ and $1p_{3/2}$ contributions in transitions to 2^- and 3^- states. We report on an attempt to disentangle these contributions using the vector analyzing powers (VAPs) of the $(\vec{d},t)$ and $(\vec{d},\tau)$ reactions on ^{17}O. We utilize the slow Q- and A-dependence of the VAPs found [2] for ℓ=1 pick-up in $(\vec{d},\tau)$ reactions on nuclei between ^{12}C and ^{28}Si, and we use VAPs of pure $1p_{1/2}$ and $1p_{3/2}$ transitions from pick-up on an ^{16}O contaminant for comparison.

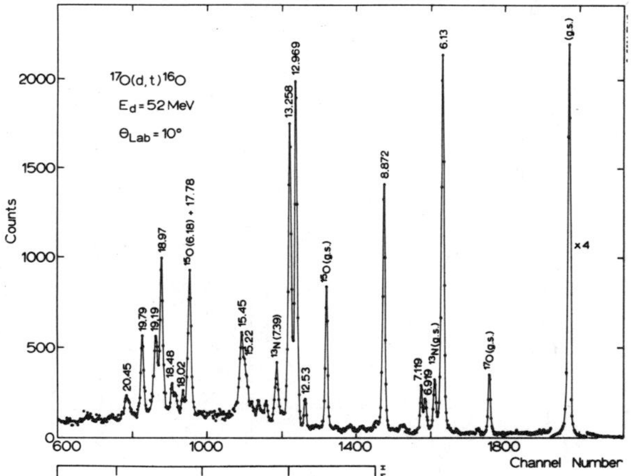

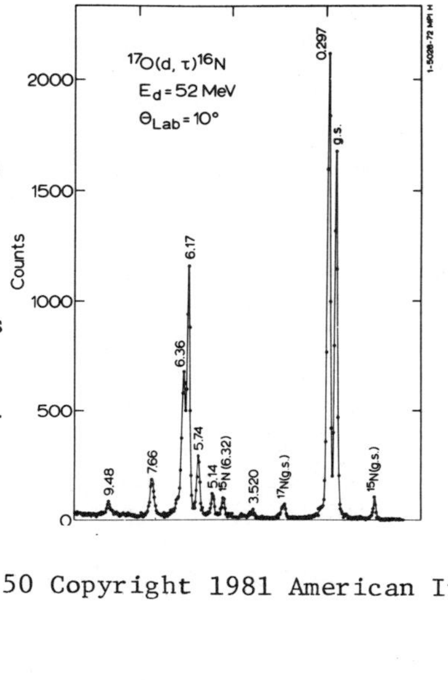

Fig. 1. Spectra of the (d,t) and the (d,τ) reactions on ^{17}O (same energy scale)

0094-243X/81/690685-03$1.50 Copyright 1981 American Institute of Physics

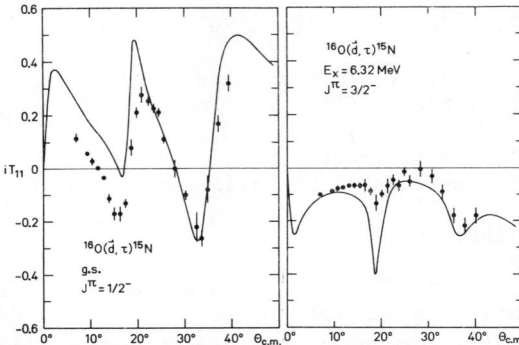

Fig. 2. Analyzing power for $1p_{1/2}$ and $1p_{3/2}$ pick-up from ^{16}O compared to DWBA calculations.

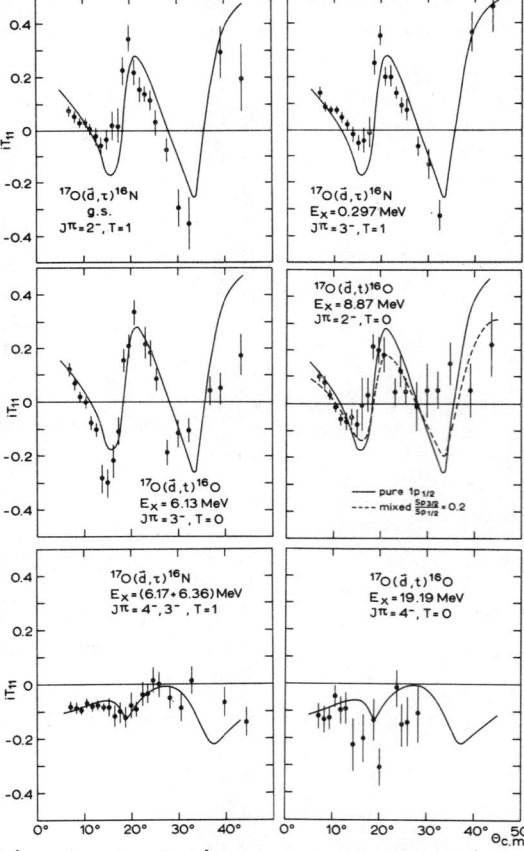

Fig. 3. Analyzing power of some selected states compared to VAPs of pure $1p_{1/2}$ or $1p_{3/2}$ pick-up (smooth curves drawn through the data points of fig.2)

EXPERIMENT AND RESULTS

The experimental technique is very similar to that described previously [3]. We used a small gas target containing enriched $^{17}O_2$ gas. Typical spectra are shown in fig. 1. In fig. 2 measured VAPs for $1p_{1/2}$ and $1p_{3/2}$ pick-up observed in the $^{16}O(\vec{d},\tau)^{15}N$ reaction are displayed together with DWBA calculations, which show satisfactory agreement. Since the DWBA description, however, is not perfect, we use experimentally determined VAPs (smooth curves drawn through the data points of fig. 2) to deduce the degree of j-mixing in the 2^- and 3^- states of ^{16}O and ^{16}N. This is justified because of the negligible mass difference and the moderate Q-value differences. In fig. 3 we present the VAPs of some states strongly excited in pick-up reactions from ^{17}O compared to VAPs of pure $1p_{1/2}$ and $1p_{3/2}$ pick-up from ^{16}O. T=1 states: Generally the data obtained for ^{16}N states agree with those of their analog (T=1) states in ^{16}O; we, therefore, discuss only results for ^{16}N states. In detail, we observe VAPs typical for $1p_{1/2}$ pick-up in the case of the 2^- g.s. and the 3^- state at 0.297 MeV. For the 2^- state the forward angle behaviour is indicative for some small $1p_{3/2}$ admixture ($R=S(p_{3/2})/S(p_{1/2}) < 0.1$). The agreement for the 3^- state, however, cannot be improved by $1p_{3/2}$ contributions ($R<0.05$). The VAP of the highly excited group comprising the 4^- and 3^-

states at 6.17 and 6.36 MeV, respectively, shows no significant deviation from a pure $1p_{3/2}$ VAP. This is consistent with the $1d_{5/2} \cdot 1p_{3/2}^{-1}$ nature of these states [1]. T=0 states: The VAP of the 3^- state at 6.13 MeV of ^{16}O shows the typical features of $1p_{1/2}$ pick-up. Due the very pronounced structure, the agreement with the VAP of the pure $1p_{1/2}$ pick-up cannot be improved by $1p_{3/2}$ contributions (R<0.15). In contrast, the structure of the VAP of the 2^- state at 8.87 MeV in ^{16}O is obviously damped (0.1<R<0.2). The evaluation of the high-lying state at 19.79 MeV is more difficult due to the increased background. Nevertheless, the observed VAP is compatible with pure $1p_{3/2}$ pick-up as required for a 4^- state.

DISCUSSION

The strongly excited particle-hole states in ^{16}O and ^{16}N show rather pure configurations in qualitative agreement with results of the shell model calculations of Gillet and Vinh Mau [4] and Hsieh [5]. If the spectra are interpreted as $1d_{5/2} \cdot 1p_{1/2}^{-1}$ and $1d_{5/2} \cdot 1p_{3/2}^{-1}$ multiplets, they can be used to extract matrix elements of the related effective interactions as has been done in ref. [1]. Since j-mixing in the 2^- and 3^- states has proved to be small, the conclusion [1] of different $1d_{5/2} \cdot 1p_{1/2}^{-1}$ and $1d_{5/2} \cdot 1p_{3/2}^{-1}$ interactions remains unchanged.

REFERENCES

1. G. Mairle et al., Nucl. Phys. A299, 39 (1978).
2. G. Mairle et al., to be published.
3. G. Mairle et al., Nucl. Phys. A339, 61 (1980).
4. V. Gillet and N. Vinh Mau, Nucl. Phys. 54, 321 (1964).
5. S.T. Hsieh et al., Nucl. Phys. A243, 380 (1975).

TENSOR ANALYSING POWER OF (d,³He) REACTIONS AND THE D-STATE OF ³He

F. Entezami, K.S. Dhuga, O. Karban, J.M. Nelson and S. Roman
Department of Physics, University of Birmingham, England

ABSTRACT

The tensor analysing powers of (d,³He) reactions on ^{19}F, ^{45}Sc and ^{63}Cu have been measured at 12.4 MeV incident deuteron energy. The results were compared with predictions of DWBA theory where the amount of the D-state component in the ³He wave function, represented by the parameter D_2, was varied. The best-fit value of $D_2 = -0.21$ fm^2 obtained is consistent with the corresponding value for the triton and is in a good agreement with recent theoretical estimates of the D-state of the tri-nucleon system based on calculations of the relative motion of the N+d clusters using realistic wave functions and on a solution of Faddeev equations with the Reid soft-core potential.

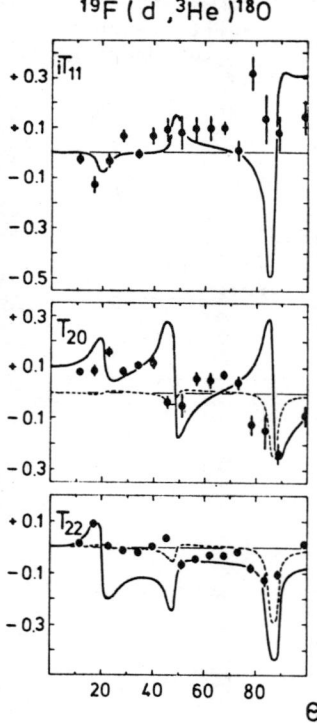

The measurements were carried out using the 12.4 MeV tensor-polarized deuteron beam from the atomic-beam ion source on the University of Birmingham Radial Ridge Cyclotron. The experimental method involved rapid three-way switching of the polarization state of the beam by means of the high frequency transitions at the ion source and azimuthal rotation of the scattering chamber by 90°. The beam polarization was monitored continuously in a polarimeter placed upstream with respect to the scattering chamber and using the ^{12}C(d,p$_0$)^{13}C reaction as an analyser. The "left-right" and "up-down" asymmetries were determined in separate runs for each scattering angle. The reaction products were detected in four silicon detector telescopes, spaced 10° apart and mounted on two rotatable arms, set at symmetric angles ±θ with respect to the incident beam direction.

The results obtained for the ^{19}F(d,³He)^{18}O g.s., 1/2$^+$, transition are shown in Fig.1 together with results of

Fig.1 The analysing powers of the ^{19}F(d,³He)^{18}O g.s. reaction. The solid and dashed lines are DWBA predictions with and without the ³He D-state contribution, respectively.

DWBA calculations including the ^{3}He D-state contributions[1]. Firstly, the role of the distorting optical-model potentials was investigated using various parameter sets available in the literature[2] for both the deuteron and ^{3}He channels. It was found that the predictions are very sensitive to the diffuseness of the ^{3}He spin-orbit potential affecting the analysing power for small angles, where smaller a_{so} values produced better fits to the data. This finding is consistent with the small ^{3}He a_{so} values determined from the elastic scattering of polarized ^{3}He in a number of experiments in this laboratory. In general, however, the DWBA predictions were more sensitive to the choice of deuteron potential.

Following the investigation of the role of the o.m. potentials, DWBA calculations were carried out varying the D_2 parameter, which determines the D-state to S-state ratio of the ^{3}He wave function. The calculations showed that $D_2 = (-0.21\pm0.03)$fm^2 was favoured for all three reactions.

The results obtained for the ^{45}Sc(d,^{3}He)^{44}Ca (7/2$^-$) and ^{63}Cu(d,^{3}He)^{62}Ni (3/2$^-$) ground state reactions are shown in Fig.2.

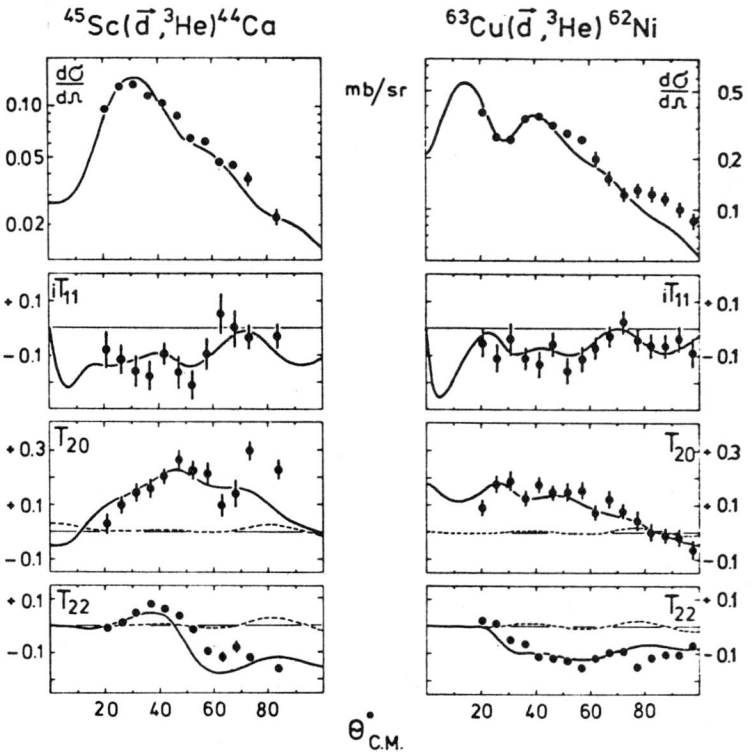

Fig.2 Differential cross sections and analysing power of the ^{45}Sc(d,^{3}He)^{44}Ca and ^{63}Cu(d,^{3}He)^{62}Ni g.s. reactions. The solid and dashed lines are DWBA predictions with and without the ^{3}He D-state contribution, respectively.

together with results of DWBA calculations including the ^{3}He D-state contributions. The predictions shown in the figures are for $D_2 = -0.21$ fm^2. This value is in agreement with the value of $D_2 = -0.22$ fm^2 obtained for the ^{27}Al(d,^{3}He)^{26}Mg reaction in this laboratory[3] and with the corresponding value for (d,t) reactions on ^{118}Sn and ^{208}Pb ($D_2 = -0.24$ fm^2) obtained by Knutson et al.[4], but in disagreement with the results of Brandan and Haeberli[5] and Knutson et al.[6] for ^{64}Zn and ^{63}Cu ($D_2 = -0.37$ and -0.339 fm^2 respectively). All experiments refer to incident deuteron energy between 12 and 13 MeV, except for the measurements on ^{63}Cu[5] which was at 9 MeV.

The ^{3}He D_2 value of the present work is in good agreement with recent theoretical predictions for the tri-nucleon system of Santos et al.[7] using a perturbation calculation with various triton wave functions, and of Kim and Muslim[8] calculations using wave function obtained from a solution of Faddeev equations with the Reid soft core potential.

REFERENCES

1. J.D. Harvey and F.D. Santos, DWCODE, Univ. of Surrey, 1970.
2. C.M. Perey and F.G. Perey, Nucl. Data Tables, 17 (1976).
3. S. Roman et al., Nucl. Phys. A289, 269 (1977)
4. L.D. Knutson et al., Phys. Rev. Lett. 31, 329 (1973).
5. M.E. Brandan and W. Haeberli, Nucl. Phys. A287, 213 (1977).
6. L.D. Knutson et al., Phys. Lett. 85B, 209 (1979).
7. F.D. Santos et al., C19, 238 (1979).
8. Y.E. Kim and Muslim, Phys. Rev. Lett. 42, 1328 (1979).

SPIN DETERMINATION OF DEEPLY-BOUND HOLE STATES FROM ($\vec{d}$,³He) REACTIONS

A. Stuirbrink, K.T. Knöpfle, G. Mairle, H. Riedesel, K. Schindler and G.J. Wagner
Max-Planck-Institut für Kernphysik, D-69 Heidelberg, W.-Germany

V. Bechtold and L. Friedrich
Kernforschungszentrum Karlsruhe, IAK, D-75 Karlsruhe, W.-Germany

ABSTRACT

From the differential cross sections and vector-analyzing powers of the ($\vec{d}$,³He) reactions on ⁹⁰Zr and ¹⁴⁵Sm at 52 MeV spins of deeply-bound hole states have been determined for the first time.

1. INTRODUCTION

While neutron pick-up reactions have been used extensively to study deeply bound hole states in heavy nuclei[1] deeply-bound proton hole states so far have only been investigated in the ¹⁴⁴,¹⁴⁸,¹⁵²Sm (d,³He) reactions at 52 MeV[2]. Previously, we have shown[3] that in light nuclei the angular momentum l and the spin j of the picked-up proton in a ($\vec{d}$,³He) reaction may be determined from the differential cross section $\sigma(\theta)$ and the vector-analyzing power $iT_{11}(\theta)$. The purpose of the present work is (i) a search for deeply-bound hole states in ⁸⁹Y; (ii) a spin determination of deeply-bound hole states in ⁸⁹Y and ¹⁴³Pm.

2. RESULTS

Figs. 1 and 2 show the ($\vec{d}$,³He) spectra on ⁹⁰Zr and ¹⁴⁴Sm targets obtained in maxima of $\ell=3$ and $\ell=4$ angular distributions, respectively. Assumed background shapes are indicated. In addition to previously known[4,5] low-lying hole states both spectra show "giant-resonancelike" structures (shaded areas) peaking near 6.8 MeV in ⁸⁹Y and 5.0 MeV in ¹⁴³Pm.

Figs. 3 and 4 show $\sigma(\theta)$ and $iT_{11}(\theta)$ for all groups identified in ⁸⁹Y and for a known 7/2⁺ state and the deeply-bound hole state in ¹⁴³Pm. In ⁸⁹Y we have also evaluated individually the 5.0 and 6.8 MeV "fine structure peaks" by subtracting a straight background connecting the minima at the bases. Local, zero-range DWBA calculations were performed (Fig. 3,4). For ¹⁴⁴Sm they are identical to those of ref.[2]. The optical potentials for ⁹⁰Zr were taken from ref.[6](deuteron set 2a) and ref.[7] (³He, set V). A bound state potential with $r_o=1.2$fm and $a=0.65$ fm was used. Table 1 contains the results.

3. DISCUSSION

The data presented here provide a unique 7/2⁻ and 9/2⁺ assignment to the dominant fraction of the strength concentrated in the observed deeply-bound hole states in ⁸⁹Y and ¹⁴³Pm, respectively.

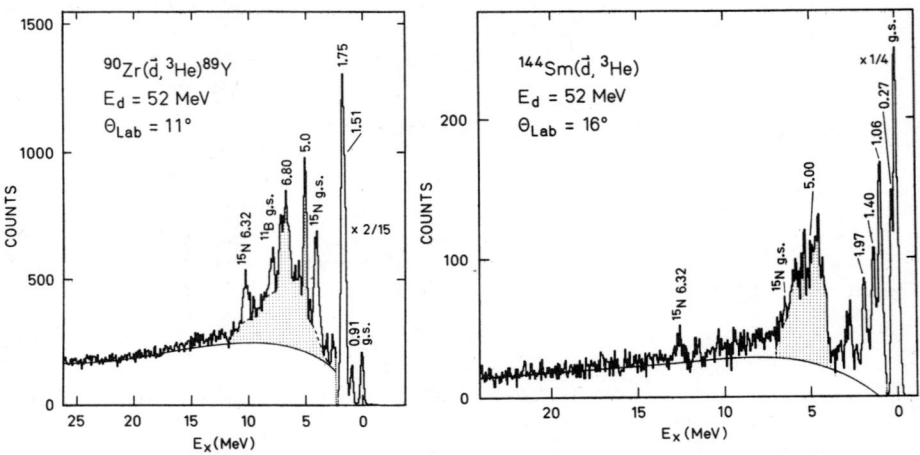

Fig. 1. Spectrum of the ^{90}Zr($\vec{d}$,^{3}He)^{89}Y reaction.

Fig. 2. Spectrum of the ^{144}Sm($\vec{d}$,^{3}He)^{143}Pm reaction.

Both, for the low-lying and for the deeply-bound hole states the spectroscopic factors exhaust the shell model strengths within the errors. This is in contrast to neutron pick-up reactions[1] where the deeply-bound hole states typically carry only 50% of the shell model strengths.

The exhaustion of the total strength allows a meaningful determination of the spin-orbit splitting ε_{so} defined as the difference of the centroid separation energies for $j=l+1/2$ and $j=l-1/2$ orbitals. The results are given in table 2. For comparison we add ε_{so} (i) as implied by the bound state potential in the DWBA calculations ($\lambda=25$); (ii) as obtained from a semi-empirical formula for single particle energies; and (iii) from a Brueckner-Hartree-Fock calculation[10]. While the empirical values (i) and (ii) are roughly correct, the fundamental approach (iii) obtains the correct order of magnitude only from the spin saturated shells (ss) but a nearly vanishing ε_{so} when the negative corrections from the non-spin-saturated shells (nss) are included. In view of this obvious difficulty of many-body theory to reproduce the spin-orbit splitting further measurements of this fundamental nuclear property using the method presented here are in order.

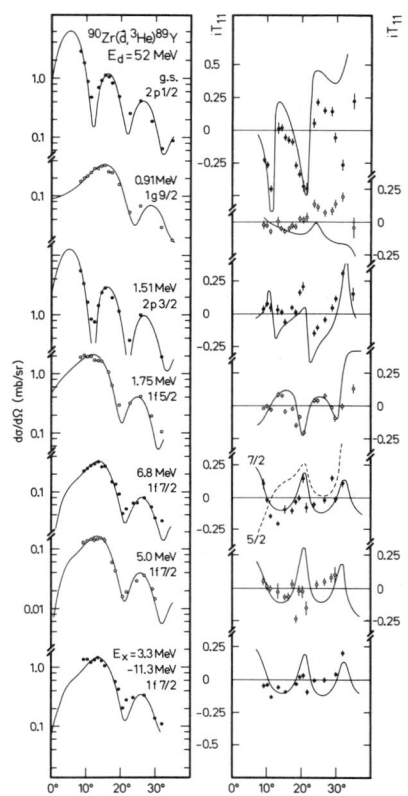

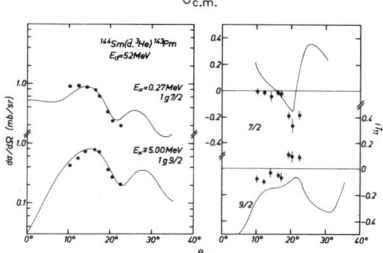

Tab.1 Summary of spectroscopic results

$^A Z$	E_x (MeV)	$n\ell j$	$C^2 S$
^{89}Y	0	2p1/2	1.8
	0.91	1g9/2	1.25
	1.51	2p3/2	3.9
	1.75	1f5/2	8.9
	5.0	1f	0.8
	6.8	1f7/2	2.1
	3.3–11.3	1f7/2	9.2
^{143}Pm	0	2d5/2	3.5
	0.27	1g7/2	8.6
	0.96	1h11/2	1.9
	1.40	2d3/2	0.54
	1.97	2p	0.4
	3.9–6.9	1g9/2	9.4

Tab.2 Spin-orbit Splitting ε_{so} (MeV)

$^A Z$	$n\ell$	exp	W.-S.	semi-emp.[9]	BHF[10] ss + nss
^{90}Zr	1f	5.1	4.0	6.1	3.9–2.5
^{144}Sm	1g	4.7	4.1	5.7	3.9–3.9

Fig. 3 Differential cross sections and vector-analyzing powers for the ^{90}Zr($\vec{d}$,^{3}He) reaction together with DWBA results.

Fig. 4 As Fig.3 but for the ^{144}Sm($\vec{d}$,^{3}He) reaction.

REFERENCES
1. E. Gerlic et al., Phys. Rev. C21 (1980) 124 and refs. therein.
2. P. Doll et al., Phys. Lett. 82B (1979) 357.
3. V. Bechtold et al., Phys. Lett. 72B (1977) 169.
4. B.M. Preedom, E. Newman, and J.C. Hiebert, Phys.Rev.166(1968)1156.
5. B.H. Wildenthal, E. Newman, and R.L. Auble, Phys.Rev.C3(1971)1199.
6. G. Mairle et al., Nucl. Phys. A339 (1980) 61.
7. E.F. Gibson et al., Phys. Rev. 155 (1967) 1194.
8. G.J. Wagner, Invited paper, Proc. Int. Symp. on Highly Excited States in Nuclear Reactions, Osaka (1980).
9. M.A.K. Lohdi and B.T. Wack, Phys. Rev. Lett. 33 (1974) 431.
10. R.R. Scheerbaum, Nucl. Phys. A257 (1976) 77.

THE (t,d) REACTION ON THE Ni ISOTOPES WITH POLARIZED TRITONS

E. R. Flynn, J. A. Cizewski, Ronald E. Brown, R. A. Hardekopf and
J. W. Sunier
University of California, Los Alamos Scientific Laboratory
Los Alamos, New Mexico 87545

ABSTRACT

The ($\vec{t}$,d) reaction has been measured on targets of ^{58,60,62,64}Ni with 17 MeV polarized tritons. Spectroscopic factors, angular momentum and total spin transfer were obtained from the differential cross section and A_y values of levels up to 3.5 MeV in excitation energy. The present (t,d) measurement enables a better description of the $9/2^+$ and $5/2^+$ states which show significant shell crossing effects as a function of increasing neutron number.

The (t,d) reaction favors a higher angular momentum transfer than does the (d,p) reaction and thus can be used more easily to study higher spin states. In addition, the (t,d) reaction has strongly absorbed particles in both entrance and exit channels and thus produces a sharp diffraction pattern in the measured angular distributions which is well described by distorted wave (DW) theory. It has been shown[1] that this strong absorption characteristic also yields significant values of analyzing powers (A_y) for a polarized triton beam, and that J^π values may be easily extracted from these data.

The Ni isotopes represent an extremely interesting nuclear region where the ($\vec{t}$,d) reaction is a very useful tool because of the large range of angular momenta associated with the particle states. As the shell fills in going away from N=28, the $g_{9/2}$ orbital drops rapidly due to pairing forces and produces a deformation tendency in the heavier Ni isotopes as suggested by Hartree-Fock calculations. This tendency also produces a major shell crossing of other orbitals, principally the $d_{5/2}$. Thus significant deviations occur from expected shell model spacing of the single particle orbitals.

The experiments were done using a 17 MeV polarized triton beam of average intensity 45 na and polarization 0.75 at the LASL Van de Graaff facility. Metallic Ni targets of ∽ 300 µgm/cm^2 thickness were bombarded and the resulting reaction deuterons were detected by a helical focal plane detector in a Q3D magnetic spectrometer.[2] The average resolution was 15-18 keV as caused by the target thickness. Cross sections as small as 2 µb/sr were measured using the full Q3D solid angle (14.3 msr). The polarization was measured at the beginning and end of each run and the spin flipped at the source for each angle. The entire procedure was computer controlled.

---

* Work performed under the auspices of the U.S. Department of Energy.

Approximately 25-30 levels were measured in ^{59,61,63,65}Ni up to an excitation energy of ~3.5 MeV. Typical analyzing powers for $1/2^-$, $3/2^-$, $9/2^+$ and $5/2^+$ states in ^{63}Ni are shown in Fig. 1 where they are compared with DW calculations. These calculations were done using published triton parameters (with an added spin-orbit potential) and deuteron parameters. The DW results reasonably describe the A_y data and fit the differential cross sections very well. A slight phase shift between data and calculation is noted in this figure but no adjustments of DW parameters were made to account for this. The principal sensitivity is in the deuteron parameters.

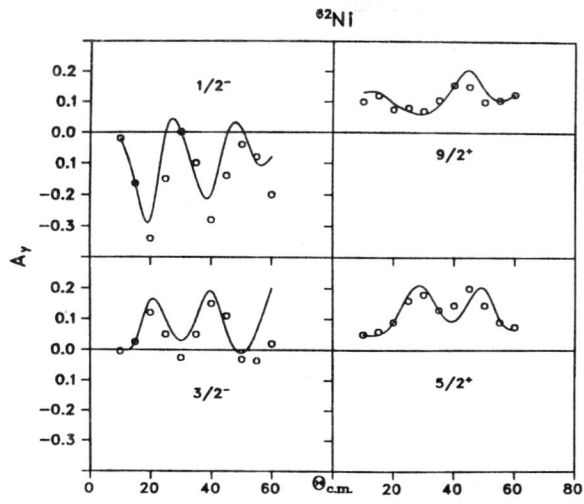

Fig. 1. Typical values for the ^{62}Ni($\vec{t}$,d) reaction.

Table 1 contains the information obtained in the present experiment for the ^{62}Ni($\vec{t}$,d)^{63}Ni reaction; other isotopes will be described in a later report. This table also contains previous results[3] for comparison. The spectroscopic factors obtained in the present work compare well with other results except for the $9/2^+$ state. This discrepancy exists for all the targets when comparing the (t,d) and (d,p) data. However, the (t,d) results give an excellent fit to A_y and fit $d\sigma/d\Omega$ better than do the (d,p) results. The new A_y data confirm most of the suggested spin values and assign several new ones.

Examination of the systematic trend of the $9/2^+$, $5/2^+$ and $1/2^+$ major fragments shows a trend where the $9/2^+$ state drops steeply in excitation energy between ^{59}Ni and ^{63}Ni from over 3 MeV to near 1 MeV and then levels out. This fragment contains ~40% of the total $g_{9/2}$ strength. This trend deviates from that suggested by shell model calculations.[4] The $5/2^+$ strength does not drop as sharply but does drop from above 3.5 MeV in ^{59}Ni to below 2 MeV in ^{65}Ni with about 1/6 of the total strength in one level. This represents a considerable

Table I. ^{62}Ni($\vec{t}$,d) Results

Level	Ex (keV)	Previous Results Assumed J^π	(2J+1) S	ℓ Transfer	Present Results J^π	(2J+1) S
1	0	$1/2^-$	0.85	1	$1/2^-$	0.74
2	89	$5/2^-$	3.40	3	$5/2^-$	3.42
3	155	$3/2^-$	1.15	1	$3/2^-$	0.96
4	518	$3/2^-$	0.32	1	$3/2^-$	0.27
5	(905)				$(9/2^+)$	
6	1001	$1/2^-$	0.82	1	$1/2^-$	0.55
7	(1064)			(1)	$(1/2^-)$	0.008
8	(1145)			(1)	$(3/2^-)$	0.014
9	(1256)					
10	1294	$(9/2^+)$	6.72	4	$9/2^+$	4.20
11	(1324)	$3/2^-$		1	$3/2^-$	0.095
12	(1557)			3	$5/2^-$	0.20
13	(1720)					
14	1787	$5/2^-,7/2^-$				
15	1899	$(5/2,7/2^-)$		1	$(3/2^-)$	0.030
16	2149	$1/2,3/2^-$		1	$(3/2^-)$	0.025
17	2297	$5/2^+$	1.99	2	$5/2^+$	1.37
18	2346			2	$5/2^+$	0.026
19	2519	$(9/2^+)$	2.58	4	$9/2^+$	2.04
20	2700	$1/2^+$	0.10	1	$1/2^-$	0.10
21	2822				$\ell=1/2$	
22	2953	$1/2^+$	0.23	0	$1/2^+$	0.11
23	3010				$\ell=1/2$	
24	3092		0.04	1	$1/2^-$	0.052
25	3180	$5/2,7/2^-$		(3)	$(7/2^-)$	0.17
26	3283	$5/2^+$	0.43		$\ell+1/2$	
27	3340	$5/2,7/2^+$		2	$5/2^+$	0.65
28	3430				$\ell+1/2$	
29	3520	$3/2,5/2^+$		2	$5/2^+$	0.61

washing out of the shell gap at N=50 and is characteristic of a nucleus going deformed. The low spin $1/2^+$ states are not as perturbed.

The present results have confirmed or added to our knowledge of the spin values in the odd Ni nuclei as well as indicating the trend toward deformation of the heavier Ni isotopes.

REFERENCES

1. E. R. Flynn, Ronald E. Brown, F. D. Correll, D. L. Hanson, and R. A. Hardekopf, Phys. Rev. Letts. <u>42</u>, 626 (1979).
2. E. R. Flynn, S. D. Orbesen, J. D. Sherman, J. W. Sunier and R. Woods, Nucl. Instrum. <u>128</u>, 35 (1975).
3. I. M. Turkiewicz, P. Beuzit, J. DeLaunay and J. P. Fouan, Nucl. Phys. <u>A143</u>, 641 (1970).
4. See e.g. C. J. Veje quoted in M. Bohr and B. Mottleson, <u>Nuclear Structure</u>, W. A. Benjamin (1969) p. 239.

MEASUREMENTS OF MASSES AND SPINS OF NEUTRON RICH NUCLEI BY THE $(\vec{t},\alpha)$ REACTION

F. Ajzenberg-Selove
University of Pennsylvania, Philadelphia, PA 19104

E. R. Flynn, Ronald E. Brown, J. A. Cizewski and J. W. Sunier
University of California, Los Alamos Scientific Laboratory*
Los Alamos, New Mexico 87545

ABSTRACT

The $(\vec{t},\alpha)$ reaction has been used to study neutron rich nuclei near A=100 to measure several masses more accurately as well as assign spin values to a variety of states. The nuclei studied are ^{109}Rh, ^{103}Tc, ^{99}Nb and ^{95}Y.

The $(\vec{t},\alpha)$ reaction using polarized tritons is known to give very large analyzing powers (A_y) over a range of nuclei.[1] This feature has permitted extensive surveys of proton hole states in the rare earths with many new spins assigned because of the unique character of the A_y values[2]. An additional feature of this reaction is that when performed on a neutron rich target, it reaches a product which is even more neutron rich. Indeed most of the residual nuclei for such studies are known only from β-decay. The competing reaction to the (t,α) studies is the $(d,^3He)$ reaction which suffers from a negative Q-value instead of the high positive Q-value of the $(\vec{t},\alpha)$ reaction. This has previously restricted such studies to cyclotrons. The use of the polarized triton beam and its resulting distinctive A_y values has now circumvented the principal problem of the (t,α) reaction, the rather featureless angular distributions.

In this study, targets of ^{110}Pd, ^{104}Ru, ^{100}Mo and ^{96}Zr of 200-300 μgm/cm^2 thickness were bombarded with a 17 MeV beam of polarized tritons with intensity averaging 45 na and polarization of 0.75. The reaction product α-particles were detected in a helical focal plane detector of a Q3D spectrometer.[3] The energy resolution was dominated by target thickness effects and varied from 20 keV to 30 keV as a result. The mass identification ability of the detector system was excellent, permitting very clean α-spectra and thus excellent mass determination. Targets of ^{60}Ni and ^{62}Ni were used as calibration targets because of the known masses and similar Q-values.

A spectrum of the reaction ^{110}Pd$(\vec{t},\alpha)^{109}$Rh is shown in Fig. 1 and illustrates the typical quality of the data. As stated earlier, the analyzing powers can be quite large, -0.8 being observed, and this makes spin assignments straightforward, if a direct reaction can be assumed. The first two states of ^{109}Rh are illustrated in Fig. 2

---
*Work performed under the auspices of the U.S. Department of Energy.

0094-243X/81/690697-03$1.50 Copyright 1981 American Institute of Physics

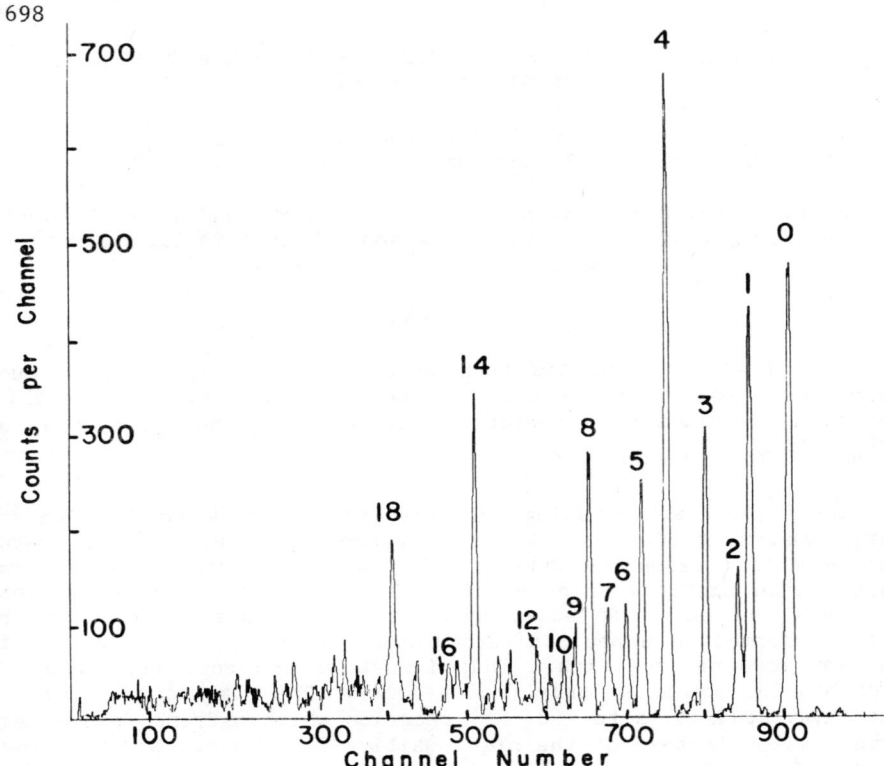

Fig. 1. Spectrum of the ^{110}Pd$(\vec{t},\alpha)^{109}$Rh reaction at 10°. The triton spin was up.

where the A_y values are given as a function of angle. The curves are a result of Distorted Wave (DW) calculations following the procedures of Ref. 1. Indeed the data shown in Fig. 2 very closely resemble the data taken on Zr of Ref. 1. These results thus establish the spins of the ground and first excited states of ^{109}Rh to be 9/2$^+$ and 1/2$^-$ respectively. The mass defect measurement of ^{109}Rh is established from the measured Q-value of 8.99 ± 0.03 MeV to be Δm = -84,800 ± 40 keV as compared to a prediction based on systematics[4] of -85,110 keV. Part of the present uncertainty is due to the ±20 keV error in the ^{110}Pd mass.

The ^{103}Tc results are somewhat different in that the ground and first excited states are rather weak and appear to be formed from a weak coupling multiplet. The mass defect given by the ground state Q-value of 9.05 ± .03 MeV is Δm = -84,724 ± 30 keV. This is to be compared with -84,910 ± 100 given in Ref. 4. The discrepancy considerably exceeds the error given from the β-decay work. The spin of

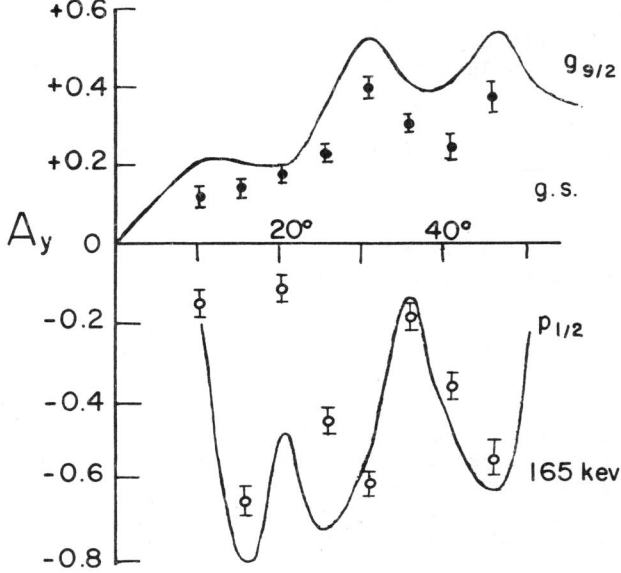

Fig. 2. Examples of A_y values for the ground and first excited states in the $^{110}Pd(\vec{t},\alpha)^{109}Rh$ reaction. The lines represent DW calculations.

the ground state is not certain because it is weakly populated, and a careful analysis must be made. The strong $9/2^+$ and $1/2^-$ states expected in this region lie at 141 keV and 178 keV, respectively, and may be assigned by their similarity in A_y values to Fig. 2.

The $^{100}Mo(\vec{t},\alpha)^{99}Nb$ reaction confirms the suggested ground state spin assignment of $9/2^+$ and places the $1/2^-$ spin as the first excited state. For the $^{96}Zr(\vec{t},\alpha)^{95}Y$ case these spins are reversed with the $1/2^-$ being the ground state.

The present results indicate that the $(\vec{t},\alpha)$ reaction is a valuable tool in measuring both the masses and spins of neutron rich nuclei.

REFERENCES

1. E. R. Flynn, R. A. Hardekopf, J. D. Sherman, J. W. Sunier and J. P. Coffin, Phys. Rev. Letts. 36, 79 (1975).
2. D. G. Burke, G. Lovhoiden, E. R. Flynn and J. W. Sunier, Nucl. Phys. A318, 77 (1979) and references therein.
3. E. R. Flynn, S. D. Orbesen, J. D. Sherman, J. W. Sunier and R. Woods, Nucl. Instrum. 128, 35 (1975).
4. A. H. Wapstra and K. Bos, Atomic Data and Nucl. Data Tables 19, 215 (1977).

THE ^{194,196,198}Pt$(t,\alpha)^{193,195,197}$Ir REACTIONS WITH POLARIZED TRITONS

J. A. Cizewski, E. R. Flynn, J. W. Sunier, Ronald E. Brown
University of California, Los Alamos Scientific Laboratory
Los Alamos, New Mexico 87545

D. G. Burke
Physics Dept., McMaster University, Hamilton, Ont., Canada, L8S 4K1

ABSTRACT

The ^{194,196,198}Pt$(\vec{t},\alpha)^{193,195,197}$Ir reactions have been measured. Angular distributions of cross sections and analyzing powers for levels up to ∿2.5 MeV in each residual nuclide have been obtained, and comparisons with DWBA predictions permit spins, parities and pickup spectroscopic strengths to be determined. The results are being analyzed with the aim of testing the existence of supersymmetric structures in nature.

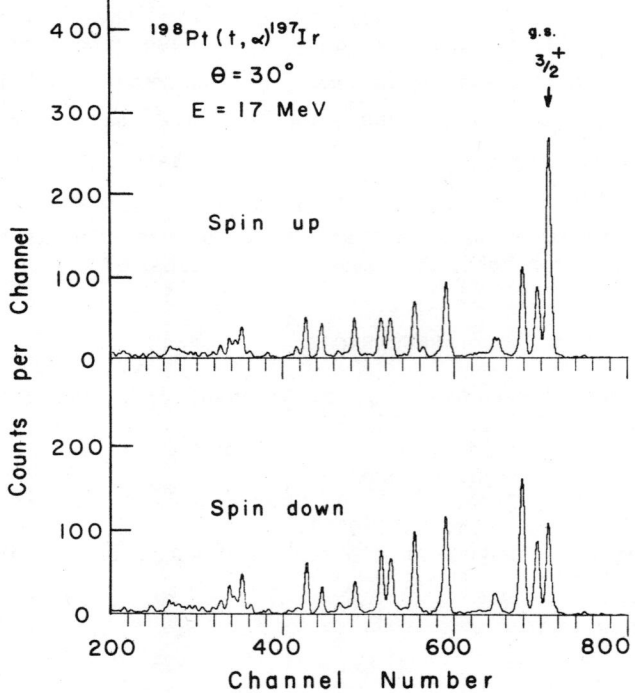

Fig. 1. Spectra from the ^{198}Pr$(\vec{t},\alpha)$ reaction at θ=30°.

*Work performed under the auspices of the US Department of Energy.

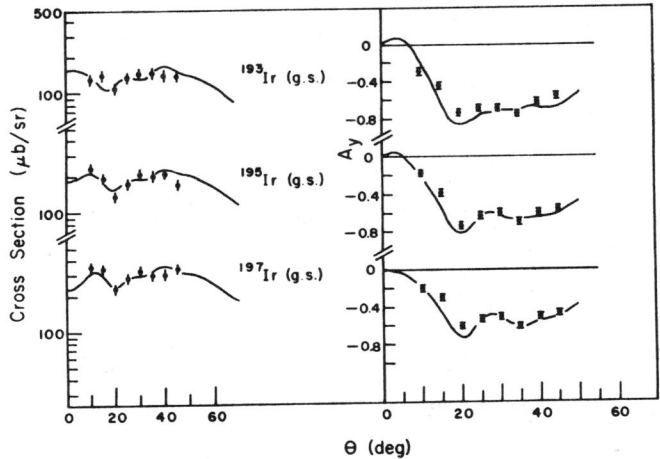

Fig. 2. Cross sections and analyzing powers for the three ground state transitions.

INTRODUCTION

The levels at low excitation energies in ^{193}Ir have been studied by several reaction and decay processes but very little information existed about the properties of the excited states in ^{195}Ir and nothing was known about the levels of ^{197}Ir. Iachello[1] has pointed out the need for more detailed knowledge of states in odd-mass nuclei in the Os-Pt transitional region. These considerations prompted the present study with the $(\vec{t},\alpha)$ reaction, which has been well established as a powerful spectroscopic tool for the study of proton hole states (e.g. Refs. 2,3).

EXPERIMENTAL PROCEDURE

The ^{194,196,198}Pt$(\vec{t},\alpha)^{193,195,197}$Ir reactions have been studied with 17 MeV polarized tritons from the Tandem Van de Graaff accelerator of the Los Alamos Scientific Laboratory. The reaction products were analyzed with a Q3D magnetic spectrometer and detected with a position-sensitive helical-cathode proportional counter. The isotopically enriched targets were prepared with a sputtering technique and had thicknesses of ∼120 μg/cm². At each angle one spectrum was recorded with the incident triton spin "up" and another with the spin "down." As typical results, the spectra from the ^{198}Pt$(\vec{t},\alpha)$ reaction at θ=30° are shown in Fig. 1, where the effects of the large analyzing powers for several of the states can be readily seen. The resolution was typically ∼ 20 keV (FWHM).

Angular distributions of cross sections and analyzing powers for the ground states of ^{193,195,197}Ir are shown in Fig. 2. The solid curves are DWBA predictions for $I^\pi = 3/2^+$ using the optical parameters of ref. 3. The ^{193}Ir ground state has long been known to be $3/2^+$, and that of ^{195}Ir was shown to be $3/2^+$ in a recent ^{193}Ir(t,p) experiment[4]. The results in Fig. 2 show that the ^{197}Ir ground state is also $3/2^+$. Similar comparisons with DWBA predictions provide spin and parity assignments for many levels in each of the nuclides studied.

DISCUSSION

An interesting feature of the data shown in Fig. 2 is that the ground state cross section is seen to increase significantly as the neutron number increases for the three cases studied. When spectroscopic factors, S, are extracted using

$$\frac{d\sigma}{d\Omega} = NS \left(\frac{d\sigma}{d\Omega}\right)_{DWBA} \tag{1}$$

with N=23, one obtains values of S = 1.9, 2.6 and 3.9 for the ground state transitions with targets of ^{194,196,198}Pt, respectively. The uncertainties in these strengths may be of the order of 30% (Ref. 3). The value of S = 1.9 for the ^{193}Ir gound state is comparable to the strength expected for the 3/2 $3/2^+$[402] Nilsson state, which has previously been considered a likely interpretation of the ground state. On the other hand, the larger strength observed for the ^{197}Ir ground state is almost equal to the value of 4 expected for pickup from a filled $d_{3/2}$ shell, even though the Z=78 targets would be expected on the basis of either a spherical or deformed shell model to have the $d_{3/2}$ orbitals only half filled. The observed strength is in fact comparable to that reported in the ^{208}Pb($\vec{t},\alpha$) study[2], for which the target could be expected to have the $d_{3/2}$ shell filled. Analysis of the data is continuing with the particular aim of comparing observed spectroscopic strengths for positive-parity states to the predictions of a supersymmetry scheme.

REFERENCES

1. F. Iachello, Phys. Rev. Lett. **44**, 772 (1980), and private communication.
2. E. R. Flynn, R. A. Hardekopf, J. D. Sherman, J. W. Sunier and J. P. Coffin, Nucl. Phys. **A279**, 394 (1977).
3. D. G. Burke, G. Lovhoiden, E. R. Flynn and J. W. Sunier, Phys. Rev. **C18**, 693 (1978).
4. J. A. Cizewski, E. R. Flynn, R. E. Brown and J. W. Sunier, to be published.

APPLICATION OF ($\vec{^3\text{He}}$,d) AND ($\vec{^3\text{He}}$,α) REACTIONS IN SPECTROSCOPY

O. Karban, A.K. Basak, G.C. Morrison, J.M. Nelson and S. Roman
Department of Physics, University of Birmingham, England

ABSTRACT

Angular distributions of analysing powers of one-nucleon transfer reactions induced by 33 MeV polarized ^{3}He beam show characteristics which can be unambiguously associated with the total angular momentum of the transferred nucleon. Over 120 transitions were observed with 19 target nuclei from A=4 to 58 involving single-particle orbits from $1s_{1/2}$ to $1g_{9/2}$. Such experiments can be used for a model-independent determination of spins and configurations of nuclear states.

The j-dependence of analysing powers (AP) measured in one-nucleon transfer reactions with a 33 MeV polarized ^{3}He beam was first observed in the 1p shell[1] and later also in the 1d and 2p shells[2,3]. During the past 4 years polarization effects in interaction of ^{3}He projectiles with nuclei were studied on 19 targets with A=4 to 58 and in all cases data were taken for both proton stripping and neutron pick-up reactions. In total, over 120 angular distributions of analysing powers were obtained involving momentum transfer of nucleons in the $1s_{1/2}$ to $1g_{9/2}$ shells. This amount of data made it possible to establish some regularities in behaviour of the analysing powers, in particular their systematic dependence on the j-value of the transferred nucleon.

The pattern of the j-dependence can be deduced by comparing the analysing powers for transitions where the j-value is well known and unique. In the 1p shell such "reference" curves can be obtained e.g. from the ^{12}C(^{3}He,α)^{11}C and ^{11}B(^{3}He,d)^{12}C reactions. Here the AP behaviour in the 15°<θ_{cm}<60° angular range, confirmed on a number of transitions, is as follows: in the (^{3}He,α) reaction, the AP for j=ℓ+1/2 is negative while for j=ℓ-1/2 it is about twice as large and positive; in the (^{3}He,d) reaction both j=ℓ±1/2 transfers have generally negative AP but for j=ℓ-1/2 there is an extra minimum (oscillation) around 20°.

An example of application of this j-dependence is shown in Fig.1 where the AP of the ^{17}O(^{3}He,α)^{16}O reaction populating the 12.97 + 13.25 MeV doublet and the 18.48 MeV state are compared with AP of the "reference" transitions to ^{15}O$_{g.s.}$ (1/2$^-$) and ^{15}O*6.18 (3/2$^-$) respectively (represented by smooth lines drawn through experimental points[3]). These momentum transfers indicate that the configuration of the 12.97 and 13.25 MeV states is ($1d_{5/2}1p_{1/2}^{-1}$) and that of the 18.48 MeV state is ($1d_{5/2}1p_{3/2}^{-1}$) in agreement with assumptions of ref.[4].

Analysing powers for the 3/2$^+$ and 5/2$^+$ transitions in the 1d shell are compared in Fig.2 where solid lines were drawn through the data points to guide the eye. The AP behaviour in both

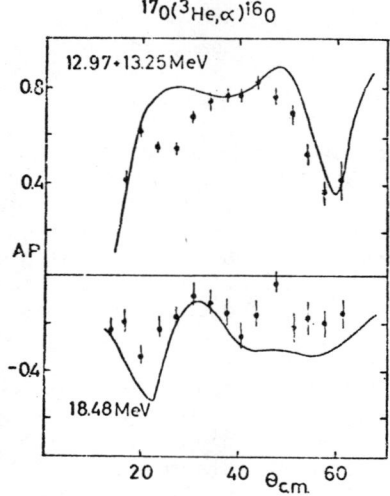

Fig.1 Analysing powers of the $^{17}O(^3He,\alpha)^{16}O$ reaction leading to the 12.97 + 13.25 and 18.48 MeV states compared with AP of the $^{16}O(^3He,\alpha)^{15}O$ reaction to the ground and 6.18 MeV states (solid lines).

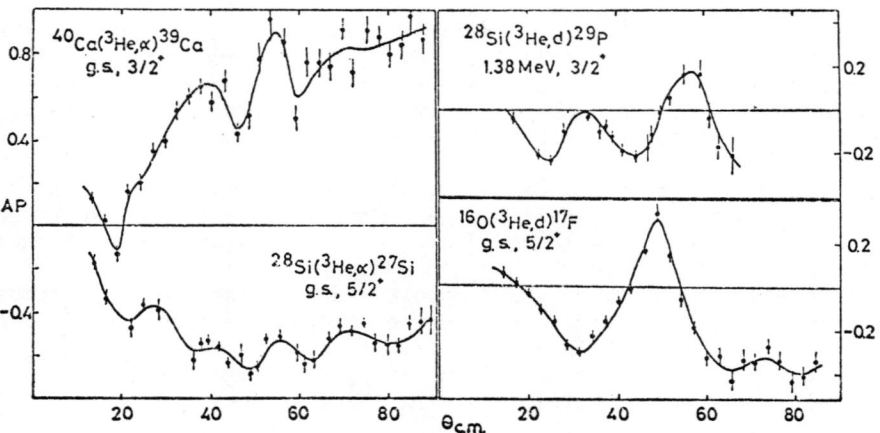

Fig.2 Comparison of the analysing powers of the $3/2^+$ and $5/2^+$ transitions in the 1d shell.

reactions is essentially the same as described above for the 1p shell. The main features of the $(^3He,\alpha)$ reaction analysing powers, namely the large values which are practically constant with angle but opposite in sign for $j=\ell\pm1/2$ together with an apparent ℓ-independence of the effect, suggest that the reactions are dominated by some common simple process. A classical model of the $(^3He,\alpha)$ reaction which can account for these features has been presented in ref[3].

In the 1f and 2p shells the $(\vec{^3He},d)$ reaction AP have a j-dependence which is similar to that in the $(\vec{d},p)$ reactions, i.e. the two angular distributions oscillate practically in opposite phase^2. The $f_{7/2}$ AP in the (^{3}He,α) reaction has the expected negative sign but not enough data have been accumulated for the $f_{5/2}$ and 2p-shell transfers. The ($\vec{^3\text{He}}$,d) reaction has been recently used[5] to deduce spins of some analog and anti-analog states in ^{55}Co.

REFERENCES

1. O. Karban et al., Nucl. Phys. A269 312 (1976).
2. S. Roman et al., Nucl. Phys. A284 365 (1977).
3. Y.-W. Lui et al., Nucl. Phys. A333 221 (1980).
4. G. Mairle et al., Nucl. Phys. A299 39 (1978).
5. O. Karban et al., Ann.Prog.Rep. Birmingham, 1980, unpublished.

REACTION MECHANISM STUDIES WITH POLARIZED ^{3}He

O. Karban, A.K. Basak, and S. Roman
Department of Physics, University of Birmingham, England

ABSTRACT

Strong sensitivity of the (^{3}He,α) reaction analysing power to the transferred angular momentum j can be used to study reaction mechanisms. Examples of analysing powers measured on ^{7}Li, ^{13}C and ^{24}Mg targets show the effect of j-transfers which are not allowed by a direct reaction selection rules, indicating that a contribution of two-step processes are involved.

In a direct one-nucleon transfer reaction on a zero-spin target the transferred total angular momentum is defined by the spin of the final nucleus. On a non-zero spin target the initial and final spins usually allow several j-values to participate in the reaction but these are restricted by shell-model considerations. This latter argument also applies to some "shell-model forbidden" transitions, e.g. the ^{24}Mg(^{3}He,α) reaction leading to the 7/2$^+$ state at 2.05 MeV in ^{23}Mg. In order to explain such reactions, more complicated mechanisms have to be taken into account, where the transferred momenta are not restricted by the direct reaction selection rules. Measurements of analysing powers (AP) in these reactions should be well suited to study the reaction mechanism due to their sensitivity to the total angular momentum transfer, in particular the (^{3}He,α) reaction where large polarization effects are observed[1]. A deviation from the expected AP could indicate a presence of multi-step processes.

Two examples of such a situation have been observed in the 1p shell, namely in the ^{7}Li(^{3}He,α)^{6}Li*$^{2.184}$(3$^+$) and ^{13}C(^{3}He,α)^{12}C*$^{4.44}$(2$^+$) reactions. In both cases the transfer is expected to be a pure j=3/2$^-$ for which the AP has a deep minimum at forward angles. As can be seen in Fig.1 this minimum is missing in the AP angular distributions. The dashed lines in the figure were drawn through the "reference" data points of the ^{7}Li(^{3}He,α)^{6}Li*$^{3.56}$(0$^+$) and ^{12}C(^{3}He,α)^{11}C$_{g.s.}$(3/2$^-$) reactions respectively. It is well known that inelastic scattering to the first excited states of ^{6}Li and ^{12}C is strong and the corresponding coupling to the ground state makes it possible for the transfer reaction to proceed also in two steps. Some preliminary calculations using the code[2] CHUCK3 were performed for both reactions with spectroscopic amplitudes taken from ref.[3] and deformation parameters β(^{6}Li) = 0.7 and β(^{12}C) = 0.5. The AP predicted by the 2-step path alone are shown as dotted lines while AP due to coherent mixture of direct + 2-step processes are represented by solid lines in Fig.1. The calculated two-step AP has positive analysing powers and it is apparently responsible for the observed "dilution" of the 3/2$^-$ transfer AP. The 2.184 MeV transition is well described while interference effects overestimate the positive AP contribution to the 4.44 MeV ^{12}C data.

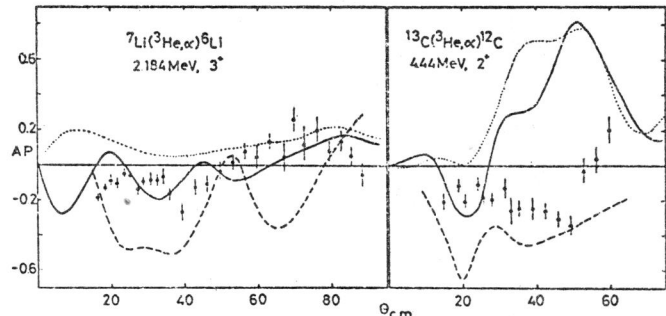

Fig.1. Analysing powers of the (^{3}He,α) reaction leading to the 2.184 MeV (3$^+$) state in ^{6}Li and 4.44 MeV (2$^+$) state in ^{12}C compared with "reference" data for pure 3/2$^-$ transfer (dashed lines) and predictions of coupled-reaction calculations.

Three states populated in the ^{24}Mg(^{3}He,α)^{23}Mg reaction have analysing powers inconsistent with a direct transfer, namely the ground state (3/2$^+$), 2.05 MeV (7/2$^+$) and 3.79 MeV (3/2$^-$) states. From the empirical j-dependence of the (^{3}He,α) analysing powers[1,4] it can be concluded that the ground state transition involves both j=3/2$^+$ and 5/2$^+$, the 2.05 MeV AP has a signature of the 5/2$^+$ transfer and a 1/2$^-$ transfer contributes to the population of the 3.79 MeV state. Collective-model calculations[5] of spectroscopic amplitudes of direct and two-step processes via the 1.37 MeV (2$^+$) state in ^{24}Mg are in agreement with these conclusions. Predictions of CHUCK3 using spectroscopic amplitudes of ref.[5] are compared with the data in Fig.2. Although details of the AP angular distributions were not reproduced the main features are well described by the coupled-reaction calculations.

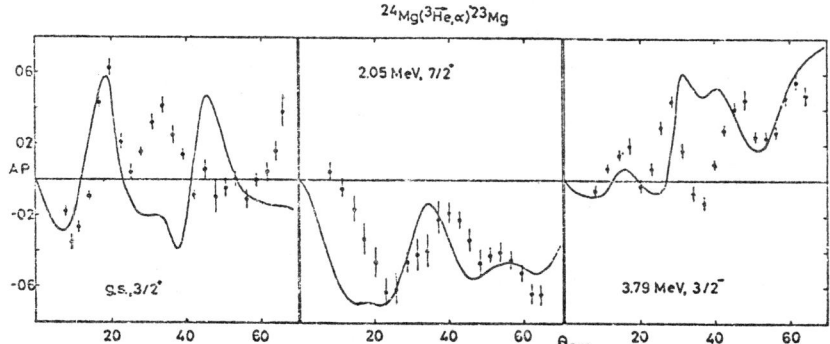

Fig.2. Analysing powers of the (^{3}He,α) reaction populating the ground, 2.05 and 3.79 MeV states in ^{23}Mg compared with coupled-reaction calculations.

Optical-model parameters derived from fitting the ^{3}He - ^{24}Mg elastic scattering data and deformation parameter $\beta = 0.45$ were used in the calculations. In the case of a strong coupling the optical potential has to be re-adjusted since the inelastic channels are accounted for explicitly. It is expected that such a procedure would lead to an improved fit to the reaction data.

REFERENCES

1. Y.-W. Lui et al., Nucl. Phys. A333, 221 (1980).
2. J. Comfort, code CHUCK3, private communication.
3. S. Cohen and D. Kurath, Nucl. Phys. A101, 1 (1967).
4. O. Karban et al., Nucl. Phys. A269, 312 (1976);
   F. Entezami et al., this Proceedings.
5. R.O. Nelson and N.R. Robertson, Phys. Rev. C6, 2153 (1972).

UNNATURAL PARITY TRANSITION IN (p,t) REACTION

Y. Toba, Y. Aoki, H. Iida, S. Kunori, K. Nagano and K. Yagi
Institute of Physics and Tandem Accelerator Center
The University of Tsukuba, Ibaraki 305, Japan

ABSTRACT

Analyzing powers for ^{208}Pb(p,t)^{206}Pb and ^{62}Ni(p,t)^{60}Ni reactions leading to unnatural parity 3^+ states at 1.34 MeV (for ^{206}Pb) and 2.63 MeV (for ^{60}Ni) were measured at E_p=22.0 MeV. For the ^{208}Pb(p,t)^{206}Pb(3^+) reaction, the cross section was fitted well by a two-step (p,d,t) calculation, but the analyzing power was not so well. Effects of other reaction mechanisms, e.g. (p,p',t) and (p,d*,t) processes, were estimated and found to be small.

EXPERIMENTAL RESULTS AND ANALYSES

In an unnatural parity (p,t) transition, a one-step process is forbidden within the limits of the conventional zero-range DWBA analysis. So far, two-step processes have been considered straightforwardly to explain several unnatural parity transitions.[1] On the other hand, the one-step process has been shown by Nagarajan et al.[2] to have as large effect as to explain the experimental cross section data of the ^{208}Pb(p,t)^{206}Pb(1.34 MeV, 3^+) reaction at E_p=35 MeV, if it is estimated with an exact finite-range (EFR) DWBA calculation using a realistic triton wave function.

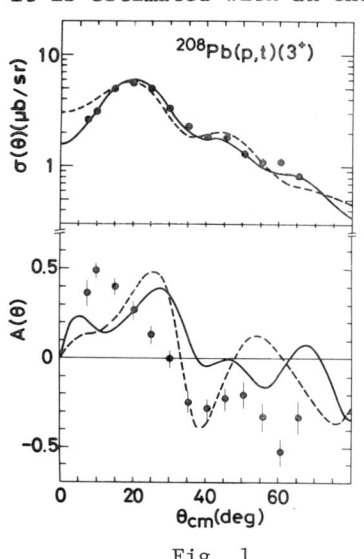

Fig. 1

To clarify the reaction mechanism for an unnatural parity (p,t) transition, analyzing power measurements of ^{208}Pb(p,t)^{206}Pb(1.34 MeV, 3^+)[3] and ^{62}Ni(p,t)^{60}Ni(2.63 MeV, 3^+) reactions were done using a 22.0 MeV polarized proton beam. The experimental data are shown in Figs. 1 and 2. Cross sections are small in magnitude by 2- to 3- order compared to natural parity transitions (e.g. 0^+_g, 2^+_1, 4^+_1, etc.).

For the ^{208}Pb(p,t)^{206}Pb(3^+) reaction, calculations were done assuming (p,d,t) sequential transfer process. The wave function of the 3^+ state can be assumed to have a pure configuration $(p_{1/2}^{-1} f_{5/2}^{-1})$.[4] Therefore, the $1/2^-$ ground

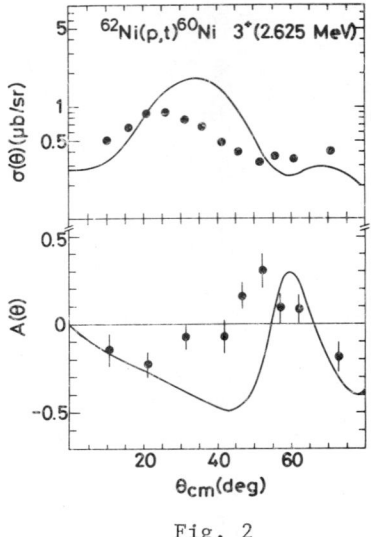

Fig. 2

state and the 5/2⁻ state (0.57 MeV) in ^{207}Pb are taken into account as intermediate channels. Solid and dashed lines correspond to the results of the calculations using two different optical potential sets, respectively. The experimental cross section is well reproduced. On the other hand, fitting of the calculation for the analyzing power is not so good. Finite-range calculations for the (p,d,t) process were also tried and the resultant analyzing power was found to be almost unchanged.

Effects of other reaction mechanisms, e.g. (p,p',t) and (p,d*,t), on the ^{208}Pb(p,t)^{206}Pb(3⁺) reaction were also estimated. The contribution of an inelastic multistep process via the 3⁻ state (2.61 MeV) in ^{208}Pb is shown in Fig. 3 as dash-dotted line. The calculated cross section is much smaller than the experimental one especially at forward angles. Deuteron break-up effect on the (p,d,t) process was estimated following the calculation method by Hashimoto.[5] Dashed and dash-dotted lines in Fig. 4 show the result of the calculation corresponding to the ^{3}S and ^{1}S states of the (p,n) system in the intermediate channel. Node=n in the figure means the Weinberg state with n nodes as the (p,n) system. Therefore, the dashed line of Node=0 corresponds to the normal (p,d,t) process. The optical potential for the (p,n) system is obtained from a folding method. It can be mentioned that contribution of the (p,d*,t) process is small compared to the normal (p,d,t) process (^{3}S, Node=0).

Recently, Igarashi and Kubo have performed the EFR (p,t) and (p,d,t) calcula-

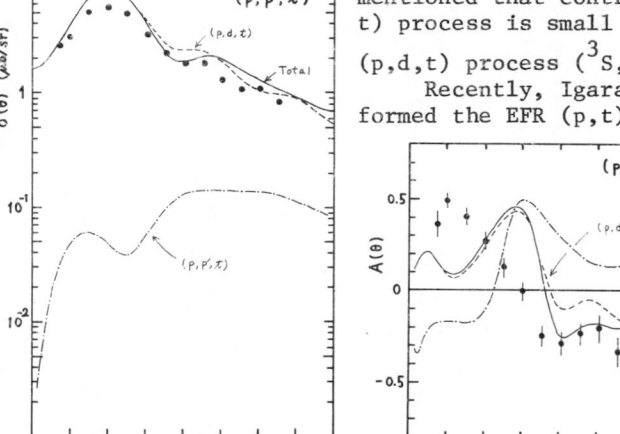

Fig. 3

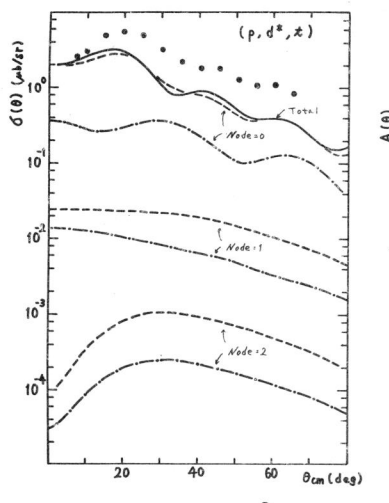
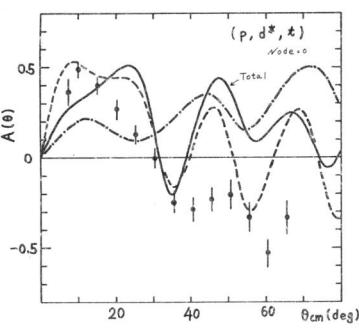

Fig.4

tions for the ^{208}Pb(p,t)^{206}Pb(3^+) reaction using realistic deuteron and triton (Sasakawa and Sawada) wave functions and found that the one-step term is small against the calculation of Nagarajan et al.[2] and the two-step term is larger than the one-step one by about one order in magnitude of cross sections.[6]

REFERENCES

1. J. D. Burch, M. J. Schneider and J. J. Kraushaar, Nucl. Phys. A299, 117 (1978) and references therein.
2. M. A. Nagarajan, M. R. Strayer and M. F. Werby, Phys. Lett. 47B, 237 (1973).
3. Y. Toba, Y. Aoki, S. Kunori, K. Nagano and K. Yagi, Phys. Rev. C20, 1204 (1979).
4. W. W. True and K. W. Ford, Phys. Rev. 109, 1675 (1958).
5. N. Hashimoto, Prog. Theor. Phys. 59, 804 (1978).
6. M. Igarashi and K.-I. Kubo, in Proceedings of the Tsukuba Symposium on Polarization Phenomena in Nuclear Reactions edited by K. Yagi, (NSSRP, The University of Tsukuba, 1980) p. 182; Contribution to this Conference.

MULTISTEP PROCESSES IN THE REACTION ^{116}Sn $(\vec{p},t)$ AT 25.1 MeV

W.F. Feix, J.H. Polane and P.J. van Hall,
Cyclotron Laboratory, Eindhoven University of Technology
Eindhoven, The Netherlands.

The contributions of two successive one-neutron transfer steps in the (p,t) reaction mechnism, especially in forbidden transitions, have been studied by several authors. Our purpose was to investigate the influence of these processes upon the allowed transitions to the ^{114}Sn 0^+gs and 2^+ (1.30 MeV) states. Moreover, we included a search for inelastic scattering effects in the transfer process into the analysis.

The ^{116}Sn $(\vec{p},t)^{114}$Sn $0^+, 2^+$ measurements were performed at the Eindhoven A.V.F. Cyclotron and included the simultaneous record of the one-neutron transfer channels ^{116}Sn $(\vec{p},d)$ ^{115}Sn $1/2^+$, $5/2^+$ as well. The zero-range calculations, including non-locality corrections, were carried out with the code CHUCK2[1]. The full coupling scheme was the following (inelastic proton scattering was found to give only minor corrections to our results):

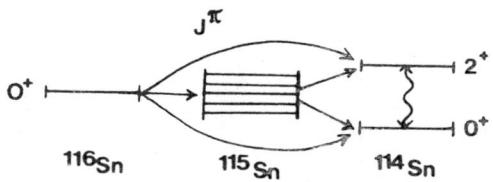

We calculated the spectroscopic amplitudes for both one and two-neutron transfer steps according to the pairing model of vibrational states[2] from experimentally determined occupation amplitudes[3] and from the BCS wave functions of Gillet[4]. The microscopic two-neutron formfactors were constructed according to the Bayman-Kallio method, taking into account the five valence shells of ^{115}Sn. Inelastic scattering amplitudes were assembled with the collective model (β_2^{triton} = 0.114). For the zero-range normalizations of the direct processes we took (in units of MeVfm$^{3/2}$): $N_o(p,d)$ = +122.5, $N_o(d,t)$ = -225., and $N_o(p,t)$ = -905. The one-neutron transfer values are widely used, whereas $N_o(p,t)$ is the result of a procedure adjusting the calculated magnitude of the ground state transition to the experimental value within the coupling scheme given above.

Fig. 1a shows up the angular distributions of the differential cross sections (normalized to the experimental ones) and the analyzing power, calculated from our most complex coupling scheme with two different sets of optical model parameters. It turns out that the proton and deuteron potentials PM, DM, achieved from a ^{116}Sn (p,d) analysis[5], together with an average triton potential TB[6] provide a good description of our data within the reaction model assumed. The choice of the proton and triton potentials PF, TF[7], successful in the limit of direct processes, can be excluded here. Moreover, this view

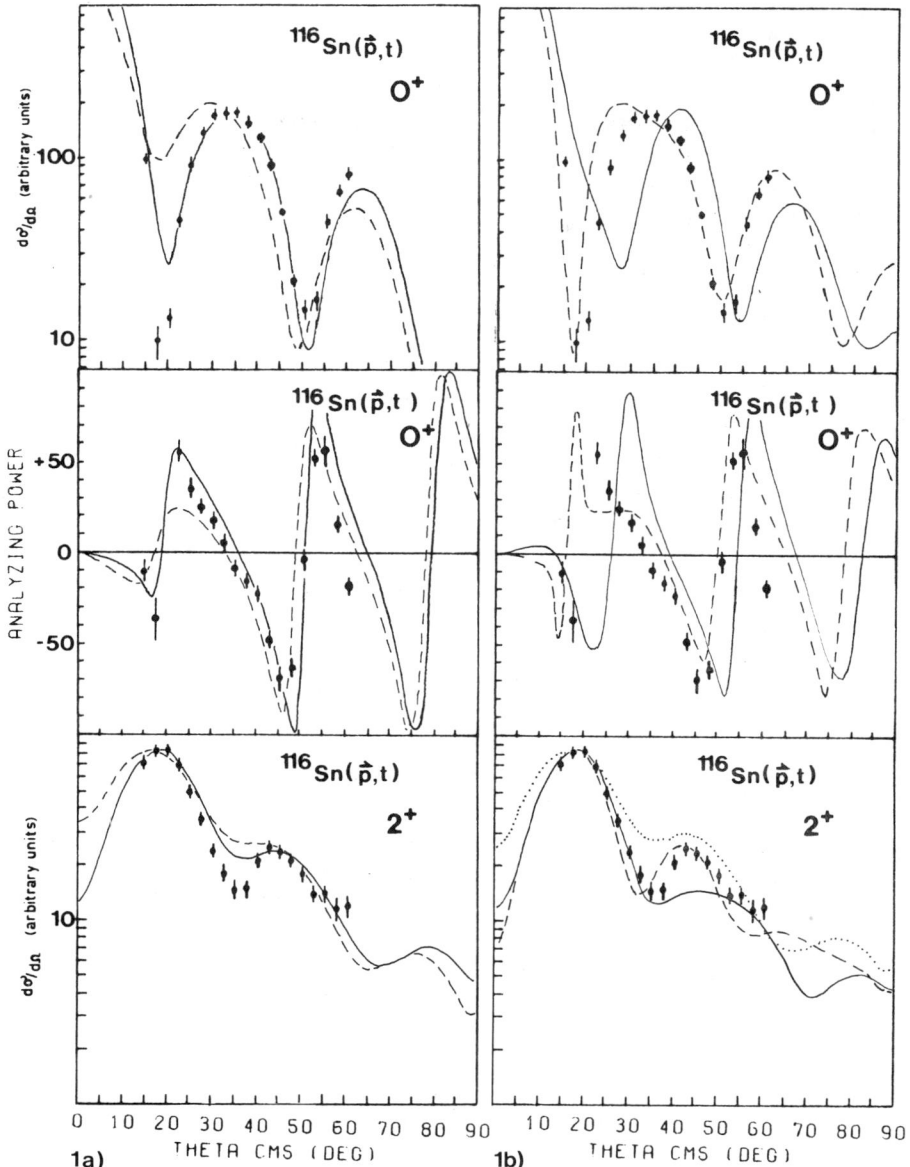

Fig. 1a Full calculation, full line : PM, DM, TB potentials, dashed line : PF, DM, TF potentials.

Fig. 1b Calculations of part processes, full line : direct steps only, dashed line : sequential steps only, dotted line : $(p,t_o)(t,t')$.

is stressed if we look at the results for the simulaneously calculated one-neutron transfer channels (fig. 2), which show a preference for the PM potential as well.

Fig. 1b presents the results of calculations (potentials PM, DM, TB) taking only part processes of the total coupling scheme into account. It is seen that the direct or sequential transfer alone cannot describe the 0^+ data. The sequential pick-up reproduces the 2^+ angular distribution, whereas the direct step or the $(p,t_o)(t,t')$ process do not.

Concerning the magnitudes we conclude from our calculations that for the groundstate transition direct and sequential processes have the same order of magnitude, whereas the direct transfer and the $(p,t_o)(t,t')$ process do so for the 2^+ transition. In the full calculation we reproduce the experimentally observed ratio $d\sigma(0^+):d\sigma(2^+)$ and the absolute magnitudes, which is not the case for the direct or sequential processes alone.

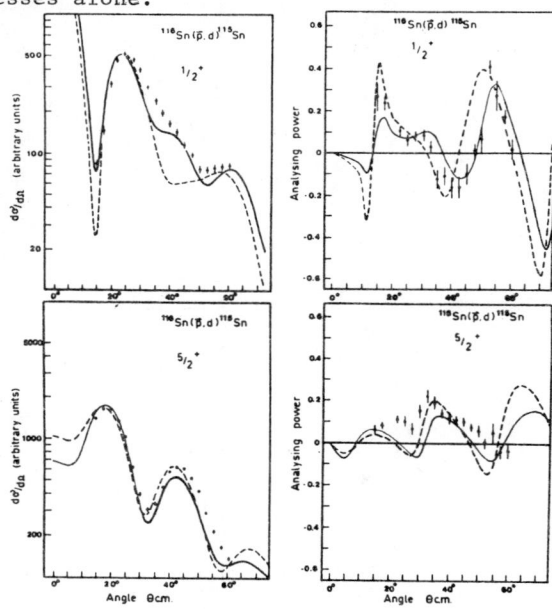

Fig. 2   *One-neutron transfer channels, full line : PM, DM potentials, dashed line : PF, DM potentials.*

REFERENCES

[1] P.D. Kunz, Univ. of Colorado, unpublished.
[2] S. Yoshida, Phys. Rev. 123 (1961) 2122; Nucl. Phys. 38 (1967) 380
[3] S.Y. Van der Werf et al. Nucl. Phys. A289 (1977) 141
[4] E.J. Schneid et al. Phys. Rev. 156 (1962) 1316
[5] V. Gillet et al, Le Journal de Phys. 37 (1976) 189
[6] B. Mayer et al, Nucl. Phys. A177 (1971) 205
[7] H.W. Barz, Nucl. Phys. A122 (1968) 625
[8] D.G. Fleming et al, Nucl. Phys. A157 (1970) 1

REACTION MECHANISMS OF THE ^{208}Pb(p,t)^{206}Pb (3^+ Ex=1.34MEV) REACTION

M. Igarashi
Department of Physics, Tokyo Medical College,
Shinjuku, Shinjuku-ku, Tokyo 160

K.-I. Kubo
Department of Physics, Faculty of Sience,
Tokyo Metropolitan University, Fukazawa, Setagaya-ku, Tokyo 158

ABSTRACT

The roles of the one-step and the sequential (two-step) transfer process in the ^{208}Pb(p,t)^{206}Pb reaction leading to the unnatural parity state 3^+ (Ex=1.34Mev) were investigated. The exact evaluations of both processes were carried out employing the realistic deuteron and triton wave functions and the realistic interaction which causes the transfers of each step. The sequential transfer process is dominant in this reaction and the data of analyzing power comfirm this fact.

INTRODUCTION

The (p,t) transitions to unnatural parity states of some even-even nuclei are completely forbidden in simple zero-range DWBA. Finite-range DWBA calculation can allow for such transitions. The analysis of ^{22}Ne(p,t)^{20}Ne (2^-) reaction by Bayman and Feng,[1,] where they used a more realistic variational wave function, predicted much smaller cross section than yhe observed ones. On the other hand, de Takascy,[2,] and Charlton [3,] showed that the cross section ^{208}Pb(p,t)^{206}Pb (3^+ 1.34Mev) at 35Mev could be well reproduced in magnitude and angular distribution by 2-nd oder DWBA calculation assuming the p-d-t process. This paticular reaction on ^{208}Pb is especially suited for the investigation of p-d-t process because there is no significant contribution of multi-step processes via inelastic channels. According to a paper by Nagarajan et al.[4,] however, finite-range DWBA with realistic triton wave function of Strayer and Sauer [5,] can reproduce the mentioned cross section very well. The agreement with the experimental cross section was so good as to cast doubt on the interpretation in terms of the sequential transfer mechanism.

The aim of this paper is to settle this conflict between the interpretations, one-step vs. two-step mechanisms. The exact finite-range calculation of both mechanisms was done with the consistent assumptions with one another.

NUMERICAL CALCULATION

The triton wave function employed was obtained by Sasagawa and Sawada [6,] as a solution of the three body Faddeev equation with Reid soft core potentials. The binding energy of triton is 6.375Mev. The probability of the S, S' and D states are the following values: P(S)= 90.08%, P(S')=1.92% and P(D)=8.00%. The deuteron wave function obtained by Reid [7,] as a solution of the two-body Hamiltonian with Reid soft core potentials has also the D-state component. The same potentials were used for the reaction interaction which cause the transfer of each step. The form factor of the p-t process was calculated by generalized Bayman-Kallio method [8]. $(\phi_d|V_{pn}+V_{nn}|\phi_t)$, $(\phi_d|\phi_t)$ and $(\phi_p|V_{np}|\phi_d)$ are composed of $l=0$ (S-state) and $l=2$ (D-state) components. The tensor and vector analyzing powers of the ^{208}Pb(d,t) ^{207}Pb reactions[9] are well reproduced using this form factor but the magnitudes of the cross sections are reduced about factor 2.2. Exact finite-range calculations of both processes were carried out by the code TWOFNR.

RESULTS AND DISCUSSION

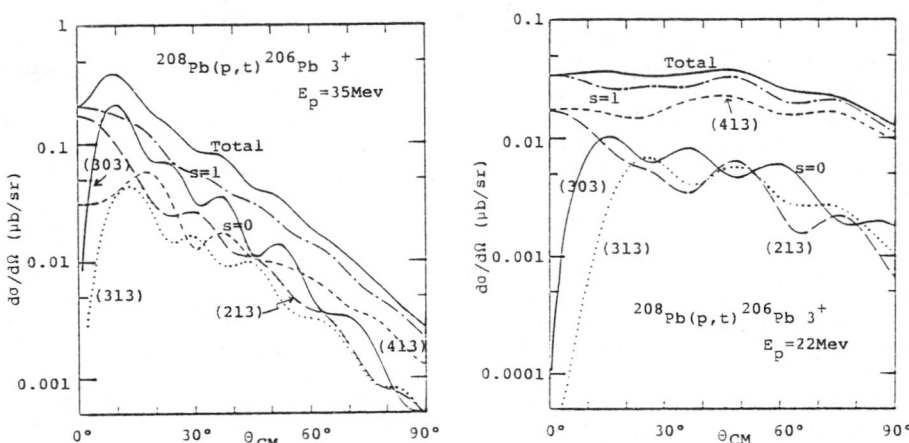

Fig. 1. Each contribution of p-t process.   Fig. 2. Same as Fig. 1

Each contribution of the transfer angular momentum component (lsj) for the p-t process are shown in Fig. 1 and 2. The p-t process depends on incident energy because of the angular momentum miss match condition. The magnitude of the cross sections are reduced about factor 30 from the result of Nagarajan et al.

The sequential transfer process plays the main role in this reaction as shown in Fig. 3 and 4. (the notations of Fig. 3 and 4 are same as Fig. 3.) The D-state effects can not be neglected in the evaluation of the p-d-t process leading to the unnatural parity state. The theoretical cross sections of the mentioned reactions are smaller than the observed ones by factor 3 ~ 4. The analyzing power data [10] confirm confirm that the sequential transfer process plays the main role in the unnatural parity transition as shown in Fig. 5.

Now we are trying to investigate the discrepancy of the theoretical absolute magnitudes of (d,t) and (p,t) reactions from the experimental results. Triton wave function has not correct binding energy. This fact may come from partly the ambiguity of nucleon- nucleon interaction and partly the other 3-body forces. We are trying to use various triton wave functions, which are the solutions of different nucleon- nucleon interactions.

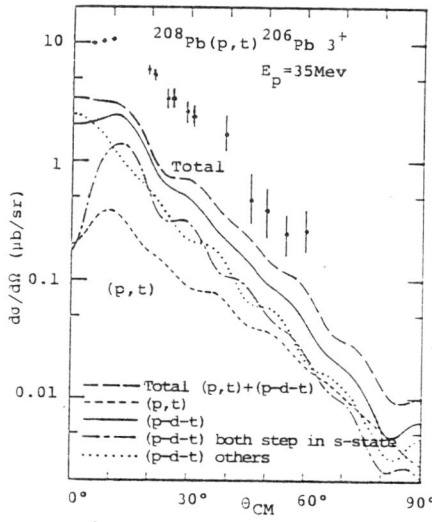

Fig. 3.

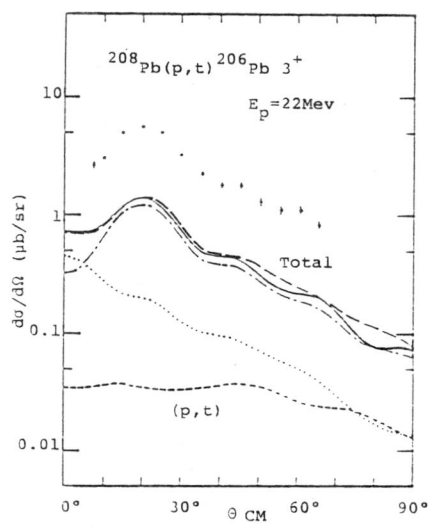

Fig. 4.

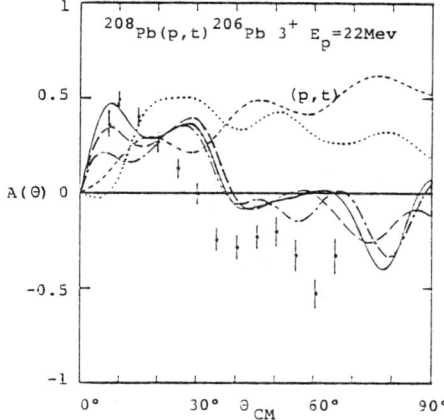

Fig. 5.

REFERENCES

1. B. F. Bayman et al., Nucl. Phys. A205 (1973) 513
2. N. B. de Takascy, Phys. Rev. Lett. 31 (1973) 1007
3. L. A. Charlton, Phys. Rev. C14 (1976) 506
4. M. A. Nagarajan et al., Phys. Lett. 68B (1977) 421
5. M. R. Strayer et al., Nucl. Phys. A231 (1974) 1
6. T. Sasakawa et al., Phys. Rev. C19 (1979) 2035
7. R. V. Reid, Ann. Phys. 50 (1968) 411
8. B. F. Bayman et al., Phys. Rev. 156 (1967) 1121
   C. W. Wong et al., Nucl. Phys. A183 (1972) 210
9. H. S. Liers et al., Phys. Rev. Lett. 26 (1971) 261
   L. D. Knutson et al., Phys. Rev. Lett. 35 (1975) 1570
10. Y. Toba et al., Phys. Rev. C20 (1979) 1204

CROSS SECTION AND ANALYZING POWER IN THE ^{206}Pb($\vec{t}$,p) ^{208}Pb(4^-) REACTION*

W. P. Alford
The University of Western Ontario

R. N. Boyd
Ohio State University

E. Sugarbaker and F. deBoer
University of Colorado

Ronald E. Brown and E. R. Flynn
Los Alamos Scientific Laboratory

Cross sections and analyzing powers have been measured in the ^{206}Pb($\vec{t}$,p) ^{208}Pb reaction leading to the 4^-, 5^-, states of the ($p_{1/2}^{-1}$ $g_{9/2}$) doublet. Standard DWBA calculations reproduce the cross section but not the analyzing power for the 5^- state. The opposite is true for a sequential transfer (t,d) (dp) calculation for the 4^- state.

Considerable effort has been directed at understanding the forbidden transition (1) in the ^{208}Pb(p,t) ^{206}Pb(3+) reaction. Theoretical analyses have been able to reproduce the measured cross section assuming either a two-step sequential transfer reaction mechanism (2,3) or a single step process taking account of finite range effects and using a realistic wave function for the triton (4). Recently Toba et al (5), have also measured the analyzing power in this reaction at 22 MeV and have concluded that the data cannot be reproduced by a sequential transfer calculation. Thus the importance of two-step processes in this reaction remains in question. We have measured the cross section and analyzing power in the ^{206}Pb($\vec{t}$,p) ^{208}Pb reaction to the 4^- state at 3.475 MeV in an effort to determine the mechanism of such "forbidden" two-nucleon transfers.

Measurements were carried out at an energy of 17 MeV using the LASL polarized triton beam. Protons were detected with a helix counter in the Q3D magnetic spectrometer (6). Data were taken in 5° steps from 10° to 65° using the full angular aperture (14.3 msr) of the spectrometer. Only the results for transitions to the 5^-(3.198 MeV) and 4^-(3.475 MeV) states are reported here, though data were obtained for states between 2.5 and 4.9 MeV.

*Work supported by the U. S. Department of Energy.

Measured cross sections and analyzing powers for the two states of interest are shown in Fig. 1. Error bars on the data points indicate statistical uncertainties only. The uncertainty in absolute cross sections is estimated to be less than 20%.

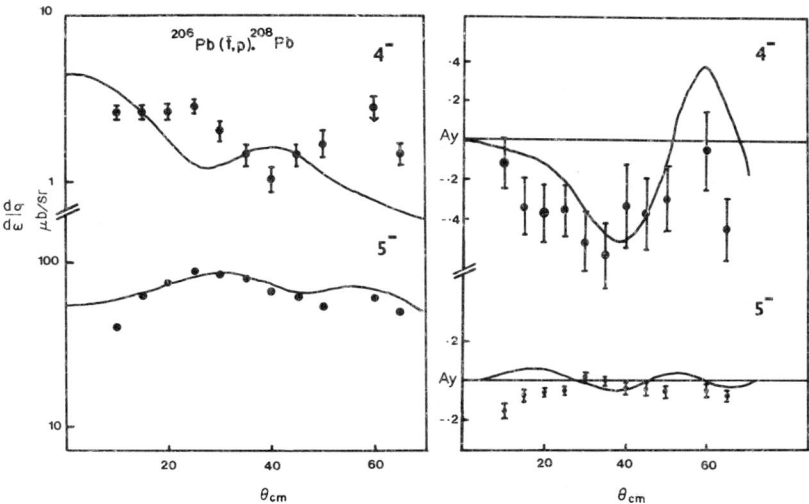

Figure 1 - Cross sections and analyzing powers for the 5^-(3.198 MeV) and 4^-(3.475 MeV) states in ^{208}Pb. Bars represent uncertainties due to counting statistics. The curves are the result of calculations described in the text.

The dominant component (7,8) of the wavefunctions of these states is $(p_{1/2}^{-1}\, g_{9/2})_\nu$. Thus our initial analysis of these results has assumed that these states are the two members of the $(p_{1/2}^{-1}\, g_{9/2})_\nu$ 4^-, 5^- doublet.

Reaction calculations were carried out using the code CHUCK (9) with optical model potentials shown in table I. All calculations assumed a zero range interaction. For the transition to the 5^- state, a single-step reaction was assumed with a form factor $(p_{1/2}\, g_{9/2})$ coupled to the $(p_{1/2})^{-2}$ component of the ^{206}Pb ground state. The transition to the 4^- state was assumed to proceed via a sequential (t,d) (d,p) reaction.

These calculations provide reasonable agreement with the measured angular distribution for the L=5 transition. The magnitude of the measured analyzing power is small, in agreement with the calculations, but the angular variation does not agree well. Similar results have been observed for large-L transitions in lighter nuclei (10). In contrast, for the 4^- state the theory predicts the general magnitude of the cross section, but does not reproduce the angular dependence, while the theory shows fairly good agreement with the analyzing power data.

Possible reasons for the above discrepancies are two step contributions to the cross-section for the 5^- state, and additional contributions to both transitions from small components of the wave functions. Calculations are in progress to test these possibilities.

TABLE 1

OPTICAL PARAMETERS USED IN REACTION CALCULATIONS

	V MeV	r_o fm	a_o fm	W_V Me	W_D MeV	r_I fm	a_I fm	V_{so}	r_{so}	a_{so}	r_c
p	-60.45	1.12	.78	0	7.1	1.32	.59	-6.2	1.0	.75	1.3
d	-98.1	1.10	.82	0	15.9	1.32	.71	-2.8	.98	1.0	1.3
t	-165	1.16	.752	-16.4	0	1.5	.82	06.0	1.16	.752	1.3

REFERENCES

1. W. A. Lanford and J.B. McGrory, Phys. Lett. <u>45B</u> 238 (1973).
2. N. B. deTakacsy, Phys. Rev. Lett. <u>31</u> 1007 (1973).
3. V. Managoli and D. Robson, Nucl. Phys. <u>A252</u> 354 (1975).
4. M. A. Nagarajan, M. R. Strayer and M. F. Weby, Phys. Lett <u>68B</u> 421 (1977).
5. Y.Toba, Y.Aoki, S.Kunori, K.Nagano and K.Yagi, Phys. Rev. <u>C20</u> 1204 (1979).
6. E.R. Flynn, S.Orbesen, J.D. Sherman, J.W. Sunier and R.Woods, Nucl. Inst. <u>128</u> 35 (1975).
7. P.Richard, P.vonBrentano, H.Weiman, W.Wharton, W.G. Weitkamp and W.W.McDonald, Phys. Rev. <u>183</u> 1007 (1969).
8. W.W.True, C.W. Ma and W.T. Pinkston, Phys. Rev. <u>C3</u> 2421 (1971)
9. P.D. Kunz, private communication.
10. W.P. Alford, R.N.Boyd, E.Sugarbaker, D.L.Hanson and E.R.Flynn Phys. Rev. <u>21C</u> 1203 (1980).

SPIN-PARITY DETERMINATIONS FROM THE ^{42}Ca($\vec{d},\alpha$)^{40}K REACTION NEAR 0°

Shang Ren-cheng*, J.A. Kuehner, A.A. Pilt,
M.A.M. Shahabuddin and A. Trudel
Tandem Accelerator Laboratory, McMaster University,
Hamilton, Ontario, Canada L8S 4K1

ABSTRACT

The tensor analyzing powers T_{20} have been measured for 24 states of ^{40}K below 5 MeV following the ^{42}Ca($\vec{d},\alpha$)^{40}K reaction, leading to many new spin-parity determinations.

INTRODUCTION

The knowledge of spins and parities of states in ^{40}K would give important information on the structure of this nucleus, and furthermore, plays an important part in understanding the structure of all the isotopes of potassium. The spin-parity combinations of many states of ^{40}K are unknown, especially, above 3 MeV excitation energy. Up to now, few investigations of ^{40}K via the (d,α) reaction have been carried out. Marinov et al[1] and Kolata et al[2] have studied only some low-lying levels of ^{40}K from the ^{42}Ca(d,α)^{40}K reaction. Polarized deuteron beams seem not to have been used in such experiments.

In this experiment, α particles from the ^{42}Ca($\vec{d},\alpha$)^{40}K reaction were observed near θ_α = 0°. Kuehner et al[3] have shown that for an even-even target nucleus, the measurement of tensor analyzing powers T_{20} near 0° allows unambiguous, model-independent natural and unnatural parity assignments for the energy levels of the residual nucleus to be made. Furthermore, $J^\pi = 0^-$ levels could also be identified.

As shown by Petty[4], the larger the number of incident beam energies the more confidence one would have in making the assignment. For 3 different measurements of T_{20}, one can determine unnatural or natural parity, or $J^\pi = 0^-$, with 99.9% confidence.

The present measurements were carried out at small angles ($\theta = 4°$), instead of 0°. This makes T_{20} weakly model dependent, but since the effects can be estimated[4], one can still make parity assignments with confidence.

EXPERIMENTAL PROCEDURE

A polarized deuteron beam was obtained from the McMaster University Lamb-shift polarized ion source[5] and FN tandem accelerator. The outgoing α particles were momentum analyzed with an Enge split-pole magnetic spectrograph and detected with a resistive wire

---

* On leave from Tsinghua University, Beijing, People's Republic of China

gas proportional counter mounted on the focal plane. Measurements were taken at θ= 4° (Lab). The experiments were carried out at beam energies of 7, 7.5, 8 and 9 MeV, but all levels were not observed at each energy. Since the counter accepts only 20 cm of the focal plane, states above 2.5 MeV excitation were studied separately at a different magnetic field setting.

The thickness of targets of ^{42}Ca was 40 μg/cm , using enriched ^{42}Ca (95%) evaporated onto 10 μg/cm^2 carbon backing. Targets of ^{40}Ca were used to subtract the peaks of $^{40}Ca(d,\alpha)^{38}K$ reaction from the ^{42}Ca spectra and to set energy calibration points. WO_2 targets were used to measure the fractional beam polarization. Further details have been given previously[6].

RESULTS

The energy resolution obtained was better than 20 keV. About 70 peaks were observed, three of them from ^{40}Ca, one from ^{16}O, five are unknown, all others are presumed to be from ^{42}Ca. Because of the good resolution, all levels except a few doublets were separated.

Twenty-four selected spin-parity assignments of ^{40}K levels are listed in table 1. Only those assignments which were confirmed from at least three different beam energies are listed. Nonlinear Gaussian fitting routines were used to separate some partially overlapping levels, but additional errors were thereby introduced. These data are denoted by a star in the table.

The 1.6437 MeV and 3.6303 MeV levels were not observed at any energy. The former is probably forbidden by isospin selection rules. The reason for the non-observation of the 3.6303 MeV level is not known.

Table I. Spin-parity determinations to selected levels in ^{40}K.

E*[a] (MeV)	J^π [a]	$<T_{20}>$ (at 4°)	Parity[b]	J^π (adopted)
0.0	4^-	-0.57 ± 0.25	U	4^-
0.0296	3^-	0.67 ± 0.28	N	3^-
0.8001	2^-	-0.47 ± 0.22	U	2^-
1.9590	2^+	0.71 ± 0.18	N	2^+
2.2605	3^+	-1.01 ± 0.17	U	3^+
2.3976	4^-	-0.87 ± 0.12	U	4^-
2.5428	7^+	-0.45 ± 0.14	U	7^+
2.6258	$(0-2)^-$	-1.50 ± 0.24	0^-	0^-
2.9866	$1^- - 4^-$	-0.01 ± 0.23	U	$2^-, 4^-$
3.0284	$(2^-, 3)$	-0.04 ± 0.18	U	$2^-, 3^+$

Table 1 (Continued)

E^{*a} (MeV)	$J^{\pi\,a}$	$<T_{20}>$ (at 4°)	Parity[b]	J^π (adopted)
3.3936	$(0-3)^-$	0.17 ± 0.22	U	2^-
3.4390	$(3-5)^+$	-0.45 ± 0.18	U	$3^+, 5^+$
3.4868	$(2,3)^-$	0.25 ± 0.13	U	2^-
3.5991	$(1-3)^-$	-0.47 ± 0.23	U	2^-
3.6636		0.21 ± 0.14	U	
3.7379		$-0.18 \pm 0.58*$	U	
3.8216	$(2,3)^-$	0 ± 0.20	U	2^-
3.8686	$(2,3)^-$	$0.33 \pm 0.82*$	U	2^-
4.1046	$(0-3)^-$	0.07 ± 0.15	U	2^-
4.352		-0.21 ± 0.26	U	
4.4646	$(0-2)^-$	0.42 ± 0.16	U(N)	$2^-(1^-)$
4.5373		$-0.04 \pm 0.75*$	U	
4.5868	$(0-3)^-$	-0.11 ± 0.14	U	2^-
4.6658	$(1-3)^-$	-0.22 ± 0.11	U	2^-

a   Reference 7

b   N : natural; U : unnatural

REFERENCES

1. A. Marinov, Ch. Drory, J. Burde, E. Navon and Sh. Morderchai. Nucl. Phys. A168, 267 (1971).
2. J.J. Kolata, P. Shapiro and L.S. August, Phys. Lett. 32B 277 (1979).
3. J.A. Kuehner, P.W. Green, G.D. Jones and D.T. Petty, Phys. Rev. Lett. 35, 423 (1975).
4. D.T. Petty, thesis, McMaster University, 1976 (unpublished).
5. J.W. McKay, thesis, McMaster University, 1975 (unpublished).
6. S. Angelo, A.A. Pilt and J.A. Kuehner, to be published.
7. P.M. Endt and C. v.d. Leun, Nucl. Phys. A310, 1 (1978).

INVESTIGATION OF THE REACTION ^{14}N(d,α)^{12}C WITH VECTOR POLARIZED DEUTERONS *)

W. Kretschmer, G. Pröbstle[+], and W. Stach
Physikalisches Institut der Universität Erlangen-Nürnberg, W.-Germany

We have investigated the ^{14}N($\vec{d}$,α)^{12}C reaction[1] with vector polarized deuterons to get some information about the transferred angular momenta L and J. The measurement was performed at E_d=10 MeV with the vector polarized beam of the Lamb-shift source and a windowless, high density gas target[2] with a target thickness of 0.3 mg/cm³ for an entrance pressure of 20 bar. The reaction products were identified with eight symmetrically arranged ΔE-E telescopes (50μ, 2000μ), the beam polarization was monitored in a ^{12}C-polarimeter. The analysis was performed with the codes DWUCK and CHUCK assuming a deuteron-cluster transfer. The deuteron optical potential was obtained from a fit to the elastic scattering and was also used for the description of the (d,p) reactions on ^{14}N [3], the α-potential was taken from Scott et al[4] (with U_R=210 MeV instead of 200 MeV).

The vector analyzing power (VAP) of the transitions to the 0^+-states in ^{12}C (E_x=0.0 and 7.65 MeV) is shown in fig. 1 compared with DWUCK calculations. In both cases the total angular momentum transfer is unique J=1, but the orbital angular momentum transfer L can be 0 or 2. The ground state transition is a clear L=2 J=1 transfer (fig.1)

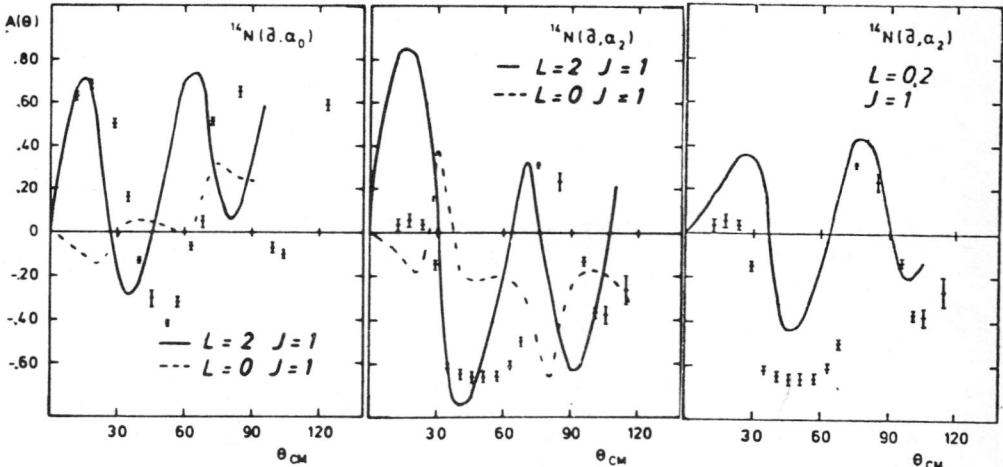

Fig. 1. $A_y(\theta)$ for the transitions to the $J^\pi=0^+$-states in ^{12}C compared with calculations for pure transfers (a and b) and for a coherent mixing (1:1) of J=1, L=0 and L=2 (c).

in accordance with ref. 5 whereas the (d,α$_2$) transition may be better described by a coherent mixing of both L=0 and L=2 (fig.1c).

The transitions to the first (E_x=4.43 MeV, $J^\pi=2^+$), fourth (E_x=10.8 MeV, $J^\pi=1^-$) and fifth (E_x=11.8 MeV, $J^\pi=2^-$) excited states in ^{12}C are shown in fig. 2. The (d,α$_1$) transition (fig. 2a) is best described by an L=2 J=2 transfer, in accordance with higher energy

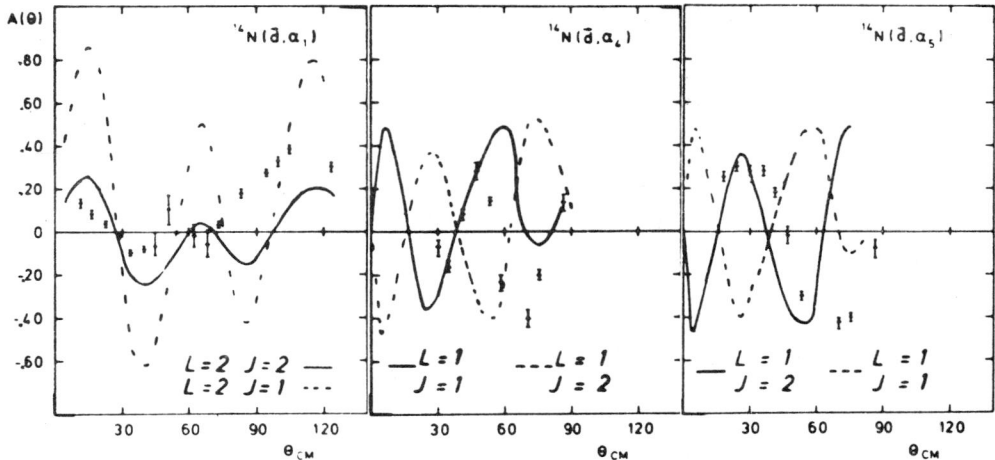

Fig. 2. $A_y(\theta)$ for transitions to the $J^\pi = 2^+$, 1^-, and 2^--states in ^{12}C compared with DWBA calculations.

data, a coupled channels calculation $^{14}N(g.s) \longrightarrow ^{12}C(g.s.) \longrightarrow ^{12}C(2^+)$ completely fails to reproduce the data. The transitions to both negative parity states are described by an L=1 J=1 transfer (fig.2b) and an L=1, J=2 transfer (fig.2c), although there are some discrepancies between experiment and calculation.

The transition to the third (E_x=9.63 MeV, J^π=3$^-$) excited state in ^{12}C is shown in fig.3 compared with different calculations. Out of the four possible transfers only those with L=2, J=2,3 (fig.3a) show the correct sign for $A_y(\theta)$, but the absolut values are wrong for these pure transfers.

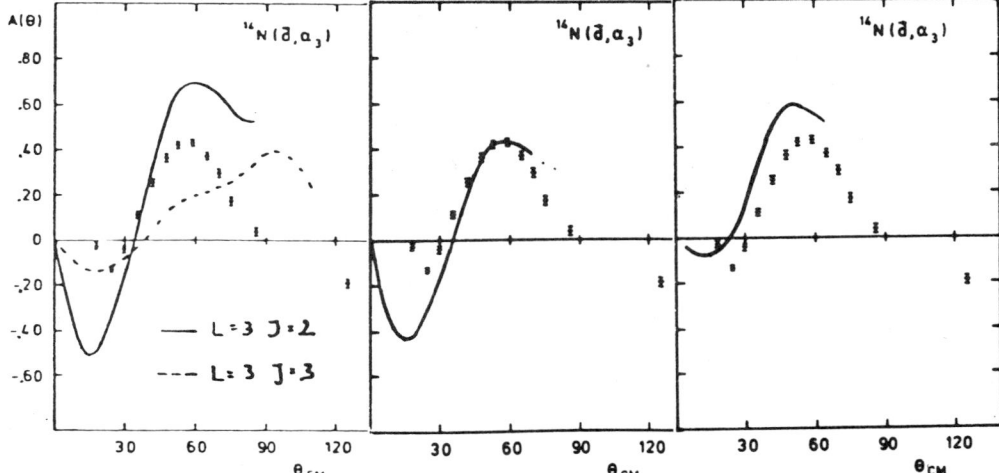

Fig. 3. $A_y(\theta)$ for the transition to the J^π=3$^-$ state in ^{12}C. (a) pure transfers L=3, J=3, and L=3, J=2 (b) coherent mixing of both transfers (c) coherent mixing of one and two step processes.

A better description of the data is obtained by a coherent mixing of different one step transfers ((L=3, J=3):(L=3,J=2)= -0.5: :0.9) or by a coherent mixing of one- and two-step processes ($[^{14}N(g.s) \rightarrow ^{12}C(g.s) \rightarrow ^{12}C(3^-)] : [^{14}N(g.s) \rightarrow ^{12}C(3^-)]$ = 1:10).

1. G. Pröbstle, diplom thesis, Erlangen 1978
2. G. Bittner, W. Kretschmer, and W. Schuster, Nucl. Instr. and Meth. 161, 1 (1979)
3. W. Kretschmer, G. Pröbstle, and W. Stach, Nucl. Phys. A333, 13 (1980)
4. D.K. Scott, P.M. Portner, J.M. Nelson, A.M. Mitchell, N.S. Chant, D.G. Montague, and K. Ramavataram, Nucl. Phys. A141, 497 (1970)
5. E.J. Ludwig, T.B. Clegg, P.G. Ikossi, and W.W. Jacobs, in Proc. IV Int. Symp. Polarization Phenomena in Nuclear Reactions, ed. by Grüebler and V. König (Birkhäuser, Berlin, 1975) p. 687

*) Work supported by the Deutsche Forschungsgemeinschaft
+) permanent address: Institut für Nuklearforschung, Villigen (SIN), Switzerland

INVESTIGATION OF THE REACTION ^{24}Mg$(d,\alpha)^{22}$Na WITH VECTOR POLARIZED DEUTERONS *)

W. Kretschmer, E. Heitz, C. Glashausser[+], A.B. Robbins[+], J. Duder[++], D. Melnik[+++]
Physikalisches Institut der Universität Erlangen-Nürnberg, W.-Germany

The investigation of (d,α)-reactions is a useful spectroscopic tool for the study of the levels in the residual nucleus. In contrary to (d,p)-reactions the L and J of the transferred n,p pair is not unique for 0^+ target nuclei, which complicates the analysis of this type of reaction.

We have measured the vector analysing power (VAP) of the (d,α) reactions on ^{24}Mg at E_d=15MeV to improve the insight of this reaction, which has been investigated with unpolarized deuterons by Schneider and Olness[1]. The VAP is much more sensitive[2] to the transferred angular momenta L and J than the differential cross section and so details of the reaction mechanism like coherent mixing of different L,J-transfers or two step processes can be investigated more reliable[3]. The measurements were performed at Rutgers University with the polarized atomic beam source and the 8 MeV tandem accelerator. The polarized beam was collimated through two slit pairs and an antiscattering tube to the target, the beam polarization was continuously monitored in a ^{4}He-polarimeter. The outgoing α-particles were detected in symmetrically arranged 300μ surface barrier detectors, for cross section normalization two 3000μ detectors were placed at ±6°. The analysis was performed with the codes DWUCK 4 and CHUCK, mostly a deuteron-cluster transfer was assumed, in some cases the separate transfers of a n,p pair from the s,d shell with theoretical coefficients given by Wildenthal[4]. The optical potentials for these calculations are very similar to those used in ref. 2 for the (d,α)-reaction on ^{28}Si.

The results for the transitions to the three lowest positive parity states in ^{22}Na are shown in fig. 1: the ground state transfer is well described by L=2 J=3, whereas L=4 J=3 can be ruled out completely. The transition to the first $J^\pi=1^+$ state is described by a predominant L=0, J=1 transfer, the possible L=2, J=1 admixtures will be discussed in more detail in another contribution[5] to this conference concerning the tensor analysing powers. The transition to the $J^\pi=5^+$ state is very favoured, since for this self-conjugate target nucleus the neutrons and protons picked up from the same shell preferentially populate states with maximum J (which is 5^+ for p and n from the $1d_{5/2}$ shell). In this situation the transfer is rather unique L=4, J=5 and indeed the corresponding DWUCK curve reproduces the data quite well (the agreement was improved by an artificially increased Q-value which gives a better angular momentum matching).

Three other transitions are shown in fig. 2: the transition to the $J^\pi=1^+$ state at E_x = 3.944 MeV is predominately a L=2, J=1 transfer with some admixtures[5] of L=0, J=1. The dominant transfer to the $J^\pi=2^-$ state at E_x = 2.572 MeV is L=3, J=2 with some admixture of L=1, J=2. The transition to the $J^\pi=4^+$ state is unique L=J=4 if it is a one step process. Since for this L=J transfer the VAP is small at

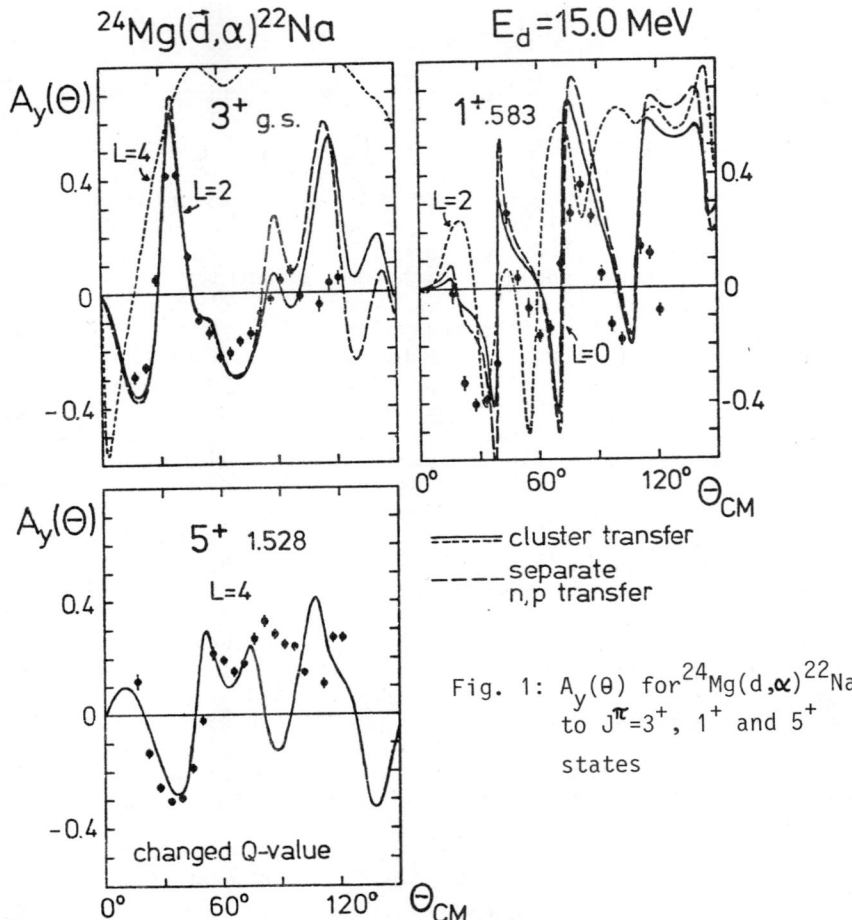

Fig. 1: $A_y(\theta)$ for $^{24}Mg(d,\alpha)^{22}Na$ to $J^\pi = 3^+$, 1^+ and 5^+ states

forward angles irrespective of the optical potentials and the data show a marked minimum at 60° similar to the L=4, J=5 transfer, a two-step process was tried with an excitation of the 2^+ level in ^{24}Mg (with β=0.42) and a successive L=4 J=5 transfer to the 4^+ state in ^{22}Na. At least in the forward angular region the VAP is better described thus supporting this reaction mechanism.

The other transitions are more complicated, the calculation of different mixtures and two step processes is in progress.

*) Work supported by the Deutsche Forschungsgemeinschaft and the US National Science Foundation
+) Rutgers University, New Brunswick, New Jersey
++) University of Auckland, New Zealand
+++) Bell Laboratories, Murray Hill, New Jersey

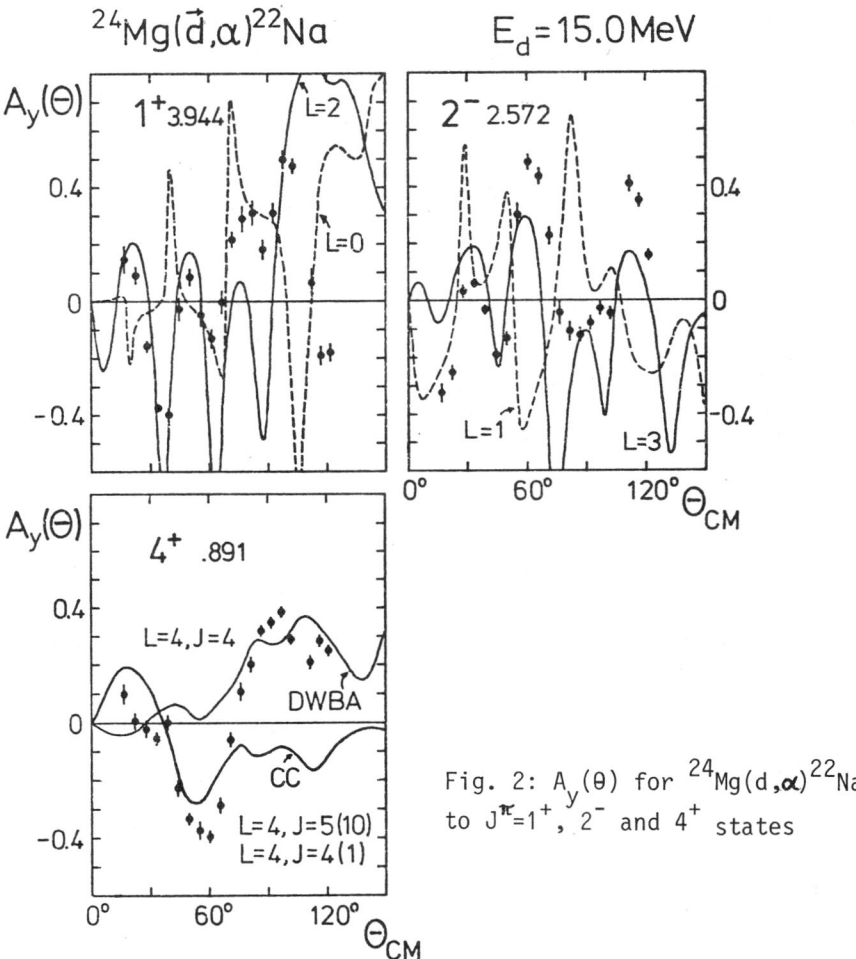

Fig. 2: $A_y(\theta)$ for ^{24}Mg(d,α)^{22}Na to $J^\pi = 1^+$, 2^- and 4^+ states

1. M.J. Schneider and J.W. Olness, Phys. Rev. C13, 1392 (1976)
2. E.J. Ludwig, T.B. Clegg, W.W. Jacobs, and S.A. Tonsfeldt, Phys. Rev. Lett. 40, 441 (1978)
3. W. Kretschmer, G. Pröbstle, and W. Stach, contribution to this conference
4. B.M. Preedom and B.H. Wildenthal, Phys. Rev. C6, 1633 (1972)
5. W. Kretschmer, E. Heitz, C. Glashausser, B.A. Robbins, J. Duder, and D. Melnik, contribution to this conference

TENSOR ANALYZING POWERS IN THE REACTION $^{24}Mg(d,\alpha)^{22}Na$ *)

W. Kretschmer, E. Heitz, C. Glashausser +), A.B. Robbins +), J.C.
Duder ++), and D. Melnik +++)
Physikalisches Institut der Universität Erlangen-Nürnberg, W.-Germany

The sensitivity of the vector-analyzing power in (d,α)-reactions to the transferred angular momenta L and J has been demonstrated in other contributions to this conference[1]. The purpose of the present investigation was the question whether one can get some new information from the tensor analyzing powers (TAP) in (d,α)-reactions, especially whether the tensor potential for the entrance channel and the deuteron D-state are essential for the description of the TAP.

We have measured the tensor analyzing powers T_{20}, T_{21}, and T_{22} for the reaction $^{24}Mg(d,\alpha)^{22}Na$ at E_d = 15 MeV. The measurement was performed at Rutgers University with the polarized atomic beam ion source. The tensor polarization of the beam was continuously monitored with a ^{3}He polarimeter behind the target chamber. The measurement of the tensor observables was performed as described in ref. 2.

The analysis was done with the code CHUCK, modified by H.Clement[3] for the inclusion of the tensor potential. The TAP for the transition to the ground state with $J^\pi=3^+$, to the $J^\pi=5^+$ state at 1.528 MeV and to the $J^\pi=2^-$ state at 2.572 Mev are shown in fig. 1. Especially at forward angles all three TAP of the ground state transition are described rather well with a L=2 J=3 transfer. A comparison of the calculations with and without tensor potential shows only small differencies, the most prominent effect may be seen at backward angles for $T_{20}(\theta)$, where the strongly negative values are reproduced by the curve with tensor potential. The other allowed transfer with L=4 J=3, which has already been ruled out by the vector analyzing power measurement[1], shows a smooth behaviour and completely fails to describe the data. The other rather pure transition to the $J^\pi=5^+$ state shows for a L=4 J=5 transfer a good agreement for $T_{21}(\theta)$ and $T_{22}(\theta)$ (at least in the structure), whereas T_{20} is only poorly described. For this L=4 transfer the Q-value has been increased by 5 MeV to get a better angular momentum matching. The transition to the 2^- state is best described by a coherent mixing of L=3 J=1 and L=1 J=1 with a ratio of 3:1, but the agreement is only qualitative.

Finally the TAP of the two transitions to $J^\pi=1^+$ states are shown in fig. 2. From VAP measurements[1] both transitions are supposed to be mixed with a dominant L=0 J=1 transfer to the E_x=0.583 MeV state and a dominant L=2 J=1 transfer to the E_x=3.944 MeV state, both shown as dashed lines in fig. 2. In both cases this dominant transfer completely fails to reproduce the data, whereas the mixed transfers ((L=0):(L=2)=3:1) for the first transition, (L=0):(L=2) = 1:2 for the second transition) give a better description at least of the structure of the angular distribution. It should be mentioned, however, that the description of the first 1^+ transition is only qualitative and so a more complicated mechanism may be possible.

It has been shown here that the tensor analyzing powers are also very sensitive to the transferred angular momenta and therefore are very useful for the disentangling of different L,J transfers.

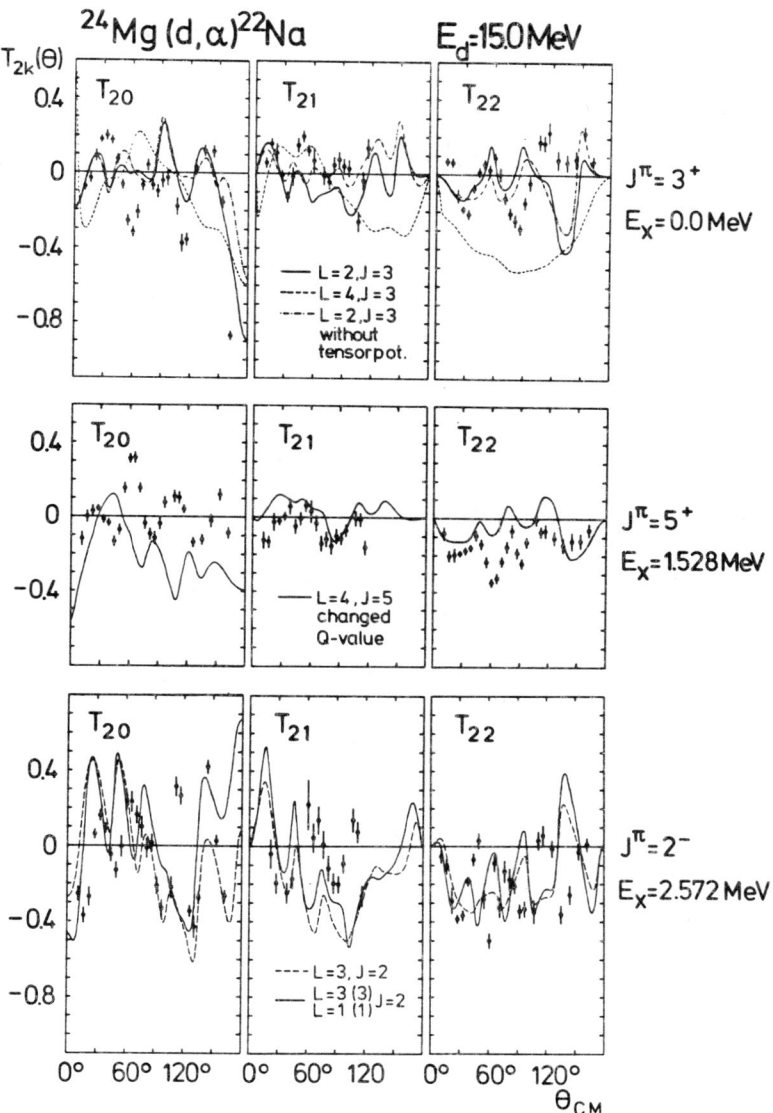

Fig. 1: Tensor analyzing powers for ^{24}Mg(d,α)^{22}Na to $J^\pi = 3^+$, 5^+ and 2^- states

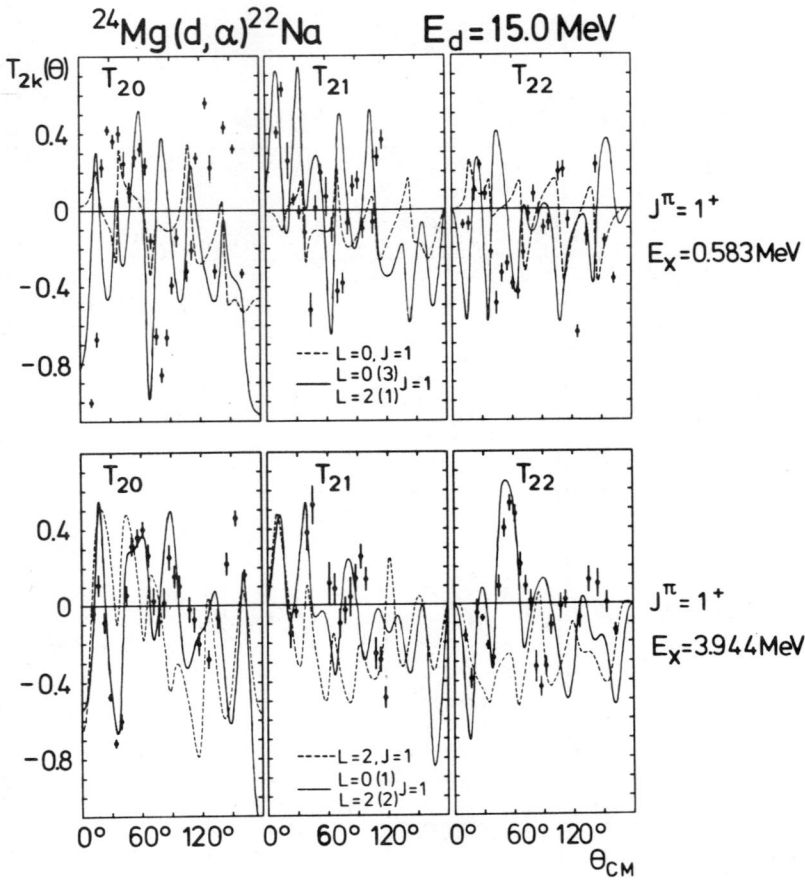

Fig. 2: Tensor analyzing powers for ^{24}Mg(d,α)^{22}Na to states with $J^\pi = 1^+$.

1. W. Kretschmer, E. Heitz, C. Glashausser, B.A. Robbins, J.C. Duder, and D. Melnik, contribution to this conference
2. F.T. Baker, S. Davis, C. Glashausser and A.B. Robbins, Nucl. Phys. A233, 409 (1974)
3. H. Clement, University of Munich, private communication

+) Rutgers University, New Brunswick, New Jersey
++) University of Auckland, New Zealand
+++) Bell Laboratories, Murray Hill, New Jersey
*) Work supported by the Deutsche Forschungsgemeinschaft and the US National Science Foundation

INVESTIGATION OF THE REACTION ^{27}Al$(d,\alpha)^{25}$Mg WITH VECTOR POLARIZED DEUTERONS *)

W. Kretschmer, E. Heitz, C. Glashausser+), A.B. Robbins+), J.C. Duder+)
D. Melnik+++)
Physikalisches Institut der Universität Erlangen-Nürnberg, W.-Germany

The vector analysing power (VAP) in (d,α)-reactions has proved to be essential for the extraction of the transferred angular momenta L and J [1,2], since its sign and magnitude at forward angles depend on J=L-1, L, L+1. In this contribution we discuss the (d,α)-reaction on a target nucleus with high spin $J^\pi=5/2^+$, where both L and J are not unique and cannot be determined from a cross section measurement alone.

We have measured the reaction ^{27}Al$(d,\alpha)^{25}$Mg at E_d=15 MeV with the polarized atomic beam source at Rutgers University as described in ref. 2. In contrast to a lower energy (E_d=11.2 and 11.8 MeV) polarization measurement in Birmingham[3] the reaction is mostly direct at this higher energy and so an analysis in the DWBA formalism seems to be adequate. It was performed with the code DWUCK4 with the optical potentials used in ref. 1 for the ^{28}Si$(d,\alpha)^{26}$Al reaction.

Typical examples are shown in fig. 1 and 2 for transitions to positive parity states with spins ranging from 1/2+ to 9/2+. The transitions to the $J^\pi=1/2^+$ states seem to be a predominant L=2, J=3 transfer with a possible L=2, J=2 admixture for the first one. The L=4 J=3 transfer can be ruled out completely. The transition to the $J^\pi=9/2^+$ state is best described by a L=4 J=5 transfer, whereas the ground state transition to the $J^\pi=5/2^+$ state is a dominant L=2 J=2 transfer in accordance with the analysis of Nemets et al.[4].

The $J^\pi=3/2^+$ states shown in fig. 2 are populated by different transfers: the E_x=0.975 MeV state by a dominant L=2 J=3 transfer with possible admixtures of L=2 J=1,2, the E_x=2.801 MeV state by a L=2 J=3 transfer and the E_x=4.359 MeV state by a L=0 J=1 transfer.

The transitions to the $J^\pi=7/2^+$ states are supposed to be best described by mixed transfers: the E_x=1.612 MeV state by a mixing of L=0 J=1 and L=4 J=5, the E_x=2.738 MeV state by a mixing of L=4 J=5, L=0 J=1 and L=2 J=2, the E_x=5.012 MeV state by a mixing of L=0 J=1 and L=4 J=5 transfers.

It should be mentioned that the curves shown in figs. 1 and 2 are calculated for pure transfers and that the other allowed L,J transfers, not shown in the figures, are ruled out by the measurement. The calculation of coherent constructive or destructive mixtures of different L,J transfers is in progress and will be published elsewhere.

1. E.J. Ludwig, T.B. Clegg, W.W. Jacobs, and S.A. Tonsfeldt, Phys. Rev. Lett. **40**, 441 (1978)
2. W. Kretschmer, E. Heitz, C. Glashausser, B.A. Robbins, J.C. Duder, and D. Melnik, contribution to this conference
3. Y. Takeuchi, J.A.R. Griffith, O. Karban, and S. Roman, Nucl. Phys. **A220**, 589 (1974)

4. O.F. Nemets, Yu.S. Stryuk, A.M. Yasnogorodsky, in Proc. IV. Int. Symp. Polarization Phenomena in Nuclear Reactions, ed. by W. Grüebler and V. König (Birkhäuser, Berlin, 1975) p. 689

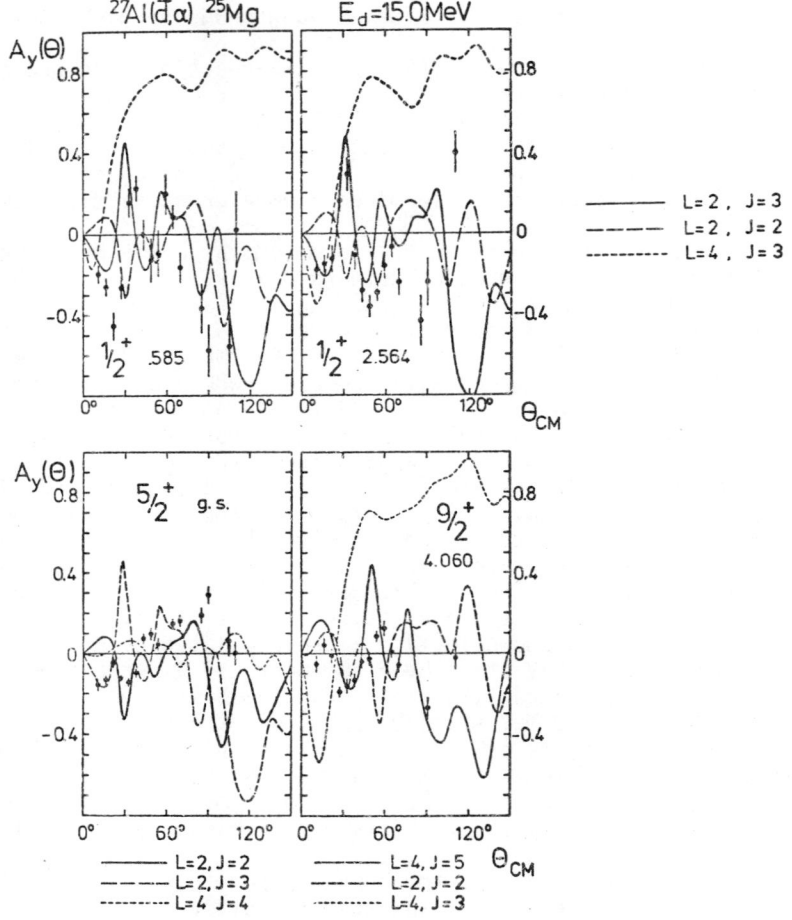

Fig. 1. $A_y(\theta)$ for ^{27}Al(d,α)^{25}Mg for transitions to states with J^π=1/2+, 5/2+, and 9/2+ compared with DWBA calculations.

+) Rutgers University, New Brunswick, New Jersery
++) University of Auckland, New Zealand
+++) Bell Laboratories, Murray Hill, New Jersey
*) Work supported by the Deutsche Forschungsgemeinschaft and the US National Science Foundation

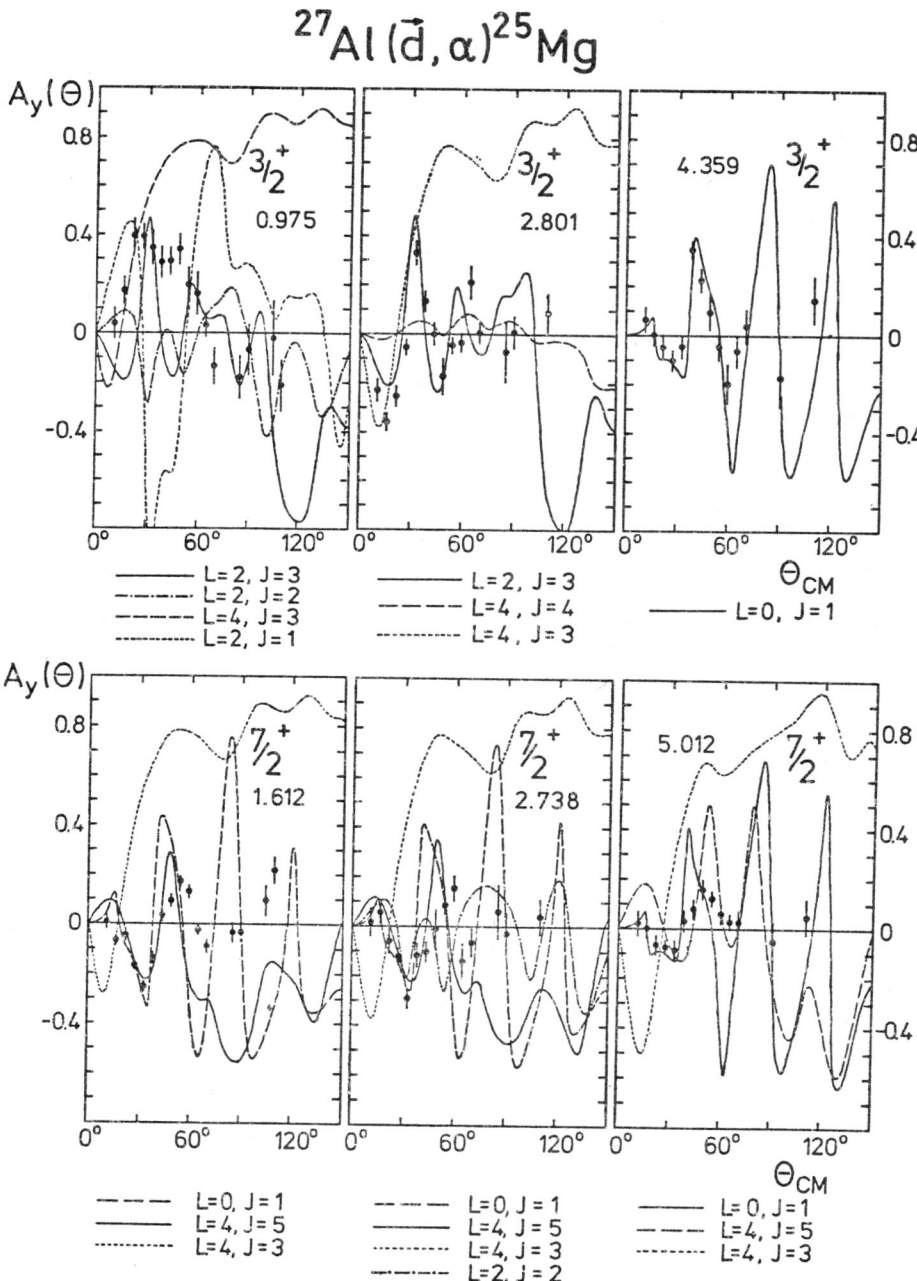

Fig. 2. $A_y(\theta)$ for ^{27}Al(d,α)^{25}Mg for transitions to $J^\pi = 3/2^+$ and $7/2^+$ states.

TENSOR ANALYZING POWER MEASUREMENTS FOR (d,α) ON s-d SHELL NUCLEI

Y. Tagishi [*], T. B. Clegg, E. J. Ludwig,
S. A. Tonsfeldt and J. F. Wilkerson [+]
Department of Physics and Astronomy, University of North Carolina,
Chapel Hill, NC 27514 and
Triangle Universities Nuclear Laboratory, Durham, NC

ABSTRACT

In previous studies of the (d,α) reactions on s-d shell nuclei, it was found that the angular distributions of vector analyzing power depend distinctly on the orbital angular momentum L and total angular momentum J transferred. We report that the angular distribution pattern of the tensor analyzing power has a considerable sensitivity of a different nature to differences in the J-transfer.

Cross sections and vector and tensor analyzing powers, iT_{11}, T_{20} and T_{22} were measured at 16 MeV for $^{32}S(d,\alpha)^{30}P$, $^{36}Ar(d,\alpha)^{34}Cl$ and $^{38}Ar(d,\alpha)^{36}Cl$. These target nuclei were chosen so that well-separated states in the residual nuclei would be reached via the (d,α) reaction. The final states provided examples of J=L-1, L and L+1 transfers. Previously, only iT_{11} was measured.[1]

The experiment was performed with a 16 MeV polarized deuteron beam from the TUNL accelerator using a Lamb-shift polarized ion source. Targets of H_2S, ^{36}Ar and ^{38}Ar gas were contained in a cell at pressures of 1 atm or less. The $^3He(\vec{d},p)^4He$ reaction was used to monitor the polarization of the incident deuteron beam. The DWBA calculations of the T_{22} and T_{20} for J transfers of 1, 2, and 3 are shown in Fig. 1. The T_{22} for the J=1 (=L-1) and J=3 (=L+1) transfers show predominantly negative values while J=2 (=L) transfer has

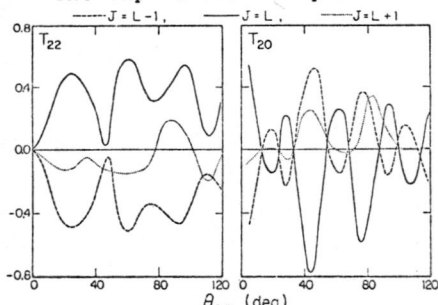

Fig.1. The T_{20} and T_{22} angular distributions for $^{38}Ar(d,\alpha)^{36}Cl$ from DWBA calculations for L=2; J=1, 2, and 3.

positive analyzing power values. Such a difference of the sign of

----

[*] On leave from the University of Tsukuba, Ibaraki 305, Japan
[+] Work supported in part by U. S. Department of Energy.

0094-243X/81/690736-03$1.50 Copyright 1981 American Institute of Physics

T_{22} between J=L±1 and J=L transfers is discussed later.

Typical experimental results of the tensor analyzing powers, T_{22} and T_{20}, for different L and J transfer cases are shown in Fig. 2.

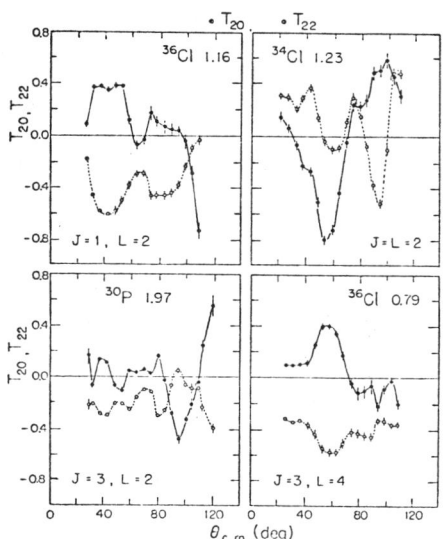

Fig.2. Measured T_{20} and T_{22} values for $^{32}S(d,\alpha)^{30}P(1.97, L=2, J=3)$, $^{36}Ar(d,\alpha)^{34}Cl(1.23, L=2, J=2)$ and $^{38}Ar(d,\alpha)^{36}Cl$ (0.79, L=4, J=3; 1.16, L=2, J=1). Lines are drawn to guide the eye.

For $J^\pi = 3^+$ and $J^\pi = 1^+$ states, two L values, L=J−1 and L=J+1, are allowed. For each of these cases the experimental distributions for cross section and iT_{11} compare quite closely with DWBA calculations made assuming a single L value is involved in the transfer[1,2]. As shown in Fig. 2., the T_{22} angular distributions are predominantly negative in value for cases of J=L±1 transfers, while for J=L transfer the T_{22} patterns are more oscillatory and more predominantly positive at forward angles. These patterns are typical of a large number of transitions studied. A more surprising feature for the measured tensor analyzing powers is that the T_{20} angular distributions for J=L±1 transfers are out of phase with the T_{22} distributions. The corresponding pattern for J=L transfer do not exhibit these relations nor do the DWBA calculations for these tensor analyzing powers (see Figs. 1 and 2).

The observed sign rule for T_{22} is well understood from the definition of T_{22}.

The T_{22} is given with a purely aligned beam of deuterons (no m=0 deuterons) as following:

$$T_{22} = (\sigma_x - \sigma_y)/\sqrt{3}\sigma_0 ,$$

where σ_0 is the cross section with unpolarized beam and σ_x and σ_y are cross sections with a purely aligned beam of deuterons along the x- and y-axis, respectively. The y-axis is normal to the reaction plane and x-axis is normal to the incident beam direction and y-axis. Since outgoing alpha particles have zero spin, the spin transfer S=1 is anti-parallel to the incoming deuteron spin. When σ_y is measured, S is along y; L and S add or subtract, and a

$J=L\pm 1$ deuteron cluster tends to be transferred. A larger σ_y compared to σ_x then implies a negative T_{22}. When σ_x is measured, S is along x; L and J are perpendicular, and there will be more probability for transferring a $J=L$ deuteron cluster. A larger σ_x compared to σ_y then implies a positive T_{22}. This simple vector-model coupling argument is not possible for T_{20} and T_{21}.

Other quantities particularly sensitive to the J transfer are σ_x/σ_0 and σ_y/σ_0 as shown in Fig. 3. The σ_x/σ_0 and σ_y/σ_0 are obtained from the measured T_{20} and T_{22} as following:

$$\sigma_x/\sigma_0 = 1 + \sqrt{3}T_{20}/2 - \sqrt{2}T_{22}/4 \quad , \quad \sigma_y/\sigma_0 = 1 - \sqrt{3}T_{20}/2 - \sqrt{2}T_{22}/4 \; .$$

It can be seen from Fig. 3 that σ_y is larger than σ_x for $J=L\pm 1$ transfers (3^+ and 1^+ transitions). For the $J=L$ transfer (2^+

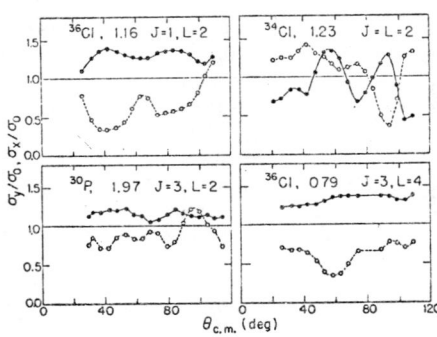

Fig.3. The ratio of the cross sections for aligned beams σ_x and σ_y to that for an unpolarized beam, σ_0 for the transitions shown in Fig.2. The solid (open) points are for σ_x/σ_0 (σ_y/σ_0). Lines are drawn to guide the eye.

transition) σ_x is larger than σ_y at forward angles. Similar patterns resulted for all final states studied in these nuclei.

In conclusion, the angular distribution patterns of T_{22} in (d,α) reactions clearly distinguish between $J=L$ and $J=L\pm 1$ transfers in an understandable way. For the $J=L\pm 1$ transfer cases studied the patterns for T_{22} are out of phase with those for T_{20} although this effect is not predicted by the DWBA calculations performed to date. The information provided by tensor analyzing power distributions compliments that provided by vector analyzing powers which readily distinguishes between $J=L+1$ and $J=L-1$ transfers.

REFERENCES

1. E. J. Ludwig, T. B. Clegg, W. W. Jacobs and S. A. Tonsfeldt, Phys. Rev. Lett. <u>40</u>, 441(1978).
2. Y. Tagishi, T. B. Clegg, J. F. Wilkerson, E. J. Ludwig, S. A. Tonsfeldt and W. Wylie, Bull. Am. Phys. Soc. <u>25</u>, 555(1980).

SPECTROSCOPY OF STRETCHED CONFIGURATIONS WITH THE $(\vec{d},\alpha)$ REACTION AT 52 MeV

G. Mairle, Liu Ken Pao, K.T. Knöpfle, H. Riedesel,
K. Schindler, G.J. Wagner
Max-Planck-Institut für Kernphysik, 6900 Heidelberg, W.-Germany

V. Bechtold, J. Bialy, L. Friedrich
Kernforschungszentrum Karlsruhe, IAK, 7500 Karlsruhe, W.-Germany

ABSTRACT

The reactions $^{12}C(\vec{d},\alpha)^{10}B$, $^{18}O(\vec{d},\alpha)^{16}N$ and $^{34}S(\vec{d},\alpha)^{32}P$ have been investigated at E_d = 52 MeV. The observed vector analyzing powers are pronounced ($|i\,T_{11}| \leq 0.9$). They exhibit characteristic patterns for final spins J=L, J=L±1 and provide spin determinations at least for states of unique L-transfer. DWBA calculations assuming deuteron-cluster pick-up reproduce qualitatively the observed effects. The method has been applied to locate states of maximum spin in $^{16}N(3^+,4^-)$ and in $^{32}P(5^+)$.

INTRODUCTION

Spectroscopic investigations by means of the (d,α) reaction are usually limited, because the observed angular distributions are unstructured and the a priori knowledge of suitable two-particle form-factors is required to describe them in the framework of the DWBA. It has been shown [1,2], however, that analyzing powers of $(\vec{d},\alpha)$ reactions are sensitive as to the coupling of the deuteron spin to the transferred angular momentum L. This allows spin determinations at least for states excited with an unique L-value. The method, therefore, is excellently suited to study stretched two particle configurations.

RESULTS

The experimental set-up has been described in detail in ref.[3]. For the measurements we used C_3H_8, $^{18}O_2$ (99% enriched) and $H_2^{34}S$ (91% enriched) gases, which were kept in a gas cell (7 cm ⌀) at a pressure of 450 mbar. The α-particles were detected in ΔE-E telescopes, each consisting of 200 μm ΔE and 1500 μm E silicon surface barrier counters. Local, zero-range DWBA calculations were performed assuming the pick-up of a deuteron cluster. The optical potentials for the entrance channel were determined from the analysis of elastic $\vec{d}$-scattering[3].

In fig. 1 the analyzing powers of some states observed in the $^{18}O(\vec{d},\alpha)^{16}N$ reaction are shown. For states of known spin J and unique L they show consistently large negative values at forward angles for J=L+1, large positive values for J=L-1 and small positive values for J=L. The measured analyzing powers for negative parity states at 0.0, 0.297 and 6.17 MeV are consistent with the spins 2^-, 3^- and 4^-, respectively, if L=3 transfer is assumed in the DWBA cal-

culations. For the g.s. transition L=1 is also allowed; L=3 is
unique for 3^- and 4^- states in the frame of the simple shell model.
A spin 4^- for the 6.17 MeV state is in agreement with results of ^{17}O
(d,τ), ^{17}O$(\vec{d},\tau)$ and ^{16}O(p,p') experiments [4, 5, 6]. The analyzing
power for the 3^+ state at 3.96 MeV (unique L) is in perfect agreement
with a DWBA calculation assuming L=2. For the 1^+ states at 3.355 MeV
and 4.319 MeV a mixture of L=0 and L=2 would be consistent with the
data.

The analyzing powers observed in the ^{12}C$(\vec{d},\alpha)^{10}$B and ^{34}S$(\vec{d},\tau)^{32}$P are not shown for brevity. At least for states of maximum spin
there is a clear-cut situation with negative values of the analyzing
power at forward angles. This is e.g. the case for the 3^+ states at
0.0 and 4.77 MeV in ^{10}B (L=2) and for strongly excited states at
7.96 and 8.09 MeV in ^{32}P which are described by DWBA calculations
assuming L=4 pick-up, with the consequence of a spin 5^+ for these
to-date unknown states.

From a shell model point of view, the considered states are
characterized by their stretched configurations:

^{10}B: $(1p_{3/2}^{-2})_{J=3^+,T=0}$

^{16}N: $(1d_{5/2}\ 1p_{3/2}^{-1})_{J=4^-,T=1}$ and

$|(1d_{5/2}^2)_{0,1}\ (1p_{3/2}^{-2})_{3,0}|_{J=3^+,T=1}$

^{32}P: $|(1d_{3/2}^2)_{0,1}\ (1d_{5/2}^{-2})_{5,0}|_{J=5^+,T=1}$

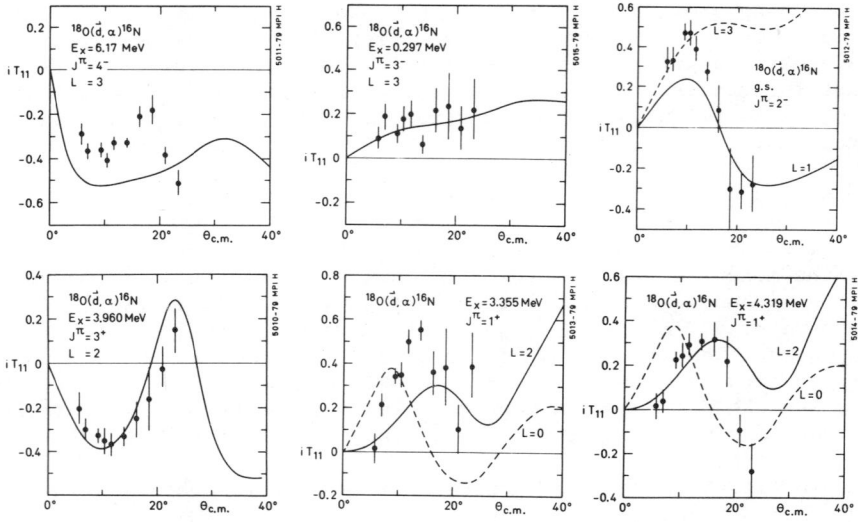

Fig. 1. Vector analyzing powers measured in the ^{18}O$(d,\alpha)^{16}$N reaction compared to DWBA calculations assuming pick-up of a deuteron cluster of given L.

The energies of these states represent a valuable information on effective interactions and hence for shell model calculations.

REFERENCES

1. J.D. Crossairt et al., Nucl. Phys. A287, 13 (1977).
2. P.G. Ikossi et al., Nucl. Phys. A297, 1 (1978).
3. G. Mairle et al., Nucl. Phys. A339, 61 (1980).
4. G. Mairle et al., Nucl. Phys. A299, 39 (1978).
5. G. Mairle et al., Contribution to this Conference.
6. R.S. Henderson et al., Australian J. of Phys. 32, 411 (1979).

THE ^{46,48}Ti$(\vec{p},\alpha)^{43,45}$Sc REACTION

R.N. Boyd, S.L. Blatt, T.R. Donoghue and H.J. Hausman,
The Ohio State University, Columbus, OH 43210;

E. Sugarbaker,
University of Colorado, Boulder, CO 80302;

S.E. Vigdor,
Indiana University, Bloomington, IN 47401

ABSTRACT

The $(\vec{p},\alpha)$ reaction has been studied at E_p = 79.2 MeV. Strong selectivity of final states, especially of high spin states, and large analyzing powers are observed. DWBA reaction calculations are found to give good representations of the general features of the data.

INTRODUCTION

The present $(\vec{p},\alpha)$ reaction study has been conducted with two primary goals in mind. Firstly, a number of questions exist as to the (p,α) reaction mechanism. It was hoped that the detail provided by both cross section and analyzing power data would help to clarify that mechanism. Secondly, the momentum matching at medium energies for this reaction should favor high spin final states. Thus this reaction study should help to identify such states in the final nuclei.

EXPERIMENTAL DETAILS

The polarized proton beam of the Indiana University Cyclotron Facility (IUCF) was used in this study. Typical beam intensities of the dispersion matched beam were 200 nA at the proton energy of 79.2 MeV. The beam polarization (typically about 0.7) was measured in a Helium polarimeter located between the injector and mainstage cyclotrons. The targets were about 1.0 mg/cm^2 thick. The 20% ^{48}Ti impurity in the ^{46}Ti target necessitated accounting for the ^{45}Sc states which appeared in the ^{43}Sc spectrum. Identification and analysis of the reaction products were accomplished with the IUCF QDDM spectrograph and focal plane detection system, consisting of a wire helix followed by two plastic scintillation detectors. Typical resolution achieved was 70 to 80 keV FWHM.

DISCUSSION

While differential cross section and analyzing power data were obtained for many states in ^{43}Sc and ^{45}Sc, this report will focus on

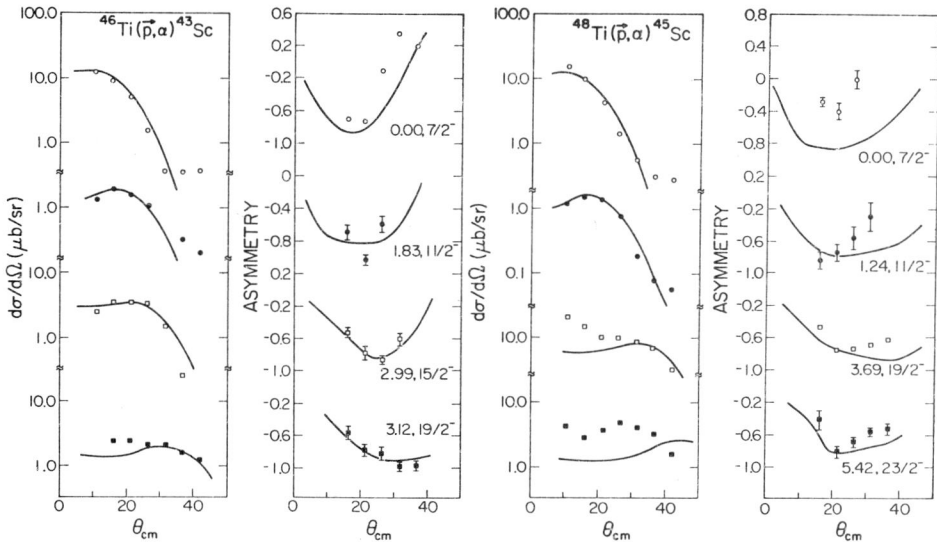

Fig. 1. Data for the ^{46}Ti$(\vec{p},\alpha)$ reaction.

Fig. 2. Data for the ^{48}Ti$(\vec{p},\alpha)$ reaction.

several of the most interesting ones. Their angular distributions are shown in Figures 1 and 2.

Noteworthy in these data is the fact that the largest cross sections, except for the most forward angles, are those for the high spin states[1,2]. In particular, the 19/2⁻ states (the maximum angular momentum transfer which can result from pickup of a $\nu f_{7/2}^2 \pi f_{7/2}$ configuration in a one step process) have the largest cross sections seen, while states with similarly high spins, e.g., 15/2⁻ and 23/2⁻ or 21/2⁻, are also strongly excited. Secondly the observed analyzing powers are very large for many of the states.

The curves shown in conjunction with the data are the results of DWBA calculations. Typical optical parameters[3,4] were used in these calculations, along with a bound state (for a mass 3 particle) of $(r_0, a) = (1.3, 0.65)$. The number of nodes N for the mass 3 cluster was given by $2N + L_{tr} = 9$. Those calculations are seen to do quite well in representing the shapes of the angular distributions. The 5.42 MeV state in ^{45}Sc has been assigned[2] a spin of 21/2⁻ or 23/2⁻. The curves shown in conjunction with those data result from assuming that state is populated by a two step process involving $J_{tr} = 19/2$ triton transfer and $L_{tr} = 2$ inelastic excitation. The only adjustable parameter in the calculation, assuming the same scaling factor as for the 19/2⁻ state, is the β_2 for the inelastic excitation: it was taken to be 0.2. If the spin is assumed to be

$23/2^-$, the cross section is in reasonable agreement with the observed magnitude. However, if the $21/2^-$ assignment is assumed, the cross section is predicted to be a factor of 20 to 50 too small. Thus the $23/2^-$ assignment appears to be strongly favored.

Numerous other high spin states were also populated in this reaction study. These are being investigated to learn as much as possible about their spins. Microscopic form factor calculations are also proceeding so as to determine the magnitude of the primary component for the wave function of each of the states seen.

REFERENCES

1. A.R. Poletti, E.K. Warburton, J.W. Olness, J.J. Kolata and Ph. Gorodetzky, Phys. Rev. C13, 1180 (1976).
2. P.G. Bizzeti, A.M. Bizzeti-Sona, M. Bucciolini, R. Huber, W. Kutschera, H. Morinaga, R.A. Ricci and C. Signorini, Nuovo Cim. 26A, 25 (1975).
3. F.D. Becchetti, Jr. and G.W. Greenlees, Phys. Rev. 182, 1190 (1969).
4. H. Rebel, G. Hauser, G.W. Schweimer, G. Nowicki, W. Wiesner and D. Hartmann, Nucl. Phys. A218, 13 (1974).

ASYMMETRIES IN (^{3}He,^{7}Be) CROSS SECTIONS AND SPIN-ORBIT COUPLING[+)]

P. Lezoch, H.-J. Trost, Md.A. Rahman and U. Strohbusch
Zyklotron Labor, I. Institut für Experimentalphysik
D-2000 Hamburg 50, Fed. Rep. of Germany

This paper is concerned with the differences consistently observed in (^{3}He,^{7}Be($3/2^-$)$_{g.s.}$) and (^{3}He,^{7}Be($1/2^-$)$_{0.429}$) experimental angular distributions. Due to the α-^{3}He cluster configuration of the relevant ^{7}Be states (2p3/2 and 2p1/2 state of relative motion), it is agreed that the (^{3}He,^{7}Be) reaction proceeds via α-pickup mechanism. Corresponding DWBA calculations, however, failed to reproduce the $3/2^-$ - $1/2^-$ asymmetries. For a given final nucleus state they essentially yield the same angular distribution patterns for both ^{7}Be states, contrary to the experimental evidence.

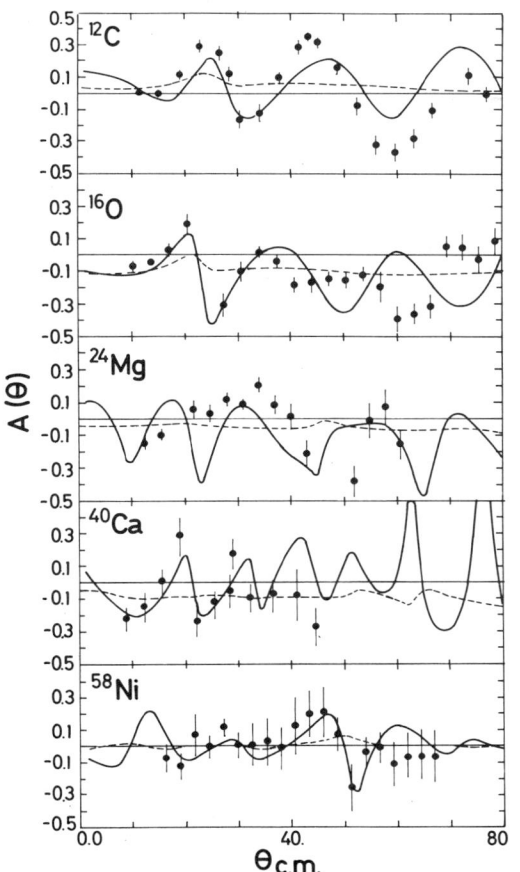

In order to study this in more detail we have extended the FRDWBA code LOLA to include spin-orbit coupling (neglected so far in these calculations). The figure displays DWBA predictions and experimental data for the asymmetry defined as

$$A(\theta) = \frac{\sigma_{3/2}(\theta) - 2\sigma_{1/2}(\theta)}{\sigma_{3/2}(\theta) + 2\sigma_{1/2}(\theta)}$$

The experimental points have been obtained at this laboratory at 41 MeV, except for the ^{40}Ca data (70 MeV) which are taken from ref. 1

The results of the calculations are as follows:
(i) $A(\theta)$ is essentially constant with no spin-orbit interaction (dashed curves).
(ii) Drastic changes are observed when spin-orbit potential is switched on in the ^{7}Be channel, even with the relatively small strength parameter $V_{SO} \approx 1.5$ MeV.
(iii) Introducing spin-orbit interaction in the ^{3}He

entrance channel alone already produces strong asymmetries. They are found to be mediated by the polarization of the transferred orbital momentum which in turn is caused through momentum matching conditions.

The solid curves show the combined effect with spin-orbit coupling in both channels simultaneously. Despite the deficiencies in the details of the fits, the results suggest spin-orbit interaction to represent the major reason for the asymmetries observed experimentally.

REFERENCES

+) Supported by Bundesministerium für Forschung und Technologie, Fed. Rep. of Germany
1. W.F. Steele et al., Nucl. Phys. A266, 445 (1976)

PHOTONUCLEAR REACTIONS WITH LINEARLY POLARIZED PHOTONS

K. Wienhard,[*] K. Ackermann, K. Bangert, U. E. P. Berg, C. Blässing, K. Kobras, W. Naatz, D. Rück, R. K. M. Schneider and R. Stock

Institut für Kernphysik, Universität Giessen, D-63 Giessen, Germany

A linearly polarized photon beam was used to study $^{16}O(\vec{\gamma},p)$ and to determine M1- and E1-strength in s-d shell nuclei via Nuclear Resonance Fluorescence (NRF)-scattering. Linearly polarized photons from off-axis electron bremsstrahlung were selected by a narrow lead collimator. The $d(\gamma,p)$-reaction was used to measure and to continuously monitor the polarization. Fig. 1 shows the measured polarization as a function of photon energy for 30 MeV electrons hitting a 50 μm thick aluminum radiator and an off-axis angle of $1.4°$.

THE $^{16}O(\vec{\gamma},p)$ ^{15}N-REACTION

From studies of photonuclear reactions with linearly polarized photons one learns: (1) the parity of the absorbed radiation in an unambiguous and model independent way, (2) the decay amplitudes of overlapping resonances with different parities, and (3) the angular distribution for a dipole transition of known parity from a polarized measurement at only one angle. The $^{16}O(\gamma,p)$ reaction was studied at bremsstrahlung endpoint energies of 22 and 30 MeV with four proton detectors at $\theta = 90°$ symmetrically around the beam and with two orthogonal directions of the photon polarization to cancel experimental asymmetries.

For a pure E1-transition the analyzing power at $\theta = 90°$ is $A(90°) = 3 a_2/(2-a_2)$ where a_2 is the coefficient of the unpolarized distribution.[1] The lower part of Fig. 2 depicts the measured analyzing power as error bars together with values calculated from a_2 coefficients in the literature[2,3,4] assuming pure E1-absorption. This gives direct evidence that all these resonances (except the one at E_x = 16.2 MeV) are predominantly E1. The only previous measurement with linearly polarized photons[5] suggested dominant E1-absorption for the 17.3 MeV region. As can be seen from the top of Fig. 2 there are actually two resonances at this energy. For the weaker one at 17.15 MeV an M1 assignment was proposed.[3] In our experiment, this state is not resolved from the stronger neighboring 1^--state at 17.3 MeV.

A clear deviation of the analyzing power from the pure E1-value is obvious at the 16.2 MeV resonance. At energies below and above the 16.2 MeV resonance the measured analyzing power agrees with the values calculated for E1-excitation. The large negative a_2 coefficient observed at the 16.2 MeV resonance excludes $J^\pi = 2^+$ and therefore gives direct evidence that this state has $J^\pi = 1^+$ and overlaps with a broad E1-resonance. This confirms previous (e,e')[6] and

---

[*]Present address: HEPL, Stanford University, Stanford, CA 94305.

0094-243X/81/690747-03$1.50 Copyright 1981 American Institute of Physics

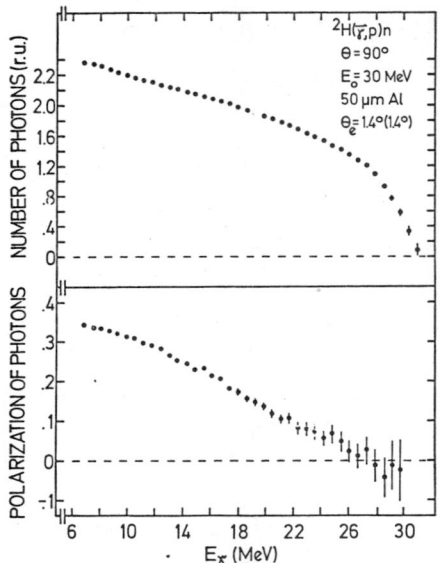

Fig.1 Intensity and Polarization for 30 MeV bremsstrahlung.

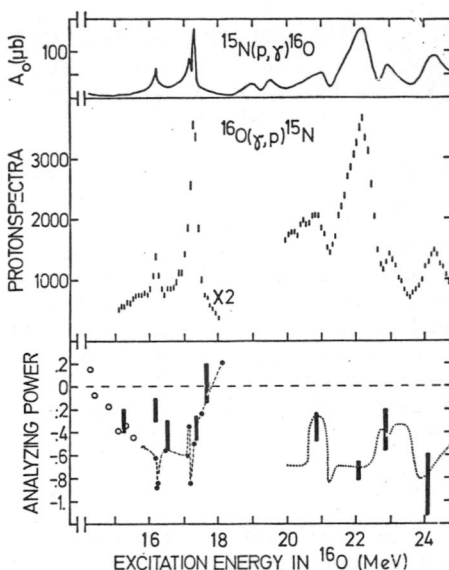

Fig.2 The $^{16}O(\gamma,p_o)$ cross section^2 with the measured proton spectra and the measured (error bars) and the calculated $A(90°)$ assuming only E1 excitation.

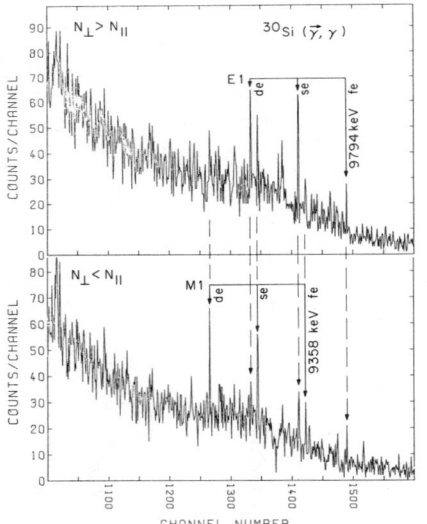

Fig.3 Ge(Li) spectra with the photon polarization perpendicular and parallel to the scattering plane for NRF scattering from ^{30}Si.

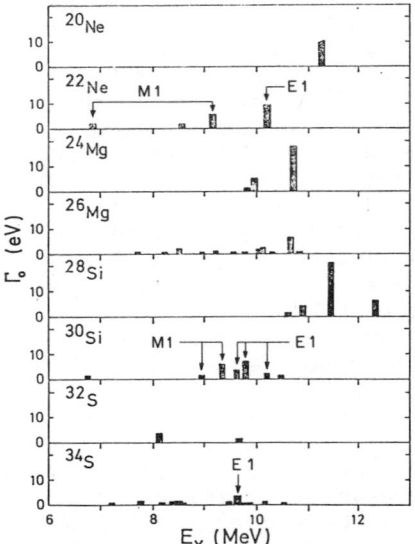

Fig.4 Dipole strength in s-d shell nuclei.

$(\vec{p},\gamma)^3$ studies. However, the magnitude of the analyzing power provides new information about the decay amplitudes of this 1^+ state. For overlapping E1 and M1 states the analyzing power at $\theta = 90°$ is given by[1] $A(90°) = [3 a_2(E1) - 3 a_2(M1)]/[2 - a_2]$ where $a_2 = a_2(E1) + a_2(M1)$ and $a_2(E1)$ and $a_2(M1)$ designate the E1 and M1 contribution to the a_2 coefficient. From the energy averaged ($\pm$ 100 keV) measured value $A(90°) = -0.21 \pm 0.09$ together with the energy averaged value[2,3] of $a_2 = -0.8 \pm 0.1$ we obtain $a_2(E1) = -0.5 \pm 0.1$ and $a_2(M1) = -0.3 \pm 0.1$. M1-absorption in ^{16}O is associated with ^1P and ^3P-waves for proton decay to the ground state of ^{15}N and $a_2(M1) = -|^1P|^2 + 0.5 |^3P|^2$. We derive $|^1P|^2 + |^3P|^2 = 0.4 \pm 0.1$ by taking the ratio of the peak cross section of the 16.2 MeV resonance to the underlying broad E1-resonance from the total cross section[2] and averaging over our experimental energy resolution using resonance parameters[3] for the 16.2 MeV resonance. Together with $a_2(M1) = -0.3 \pm 0.1$ we obtain $|^3P|^2 = 0.07 \pm 0.13$ and $|^1P|^2 = 0.33 \pm 0.10$. This shows that the 16.2 MeV 1^+ state in ^{16}O decays preferentially by channel spin $S = 0$ to the ground state of ^{15}N.

In the 30 MeV data, protons primarily from transitions to the $3/2^-$ third excited state in ^{15}N contributed to the lower energy part of the spectrum.[7] The total $^{16}O(\gamma,p_3)$-cross section has been measured,[8] but not the angular distribution. Protons from the (γ,p_3) decay of the main peak of the ^{16}O GEDR at 22.2 MeV were identified and their analyzing power measured. Fig. 2 shows that at $E_x = 22.2$ MeV E1-absorption is dominant. Therefore we can derive the a_2-coefficient for the $^{16}O(\gamma,p_3)$ reaction from the measured value of $A(90°)$.

From $A_{p_3}(90°) = -0.7 \pm 0.2$ we obtain $a_2 = -0.63 \pm 0.23$. This shows that in the main peak of the ^{16}O GEDR the (γ,p_3) angular distribution is also strongly peaked around $90°$.

NUCLEAR RESONANCE FLUORESCENCE SCATTERING

With linearly polarized bremsstrahlung E1 and M1 excitation can easily be separated in NRF scattering (Fig. 3). The systematic study of dipole excitations in $T = 0$ and $T = 1$ s-d shell nuclei shows that in ^{22}Ne, ^{30}Si and ^{34}S strong E1 levels are present in the region of the magnetic dipole resonance (Fig. 4).

REFERENCES

1. L. W. Fagg, S. S. Hanna, Rev. Mod. Phys. **31**, 711 (1959).
2. E. D. Earle and N. W. Tanner, Nucl. Phys. **A95**, 241 (1967).
3. K. A. Snover et al., Phys. Rev. Lett. **43**, 117 (1979).
4. W. J. O'Connell and S. S. Hanna, Phys. Rev. **C17**, 892 (1978).
5. K. Shoda, J. Phys. Soc. Japan **16**, 1841 (1961).
6. M. Stroetzel and A. Goldmann, Z. Physik **233**, 245 (1970).
7. Y. S. Horowitz, et al., Nucl. Phys. **A151**, 161 (1970).
8. J. T. Caldwell, et al., Phys. Rev. Lett. **19**, 447 (1967).

ANTISYMMETRIZATION EFFECTS IN DEUTERON-NUCLEUS ELASTIC SCATTERING

J. A. Tostevin, M. H. Lopes and R. C. Johnson,
Department of Physics, University of Surrey, Guildford, U.K.

ABSTRACT

An approximate local energy dependent prescription for the deuteron optical potential is presented which includes the effects of both the deuteron D-state and antisymmetrization. A large T_p tensor potential is obtained which has a strong effect upon calculated reaction observables at energies of current experimental interest.

Under the assumption that the target nucleus is described by a single determinant, Φ, of single particle states, and is not excited during the reaction, the fully antisymmetrized wavefunction for the d+A system is

$$\Psi(n,p,A) = \mathcal{A}\{\psi(n,p)\Phi\}$$

where $\mathcal{A}$ is the antisymmetrization operator for A+2 nucleons. The state $\psi(n,p)$, which describes the motion of the incident n-p pair, then satisfies an equation of the Bethe-Goldstone type, namely $\{E - H\}\psi(n,p) = 0$, where $H = K(r) + K(R) + V(HF) + QV(np)Q$ and $(Q - 1)\psi(n,p) = 0$. Here, Q projects off all nucleon states occupied in Φ, V(HF) is the sum of the Hartree-Fock potentials for the n and p-nucleus systems, V(np) is the n-p interaction and $K(r)[K(R)]$ the relative [centre of mass] kinetic energy operator for the n-p pair.

Decomposing $H = h + V(HF) + V(PP)$, we include in $V(PP) = QV(np)Q - V(np)$ those Pauli Principle effects normally neglected in the 3-body Hamiltonian H. The deuteron optical potential W may then be written formally as

$$W = \langle\phi(d)|U|\phi(d)\rangle$$

where, if Q(d) projects off the deuteron ground state wavefunction $\phi(d)$,

$$U = \{V(HF) + V(PP)\}\left\{1 + \frac{Q(d)}{E-h} U\right\}.$$

Neglecting strong nuclear breakup effects due to V(HF), we solve the integral equation for U to first order in V(HF). However, Pauli breakup effects through the interaction V(PP) will be included to all orders. Our approximation for the deuteron optical potential is therefore

$$W = \langle\chi|V(PP)|\phi(d)\rangle + \langle\chi|V(HF)|\chi\rangle, \qquad (1)$$

$$|\chi\rangle = |\phi(d)\rangle + \frac{Q(d)}{E-h} V(PP)|\chi\rangle.$$

Clearly the second term of W comprises a modified folding model in which the nucleon-A interactions are averaged over the n-p relative motion in the state $|\chi\rangle$. In the calculations presented below V(HF) is replaced by phenomenological n- and p-A optical potentials[1] corresponding to $\frac{1}{2} E_d$. In the limit V(PP) = 0, W reduces to the Watanabe form, which the present calculation corrects for those Pauli effects associated with the composite nature of the projectile.

The state $|\chi\rangle$ is very complicated in general. It can however be approximated by a modified deuteron internal wave function at each C.M. position. This is possible because at both high and low incident deuteron energies, for those states most strongly coupled by V(PP) to $\phi(d)$,

$(E - h)^{-1} V(PP)$

$\approx (-\varepsilon - K(r) - V(np))^{-1} V(PP)$.

Here ε is the deuteron binding energy. The folding for W is now evaluated using the local Fermi gas approximation[2] for the Pauli operator Q and a separable model for V(np)[3] which includes the n-p tensor force. Both terms of equation (1) produce forces of the T_p type, which are illustrated in Figure 1 for scattering from ^{58}Ni at 80 MeV. The first term, which has the form of an energy shift, produces the attractive interaction indicated by the dash-dot curve. This interaction has the same origin as that of Ioannides and Johnson[2] and is of similar shape and magnitude. The remaining curves show the results when the folding term of W, not considered in ref. 2, is included. This term is seen to dominate, producing an attractive real T_p force of strength $\approx$ 10 MeV. For the same reaction, figure 2 shows the correction to the real central Watanabe potential due to the inclusion of V(PP). All other potential formfactors are very close to the corresponding Watanabe predictions. In figure 3 we present calculations of the observables for this reaction using an optical potential calculated via : (i) the Watanabe (solid curves) and (ii) present Modified folding procedures (dashed curves).

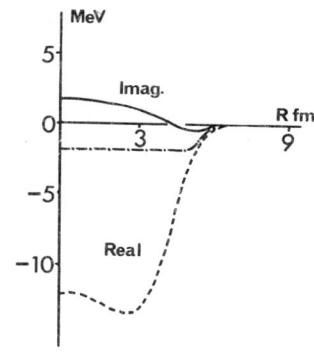

Fig. 1   Calculated T_p tensor forces.

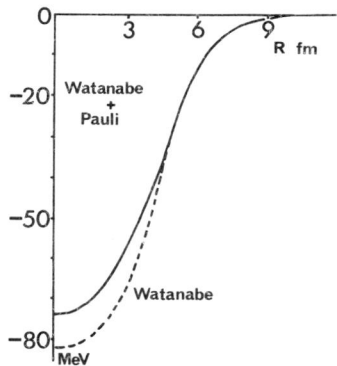

Fig. 2   Calculated central potentials.

We see interesting effects due to the inclusion of V(PP) in an energy region where data is now becoming available[4].

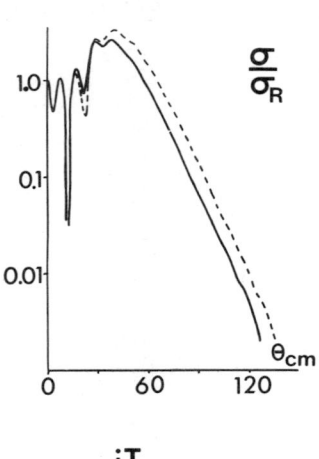

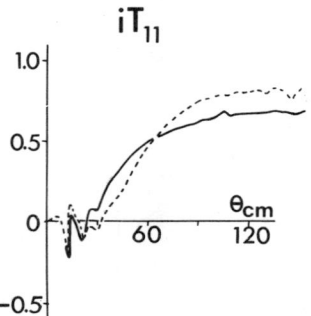

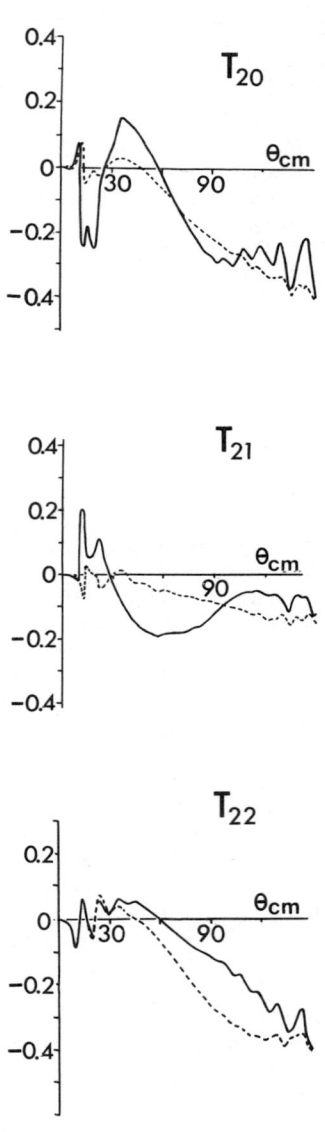

Fig. 3 Calculated observables for the reaction ^{58}Ni (d,d) using the Watanabe and Modified folding model potentials.

REFERENCES

1. F.D. Bechetti and G.W. Greenlees, Phys. Rev. 182, 1190 (1969).
2. A.A. Ioannides and R.C. Johnson, Phys. Rev. C17, 1331 (1978).
3. Y. Yamaguchi and Y. Yamaguchi, Phys. Rev. 95, 1635 (1954).
4. E.J. Stephenson et al., This Conference paper No. 2.33.

AUTHOR INDEX

R. Abegg..............188, 1205, 1287, 1317
K. Ackermann 747
G. S. Adams 146, 554
E. G. Adelberger.....1367
M. Ahmed 389
F. Ajzenberg-Selove..697
D. M. Alde 1449
W. P. Alford 682, 718
M. S. A. L. Al-Ghazi.1290
J. C. Allred 169
M. R. Anastasio 200
H. J. Andrä 1037
K. Aniol 158
L. E. Antonuk 188, 568
Y. Aoki 529, 611, 614, 709
E. Aprile 143, 161, 955
M. Araki 194
R. A. Arndt 179
W. Arnold 882, 1258
J. Arvieux 1201
E. G. Auld 93, 550
O. Avila 1332
V. G. Avrigeanu 1085

A. D. Bacher 220, 457, 632, 1455
N. L. Back 505
F. T. Baker 554
S. D. Baker 169
R. Balzer 1439
K. Bangert 747
M. D. Barker 137, 877, 931
M. Barlett 135
J. M. Barnwell 502
A. K. Basak 587, 590, 593, 659, 703, 706, 1284
B. Bauer 916, 979
G. Baum 785
V. Bechtold 685, 691, 739
R. Beckmann 964, 1308
D. A. Bell 152
H. Berg 882, 1258
U. E. P. Berg 747
U. Berghaus 1305
M. R. Bergstrom 933
M. Bernas 1120
K. Bethge 1085
P. R. Bevington 149, 958
H. C. Bhang 505, 511
T. S. Bhatia 123, 126, 129, 132, 166, 871, 874, 949
J. Bialy 739
L. C. Biedenharn.....1

J. A. Bieszk.....641
J. Birchall......973, 1237, 1263, 1290
G. S. Blanpied...135
C. Bläsing.......747
S. L. Blatt......742
E. Bleszynski....556
M. Bleszynski....556, 1217, 1220
C. O. Blyth......1284
D. Boal..........158
H. G. Bohlen.....1117
J. G. Boissevain.949
B. E. Bonner.....123, 126, 129, 132, 135, 149, 166, 871, 874, 949, 956, 958
R. Böttger 916, 919, 970, 979, 991, 1096, 1099, 1102, 1105, 1108
T. J. Bowles 939
J. D. Bowman.....1449
R. N. Boyd 373, 682, 718, 742
D. M. Brink......1105
W. Broermann.....925
F. D. Brooks.....656
J. C. Brown......1466, 1469
R. E. Brown......496, 694, 697, 700, 718, 1240, 1243, 1246, 1249, 1252, 1263, 1314
H. Brückmann.....1305
J. Bruckshaw.....890
J. A. Buchanan...152, 169
W. Burgmer.......1299
D. G. Burke......700
B. L. Burks......430, 467, 647
R. C. Byrd.......398, 401, 404, 407, 413, **988**, 1272, 1275, 1323, 1326, 1472, 1475, 1478, 1481
J. Bystricky 177, 178

R. R. Cadmus, Jr.647
J. R. Calarco ...596, 600
M. M. Calkin.....152
R. Caloi.........175, 928
C. P. Cameron....602, 605
J. A. Cameron....679, 682
J. M. Cameron....161, 188, 454, 1205, 1287
J. Carroll.......1217
L. Casano........175, 928
G. Caskey........877, 931
E. P. Chamberlin.887
P. Chatelain 163, 1232
A. Chisholm......1488
Y. H. Chiu 1311
J. A. Cizewski ..694, 697, 700
H. W. Clark 1466, 1469

N. M. Clarke 502
G. Clausnitzer 882, 1258
T. B. Clegg 398, 401, 404, 407, 413, 415, 430, 467, 647, 736, 884, 944, 961, 1126, 1272, 1323, 1326, 1472, 1478

H. Clement 473, 476, 571, 573, 576, 579, 664, 667, 670
J. M. Clement 152, 169
J. E. Clendenin 936
M. Clover 1117
P. C. Colby 137
J. C. Collins 484
J. R. Comfort 547
H. E. Conzett 1195, 1263, 1422, 1442, 1452
D. W. Cooke 1057
M. Copel 169
W. D. Cornelius 149, 958, 994, 1223, 1335, 1461
F. D. Correll 496, 1240, 1243, 1246, 1249, 1252, 1314
S. R. Cotanch 1481
J. Côté 114
D. G. Crabb 11
J. G. Cramer 505
G. M. Crawley 632

W. W. Daehnick 484, 487
J. M. Daniels 1266
S. E. Darden 1332
J. F. J. Dautzenberg 517, 520, 523, 620
A. R. Davis 939
F. deBoer 718
M. de Jong 890, 973, 1237, 1290
W. Del Bianco 175
P. P. J. Delheij 947
M. P. De Pascale 175, 928
R. de Swiniarski 541
R. Detomo, Jr. 1466, 1469
R. de Tourreil 114
D. W. Devins 635
S. K. Dhawan 933
K. S. Dhuga 656, 688
F. Diaf 1120
H. Dobiasch 1260
P. Doleschall 1208, 1211, 1246, 1249, 1252
T. R. Donoghue 742, 1466, 1469
W. H. Dragoset 152
W. Drenckhahn 644, 653
W. Dreves 916, 925, 970, 979, 991, 1085, 1096, 1099, 1102, 1105, 1108
L. J. Dries 1469
I. M. Duck 172
J. C. Duder. 727, 730, 733, 1488
J. S. Dunham 941

D. DuPlantis....1455

E. D. Earle.....1436
J. A. Edgington.84
P. Egelhof......916, 919, 970, 979, 991, 1085, 1096, 1099, 1102,
                                                    1105, 1108
D. Ehrlich......893
H. Einenkel.....904
C. Eisenegger...955
E. Elbaz........499, 1083, 1091
M. Elbel........925
K. Elsener......1211, 1296
F. Entezami.....587, 659, 688
M. B. Epstein...1205, 1287

B. Fabbro.......1120
A. Faessler 200
B. Favier.......163
C. Fayard.......1202
J. M. Feagin....1126
L. Federici 175, 928
A. Feigel.......418, 644, 653
W. F. Feix 517, 620, 712
M. S. Feld......804
M. Fernandez....1332
D. Fick.........916, 919, 925, 970, 979, 991, 1085, 1096, 1099, 1102,
                                                    1105, 1108
E. Finckh.......418, 644, 653
F. W. K. Firk...389, 1311
R. Fischer......1260
G. A. Fisher....596, 600
C. Fitzpatrick .602
C. E. Floyd 398, 401, 404, 407, 884, 1272, 1323, 1326, 1472, 1478

E. R. Flynn.....270, 682, 694, 697, 700, 718
R. A. Fong-Tom .933
F. Foroughi 163, 1232
C. C. Foster....481, 484, 547
J. L. Foster ...1332
K. H. Frank 427
S. J. Freedman .939
R. Frick 473, 571, 573, 576, 579, 664, 667, 670, 893
L. Friedrich....685, 691, 739
D. L. Friesel...484, 632
D. Fröhling.....1308
S. Frullani.....175, 928
T. Fujisawa 535, 673, 901, 906
Y. Fujita.......626
M. Fujiwara 626
T. Fukada.......1111
S. Fukumoto 879

K. Furuno 1226
M. Furić 152
H. Furutani 1269

G. Gademann 644, 653
C. A. Gagliardi . 939
S. Gales 632
C. S. Galovich . . 941
R. Garrett 1488
G. T. Garvey 939
M. Gazzaly 146, 1234
J. Geenen 982
E. Gerlic 632
W. R. Gibbs 132
G. Giordano 175, 928
B. Girolami 175, 928
C. Glashausser . . 554, 727, 730, 733
G. Glass 123, 126, 129, 132, 166, 871, 874, 949
D. Gola 1299
D. A. Goldberg . . 481, 484
A. M. Gonçalves . 1123
C. D. Goodman . . . 547
C. A. Gossett . . . 650, 877, 931
A. Goto 445
C. R. Gould 944
G. Graw 473, 571, 573, 576, 579, 664, 667, 670, 893, 1027

R. G. Green 913
S. J. Greene 956
L. G. Greeniaus . 158, 1205, 1287, 1317
R. J. Griffiths . 502
M. Grimm 554
G. Gruber 1085
W. Grüebler 848, 899, 1208, 1211, 1214, 1293, 1296
G. W. Guest 913
C. Günther 638
Y.-N. Guo 933
D. P. Gurd 1317
I. Gusdal 890, 973
P. P. Guss 398, 401, 404, 407, 1272, 1275, 1323, 1326, 1472, 1478

D. G. Haase 944
E. Habib 679
R. W. Hackenburg 169
W. Haeberli 137, 641, 877, 931, 1340, 1439
B. Haesner 1260
M. Haji-Saeid . . . 1217, 1234
J. R. Hall 484
I. Halpern 511
J. W. Hammer 392, 395

A. D. Hancock 169
S. S. Hanna 596, 600, 662, 941, 1129, 1492
R. A. Hardekopf ..496, 694, 896, 1240, 1243, 1246, 1249, 1252, 1314

W. C. Hardy 913
T. Hasegawa 535, 673, 901, 906
D. K. Hasell 1205, 1287
K. Hatanaka 140, 437, 440, 445, 448, 451, 478, 544, 562, 581,
                  952, 967, 1229, 1255, 1464

R. Hausammann 143, 161, 955
H. J. Hausman 742
E. Heer 143, 161, 955
W. Heeringa 770, 1260
C. Heinrich 1299
L. Heins 197
E. Heitz 424, 427, 727, 730, 733
H. J. Helten 1299
R. Henneck 1439
R. Hess 143, 161, 955
B. P. Hichwa 1332, 1455
J. C. Hiebert 120, 123, 126, 129, 132, 166, 871, 874, 949, 1223,
                  1278, 1281, 1335, 1461
T. F. Hill 1105, 1108
A. Hintermann 1068
C. M. Hoffman 1449
E. W. Hoffman 149, 958
G. W. Hoffmann ...135, 554
K. W. Hoffmann ...392, 395
H. M. Hofmann ...1458
J. Hoftiezer 163
B. Hoistad 135, 146
K. Holinde 197, 200
C. L. Hollas.....123, 126, 129, 132, 149, 166, 871, 874, 949, 956,
                  958
U. Holm 964, 1308
R. J. Holt 1180
K. Hosono 544, 559, 629, 967, 1255
H. Hübel.........638
D. W. Hudgings...985
J. W. Hugg 662, 1129
V. W. Hughes 933
E. V. Hungerford.169
D. A. Hutcheon ..158, 188, 454, 568, 1205, 1317
E. Huttel........1258
C. Hwang 169

M. Igarashi 715
G. Igo 146, 554, 1157, 1217, 1220, 1234
H. Iida 611, 709
H. Ikegami 626

K. Imai 140, 437, 478, 581, 626, 952, 1229, 1464
G. Ingold 1117
L. Ingrosso 175
H. Ingwersen 919
F. Irom 1220, 1234
M. Ishikawa 1226
T. Itahashi 1255
K. Ito 879
S. Iversen 1338
K. Iwatani 673

S. Jaccard 143, 161, 163, 904, 1439
W. W. Jacobs 457, 484, 635, 1455
Ch. Jacquemart . . 1439
J. J. Jarmer 126
N. Jarmie 1243, 1246, 1249, 1252, 1314
A. J. Jason 985
B. Jenny 899, 1208, 1211, 1214, 1293, 1296
H. Jeremie 175
K. A. Johns 152
R. C. Johnson . . . 750
P. L. Jolivette . 1332, 1455
G. Jones 550
W. P. Jones 484, 635
J. Jordan 421, 424, 427
T. Joyce 1217
D. M. Judd 169

Y. Kadota 544, 629
S. Kailas 484
M. Kaitchuk 484
M. Kaletka 1338
W. Kamke 925
J. E. Kammeraad . 641
O. Karban 587, 590, 593, 659, 688, 703, 706, 913
J. Kasagi 623, 632
I. Katayama 626
S. Kato 544, 581, 629, 952
J. Kawa 1111
N. Kishida 535, 623, 673
H. Kitazawa 602
P. Kitching 454, 568
H. O. Klages 1260
S. S. Klein 517, 520, 523, 532, 620
R. Klem 1217
W. Kloet 1132
K. T. Knöpfle . . . 685, 691, 739
J. W. Knowles . . 1436
L. D. Knutson . . . 641, 650, 1455
S. Kobayashi 440, 445, 448, 451, 581, 952
A. M. Kobos 432

K. Kobras 747
I. Koenig 916, 919, 970, 979, 1096, 1099, 1102
Y. Koike 194, 1299, 1302
D. Kollewe 392, 395
V. V. Komarov 191
K. Kondo 933
M. Kondo 544, 629, 967, 1255
V. König 899, 1208, 1211, 1214, 1293, 1296
F. Konopasek 890
H.-G. Körber 1308
S. Kossionides 916, 919
J. Krabbenborg 514
K. Krämer 1258
W. Kratschmer 392, 395
W. Kretschmer 421, 424, 427, 724, 727, 730, 733, 1329
K.-I. Kubo 715
T. Kubo 535, 623, 673
C. Kubota 879
J. A. Kuehner 721
S. Kunori 529, 611, 614, 709
P. D. Kunz 361
P. M. Kurjan 596, 600

M. Lacombe 114
J. M. Lambert 1252
S. K. Lamoreaux 511
G. H. Lamot 1202
J. Lang 1439
P. Lara 1305
R. M. Larimer 1263
R. D. Ledford 605
F. Lehar 177, 178
C. Lechanoine-LeLuc . 143, 161, 177, 178, 955
H. Lenske 565
W. Leo 143, 161, 955
J. D. Lesikar 152
H. Lettau 1117
P. M. Lewis 587, 590, 593
P. Lezoch 745
R. P. Liljestrand .. 158, 188, 454, 1205, 1317
A. Lindner 470
P. Lisowski 947
P. M. Lister 656
K. P. Liu 739
N. Lockyer 1449
H. Löh 421, 424, 427
B. Loiseau 114
G. J. Lolos 550
M. H. Lopes 750
W. G. Love 554
E. J. Ludwig 413, 415, 430, 467, 647, 736, 961, 1126

R. N. MacDonald 454
R. Machleidt........200
G. Mack 404, 410, 1275, 1323
R. S. Mackintosh....432
J. J. Madsen........909
K. Maeda 559
G. Mairle...........685, 691, 739
Y. Makdisi..........1217
S. Mango............143, 161
D. J. Margaziotis...1205, 1287
M. J. Marolda 120
M. Marshak..........1217
L. Mathelitsch......185
E. L. Mathie........550
G. Matone...........175, 928
N. Matsuoka.........544, 559, 562, 581, 629, 952, 967, 1255
T. Matsusue 140, 437, 478, 1229, 1464
M. Mattioli 175, 928
G. Maughan 1266
D. G. Mavis.........608, 877, 931, 941
A. D. May 1266
B. W. Mayes 169
R. H. McCamis.......158, 1266
J. B. McClelland....146, 1217, 1220
D. K. McDaniels 454
A. B. McDonald 1358, 1436
W. J. McDonald......188, 454, 568
J. McGill...........135
A. McIlwain 890, 973
J. S. C. McKee......973, 1237, 1290
R. D. McKeown.......939
J. L. McKibben......830
M. W. McNaughton....126, 132, 149, 169, 818, 956, 958
J. A. McNeil 461
D. Melnik 727, 730, 733
J. P. M. G. Melssen 508, 532, 538
F. Merz 573, 664
E. Merzbacher 1126
U. Meyer-Berkhout ..893
H.-O. Meyer.........457, 1455
J. Meyer 499, 1083, 1091
H. E. Miettinen.....152
C. A. Miller 188, 454, 568, 1205, 1317
D. W. Miller 547, 635
T. Minamisono 1114
R. E. Mischke 1449
S. M. Mitchell......884
I. Miura 1111
S. Miyashita 933
G. L. Moake 547, 1455
K.-H. Möbius 916, 919, 970, 979, 991, 1096, 1099, 1102, 1105, 1108

D. R. Moffett 1449
W. H. L. Moonen 432, 434, 514
G. L. Morgan.........947
Y. Mori 879
S. Morinobu 626
Z. Moroz............916, 970, 979, 1096, 1099, 1102, 1105, 1108
C. L. Morris.........956, 1217
G. C. Morrison 593, 703
G. A. Moss..........158, 1205, 1287, 1317
J. M. Moss 334, 554
B. Mossberg 1217
S. Motonaga.........901, 906
S. N. Mukherjee 490
N. C. Mukhopadhyay...1068
T. A. Mulera.........152
Th. Müller 638
G. Murillo 1332
D. E. Murnick 804
R. K. Murphy 398, 401, 404, 407, 884, 988, 1272, 1323, 1326,
1472, 1478
G. S. Mutchler 152, 169, 172
H. Müther 200
B. Myslek-Laurikainen.939

W. Naatz 747
S. Nagamachi........559, 629, 1255
K. Nagano...........529, 611, 614, 709
A. Nagao 535
Y. Nagashima 1226
D. E. Nagle 1449
M. Nakamura 440, 445, 448, 451, 478, 581, 952
I. Nakano 933
S. Nakayama 1111
H. Nann 1338
S. Nath 120
F. Naulin 1120
G. C. Neilson 454, 568
J. M. Nelson 656, 688, 703
C. R. Newsom.......123, 126, 129, 166, 871, 874, 949
K. Nagano
G. J. Nijgh 517, 520, 523, 532, 620
K. Nisimura 140, 437, 478, 1229, 1464
J. A. Niskanen......62
Y. Nojiri 1114
J. M. Normand 997
T. Noro.............440, 445, 448, 451, 478, 544, 581, 629, 952
L. C. Northcliffe ...111, 120, 123, 126, 129, 132, 166, 871, 874,
949, 1223, 1278, 1281, 1335, 1461
H. Noya 493
Ch. Nussbaum 163, 1232
C. Nuytten 982

I. Oelrich........576
H. Offermans......514
H. Ogata.........1111
H. Ogawa.........952
K. Ogino.........544, 629
S. Oh............890, 973
G. G. Ohlsen.....496, 1240, 1243, 1246, 1249, 1252, 1314
H. Ohnishi.......493
H. Ohnuma........623, 673, 676
F. Ohtani........440, 445, 448, 451, 1255
K. Okada.........544, 559, 562, 629
A. Okihana.......1229, 1404
N. T. Okumusoğlu..1237, 1266, 1284, 1290
W. C. Olsen 454, 568
Y. Onel..........143, 161, 955
R. F. Oppenheim ..933
A. Osman 1094
H. Ossenbrink....1117
H. Oswald........1299
F. Otani.........952

E. F. Parker.....909
G. Pasquariello...175, 928
H. Pattyn........982
G. Pauletta......146, 1234
P. Pelfer 175, 928
V. Penumetcha....554
G. P. Pepin 166, 169
E. A. Peterson...1217
D.-L. Pham.......541, 1088
G. C. Phillips ...152, 169
J. Philpott 1144
P. Picozza 175, 928
A. A. Pilt.......679, 721
L. S. Pinsky.....169
P. Pirès 114
E. Plagnol 1120
W. Plessas.......182, 185
R. Pogson........890
J. H. Polane.....517, 520, 523, 617, 620, 712
E. Poldi 175, 928
E. C. Pollacco ...590
Yu. V. Popov.....191
A. M. Popova 191
O. J. Poppema ...517, 520, 523, 532, 620
M. Potakar.......605
J. M. Potter 1449
F. Pougheon 1120
W. B. Powell.....913
C. Y. Prescott...1400
D. Presinger 916, 919

G. Pröbstle.......724
D. Prosperi.......175, 928

P. A. Quin........464, 650, 866, 877, 931, 1455

A. Rahbar.........1234
Md. A. Rahman.....745
J. Ramirez........1332
R. D. Ransome.....123, 126, 129, 132, 149, 166, 871, 874, 949, 956, 958
J. R. Rapaport....547
D. Rapin..........143, 161, 955
G. H. Rawitscher..490
L. Ray............295, 454
J. Raynal.........584, 997
W. Reichert.......1439
F. Riess..........664
E. Renner.........1329
H. Riedesel.......685, 691, 739
S. Riedhauser.....931
P. J. Riley.......123, 126, 129, 132, 149, 166, 871, 874, 949, 958

A. S. Rinat.......1201
T. C. Rinckel.....1466, 1469
E. L. Rios 887
A. B. Robbins.....727, 730, 733
N. R. Roberson....598, 602, 605
J. B. Roberts 31, 152
R. G. H. Robertson 939
F. Roesel.........991
S. Roman..........282, 587, 590, 593, 659, 670, 688, 703, 706, 913

T. A. Romanowski..1449
Th. Roser.........1446
E. Rost...........361
G. Rotbard........1120
P. Roussel........1120
B. Rowedder.......418
G. Roy............158, 1317
M. C. Rozak.......1332
D. Rück...........747
K. Ruddick........1217
K. Rüskamp........418, 644, 653
J. F. A. G. Ruyl..514

A. Sagle 1217
T. Saito..........140, 437, 544, 629, 967, 1229, 1255, 1464
H. Sakaguchi 440, 445, 448, 451, 581, 952, 1255
H. Sakai..........559, 562
H. Sakamoto.......440, 445, 448, 451, 478, 581, 952
J. Sanada.........1226

F. D. Santos..........571, 670, 1123
M. Sawada............1226
M. Sawamoto..........155
C. Schaerf...........175, 928
H. J. Scheerer.......576
H. Scheuring.........418
H. Paetz gen Schieck.638, 1299
P. Schiemenz.........473, 571, 573, 576, 579, 664, 667, 670, 893
J. Schimizu..........1226
K. Schindler.........685, 691, 739
G. Schleussner.......392, 395
P. A. Schmelzbach....848, 899, 1208, 1211, 1214, 1293, 1296
R. K. M. Schneider...747
U. Schneidereit......418
R. Schuch............916
K. P. Schüler........933, 936
P. F. Schultz........909
W. Schuster..........424, 427
D. Schütte...........197
P. Schwandt..........454, 457, 481, 484, 496, 547, 1249, 1252, 1455

P. Schwarz...........1260
C. Schweizer.........1208, 1211, 1293, 1296
A. Scott.............554
M. L. Seely..........933
R. E. Segel..........547
N. Seichert..........473, 571, 573, 576, 579, 664, 667, 670
S. Seki..............1226
M. Sekiguchi.........535, 901
K. K. Seth...........1338
M. A. M. Shahabuddin.721
R.-C. Shang..........721
J. R. Shepard........361
D. M. Sheppard.......454
H. S. Sherif.........454, 1317
A. Shimizu...........967, 1255
H. Shimizu...........140, 437, 478, 544, 559, 562, 581, 629, 952,
                                                              1229, 1464
J. Shirai............140, 437, 478, 1229
J. E. Simmons........123, 126, 129, 132, 166, 871, 874, 949
M. Simonius 1439, 1446
I. Slaus.............1252
R. J. Slobodrian.....797
K. A. Snover.........321
P. A. Souder.........933
J. Soukup............1317
J. Sowinski..........608, 877, 931, 1455
K. Sparks............598
E. Speller...........392, 395
F. Sperisen..........1208, 1211, 1214, 1296

H. Spinka........149
K. Spitzer.......427
W. Stach.........421, 424, 427, 724
E. Steffens 916, 919, 925, 970, 979, 991, 1001, 1085, 1096,
                                                1099, 1102, 1105, 1108
E. J. Stephenson.481, 484, 1455
A. W. Stetz......158, 568, 1205, 1287
R. Stingl........424
G. M. Stinson....454, 568, 1317
S. J. St. Lorant.933
R. Stock.........747
U. Strohbusch....745
C. E. Stronach...188, 454
A. Stuirbrink....691
E. Sugarbaker....682, 718, 742
K. Sugimoto......1114
T.-H. Sun........473, 571, 573, 576, 664, 667, 670
J. W. Sunier.....682, 694, 697, 700
J. P. Svenne.....1237
E. C. Swallow....1449
L. W. Swensen....454
W. Swenson.......554

Y. Tagishi.......430, 467, 736
A. Takagi........879
N. Takahashi.....1016, 1114
R. Takashima.....1229, 1464
R. L. Talaga.....1449
T. Tamura........565
K. H. Tanaka.....1114
M. Tanaka........1111
M. Tanifuji 493
R. B. Taylor.....550
H. A. Thiessen...956, 1234
W. J. Thompson...204, 407, 413, 415, 430, 467, 944, 1126
D. R. Tilley.....598, 602, 605
J. R. Tinsley....188, 454
W. B. Tippens....123, 126, 129, 132, 166, 871, 874, 949
Y. Toba..........529, 611, 614, 709
S. A. Tonsfeldt..736, 961
W. Tornow........117, 404, 407, 410, 1275, 1320, 1323
K. Toshioka 149, 958
J. A. Tostevin...750
T. A. Trainor....505, 511
P. A. Treado.....1252
P. Tröger........418
H.-J. Trost......745
A. Trudel........721
G. Tungate 401, 916, 970, 979, 991, 1096, 1099, 1102, 1105,
                                                1108

J. D. Turner........598
S. E. Turpin.......152
A. Turrin..........976

T. Udagawa.........565
N. Ueda............535, 673, 901, 906
T. Ueda............194
J. Ulbricht........137, 931, 964, 1296
E. A. Umland.......169, 172
P. Urbainsky......424, 427

D. Vandeplassche...982
O. B. van Dyck.....149, 958, 985
P. J. van Hall.....432, 434, 508, 514, 517, 520, 523, 526, 532, 538,
                                                        617, 620, 712
L. Vanneste........982
M. Vanni...........928
W. T. H. van Oers..1205, 1263, 1266, 1287
E. Van Walle.......982
R. L. Varner.......430, 467, 884
P. J. T. Verheijen.1266
B. J. VerWest......179
S. E. Vigdor.......457, 742, 1429, 1455
R. Vinh Mau........114
B. J. vom Feld.....638
R. Von Lintig......505
W. von Oertzen 1117
P. von Rossen......1263, 1442
U. von Rossen......1442
H. Vuillème........1232

T. Wada............673
G. J. Wagner.......685, 691, 739
S. Wakaizumi 155
Y. Wakuta..........535
P. L. Walden 550
R. L. Walter 117, 344, 398, 401, 404, 407, 413, 944, 988, 1272,
                   1275, 1320, 1323, 1326, 1472, 1475, 1478, 1481

A. T. M. Wang 146, 1234
M. Wangler.........644, 653
M. B. Wango........424, 427
S. D. Wassenaar ...434, 508, 517, 520, 523, 526, 532, 538, 620

J. Watson 1263
Ch. Weddigen.......163, 1439
W. G. Weitkamp.....511
H. R. Weller.......308, 598, 602, 605
S. A. Wender 398, 404, 407, 884, 1272, 1323, 1326, 1472, 1478

R. E. White 1208
J. Whittaker........1217
C. A. Whitten, Jr. .146, 556, 1220, 1234
K. Wick.............1305
K. Wienhard.........747
J. Wilczynski.......1260
J. F. Wilkerson 415, 430, 467, 736, 961, 1126
H. B. Willard.......149
H. E. Williams......887
T. M. Williams 169
W. L. Williams 1384
A. Willis 158
N. Willis...........158
H. Wilson 1205, 1317
A. Winnacker 933
H. Winter...........922
T. Wise 877
S. W. Wissink.......941
R. M. Woloshyn 158
W. H. Wong..........464
B. E. Wood..........554
W. F. Woodward 120
H. L. Woolverton....120, 1278, 1281
E. Woye.............401, 404, 410, 1275, 1320, 1323
A. Wriekat..........146
M. E. Wright........884
W. R. Wylie 884, 1272, 1323

K. Yagi 254, 529, 611, 614, 709
S. Yamada 901, 906
T. Yamazaki 626
A. M. Yasnogorodsky.1484
M. Yasue 535, 623
R. L. York 149, 887, 958, 994, 1223, 1278, 1281, 1335, 1461
M. I. Youssef.......1094

W. Zahn.............1458
H. Zankel...........1413
B. Zeitnitz.........1260
L. Zemlo............653
P. Zupranski 970, 991, 1080

AIP Conference Proceedings

No.24	Magnetism and Magnetic Materials - 1974 (20th Annual Conference, San Francisco)	75-2647	0-88318-123-1
No.25	Efficient Use of Energy (The APS Studies on the Technical Aspects of the More Efficient Use of Energy)	75-18227	0-88318-124-X
No.26	High-Energy Physics and Nuclear Structure - 1975 (Santa Fe and Los Alamos)	75-26411	0-88318-125-8
No.27	Topics in Statistical Mechanics and Biophysics: A Memorial to Julius L. Jackson (Wayne State University, 1975)	75-36309	0-88318-126-6
No.28	Physics and Our World: A Symposium in Honor of Victor F. Weisskopf (M.I.T., 1974)	76-7207	0-88318-127-4
No.29	Magnetism and Magnetic Materials - 1975 (21st Annual Conference, Philadelphia)	76-10931	0-88318-128-2
No.30	Particle Searches and Discoveries - 1976 (Vanderbilt Conference)	76-19949	0-88318-129-0
No.31	Structure and Excitations of Amorphous Solids (Williamsburg, VA., 1976)	76-22279	0-88318-130-4
No.32	Materials Technology - 1975 (APS New York Meeting)	76-27967	0-88318-131-2
No.33	Meson-Nuclear Physics - 1976 (Carnegie-Mellon Conference)	76-26811	0-88318-132-0
No.34	Magnetism and Magnetic Materials - 1976 (Joint MMM-Intermag Conference, Pittsburgh)	76-47106	0-88318-133-9
No.35	High Energy Physics with Polarized Beams and Targets (Argonne, 1976)	76-50181	0-88318-134-7
No.36	Momentum Wave Functions - 1976 (Indiana University)	77-82145	0-88318-135-5
No.37	Weak Interaction Physics - 1977 (Indiana University)	77-83344	0-88318-136-3
No.38	Workshop on New Directions in Mossbauer Spectroscopy (Argonne, 1977)	77-90635	0-88318-137-1
No.39	Physics Careers, Employment and Education (Penn State, 1977)	77-94053	0-88318-138-X
No.40	Electrical Transport and Optical Properties of Inhomogeneous Media (Ohio State University, 1977)	78-54319	0-88318-139-8
No.41	Nucleon-Nucleon Interactions - 1977 (Vancouver)	78-54249	0-88318-140-1
No.42	Higher Energy Polarized Proton Beams (Ann Arbor, 1977)	78-55682	0-88318-141-X
No.43	Particles and Fields - 1977 (APS/DPF, Argonne)	78-55683	0-88318-142-8
No.44	Future Trends in Superconductive Electronics (Charlottesville, 1978)	77-9240	0-88318-143-6

No.	Title		
No. 45	New Results in High Energy Physics - 1978 (Vanderbilt Conference)	78-67196	0-88318-144-4
No. 46	Topics in Nonlinear Dynamics (La Jolla Institute)	78-057870	0-88318-145-2
No. 47	Clustering Aspects of Nuclear Structure and Nuclear Reactions (Winnepeg, 1978)	78-64942	0-88318-146-0
No. 48	Current Trends in the Theory of Fields (Tallahassee, 1978)	78-72948	0-88318-147-9
No. 49	Cosmic Rays and Particle Physics - 1978 (Bartol Conference)	79-50489	0-88318-148-7
No. 50	Laser-Solid Interactions and Laser Processing - 1978 (Boston)	79-51564	0-88318-149-5
No. 51	High Energy Physics with Polarized Beams and Polarized Targets (Argonne, 1978)	79-64565	0-88318-150-9
No. 52	Long-Distance Neutrino Detection - 1978 (C.L. Cowan Memorial Symposium)	79-52078	0-88318-151-7
No. 53	Modulated Structures - 1979 (Kailua Kona, Hawaii)	79-53846	0-88318-152-5
No. 54	Meson-Nuclear Physics - 1979 (Houston)	79-53978	0-88318-153-3
No. 55	Quantum Chromodynamics (La Jolla, 1978)	79-54969	0-88318-154-1
No. 56	Particle Acceleration Mechanisms in Astrophysics (La Jolla, 1979)	79-55844	0-88318-155-X
No. 57	Nonlinear Dynamics and the Beam-Beam Interaction (Brookhaven, 1979)	79-57341	0-88318-156-8
No. 58	Inhomogeneous Superconductors - 1979 (Berkeley Springs, W.V.)	79-57620	0-88318-157-6
No. 59	Particles and Fields - 1979 (APS/DPF Montreal)	80-66631	0-88318-158-4
No. 60	History of the ZGS (Argonne, 1979)	80-67694	0-88318-159-2
No. 61	Aspects of the Kinetics and Dynamics of Surface Reactions (La Jolla Institute, 1979)	80-68004	0-88318-160-6
No. 62	High Energy e^+e^- Interactions (Vanderbilt, 1980)	80-53377	0-88318-161-4
No. 63	Supernovae Spectra (La Jolla, 1980)	80-70019	0-88318-162-2
No. 64	Laboratory EXAFS Facilities - 1980 (Univ. of Washington)	80-70579	0-88318-163-0
No. 65	Optics in Four Dimensions - 1980 (ICO, Ensenada)	80-70771	0-88318-164-9
No. 66	Physics in the Automotive Industry - 1980 (APS/AAPT Topical Conference)	80-70987	0-88318-165-7
No. 67	Experimental Meson Spectroscopy - 1980 (Sixth International Conference, Brookhaven)	80-71123	0-88318-166-5
No. 68	High Energy Physics - 1980 (XX International Conference, Madison)	81-65032	0-88318-167-3
No. 69	Polarization Phenomena in Nuclear Physics - 1980 (Fifth International Symposium, Santa Fe)	81-65107	0-88318-168-1
No. 70	Chemistry and Physics of Coal Utilization - 1980 (APS, Morgantown)	81-65106	0-88318-169-X

RAYMOND H. FOGLER LIBRARY
DATE DUE

BOOKS ARE SUBJECT TO RECALL AFTER TWO WEEKS